5

# The Dictionary of Butterflies and Moths in color

# The Dictionary of Butterflies and Moths in color

Allan Watson Paul E.S. Whalley
With an Introduction by
W. Donald Duckworth
AMERICAN CONSULTANT EDITOR

Exeter Books
NEW YORK

Published in USA 1983
by Exeter Books
Distributed by Bookthrift
Exeter is a trademark of Simon & Schuster, Inc.
Bookthrift is a registered trademark of Simon & Schuster, Inc.
New York, New York



ISBN 0-671-06141-0

Printed in Hong Kong

# Contents



All common names are fully cross-referenced in the dictionary section.

# Preface

In this book we have illustrated examples of nearly every currently accepted family of the insect group Lepidoptera (moths and butterflies), from minute and insignificant clothes moths (Tineidae) to giant emperor moths (Saturniidae) and the brilliantly coloured birdwing butterflies (*Ornithoptera* and *Troides*). The few exceptions are a handful of 'microlepidoptera' families such as the Lathrotelidae, a recently described family based on two specimens from Rapa Island in the South Pacific. Species have been selected from every major geographical region, so that the plates should give a good idea of the vast array of shapes, colours and patterns found among the 165,000 or so Lepidoptera of the world. Some species chose themselves: the largest and smallest moths in the world (*Attacus*, *Thysania* and *Nepticula*), the valuable birdwing butterflies, the 'Vampire' moth (*Calpe*), the only known killer (*Lonomia*), the hawkmoth whose existence Charles Darwin predicted (*Xanthopan*) and other famous or infamous species. Many of the figured specimens illustrate biological phenomena such as mimicry and warning coloration (see below), protective coloration (*Phalera*) and sexual dimorphism (*Orgyia*). Other species have been illustrated or mentioned because of their importance as pests of food-crops, forests, orchards, plantations and stored foods (especially Noctuidae, Pyralidae, Lymantriidae and Sphingidae) or their usefulness to man (silk-moths of the families Bombycidae and Saturniidae), or because they have unusual life-histories (the aquatic Nymphulinae) or feeding habits (the predatory Hawaiian Geometrid, *Eupithecia*, and various species of Lycaenidae). Some 'non-species' are mentioned: the legendary Queen Victoria moth and the hand-painted 'humbug' (*Gonepteryx eclipsis*). Many species are illustrated for the first time.

There are two main reasons why we have not illustrated a single egg, caterpillar or chrysalis. Firstly only a small proportion of the world's species are known from these early stages, and secondly, we should need more space than we have available in these pages because of the variety in shape and colour of caterpillars. Even within one species the various instars (stages) of the caterpillar, which moults (changes its skin) several times, may differ greatly from one another – for example the Alder moth, *Acronicta alni*.

## *Mimicry*

Some of the most dramatic examples of proved and supposed models and mimics are illustrated in the plates. The phenomenon of mimicry has been the subject of controversy since 1862 when H. W. Bates, drawing on observations he made during eleven years in the rain-forests of the Amazon, first drew attention to the fact that various species of butterfly and moth which fly together in the tropics resemble each other in surprising detail. He suggested that the reason for this was that one species of each similarly patterned complex of species was noxious to predators and that the remaining palatable species were mimics, deceiving their enemies into ignoring them by means of the protective umbrella of the model's warning colour pattern – a situation now called Batesian mimicry.

Little effort was directed at first towards finding out whether the supposed warningly coloured species were in fact unpalatable to birds and other predators, but in recent years experiments have confirmed that many moths and butterflies are highly toxic or distasteful to birds, bats and other vertebrates. The chemical basis of this toxicity has also been studied and a formidable battery of cardenolides (heart poisons), alkaloids (the group of chemicals which includes strychnine) and hydrogen cyanide has been found by Dr M. Rothschild and her co-workers and by others in the tissues of various species of Arctiidae, Hypsidae, Ctenuchidae, Zygaenidae, Heliconiinae, Ithomiinae, Danainae, Acraeinae and those species of Papilionidae which feed on plants of the family Aristolochiaceae.

One of the best known mimetic complexes illustrated is that between the distasteful *Asclepia*-feeding *Danaus chrysippus* and the presumably palatable *Limenitis archippus*. It also includes a species of the probably unpalatable Hypsid moth genus *Phaegorista*, which is a partner rather than a mimic of *Danaus chrysippus*; the term Müllerian mimicry is commonly applied to this kind of partnership. It sometimes happens that a species of a Müllerian complex, although generally unpalatable to predators, is quite palatable to a particular predator, so that the same moth or butterfly can be both a Batesian mimic and a Müllerian partner. A further complication discussed under *Danaus plexippus* occurs when some specimens of a generally unpalatable species are consistently palatable to most predators.

The sources of the poisonous substances found in unpalatable Lepidoptera are usually the caterpillars' foodplants, which include species of the toxic families Asclepiadaceae, Apocynaceae, Aristolochiaceae, Solanaceae and Passifloraceae. The caterpillars are able to extract the plant toxins and pass them on, via the chrysalis, to the adult moth or butterfly which advertises its noxious qualities with bright colours or a conspicuous colour pattern. Adult and caterpillar colour patterns often include red, yellow and orange, colours which are within the optimal range of a bird's vision. An indication of the virulence of some of these poisonous plants can be gathered from the fact that species of *Crotalaria* (Leguminosae), which nourish various species of the Arctiid genera *Utetheisa* and *Amphicallia*, have been known to kill cattle when eaten by them. A few species are able to manufacture poisons themselves without the need for a poisonous food source. For example the deterrent produced by some species of Zygaenidae is hydrogen cyanide, a generally lethal chemical.

Mimicry of an acoustic and unphotographable kind is practised by both sexes of many Arctiid moths, although some are both visual and acoustic mimics as the time of day demands. Sounds produced by males alone probably serve a different function (see *Setina* and *Pemphegostola*).

## *Fossils*

There are relatively few fossil Lepidoptera preserved in the rocks, compared with some of the other insect orders. This cannot be entirely due to their fragility since many apparently more fragile insects have a long fossil history. The following list gives an indication of the time scale we are discussing:

Man first appeared about 3 million years ago
Modern mammals arose around 40 million years ago
First evidence of Lepidoptera, around 100 million years ago
Last Dinosaurs disappeared around 135 million years ago
Plants, which subsequently became coal, flourished around 350 million years ago
Earliest land plants appeared about 440 million years ago.

The earliest known fossils which are accepted as lepidopterous are a few scales from a wing found in fossil resin from the Middle Cretaceous (around 100 million years ago) in France. These were discovered in 1973 and until then the earliest was the head of a caterpillar preserved in amber from Canada. Unlike European amber, which is mostly Oligocene (about 40 million years old), the Canadian amber is Cretaceous and is estimated at around 70–100 million years old.

There is a tantalizing gap of some 60 million years in the fossil record of the Lepidoptera after the Cretaceous – then, suddenly, fossilized adult butterflies and moths appear in the geological record. These, particularly the fossils in the Baltic amber, about 40 million years old, are recognizable as some of the ancestors of the present day families, or at least very close to them. By the Miocene period, roughly 25–12 million years ago, the few lepidopterous fossils are all recognizable as representative of present day genera, and in some cases perhaps even present day species. However, fossil Lepidoptera have not been well studied and the picture is by no means clear.

We believe that Lepidoptera, and another present day insect order, Trichoptera (Caddis-flies), evolved from a common ancestor. In the absence of fossil evidence we can only speculate on their subsequent evolution. It has always been suggested that, because of their close relationship with the flowering plants, these and Lepidoptera evolved at the same time. Flowering plants began their major evolutionary 'explosion' in the middle of the Cretaceous period, roughly 80 million years ago. With the increase of flowering plants making available a vast new source of food for lepidopterous insects, the latter too began a period of 'rapid' evolution. This does not preclude the possibility that as a group, the Lepidoptera arose at an earlier period; in fact, it is very likely that the earliest Lepidoptera existed long before the present known fossil record. They may have fed on plant juices with their proboscis (if they had one!) and with the 'arrival' of flowers were able to make greater use of this new source of food. Lepidoptera-like insects may well have been around quite early in the evolution of insects.

Evolution was formerly regarded as a process which needed a long period of time for changes to take place, but it now seems likely that quite striking evolutionary changes can take place in a relatively short time, in hundreds and certainly thousands, rather than millions, of years.

Butterflies and moths are remarkably alike morphologically, compared for example with beetles, but nevertheless present an amazing array of specialized shapes, colours and habits, many of which we portray or describe in this book.

## *Life History and Biology*

It may be surprising to learn how little is known about most species of moths and many butterflies. A fair amount has been published about the life histories of the butterflies and larger moths of the Northern Hemisphere, but very little about the species found in most other parts of the world. This is especially true of the tropical regions of the New World and of S.E. Asia and Australasia. In these areas probably less than half the species have so far been described, and these almost invariably from the adult insect, whereas the egg, caterpillar and chrysalis are completely unknown. Even in Europe new species are discovered from time to time; for example, the large Brahmaeid *Acanthobrahmaea europaea* was described as recently as 1963. Unravelling the life history of a species can be difficult because the only certain way to correctly associate the various stages is to follow the breeding cycle in the laboratory. You might start, perhaps, with a female, hopefully fertilized by a male of the same species (hybrids do occur). Before the eggs hatch you must gather food for the caterpillars. Perhaps they will be wood-borers (e.g. Cossidae and Sesiidae), root feeders (some Noctuidae), carnivorous (*Megalopalpus*), aquatic (*Nymphula*), inhabitants of ants' nests or dependent on epiphytic orchids and bromeliads (Castniidae). It is probably easier to start with caterpillars found on their foodplant and hope that neither they nor the resulting chrysalids are parasitized by an ichneumon wasp or a parasitic fly. The life-cycle may take a few weeks to complete, but if your subject is a Cossid, you may have to wait a year or more, especially in temperate latitudes. The particular importance of the caterpillar stage to us and our environment is that this is the feeding stage of a species. A single caterpillar can eat a surprising amount of food; an infestation of the Gypsy moth, *Lymantria dispar*, in the east United States can defoliate a tree in a few days, and acres of forest in as short a time if the outbreak is severe. Scientific 'firsts' await anyone patient enough to discover the foodplants of, for example, nearly any species of Nemeobiid butterfly in the American tropics and to record and collect the various stages on colour film and in preservative( eighty per cent alcohol is best).

## *Names*

To overcome the language barrier, a 'language' based on Latin and Greek was devised by zoologists to form a common medium for their work. It is based on the standard use of Latin by the early writers of the Middle Ages. This system has worked for some 200 years, but because of the large number of animal species known (around 1,500,000), each with its own name, the system is creaking! However, with the advent of computer storage of information there is still hope.

The name is the peg on which all the information about the species is hung, hence the vital need for an international name for each insect. Over the years a system of rules has been evolved, not without struggle, which has been crystallized in the International Code for Zoological Nomenclature. These rules are supervised by an august body, with representatives from most countries or spheres of influence in the world. It is their function to arbitrate in disputes over the names of animals and to keep an eye on the procedures used in naming new species. A similar set of rules applies to plants.

How is an insect named? Let us assume that a butterfly from a remote area cannot be identified, either by comparison with existing named specimens, or from the vast literature where insects have been described and named. Assuming that the new species can at least be recognized as a member of an existing genus, a publication is produced in which it is described, illustrated and compared with its most closely allied relatives. At some point a name must be chosen for the new species. This is generally a name which helps to describe it, for example *subnitida* – slightly shining, or refers to the place where it was found, e.g. *hawaiiensis.* The rule to be followed is that the name selected must never have been used in combination with its particular genus. It does not matter if the specific name is already used in another genus – a glance at this book will show how many specific names are the same. Thus all species have a binomial name; the genus, which always starts with a capital letter, and the species which starts with a small letter. These two names are generally printed in italics. When the new name is published, following the rules, it is available for use. One of the specimens examined by the describer is called the holotype (or type); this is the specimen which is re-examined if there is any question of the correct interpretation of the species. When the species is subsequently referred to in print, the generic and specific names are followed by the name of the person who described it.

*Papilio machaon* Linnaeus shows us that the species, *machaon*, belongs to the genus *Papilio*, and was described by Linnaeus. This genus contains a number of species which are fairly closely related. Thus *Papilio alexanor* Esper is a related species of Swallowtail (*Papilio*), described by Esper. A generic name may be followed by an author, when it is used on its own. This indicates the first person who recognized the particular generic grouping, for example, *Papilio* Linnaeus. These scientific names are international and are used in all biological publications. There are many problems in finding the correct name for an animal, so much so that an entire journal is devoted to the *names* of animals (*Bulletin of Zoological Nomenclature*).

Does it really matter what an insect is called by the Australians, Chinese or Russians? Not only is it a matter of convenience to scientists internationally but it can be financially important. For example, if reliable control measures for a particular pest are discovered in one country, it is obviously desirable that they should be applied in other countries where the insect is also a pest. First and foremost it is vital to know if one is dealing with the same pest; hence the importance of applying the same name to the same insect. Unfortunately, using the same name is no guarantee that the same insect is involved. If the species is misidentified then more problems arise. This is where the identification services and museums help in checking the identification against reliably determined specimens (often the holotype of the species in question). Recently it has been shown that species which were believed to be world-wide pests, involve several similar-looking species. The corn-borer moth, *Ostrinia nubilalis*, was thought to be a world-wide pest, but differing results were obtained in different countries from similar control measures and it has now been shown that there were twenty different species involved under the name *Ostrinia nubilalis*.

## *Sequence of Illustrations*

This approximately follows the sequence in the Lepidoptera section of *Insects of Australia*, 1970, ed. D. Waterhouse, but with some additions and modifications. Generally the more primitive families are at the beginning with the supposedly more advanced at the end of the work. As with any linear arrangement of a complicated multidirectional evolutionary process, the actual sequence is somewhat arbitrary and the arrangement of an evolutionary 'tree' in the Lepidoptera is subject to much argument and disagreement.

Under each family entry (Lycaenidae, Arctiidae etc.) we have listed the genera which have one or more species illustrated or have some text data. The type-genus of the family, on which the family name is based is usually included, e.g. *Lycaena* under Lycaenidae and *Arctia* under Arctiidae. An attempt has been made to keep together the illustrations of the species of a given genus but problems of size and shape have sometimes made this impossible.

Nearly all the currently recognized families of Lepidoptera are illustrated here, apart from new species or genera, which are being discovered continually; even new families of Lepidoptera have been described since this work was started, while research into the status of some existing families has resulted in them being amalgamated with others. The systematic arrangement of the families is given below.

---

LEPIDOPTERA (Butterflies and moths)

MICRO-MOTHS (Microlepidoptera)

| | | |
|---|---|---|
| Micropterigidae | Acrolophidae | Blastobasidae |
| Agathiphagidae | Ochsenheimeriidae | Xyloryctidae |
| Eriocraniidae | Lyonetiidae | Stenomidae |
| Neopseustidae | Hieroxestidae | Symmocidae |
| Lophocoronidae | Gracillariidae | Gelechiidae |
| Mnesarchaeidae | Phyllocnistidae | Strepsimanidae |
| Prototheoridae | Sesiidae | Copromorphidae |
| Palaeosetidae | Glyphipterigidae | Aluctidae |
| Hepialidae | Douglasiidae | Carposinidae |
| Nepticulidae | Heliodinidae | Heterogynidae |
| Opostegidae | Yponomeutidae | Castniidae |
| Incurvariidae | Epermeniidae | Zygaenidae |
| Heliozelidae | Schreckensteiniidae | Chrysopolomidae |
| Tischeriidae | Coleophoridae | Megalopygidae |
| Cossidae | Agonoxenidae | Cyclotornidae |
| Dudgeoneidae | Elachistidae | Epipyropidae |
| Metarbelidae | Scythrididae | Dalceridae |
| Tortricidae | Stathmopodidae | Limacodidae |
| Cochylidae | Oecophoridae | Hyblaeidae |
| Pseudarbelidae | Ethmiidae | Thyrididae |
| Arrhenophanidae | Lecithoceridae | Tineodidae |
| Somabrachyidae | Parametriotidae | Oxychirotidae |
| Deuterotineidae | Momphidae | Lathrotelidae |
| Psychidae | Metachandidae | Pyralidae |
| Compsoctenidae | Anomologidae | Pterophoridae |
| Tineidae | Pterolonchidae | |

BUTTERFLIES

| | | |
|---|---|---|
| Hesperiidae | Pieridae | Lycaenidae |
| Megathymidae | Nymphalidae | Nemeobiidae |
| Papilionidae | Libytheidae | |

MACRO-MOTHS (Macrolepidoptera)

| | | |
|---|---|---|
| Drepanidae | Lasiocampidae | Sphingidae |
| Cyclidiidae | Anthelidae | Dioptidae |
| Thyatiridae | Eupterotidae | Notodontidae |
| Geometridae | Mimallonidae | Thyretidae |
| Apoprogenidae | Apatelodidae | Lymantriidae |
| Uraniidae | Bombycidae | Ctenuchidae |
| Epicopeidae | Lemoniidae | Arctiidae |
| Epiplemidae | Carthaeidae | Hypsidae |
| Axiidae | Oxytenidae | Nolidae |
| Sematuridae | Cercophanidae | Noctuidae |
| Callidulidae | Saturniidae | Agaristidae |
| Pterothysanidae | Brahmaeidae | |
| Endromidae | Ratardidae | |

# Introduction

The preparation of an adequate introduction to a book as ambitious in its coverage as this one is no easy task. While there have been numerous popular books dealing with butterflies or moths from either a restricted area or selected examples from various parts of the world, I believe the present work represents a new plateau in efforts to gather together information concerning the entirety of the Order Lepidoptera from a world standpoint. The cross-indexed, alphabetical format allows ready access to the enormous amount of information contained in the text and the magnificent photographs illustrating virtually every presently recognized family are without peer in the existing literature.

Two aspects of this book, in my opinion, are especially noteworthy. Due to the broad coverage of the text, from well-known groups such as butterflies to the most obscure moth families, it is possible to not only learn what is known but to also perceive how much more there is to be discovered. In many more instances than is generally recognized, our current knowledge of many species consists of little more than a description and its accompanying name assignment based on a few museum specimens. Not only is information on the life histories and habits yet to be gathered but frequently only one sex of the adult is known! Thus, this book is a handy reference to what is known, as well as a useful guide for determining areas and groups in desperate need of additional studies.

Secondly, the extensive use of photographs from nature greatly enhances, it seems to me, the dynamic nature and the breathtaking beauty of this extraordinary group of animals. Many people are aware of the beauty and drama of some of the larger butterflies and moths; however, the photographs in this book extend that awareness well beyond the ordinary. Even the photographs of the flawless set specimens from the collection of the British Museum (Natural History) pale in comparison to the living organisms captured in natural poses in their native habitats. This contrast may also indicate a basic shift in interest which I believe is occurring in the study and appreciation of butterflies and moths by enthusiasts of all types – from collecting and admiring dead specimens as inanimate objects in containers to observing and studying the living organisms in relation to their natural environment. Additional comments on this subject are included in the final section of this discussion.

My colleagues, Watson and Whalley, have provided in the Preface a series of articles containing information pertinent to users of this book. I, in turn, will touch upon some general features of butterflies and moths which seem appropriate for inclusion, and also briefly describe several activities which provide not only opportunities for pleasure, recreation and learning for anyone interested in pursuing more knowledge of Lepidoptera; but also provide a very real opportunity to add significantly to the gathering of new information on their habits.

The Order Lepidoptera consists of those insects popularly known as the butterflies and moths. While the butterflies as a group are better known to the public because of their usually bright colours and daytime activity, they represent in fact the minority group within the Order. Outnumbering the butterflies in numbers of species by approximately 10 to 1, the largely night-flying moths offer a marked contrast in the diversity of sizes, habits, patterns and colours.

Ranging in size from tiny leaf-mining moths with a wingspan of a few millimeters to giant emperor moths and birdwing butterflies, the Lepidoptera comprises one of the largest orders of insects. With approximately 165,000 species described thus far, and new species being recognized and defined regularly, the success of the group on a world basis is readily apparent.

The name Lepidoptera is derived by combining the Greek words for "scale" and "wing," thus signalling one of the distinguishing features of the organisms making up the group – four wings covered with minute scales. In addition, the remainder of the body is also covered with scales of various types, and it is through the medium of these scales that the great array of colours and patterns so notable in the group is developed.

The colours of adult butterflies and moths are the result of two mechanisms operating individually or in combination: the physical structure of the scales and the presence of pigments. Each scale consists of an upper and lower surface with a space in between. The lower surface is usually smooth but the upper has a series of longitudinal and transverse ridges.

Structural colours are usually recognized as metallic or iridescent effects which vary with the angle of view. These colours are due to the interaction between light and the minute structure of the scales which reflect certain parts of the light and absorb others. These colours are permanent and do not diminish after death.

Colours due to chemical pigments are deposits in the upper and lower surfaces of the scales or in the cavity between. Pigmentary colours are maintained after death but may lose some of their intensity and gradually fade with time.

Another characteristic feature of the group is the development of the mouthparts into a tubular structure through which liquid food may be drawn. This structure is commonly called the proboscis and the success of the group is frequently attributed to its development. The ability of most adult Lepidoptera to obtain and utilize the carbohydrates contained in nectar, which can be converted to and stored as fats, became a major asset with the rise and spread of flowering plants. The moth family Micropterigidae, the most primitive of contemporary Lepidoptera, retains chewing mouthparts which are adapted to grind vegetable matter such as pollen, and in some groups the proboscis is reduced and nonfunctional. In the latter, the adults utilize food stored in the body tissues during the larval stage for the essential activities of mating and egg laying and the adult life span is understandably short. In some groups, notably the Sphingidae or hawk moths, the proboscis is extremely well developed and frequently exceeds the length of the body.

Having touched upon two of the more obvious characteristics which distinguish butterflies and moths from other insects, the next logical question is "How are the two distinguished from each other?" As indicated in the Preface, this book uses a system which divides the Lepidoptera into three groups. The moths are separated into micros and macros, and the butterflies are distinguished as a third group. This arrangement is admittedly superficial and although it has been in existence for many years, it does not afford a classification which adequately reflects relationships within the Lepidoptera. However, since this book is not intended to be a treatise on evolution and classification but rather to provide abbreviated information over a broad representative sample of the world's Lepidoptera fauna, the simplicity of the system is convenient. The Glossary provides the authors' definitions of the groups, and I will only add a reminder that any such criteria are generalizations and numerous exceptions exist in nature. As with any active field of inquiry the details of the classification within the Lepidoptera are the subject of debate among workers, and as new studies are completed and new facts brought to light the system is altered and refined.

It now seems appropriate to briefly explore some of the more dynamic aspects of butterflies and moths, especially to emphasize the obvious fact that they are living organisms which play a variety of roles in the ecosystem around us. As is the case for all animals, butterflies and moths are faced with a major problem – their survival. Their entire makeup, from structure to behaviour and life history to internal chemistry, is adjusted to allow them to locate food and suitable mates, tolerate adverse climatic conditions and locate more favourable ones, and, as species, to endure the attacks of predators, parasites and disease.

Virtually all insects maintain their population levels by producing large numbers of young, only a few of which reach maturity. Since survival of only two offspring from a given female is required for maintenance of constant population levels it would seem that large numbers alone would suffice. This is not the case however, for the numerous perils faced from egg to adult are such that diverse protective mechanisms have evolved which additionally contribute to the successful completion of the life cycle. That the adjustments to the physical and biological environment are never complete is evidenced by the periodic increases and decreases in numbers of any given species.

One of the most common survival characteristics observed in the Lepidoptera is the use of colour and pattern to influence the behaviour of potential predators. Many species have coloration which results in their blending into their background and escaping detection by predators. These "cryptic" species usually have behavioural traits and other structural modifications which contribute to their camouflage. Numerous species are distasteful to birds, lizards and other animals and frequently display bright warning colours and obvious behaviour which serve as a signal to experienced predators that consumption will result in unpleasant consequences. The colours serve to reinforce learning in inexperienced animals that suffer the consequences of attacking an unpalatable species.

Many species have circular markings on their wings which are generally termed "eyespots." The survival advantage of these markings is as yet not clearly understood. It has been suggested that they may sufficiently resemble vertebrate eyes to deter attack by small predators. Many moths have such spots on their hindwings and in the normal resting position they are concealed by cryptically patterned forewings. When disturbed, these moths suddenly move the covering wings forward exposing the eyespot, presumably startling the predator, at least momentarily. Alternatively, some workers have proposed that the eyespots serve to focus the attention of the predator on the least damaging area, the wing, and in this way the butterfly or moth has an improved chance to escape. Careful observation of behaviour in nature is needed to improve our understanding of the function of eyespots.

Mimicry has been discussed in the Preface, and I will only add a few comments. The presence of distasteful butterflies and moths in an area, together with the tendency of predators to visually generalize a recognition of their bright colours and obvious behaviour, creates an opportunity for increased survival of palatable species. The tendency of predators to avoid edible species if they look similar to distasteful ones has resulted in numerous mimetic associations frequently involving numbers of species belonging to several different families. Apparently the ability to distinguish between a good resemblance and a poor one by predators is sufficient to provide an increased survival advantage to any individual with even a small improvement in similarity. Thus, each improvement through genetic change increases the survival chances of its bearer and a gradual increase in similarity between model and mimic occurs.

One of the more dynamic aspects of all living things is the variation which occurs between individuals of the same species. This variation is the outward indication that virtually no organism has a genetic constitution identical to that of any other organism. Among butterflies and moths variation between individuals of the same species may be slight and difficult to detect, or they may be so great that they appear to be a different species. Some types of variation are due to the environmental influences exerted on organisms with uniform genetic makeups. Thus, the spring generation may vary in appearance from the summer or fall generation. In tropical regions the rainy-season generation may differ from those of the dry season.

Where populations of a species are isolated from one another over great distances geographical variation becomes evident. This type of variation is, in fact, virtually always present to one degree or another inasmuch as it results from the spatial effect of a species' range. Natural selection tends to favour changes which better adapt populations to local environments, whereas species stability is maintained by the gradual flow of genes between populations. Thus, the individuals from either extreme of the total range of the species may differ markedly from each other. When there is a gradual change in characteristics from one end of the range to the other, the term "cline" is used and where sudden changes in characteristics occur within a cline these are called "steps." Steps in a cline are usually due to major physical barriers such as mountains or large bodies of water or zones in which food plants do not occur. These barriers serve to reduce gene flow between populations on either side of the barrier and increase differences in some characteristics.

Sexual variation occurs in many species of butterflies and moths and is used to indicate any distinction between the sexes resulting from the differing complement of genes in the sex chromosomes. Sexual variation may be so slight as to require examination of the genitalia to distinguish between males and females. On the other hand, in many species the differences are so great that they appear to be different species. On many occasions this has led to the two sexes being described as separate species and only later, through rearing or field observations, has the true relationship been discovered.

Another type of variation which occurs in butterflies and moths is one in which two or more strikingly different forms of a species occur together in the same habitat. These major differences are generally controlled by a single gene and the term polymorphism is used to describe their occurrence in a population. Polymorphism occurs as a result of a mutation in one of the genes in an interbreeding population which, because it affords some advantage, increases in frequency. In butterflies and moths the most obvious form of polymorphism is that in which two or more colour forms occur together. Although the relative proportions of the polymorphs may vary with population size, seasonal conditions and geographic region, the tendency is for them to remain relatively constant. A species may also be polymorphic in one part of its range and not in another.

Obviously variation is of paramount importance in the survival of organisms. It is the outward manifestation of a dynamic gene pool which is constantly generating individuals with slightly differing characteristics. If any of the characteristics better enable the individual bearing them to survive and prosper, it will be favoured by natural selection and increase in occurrence through time in the population. This provides a constant mechanism for strengthening the gene pool of the species and providing flexibility to cope with changing environmental conditions.

Fundamental to the welfare of any sexual organism is the location of a suitable mate. Once found, acceptance must usually be established through courtship, and mating achieved by transmission of genetic material from male to female. In most butterflies and moths the depositing of the eggs on or near to a suitable food source for the newly hatched caterpillars completes the process and terminates the adult's contribution to succeeding generations.

The location of a suitable mate is generally accomplished in two basic ways in butterflies and moths. If the species is day-flying, visual recognition usually plays a primary role in mate seeking. Frequently, the females localize in the vicinity of the larval food plant and the males presumably seek out these areas for courtship. In other instances, males and females congregate in certain areas such as hilltops where courtship and mating occurs. Once visual recognition is achieved courtship begins. Very little information resulting from observations in nature exists on this activity in butterflies and moths, and much more information is needed. Undoubtedly, numerous elements involving sight, touch, aroma and sound are involved in the preliminary activities to mating. In many species the males possess scent-producing organs. These occur in various forms from modified scales or pouches on the wings to specialized structures on the abdomen. Presumably, receptivity on the part of the female is at least partially influenced by the scents produced by these organs.

In night-flying butterflies and moths, the location of a suitable mate visually is complicated by the reduction in available light. In many species smell becomes the primary means for bringing the two sexes together. In moths the virgin female of many species produces a powerful sex scent (pheromone) that is capable of attracting males from distances of hundreds if not thousands of yards. Again, once the male and female are in close proximity, information on specific courtship behaviour is very limited.

Once mating has been achieved egg laying usually begins almost immediately. Of the many factors involved in the attraction of a female to a particular plant and subsequent laying of eggs, the aroma resulting from essential oils in the plant are believed to be especially important. Physical factors such as suitable temperatures, sunlight or darkness, colour and texture of the material to receive the egg, etc., are undoubtedly important in setting the scene for oviposition. Once egg laying begins the manner in which they are deposited is frequently characteristic. The eggs may be laid individually or in clusters. In some groups (Incurvariidae, Eriocraniidae) the eggs are inserted into plant tissues which the females pierce with their ovipositors. In others (Hepialidae) the eggs are scattered at random during flight or in loose masses on the ground.

The degree of dependence a given species has on a particular food source varies from very specific to rather general. In general, the majority appear to require a specific plant or group of related plants for the caterpillars to survive. Thus, the welfare of the insect is to a greater or lesser degree linked to the welfare of its host plants. If a butterfly or moth can complete its life cycle only on a particular species of plant and that plant declines or disappears in the habitat, then a similar decline or loss of the insect occurs. Fluctuations in numbers of both plants and animals in the environment occurs continually in nature and has done so through time. As mentioned previously, such changes are the result of the constant interactions between organisms and their physical and biological environments. These are normally gradual and only major climatic or biological disruptions result in dramatic changes. In many parts of the world just such a major disruption is now occurring. This will be discussed in the following few paragraphs.

Much has been written in recent years about the enormous increase in human populations and the pressures placed on our natural resources as a result of this growth. The constantly rising demands, which have accelerated in recent years, for food, housing, factories, etc., have resulted in a corresponding loss of open spaces and natural habitats. In addition, the introduction into the environment of toxic by-products of our technology, along with broad application of chemical pesticides, has degraded many biological communities. In many parts of the world these events have reached sufficient magnitude to exert a disruptive effect of major proportions on species of plants and animals. In some instances, entire habitats have disappeared or are in danger of doing so. Obviously, any plants or animals that live only in these habitats have also disappeared or are threatened.

In many places there is a growing concern for the conservation and preservation of natural habitats and the organisms that populate them. Efforts to regulate the use of land and to evaluate the impact of major habitat disturbances are being exerted by many governmental bodies, and laws to protect and preserve threatened and endangered species are becoming more widespread each year.

As butterflies and moths are primarily dependent on plants as a food source both as caterpillars and adults, the broad-scale disruption of habitats which reduces or eliminates plant species or communities is especially harmful. In addition, our knowledge of the distribution, life history and host plant associations of many species of Lepidoptera is so incomplete that it is difficult to determine the degree of danger in advance. A number of European countries have extensive programmes aimed at stimulating the study of plants and animals and the mapping of their distributions. These programmes are designed to encourage the accumulation of information by all interested individuals and groups to allow for more effective management of resources and the identification as early as possible of species that are endangered.

In the United States, a recent report on endangered and threatened plant species compiled by the Smithsonian Institution lists 100 species of plant recently extinct or possibly extinct, 761 species endangered and 1238 species threatened. The appearance of this report serves as a signal that many species of butterfly and moth are also potentially threatened through destruction or alteration of their habitats. If we are to preserve for future generations a truly significant part of our outdoor heritage, it is necessary that efforts to preserve habitats be increased and encouraged and additional efforts to gather information about the organisms that live in them be accelerated. The final section of this introduction provides suggestions of ways in which anyone interested in butterflies can participate.

It is probable that two of our oldest traits are collecting objects which interest and intrigue us and an awareness of things other than ourselves, both animate and inanimate, which exist in the world around us. While the beginnings of these interests are lost in antiquity and their cultural expression is as varied as our history, they have continued through time and provided the fuel for numerous activities which have given both personal pleasure and gratification as well as contributed to the common pool of knowledge. Our museums trace their beginnings to the "cabinets of curiosities" which flourished in the seventeenth and eighteenth centuries and were simply outgrowths of the collecting urge elaborated by increased wealth and an expanding awareness of distant places. In like fashion, the various branches of knowledge which explore the physical and biological world around us owe their origins to individuals whose curiosity about things observed in nature extended beyond a passing fancy. Today, as our knowledge and our institutions continue to grow at an ever increasing rate, it would seem that the role of the individual enthusiast, armed only with his interest in nature and the pleasure gained from learning, would have a small role to play in acquiring new knowledge or collecting more objects. In fact, this is not the case – just as discovering one new fact generally suggests numerous new questions to be explored, the vast accumulation of knowledge to date has only served to expand the opportunities for professional and amateur alike to explore new and rewarding avenues together.

In the biological world many groups of organisms have become better known to us through the combined interest and efforts of both professional researchers and enthusiastic amateurs of all ages. Prominent among these various groups of organisms is the insect Order Lepidoptera – the butterflies and moths. These strikingly varied and beautiful animals are an obvious part of our environmental heritage and their presence and impact is documented in numerous ways by poets and philosophers as well as scientists and nature lovers.

Today there is a growing awareness of the ease with which our activities can disrupt and destroy the delicate balances in nature. At the same time there is an accelerated interest in both learning more about the environment and enjoying the wonders of the natural world. More than ever before, there is the urge, need and opportunity for the application of our energies to collect information and gain knowledge and pleasure from the world around us. In this regard, butterflies and moths afford a major leisure-time resource which should, and in my opinion will, rival that now provided by birds to millions of people the world over. The implication is inherent that the collecting urge mentioned earlier must be satisfied in ways other than simply the accumulation of dead specimens – and properly so for several reasons.

While the collection of specimens is universally recognized as a fundamental necessity for the proper conduct of various types of research studies, it is not necessarily the best approach, at least in the traditional sense, to gaining a genuine appreciation of the wonders and intricacies of living organisms. In addition, our knowledge of butterfly and moth behaviour and their life histories trails far behind our inventory of numbers of kinds which exist. Thus, the opportunities for contributing to the accumulation of new knowledge as well as the personal satisfaction of observing the details of another living organism are greatest if specimen-gathering is made a secondary activity.

One valuable and challenging alternative to specimen collecting is gaining in popularity – the photographing of butterflies and moths. A brief examination of the photographs in this book will quickly demonstrate the advantages of collecting specimens on film as compared to the more traditional accumulation of set specimens in a drawer. On the one hand is a record of the living organism in a natural position engaged in some facet of its behaviour. On the other is a series of dead objects in unnatural positions more suggestive of trophies or art objects. A single specimen may be photographed any number of times by one or more collectors and each picture may reveal a different aspect of its behaviour, whereas it can be captured and killed only once. In addition, the colours of living specimens, especially the ones due to chemical pigments, are usually more vivid than they are after death, and with the technique available in modern photography these colours in most cases last longer on film than on specimens in a drawer.

It should also be emphasized that the entire life history of any species of butterfly or moth can be recorded on film, so the building up of a photo album or slide collection, especially if accompanied by written notes on observations of habits, is not only an exciting and challenging outdoor activity but also affords a greater degree of insight and understanding of nature than ever provided by collecting specimens. The various stages in the life cycle of butterflies and moth – egg, caterpillar, pupa, adult – provide virtually a limitless variety of challenging puzzles to be solved and recorded. Coupled with the numerous behavioural aspects to be observed, such as feeding, egg laying, mating, cocoon building, etc., the range of activities for exploring and learning are further increased.

A fascinating alternative to specimen capture is the rearing of butterflies and moths in enclosures which permit more direct and dependable observation of the various stages. An attractive aspect of breeding Lepidoptera is that it can be pursued during all seasons of the year, an important consideration in those parts of the world where winter months would otherwise interrupt activities. The basic requirements for this rewarding hobby involve acquiring the eggs, caterpillars or pupae or obtaining eggs from a living female and keeping them in appropriate containers as they grow and change. Provision must be made for providing fresh food and proper temperature and humidity levels. If the host plant is known for the species being reared, it is easy to provide the caterpillars with fresh foliage. However, if the host plant is not known or if the larvae feed on other parts of the plant, such as borers in stalks or roots, it is possible to use a mixture of ingredients to form a substitute food source. A number of generalized diets have been developed for rearing Lepidoptera in recent years and their use increases the opportunities for rearing species whose life histories are unknown. These diets consist of dried, ground plant material (such as lima beans), added to ground soy bean, wheat germ, yeast and vitamin C. These materials are mixed with water to obtain a moist, firm consistency or dispersed in agar and water to obtained a jelly-like consistency. Antibiotics and antifungal agents are added to retard spoilage and reduce the buildup of disease organisms. Once mixing diets is mastered, special variations can be produced for trial feedings and rearing experiments. Attempting to solve the special problems such as those presented by caterpillars which bore in wood provides numerous opportunities for innovation. Success is likely to provide previously unknown information and new techniques.

The pleasure and challenge of developing techniques and procedures for breeding and rearing butterflies and moths is exceeded only by the wonder of observing close at hand the remarkable transformations which take place as the insects grow and develop. Also, the colours and forms of the various immature stages are often as exciting and visually pleasing as the more familiar adults.

Now that some of the alternatives to specimen collecting have been discussed it seems appropriate to return to that activity and examine it in terms of its significance to the study of Lepidoptera. The careful collecting of actual specimens is not only appropriate for documentation of identity, but studies of distribution, taxonomy and genetics depend in large part on preserved organisms for various kinds of information.

Perhaps the single most important function for the preserved specimen is its role in determining species identity. As described in the Preface, the name of an organism is the key to all the information known about it. Thus, a fundamental requirement for the serious study of butterflies and moths is the accumulation of accurately identified specimens which are properly prepared, preserved and documented. Once this has been accomplished all observations dealing with that species can be properly associated with the correct name and the preserved specimen serve as a voucher as well as a source for structural and distributional information.

Above all else, the collecting of specimens should be conducted with a serious goal in mind and not merely to acquire objects of interest. If collecting specimens is to be pursued for worthwhile purposes, it is important that the proper techniques for preparing the material be learned and that each specimen be labelled with at least the exact location of capture, date of capture, and name of the person who made the capture. A carefully assembled personal collection developed with adequate field notes is not only a source of personal satisfaction but becomes a valuable part of the scientific record and should be transferred to a museum or university collection when it has fulfilled the collector's purpose. Adhering to these principles will insure a more enjoyable and meaningful leisure time activity and at the same time contribute to the advancement of knowledge.

It is beyond the scope of this brief introduction to explore the details of the various techniques and procedures mentioned in the preceding paragraphs. Many of the books listed in my colleagues' bibliography on the following two pages describe these activities in detail, as well as many others. A visit to the nearest public or school library will generally result in the discovery of additional publications on the subject. It will become apparent that the popular literature on butterflies, especially field guides, is vastly greater than on moths. This is to be expected inasmuch as moths are much more numerous than butterflies and, due to their largely nocturnal habits, have received less attention from both professionals and amateurs. This situation is changing however, and new publications treating the moths of various areas, such as North America, are appearing frequently.

Above all else, it is important to develop an appreciation for butterflies and moths in nature. Whether it is in the back yard of your home or a wilderness area, begin to perceive the presence of these interesting animals by adjusting to the somewhat smaller scale of their activity and actively observing their behaviour. Once this awareness has been achieved an entirely new world of intellectual and aesthetic satisfaction becomes available, and its exploration is limited only by each person's energy and enthusiasm.

W. DONALD DUCKWORTH

# Acknowledgments and Bibliography


## PREFACE ACKNOWLEDGMENTS

Our grateful thanks are due to our families whose patience and help enabled the manuscript to be completed on time; especially Mary Whalley and Vivienne Watson. Timothy Watson collected one of the illustrated butterflies. Our colleagues in the Museum have helped with some problems; Harry K. Clench, Carnegie Museum, read part of the Lycaenid draft and W. Donald Duckworth, Smithsonian Institution, commented on the final manuscript.

Curigwen Lewis and Tom Wellsted of George Rainbird Ltd have been a constant source of editorial wisdom, and George Sharp has been our expert adviser on matters to do with the illustrations.

We are grateful to the photographers whose talents are demonstrated by the many beautiful illustrations of living and set specimens.

We should like to thank the Trustees of the British Museum (Natural History) who gave permission for this work to be undertaken; Dr Paul Freeman who supported our project and Pamela Gilbert who was frequently able to help with bibliographical problems.

Our helpers corrected many errors and ambiguities but are in no way responsible for any that remain.

## PHOTOGRAPHIC ACKNOWLEDGMENTS

*Ardea Photographics:* Hans & Judy Beste: 316; Elizabeth Burgess: 110, 329; Eric Lindgren: 255; Peter Steyn: 332.

*Dr J. D. Bradley:* 17, 271, 321, 393.

*Professor K. S. Brown:* 90, 95, 132–3, 135–41, 144–5, 147, 149, 154–5, 158–9, 218–9, 222, 224, 232–5, 264.

*Los Angeles County Museum (Dr C. Hogue):* 128, 210, 226, 254, 319, 328.

*G. E. Hyde:* 4, 10, 13, 18, 19, 21–2, 24–5, 29, 31, 52–3, 55, 59, 61, 71–2, 78, 81, 83–7, 92–3, 98, 102, 108, 109, 116–17, 121–5, 130–31, 171, 173, 180–2, 186–7, 189, 193–200, 202, 220, 238–9, 245, 247–53, 257–9, 267, 269–70, 273–9, 281–3, 285–7, 289–90, 292–3, 295, 297–301, 304, 306–13, 320, 322–5, 330, 334–6, 339, 341, 345, 347–8, 350–3, 355–6, 358–9, 361, 366, 370–5, 377–9, 384–5, 388, 394, 396–8, 400, 401, 403–5.

*N. Macfarland:* 305.

*J. L. Mason:* 36, 74, 82, 89, 113–14, 126, 129, 143, 148, 156–7, 164, 166, 178, 183–4, 201, 206, 207, 211–14, 218, 221, 223, 225, 227–8, 261–2, 315, 349, 380, 383.

*Harold Oldroyd:* 3, 75, 118, 203, 204, 256, 387.

*Robert Smiles:* 120, 165, 167–70, 176–7, 191, 205, 302, 303, 346, 360, 395.

*Dr John R. G. Turner:* 146.

*M. W. Tweedie:* 2, 9, 12, 15, 16, 20, 23, 26, 30, 32, 35, 40, 48, 50, 57–8, 60, 62, 64–70, 73, 76–7, 88, 172, 175, 179, 190, 192, 246, 280, 284, 288, 291, 354, 365, 386.

*Dr L. Vári:* 338.

*Allan Watson:* 107.

All other photographs were taken by Peter York of set specimens from the British Museum (Natural History).


## BIBLIOGRAPHY

The following are references to some of the most useful works for the identification of Lepidoptera. Most of these deal solely with butterflies, but moths are now receiving better treatment in the literature. The only publication which attempts to cover the world fauna of both butterflies and moths is 'Seitz', but its volumes are expensive.


D'Abrera, B., 1971. *Butterflies of the Australian Region,* Lansdowne, Melbourne.

Amsel, H. G., Gregor, F., and Reisser, H. (eds). *Microlepidoptera Palaearctica* Vol. 1 Crambidae, S. Bleszynski, 1965; Vol. 2 Ethmiidae, K. Sattler, 1967; Vol. 3 Cochylidae, J. Razowski, 1970. (Eventually this work will cover all the microlepidoptera of Europe and temperate Asia.)

Barcant, M., 1970. *Butterflies of Trinidad and Tobago,* Collins, London.

Beirne, B. P., 1952. *British Pyralid and Plume Moths,* Frederick Warne, London.

Brown, F. M. and Heineman, B., 1972. *Jamaica and its Butterflies,* Classey Ltd, London.

Clarke, J. F. G., 1955–70. Catalogue of the type specimens of Microlepidoptera in the British Museum (Natural History) described by Edward Meyrick. Vols 1–8, British Museum (Natural History), London.

Common, I. F. B., and Waterhouse, D. F., 1972. *Butterflies of Australia,* Angus & Robertson, Sydney.

Corbet, A. S., and Pendlebury, H. M., 1956. *The Butterflies of the Malay Peninsula,* 2nd Ed., Oliver & Boyd, Edinburgh.

Dominick, R. B. et al (eds), 1971– . *The Moths of North America,* Classey Ltd, London. (A series in production intended to cover all the North American species. Several parts are already published.)

Ehrlich, P. R., and Ehrlich, A., 1961. *How to know the Butterflies* (of the USA), Wm Brown & Co., Iowa.

Forbes, W. T. M., 1923–1960. *Lepidoptera of New York and Neighboring States,* Vols 1–4, Cornell University, Ithaca.

Forster, W., and Wohlfahrt, T. A., 1954–74. *Die Schmetterlinge Mitteleuropas,* W. Keller & Co., Stuttgart.

Fox, R. M., Lindsey, A. W., Clench, H. K., and Miller, L. D., 1965. *The Butterflies of Liberia,* American Entomological Society, Philadelphia.

Gifford, D., 1965. *The Butterflies of Malawi,* Malawi Society, Blantyre.

Hannemann, H. J., 1964. *Die Tierwelt Deutschland, Kleinschmetterlinge oder Microlepidoptera,* Gustav Fischer, Jena.

Higgins, L. G., and Riley, N., 1974. *Field Guide to the Butterflies of Britain and Europe,* 2nd Ed., Collins, London.

Holland, W. J., 1903. *The Moth Book,* Doubleday Page and Co., New York.

Howarth, T. G., 1973. *South's British Butterflies,* Frederick Warne, London.

Hudson, G. V., 1928. *The Butterflies and Moths of New Zealand,* Ferguson & Osborn, Wellington.

Hudson, G. V., 1939. *A Supplement to the Butterflies and Moths of New Zealand,* Ferguson & Osborn, Wellington.

Inoue, H., et al, 1971. *Iconographia Insectorum Japonicorum* Vol. 1 Lepidoptera, Hokuryukan Ltd, Tokyo.

Kirby, W. F., 1903. *The Butterflies and Moths of Europe,* revised Ed., Cassells & Co., London.

Klots, A. B., 1951. *A Field Guide to the Butterflies (of the USA),* Houghton Mifflin and Co., Cambridge, Mass.

Lewis, H. L., 1974. *Butterflies of the World,* Harrap, London.

Meyrick, E., 1928. *Handbook of British Lepidoptera,* Macmillan & Co., London.

Pinhey, E. C. G., 1949. *Butterflies of Rhodesia,* Rhodesia Scientific Association, Salisbury.

Pinhey, E. C. G., 1965. *Butterflies of South Africa,* Nelson, Johannesburg.

Pinhey, E. C. G., 1972. *Emperor moths of South and South Central Africa,* Struik, Cape Town.

Seitz, A. (ed.), 1907–1935. *The Macrolepidoptera of the World,* Vols 1–16, Kernen, Stuttgart. (English, French and German editions). (New volumes of this work are still being produced. The first sixteen now cost over £1000, $2200.)

South, R., 1961. *The Moths of the British Isles,* new ed., Frederick Warne, London.

van Son, G., 1949–63. *The Butterflies of Southern Africa,* Transvaal Museum, Pretoria.

Vári, L., 1961. *South African Lepidoptera* Vol. 1 Lithocolletidae, pp. 1–238, Transvaal Museum, Pretoria.

Waterhouse, D. F., (ed.) 1970. *The Insects of Australia,* C.S.I.R.O., Canberra.

Woodhouse, L.G. O., 1952. *The Butterfly Fauna of Ceylon,* 2nd ed., Colombo.

Williams, J. G., 1969. *A Field Guide to the Butterflies of Africa,* Collins, London.

Wynter-Blyth, M. A., 1957. *Butterflies of the Indian Region,* Bombay Natural History Society, Bombay.

Books published before 1960, and even some of those published since, may include scientific names not in current use. These can be checked against more recent check-lists of scientific names which appear from time to time in entomological periodicals.

# The Plates

SYMBOLS USED IN THE CAPTIONS

♂ male

♀ female

◒ to indicate that the undersurface of the wing is illustrated in the set specimens

1

2

3

4

5

6

**1**

a *Micropterix schaefferi*

b *Opostega salaciella*

c *Nepticula aurella*

d *Stigmella ulmivora*

e *Mnesarcha loxoscia*

f *Antispila pfeifferella*

g *Eriocrania sparrmannella*

h *Lophocorona pediasia*

**2**

*Micropterix calthella*

**3**

*Adela reaumurella*
Golden Long-horn

**4**

*Nemophora degeerella*
Long-horned Moth

**5**

a *Somabrachys albinervis*

b *Trichophaga tapetzella*
White-tipped Clothes Moth, Tapestry OR Carpet Moth

c *Phalacropterix apiformis*

c *Tineola bisselliella*
Common OR Webbing Clothes Moth

e *Lampronia oehlmanniella*

f *Tegeticula yuccasella*
Yucca Moth

g *Nemapogon granella*
European Grain Moth

h *Prototheora petrosoma*

j *Ochsenheimeria mediopectinellus*

**6**

a *Tischeria ekebladella*

b *Phyllocnistis saligna*

c *Lyonetia latistrigella*

d *Phyllonorycter cavella*

e *Phyllonorycter lautella*

f *Oinophila v-flava*

**7**

a *Zelotypia stacyi*
Bent-wing Swift Moth

b *Charagia daphnandra*

c *Leto venus*

**8**

a *Xyleutes affinis*

b *Dudgeonea actinias*

c *Allostylus caerulescens*

d *Xyleutes eucalypti*

**7**

**8**

9

10

11

12

13

14

**9**

*Acleris variegana*

**10**

*Hepialus fusconebulosa*
Map-winged Swift

**11**

a *Arrhenophanes perspicilla*

b *Pseudatteria volanica*

c *Oleuthreutes mniochlora*

d *Metarbela triguttata*

e *Compsoctena primella*

f *Chresmarcha delphica*

**12**

*Tortrix viridana*
Green Oak Tortrix

**13**

*Acleris comariana*
Strawberry Tortrix Moth

**14**

a *Zacorisa holantha*

b *Commophila aeneana*

c *Zacorisa toxopei*

d *Cydia saltitans*
Jumping Bean Moth

e *Cydia pomonella*
Apple Codlin Moth or Apple Maggot

f *Cydia ninana*

g *Cerace xanthocosma*

h *Cydia egregiana*

15
*Epinotia stroemiana*

16
*Archips oporana*

17
*Epiblema foenella*

18
*Rhyacionia buoliana*
Pine-shoot Tortrix
OR European
Pine-shoot Moth

19
*Cydia pomonella*
Apple Codlin Moth
or Apple Maggot

20
*Pseudargyrotoza conwagana*

15

16

17

18

19

20

21

22

23

24

25

26

**21**
*Agapeta zoegana*

**22**
*Agapeta hamana*

**23**
*Croesia forsskaleana*

**24**
*Psyche casta*

**25**
*Cossus cossus*
Goat Moth

**26**
*Phyllonorycter oxyacanthae*

**27**

a *Alcathoe autumnalis*

b *Conopia chrysophanes*

**28**

a *Sesia apiformia*
Hornet Moth

b *Melittia gloriosa*
Manroot OR Big-root Borer

**29**

*Bembecia scopigera*
Six-belted Clearwing

**30**

*Anthophila fabriciana*

**31**

*Yponomeuta cagnagella*

**32**

*Ypsolophus dentella*
Tooth-streak Hooked Tip Moth

27

28

29

30

31

32

33

34

**33**

a *Imma saturata*

b *Imma racemosa*

**34**

a *Opogona sacchari*

b *Burlacena vacua*

c *Glyphipteryx lathamella*

d *Mictopsichia durranti*

e *Tebenna bjerkandrella*

f *Hilarographa excellens*

g *Glyphipteryx lineella*

h *Glyphipteryx equitella*

j *Atychia appendiculata*

k *Choreutis dolosana*

l *Hypertropha desumptana*

m *Tortyra divitiosa*

35

36

**35**

*Endrosis sarcitrella*
White-shouldered Clothes OR House Moth

**36**

*Hofmannophila pseudospretella*
Brown House OR False Clothes Moth

37

38

**37**

a *Atteva monerythra*

b *Atteva mathewi*

c *Comocritus olympia*

d *Atteva pastulella*
Ailanthus Webworm

e *Atteva megalastra*

**38**

a *Lactura suffusa*

b *Anticrates tridelta*

c *Lactura coleoxantha*

d *Lactura callopisma*

e *Trisophista doctissima*

**39**

a *Ethmia hilarella*

b *Thalamarchella alveola*

c *Ethmia lineatonotella*

d *Timyra cingalensis*

e *Ethmia aurifluella*

f *Ethmia bipunctella*

**40**

*Alabonia geoffrella*

**41**

a *Epermenia pontificella*

b *Eretmocera chrysias*

c *Argyresthia pruniella*

d *Douglasia anchusella*

e *Eretmocera laetissima*

f *Prays citri*

g *Eretmocera fuscipennis*

h *Anticrates autobrocha*

j *Plutella xylostella* Diamond Back Moth

**42**

a *Uzucha humeralis*

b *Loxotoma elegans*

c *Cryptophasa nephrosema*

d *Stenoma sequitierta*

e *Cyanocrates grandis*

f *Antaeotrica griseana*

**43**

a *Pterolonche pulverentella*

b *Scythris grandipennis*

c *Gelechia ophiaula*

d *Schreckensteinia festaliella*

e *Phthorimaea operculella* Potato Tuber Moth or Tobacco Splitworm

f *Metachanda citrodesma*

g *Gelechia rhombella*

h *Symmoca signella*

j *Aeolanthus sagulatus*

k *Physoptilia pinguivora*

l *Aphthonetus mediocris*

m *Hyposmocoma arenella*

**44**

a *Sphaerelictis dorothea*

b *Cosmopteryx scribaiella*

c *Trachydora leucobathra*

d *Synploca gumia*

e *Stathmopoda pedella*

f *Holocera iceryaeella*

g *Coleophora virgatella*

39

40

41

42

43

44

**44 contd.**

h *Sathrobrota rileyi*

j *Scythris cuspidella*

k *Agonoxena argaula*

l *Elachista regificella*

**45**

a *Eterusia repleta*

b *Cyclosia midamia*

c *Aglaope bifasciata*

d *Gymnautocera rhodope*

e *Erasmia pulchella*

f *Campylotes desgodinsi*

g *Erasmia sanguiflua*

h *Cyclosia curiosa*

j *Cyanostola diva*

k *Zegara zagraeoides*

l *Riechia acraeoides*

m *Gazera linus*

n *Neocastnia nicevillei*

p *Synemon sophia*

q *Enicospila marcus*

r *Amauta cacica*

s *Castnia licus*
Giant Sugarcane-borer

**46**

a *Zygaena trifolii* Five-spot Burnet Moth

b *Himantopterus fuscinervis*

c *Himantopterus dohertyi*

**47**

a *Heterogymna pardalota*

b *Alucita dohertyi*

c *Meridarchis trapeziella*

d *Corpromorpha tetracha*

**48**

a *Trosia bicolor*

b *Chrysopholoma similis*

c *Dalcera abrasa*

d *Megalopyge lanata*

**49**

*Zygaena occitanica* Provence Burnet Moth

**50**

*Zygaena filipendulae* Six-spot Burnet Moth

**46**

**47**

**48**

**49**

**50**

**51**

**52**

**53**

**54**

**55**

**51**

a *Hepialodes follicula*

b *Belenoptera cancellata*

**52**

*Adscita statices*
Forester

**53**

*Catoptria pinella*
Pearl Grass-veneer Moth

**54**

a *Heterogynis pennella*

b *Cyclotorna monocentra*

c *Epipomponia nawi*

d *Coenobasis amoena*

e *Parasa hilarata*

f *Chrysamma purpuripulcra*

**55**

*Pterophorus pentadactyla*
White Plume Moth

**56**

a *Galleria melonella*
Greater Wax Moth

b *Dichocrosis zebralis*

c *Palpita unionalis*

d *Rhodoneura zurisana*

e *Ephestia kuehniella* ♂
Mediterranean Flour Moth

f *Apomyelois bistriatella*

g *Filodes fulvidorsalis*

h *Acentria nivea*
Aquatic OR False Caddis Water-veneer Moth

j *Maruca testulalis*
Bean Pod-borer

k *Plodia interpunctella*
Indian Meal Moth

l *Pempelia formosa*

m *Hyblaea sanguinea*

n *Pyrausta daphalis*

p *Etiella zinckenella*
Lima-bean Pod-borer

q *Uresiphita limbalis*

r *Ephestia elutella*
Cacao OR Tobacco Moth

s *Ephestia kuehniella* ♀
Mediterranean Flour Moth

t *Rhodoneura limatula*

u *Crambus occidentalis*

v *Oncocera semirubella*

w *Nomophila noctuella*
Celery Webworm OR Rush Veneer Moth

x *Herdonia osacesalis*

y *Anania funebris*

z *Lineodes peterseni*

aa *Aulacodes splendens*

bb *Stericta flammealis*

cc *Thyris lugubralis*

dd *Agdistis bennetti*

ee *Castoblastis cactorum*
Cactus Moth

ff *Epipagis pictalis*

57

58

59

60

61

62

**57**

*Crambus perlella*

**58**

*Chrysoteuchia culmella*
Garden Grass Veneer Moth

**59**

*Agriphila tristella*

**60**

*Aphomia sociella*
Bee Moth

**61**

*Hypsophygia costalis*
Gold-fringe, Clover Hayworm OR Clover Webworm

**62**

*Pyralis farinalis*
Meal Moth

**63**

a *Sindris magnifica*

b *Arctioblepis rubida*

c *Midila quadrifenestrata*

d *Daulia aurantialis*

e *Cadarema sinuata*

f *Hypsidia erythropsalis*

g *Dracaenura stenosoma*

h *Orybina flaviplaga*

j *Siga liris*

k *Diaphania superalis*

l *Neadeloides cinerealis*

m *Parotis laceritalis*

n *Plagerepne torquata*

p *Ramphidium gigantalis*

q *Tamyra penicillana*

r *Ethopia roseilinea*

64

65

66

67

68

69

70

**64**

*Eudonia mercurella*

**65**

*Scoparia arundinata*

**66**

*Cataclysta lemnata*
Small China-mark

**67**

*Parapoynx stagnata*
Beautiful China-mark

**68**

*Nymphula nympheata*
Brown China-mark

**69**

*Pyrausta purpuralis*

**70**

*Parapoynx stratiotata*
Ringed China-mark

71
*Eurrhypara hortulata*
Small Magpie Moth

72
*Pleurotypa ruralis*
Mother-of-Pearl

73
*Anania funebris*

74
*Capperia britanniodactyla*

75
*Syngamia florella*

71

72

73

74

75

76

77

8

79

81

76

*Emmelina monodactyla*

77

*Amblyptilia punctidactyla*

78

*Platyptilia ochrodactyla*
Ochreous Plume Moth

79

a *Euschemon rafflesia*
Regent Skipper

b *Megathymnus yuccae*
Yucca Skipper

80

a *Abantis paradisea*
Paradise Skipper

b *Coeliades forestan*
Striped Policeman

c *Argopteron aureipennis* ◒
Gold Skipper

d *Allora doleschalli*
Peacock Awl

81

*Carterocephalus palaemon*
Arctic OR Chequered Skipper

82

*Spialia rebeli*

83

*Thymelicus acteon*
Lulworth Skipper

84

*Thymelicus lineola*
Essex or European Skipper

85

*Erynnis tages*
Dingy Skipper

86

*Pyrgus malvae*
Grizzled Skipper

87

*Ochlodes venatus*
Large Skipper

82

83

84

85

86

87

88

89

o

91

93

**88**

*Muschampia tessellum*
Tessellated Skipper

**89**

*Metisella orientalis*

**90**

*Mimionades versicolor*

**91**

*Ornithoptera priamus poseidon*

**92**

*Papilio machaon*
Swallowtail

**93**

*Iphiclides podalirius*

**94**

a *Baronia brevicornis*

b *Archon appollinus*
False Apollo

c *Hypermnestra helios*

d *Parnassius eversmanni*

e *Parnassius hardwickei*
Common Blue Apollo

f *Parnassius charltonius*
Regal Apollo

g *Serecinus telamon*

h *Luehdorfia japonica*

j *Allancastria cerisyi*
Eastern Festoon

k *Zerynthia rumina*
Spanish Festoon

l *Parnassius clodius*

m *Lamproptera meges*
Green Dragontail

n *Parnassius phoebus*
Small Apollo

**94**

**95**

**96**

**95**

*Parides tros*

**96**

a *Teinopalpus imperialis* Kaiser-I-Hind Butterfly

b *Bhutanitis thaidina*

c *Parides antenor*

**97**

- a *Eurytides protesilaus* Swordtail
- b *Euryades duponchelii*
- c *Graphium macleayanus* Macleay's Swallowtail
- d *Eruytides crassus*
- e *Eurytides pausanias*
- f *Eurytides marchandii*
- g *Graphium codrus*

**98**

*Papilio demetrius*

**99**

a *Graphium cloanthus*

b *Graphium eurypylus*
Pale Green Triangle Butterfly

c *Graphium agamemnon*
Tailed Jay OR Green-spotted Triangle Butterfly

d *Graphium cyrnus*

e *Graphium antiphates*
Fivebar Swordtail

f *Graphium colonna*
Mamba OR Black Swordtail

g *Papilio clytia*
Common Mime

h *Papilio cresphontes*
Orange Dog, Giant Swallowtail

j *Papilio paradoxa*
Great Blue Mime

k *Papilio slateri*

**99**

a *Papilio zalmoxis* Giant Blue Swallowtail

b *Papilio antimachus* African Giant Swallowtail

c *Troides prattorum*

**101**

*Papilio troilus, Papilio glaucus (3)*
Spicebush Swallowtail, Tiger Swallowtail

**102**

*Parnassius apollo*
Apollo

**101**

**102**

**103**

a *Papilio anactus*
Dingy Swallowtail

b *Papilio polymnestor*

c *Papilio bianor*

d *Papilio paris*
Paris Peacock

e *Papilio demoleus*
Chequered OR Lime Swallowtail

f *Papilio blumei*

g *Papilio ophidicephalus*
Emperor Swallowtail

**104**

a *Ornithoptera allottei*

b *Ornithoptera alexandrae* Queen Alexandra's birdwing

c *Troides trojana*

**105**

a *Papilio montrouzieri*

b *Papilio dardanus* ♂
Mocker Swallowtail

c *Graphium ridleyanus*
Acraea Swallowtail

d *Papilio dardanus* ♀
Mocker Swallowtail

e *Papilio rex*
Regal Swallowtail

f *Papilio dardanus* ♀
Mocker Swallowtail

g *Papilio laglaizei*

h *Dabasa payeni*

**106**

a *Papilio zagreus*

b *Papilio nobilis*

c *Papilio torquatus* ♀

d *Papilio aegeus* Orchard Swallowtail

e *Papilio torquatus* ♂

f *Papilio euterpinus*

g *Papilio hesperus*

**107**

a *Ornithoptera victoriae* Queen Victoria's Birdwing

b *Ornithoptera paradisea* Paradise OR Tailed Birdwing

c *Ornithoptera croesus*

d *Ornithoptera aesacus*

e *Ornithoptera goliath* Goliath Birdwing

f *Troides helena* Common Birdwing

**108**

*Graphium sarpedon*
Blue Triangle OR
Common Bluebottle
Butterfly

**109**

*Zerynthia polyxena*
Southern Festoon

**110**

*Eurytides marcellus*
Zebra Swallowtail

**108**

**109**

**110**

a *Cressida cressida*
Big Greasy Butterfly

b *Parides philoxenus*
Common Windmill

c *Parides coon*
Common Clubtail

d *Pachliopta aristolochiae*
Common Rose

e *Pachliopta hector*
Common Rose

f *Parides priapus*

a *Parides ascanius*
Ascanius Swallowtail

b *Parides aeneas*

c *Parides hahneli*

d *Parides zacynthus*

e *Parides neptunus*

f *Battus streckerianus*

g *Battus laodamas*

h *Battus philenor*
Pipe-vine OR Blue
Swallowtail

113

114

115

116

117

118

113

*Papilio phorcas*
Green-patch Swallowtail

114

*Papilio zoroastres*

115

*Graphium weiskei*
Purple Spotted Swallowtail

116

*Leptidea sinapis*
Wood White

117

*Pieris brassicae*

Large OR Cabbage White

118

*Gonepteryx rhamni*
Brimstone

**119**

a *Pereute leucodrosime*

b *Catastica uricoecheae*

c *Archonias bellona*

d *Cepora aspasia*

e *Pontia callidice*
Peak, Checkered OR Common White

f *Ixias reinwardti*

g *Anaphaeis java* ♀
Caper White

h *Colias hyale*
Pale Clouded Yellow

j *Anaphaeis java* ♂
Caper White

k *Colias crysotheme*
Lesser Clouded Yellow

l *Colotis eris*
Banded Gold Tip

m *Colotis zoe*
Purple Tip

n *Catopsilia pomona*
Lemon Migrant

p *Ixias vollenhoferi*

q *Colotis danae*
Crimson Tip

r *Appias nero*
Orange Albatross

s *Enantia licina*
White Dismorphia

t *Dismorphia amphione*
Tiger Pierid

120

*Catopsilia florella*
African Migrant

**121**

*Colias crocea*
Clouded Yellow

**122**

*Pieris rapae*
Small OR Garden
White Cabbage
Butterfly OR
Imported Cabbageworm

**123**

*Pieris napi*
Green-veined OR
Mustard White

**121**

**122**

**123**

*Anthocharis cardamines*
Orange-tip

**125**
*Aporia crataegi*
Black Veined White

**126**
*Amauris niavius*
Friar Butterfly

125

126

**127**

a *Danaus gilippus* Queen Butterfly

b *Danaus chrysippus* Plain Tiger OR African Monarch

c *Amauris albimaculata* Layman Butterfly

d *Danaus limniace* Blue Tiger OR Blue Monarch

e *Danaus schenckii*

f *Danaus sita* Chestnut Tiger

g *Danaus pumila*

h *Phoebus avellaneda* ♂

j *Phoebus avellaneda* ♀

k *Delias belisama*

l *Phoebus philea* Orange-barred Sulphur OR Yellow Apricot

m *Phoebus rurina*

**128**

*Danaus plexippus*
Monarch, Milkweed or Wanderer Butterfly

**129**

*Danaus plexippus*
Monarch, Milkweed or Wanderer Butterfly

**128**

**129**

**130**

a *Ideopsis vitrea*

b *Ideopsis daos*

c *Lycorea ceres*
Large Tiger

d *Euploea leucostictos*
Blue-branded King Crow

e *Euploea mulciber*
Striped Blue Crow

f *Euploea core kalaona*
Common Indian Crow, Australian Crow OR Oleander Butterfly

g *Idea idea*

h *Euploea diocletiana*
Magpie Crow

**131**

- a *Tithorea harmonia* Tiger Butterfly
- b *Elzunia pavonii*
- c *Melinaea mneme*
- d *Olyras insignis*
- e *Scada karschina*
- f *Hypothyris oulita*
- g *Thyridia confusa*
- h *Tellervo zoilus*
- j *Napeogenes larilla*
- k *Aeria agna*
- l *Oleria synnova*
- m *Mechanitis egaensis*
- n *Napeogenes corena*
- p *Greta quinta*

132

133

134

**132**
*Heliconius nattereri*

**133**
*Hypoleria oreas*

**134**
a *Melinaea lilis imitata*
b *Melinaea lilis messatis*

**135**
*Heliconius erato*
Small Postman

**136**
*Heliconius erato*
Small Postman

**137**
*Heliconius hermathena*

**138**
*Dryas iulia delila*

**139**
*Heliconius charitonius simulator*
Zebra

**140**
*Eueides lybia olympia*

135

136

137

138

139

140

141

*Heliconius astraea rondonia*

a *Heliconius ethillus*
Rare Tiger

b *Heliconius ricini*

c *Heliconius doris*
Doris Butterfly

d *Heliconius wallacei*
Blue Grecian

e *Eueides isabella*

f *Heliconius doris*
Doris Butterfly

g *Heliconius anderida*

h *Heliconius vulcanus*

j *Podotricha telesiphe telesiphe*

k *Dryadula phaetusa*
Caroni Flambeau

143

144

145

146

47

143

*Agraulis vanillae vanillae*
Gulf Fritillary
OR Silver-spotted
Flambeau

144

*Heliconius sapho*

145

*Heliconius erato cyrbia*
Small Postman

146

*Heliconius melpomene*
Postman

147

*Heliconius melpomene cytherea*
Postman

**148**

*Agraulis vanillae vanillae* ◒
Gulf Fritillary OR
Silver-spotted
Flambeau

**149**

*Philaethria werneckei*

148

149

**150**

*Dynastor napoleon*

**151**

*Dynastor napoleon* ◒

**152**

*Caligo prometheus*

**153**

*Caligo prometheus* ◒

50

151

2

153

154
*Dismorphia melia*

155
*Caligo idomeneus*
Owl Butterfly

156
*Acraea uvui*
Tiny Acraea

157
*Acraea bonasia*

158
*Morpho aega*

159
*Cithaerias aurora*

154

155

156

157

158

159

**160**

a *Taenaris schoenbergi*

b *Hyantis hodeva*

c *Zeuxida amethystus*
Saturn Butterfly

d *Kallima inachus*
Orange Oak Leaf OR
Indian Leaf Butterfly

**161**

a *Morpho portis*
Sky Blue Morpho

b *Morpho zephyrites*

c *Morpho cypris*
Blue Morpho

160

161

a *Morpho rhetenor*
Blue Morpho

b *Morpho laertes*

c *Morpho menelaus*
Blue Morpho

d *Morpho didius*

e *Morpho deidamia*

f *Morpho perseus*

**163**

a *Acraea chilo*

b *Araschnia levana*
Map Butterfly

c *Vagrans egista*
Australian Vagrant Butterfly

d *Bematistes aganice*
Wanderer

e *Chlorosyne nycteis*
Silvery Checkerspot

f *Speyeria cybele*
Great Spangled Fritillary

g *Clossiana frigga*
Saga OR Frigga's Fritillary

h *Nymphalis vau-album*
False Comma OR Compton Tortoiseshell

j *Phyciodes tharos*
Pearl Crescent

k *Siderone galanthis*

l *Acraea cephus*

m *Actinote ozomene*

n *Polygonia interrogationis*
Question Mark

**164**

*Morpho peleides* ◒

**165**

*Brintesia circe*
Great Banded
Grayling

**166**

*Morpho peleides*

**167**

*Pyronia cecilia*
Southern Gatekeeper

**164**

**165**

**166**

**167**

168

169

170

171

72

173

**168**

*Pyronia tithonus*
Gatekeeper

**169**

*Coenonympha pamphilus*
Small Heath

**170**

*Aphantopus hyperantus*
Ringlet

**171**

*Coenonympha tullia*
Large Heath

**172**

*Coenonympha pamphilus*
Small Heath

**173**

*Erebia epiphron*
Mountain Ringlet

a *Haetera macleannania*

b *Mycalesis terminus*
Orange Bush Brown

c *Elimniopsis lise*

d *Cithaerias esmeralda*

e *Cepheuptychia cephus*
Blue Night Butterfly

f *Cepheuptychia cephus* ◒
Blue Night Butterfly

g *Erebia rossii*
Ross's Alpine Butterfly

h *Mandarina regalis*

j *Melanitis leda*
Evening Brown OR
Twilight Brown

k *Meneris tulbaghia*
Mountain Beauty OR
Mountain Pride

l *Magneuptychia tricolor*

m *Tarsocera southeyae*

n *Argyrophorus argenteus*
Silver Butterfly

p *Enodia portlandia*
Pearly Eye Butterfly

q *Orinoma damaris*

r *Lethe europa*
Bamboo Tree Brown

175

176

177

178

179

180

175
*Coenonympha leander*
Russian Heath

176
*Lasiommata megera*
Wall Butterfly

177
*Maniola jurtina*
Meadow Brown

178
*Junonia stygia*

179
*Melanargia larissa*
Balkan Marbled White

180
*Melanargia galathea*
Marbled White

**181**

*Erebia aethiops*
Scotch Argus

**182**

*Pararge aegeria*
Speckled Wood

**183**

*Junonia sophia*
Little Commodore

**184**

*Junonia westermanni*
Blue-spot Commodore

181

182

183

184

**185**

- a *Libythea celtis* Nettle-tree OR European Beak Butterfly
- b *Junonia orithya* Blue Pansy OR Blue Argus
- c *Junonia rhadama*
- d *Precis octavia*
- e *Antanartia schaenia* Long Tailed Admiral
- f *Junonia lavinia* Buckeye
- g *Precis octavia*
- h *Symbrenthia hypselis* Himalayan Jester Butterfly
- j *Junonia villida* Meadow Argus
- k *Symbrenthis hypselis* ◒ Himalayan Jester Butterfly
- l *Terinos clarissa*
- m *Cethosia myrina*

**186**

*Issoria lathonia*
Queen of Spain Fritillary

**187**

*Mellicta athalia*
Heath Fritillary

**188**

*Pierella hyceta*

**186**

**187**

**188**

**189**

*Euphydryas aurinia*
Marsh Fritillary

**190**
*Melitaea didyma*
Spotted Fritillary

**191**
*Melitaea cinxia*
Glanville Fritillary

**192**
*Pandoriana pandora*
Cardinal Butterfly

**193**
*Fabriciana adippe*
High Brown
Fritillary

**194**
*Argynnis paphia*
Silver-washed
Fritillary

190

191

192

193

194

**195**

*Clossiana selene*
Silver-bordered OR
Small Pearl-bordered
Fritillary

**196**

*Aglais urticeae*
Small Tortoiseshell

**196**

97

98

197

*Nymphalis antiopa*
Mourning Cloak, White-bordered OR Camberwell Beauty

198

*Inachis io*
Peacock

**199**

*Apatura iris*
Purple Emperor

**200**

*Cynthia cardui*
Painted Lady

**199**

**200**

201

202

203

204

205

206

207

**201**
*Clossiana titania*
Purple Lesser OR
Titania's Fritillary

**202**
*Nymphalis polychloros*
Large Tortoiseshell

**203**
*Vanessa atalanta*
Red Admiral

**204**
*Polygonia c-album*
Comma

**205**
*Polygonia c-album* ◒
Comma

**206**
*Euphaedra medon*

**207**
*Anartia amathea*
Coolie Butterfly

a *Hypolimnas dexithea*

b *Hypolimnas bolina*
Common or Great Eggfly

c *Hypolimnas pandarus*

d *Kallima jacksoni*
Jackson's Leaf Butterfly

e *Salamis anacardii*
Clouded Mother-of-pearl

f *Laringa castelnaui*

g *Stibochiona coresia*

h *Hypolimnas salmacis*

j *Hypolimnas misippus* ♀
Six Continent Butterfly, Danaid Eggfly or Diadem Butterfly

k *Hypolimnas alimena*
Blue-banded Eggfly

a *Hamadrayas feronia*
Cracker

b *Hamadryas arethusia*
Queen Cracker

c *Nessaea hewitsonii*

d *Nessaea obrinus*

e *Eunica alcmena*

f *Nessaea obrinus* ◒

g *Amnosia decora*

h *Panacea procilla*

j *Hamadryas velutina*

k *Baeotus baeotus*

*Metamorpha stelenes*
Malachite OR Bamboo
Page Butterfly

211

212

213

214

**211**

*Kallima rumia*
African Leaf Butterfly

**212**

*Kallima rumia* ◒
African Leaf Butterfly

**213**

*Salamis temora*
Blue Salamis

**214**

*Salamis temora* ◒
Blue Salamis

**215**

*Callicore sorona*

215

a *Batesia hypochlora*

b *Neptis sappho*
Common Glider OR
Common Sailer

c *Pantoporia eulimene*

d *Euthalia adonia* ♀

e *Limenitis arthemis*
White Admiral OR
Purple Banded
Butterfly

f *Euthalia adonia* ♂

g *Cymothoe sangaris*
Blood Red Cymothoe

h *Doxocopa lavinia*

j *Marpesia iole*

k *Tanaecia pelea*
Malay Viscount
Butterfly

l *Cymothoe coccinata*

m *Marpesia petreus*
Ruddy Dagger Wing
OR Tailed Flambeau

n *Limenitis archippus*
Viceroy OR Mimic

p *Asterocampa clyton*
Tawny Emperor

q *Parthenos sylvia*
Clipper Butterfly

a *Panacea prola*

b *Asterope rosa* ◒
Lilac Nymph

c *Asterope rosa*
Lilac Nymph

d *Panacea prola* ◒

e *Callicore astarte* ◒
'B.D.' Butterfly

f *Myscelia cyaniris*

g *Biblis hyperia*
Red Rim

h *Callicore sorona*

j *Myscelia orsis*

k *Dynamine mylitta* ♀
Large Dynamine Butterfly

l *Callicore cynosura* ◒

m *Eunica sophonisba* ♂

n *Diaethria aurelia* ◒
'89' Butterfly

p *Dynamine mylitta* ♂
Large Dynamine Butterfly

q *Eunica sophonisba* ♀

**218**

*Neptis melicerta*

**219**

*Agrias claudia godmani*

**220**

*Ladoga camilla*
White Admiral

**221**

*Diaethria clymena*
Figure of Eight OR
Jewel Butterfly

218

219

220

221

222
*Anaea otrere*

223
*Salamis parhassus*
Mother-of-pearl Butterfly

224
*Marpesia coresia*
Waiter Butterfly

225
*Cymothoe herminia johnstoni*

226
*Colobura dirce*
Zebra Butterfly

227
*Charaxes pollux*
Black Bordered Charaxes

228
*Euphaedra eleus*
Orange Forester

222

223

224

225

226

227

228

**229**

a *Catonephele numilea* ♂
Grecian Shoemaker

b *Catonephele numilea* ♀
Grecian Shoemaker

c *Marpesia marcella*

d *Marpesia camillus*
African Porcelain OR
African Map Butterfly

e *Cyrestis nivea*

f *Epiphile hubneri*

g *Catonephele acontius* ♂
Orange-banded
Shoemaker

h *Catonephele acontius* ♀
Orange-banded
Shoemaker

j *Perisama vaninka* ◓

k *Perisama vaninka*

l *Callisthenia salvinii* ◓

m *Callisthenia salvinii*

n *Callisthenia markii*

p *Perisama oppelii*

q *Perisama bonplandi* ◓

r *Perisama oppelii* ◓

s *Diaethria clymena*
Figure of Eight OR
Jewel Butterfly

t *Callicore mionina*

u *Callicore maimuna*
Eighty-eight
Butterfly

a *Euthalia patala*
Grand Duchess

b *Anaea panariste*

c *Prepona praeneste*

d *Anaea andria*
Goat-weed Butterfly

e *Euthalia durga*
Blue Duke

f *Sasakia charonda*

g *Anaea tyrianthina*

h *Anaea nessus*

j *Agrias claudia*

**231**

a *Cymothoe lucasi*

b *Metamorpha epaphus*

c *Euxanthe tiberius*

d *Rhinopalpa polynice*
Wizard

e *Symphaedra nais*
Baronet Butterfly

f *Neurosigma doubledayi*

g *Rhinopalpa polynice* ◒
Wizard

h *Pseudacraea boisduvalii*
Trimen's False Acraea

j *Euphaedra spatiosa*

k *Yoma sabina*
Lurcher

l *Euphaedra eleus*
Orange Forester

232
*Hypna clytemnestra forbesi*

233
*Doxocopa agathina vacuna*

234
*Prepona deiphile*

235
*Siderone marthesia*

232

233

234

235

**236**

a *Hestinalis assimilis*

b *Prepona meander*
Branded King
Shoemaker

c *Consul hippona*

d *Polyura eudamippus* ♀
Great Nawab

e *Polyura eudamippus* ♂
Great Nawab

f *Polyura pyrrhus*
Tailed Emperor

g *Coenophlebia archidona*

h *Polyura dolon*
Stately Nawab

**237**

a *Charaxes ameliaea*

b *Charaxes bohemani*
Large Blue Charaxes

c *Charaxes eupale*
Green Charaxes

d *Zingha zingha*

e *Charaxes jasius*
Two-tailed Pasha

f *Euphaedra eusemoides*

g *Euphaedra themis*

h *Euphaedra neophron*
Gold-banded Forester

j *Euphaedra francina*

38

239

40

241

2

243

**238**

*Lycaena dispar batavus*
Dutch Large Copper

**239**

*Lycaena phlaeas*
Small Copper

**240**

a *Aurea aurea* ♂

b *Iolaus lalos* ♀
Pale Sapphire

c *Aurea aurea* ♀

d *Iolaus lalos* ♂
Pale Sapphire

**241**

*Liphyra brassolis*
Moth-Butterfly

**242**

a *Aphnaeus hutchinsoni* ♂ ◒
Silver-spot

b *Danis danis* ♀ ◒
Large Green-banded Blue

c *Aphnaeus hutchinsoni* ♂
Silver-spot

d *Danis danis* ♂
Large Green-banded Blue

**243**

a *Eumaeus atala* ♂ ◒
Atala Butterfly

b *Theorema eumenia* ♀

c *Eumaeus atala* ♂
Atala Butterfly

d *Theorema eumenia* ♂

a *Pentila tropicalis*
Spotted Buff

b *Mimacraea marshallii*
Marshall's Acraea Mimic

c *Poritia erycinoides*
Blue Gem

d *Feniseca tarquinius*
Harvester

e *Thecla coronata*

f *Evenus regalis* ◒
Bank-note Blue

g *Narathura micale*
Common Oak Blue

h *Hypochrysops apelles* ◒
Copper Jewel

j *Hypochrysops theon* ◒

k *Curetis thetis*
Indian Sunbeam

l *Ogyris ianthis*
Sydney Azure

m *Amblypodia acron*

n *Jalmenus evagoras*
Common Imperial Blue

p *Loxura atymnus*
Yamfly

q *Myrina silenus*
Figtree Blue

r *Neomyrina hiemalis*
White Imperial

s *Chrysozephyrus ataxus*
Wonderful Hairstreak

t *Poecilmitis thysbe*
Opal-centred Copper

u *Ogyris aenone*

v *Tajuria cippus*
Peacock Royal Butterfly

w *Suasa lisides*
Red Imperial

x *Jacoona anasuja*

y *Hypolycaena danis*
Black and White Tit

z *Zeltus amasa*
Fluffy Tit

aa *Virachola perse*

bb *Bindhara phocides*
Plane Butterfly

cc *Rapala jarbus*
Common Red Flash

245

**245**

*Plebejus argus*
Silver-studded Blue

**246**

*Palaeochrysophanus hippothoe*
Purple-edged Copper

**247**
*Cupido minimus*
Little Blue

**248**
*Lampides boeticus*
Lucerne Blue

**249**
*Celestrina argiolus*
Holly Blue OR Spring Azure

**250**
*Lysandra coridon*
Chalkhill Blue

**251**
*Polyommatus icarus*
Common OR Violet Meadow Blue

**252**
*Aricia agestis*
Brown Argus

247

248

249

250

251

252

253

254

255

256

**253**
*Wagimo signata*

**254**
*Philotes battoides*

**255**
*Hypochrysops pythias*

**256**
*Callophrys rubi*
Green Hairstreak

**257**
*Thecla betulae*
Brown Hairstreak

**258**
*Quercusia quercus*
Purple Hairstreak

**259**
*Strymonidia w-album*
White Letter
Hairstreak

**260**
- a *Freyeria trochylus*
  Grass Jewel
- b *Brephidium exilis*
- c *Syrmatia dorilas*
  White-spotted Tadpole
  Butterfly
- d *Calephelis virginiensis*
  Little Metalmark

**261**
*Phylaria cyara* (2), *Phylaria heritsia*

**262**
*Uranothauma falkensteini*

257

258

259

260

261

262

**263**

a *Mithras hemon*
Black-backed Blue

b *Rekoa meton*
Mother-of-pearl

c *Atlidas halesus*
Great Purple Hairstreak

d *Japonica saepestriata*

e *Lycaena alciphron*
Purple-shot Copper

f *Lycaena thetis*
Golden Copper

g *Heliophorus androcles*
Green Sapphire

h *Heliophorus tamu* ◒

j *Jamides alecto*
Metallic Cerulean

k *Drupadia ravindra*
Common Posy

l *Philotes sonorensis*
Sonora Blue

m *Castalius rosimon*
Common Pierrot

n *Glaucopsyche alexis* ◒
Green-underside Blue

p *Euselasia euriteus*

q *Euselasia euriteus* ◒

r *Euselasia issoria* ◒

s *Cremna actoris*
Speckled Black Nymph

t *Heliocopis endymion*
Six-tailed Helicopsis

u *Menander menander*
Blue Tharops

v *Menander hebrus*

w *Uraneis ucubris*

x *Hyphilaria parthenis*

y *Semomesia capanea*

z *Mesosemia mevania*

aa *Lymnas pixe*

bb *Lymnas xarifa*
Orange-tipped Underleaf Butterfly

cc *Hermathena candidata*
White Erycid

dd *Alesa prema*
Green Dragon

**264**

*Heliocopis acis*

**265**

a *Stalachtis calliope*

b *Stalachtis calliope*

**266**

a *Mesenopsis albivitta*

b *Stalachtis phaedusa*

c *Mesene margaretta*

d *Esthemopsis clonia*

**267**

*Hamearis lucina*
Duke of Burgundy Fritillary

**264**

**265**

**266**

**267**

**268**

a *Ancycluris formosissima* ◒

b *Ancycluris formosissima*

c *Rhetus arcius*

d *Ithomeis eulema*

e *Ithomeis corena*

f *Chorinea faunus*
Bee Butterfly

g *Stalachtis euterpe*

h *Amarynthis meneria*

j *Caria mantinea*

k *Esthemopsis thyatira* ♀

l *Esthemopsis thyatira* ♂

m *Nymphidium onaeum*
Yellow-banded Nymphidium

n *Lasaia sessilis*

p *Polystichtis siaka*

q *Theope pieridoides*
White Theope Butterfly

r *Styx infernalis*

s *Echenais alector*
White-tailed Erycid

t *Lyropteryx apollonia* ♂

u *Lyropteryx apollonia* ♀

v *Apodemia nais*
Nais Metalmark

**269**

*Habrosyne pyritoides*
Buff Arches

**270**

*Cilix glaucata*
Chinese Character

**271**

*Thyatira batis*
Peach-blossom

**269** C. J.

**270**

**271** J.

**272**

a *Pterothysanus noblei*

b *Hibrildes novax*

c *Dysphania cuprina*

d *Callidula lunigera*

e *Epiplema himala*

f *Apoprogones hesperistis*

g *Macrauzata maxima*

h *Epicmelia theresiae*

j *Habrosyne scripta*

k *Oreta singapura*

l *Tridrepana flava*

m *Oreta rosea*

n *Axia margarita*

p *Cyclidia dictyaria*

q *Erebomorpha fulguritia*

r *Drepana falcataria*
Pebble Hook-tip

s *Catocalopsis medinae*

t *Percnia felinaria*

u *Pterodecta felderi*

v *Carpella districta*

**273**

*Geometra papilionaria*
Large Emerald

**274**

*Colostygia pectinataria*
Green Carpet

**275**

*Rhodometra sacraria*
Vestal Moth

**276**

*Camptogramma bilineata*
Yellow-shell Moth

**277**

*Catarhoe rubidata*
Flame Moth

**278**

*Idaea muricata*
Purple-bordered Gold

**273**

**274**

**275**

**276**

**277**

**278**

279

*Rheumaptera hastata*
Spear-marked OR Argent and Sable Moth

280

*Biston strataria*
Oak Beauty

281

*Rheumaptera undulata*
Scallop Shell

279

280

281

282

*Operophtera brumata* ♀
Winter Moth

283

*Operophtera fagata* ♂
Northern Winter Moth

284

*Eulithis pyraliata*
Barred Straw Moth

285

*Cleora cinctaria*
Ringed Carpet

286

*Thera juniperata*
Juniper Carpet Moth

287

*Chesias rufuta*
Broom-tip

282

283

284

285

286

287

88

289

90

291

92

293

**288**

*Eupithecia centaureata*
Lime-speck Pug Moth

**289**

*Eupithecia venosata*
Nettle Pug

**290**

*Perizoma sagittata*
Marsh Carpet

**291**

*Plagodis dolabraria*
Scorched Wing

**292**

*Odezia atrata*
Chimney-sweeper

**293**

*Venusia cambrica*
Welsh Wave OR Cambric Wave Moth

a *Angerona prunaria*
Orange Moth

b *Agathia arcuata*

c *Euchlaena irraria*

d *Erannis defoliaria*
Mottled Umber

e *Ophthalmophora branickiaria*

f *Rhodophthitus simplex*

g *Osteosema sanguilineata*

h *Plutodes discigera*

j *Corymica vesicularia*

k *Anisozyga stellifera*

l *Amnemopsyche sanguinaria*

m *Casbia fasciata*

n *Stamnodes danilovi*

p *Edule ficulnea*

q *Erateina lineata*

r *Pityeja histrionaria*

s *Erateina julia*

t *Phyle arcuosaris*

u *Rhodophthitus roseovittata*

v *Agathia succadanea*

w *Xanthabraxas hemionata*

x *Chorodnodes rothi*

y *Iotaphora admirabilis*

295

*Epione paralellaria* ♀ ♂
Dark Bordered Beauty

**296**

- a *Milionia pericallis*
- b *Milionia grandis*
- c *Milionia elegans*
- d *Milionia basalis*
- e *Milionia euchromozona*
- f *Milionia obiensis*
- g *Milionia paradisea*
- h *Milionia exultans*
- j *Milionia aroensis*
- k *Milionia aglaia*
- l *Milionia ovata*
- m *Milionia callima*
- n *Milionia plesiobapta*
- p *Milionia coeruleonitens*
- q *Milionia lamprima*
- r *Arycanda emolliens*
- s *Callioratis milliaria*
- t *Arycanda vinaceostriga*
- u *Callipia paradisea*
- v *Eumelea fumicosta*
- w *Gonora hyelosioides*

**296**

**297**

*Crocallis elinguaria*
Scalloped Oak

**298**

*Ennomos autumnaria*
Large Thorn Moth

**299**

*Ematurga atomaria*
Common Heath

**300**

*Campaea margaritata*
Light Emerald

97

298

9

300

**301**

*Abraxas grossulariata*
Magpie OR Currant Moth

**302**

*Biston betularia*
Pepper-and-salt Moth
OR Peppered Moth

**303**

*Biston betularia*
Pepper-and-salt Moth
OR Peppered Moth

**304**

*Abraxas sylvata*
Clouded Magpie

**301**

**302**

**303**

**304**

305

306

307

308

**305**

*Thalaina angulosa*

**306**

*Ourapteryx sambucaria*
Swallow Tail Moth

**307**

*Selidosema brunnearia*
Bordered Grey

**308**

*Erannis defoliaria* ♀
Mottled Umber

**309**

*Opisthograptis luteolata*
Brimstone Moth

**310**

*Pseudopanthera macularia*
Speckled Yellow

**311**

*Peribatodes rhomboidaria*
Willow Beauty

**312**

*Bupalis piniaria*
Bordered White OR
Pine Looper

309

310

311

312

313

*Selenia tetralunaria*
Purple Thorn

**314**

a *Lyssa patroclus*

b *Alcides agathyrus*

c *Urania sloanus*

d *Coronidia orithea*

e *Chrysiridia riphearia*

f *Epicopeia polydora*

g *Nothus lunus*

315

16

315
*Urania leilus*

316
*Alcides zodiaca*

**317**

- a *Alcides aurora*
- b *Cyphura pardata*
- c *Strophidia fasciata*
- d *Anthela nicothoe*
- e *Bombyx mori* The Common Silk-moth, Silkworm
- f *Carthaea saturnioides*
- g *Bombyx mandarina*
- h *Phiala cunina*
- j *Acrojana rosacea*

**318**

a *Oxytenis peregrina*

b *Asthenidia lactucina*

c *Spiramiopsis comma*

d *Sabalia barnsi*

e *Lemonia dumi*

f *Cercophana venusta*

g *Epia muscosa*

h *Hygrochora torrefacta*

j *Ratarda furvivestia*

k *Cicinnus despecta*

l *Lacosoma valeria*

m *Grammodora nigrolineata*

n *Dendrolimus pini*
Pine Lappet Moth

p *Mirina christophi*

**319**
*Tolype glenwoodi*

20

321

22

323

4

325

320

*Endromis versicolora*
Kentish Glory

321

*Gastropacha quercifolia*
Lappet Moth

322

*Malacosoma castrensis*
Ground Lackey

323

*Macrothylacia rubi*
Fox Moth

324

*Philudoria potatoria*
Drinker

325

*Lasiocampa quercus*
Oak Eggar Moth

**326**

*Athletes steindachneri*
Steindachner's Emperor

**327**

*Attacus atlas*
Atlas Moth

326

327

328

329

330

328

*Hyalophora euryalus*
Ceanothus or
California Silkmoth

329

*Callosamia promethea*
Promethea Moth

330

*Antheraea polyphemus*
Polyphemus Moth

**331**

a *Hemileuca electra*

b *Micragone ansorgei*
Ansorge's Prince Moth

c *Ludia orinoptena*

d *Hemileuca hera*
Sagebrush Sheep-moth

e *Hemileuca tricolor*
Tricolor Buck

f *Holocerina angulata*

g *Vegetia dewitzi*
Dewitz's Princeling

h *Saturnia mendocino*

j *Eubergia boetifica*

k *Aglia tau*
Tau Emperor

l *Goodia kuntzei*
Lunar Prince

m *Automeris io*
Io Moth

n *Pseudaphelia apollinaris*
Apollo Moth

**331**

332

*Argema mimosae*
African Moon-moth

**333**

a *Ubaena fulleborniana*
Mourning Emperor

b *Samia cynthia*
Ailanthus Silk-moth,
Cynthia Moth

c *Eudaemonia brachyura*

d *Bunaea alcinoe*
Common Emperor

e *Citheronia splendens*

f *Eacles oslari*

334

335

336

334

*Saturnia pavonia*
Emperor Moth

335

*Rhodinia fugax*
Squeaking Silk-moth

336

*Actias selene*
Indian Moon-moth

**337**

a *Antheraea pernyi*
Chinese Oak Silk-moth

b *Graellsia isabellae*

c *Rothschildia zacateca*

d *Heliconisa pagenstecheri*

e *Loxolomia serpentina*

f *Lonomia achelous*

338

339

**338**

*Eochroa trimeni*
Roseate Emperor

**339**

*Hyalophora cecropia*
Cecrophia or Robin Moth

**340**

- a *Actias maenas*
- b *Acanthobrahmaea europaea*
- c *Saturnia pyri* Great Peacock Moth, Viennese Emperor
- d *Brahmaea wallichii*
- e *Calliprogonos miraculosa*
- f *Bunaeopsis jacksoni* Jackson's Emperor
- g *Dactyloceras widenmanni*

**340**

**341**
*Hemaris fuciformis*

**342**
*Acherontia atropos*
Death's Head Hawk-moth

**343**
*Cechena mirabilis*

41

42

343

a *Daphnis nerii*
Oleander Hawk-moth

b *Arctonotus lucidus*

c *Rethera komarovi*

d *Macroglossum stellatarum*
Humming-bird Hawk-moth

e *Microsphinx pumilum*
Bedstraw Hawklet

f *Xylophanes pluto*

g *Euchloron megaera*
Verdant Hawkmoth

h *Erinnyis ello*

j *Sphinx ligustri*
Privet Hawk-moth

k *Rhodosoma triopus*

l *Agrius cingulatus*
Sweet Potato Hornworm

m *Manduca quinquemaculata*
Tomato Hornworm

n *Hyles calida*

p *Hyles lineata*
Striped Hawk-moth OR
White-lined Sphinx

q *Hippotion celerio*
Silver-striped Hawk-moth

r *Tinostoma smaragditis*
Green Sphinx of Kauai

s *Sphecodina abboti*
Abbot's Sphinx Moth

t *Callionima parce*

u *Manduca ochus*

v *Proserpinus juanita*

45

346

347

48

349

o

351

**345**

*Agrius convolvuli*
Convulvulus Hawk-moth

**346**

*Mimas tiliae*
Lime Hawk-moth

**347**

*Smerinthus ocellata*
Eyed Hawk-moth

**348**

*Deilephila elpenor*
Elephant Hawk-moth

**349**

*Laothoe populi*
Poplar Hawk-moth

**350**

*Sphinx pinastri*
Pine Hawk-moth

**351**

*Hyles euphorbiae*
Spurge Hawk-moth

**352**

*Leucodonta bicoloria*
White Prominent

352

53

**354**

55

**356**

**353**

*Stauropus fagi*
Lobster Moth

**354**

*Pterostoma palpina*
Pale Prominent

**355**

*Phalera bucephala*
Buff-tip

**356**

*Cerura vinula*
Puss Moth

**357**

a *Tarsolepis sommeri*

b *Furcula rivera*

c *Moresa magniplaga*

d *Thaumetopoea processionea*

e *Danima banksiae*
Banksia Moth

f *Uropyia meticulodina*

g *Anaphe barrei*

h *Epicoma melanospila*

j *Spatalia argentina*

k *Galona serena*

l *Pheosia rimosa*

m *Automolis meteus*

n *Euproctoides acrisia*

p *Euproctis chrysorrhoea*
Brown-tail

q *Perina nuda*

r *Cozola collenettei*

s *Lymantria dispar*
Gypsy Moth

t *Lymantria monacha*
Black Arches

358

359

60

361

62

363

**358**

*Euproctis similis*
Yellow-tail OR
Gold-tail (U.S.)

**359**

*Orgyia antiqua* ♂ ♀
Vapourer OR Rusty
Tussock

**360**

*Dasychira pudibunda*
Pale Tussock

**361**

*Leucoma salicis*
White Satin OR
Satin Moth

**362**

a *Dioptis egla*

b *Scea steinbachi*

c *Josia lativitta*

**363**

*Chliara croesus*

a *Ctenucha virginica*
Virginian Ctenucha

b *Euchromia lethe*

c *Chrysocale ignita*

d *Dasysphinx torquata*

e *Gymnelia gemmifera*

f *Cyanopepla pretiosa*

g *Napata walkeri*
Barred-tip Yellow

h *Cosmosoma auge*

j *Setina irrorella*
Dew Moth

k *Eilema complana*
Scarce Footman

l *Chionaema scintillans*

m *Damias esthla*

n *Chionaeama moupinensis*

p *Spiris striata*
Feathered Footman

q *Premolis semirufa*

r *Utetheisa bella*
Beautiful Utetheisa
OR Bella Moth

365

366

367

**365**

*Syntomis phegea*

**366**

*Utetheisa pulchella*
Crimson Speckled Moth

**367**

a *Pompilopsis tarsalis*

b *Phaeosphecia opaca*

c *Lycomorpha pholus*

d *Pseudosphex crabronis*

e *Horama oedippus*

f *Macrocneme thyra*

g *Didasys belae*
Double-tufted Wasp Moth

h *Correbidia assimilis*
Yellow-banded Beetle Mimic

j *Callisthenia plicata*

**368**

a *Eucharia casta*

b *Amastus adela*

c *Diacrisia breteaudeaui*

d *Diacrisia metelkana*
Swamp Tiger Moth

e *Graphelysia strigillata*

f *Aemilia ambigua*
Red-banded Aemilia

g *Eupseudosoma involuta*
Snowy Eupseudosoma

h *Rhipha pulcherrima*

j *Amaxia chaon*

k *Viviennea moma*

l *Bertholdia trigona*

m *Ormetica tanialoides*

n *Pyrrharctia isabella*
Banded Woollybear

p *Rhodogastria caudipennis*

q *Parathyris semivitrea*

r *Halysidota schausi*
Pale Tussock

s *Halysidota caryae*
Hickory Tussock

t *Apocrisias thaumasta*

**369**

a *Pericallia matronula*
Large Tiger Moth

b *Phragmatobia fuliginosa*
Ruby Tiger Moth

c *Apantesis virgo*
Virgin Tiger Moth

d *Arachnis zuni*
Zuni Tiger Moth

e *Hyphantria cunea*
Fall Webworm

f *Chlanidophora patagiata*

g *Ecpantheria decora*

h *Ammobiota festiva*

j *Estigmene acrea*
Saltmarsh Caterpillar

k *Platarctia parthenos*
St Lawrence Tiger Moth

l *Amsacta lactinea*

m *Cymbalophora pudica*
Festooned Tiger Moth

n *Acerbia alpina*

p *Arctia flavia*
Yellow Tiger Moth

q *Diacrisia purpurata*

**370**
*Ryria jacobaeae*

**371**
*Diacrisia sannio*
Clouded Buff

**372**
*Diaphora mendica* ♀ ♂
Muslin Moth

**373**
*Arctia villica*
Cream-spot Tiger Moth

**374**
*Arctia caja*
Garden Tiger Moth

**375**
*Parasemia plantaginis*
Wood Tiger Moth

**370**

**371**

**372**

**373**

**374**

**375**

**376**

a *Dysschema mariamne*

b *Composia credula*

c *Dysschema tricolor*

d *Hypocrita aletta*

e *Calodesma maculifrons*

f *Phaloesia saucia*

g *Hypocrita dejanira*

h *Gnophaela arizonae*

j *Ephestris melaxantha*

k *Phaegorista similis*

l *Camptoloma interiorata*

m *Anaxita drucei*

n *Chetone histrio*

p *Chetone angulosa*

**377**
*Spilosoma lubricipeda*
White Ermine Moth

**378**
*Callimorpha dominula*
Scarlet Tiger Moth

**379**
*Euplagia quadripunctaria*
Jersey Tiger Moth

**380**
*Amphicallia pactolicus*

377

378

379

380

381

**381**

a *Psephea orbicularis*

b *Mesenochroa guatelmalteca*

c *Chetone phyleis*

d *Euplocia membliaria*

e *Hyalurga fenestra*

f *Notophyson heliconides*

g *Hyalurga osiba batesi*

h *Calodesma melanochroia*

j *Sagaropsis brevifasciata*

k *Darna colorata*

382

**382**

a *Apantesis quenseli*

b *Holomelina lamae*

c *Miltochrista miniata* Rosy Footman

d *Ocnogyna leprieuri* ♂

e *Ocnogyna leprieuri* ♀

f *Chrysochlorosia splendida*

**383**
*Diphthera festiva*

**384**
*Dypterygia scabriuscula*
Bird's Wing Moth

**385**
*Euclidia glyphica*
Burnet Companion

**386**
*Grammodes stolida*

**387**
*Autographa gamma*
Silver-Y Moth

**388**
*Cerapteryx graminis*
Antler Moth

383

384

385

386

387

388

**389**

*Thysania agrippina*

**390**

a *Nola albula*
Kent Black Arches

b *Cydosia nobilitella*

c *Mazuca amoena*

d *Emmelia trabealis*
Spotted Sulphur

e *Acontia aprica*
Bird-dropping Moth

f *Eublemma ostrina*
Purple Marbled Moth

g *Earias insulana*
Egyptian Bollworm OR
Cotton Spotted
Bollworm

h *Spodoptera litura*
Oriental Leafworm

j *Spodoptera exigua*
Beet OR Lesser
Armyworm, Pig-weed
Caterpillar OR
Small Mottled
Willow Moth

k *Heliothis zea*
Corn Earworm, Cotton
Bollworm OR Tomato
Fruitworm

9

o

a *Mazuca strigicincta*

b *Ochropleura praecox*
Portland Moth

c *Baorisa hieroglyphica*

d *Calamia tridens*
Burren Green

e *Agrotis infusa*
Bogong Moth

f *Agrotis ipsilon*
Dark Sword-grass OR
Black Cutworm

g *Peridroma saucia*
Pearly Underwing OR
Variegated Cutworm

h *Mamestra brassicae*
Cabbage Moth

j *Amathes c-nigrum*
Setaceous Hebrew
Character OR Spotted
Cutworm

k *Apsarasa radians*

l *Gonodonta distincta*

m *Eosphoropteryx thyatiroides*
Peach Blossom Plusia

n *Nacna malachitis*

p *Harrisimemna trisignata*
Harris's Three Spot

q *Argyritis argentea*

r *Xanthia flavago*
Pink-barred Sallow
Moth

s *Anarta cordigera*
Small Dark Yellow
Underwing

t *Phlogophora iris*

u *Plusia balluca*

**392**

a *Ascalapha odorata*
Black Witch

b *Catocala relicta*
White Underwing

c *Miniophyllodes aurora*

d *Othreis fullonia*
Tropical Fruit-piercer

e *Catocala fraxini*
Clifden Nonpareil

f *Phyllodes floralis*

g *Epicausis smithi*

**393**
*Acronycta psi*
Grey Dagger

**394**
*Catocala nupta*
Red Underwing

**395**
*Scoliopteryx libatrix*
Herald, Scalloped OR European Fruit-piercing Moth

**396**
*Plusia festucae*
Gold Spot

**397**
*Spodoptera littoralis*
Egyptian Cotton Leafworm OR Mediterranean Brocade Moth

**398**
*Plusia chrysitis*
Burnished Brass

**393**

**394**

**395**

**396**

**397**

**398**

**399**

a *Amphipyra pyramidoides* American Copper Underwing

b *Capillamentum nigrofasciatum*

c *Daphoenura fasciata*

d *Eudaphnaeura splendens*

e *Noctua pronuba* Large Yellow Underwing

f *Cucullia verbasci* Mullein Moth

g *Zale lunata*

h *Calpe eustrigata* Vampire OR Blood-sucking Moth

j *Erocha leucodisca*

k *Weymeria athene*

l *Choerapais jucunda*

m *Platagarista tetrapleura*

n *Agarista agricola*

p *Copidryas gloveri* Glover's Purslane

q *Phalaenoides glycine* Vine Moth

r *Cocytia durvillii*

**400**

*Mythimna unipuncta*
Armyworm, White-speck
OR American Wainscot
Moth

**401**

*Pseudoips fagana*
Green Silver Lines

**402**

- a *Pemphigostola synemonistis* ♀
- b *Pemphigostola synemonistis* ♂
- c *Alypia octomaculata* Eight-spotted Forester
- d *Hecatesia fenestrata* Whistling Moth
- e *Musurgina laeta*
- f *Psychomorpha epimenis* Grape-vine Epimenis

**403**

*Bena prasinana*
Scarce Silver Lines

**404**

*Nonagria typhae*
Bulrush Wainscot

**405**

*Panolis flammea*
Pine Beauty

400

401

402

403

404

405

# The Dictionary

## GLOSSARY

ABDOMEN: the hindpart of the body, without external processes in the lepidoptera but often with tufts or brushes of modified scales (e.g. *Alcides aurora*, q.v.)

ANAL ANGLE (of the wing): the rear, outer corner of the forewing. When the wings are spread, the apex (q.v.) is at the front of the forewing, the anal angle is the one behind this. Also used for the inner angle of the hindwing.

ANTENNAE: the pair of processes on top of the head of the adult moth or butterfly, present, but inconspicuous, in caterpillars. There are many different shapes of antenna. Those of butterflies usually have club-shaped tips, while those of moths are usually not clubbed. In moths they may be variously modified; for example, they may have extensive lateral processes and look feathery (pectinate) or may be swollen at the tips, like those of butterflies, particularly in some day-flying moths (e.g. Agaristidae, Callidulidae, Castiniidae and Zygaenidae).

APEX (of the wing): the tip of the forewing, furthest from the body when the wings are spread out. The apex is hooked or pointed in some species.

BIOLOGICAL CONTROL: the use of living animals (or plants) to control a plant or animal which has become a pest. A pest is often serious in a country to which it has been accidentally introduced, without the predators or parasites which normally keep it under control in its native land. These enemies of the pest can be introduced in an effort to control it. Biological control, when carefully carried out, is less dangerous to man and his environment than conventional insecticides, and is usually cheaper in the long run if effective.

BUTTERFLIES: mainly day-flying Lepidoptera, generally without a bristle-coupling mechanism for the wings (frenulum, q.v.) and usually with clubbed antennae.

CATERPILLAR: the crawling, post-egg, stage of butterflies and moths is popularly called a caterpillar. This, in common with many other immature insects, is also called a larva. The caterpillars of clothes moths and other micro-moths (q.v.) are also called 'grubs', 'maggots' or 'worms'. The term moth-eaten refers to the damage done by the caterpillar, not by the moth. Rarely is there any noticeable difference between the sexes in the caterpillar stage.

CELL: this is an enclosed area between the veins (q.v.) from the base of the fore- or hindwing to about the middle of the wing. A number of veins in the wing arise from the cell.

CHRYSALIS: this is the stage in the life history between the caterpillar and the adult moth or butterfly. After the final moult of the caterpillar, the chrysalis (pupa or 'resting stage') is produced. It is very active biologically as it is in the chrysalis stage that the caterpillar undergoes transformation (metamorphosis) into the moth or butterfly. Pupae of moths are frequently protected by a cocoon and are often buried in the soil, but in butterflies the chrysalis usually lacks a cocoon.

COSTA: this is the front margin or leading edge of the wing when in the spread position or in flight.

CRYPTIC: this describes a colour pattern or behavioural trait which tends to allow an animal to blend into its surroundings (see for example, the bark-like pattern of some of the illustrated moths and the remarkable dead leaf appearance of the Lappet, *Gastropacha quercifolia*).

DIMORPHISM: two forms. Frequently used in connection with males and females of the same species whose patterns are different (sexual dimorphism).

DISRUPTIVE: this word describes a bold colour pattern whose effect is to break-up the shape of an animal; see for example, *Apsarasa radians*.

DISTAL: furthest from the body (opposite of proximal).

EYE: the large compound eyes made up of many separate facets (ommatidia) can be seen on each side of the adult head. Some species have two or more ocelli (simple eyes) on top of the head. Caterpillars have ocelli on each side of the head, usually five, but never have compound eyes like the adult.

EYESPOT: nothing to do with the eyes but the name given to the concentric rings of colour, often with black or white centres, which are commonly found on the wings of some species. Many of the rings resemble eyes and are supposed to frighten possible predators, particularly small birds (see for example *Automeris*).

FAMILY: the group name for a collection of genera (q.v.) which are believed to be related to each other. A family may be divided into subfamilies, which are groups of genera within the family thought to be more closely related to each other than to others in the same family. Family names can be recognized by the ending, *idae*, e.g. Pyralidae, Nymphalidae. Subfamily names always end with *inae*, e.g. Crambinae, Heliconiinae.

FORMS: specimens which differ in colour, pattern or size from the more typical specimens are called forms. The form name is not governed by the rules of zoological nomenclature. (See season forms, sexual dimorphism and variation.)

FRASS: the small black lumps which are the excreta of the caterpillar. Often confused with eggs which are quite different and only produced by the adult female butterfly or moth.

FRENULUM: the bristle from the hindwing which catches in the retinaculum, or hook, on the underside of the forewing; this is found in most moths but is absent in the majority of butterflies. Usually the frenulum is single in the male and multiple in the female, but there are many exceptions. The frenulum and retinaculum form a wing-locking mechanism which enables the fore- and hindwings to beat as one. (See also jugum.) In butterflies, the forewing and hindwing overlap to a greater extent which helps them to function together.

GENUS: the group into which allied species are placed. There is often more dispute about the correct genus for a species of Lepidoptera than about the identity of the species itself. Genera can be divided into subgenera, which are groups of species within the genus whose affinities are thought to be closer to each other than to others in the same genus. Generic names always start with a capital letter; species with a small one. Normally the generic name precedes the species name in any combination.

HAIR-PENCIL: this is a group of long hair-like scales found on various parts of the body, legs or wings of male moths and butterflies. Hair-pencils are normally assumed to be scent distributors and have been proved to be so in several species.

HAUSTELLUM: see proboscis.

HOLOTYPE: one of the specimens used in the original description is designated the holotype, sometimes contracted to 'type'. All subsequent discussion about the identity of the species and the use of the name attached, are based on this specimen. The other specimens in the original series are called paratypes.

JUGUM: the wing-coupling lobe at the base of the forewing of the more primitive families of moths (like Hepialidae). This overlaps the hindwing and helps the fore- and hindwing to beat together.

LABIAL PALPS: see palps.

MACRO-MOTHS, MACROS: the larger moths, for example the species in the families Noctuidae, Geometridae, Sphingidae, etc. (q.v.). Generally these are the most popular groups which have been most studied. The term macro-moth, often contracted to 'macros', contrasts the term micro-moth (q.v.). The butterflies, together with the macros are sometimes collectively termed macrolepidoptera. These terms are traditional and have been used for many years as broad divisions of the Lepidoptera.

METAMORPHOSIS: the change of state from the caterpillar through the chrysalis to the adult butterfly or moth.

MICRO-MOTHS, MICROS: a long-standing, but somewhat arbitrary, typological division of moths which include the more primitive and generally smaller species of Lepidoptera. Some species of micros are larger than many macros but the majority are smaller. The micros are generally less well known but include a few widely known species, like the clothes moths, which are closely associated with man.

MOTHS: see macro-moths and micro-moths.

NOMINATE SUBSPECIES: the first named of any subspecies is called the nominate one and bears the same name as the species, e.g. *caja caja* or *toxopei toxopei*.

OCELLI: these are simple eyes. They are sensitive areas sometimes showing on the top of the head of the moth or butterfly as raised spots. The caterpillar does not have a compound eye (see eye) but has ocelli on the side of the head.

ORIENTAL REGION: a term used by zoogeographers (students of animal distribution) for an area which includes India, Sri Lanka, Pakistan, Southern China, South East Asia, south eastwards as far as Sulawesi (Celebes), and usually the Solomon Islands and New Guinea but excluding Australia.

OSMATERIUM: see Papilionidae.

PALPS: in caterpillars there are usually two pairs at the front of the head. They are short segmented structures associated with feeding. In adults there is a pair of maxillary palps which are generally small and not easily seen, and a pair of labial palps which are always present and can be of various shapes and sizes.

PARASITE: in zoology this is an animal which lives inside or more rarely outside another animal, its 'host', from which it obtains its nourishment. The most important parasites of Lepidoptera are the larvae of certain wasps and flies.

PARATYPE: specimens in the original series, other than the holotype (q.v.), that were examined by the describer of the species.

PECTINATE: this refers to the antennae (q.v.) of moths when they have numerous side branches and appear feathery.

POLYMORPHIC: occurring in several different forms. Solely female polymorphism occurs in some mimetic butterflies, for example *Papilio dardanus*. Here the males are all similar but the females occur in several different colour forms which can be strikingly different from each other.

PREDATOR: this is an animal which searches for, attacks and eats other animals. The daytime predators of Lepidoptera include birds, lizards and spiders; the most important nocturnal predators are probably bats.

PROBOSCIS: the coiled feeding tube under the head of the adult butterfly or moth (also called the tongue or haustellum) through which it sucks its food. In some families the proboscis has scales on it, in others it is smooth. The proboscis of fruit-piercing moths has small teeth near the apex. It is reduced in size or absent in many species, which are consequently unable to feed. The proboscis can be uncoiled when the insect is feeding and in some species is of considerable length to enable the insect to reach nectar in elongate flowers like honeysuckle *(Lonicera)*.

PROXIMAL: nearest to the body; contrast distal.

SCALES: the dust-like particles which rub off the wings of butterflies and moths when they are handled. The scales are important in pattern, colour and for flight and may be modified in various ways, for example scent scales. The colours of iridescent scales are often the result of their shape. Scales are found all over the body, wings and legs of adult butterflies and moths but not caterpillars. Although some other insects have scales on the wings, the scales of Lepidoptera are unique in shape and the tasks they perform.

RARE: the use of this word indicates rarity in collections, not necessarily rarity in nature, although this may also be the case.

SEASONAL FORMS: as the term implies, this refers to butterflies or moths which are on the wing at a particular time of year and differ in some way from those broods or generations of the same species which appear at a different season of the year. In the tropics, wet season forms, which fly during the rains, may differ considerably from dry season forms (see *Precis octavia*). In the temperate regions there are often marked differences between the spring and summer broods.

SEXUAL DIMORPHISM: mention has been made in the text if the two sexes differ greatly from each other in coloration, colour pattern or wingshape; if no mention is made, it can be assumed that there are only small differences between males and females. The butterfly *Papilio dardanus* is one of the better known examples of sexual dimorphism (and of mimicry) and is illustrated in the plates together with other equally striking dimorphic species of both butterflies and moths.

SPECIES: the name given to an identifiable interbreeding group of similar individuals. There are many disputes about the definition of a species. Species may be further divided into subspecies, often geographically separated, slightly different members of one species. Species are made up of individuals (specimens) which form populations. The study of variation in species is outside the scope of this book but see *Aglia tau* and *Arctia caja* for example. Specimens of one species may vary slightly from one another and those slightly different from the majority are called forms. These may be seasonal (q.v.) or may differ for other reasons. Species names always start with a lower-case letter; in contrast generic names start with a capital letter.

SUBSPECIES: a division of a species (q.v.). A species may be separated into a number of isolated populations, e.g. on islands, which differ slightly from one another. There are many other isolating mechanisms which may give rise to slight differences between populations of one species. These, when recognizably distinct, are often given subspecific names.

SUPERFAMILY: this is a supposedly natural grouping of families. The name of a superfamily ends in 'oidea', for example Papilionoidea, the superfamily to which most families of butterflies belong.

SYNONYMS: these are names by which the species or genus has been known in the past. Generally they have been proposed at a later date than the currently accepted name. An example of a generic synonym is *Crenis*, given in the text as *Asterope* Hübner syn. *Crenis*. A specific synonym appears in a similar way, e.g. *pandion* Geyer syn. *A. lalage* under *Appias*. Synonyms are only given in this work when the synonym has been recently used as the name for the genus or species.

TEGULA: scale covered projections on each side of the thorax, which cover the base of the wings.

THORAX: the middle part of the body from which the legs, and wings arise and where the wing-operating muscles are housed.

TIBIA: a segment of the leg of an insect.

TRIBE: in zoology this is a subdivision of a subfamily. The name of a tribe ends in 'ini', for example Troidini, the tribe of Papilionid butterflies to which the birdwings belong.

TYPE: see holotype.

VARIATION: no two individuals of a species are absolutely identical. The differences between them are usually so slight that it is not noticeable. In other instances, the variation may be so great that it is difficult to believe they represent the same species (see *Papilio dardanus* or *Orgyia antiqua*). Variation within one species may be caused by a variety of factors, including time of year, at which the insect emerges (seasonal forms q.v.) the area in which they live (geographical variations) and their sex (sexual dimorphism).

VEINS, VENATION: these are the hollow struts which can be seen across the wings. They run either along the wing or as short cross-veins. The veins enclose an area in each wing called the cell. Recent research has shown that there is a circulation of blood in these veins. After a butterfly or moth emerges from the chrysalis the wings are crumpled and unexpanded; blood pumped along the veins gradually expands the wing into its flat smooth shape.

WINGSPAN: the measurements given under the headings of species give an idea of the range of size of a particular species from one wingtip to the other. Normally the smallest specimen is a male and the largest a female. Most of the measurements were taken from specimens in the British Museum (Natural History).

**Abantiades** Herrich-Schäffer (HEPIALIDAE)
There are some 20 species in this Australian genus. The caterpillars feed on the roots of *Eucalyptus*. In spite of their large size the moths come in the grouping popularly termed micro-moths.

**hydrographus** Felder
Australia 90–180 mm. (3·54–7·09 in.)
This is a very large Hepialid, with grey brown patterned forewings and white streaks in the middle of the wing. The whole forewing is covered by an intricate swirling pattern.

**Abantis** Hopffer (HESPERIIDAE)
SKIPPER BUTTERFLIES
A genus of skipper butterflies found in Africa south of Sahara. They are brightly coloured species, unlike many of the more sombre coloured relatives in the family.

**paradisea** Butler PARADISE SKIPPER
East to South Africa 36–50 mm. (1·42–1·97 in.)
The black forewings have many hyaline spots, tinged with yellow at the base. The hindwings have large hyaline areas, edged by the black veins or with a broad border. This is one of the more colourful skippers found in the bushveldt and rocky hills in South Africa and from Mozambique to E. Africa. The caterpillar feeds on *Cola* (Sterculaceae) and *Hibiscus* (Malvaceae). There are several similar looking species in other parts of Africa. **80**a

ABBOT'S SPHINX see **Sphecodina abbotii**
**abbotii** see **Sphecodina**

**Abisara** Felder (NEMEOBIIDAE) JUDY BUTTERFLIES
There are about 15 species in this African and tropical Asian genus of butterflies. They are chiefly brown in colour, some with a slight purple sheen to the wings, and some with a short hindwing tail. The known foodplants are species of Myrsinaceae.

**echerius** Stoll PLUM JUDY
Sri Lanka, India, Burma, Thailand, China, the Philippines 35–44 mm. (1·38–1·73 in.)
This is a common, purplish brown (male) or brown (female) species which flies along forest trails and margins, invariably during the late afternoon and evening. The brown pattern of the under surface is partly represented on the upper surface of the female wings. Numerous subspecies of this widely distributed butterfly have been described.

**abrasa** see **Dalcera**

**Abraxas** Leach (GEOMETRIDAE)
CURRANT OR MAGPIE MOTHS
A genus of moths with many species, most of which are brightly coloured. Many are known to be distasteful to birds and other predators.

**grossulariata** Linnaeus MAGPIE OR CURRANT MOTH
Europe, through W. & C. Asia to Japan 36–43 mm. (1·42–1·69 in.)
A brightly coloured moth, known to many gardeners growing currants or gooseberries *(Ribes)*. These moths are common in the fruit bushes and when disturbed during the day flap lazily out of them. At times the caterpillars can cause serious damage. They are almost as conspicuous as the moths, being white with black transverse spots on the back and an orange-yellow stripe on each side, spotted above and below with black. The caterpillars are in the bushes in May and June. Many varieties of the moth have been named. Some of the rarest ones are valuable to collectors and sell at prices which reflect their rarity. They are distasteful to birds and their bright colour is believed to be a warning sign. The moths have been used for genetic studies on inheritance of character. **301**

**sylvata** Scopoli CLOUDED MAGPIE
Europe including Britain, Asia including Japan 36–45 mm. (1·42–1·77 in.)
The Japanese specimens belong to a distinct subspecies; the illustration is of the European subspecies. The bluish white caterpillar has 11 black lines along the body; it lives on elm *(Ulmus)*, although it has been found feeding on other trees. The adults, which fly in June and July, vary considerably over the whole range and, apart from the Japanese subspecies, others have been described from different localities. **304**

**abrota** see **Ogyris**

**Acanthobrahmaea** Sauter (BRAHMAEIDAE)
This genus was erected in 1967 to accommodate a single moth species, *A. europaea*, which was previously placed in the genus *Brahmaea*.

**europaea** Hartig
S. Italy 50–65 mm. (1·97–2·56 in.)
This species was described as recently as 1963, by Count Hartig, from a single S. Italian specimen. That such a large moth had remained undiscovered for so long is an indication of the lack of collecting in this part of Italy, rather than the scarcity of this species. It is the only representative of the family Brahmaeidae in Europe. **340**b

**Acanthosphinx** Aurivillius (SPHINGIDAE)
There is only one species in this genus of hawk-moths.

**guessfeldti** Dewitz WIDOW SPHINX
Tropical Africa 120–140 mm. (4·72–5·51 in.)
The large, pointed spurs on the legs of the adult are capable of inflicting unpleasant scratches, and the strong scent of this moth is thought to be repellent to potential enemies. The forewing has a contorted pattern of yellowish green, brownish grey and greenish brown which is apparently either cryptic or disruptive. The hindwing is brownish grey, with a yellowish green outer margin. The caterpillar is unknown.

**Acentria** Stephens: syn. *Acentropus* (PYRALIDAE)
AQUATIC MOTHS
This small European genus of Pyralid moths is currently placed in the subfamily Schoenobiinae. It contains species where the moth itself can swim under water, almost unknown in other genera, although there are many species whose caterpillars are aquatic. When they were first discovered in the 18th century they were not recognized as moths and were described as Caddis-flies *(Trichoptera)*. It was not until many years later that their true identity was recognized.

**nivea** Olivier AQUATIC OR FALSE-CADDIS WATER-VENEER MOTH
Europe including Britain, N. America 10–16 mm. (0·39–0·63 in.)
This moth lives near fresh water ponds. The caterpillar feeds under water on species of *Chara*, *Potamogeton* and other water plants. There are two kinds of females, one is similar to the illustration, with long wings; in the other the wings are reduced. It is the latter which lives in the water. The moth is widely distributed in Europe but because of its small size is not well known. The caterpillar, living under water, obtains its oxygen from the air spaces in plants but also, it is believed, by diffusion of oxygen through the skin. It pupates under water in a cocoon attached to a water plant and the moth apparently emerges under water. It can be found from July to September although it is believed that the individual life of each moth is short, possibly only 2 days. The moths are active at dusk and during the night when they fly round the surfaces of ponds and stagnant streams. **56**h

**acerata** see **Acraea**
**acerbella** see **Epichoristodes**

**Acerbia** Sotavalta (ARCTIIDAE)
Three species are assigned to this genus: the circumpolar *A. alpina*, dealt with below, *A. seitzi* Bang-Haas, found in central Russia, and *A. kolpakofskii* Alpheraky in W. China. All 3 have the same type of adult colour pattern.

**alpina** Quensel 40–49 mm. (1·57–1·93 in.)
Scandinavia, U.S.S.R. (Siberia and high altitudes in N. Mongolia), U.S. (Alaska), Arctic Canada
This is one of the rarest species of tiger-moths in Europe or N. America. The only good series of *A. alpina* have been collected in N. Mongolia in the regions around Lake Baikal. There had been no report of *A. alpina* between the time of its first discovery in 1799, by an Italian traveller Guiseppe Acerbi, and the 1920s when a specimen was seen but not taken. Not until 1962 was a second European specimen captured, by the Finnish zoologist Olavi Sotavalta. N. American records date from 1914 and 1916 and account for a meagre 5 specimens. **369**n

**achelous** see **Lonomia**

**Acherontia** Laspeyres (SPHINGIDAE)
There are about 5 species in this genus of hawk-moths. They are found in Africa, Europe and Asia, as far east as Japan, and in S.E. Asia from Sri Lanka and India to the Moluccas. All the species have a skull-shaped marking on the upper surface of the thorax.

**atropos** Linnaeus DEATH'S-HEAD HAWK-MOTH
Europe, N. Asia, Africa 102–140 mm. (4·02–5·51 in.)
This large moth gets its name from the skull-like markings on top of the thorax, a characteristic which together with its ability to emit a series of squeaks when handled have combined to give it an awesome but undeserved reputation. An 18th-century name for *A. atropos*, the 'Bee Tyger Hawk Moth', refers to the fact that the adult moth will visit bee-hives to feed on the honey: the squeaking or hissing sound it makes is said to sound like the pre-swarming sound of a queen bee and this might give some protection to the intruding moth. The huge green or purple caterpillar, often over 120 mm. (4·7 in.) in length, is equally fearsome in appearance, but is also harmless to man. It feeds on potato *(Solanum)*, tea-tree *(Lycium)*, woody nightshade *(Solanum)* and the highly poisonous deadly nightshade *(Atropa)*. **342**

**achine** see under **Colotis antevippe**
**achlora** see under **Antinephele maculifera**

**Achroia** Hübner (PYRALIDAE) HONEY MOTHS
A genus of Pyralid moths with 2 species. One of these is practically world wide in distribution, and has mostly been spread accidentally by man.

**grisella** Fabricius HONEY OR LESSER WAX MOTH
Worldwide 15–22 mm. (0·59–0·87 in.)
This Pyralid moth has a shiny yellow head and yellowish brown wings, more or less unmarked. This pattern separates it from the other species in the wax-moth subfamily (Galleriinae) to which it belongs. The greyish white caterpillar has a dark line along the back. It is very destructive in old bee-hives but also feeds on a variety of dried foods. With improved hygiene in beehives, this species seems to have become less common in recent years.

**acidula** see **Caloptilia**
**acis** see **Helicopis**

**Acleris** Hübner (TORTRICIDAE) BELL-MOTHS
There are upwards of 500 species in this genus of moth, mostly in Europe, Asia and N. America but also with a few Indian and one or two African species.

**comariana** Lienig & Zeller
STRAWBERRY TORTRIX MOTH
Europe including Britain, N. America 15–16 mm. (0·59–0·63 in.)
At times the caterpillar of this moth can be a serious pest of strawberries *(Fragaria)*. After hatching from the eggs, the caterpillars feed on the young leaves of the strawberry plants. Later they feed on the unopened flowers with the result that attacked flowers either do not form fruit or only form distorted ones. The caterpillars bind several leaves together where they live in webs. If touched they wriggle quickly, usually backwards, in a way characteristic of Tortricid caterpillars. The moth has grey, or reddish grey forewings with a characteristic triangular brown or black marking on the front margin of the forewing. This species was first reported in N. America in 1924 when it was accidentally imported in azaleas *(Rhododendron)* from Europe. **13**

**cristana** Denis & Schiffermüller
Europe including Britain, temperate Asia including Japan 18–21 mm. (0·71–0·83 in.)
A species remarkable for the variation in wing color-

ation. A recent publication showed that there were nearly 120 named forms in Britain alone, differing slightly in colour or pattern in some cases, but dramatically different in others. The caterpillar feeds on Blackthorn *(Prunus spinosa)* and possibly on Hawthorn *(Crataegus)*. There is only one brood a year with the moth out from August to October. They hibernate over the winter and those which survive lay their eggs in spring. The moth generally has a grey-white head and thorax, a slightly wavy anterior margin to the forewing which varies from black with white to reddish brown, and with a yellow streak on the posterior margin. The hindwings do not vary and are generally grey-brown or yellow-brown. This moth was first described in 1775 from specimens collected near Vienna, Austria.

**hastiana** Linnaeus
Europe, temperate Asia, 18–21 mm.
N. America (0·71–0·83 in.)
The pale green caterpillar feeds on willow *(Salix)*. The adult moth has many colour variations, usually grey-brown or purplish brown with patterns of darker marks. This is a species on which breeding work would provide an interesting approach to the genetics of the many colour forms which are known.

**variegana** Denis & Shiffermüller
Europe including Britain, 13–16 mm.
N. America (0·51–0·63 in.)
Typical in appearance of the species popularly called Bell-moths from their shape when at rest, this is common in Europe. The caterpillar is pale green or yellowish with a darker green or red-brown line along the back. It feeds on hawthorn *(Crataegus)*, blackthorn *(Prunus)*, rose *(Rosa)* and other related plants. The caterpillars tie the leaves together with silk and if disturbed run backwards quickly and drop off the plant, hanging by a silken thread. They climb back up this thread when the danger is past. In N. America this species has been collected in the Pacific N.W. and California. **9**

**acanthea** see **Euthalia**

**Acontia** Ochsenheimer (NOCTUIDAE)
Over 150 species are placed in this almost cosmopolitan genus of small moths.

**aprica** Hübner BIRD-DROPPING MOTH
U.S., Canada 23–26 mm. (0·91–1·02 in.)
This is one of many species of its genus which, when at rest, resemble bird-droppings. The greenish grey caterpillar is marked with white and black spots; it feeds on hollyhock *(Althaea)*. There are normally 2 broods per year. **390**e

**acontius** see **Catonephele**

**Acraea** Fabricius: syn. *Pareba* (NYMPHALIDAE)
These butterflies belong to the widespread subfamily Acraeinae, well represented in Africa with a few species in the tropics of the Orient and one genus, *Actinote*, in South America. In Africa *Acraea* is probably one of the largest genera, with some 150 species. Most species are polymorphic, with different forms in different seasons. They occur everywhere in Africa, from the rain-forests to the desert regions. All the species are distasteful to vertebrate predators and so are left alone by birds and small mammals. Many, because they are distasteful, are imitated by other more palatable species (mimicry) and a number of species of *Acraea* themselves are similar, deriving mutual protection from their similar appearance (Müllerian mimicry). Acraeas are generally slow flying butterflies, often coloured in shades of brown, red, orange and yellow. The caterpillars of *Acraea* generally feed on Compositae and toxic plants of the family Passifloraceae.

**acerata** Hewitson
Africa 28–38 mm. (1·1–1·5 in.)
This butterfly is common from north of the Zambesi to the wetter parts north of the equator. The caterpillar feeds on sweet potato *(Ipomoea)* and at times can be a locally important pest on this crop. The adult is pale yellow to orange-brown with a broad black border and black in the cell and basal half of the forewing, with only a black border on the base of the hindwing. The extent of the black on the hindwing varies in the different subspecies. The specimen illustrated is from Uganda. The butterfly is out in most months of the year.

**andromache** Fabricius GLASSWING
New Caledonia, New Georgia,
Indonesia, Sulawesi (Celebes), 50–60 mm.
New Guinea, Samoa, Fiji, Australia (1·97–2·36 in.)
This is the only Australian representative of the genus. It is a typical *Acraea* in shape and colour but has almost transparent forewings, with the underside spots showing through and a pale, black-spotted hindwing with a dark marking. The caterpillars feed on species of *Passiflora* (Passifloraceae) and are gregarious when small. Several subspecies have been described in different parts of its range.

**anemosa** see under **Hibrildes norax**

**bonasia** Fabricius
Africa 32–42 mm. (1·26–1·65 in.)
This species is found in the forests from Sierre Leone to Zaire, Zambia, E. Africa and in Ethiopia. Several subspecies have been described. It is very similar to a closely allied species *A. sotikensis* and possibly they are all part of one species. The specimen illustrated was photographed in Uganda. **157**

**cephus** Linnaeus
W. & C. Africa 48–50 mm. (1·89–1·97 in.)
This is primarily a forest species which ranges from Sierra Leone through Zaire into the forests of W. Uganda. The female is larger than the male. In Uganda a distinct form, which lacks the red bar below the apex of the wing, occurs. **163**l

**chilo** Godman
E. Africa, Ethiopia, 50–65 mm.
S. Arabia (1·97–2·56 in.)
The male is illustrated. The female has completely transparent wings, with a few spots on the hindwing. One subspecies is found in N. Kenya, and another in N.E. Tanzania and E. Kenya. The transparent-winged female is an unusual butterfly with a most delicate look as it flies about. **163**a

**egina** Cramer ELEGANT ACRAEA
Africa 62–96 mm. (2·44–3·78 in.)
This species is found throughout the forests of tropical Africa as far south as Rhodesia and Mozambique. Several subspecies have been described. It is broadly similar to *Pseudacraea boisduvalii* (figure **231**h) but has some differences in detail; for example it has a plain black margin to the hindwing. It is thought that *A. egina* is the distasteful model (since all *Acraea* species have proved to be distasteful to birds and small mammals) and that *P. boisduvalii* is the palatable mimic, which deceives the predators and derives an advantage by resembling the distasteful species. It is possible that both are distasteful and together derive benefit from this; predators have a greater chance of 'learning' the pattern and leaving both species alone (Müllerian mimicry).

**natalica** see under **Hibrildes norax**

**uvui** Grose-Smith TINY ACRAEA
Africa 28–34 mm. (1·1–1·34 in.)
This species, which is related to *A. bonasia*, is found in the forest of Angola, Cameroons, Zaire and into E. Africa. It is one of the smallest species of *Acraea* in E. Africa. **156**

**vesta** Fabricius YELLOW COSTER
N. India, Pakistan, Burma, 45–85 mm.
W. & S. China (1·77–3·35 in.)
Although usually yellowish brown above, some specimens are almost black. They fly between 610–2130 m. (2000–7000 ft), usually not far from the foodplant of the caterpillar, in open country and scrubland. The caterpillar has been found on several species of Urticaceae and on *Buddleia* (Loganiaceae), the latter being a family containing poisonous species of plants, from which highly toxic strychnine is extracted.

**violae** see **Telchinia violae**

**zetes** Linnaeus LARGE SPOTTED ACRAEA
Africa 60–75 mm. (2·36–2·95 in.)
A widespread and common species found in the savannah and forest edges throughout most of Africa south of Sahara. Several subspecies have been described. The wings are heavily marked with black, with a general ground-colour red. This species is a typical *Acraea* shape. The caterpillars feed on *Moducca* and *Passiflora* (Passifloraceae), and probably derive protection from the poisons which these plants contain or synthesizes its own from them.

ACRAEA,
ELEGANT see **Acraea egina**
LARGE SPOTTED see **Acraea zetes**
TINY see **Acraea uvui**
TRIMEN'S FALSE see **Pseudacraea boisduvalii**
ACRAEA
MIMIC, MARSHALL's see **Mimacraea marshalli**
SWALLOWTAIL see **Graphium ridleyanus**

**Acraeinae** (NYMPHALIDAE)
A subfamily of butterflies with mostly African species, a few in the Orient and a single genus with species in S. America. Generally distasteful and often mimicked by more palatable species, they are usually slow fliers, out in the daytime and characteristically coloured (see *Acraea*, *Actinote*). The subfamily has often been considered sufficiently different to be regarded as a family distinct from, but related to, the Nymphalidae.

**acraeoides** see **Riechia**
**acrea** see **Estigmene**
**acrisia** see **Euproctoides**

**Acrobasis** Zeller (PYRALIDAE)
A genus of Pyralid moths with many species, some of which are probably unrelated, but further research on their affinities is needed. The species in the genus range from N. America to Europe and Asia with a few species in Africa.

**consociella** Hübner
Europe including Britain 18–20 mm. (0·71–0·79 in.)
Reddish brown or purplish brown forewings with a black and white line across them are characteristic of this species. There are several similar ones in the genus, all with narrow wings and long legs typical of the subfamily. The moth is common in oak woods *(Quercus)* flying at dusk and after dark and is often attracted by light. The caterpillar is pale yellow-grey or greenish grey with dark lines along the back. It feeds on oak leaves, hibernating over winter among the dead leaves. Generally there are several caterpillars together, each living in a silk tube and making a webbing which conceals the tubes where they live. They pupate in silken, debris-covered cocoons attached to the chewed leaves.

**jugulandis** Le Baron PECAN-LEAF CASE-BEARER
U.S., Canada 14–17 mm. (0·55–0·67 in.)
The forewings are grey with white basal areas and a blackish line near the outer margins, the hindwings are smokey grey. The caterpillar feeds on buds, leaves and flowers of the pecan, walnut, hickory *(Juglandaceae)* and other trees in early spring. It can be numerous at times and may be a serious problem as a defoliator of pecan.

**Acrocercops** Wallengren (GRACILLARIIDAE)
A huge, worldwide genus of micro-moths with 300–400 species. They are mostly small moths and a number are serious pests of agricultural crops.

**cramerella** Snellen COCOA MOTH
Worldwide in cocoa growing areas 14 mm. (0·55 in.)
The caterpillars of this moth are serious pests of cocoa *(Theobroma cacao)*. They bore into the cocoa pods and make galleries which fill up with frass and debris. Although the damage to the pods cannot be seen from the outside, the beans inside are made quite worthless and the attacks can result in the loss of crops worth thousands of pounds sterling. The caterpillars leave the pods to pupate, making a flat cocoon on leaves or on the ground. Their life-cycle from egg to adult takes about 4 weeks. The adults swarm at sunset over the bushes.

**Acrojana** Aurivillius (Eupterotidae)
Four species are placed in this genus of W. African moths.

**rosacea** Butler
Nigeria, Ghana 75–104 mm. (2·95–4·09 in.)
This apparently rare moth is probably the most colourful species of its family, most of whose members are sombrely coloured. The fact that only the hindwings are brightly coloured indicates that this is probably another example of 'flash-coloration' by which the moth is able to disconcert potential predators in a 'now you see it, now you don't' display as the coloured areas are alternately revealed when in flight and hidden when at rest. **317**j

**splendida** Rothschild
Nigeria, Ghana 100–110 mm. (3·94–4·33 in.)
*A. splendida* is similar in wing-shape and pattern to the smaller, illustrated *A. rosacea* and is almost as colourful. Its wings are mainly light green above except for the front edge of the hindwing which is a beautiful rosy pink. Very few specimens exist in collections.

**Acrolophidae**
The micro-moths in this family are predominantly brown in colour. The caterpillars, popularly called burrowing webworms, feed on various grasses, bromeliads, orchids and on corn. The family is restricted to the New World. The genus *Acrolophus* was formerly included in the Tineidae.

**Acrolophus** Poey (Acrolophidae)
BURROWING WEBWORMS
These micro-moths are widespread in America. They occur in N. and S. America, the West Indian Islands and in Hawaii. They are small to medium; wingspan around 18–40 mm. (0·71–1·57 in.) and mostly brown in colour. The caterpillars attack the roots of corn *(Zea)*, orchids, bromeliads and many pasture grasses.

**popeanellus** Clemens
N., C. & S. America 25–38 mm. (0·98–1·5 in.)
The moth is reddish brown with some markings; the hindwings are yellow-brown. The male antennae are stout and deeply serrated, the female's are simple. The original specimen on which the description of the moth was first based was collected in Texas about the middle of the 19th century. The caterpillars live on the ground feeding on grass roots. The moth is common over its range except for the northern States and the extreme west of the U.S.

**acron** see **Amblypodia**

**Acronicta** Ochsenheimer (Noctuidae)
DAGGER MOTHS
Nearly all the species of this large genus have one or more dark, dagger-shaped marking on the forewing. They are chiefly northern temperate in distribution, both in the Old and New Worlds. The rather uniformly marked adults contrast with the strikingly patterned caterpillars, most of which bear several tufts of barbed hairs unlike most other species of Noctuidae. About 75 species of this genus are found in N. America.

**alni** Linneaus ALDER MOTH
Europe, temperate Asia 35–40 mm.
E. to Japan (1·38–1·57 in.)
The adult moth of this species is generally similar in colour pattern to the illustrated *A. psi*, but is variably clouded with black. The curious caterpillar, however, is quite unlike that of its relation. In the early stages it closely resembles bird-droppings; later it is marked with highly conspicuous bands of yellow and black and bears several long, clubbed hairs. Brightly coloured caterpillars are usually considered to be warningly coloured, but there is no evidence yet as to whether the caterpillars of *A. alni* are distasteful to predators in their later stages. They feed on alder *(Alnus)* but also on the foliage of many other trees and shrubs.

**americana** Harris AMERICAN DAGGER
E. Canada, U.S. 50–60 mm. (1·97–2·36 in.)
This relatively common species is similar in colour pattern to the illustrated Old Word *A. psi*. Its caterpillar feeds on the leaves of maple *(Acer)*, oak *(Quercus)* and other broad-leaved trees. Its range in the U.S. extends S. to the N. border of Texas.

**interrupta** Guenée AMERICAN GREY DAGGER
E. Canada, N. U.S. 35–40 mm.
E. of the Rockies (1·38–1·57 in.)
This species has been confused in the past with the rather similarly patterned *A. psi* of the Old World. The dark melanic form of *A. interrupta* is much more common today than at the beginning of this century, possibly as a result of increased industrial pollution. Caterpillars have been recorded from elm *(Ulmus)*, apple *(Malus)*, plum *(Prunus)* and birch *(Betula)*.

**oblinata** Smith SMEARED DAGGER
Canada, U.S. 35–50 mm. (1·38–1·97 in.)
This is a common species in E. Canada and in the U.S. E. of the Rockies S. to Florida and Texas. The brightly coloured black, yellow and red caterpillar is an occasional pest of cotton *(Gossypium)* in the S. States but feeds more commonly on smartweed *(Polygonum)* and other low-growing plants. The colour pattern of the moth is similar to that of *A. psi* (illustrated). There is a dark grey melanic form.

**psi** Linnaeus GREY DAGGER
Europe, N. Africa, 35–40 mm.
W. & C. Asia (1·38–1·57 in.)
This is a common species in much of its range. Its distinctive caterpillars are yellow above, with black-edged, red spots on each side; they feed on hawthorn *(Crataegus)*, blackthorn and plum *(Prunus)*, apple *(Malus)*, pear *(Pyrus)* and on many other deciduous trees. In the islands of the Hebrides it has been recorded from bracken *(Pteris)*. The closely related species, the Dark Dagger *(A. tridens)* has a red-backed, differently patterned caterpillar which is easily distinguished from that of the Grey Dagger, whereas the adults are nearly identical in colour pattern. **393**

**tridens** see under **A. psi**

**actea** see **Satyrus**
**acteon** see **Thymelicus**

**Actias** Leach (Saturniidae) MOON-MOTHS
Most species of this small genus of large pale green, long-tailed moths are E. Asiatic in range, but one occurs in N. America (the American Moon-moth) and another in Mexico. Their popular name refers to the distinctive moon-like markings on the wings. The adult is provided with a sharp spine at the base of the forewing with which it bores its way out of the cocoon after emergence from the chrysalis. Males, in particular, are capable of swift and vigorous flight. Other moon-moth genera are the African *Argema* and the S. European *Graellsia*.

**artemis** Bremer & Grey
E. U.S.S.R., Korea, 80–120 mm.
Japan (3·15–4·72 in.)
This is similar in general appearance to the illustrated *A. selene* but the tails of the hindwing are much shorter. The moth is on the wing in June and July. Its caterpillars feed on the foliage of a variety of broad-leaved trees.

**luna** Linnaeus AMERICAN MOON-MOTH or LUNA
Mexico, U.S., 75–90 mm.
Canada (rarely) (2·95–3·54 in.)
This is one of the moths that most small boys (American, at least) can name on sight. It is commonly cage-bred throughout the world and frequently manages to escape, or is purposely given its freedom, and as a result has been captured in several unexpected places, including the centre of London. It is similar in appearance to the illustrated Indian Moon-moth *(A. selene)* but the 'moon' mark on the forewing is connected to the front edge of the wing. The caterpillars feed on the foliage of birch *(Betula)*, walnut *(Juglans)*, aspen *(Populus)*, American Sycamore *(Platanus occidentalis)* and several other broad-leaved trees.

**maenas** Doubleday
S.E. Asia 100–120 mm. (3·94–4·72 in.)
The illustrated male of this elegant species is from the island of Sulawesi (Celebes) and represents the subspecies *A. maenas latana*. The males of other subspecies are more greenish or nearly completely brown in colour. The female of each subspecies is a generally more greenish insect, with hardly a trace of wing markings and a much shorter tail to the hindwing. **340**a

**selene** Hübner INDIAN MOON-MOTH
Sri Lanka, India, China, 80–120 mm.
Malaysia, Indonesia (3·15–4·72 in.)
Like *A. luna* this is a very popular species with breeders, whose choice could hardly be criticized; the beauty of a newly emerged moth has few parallels in the insect world. There are probably at least 2 broods in nature, and probably more in the southern parts of its range. The caterpillars are at first red, then mainly green except for the brown head and legs, and the spiny, orange tubercles. Its foodplants include *Coriaria*, *Andromeda*, *Hibiscus*, *Salix* and *Prunus*. In captivity the leaves of plum *(Prunus)*, apple *(Malus)*, pear *(Pyrus)* and hawthorn *(Crataegus)* are eaten. **336**

**actinias** see **Dudgeonea**

**Actinote** Hübner (Nymphalidae)
This genus is in the subfamily Acraeinae and is the only South American member of that subfamily. The species are warningly coloured and distasteful, and are mimicked by other palatable species which derive protection from this mimicry.

**ozomene** Godart
S. America 44–60 mm. (1·73–2·36 in.)
This species is found in Ecuador and Colombia where it is common. The pattern is copied by other species of butterflies which are palatable, but they are avoided by predators because of their resemblance to the distasteful *A. ozomene*. **163**m

**pellenea** Hübner SMALL LACE-WING
S. America – reaching 43–50 mm.
into W. Indies (1·69–1·97 in.)
This typically warning-coloured species is basically light brown with black veins and black edges to the wings. The prominent yellow-brown patch below the apex of the forewing also has black lines across. This pattern is very reminiscent of the typical wing patterns of the African species *Acraea* in the same subfamily (Acraeinae). This species is a sunshine flier, with a slow flight, and often collects together in large numbers. It is separated into subspecies in different parts of its range from Venezuela to Argentina and in the W. Indies.

**actoris** see **Cremna**
**adactylalis** see **Tineodes**

**Adela** Latreille (Incurvariidae)
LONG-HORNED MOTHS
This genus of small, often metallic coloured micro-moths is worldwide with several species in Europe and in N. America. All are characterized by extremely long antennae which they hold out in front of them while fluttering slowly around bushes. These antennae are many times the length of the body, longer in proportion than any other species of Lepidoptera. The moths frequently occur in large numbers and always attract attention.

**bella** Chambers PURPLE LONG-HORN
U.S. 14 mm. (0·55 in.)
Common, particularly in the E. central States, this species has a purple iridescence on the wings and the general colour of the wings is purplish green.

**reamurella** Linnaeus: syn. *A. viridella*
GOLDEN LONG-HORN 14–16 mm. (0·55–0·63 in.)
Europe including Britain
Can be seen in May or June hovering around hedgerows or oak trees *(Quercus)*, with the long white antennae sticking out in front. These antennae may be nearly 1 cm. (0·39 in.) long, huge in proportion to the insect. The caterpillars live in a broad, flat case of leaf fragments on the ground amongst fallen leaves. **3**

**adela** see **Amastus**
**adippe** see **Fabriciana**
**admirabilis** see **Iotaphora**

ADMIRAL,
BLUE see **Vanessa canace**
INDIAN RED see **Vanessa indica**
INDIAN WHITE see **Limenitis trivena**
LONG-TAILED see **Antanartia schaenia**
ORANGE see **Antanartia delius**
POPLAR see **Ladoga populi**
RED see **Vanessa atalanta**
SOUTHERN WHITE see **Limenitis reducta**
WHITE see **Ladoga camilla** and **Limenitis arthemis**
**Adolias** see **Euthalia**
**adolphei** see **Mycalesis**
**adonia** see **Euthalia**
**adonira** see **Dodona**
**Adonis Blue** see **Lysandra bellargus**
**adreptella** see **Carposina**

**Adscita** Retzius: syn. *Procris* (ZYGAENIDAE)
Very large genus of European and Asian day-flying moths. Most have attractive metallic colours and they have always interested collectors with a result that many subspecies and aberrations (differences of wing pattern from the normal) have been described and given names.

**statices** Linnaeus FORESTER
Europe including Britain, Turkey, Asia 26–30 mm. (1·02–1·18 in.)
The Forester moth has blue-green or bronze-green wings and numerous aberrations have been described. The caterpillar, which feeds on sorrel *(Rumex)*, is greenish white or yellow, with brown or pink sides, the dorsal line is usually pink or brown and the hairy warts are brown. The early stage caterpillar bores into the leaves from the underside, often leaving one side of the leaf uneaten. The moth is out in June. There are several related species in Europe with a similar metallic appearance. **52**

**aeacus** see **Troides**
**aega** see **Morpho**
**aegeria** see **Pararge**
**aegeus** see **Papilio**
**Aegocera tripartita** see under **Hecatesia**

**Aemilia** Kirby (ARCTIIDAE)
There are 2 species in this New World genus: *A. ambigua* (the type-species), and *A. roseata* Walker which is probably not correctly placed here. It is probably quite closely related to the huge genus *Halisidota*.

**ambigua** Strecker RED-BANDED AEMILIA
U.S., Mexico 42–49 mm. (1·65–1·93 in.)
This is a fairly common woodland species found in the Rockies of Colorado to the mountains of S. Mexico. The pale areas of the wings are sparsely scaled and partly transparent. A similarly patterned Geometrid, *Caripeta piniata* Packard, flies with *A. ambigua* in N. America and is possibly a mimic of it. **368f**

AEMILIA, RED-BANDED see **Aemilia ambigua**
**aeneana** see **Commophila**
**aeneas** see **Parides**
**aenicta** see **Polysoma**
**aenone** see **Ogyris**

**Aeolanthus** Meyrick (XYLORYCTIDAE)
Species in this genus of micro-moths occur from India to China, with one species described from Tasmania, although it may be incorrectly placed in this genus. Many of the species are richly coloured, ranging from those similar to the species illustrated to some darker coloured ones.

**sagulatus** Meyrick
India, Pakistan 15–20 mm. (0·59–0·79 in.)
The original specimens of this species were collected in Bengal. The caterpillars tie together the edges of leaves on which they feed. Very little is known of their biology but *A. sagulatus* has been reared from leaves of *Cedrela toona* (Meliaceae). **43j**

**Aeria** Hübner (NYMPHALIDAE)
There are 7 species of butterfly in this genus of the subfamily Ithomiinae. All of them are black, or brown, and yellow, and rather Dioptid-like in colour pattern. They are found in C. America and tropical S. America.

**agna** Godman & Salvin
C. America, N. America 37–52 mm. (1·46–2·05 in.)
*A. agna* is typical in pattern of the other species of its genus. Most females are less strongly marked than the illustrated male. **131k**

**eurimedia** Cramer GREEN SWEET-OIL BUTTERFLY
C. America, tropical S. America 31–45 mm. (1·22–1·77 in.)
Damp forested regions are preferred by this small butterfly. The juxtaposition of black and yellow on the wings creates an impression of green, although no green scales are present. There is some variation in the colour pattern of the wings. *A. olena* Weymer, from Brazil, is very similar but generally smaller.

**olena** see under **A. eurimedia**

**aesacus** see **Ornithoptera**

**Aethes** Billberg (COCHYLIDAE)
A large genus of micro-moths with well over 50 species in Europe and Asia and many more in N. and C. America. Only one species is known from India.

**smeathmanniana** Fabricius
Europe including Britain, Asia, N. America 14–18 mm. (0·55–0·71 in.)
The caterpillar of this widespread moth feeds on flowerheads of various plants including species of *Anthemis*, *Achillea* and *Centaurea* (Compositae). The forewings are straw-yellow with dark brown markings; the hindwings are pale grey. The moth is widespread in N. America, where it is believed to have been introduced from Europe.

**aethiops** see **Erebia** and **Syrmatia**

**Aethiopsestis** Watson (THYATIRIDAE)
The description of this genus of 3 species from Rhodesia, Tanzania and South Africa in 1965 was the first published record of the family Thyatiridae from the continent of Africa, although it was known earlier to Dr Kiriakoff, Ghent, and Dr Pinhey, Bulawayo. The oriental genus *Mimopsestis* is probably its closest relative and resembles *Aethiopsestis* in colour pattern.

**mufindiae** Watson
Tanzania 43 mm. (1·69 in.)
This is a brownish grey moth, with poorly marked hindwings but with a broad white transverse band on the forewing. The male genitalia are highly diagnostic and provide the most reliable means of separating *A. mufindiae* from the remaining two species of its genus.

**affinis** see **Danaus** and **Xyleutes**
AFRICAN
ARMYWORM see **Spodoptera exempta**
ATLAS see **Epiphora vacuna**
CARNATION WORM, SOUTH see **Epichoristodes acerbella**
COMMON WHITE see **Anaphaeis creona**
GIANT SWALLOWTAIL see **Papilio antimachus**
GRASS BLUE see **Zizeeria knysna**
LEAF see **Kallima rumia**
MAP-BUTTERFLIES see **Marpesia**
MIGRANT see **Catopsilia florella**
MONARCH see **Danaus chrysippus**
MOON-MOTH see **Argema mimosae**
PORCELAIN see **Marpesia camillus**
RINGLET see **Ypthima asterope**
SMALL WHITE see **Dixeia doxo**
SNOUT see **Libythea labdaca**
WOOD WHITE see **Leptosis alcesta**
**africana** see **Bena**
**agamemnon** see **Graphium**

**Aganais** Boisduval (HYPSIDAE)
About 9 species are placed in this genus of yellow, brown and black moths. They are found in Africa and tropical Asia. (Syn. of **Psephea**).

**orbicularis** Walker
India, S.E. Asia 67–82 mm. (2·64–3·23 in.)
The whole of the male forewing is covered with raised, modified scales, presumably scent scales. Patches of scent scales are found on the wings of many moths and butterflies, but *A. orbicularis* is possibly unique in that the whole of a wing is modified in this respect. Females (not illustrated) are quite different in colour pattern; they have mostly orange forewings, and orange-yellow hindwings, spotted with black, and with white-veined, dark brown outer marginal bands. The spiny caterpillar is banded with black and white and has a reddish brown head. **381a**

**aganice** see **Bematistes**
**aganippe** see **Delias**

**Agapeta** Hübner (COCHYLIDAE)
A European and W. Asian genus of micro-moths with about 10 known species, some strikingly coloured.

**hamana** Linnaeus
Europe including Britain, Asia 17–25 mm. (0·67–0·98 in.)
This species is similar to *A. zoegana* with bright yellow forewings but differs in not having the yellow patch enclosed in brown on the apex of the wings. The caterpillar has been recorded on *Ononis* (Leguminosae) where it rolls leaves and feeds on the seeds. **22**

**zoegana** Linnaeus
Europe including Britain 17–21 mm. (0·67–0·83 in.)
The bright yellow forewings are brown near the apex. When the wings are closed a patch of yellow with a brown spot in the middle of the forewings and brown hindwings make this a conspicuous little moth. The caterpillar feeds in the roots of *Scabiosa columbaria* (Dipsaceae). The moth is out from June to August and is common wherever the food-plant of the caterpillar grows. **21**

**Agarista** Leach (AGARISTIDAE)
There is a single species in this genus of moths.

**agricola** Donovan
Timor, New Guinea, N. Australia 54–68 mm. (2·13–2·68 in.)
The illustrated moth is a specimen of the subspecies *A. agricola agricola* from Australia. Other subspecies differ markedly in the colour pattern and mostly have less red on the hindwing. Males (not illustrated) of *A. agricola* have a less extensive pale green area at the base of the forewing. The caterpillar is banded with orange and black, and has several long, clubbed hairs on each segment. It feeds on species of the genus *Cissus*. **399n**

**Agaristidae**
This almost worldwide family of day-flying or nocturnal moths is best represented in the Old World tropics. It is absent in Europe and represented by a few species in North America. Most of the species can be distinguished from those of the closely allied Noctuidae by the antennae which are thickened towards apex and often hooked; many of them are brightly coloured. The caterpillars are mostly orange and black, and feed on members of the Grape family (Vitaceae) and Evening Primrose family (Onagraceae). The males of *Hecatesia*, *Aegocera*, *Musurgina*, *Pemphegostola*, and *Platagarista* are capable of producing sound. See *Agarista*, *Alypia*, *Choerapais*, *Cocytia*, *Copidryas*, *Erocha*, *Hecatesia*, *Musurgina*, *Pemphegostola*, *Platagarista*, *Phalaenoides*, *Psychomorpha*, *Weymeria*.

**Agathia** Guenée (GEOMETRIDAE)
The majority of the moths in this genus are bright green and have a characteristic wingshape. They are mostly found from the Indian continent through Indonesia to Australia. Some species occur on Fiji and a few have been described from Africa. Little is known of their biology.

**arcuata** Moore
Sri Lanka, N. India, Pakistan, Thailand, Indonesia, Java and Borneo 42 mm. (1·65 in.)
There is little variation in pattern over the range of this species. There are several other similarly coloured

species in the genus, with slight differences in pattern. Nothing is known of its biology. **294**b

**succadanea** Warren
Malaysia 45 mm. (1·77 in.)
This moth is found on the higher parts of Mount Kinabalu in Sabah (Malaysia). There are many related, similar looking species, all with the bright green wing colour, from Borneo to India. The underside of the wing is a pale yellowish-brown with no trace of the green. **294**v

**agathina** see **Doxocopa** and **Mylothris**

**Agathiphaga** Dumbleton (AGATHIPHAGIDAE)
Two species are known in this genus of micro-moths, one in Fiji the other in Australia. Both have caterpillars which bore into the seeds of Kauri Pine *(Agathis australis)*. (See also under Eriocraniidae.)

**queenslandensis** Dumbleton
Australia (Queensland) 13 mm. (0·51 in.)
Broadly similar externally to species of Eriocraniid, this moth is small with rather pointed fore- and hindwings. It is rather a sombre coloured species with brown markings on a grey-brown background. The caterpillar has been reared from the seeds of Kauri Pine. It pupates within the seed, the moth emerging through a small hole which was made by the caterpillar. This moth can cause up to 20 per cent loss of fertile seeds of Kauri Pine, which can be serious when new areas of this valuable timber tree are to be planted.

**Agathiphagidae**
A family of micro-moths with only one genus. This family has species in Australia and Fiji. The family is related to the Eriocraniidae and is one of the primitive families of moths, only recognized as distinct in 1967. See *Agathiphaga*.

**agathon** see **Aporia**
**agathyrsus** see **Alcides**

**Agdistis** Hübner (PTEROPHORIDAE)
A small genus of plume-moths with long and slender, undivided wings and very long legs. They are found in Europe, N. Africa, and N. America.

**bennetii** Curtis
Europe 22–25 mm.
including Britain (0·87–0·98 in.)
The moths are locally common in S. and E. England where they live in salt marshes. There are 2 broods each year, the first in June, with the second in late July and August. When at rest the moth rolls its forewings in the typical plume-moth manner and pointing slightly forward (see figure **76**). It flies after dark and is attracted to light. The eggs are laid on Sea Lavender *(Limonium humile)*, the caterpillar feeding on the leaves of this plant. It is green with many small white spots and a yellowish line along the sides. There are 2 conical projections on the first segment behind the head (prothorax) and a reddish horn on the ninth abdominal segment. Caterpillars hatching in the autumn overwinter in withered leaves, and during the winter storms survive submergence in the sea water. **56**dd

**agestis** see **Aricia**
**agestor** see **Papilio**
**aglaia** see **Delias** and **Milionia**

**Aglais** Dalman (NYMPHALIDAE)
The butterflies in this genus are found mainly in Europe and temperate Asia, but with at least one species in N. America and another reaching Kashmir. Some are well known, like the illustrated Small Tortoiseshell, because they hibernate in the adult state. They spend the winter in outbuildings or garden sheds where they may be found during the winter, or seen as they flap about on a slightly warmer winter's day.

**urticeae** Linnaeus SMALL TORTOISESHELL
Europe including Britain, 44–50 mm.
temperate Asia including Japan (1·73–1·97 in.)
This butterfly is on the wing from early spring until autumn. The adults hibernate and emerge in the warmer late winter or early spring days. There are often 2 broods a year. The caterpillars, which feed on nettles *(Urtica)*, are variable in colour ranging from bright yellow to almost black and covered with black specklings and short hairs. There is a black line down the centre of the back, bordered by a clearer line, with a yellowish line on the sides. The longer spines are yellowish with black tips. The butterfly is common over most of its range. One subspecies occurs in Japan while there are two distinct ones in Europe; several others have been described. This species is often found in houses in winter. **196**

**aglaja** see **Mesoacidalia**
**aglaope** see **Parides**
**aglea** see **Danaus**

**Aglaope** Walker (ZYGAENIDAE)
About 40 species have been described in this genus, most of them with rather transparent wings. They are known to occur in China, N. India, Burma and Formosa. They are different from most of the others in the subfamily Chalcosinae in not having iridescent metallic scales.

**bifasciata** Hope
India, Sikkim 70–80 mm. (2·76–3·15 in.)
This species varies in the amount of the dusky colour on the wings. There are related species in Java and Burma. The males and females are similar but the male antennae are feathery (pectinate) while the female's are simple. Nothing seems to be known of the life history of this species. **45**c

**Aglia** Ochsenheimer (SATURNIIDAE)
The only species in this moth genus is one of the most distinctively patterned Saturniids in temperate regions of the Old World.

**tau** Linnaeus TAU EMPEROR
Europe, 55–90 mm.
temperate Asia to Japan (2·17–3·54 in.)
This is often common in much of its range. The high degree of individual variation has produced a multitude of latin names for this species. Specimens can vary from yellow through brown to black, and the eyespots can vary considerably in size and colour. The moths are on the wing between March and June. Males are dayfliers; females seldom move during the day. The caterpillars feed from June to July on various deciduous trees including beech *(Fagus)*, lime *(Tilia)*, oak *(Quercus)* and birch *(Betula)*; they are green, with oblique, lateral stripes when fully grown. The species overwinters as a chrysalis. **331**k

**agna** see **Aeria**
**agondas** see **Elymnias**

**Agonoxena** Meyrick (AGONOXENIDAE)
This is the only genus in this family of micro-moths. They generally have narrow pointed wings with reddish stripes along them. The caterpillars feed on the leaves of palms *(Palmae)* in the Australian-New Guinea region.

**argaula** Meyrick
Fiji, Samoa 14–16 mm. (0·55–0·63 in.)
The caterpillars of this moth are pests on coconuts *(Cocos nucifera)* and other palms. Usually they are not present in sufficient numbers to cause much loss of production, but at times severe outbreaks may occur which can be damaging. The shape of this species is typical of others in the genus. **44**k

**Agonoxenidae**
A small family of micro-moths related to the Gelechiidae and found from Indonesia to Australia and in some Pacific Islands. See *Agonoxena*.

**Agraulis** Boisduval & Leconte (NYMPHALIDAE)
As in other genera of the butterfly subfamily Heliconiinae, the caterpillars of *Agraulis* feed on species of passion flower *(Passifloraceae)* and are thought to pass on plant toxins to the adult butterfly which is chemically protected from predators as a result.

**vanillae** Linnaeus GULF FRITILLARY or SILVER-SPOTTED FLAMBEAU 65–70 mm. (2·56–2·76 in.)
N. America through tropical America to Argentina
*A. vanillae* occasionally reaches British Columbia (Canada) but is generally found only in S. N. America. This is a common butterfly of open spaces where it frequently visits flowers in search of nectar. The reddish orange coloration of the upper surface of the wings is in sharp contrast with the silver spotted under surface. The illustrated specimens from Trinidad belong to the subspecies *A. vanillae vanillae*. **143, 148**

**Agriades** Hübner (LYCAENIDAE)
There are about 11 species in this genus of blue butterflies. They are found in N. America, Europe and C. Asia, E. to China.

**aquilo** Boisduval ARCTIC BLUE
Arctic Europe, Asia, Canada, 19–27 mm.
U.S. (Alaska) (0·75–1·06 in.)
Like many other Arctic moths and butterflies, this species is smaller and more hairy than related species from further south. Males of *A. aquilo* are a lustrous, bluish grey above, with brown margins; females are brown. Both sexes are greyish brown beneath, with black, brown and white markings. The caterpillar feeds on a species of *Diapensia*, a group of chiefly Arctic and alpine evergreen shrubs, on Alpine Milk-vetch *(Astragalus alpinus)* and also possibly on cranberry and its allies *(Vaccinium)*.

**Agrias** Doubleday (NYMPHALIDAE)
A huge S. American butterfly genus with species found in most of the warmer parts of the continent. The genus has a number of species which compete as 'the most beautiful species' in many books and are much sought after by collectors. They are fast, powerful fliers with a large patch of scent scales on the hindwings.

**claudia** Schulze
S. America 70–90 mm. (2·76–3·54 in.)
A widespread and very variable species with colours ranging from bright red through orange; some varieties lack the blue on the wing and about a hundred forms and subspecies have been named. The specimen illustrated in figure **230** is from Peru. The underside is red as on the forewings but the apex is white. The wing is strongly patterned with blue, black-edged eyespots round the margin of the wing and a pale silvery band. The underside varies in pattern but probably not as much as the upperside. A study of the genetics of the different colour forms to one another, would be very interesting. The specimen photographed live, **219**, is from Brazil and is a different subspecies from the set specimen. **230**j

**agricola** see **Agarista**

**Agriphila** Hübner (PYRALIDAE) GRASS-MOTHS
This genus of Pyralid Grass-moths occurs throughout the N. Hemisphere in Europe, N. America and Asia. The moths, with their long palps sticking out in front of the head in a characteristic manner, are common in pastures and meadows where the caterpillars feed on various grasses. The moths generally sit on the grass stem with the head downwards and with the wings rolled round the body giving a wedge-shaped appearance. Their short zig-zag flight through the long grass as they are disturbed is a common feature of meadows and pastures in summer.

**tristella** Denis & Schiffermüller
Europe 24–31 mm.
including Britain (0·94–1·22 in.)
This species is widespread in pastures and is known to migrate at times in large swarms. The pattern is variable, from whitish to shades of yellow-brown. If the moth is disturbed it flies rapidly before settling almost invisibly again on a grass stem. The caterpillar feeds on various species of grass (Gramineae). **59**

**agrippina** see **Thysania**

**Agrius** Hübner (SPHINGIDAE)
There are only 5 known species in this almost cosmopolitan genus of hawk-moths. The distribution of the individual species is equally extensive.

**cingulatus** Fabricius SWEET POTATO HORNWORM
S. U.S., C. and S. America 80–115 mm.
Hawaii (3·15–4·53 in.)
Black light (ultraviolet) and chemical attractant traps have been used in attempts to control this species in the S. U.S. where its caterpillars are sometimes pests of cultivated sweet potato *(Ipomoea)*. Although it is common there, and may stray as far N. as Nova Scotia, *A. cingulatus* has a centre of distribution in tropical S. America. The pink colour of the hindwing is replaced by pinkish white in a few specimens (see figure). The party uncoiled proboscis can be seen to the left of the head of the illustrated specimen. **344**l

**convolvuli** Linnaeus CONVOLVULUS HAWK-MOTH
Temperate tropical Africa, Europe, 80–120 mm.
Asia including Japan, Australia (3·15–4·72 in.)
the islands of the Pacific
This species is a migrant to Britain and continental Europe from N. Africa. Eggs laid in Europe produce moths later in the summer but these fail to overwinter or produce offspring. The blooms of honeysuckle *(Lonicera)*, *Petunia* and tobacco *(Nicotiana)* frequently attract these moths whose long proboscis (up to 7·5 cm., 3 in. long) is able to penetrate to the nectaries of these flowers. Larval foodplants include *Convolvulus*, sweet potato *(Ipomoea)* and other species of the family Solanaceae. **345**

**Agrodiaetus** Hübner (LYCAENIDAE)
There are about 20 species in this genus of butterflies. They are found chiefly in Europe, Turkey and the Middle East.

**dolus** Hübner FURRY BLUE
S.W. Europe, Italy 25–38 mm. (0·98–1·5 in.)
This butterfly derives its common name from the large area of brown scent scales on the otherwise pale blue, pale grey or white male forewing. It is found from sea level to 1830 m. (6000 ft) in hilly countryside. The foodplants of the caterpillar include lucerne *(Medicago)* and sainfoin *(Onobrychis)*. The upper surface of the female is brown. The under surface of both sexes is pale yellow-grey or yellow-brown.

**Agrotis** Ochsenheimer (NOCTUIDAE)
This is a very large, almost cosmopolitan genus whose caterpillars (cutworms) are amongst the most serious insect pests of food-crops. The typically greasy-looking, granulate caterpillar normally burrows at the base of its foodplant, emerging at night to feed. Some idea of its prevalence as a pest genus can be gathered from the fact that in Hawaii alone there are 26 endemic species.

**infusa** Boisduval BOGONG MOTH
E. Australia 35–40 mm.
(New South Wales, Queensland) (1·38–1·57 in.)
This is a pest in S.E. Australia where the variably coloured, nocturnal caterpillars feed on numerous winter crops and grasses. The adults migrate in the spring to mountains above 1450 m. (4800 ft) in the Australian Alps to gather in caves and rock crevices where they remain dormant during the hot, dry summer months. They disperse in the autumn to lower elevations. The aestivating moths are harvested and eaten by the Australian aborigines. **391**e

**ipsilon** Hufnagel DARK SWORD-GRASS or BLACK CUTWORM
Canada, U.S., C. & S. America 42–50 mm.
Europe, Asia, Australia, Hawaii (1·65–1·97 in.)
The Greek scientific name of this infamous moth refers to the Y-shaped marking on the forewing. *A. ipsilon* is one of the most destructive species of its genus in the E. U.S. and is often equally destructive in other parts of its range. Its caterpillar, which appears to prefer damp soil, attacks the roots, stem and leaves of cabbage *(Brassica)*, strawberry *(Fragaria)*, grape *(Vitis)*, corn *(Zea)*, cotton *(Gossypium)*, tobacco *(Nicotiana)* and several other plants. **391**f

**segetum** Denis & Schiffermüller TURNIP MOTH
Europe, temperate Asia, 35–40 mm.
Africa (1·38–1·57 in.)
The moth varies considerably in coloration, but is similar in colour pattern to the slightly larger *A. ipsilon* (illustrated). In Britain, its caterpillar is probably the worst pest of root crops such as turnips and swedes *(Brassica)* but it will also attack many other crops. In Africa, damage is sometimes caused to bark at the base of young coffee trees *(Coffea)*, but more serious damage is suffered by *Brassica* crops.

AILANTHUS SILK-MOTH see **Samia cynthia**
AILANTHUS WEBWORM see **Atteva punctella**

**Alabonia** Hübner (OECOPHORIDAE)
This is a small genus of micro-moths, at present including only European species, which was recently separated from species in the larger related genus *Oecophora*. Other related genera, *Esperia* for example also have colourful species in the European region.

**geoffrella** Linnaeus: syn. *A. geoffroyella*
Europe including Britain 17–21 mm. (0·67–0·83 in.)
The shape of this species at rest is unmistakable (see figure). It is a common species over much of its range. The caterpillar is found on rotting wood where it may be feeding on the fungus developing there or on the rotting wood itself. The adults are about in May and June in hedges and woodlands where their slow undulating flight is reminiscent of species of *Adela*. They fly in the early morning, particularly in sunny weather, but also may be on the wing at almost any time of the day or night. **40**

ALBATROSS,
COMMON see **Appias paulina**
ORANGE see **Appias nero**
**albimaculata** see **Amauris**
**albinervis** see **Somabrachys**
**albipennis** see **Perisama bonplandi**
**albivitta** see **Mesenopsis**
**albofasciata** see **Saturnia**
**albula** see **Nola**

**Albulina** Tutt (LYCAENIDAE)
About 12 species of blue butterflies are placed in this genus of European and temperate Asian Lycaenidae.

**orbitulus** De Prunn ALPINE ARGUS
European Alps, Norway, 24 mm.
temperate Asia (0·94 in.)
This is typically a butterfly of alpine meadows, and is rarely seen below 1680 m. (5500 ft). It is sometimes common, especially in limestone districts. The male is sky-blue above, the female generally brown. The under surface of the wings is light grey in the male, light brown in the female with some indistinct markings. The caterpillar feeds on vetches of the genus *Astragalus*.

**Alcathoe** Edwards (SESIIDAE) CLEARWINGS
All the moths in this genus are wasp mimics and since wasps are generally avoided by predators, they obtain a measure of protection from this mimicry. The wasps they resemble belong to the genus *Psammocharidae*, the Tarantula-killer wasps. The caterpillars bore in roots and stems of *Clematis* (Ranunculaceae). There are several N. American species.

**autumnalis** Engelhardt
N. America (S.E. Texas), 32–40 mm.
Mexico (1·26–1·57 in.)
This moth was not discovered until 1946 when the American entomologist who studied the group recognized it as a new species. It was raised on roots of *Clematis*. The female is generally larger than the male. **27**b

**alcathoe** see **Euploea**
**alcesta** see **Leptosia**

**Alcides** (URANIIDAE)
There are 11 species in this genus of beautifully coloured, mostly tailed, day-flying moths. They are found in N. Australia, the Solomons, New Guinea and associated islands. The ground colour of the wings is black, bluish black or greenish black; the colour of the markings include every part of the spectrum.

**agathyrsus** Kirsch
New Guinea, Aru Island 77–98 mm. (3·03–3·86 in.)
This day-flying species is supposedly the model for the butterfly *Papilio laglaizei* which matches its circling flight high above the ground. The caterpillars of the related Uraniid species *Chrysiridia ripheus* and *Urania leilus*, from Madagascar and S. America respectively, feed on poisonous species of Euphorbiaceae, a choice of food which could produce distastefulness in the adult, and, if also the food of *A. agathyrsus*, would support the hypothesis that the latter is a model for *P. laglaizei*. **314**b

**aurora** Godman & Salvin
New Guinea, 75–82 mm.
the Bismarck Archipelago (2·95–3·23 in.)
Little is known about this moth. It apparently flies by day like other species of its genus and is fairly common at times. Some collectors consider *A. aurora* to be one of the most beautifully coloured species of butterflies or moths. The modified scales visible in the illustration are present only in the male; they probably function immediately prior to mating by dispersing aphrodisiac scents (sex pheromones). **317**a

**zodiaca** Butler
New Guinea, 85–100 mm.
N. Australia (3·35–3·94 in.)
This is a day-flying, papilionid-like species of tropical rain-forests. The bronze, olive and purplish pink iridescence of the wing markings is a distinctive feature of this beautiful species. It is a frequent visitor to flowers during the day, and in the evening circles at tree canopy level. **316**

**alcinoe** see **Bunaea**
**alciphron** see **Lycaena**

**Alcis** Curtis (GEOMETRIDAE)
A very large worldwide genus of moths with many European and Asian species.

**repandata** Linnaeus MOTTLED BEAUTY
Europe including Britain, 38–46 mm.
Asia, N. India, Japan (1·5–1·81 in.)
This moth is common and is out in June and July. It is very variable in pattern and has several named forms (polymorphism). The caterpillar is pale greenish brown with a yellowish tinge. It has darker markings and a dark line across the back. It feeds on the leaves of oak *(Quercus)*, beech *(Fagus)*, blackthorn *(Prunus)* and other trees and bushes. A study of the occurrence of the different colour forms and an investigation of their frequency would be an interesting problem for a keen collector.

**alcmena** see **Eunica**
ALDER see **Acronicta alni**
**alecto** see **Jamides**
**alector** see **Echenais**

**Alesa** Doubleday (NEMEOBIIDAE)
About 4 species are commonly placed in this S. American butterfly genus. There is a striking difference in colour pattern between the sexes.

**amesis** Cramer BLACK UNDERLEAF BUTTERFLY
N. S. America 33–38 mm. (1·3–1·5 in.)
The male is black and dark grey, with dark iridescent violet at the outer margins of the wings. The female is dark yellowish brown, with larger brown stripes. Like *A. prema*, it rests on the under surface of a leaf.

**prema** Godart GREEN DRAGON
Tropical S. America 36–48 mm. (1·42–1·89 in.)
The male of this rare, large, iridescent green and blue species is one of the most richly coloured S. American species of its family. It is nevertheless well camouflaged against green foliage when at rest. The female (not illustrated) is yellowish brown, with darker brown lines and spots and with the green markings restricted to the base of the forewing and the outer margin of the hindwing. **263**dd

**aletta** see **Hypocrita**
**alexandra** see **Colias**
ALEXANDRA SULPHUR see **Colias alexandra**
ALEXANDRA'S BIRDWING, QUEEN see **Ornithoptera alexandrae**

**alexandrae** see **Ornithoptera**
**alexanor** see **Papilio**
**alexis** see **Glaucopsyche**
ALFALFA see **Colias eurytheme**
**alimena** see **Hypolimnas**
**alinda** see **Hylesia**
ALINDA see **Hylesia alinda**
**aliphera** see **Eueides**
**aliris** see **Thauria**

**Allancastria** Bryk (PAPILIONIDAE)
There is a single known species in this genus of butterflies. It is externally similar to the species of *Parnalius* in colour pattern but differs in several structural characters.

**cerisyi** Godart EASTERN FESTOON
S.E. Europe, Turkey, the Middle E. 40–58 mm. (1·57–2·28 in.)
This is a butterfly of rough ground on hillsides and mountains where it flies at elevations of up to 1220 m. (4000 ft). The male (not illustrated) is much less heavily marked than the female, especially along the margins of the wings. Its caterpillar, like most of its allies in the genus *Parnalius*, feeds on species of *Aristolochia*. It varies from yellow to reddish brown and has red tubercles. **94**j

**Allora** Waterhouse & Lyell (HESPERIIDAE) AWL BUTTERFLIES
A genus with 2 species of butterflies which are found from the Moluccas to Australia and in New Guinea and the Solomons. They are known as Awl butterflies and are related to Skippers.

**doleschalli** Felder PEACOCK AWL
Moluccas, New Guinea, Bismarck Archipelago, Solomon Is., and N. Queensland 40 mm. (1·57 in.)
This metallic blue-green, black-winged butterfly is found along the edges of rain-forests where it generally flies high up. Several subspecies have been described in different parts of its range but little seems to be known of its biology. **80**d

**Allostylus** Hering: syn. *Cossula* (COSSIDAE)
This American genus of moths contains few species and little is known of its biology. The specimen illustrated is typical of the genus.

**caerulescens** Schaus
Costa Rica 65 mm. (2·56 in.)
The biology of this species is unknown; the sexes are similar although the females are generally larger than the male. **8**c

**allottei** see **Ornithoptera**
**almana** see **Junonia**
ALMOND see **Ephestia cautella**
**alni** see **Acronicta**
**alpina** see **Acerbia**
ALPINE,
DISA see **Erebia disa**
ROSS's see **Erebia rossii**
ALPINE
ARGUS see **Albulina orbitulus**
BLUE see **Icaricia shasta**
SKIPPER see **Oreisplanus munionga**

**Alsophila** Hübner (GEOMETRIDAE) WINTER MOTHS
A widespread genus of moths across the N. Hemisphere from Europe to Japan and N. America. Many are pests, causing defoliation of fruit trees and damage to buds. In some years outbreaks occur when large numbers of their caterpillars appear while in other years the numbers are much smaller.

**pometaria** Harris FALL CANKER WORM
U.S., Canada 25–30 mm. (0·98–1·18 in.)
The forewing is light grey with two diffuse white bands and a dark spot in the middle of the wing; the hindwing grey without any mark. The female has reduced wings and cannot fly. The caterpillars, striped either green and white or brown and white, are injurious to apple *(Malus)*, elm *(Ulmus)* and other trees. The moths are out in November and December and can be found in the N.E. States of America, Utah and California, as well as S.E. Canada.

**alta** see **Neptis**
**altella** see **Nepticula**

**Alucita** Linnaeus: syn. *Orneodes* (ALUCITIDAE) MANY PLUMED MOTH
Although the moths of this genus are among the most distinctive looking of all the lepidoptera, they are not widely known and most are rather small and fragile looking. They show one of the most interesting modifications of wingshape in the family; all the species in the genus having fore- and hindwing divided into 6 parts. As a result the wings have a feathery look and the moths are sometimes called 'feather wings'. The genus is worldwide but little is known of its biology and there is still dispute amongst specialists about the relationships of the family.

**dohertyi** Walsingham
Kenya 30–40 mm. (1·18–1·57 in.)
*A. dohertyi* is known only from one part of Kenya where it has been collected in some numbers. Nothing is known of its biology. A study of the life history and biology of this species, which is much larger than any others at present known in the family, would be most interesting. **47**b

**hexadactyla** Linnaeus MANY PLUMED MOTH
Europe including Britain, Turkey 13–16 mm. (0·51–0·63 in.)
This species is common over most of its range and when noticed, causes interest because of the unusual shape of the wings, each being divided into 6 slender segments, edged with hairs. The caterpillar feeds in the flowerbuds of honeysuckle *(Lonicera)* and the moth is out in August and September, after which it goes into hibernation to re-appear in May and June. It hibernates in dense masses of creepers, outhouses or similar sites and in mild weather may be active in winter.

**montana** Cockerell
U.S., Canada 11–16 mm. (0·43–0·63 in.)
This species, which is similar in appearance to *A. hexadactyla*, flies in April and had been recorded from New York across to the W. States. It has been suggested that the caterpillar mines the leaves of Snowberry *(Symphoricarpus)* but there seems to be an element of doubt about the host plant and also whether this species is in fact quite distinct from the European one, or has been accidentally introduced. The adult has the characteristic 6 divisions of the wings.

**Alucitidae** (ORNEODIDAE) MANY PLUMED MOTHS
A small, worldwide, family immediately recognizable by its wing-shape. The fore- and hindwings are divided into 6 long lobes (rarely short lobes), with hairs round the edges. The family has generally been associated with the Pterophoridae (Plume-moths) and has only recently been moved into the Copromorphoid group, although this relationship was suggested earlier by a famous English lepidopterist, Edward Meyrick. In a few species there are 7 lobes to the hindwings. The caterpillars tunnel in flowers, buds and in some plants cause gall formation. See *Alucita*.

**alveola** see **Thalamarchella**
**alveus** see **Pyrgus**

**Alypia** Hübner (AGARISTIDAE) FORESTERS
The 14 or so species of this genus are small day-flying moths, mostly black or dark brown, with yellow forewing markings and yellow or white hindwing markings. They are found in Canada, U.S., Mexico and tropical America.

**maccullochii** Kirby
Canada N. to Alaska 25 mm. (0·98 in.)
This northern species resembles *A. octomaculata* in coloration but usually has 3 yellow spots, not 2, on the forewing.

**octomaculata** Fabricius EIGHT-SPOTTED FORESTER
E. Canada, U.S. 26–33 mm.. (1·02–1·3 in.)
Some specimens from Canada lack the basal white spot on the hindwing, but the species is otherwise well named, as the illustration shows. The caterpillar feeds on vine *(Vitis)*, woodbine *(Lonicera)*, and Virginia creeper *(Parthenocissus)*. **402**c

**wittfeldi** Edwards WITTFELD'S FORESTER
S.E. U.S. 28–36 mm. (1·1–1·42 in.)
This species is unknown outside Florida where the caterpillar has been recorded from Japanese persimmon *(Diospyros)*. It is similar in colour pattern to *A. octomaculata* (illustrated), with which it was once confused, but differs in several details of the pattern.

**amantes** see **Narathura**

**Amarynthis** Hübner (NEMEOBIIDAE)
There is one known species in this genus of butterflies.

**meneria** Cramer
Tropical S. America 36–42 mm. (1·42–1·65 in.)
There are several variations of the colour pattern illustrated on the plate. Some of the variation is geographical, but some colour forms occur throughout the range of this butterfly. The white spots are absent in some specimens, for example, while in others the red band is greatly reduced in width. Except for the red outer bands on the wings this is remarkably like the C. and S. American and S. N. American Hypsid moth *Composia credula* Fabricius. As both species occur in Guyana, there is the possibility of a mimetic association. **268**h

**amasa** see **Zeltus**

**Amastus** Walker (ARCTIIDAE)
There are over 50 species in this tropical American genus of moths. Most have partly brown wings, marked with darker transverse bands, and a partly red or orange abdomen.

**adela** Schaus
Brazil 44–52 mm. (1·73–2·05 in.)
The red, white and black *A. adela* is rather more colourful than most species of its genus. Its life history is unknown. **368**b

**Amata** Fabricius (CTENUCHIDAE)
The species of this genus have been combined at times with those of *Syntomis* and there is currently doubt about its validity as a separate genus. The caterpillars of nearly all species feed on living foliage, but *A. trigonophora* feeds on fallen leaves and flowers.

**germana** Felder
China, Formosa, Japan 24–32 mm. (0·94–1·26 in.)
The colour pattern of this species is rather similar to that of *Syntomis phegea* (illustrated), except that the abdomen is marked with consecutive bands of black and yellow.

**amathea** see **Anartia**

**Amathes** Hübner: syn. *Xestia* (NOCTUIDAE)
This is a large genus of moths represented in N. America but chiefly temperate Asian and European in distribution.

**c-nigrum** Linnaeus SETACEOUS HEBREW CHARACTER OF SPOTTED CUTWORM
Europe, temperate Asia, N. America 36–48 mm. (1·42–1·89 in.)
This is a common species in much of its range and is sometimes a pest of food-crops. Its caterpillar, a cutworm, feeds at night on little dock *(Rumex)*, chickweed *(Stellaria)*, groundsel *(Senecio)* various legumes, grasses and cultivated crops, including fruit-trees. The moth is variable in coloration and can be purplish black or brownish pink in ground colour. *A. c-nigrum* overwinters as a partly grown caterpillar. **391**j

**Amathusia** Fabricius (NYMPHALIDAE)
A genus of butterflies from S. E. Asia and the Philippines with some 40 species and subspecies. It is in the subfamily Amathusiinae. The species in the genus are mostly large, broad winged butterflies which fly in forests, usually late in the evening or early morning.

**phidippus** Johanssen PALM KING
Burma, Malaysia, Philippines, Indonesia 100–125 mm. (3·94–4·92 in.)
The brown upperside of the male's forewing has faint orange-yellow patches near the apex. The females have a more boldly marked pattern. The undersides of the

wings are pale, with longitudinal stripes and eyespots on the margins. The green caterpillar feeds on Coconut palms *(Cocos)* where it can be found on the midrib but it is difficult to see because its colouring matches the background. The pupa is generally suspended from a leaf of the foodplant. The butterfly is on the wing at dawn and again at dusk and may be seen flying rapidly round the tree tops in December and January. In the Philippines it is common all the year, and is said to fly low in banana *(Musa)* and bamboo (Graminaea) groves.

**Amathusiinae** (NYMPHALIDAE)
A group of Nymphalid butterflies, at one time thought to be a distinct family but now considered as a subfamily of the Nymphalidae. The butterflies in this subfamily are mostly rather large, broad-winged species generally flying towards dusk (crepuscular). They have often been associated with Brassoline butterflies (q.v.).

**Amathuxidia** Standinger (NYMPHALIDAE)
This genus belongs to the subfamily Amathusiinae. The species are large broad-winged butterflies, generally flying in jungle areas towards dusk.

**amythaon** Doubleday KOH-I-NOOR BUTTERFLY
Pakistan, India, S.E. Asia, Indonesia, Philippines 110–113 mm. (4·33–4·45 in.)
A dark brown butterfly with a broad band of pale blue on the forewing in the male. These bands are dark yellow in the female. The hindwing has a small lobe, like a short tail, at the hind apex. The underside is pinkish brown or pinkish blue with prominent dark lines and large eyespots. It is a jungle butterfly, generally flying towards dusk.

**Amauris** Hübner (NYMPHALIDAE)
About 25 species comprise this genus of the subfamily Danainae. They are solely African in distribution. The only other Danainae genus found in Africa is *Danaus*, a chiefly Asian group of butterflies.

**albimaculata** Butler LAYMAN BUTTERFLY
Africa S. of the Sahara (S. to Cape Province, South Africa) 60–84 mm. (2·36–3·31 in.)
Mimics of this moist forest species include female forms of *Papilio echerioides*, *P. jacksoni* and *P. dardanus*. Its caterpillars feed on species of *Cyanchum* and *Tylophora* (Asclepiadaceae). The female (not illustrated) is less dark in ground colour and has paler yellow markings. There is variation between subspecies of *A. albimaculata* in the width of the yellow band on the hindwing; the forewing spots are yellow, not white in some specimens of *A. echeria*. **127**c

**echeria** Stoll CHIEF BUTTERFLY
Africa S. of the Sahara 60–82 mm. (2·36–3·23 in.)
Some subspecies of this butterfly resemble *A. albimaculata* (illustrated) but the under surface of the abdomen is dark, not white like that of the latter. It is mimicked by females of *Papilio echerioides* and *P. jacksoni*. The first full account of the life history of *A. echeria* was published recently by G. Van Son, based on data collected by G. C. Clark, both South African lepidopterists. The foodplants are species of *Tylophora* and *Cyanchum* (Asclepiadaceae). (See also under *Graphium leonidas*.)

**niavius** Linnaeus FRIAR BUTTERFLY
Tropical & subtropical Africa 80–115 mm. (3·15–4·53 in.)
This is typically a forest species. It varies somewhat in pattern, especially in the amount of white on the hindwing. In subspecies *A. n. dominicanus* (not illustrated) much of the hindwing is white. Several palatable species mimic *niavius* in various parts of its extensive range, 2 of the better known ones being a female form of *Papilio dardanus* and a species of *Hypolimnas* (Nymphalinae). Little is known about its life history. **126**

**Amauta** Houlbert (CASTNIIDAE)
This is a small genus of about 8 large, brightly coloured, tropical C. and S. American moths. The hindwings are conspicuously marked with red, yellow, violet or blue. The sexes are alike except in *A. hodeei*.

**cacica** Herrich-Schäffer
Tropical C. & S. America 100–135 mm. (3·94–5·31 in.)
This is one of the largest of the *Amauta* species. The scales at the base of the wings are raised, elongate and weakly iridescent green. Females (not illustrated) have a slightly less pointed apex to the forewing than the male, but are similarly dark brown, with a single narrow, transverse band on the forewing and an orange transverse band and a marginal row of orange spots on the hindwing. **45**r

**hodeei** Oberthür
Colombia 100–112 mm. (3·94–4·41 in.)
This rare species is remarkable because of the sexual dimorphism in its wing markings. The general colour pattern of the female wings is as in *A. cacica* except that the inner band on the hindwing is pale yellow except the anal margin. The male, in contrast, has no markings on the forewing and those on the hindwing reduced to an orange spot at the anal margin with a trace of pale yellow anteriorly, giving the appearance of a species of the unpalatable butterfly genus *Parides*.

**papilionaris** Walker
Tropical S. America 90–125 mm. (3·54–4·92 in.)
This is similar in pattern to *A. cacica*, but the markings of the hindwings are a bright violet-blue instead of orange and the outer row of spots is absent.

**Amaxia** Walker (ARCTIIDAE)
About 15 species of tropical American moths are placed in this genus, some of them certainly incorrectly.

**chaon** Druce
C. & tropical S. America 20–50 mm. (0·79–1·97 in.)
This is typical in colour pattern of several species of *Amaxia* except for the brilliant white metallic markings at the base of the forewing. The particularly large range of wingspan measurements is noteworthy; several of the males available for study are under half the size of most of the females. **368**j

**ambica** see **Apatura**
**ambigua** see **Aemilia**
**ambigualis** see **Scoparia**

**Amblypodia** Horsfield (LYCAENIDAE)
There are about 15 species in this tropical S.E. Asian genus of butterflies. The upper surfaces of both sexes are blue; the under surfaces are patterned brown.

**anita** Hewitson LEAF BLUE
Sri Lanka, India, Burma, Thailand, Malaya, Java 45–52 mm. (1·77–2·05 in.)
The males of this species are blue and iridescent purple above, with black borders; the females dark brown, with purple markings. The under surface of the wings so closely resembles a leaf that resting butterflies become almost indistinguishable from their surroundings. This is a species of dry open woodland. The caterpillars feed on *Olax* (Olacaceae).

**acron** Hewitson
Indonesia, New Guinea 30–35 mm. (1·18–1·38 in.)
The female of this butterfly differs above from the male (illustrated) in having much broader, black marginal bands. The under surface of both sexes is very pale, bluish grey, with brownish violet bands and spots. There is little variation in colour or pattern throughout the range of this superbly coloured species. **244**m

**Amblyptilia** Hübner (PTEROPHORIDAE)
A small plume-moth genus of about 12 species. They are found all across the Northern Hemisphere and into N. Africa. Each forewing is divided into 2 lobes by a narrow groove below the apex which is generally pointed. The hindwings are each divided into 3 lobes.

**punctidactyla** Haworth
Europe including Britain 18–21 mm. (0·71–0·83 in.)
A typical long-legged plume-moth, the caterpillar of which feeds on a variety of plants including *Stachys* (Labiatae), *Aquilegia* (Ranunculaceae) and *Geranium* (Geraniaceae), mostly in the seed heads. It is purplish-pink or greenish with a grey line along the back and whitish lines along the side. It pupates on the foodplant. The moth generally flies after dusk and is out in July with a second brood in September. It is found around hedgerows, woods and is widely distributed but not generally common over its range. **77**

**ameliae** see **Charaxes**
**americana** see **Pampa**
AMERICAN
CABBAGE WEBWORM see **Hellula rogatalis**
COPPER UNDERWING see **Amphipyra pyramidoides**
DAGGER see **Acronicta americana**
GREY DAGGER see **Acronicta interrupta**
MOON-MOTH see **Actias luna**
PEACOCK see **Anartia jatrophae**
WAINSCOT see **Mythimna unipuncta**
**americana** see **Acronicta** and **Malacosoma**
**amesis** see **Alesa**
**amethystus** see **Zeuxida**
**amica** see **Nyctemera**

**Ammobiota** Wallengren (ARCTIIDAE)
There are 2 species of moths in this genus: *A. festiva* (illustrated) and *A. culoti* Oberthür from E. Siberia.

**festiva** Hufnagel: syn. *A. hebe*
C. & S. Europe and temperate W. Asia 45–53 mm. (1·77–2·09 in.)
This colourful species is generally common in much of its range. It is on the wing in May, June and July. The caterpillar, which overwinters, is covered with black hairs along the back and brown hairs at the sides; it feeds on spurge *(Euphorbia)* and other plants in April and May. **369**h

**Amnemopsyche** Butler (GEOMETRIDAE)
This genus of moths is mainly found in Africa. The species are mostly patterned yellow and black, or white and black, with rather rounded wings. Their biology is unknown.

**sanguinaria** Bethune-Baker
Angola, Uganda 35 mm. (1·38 in.)
This is an exception in the genus in being red and black. Nothing is known of its biology. **294**l

**Amnosia** Doubleday (NYMPHALIDAE)
This is a genus of butterflies found in Indonesia. Only one species has been described, but several subspecies have been named from different Indonesian Islands.

**decora** Doubleday & Hewitson
Indonesia 74 mm. (2·91 in.)
The male is illustrated. The females are brown with more white in the stripes and the hindwing has large eyespots with blue centres. The margin of the hindwing has 2 black lines just below it. The underside is brown in both sexes but is darker in the male. The female has a white band along the underside of the forewing. The butterflies are common in Java, Sumatra and Borneo. The species is variable in pattern and many subspecies have been described. **209**g

**amoena** see **Coenobasis** and **Mazuca**

**Amphicallia** Aurivillius (HYPSIDAE)
Two species are placed in this genus of moths. The caterpillars and adults of both species are orange-yellow and black in colour. There are several similarities between this genus and some European Arctiids such as *Tyria* and it is probably not correctly associated with other genera of the family Hypsidae. The species are mimicked by the Geometrid *Callioratis milliaria*.

**bellatrix** Dalman
Madagascar, the Congo Basin, E. Africa, Ethiopia, S. to Mozambique & Malawi 50–75 mm. (1·97–2·95 in.)
This gaudily coloured species is often common in much of its range. It differs from the illustrated *A. pactolicus* chiefly in the irregular shape of the 2 central dark bands on the forewing. The dark markings on some specimens can cover much of the wing, while in others the dark bands and spots may be almost entirely absent. The orange-yellow and black banded caterpillar feeds on *Crotalaria*, from which it extracts toxic alkaloids. The chrysalis is enclosed in a loosely woven cocoon containing hairs of the caterpillar.

**pactolicus** Butler
W. Africa, the Congo Basin, E. Africa 58–75 mm. (2·28–2·95 in.)
*A. pactolicus* probably occurs together with the apparently more common *A. bellatrix* in part of its range and is probably a Müllerian partner of the latter, both signalling their unpalatability to predators with a similar pattern of orange-yellow and bluish black. **380**

**amphione** see **Dismorphia**

**Amphipyra** Ochsenheimer (NOCTUIDAE)
The 50 or so species of this genus of moths are essentially temperate European and temperate Asian in distribution, but there is a minor representation in North America and Mexico and in tropical Asia.

**berbera** Rungs RUNGS' COPPER UNDERWING
Europe including Britain, N. Africa to W. Russia 48–56 mm. (1·89–2·2 in.)
Specimens of this species have existed in British collections for many years but were misidentified as *A. pyramidea* (the Copper Underwing) until the mistake was discovered in 1968 by the Swedish collector, I. Svensson. A British subspecies *A. b. svenssoni*, was described, also in 1968, by D. S. Fletcher (London). There are small differences in colour pattern between *A. berbera* and *A. pyramidea* and other diagnostic features in the male and female genitalia.

**pyramidoides** Guenée AMERICAN COPPER UNDERWING 36–50 mm. (1·42–1·97 in.)
Canada (especially S. Ontario), U.S., Mexico
The closely allied European relatives of this species are *A. pyramidea* and *A. berbera* (q.v.). Its caterpillars feed on a great variety of broad-leaved trees and shrubs and are often common in the Atlantic States. The moths fly in August and September. **399**a

**Amsacta** Walker (ARCTIIDAE)
About 40 species of moths are included in this genus. It is represented in Africa south of the Sahara, India to Japan, and in tropical S.E. Asia to New Guinea and Australia. The species are typically white, with a red front margin to the forewing, a few black spots or streaks on both wings and a red, or yellow, and black banded abdomen.

**lactinea** Cramer
India, China, Japan, Burma, S. to Java 50–56 mm. (1·97–2·2 in.)
This species has the most northerly distribution in its genus. Its wings can be entirely without black markings or quite heavily marked with black on the hindwing. The caterpillar has been reported as a pest of teak (*Tectona*) and tea (*Thea*) foliage in India. **369**l

**marginata** Donovan
Australia 38–42 mm. (1·5–1·65 in.)
This is a highly variable species in coloration and pattern. The wings are usually either red and black, or, less often, yellow and black, and usually white in ground colour. They have black markings and a red costal margin to the forewings similar to the illustrated *A. lactinea*, but are rarely pink above on the hindwing. The abdomen is either dull red or yellow, with black markings.

**amymone** see **Mestra**
AMYMONE see **Mestra amymone**
**amyntula** see under **Everes comyntas**
**amythaon** see **Amathuxidia**
**anabella** see **Cynthia**
**anacardii** see **Salamis**
**anactus** see **Papilio**

**Anaea** Hübner (NYMPHALIDAE)
LEAF-WING BUTTERFLIES
A genus of C. and S. American butterflies with a few species in the S. U.S. and in the W. Indies. Many of them are relatively common and widely distributed. Some species show strong sexual dimorphism; in others there are only slight differences in colour or pattern between the sexes. The caterpillars have been recorded on a variety of plants including *Piper* (Piperaceae), *Nectandra*, *Goeppertia* and *Camphoromoea* (Lauraceae). The behaviour of species in this genus is very similar in some aspects to that of species in the African genus *Charaxes*. The apex of the forewing of most species is distinctly hooked and there is a slender tail on the hindwing. The uppersides are usually brilliant orange-red or blue but it is from the undersides that the popular name of Leaf-wing butterflies arises. The butterflies have a fast, erratic flight and when they alight they seem to disappear as the bright upperside is suddenly replaced by their camouflaged underside. These resemble leaves in pattern and colour.

**andria** Scudder GOAT-WEED BUTTERFLY
U.S., C. America 50–65 mm. (1·97–2·56 in.)
This butterfly is found in the more southern parts of the U.S. and in C. America. The colouring is typical of the genus. The caterpillar feeds on crotons or goat-weed (species of *Croton*, Euphorbiaceae). There are 2 broods in S. U.S. but more further S. **230**d

**nessus** Latreille
S. America 50–65 mm. (1·97–2·56 in.)
The female is brown with a broad white patch on the forewing and lacks the reddish-purple of the male. This butterfly is found in Colombia, Peru, Ecuador, Bolivia and Venezuela. **230**h

**otrere** Hübner
S. America 48–52 mm. (1·89–2·05 in.)
The butterfly is blue with a purple iridescence and short tails on the hindwing. The underside is dark and very leaf-like, speckled with silvery white. This butterfly is common in the mountains of S. Brazil where it flies fast, chasing other males which come near while it is waiting for the females to appear. **222**

**panariste** Hewitson
S. America 75–82 mm. (2·95–3·23 in.)
The female has a totally different appearance from the male, having yellow hindwings with brown bases and yellow blotches on the forewing. She is basically brown, not blue like the male. The underside is very leaf-like. The species has been collected in Colombia and differs from most others in the genus in the shape of the wings. It is always regarded as one of the rarer ones in the genus. In fact, by some authorities, it is considered to be in the genus *Consul*. **230**h

**tyrianthina** Salvin & Godman
S. America 68–85 mm. (2·68–3·35 in.)
This species is known from Bolivia and Peru and is one of the largest in the genus. Nothing seems to be known of its biology. **230**g

**Anagasta** see under **Ephestia**

**Anania** Hübner (PYRALIDAE)
A small genus of Pyralid moths, mostly European and Asian in distribution but with species in N. America. The species included in *Anania* were previously in the genus *Pyrausta*.

**florella** see **Syngamia florella**

**funebris** Ström
Europe including Britain, temperate Asia, N. America 19–21 mm. (0·75–0·83 in.)
This species is common over much of its range in the Northern Hemisphere. There is some variation in colour and pattern and in some specimens the large patches on the wings are a pale yellow colour. The greenish caterpillar feeds on *Solidago* (Compositae) where it can be found in a slight web on the underside of the lower leaves of the plant. The moth is out in June and July on both sides of the Atlantic and flies in the sunshine. **56**y, **73**

**Anaphaeis** Hübner (PIERIDAE)
A widespread butterfly genus from Africa, through India and S.E. Asia to Australia and the S.W. Pacific.

**aurota** Fabricius BROWN-VEINED WHITE
Africa, Arabia, S. Asia 50–56 mm. (1·97–2·2 in.)
A white butterfly, with black-brown wing margins, white, angular or pointed spots, and a curved blackish mark on the forewing near the middle of the front edge. Underside veins are a darker brown. The females are more heavily marked with black on the upperside than the males and are a creamy yellow or whitish colour rather than pure white. This widespread butterfly is often seen on migration in large numbers and is one of the commonest species in Africa. The caterpillar feeds on species of *Capparis*, *Boscia* and *Maerua* (Capparidaceae).

**creona** Cramer AFRICAN COMMON WHITE
Africa 46–56 mm. (1·81–2·2 in.)
Similar to *A. aurota* but it lacks the black curved mark on the forewing and generally has smaller white marks on the black margin. The underside of the hindwing is yellow, or pale brown in the dry season. It is common and widespread throughout Africa and sometimes is found with other white butterflies on migration. The caterpillar feeds on species of Capparidaceae.

**java** Sparrmann CAPER WHITE
Java, New Guinea into Australia 45–55 mm. (1·77–2·17 in.)
There are several subspecies of this very common and often abundant butterfly. In N. Australia it is one of the commonest species. The male is white, heavily patterned with black on the apex and outer forewing margin, with white spots on the margin, changing to white dashes at the apex and a curved black mark over the apex of the cell. The hindwing is broadly similar. Underside the hindwing is marked with yellow and all the veins are broadly ribbed with white. The females, which are very variable in pattern, have a yellow-white upperside with a broad black band on the margin, black apex and half the hindwing from the margin black. The underside is yellow-orange on fore- and hindwing. The species migrates in vast swarms and the causes of these immense migratory flights are at present being studied in Australia. The caterpillar feeds on species of *Capparis* (Capparidaceae) and when numerous will completely defoliate the foodplant. The female underside and male upperside are shown. **119**g, j

**Anaphe** Walker (NOTODONTIDAE)
This is a genus of about 50 basically yellow or white species of moths. They are found only in Africa and Madagascar. The barbed hairs in the anal tuft of the female contain toxic irritant chemicals. Communal cocoon masses are formed by the caterpillars of some species. A rough, native silk is manufactured from the cocoons of some species.

**barrei** Mabille
Madagascar 43–58 mm. (1·69–2·28 in.)
The colour pattern of this species is typical of other Madagascan species of *Anaphe*, all of which have the dark brown margin to the forewing but may have the orange ground colour of the wings replaced by yellow or, more rarely, by white. **357**g

**Anarsia** Zeller (GELECHIIDAE)
A large genus of micro-moths with species in Europe, Africa, the Indian sub-continent and in N. America.

**lineatella** Zeller PEACH BUD MOTH or PEACH TWIG BORER
Europe, N. America 10–15 mm. (0·39–0·59 in.)
The forewings are white, dusted with grey mottling; the hindwings are pale grey with a darker fringe. The brown caterpillar feeds in twigs and buds of peach and plum (*Prunus*) in the spring, killing the twigs. Later broods bore into the fruit and eat the stone. The moth is widespread, probably having been introduced to America, where it is injurious to fruit in the S. States. At times this moth can be a serious pest in peach growing areas, and considerable sums of money are spent on control measures.

**Anarta** Ochsenheimer (NOCTUIDAE)
This is chiefly an arctic or subarctic genus of the N. Hemisphere with minor representations at high altitudes further S.

**cordigera** Thunberg SMALL DARK YELLOW UNDERWING
Alaska, N. Canada S. to Colorado, N. Europe, N. Africa 22–27 mm. (0·87–1·06 in.)
The southernmost part of the range of this small moth is at very high elevations in the Rocky Mountains of Colorado. In Europe it extends S. to the mountains of N. Scotland where the caterpillars feed on bearberry (*Arctostaphylos*) and species of *Vaccinium* (the same

foodplant genera as in N. America). *A. cordigera* flies in sunny weather, in May and June. **391**s

**Anartia** Hübner (NYMPHALIDAE)
A small genus of butterflies with species in N. and S. America. In appearance they are similar to species of *Cynthia* and are powerful fliers.

**amathea** Linnaeus COOLIE BUTTERFLY
S. & C. America, W. Indies 38 mm. (1·5 in.)
Both fore- and hindwings of this butterfly are a deep red in the middle with broad black outer margins which have white spots in them, and a black base. This combination makes it a richly coloured species. The female is browner than the male. It is widespread and often common throughout its range, frequently seen visiting flowers. The caterpillar feeds on species of Acanthaceae. **207**

**jatrophae** Linnaeus WHITE OR AMERICAN PEACOCK, BISCUIT BUTTERFLY 48–56 mm. (1·89–2·2 in.)
C. & S. America, U.S., W. Indies
This species is found in the southern parts of Florida and Texas, occasionally reaching Kansas and Massachusetts. Over the rest of its range it is common and many subspecies have been described from different areas. There is seasonal variation in the pattern. The caterpillar feeds on *Jatropha* (Euphorbiaceae), *Lippia* (Verbenaceae) and *Bacopa* (Scrophulariaceae). It tends to be an aggressive butterfly, often chasing others. Its flight is a mixture of gliding and flapping.

**anartoides** see **Orgyia**
**anasuja** see **Jacoona**
**anaxias** see **Mycalesis**

**Anaxita** Walker (HYPSIDAE)
Seven species are placed at present in this genus of C. American and tropical S. American moths. Most have a highly conspicuous pattern which includes orange, yellow and brown, or red and brown.

**drucei** Rodriguez
C. America including Mexico 73–82 mm. (2·87–3·23 in.)
This is probably a mountain species. In Guatemala, caterpillars have been collected at 3230 m. (10,600 ft) feeding on a plant called 'Motoze' which grows amongst the rocks. The caterpillars are covered with yellow and brown hair, are bright red beneath and have silvery white spots along each side. **376**m

**anchisiades** see **Papilio**
**anchusella** see **Douglasia**

**Ancyclurisa** Hübner (NEMEOBIIDAE)
There are between 10 and 20 C. American and tropical S. American species in this genus of butterflies. Most have short tails to the hindwings and are marked above with transverse bands of red or white.

**formosissima** Hewitson
Peru, Ecuador 40–44 mm. (1·57–1·73 in.)
Specimens of this species have been captured at elevations of 3050 m. (10,000 ft) and above. Both surfaces of this colourful butterfly are illustrated. **268**a, b

**anderida** see **Heliconius**
**anderschella** see **Micropterix**
**andicola** see **Leptotes**
**andria** see **Anaea**
**androcles** see **Graphium** and **Heliophorus**
**androgeus** see **Papilio**
**andromache** see **Acraea**

**Angerona** Duponchel (GEOMETRIDAE)
A genus of moths with species from Europe across Asia to Japan. The species are often common.

**prunaria** Linnaeus ORANGE MOTH
Europe, W. & C. Asia 37–46 mm. (1·46–1·81 in.)
The specimen illustrated is the subspecies *A. p. prunaria* and there are several others from different localities. The moth is common in most parts of Europe. There are 2 colour forms, both occurring over its whole range. In the common, unbanded version the male is orange and the female yellowish. In the rarer colour form the orange and yellow is restricted to the central area of the wing. The caterpillars feed on a variety of plants including hawthorn *(Crataegus)*, plum *(Prunus)*, privet *(Ligustrum)*, and beech *(Fagus)*. They generally hibernate over the winter, pupating the following spring to emerge as moths some 4 weeks later. The moth is common in woods where it may be disturbed from the undergrowth or bracken where it rests during the day. **294**a

ANGLE-SHADES see **Phlogophora meticulosa**
ANGOLA WHITE LADY SWALLOWTAIL see **Graphium pylades**
ANGOUMOIS GRAIN see **Sitotroga cerealella**
**angulata** see **Holocerina**
**angulosa** see **Chetone** and **Thalaina**

**Anisozyga** Prout (GEOMETRIDAE)
These delicate moths are particularly common in New Guinea and form a characteristic part of the fauna. There are species in the Philippines and one or two in Borneo and India. It is a large genus, with a number of the curious semi-transparent green species. In many cases the males and females of the same species have very distinct patterns (sexual dimorphism).

**stellifera** Prout
New Guinea 29 mm. (1·14 in.)
This species has only been collected on Mount Goliath in W. Irian. Nothing is known of its biology. The moth is typical of the appearance of many of the species in the genus. **294**k

**anita** see **Amblypodia**

**Annaphila** Grote (NOCTUIDAE)
This is a genus of about 20 small, highly coloured day-flying moths. Most species are somewhat restricted in distribution and rather rare in general collections. They are found in W. N. America, with a centre of distribution in California. Few foodplants are known, but *Montia* is the host of at least 2 species and *Gilia* of another.

**baueri** Rindge & Smith
U.S. (California) 20–26 mm. (0·79–1·02 in.)
This small moth has been caught at altitudes of from 430–1520 m. (1400–5000 ft). Although difficult to catch while in flight, especially on hot days, it can be taken when feeding on flowerheads or while basking in the sunshine. Between 1 pm. and 3 pm. is apparently the feeding period of the adult, which is on the wing in February and March. The hindwings of *A. baueri* are mostly yellow, the forewings cryptically coloured.

**Anomis** Hübner (NOCTUIDAE)
This is a large, almost cosmopolitan genus of moths, many of which are pests.

**argillacea** Hübner COTTON LEAFWORM
E. Canada, E. and S.W. U.S., C. & S. America 30–35 mm. (1·18–1·38 in.)
This is an essentially tropical species which spreads northwards from C. America, reaching Canada at the end of the summer, but apparently is not capable of overwintering in N. America. Millions of specimens have been reported at times in New York city. The adult moth is able to pierce soft fruits, especially peaches and plums *(Prunus)*, figs *(Ficus)*, and grapes *(Vitis)* and sometimes causes considerable damage. The caterpillar is one of the most important pests of cotton *(Gossypium)* in the U.S. The forewings of *A. argillacea* are pale brown, with a few grey and dark brown markings; the hindwings are an unmarked brownish white.

**Anomologa** Meyrick (ANOMOLOGIDAE)
A small genus of micro-moths with only 2 species, both found in South Africa. Little is known of the species, and the relationship of the genus to others of the genera related to the Gelechiidae is not known. Studies on the biology of the species and detailed studies of the anatomy and morphology of the adult are needed.

**dispula** Meyrick
South Africa 23 mm. (0·91 in.)
Only a few specimens of this species and a related one, have been found. Nothing is known of their biology.

**Anomologidae**
A family of micro-moths, which is little known and only doubtfully separated from others in the Gelechoidea. There is one genus containing 2 species, of which only a few specimens have been found. The moths are small, rather nondescript, and nothing is known of their biology. See *Anomologa*.

ANSORGE'S
LEAF see **Kallima ansorgei**
PRINCE see **Micragone ansorgei**
**ansorgei** see **Kallima** and **Micragone**

**Antaeotrica** Zeller (STENOMIDAE)
A large genus of micro-moths containing in excess of 400 species. Although the genus reaches its greatest diversity and abundance in S. America, it also occurs in C. and N. America and in the W. Indies. The moths are mostly patterned in sombre colours and some species have been cited as examples of cryptic coloration due to their resemblance to bird droppings when in the resting position. In spite of its size and abundance, little is known of the life histories of its members in the tropical portion of its range. N. American members of the genus are known to feed on various species of oak *(Quercus)*, as well as *Pyracantha, Malus, Vaccinium, Acer, Sida, Pitheoellobium* and *Arctostaphylos*. In most species the larvae feed between the leaves of the host plant; they tie these leaves together with silk.

**griseana** Fabricius
Colombia, Guyanas, Brazil, Trinidad 22–26 mm. (0·87–1·02 in.)
This is conspicuous when compared with some of the other species in the genus as it has a striking white band on the anterior margin of the forewing. Apart from this the pattern is basically like the others. Nothing is known of the biology of the species. **42**f

**antaeus** see **Cocytius**
**antalus** see **Deudorix**

**Antanartia** Rothschild & Jordan (NYMPHALIDAE)
These butterflies generally have tailed hindwings. It is a small genus, found only in Africa, Madagascar and Mauritius.

**delius** Drury ORANGE ADMIRAL
Africa 50–60 mm. (1·97–2·36 in.)
This is a woodland species found from W. Africa and the Congo forest to Uganda, Kenya and Tanzania and N. it extends into Ethiopia. The butterflies are out in all months of the year. The caterpillar feeds on *Musanga* (Moraceae). The butterfly is brownish black with a broad curved orange-red band on the forewing from the hind margin, curving across towards the base of the wing. There are a few white spots near the apex of the wing. The hindwing has a black margin, a black tail, and a broad band of orange-red across the middle of the wing. The underside is patterned with green.

**schaenia** Trimen LONG TAILED ADMIRAL
E. to South Africa 50–62 mm. (1·97–2·44 in.)
The underside is speckled with red; the upperside is blackish with an orange band across the middle of the forewing and along the upper part of the margin of the hindwing, with a few white dots below the apex of the forewing. The flight is generally rather rapid but usually of a short duration. The caterpillar feeds on *Boehmeria, Fleurya* and *Pouzolzia* (Urticaceae). There are several subspecies, the one illustrated is *A. schaenia dubi* Howarth from Kenya. **185**e

**antenor** see **Parides**

**Anteos** Hübner (PIERIDAE) MAMMOTH SULPHUR
As the common name implies, these are large, usually yellow, butterflies. Found in C. and S. America with species occasionally straying into North America. They have very powerful flight. Some species have

been reared from caterpillars found on *Cassia* (Leguminosae).

**clorinde** Godart YELLOW-SPOTTED GONATRYX
C., S. & N. America 85–90 mm. (3·35–3·54 in.)
This species is found in the West Indies and is common on flowers. The butterfly is white, with a large yellow patch on the forewing with a black spot, and a black spot in the hindwing with a little yellow round it. Basically it is rather similar to *Gonepteryx rhamni* (figure **118**). *A. clorinde* has been found as far north as Kansas but is rather rare on the extremes of its range.

**maerula** Fabricius MAERULA BUTTERFLY
C. America, N. to Texas, Florida, W. Indies 70–88 mm. (2·76–3·46 in.)
The males are yellow; the female has an orange-yellow spot on the hindwing. The large size and sharply angled wings make this yellow butterfly easily recognized. It can be distinguished by its colour and larger size from the similarly shaped relative, *A. clorinde*, in which the male is white and the female lacks the orange spot on the hindwing.

**antevippe** see **Colotis**

**Anthela** Walker (ANTHELIDAE)
There are probably between 50 and 80 species in this genus of moths. Their range is entirely Australian.

**nicothoe** Boisduval
Australia 62–94 mm. (2·44–3·7 in.)
The males of this variable species can be either brown, red or yellow in ground colour; most have 2 greyish white spots in the middle of the forewing. Females (not illustrated) are mainly yellowish grey or brown in ground colour. Its caterpillars feed usually on a species of *Acacia*, but at times may be a pest of Monterey Pine *(Pinus radiata)* in Australia. The cocoon is protected by the sharply pointed, irritant hairs of the caterpillar. **317**d

**Anthelidae**
This is a small Australian and Papuan family of about 9 genera and 100 species of moths. It is related to the families Lasiocampidae and Eupterotidae and is placed in the superfamily Bombycoidea. Most species have lost the proboscis and are incapable of feeding as adults. The caterpillars of some species are protected by dense tufts of irritant hairs. The cocoons too, may contain larval hairs and are similarly protected. See *Anthela*.

**Antheraea** Hübner (SATURNIIDAE) TUSSUR or TUSSORE SILK-MOTHS or OAK SILK-MOTHS
Many of the 30 or more species of this mainly Old World genus have been bred commercially for hundreds of years in China, India and Japan. Silk is obtained from the cocoon, which is woven by the caterpillar prior to turning into a chrysalis. One member of the genus occurs in N, America; the other 2 New World species are confined to C. America. The remaining species are found in India eastwards to Japan and Australia.

**loranthi** Lucas
N.E. Australia 120–155 mm. (4·72–6·1 in.)
This is a typical species of *Antheraea* in wingshape and pattern, and is chiefly deep yellow or orange in ground colour; it is one of 11 species of its genus found in Australia. The caterpillars are gregarious and feed on mistletoe (Loranthaceae) mostly high up on *Eucalyptus* trees.

**pernyi** Guérin-Méneville CHINESE OAK SILK-MOTH
China, S.E. Russia, Japan 100–150 mm. (3·94–5·91 in.)
Silk from the cocoons of this species (Chinese Tussore silk) is still exported from China and is woven into the material known as Shantung silk. This is slightly more irregular in texture than that produced from the cocoons of *Bombyx mori*. Unlike *B. mori*, this species also exists in the wild. There are 2 broods per year. Its caterpillar is mainly green when fully grown, with hairy tubercles; it feeds on oaks *(Quercus)* and other broad-leaved trees. **337**a

**polyphemus** Cramer POLYPHEMUS MOTH
N. America 100–128 mm. (3·94–5·04 in.)
This is probably the commonest and most widely distributed species of N. American Saturniidae. Over 50 species of broad-leaved trees and shrubs provide food for the caterpillar. In captivity, apple *(Malus)* and hawthorn *(Crataegus)* are readily eaten. Fully grown caterpillars are mainly green, with red tubercles, and are similar in general appearance to the caterpillars of the Luna moth *(Actias luna)*. Adults can vary individually in ground colour of the wings from yellow to brownish red. There are 2 broods per year in the south and a single brood in the north of its range. **330**

**yamamai** Guérin-Méneville JAPANESE OAK SILK-MOTH
Japan 115–150 mm. (4·53–5·91 in.)
This was a closely and successfully guarded species until the year 1860 when, under the threat of death, eggs were secretly exported to Europe by the French consul Duchesne de Bellecourt. Like other species of its genus the wings of this moth vary considerably in coloration between individuals and can be light yellow, brownish red or dark brown. The caterpillar is greenish yellow when fully grown, its yellow tubercles armed with black spines. It feeds mainly on oak *(Quercus)*. Attempts have been made to establish *A. yamamai* in the wild outside Japan, with varying success. It is, however, probably now established in parts of S.W. Europe.

**Anthocharis** Boisduval (PIERIDAE)
A large genus of mostly white butterflies found in temperate Asia and N. America. Many have coloured tips to the wings.

**belia** Linnaeus MOROCCO ORANGE-TIP
N. Africa, S. Europe 36–40 mm. (1·42–1·57 in.)
The general yellow colour separates this from *A. cardamines* which is otherwise similar. There are 2 subspecies, one of which occurs in Spain and Portugal, the other in North Africa. The caterpillar feeds on *Biscutella* (Cruciferae). The butterfly is out in April and May and can be found in most suitable spots on rough ground or in woodland, up to 1520 m. (5000 ft).

**cardamines** Linnaeus ORANGE-TIP
Europe including Britain, temperate Asia including Japan 36–48 mm. (1·42–1·89 in.)
This white species is common over most of its range where it is usually closely associated with the foodplants of the caterpillars. These are species of Cruciferae, including Bitter Cress *(Cardamine)* and *Sisymbrium*. The green caterpillar, which has a white line with red spots along the side, is covered with blackish hairs and closely resembles the seed pod of the foodplant, providing itself with excellent camouflage. Generally it has only one brood a year. The Japanese specimens differ slightly in pattern from the European ones and are considered a distinct subspecies. **124**

**damone** Boisduval EASTERN ORANGE-TIP
Sicily, S. Italy, Syria to Iran 36–40 mm. (1·42–1·57 in.)
This species is similar to *A. cardamines* but instead of the white background it is yellow. It differs from *A. belia* in having more green on the underside of the hindwings. The foodplant is unknown but is probably a species of Cruciferae. The butterfly is out in April-May and flies over rocky mountain slopes up to 910 m. (3000 ft). The caterpillar has yet to be found.

**Anthomyza** Swainson (HYPSIDAE)
Two species of moths are placed in this genus: *A. heliconides* (illustrated) and *A. variifasciata* Hering (W. Colombia). The former is a mimetic partner of the Ithomiine butterfly genus *Thyridia* and others. Transparency of the wings is achieved by transparency of the scales in contrast with Ithomiinae in which there is a reduction or absence of scales, or with species of Castniidae in which the scales are placed on their edges in relation to the wing surface. Replaced by *Notophyson*.

**heliconides** Swainson
C. & tropical S. America 60–92 mm. (2·36–3·62 in.)
*A. heliconides* is a member of one of the most impressive mimetic associations of tropical New World butterflies and moths (see *Thyridia*). **381**f

**Anthophila** Haworth: syn. *Simaethis* (GLYPHIPTERIGIDAE)
A genus of micro-moths most abundant in the tropics, but with species in most parts of the world.

**fabriciana** Linnaeus
Europe including Britain 11–13 mm. (0·43–0·51 in.)
A common species over most of its range, the adult moth is out in May and June and again in August to September. The whitish caterpillar, which has pale brown spots, feeds in a small web on the leaves of nettle *(Urtica)* and *Parietaria* (Urticaceae). **30**

**Anticrates** Meyrick (YPONOMEUTIDAE)
A genus of micro-moths with some 30 species from India, Indonesia, Samoa, the Philippines, Australia and with 2 species in Africa. They are mostly orange-red and yellow in colour with a distinctive pattern and reddish-brown hindwings. The undersides are less patterned. Nothing is known of their biology and few species have been reared.

**autobrocha** Meyrick
Solomon Islands, Rossel Islands, Sudest Island, Australia 15 mm. (0·59 in.)
In Australia this species occurs only in Queensland. The round yellow spots are smaller than those on related species. There is also a similar, but larger, species on New Guinea. There is still some doubt of the exact identity of specimens of this species from the Solomon Islands and whether they really are the same species as the Australian specimens. **41**h

**tridelta** Meyrick
N. India 20–24 mm. (0·79–0·94 in.)
This species has only been collected in the Khasis Hills in Assam and there is some variation in the amount of yellow on the wings in the specimens that are known. Relatively few have ever been collected and the relationship of this moth with the others in the genus is not known. Nothing is known of its biology. With these smaller and less well known species it is always probable that they are more widespread. So much of the information about distribution reflects merely the places visited by collectors and often bears little relationship to the true distribution of the species. **38**b

**antimachus** see **Papilio**

**Antinephele** Holland (SPHINGIDAE)
There are 7 species in this genus of tropical African hawk-moths.

**maculifera** Holland BEAUTIFUL ANTINEPHELE
Tropical W. Africa, the Congo Basin, Malawi, Rhodesia 40–50 mm. (1·57–1·97 in.)
Other species of this genus have a less brightly coloured abdomen. A related species, *A. achlora* Holland, for example, carries the common name Dingy Antinephele. The wings of *A. maculifera* are patterned with dark brown, black and light brown and are possibly cryptic in effect. The conspicuous bright orange-yellow abdomen contrasts sharply with the rest of the moth and could well be an example of 'flash coloration' (a phenomenon in which the coloured area is normally concealed – by the forewings in this instance – but can be revealed suddenly in an attempt to disconcert an enemy). The caterpillar of *A. maculifera* is not known.

ANTINEPHELE,
BEAUTIFUL see **Antinephele maculifera**
DINGY see under **Antinephele maculifera**
**antinorii** see **Nyctemera apicalis**
**antiochus** see **Heliconius**
**antiopa** see **Nymphalis**
**antiphates** see **Graphium**
**antiqua** see **Orgyia**

**Antispila** Hübner (HELIOZELIDAE)
A genus of micro-moths with a few N. American species, 3 or 4 European ones, and several in Japan. The wings are narrow and pointed and the moths are mostly day fliers. The tiny caterpillar is a leaf-miner.

The genus is separated from the closely allied *Heliozela* by the apparent absence of one of the veins of the forewing which can only be seen by careful examination through a stereomicroscope.

**cornifoliella** Clemens
U.S. 7·5 mm. (0·3 in.)
The forewings are dark brown with a coppery hue. Towards the apex there is a bright coppery colour with golden spot and a narrow line on the wing. This is a species which can be more reliably identified by the host plant from which it was reared, rather than by the pattern of the wing. The caterpillars mine in the leaves of Flowering Dogwood, *Cornus florida*, in September. The moths have been collected in Ohio.

**orbiculella** Kuroko
Japan 4·5 mm. (0·16–0·2 in.)
This small moth was discovered recently when the caterpillar was bred from leaves of *Ampelopsis* (Vitaceae). The forewings are pointed, dark bronze with a reddish suffusion. Some iridescent markings form bands across the wings. One generation of this moth, from egg to adult, takes a full year in Japan, the adult moths flying in June and July. The leaf mine made by the caterpillar is in the form of a blotch on the leaf of the host plant.

**pfeifferella** Hübner
Europe including Britain 8–9 mm. (0·31–0·35 in.)
The caterpillar of this small moth lives between the upper and lower surface of the leaves of *Swida*. The mines appear externally as blotches on the leaves. In Britain the eggs are laid under the leaves of Dogwood *(Swida sanguinea)*. The caterpillars pupate above the ground. **1f**

**Antistathmoptera** Tams (SATURNIIDAE)
This is a small genus of long-tailed moths similar in wingshape to *Argema* species, but differing in characters of the antennae and genitalia. They are restricted to Africa south of the Sahara.

**daltonae** Tams FROG FOOT MOTH
Tanzania, Rhodesia, Zambia, Mozambique 90–120 mm. (3·54–4·72 in.)
This species inhabits low elevation forests where it efficiently mimics dead leaves when at rest. Its popular name is based on the shape of the hindwing, with the elongate 'thumb' of a frog's foot forming the tail of the wing. The ground colour of the wings varies considerably from orange and yellow to a brownish green; fore- and hindwings both have a small transparent patch near the middle. Females have shorter tails than the male. Felling of the forest is currently endangering the existence of the subspecies *A. daltonae rectangulata* Pinhey.

ANTLER see **Cerapteryx graminis**
**anynana** see **Mycalesis**
**apachealis** see **Scoparia**

**Apamea** Ochsenheimer (NOCTUIDAE)
There is a large number of species in this genus of moths. Most are European or temperate N. American in distribution, but some species are found in most parts of the world.

**lateritia** Hufnagel SCARCE BRINDLE
Europe, temperate Asia, N. America 40–48 mm. (1·59–1·89 in.)
This is a fairly common species throughout most of its range. The exceptions include Britain, where, as its common name indicates, it is certainly scarce and is known from only one specimen. The American common name, Red-winged Hadena, refers to the earlier placement of *A. lateritia* in the genus *Hadena*. The adult is a dull reddish brown, with mostly indistinct markings of white and darker brown. The caterpillar is a cutworm and a general feeder on herbaceous plants.

**Apantesis** Walker (ARCTIIDAE)
As presently accepted, this is almost entirely a N. American and Mexican genus, with a single species in the Bahamas and 2 Arctic representatives in Europe. The European genus *Cymbalophora* Rambur is closely related to *Apantesis*. Like many other Arctiids, the species of this genus are variable in colour pattern and coloration, both individually and, in some species, sexually. Many are difficult to identify. The foodplants include milkweeds (Asclepiadaceae) from which the caterpillars are able to extract and store plant poisons and pass these on to the consequently distasteful adult moths. The bright colours of *A. virgo*, for example, are 'warning colours' and a signal to predators of distastefulness.

**quenseli** Paykull
Europe, Asia including Japan, N. America 32–38 mm. (1·26–1·5 in.)
*A. quenseli* is a good example of a species which was presumably widespread in N. latitudes at the end of the last glacial period of the Earth's history (about one million years ago) but which is now restricted to subarctic and cool temperate latitudes and to cold refuges at very high altitudes in the alps of Europe. The caterpillar feeds on low-growing plants. It possibly overwinters twice as a chrysalis before emerging as a moth. **382a**

**virgo** Linnaeus VIRGIN TIGER-MOTH
Canada, U.S. 45–60 mm. (1·77–2·36 in.)
This is a widespread and common moth in E. Canada and much of the U.S. *A. virgo* is well-known as a species distasteful to birds, and adults are capable of ejecting noxious secretions from their thoracic glands. The hairy, mostly black caterpillars feed on low-growing plants, including species of milkweeds *(Asclepias)* from which they are able to extract plant poisons and transmit these via the chrysalis to the adult. The red of the hindwings and abdomen is replaced by yellow in a few specimens. Both red and yellow are within the optimal part of a bird's colour range and are probably equally effective as a signal to a bird that the moth is unpalatable. **369c**

**Apatelodes** see **Hygrochora**

**Apatelodidae**
This is a small family of New World moths which forms part of the superfamily Bombycoidea. They have been combined at times with the Eupterotidae and Bombycidae and are probably fairly closely related to these families. Most species are tropical in distribution, but 5 occur in North America. See *Epia, Hygrochora*.

**Apatura** Fabricius (NYMPHALIDAE)
A large European and Asian genus of butterflies with a few species extending into the Oriental region as far as Sulawesi (Celebes). They are generally large butterflies and are powerful fliers.

**ambica** Kollar INDIAN PURPLE EMPEROR
India, Pakistan, Burma 65–90 mm. (2·56–3·54 in.)
This dark brown butterfly has a single white band across both wings and in the male the brown catches the light to give a strong blue iridescence. It is similar to *A. iris* except in the colour of the iridescence and like it, is a fast flier. It settles on damp sand or carrion and is found throughout the Himalayas up to about 2440 m. (8000 ft). The caterpillar feeds on *Ulmus wallichianus* (Ulmaceae).

**ilia** Denis & Schiffermüller LESSER PURPLE EMPEROR
C. & S. Europe, temperate Asia including Japan 64–70 mm. (2·52–2·76 in.)
This is very similar to *A. iris* but with the iridescence blue. It can also be separated from that species by the presence, on the upperside, of an orange ring round the black spot near the middle of the outer wing margin. *A. iris* has a round black mark, visible on the upperside but no orange ring. Both species have orange round this mark on the underside. The caterpillar feeds on poplars *(Populus)* and willows *(Salix)*. Several subspecies have been described from Europe and Asia. In the south the butterfly is out in May and June, with a second brood in August-September although further N. there is only a single brood in July.

**iris** Linnaeus PURPLE EMPEROR
Europe including Britain, temperate Asia including Japan 62–74 mm. (2·44–2·91 in.)
The iridescent mauve of this butterfly is very conspicuous as it flies in the sunlight. The female is more soberly coloured than the male having no iridescence. It is also larger with larger white markings. The adults fly around tree tops in woodlands in July and August. The Purple Emperor, besides feeding on nectar, is also attracted to carrion – a habit used by collectors who would have difficulty in catching the fast flying butterfly as it skimmed about the tree tops. A dead rabbit or any rotting animal corpse will attract it and this unpleasant habit has caused many of these beautiful butterflies to end up in the collector's box. The caterpillar feeds on Sallow *(Salix caprea)* and White Poplar *(Populus alba)*. **199**

APEFLY see **Spalgis epeus**
**apelles** see **Hypochrysops**

**Aphantopus** Wallengren (NYMPHALIDAE)
A genus of butterflies with species in Europe and Asia. It belongs to the subfamily Satyrinae. The majority of the species have caterpillars which feed on grasses.

**hyperantus** Linnaeus RINGLET
Europe including Britain, temperate Asia 40–48 mm. (1·57–1·89 in.)
This butterfly is common over most of its range but is not found in the Arctic or Mediterranean parts of Europe. It flies in June and July in grassy meadows and open woodlands, from sea level to 1520 m. (5000 ft). There are several similar related species. The caterpillar feeds on grasses and sedges, such as *Poa, Milium* and *Carex*. There is some variation in the size and form of the pattern over its range and several subspecies have been described. The popular name is most appropriate in this case and the 'ringlet' markings on the wing can be clearly seen in the figure. **170**

**Aphnaeus** Hübner (LYCAENIDAE)
This is a widespread genus of about 60 two-tailed, blue hairstreak butterflies. It is chiefly subtropical and tropical Asian in distribution but there are a few species in Africa S. of the Sahara.

**flavescens** Stempffer
Malawi, coastal Kenya 30–35 mm. (1·18–1·38 in.)
The turquoise and black upper surface of this rare butterfly contrasts with the yellow, red and silver under surface, It differs from the illustrated *A. hutchinsoni* in the much less extensive turquoise areas and the more colourful under surface of the wings. Nothing is known about its life history.

**hutchinsoni** Trimen SILVER-SPOT
E. Africa to South Africa 26–35 mm. (1·02–1·38 in.)
The life history of this species is not known, but adults are often found in *Acacia* scrub in hilly country. The popular name refers to the metallic silver spots on the under surface of the wings. Both surfaces of the male wings are illustrated. Females differ little from the male in coloration. **242a, c**

**Aphomia** Hübner (PYRALIDAE) BEE MOTHS
This large genus of Pyralid moths has species all over the world. Some have caterpillars which feed in bees' nests but most species are known only in the adult stage and their caterpillars have yet to be discovered. Most species are a dull grey-brown colour with rather rounded wings.

**sociella** Linnaeus BEE MOTH
Europe including Britain, N. America 23–25 mm. (0·91–0·98 in.)
The caterpillars feed on the honey combs in the nests of wild bees, although at times they may be pests in bee hives. When the caterpillars are about to pupate they spin tough grey papery cocoons. Since they tend to pupate in groups, masses of these cocoons are found together often in sheds or in attics which indicates that there was once a bees' nest nearby from which the caterpillars came. **60**

**Aphthonetus** Walsingham (MOMPHIDAE)
A genus of micro-moths which was placed in a family of its own (Diplosaridae) with the genus *Hyposmocoma*. These two genera are known only from Hawaii, where each has produced many species, in the case of *Hyposmocoma* probably in excess of 500 species.

**mediocris** Walsingham
Hawaii 17 mm. (0·67 in.)
This rather dull coloured species is one of many in the genus in Hawaii. As in some other groups, for example, Fruitflies *(Drosophila)* a large number of species have evolved on the islands of Hawaii in response to some unknown local factors which may include lack of competition and the existence of many suitable habitats for the species in which they can be isolated from neighbouring populations. **43**l

**Aphysoneura** Karsch (NYMPHALIDAE)
Only 2 species are known in this genus of African butterflies which is in the subfamily Satyrinae. Generally species of Satyrine butterflies are brown with prominent markings but those of this genus are exceptions.

**pigmentaria** Karsch
Africa 46–50 mm. (1·81–1·97 in.)
Very distinct from other African Satyrinae, this is a creamy-white and black butterfly. The margins of the wings are black, with spots of creamy-white near the apex of the forewing. The rest of the wings are creamy while the undersides are more strongly patterned, but paler, and a very small tail is present on the hindwings. The life history is unknown. The butterfly is found from E. Africa south to Rhodesia, generally on the edges of rain-forests and can be seen at all months of the year.

**apicalis** see **Nyctemera**
**apiformis** see **Phalacropterix** and **Sesia**

**Apolocera** Stephens: syn. *Anaites* (GEOMETRIDAE)
Most of the moths in this genus are found in Europe. Many of the species are common.

**efformata** see under **A. plagiata**

**plagiata** Linnaeus TREBLE BAR MOTH
Europe including Britain, W. Asia, N. Africa 34–38 mm. (1·34–1·5 in.)
This is a common moth throughout Europe from June to September. There are usually 2 generations a year. The caterpillars feed on St John's Wort *(Hypericum)* and those from the second brood hibernate over the winter. Several subspecies of this moth have been described, 2 of which are present in Britain, one occurring in Scotland and the other throughout Britain. In 1923 it was discovered that there were 2 species, then confused under the name *A. plagiata,* which could only be distinguished by their internal morphology. Both species have caterpillars which feed on *Hypericum*. The second is *A. efformata* Guénee.

**Apocrisias** Franclemont (ARCTIIDAE)
There is a single known species in this genus of moths. It is related to species of *Halisidota, Hypocrisias* and their allies, but differs amongst other characters in the relatively long palps.

**thaumasta** Franclemont
S.W. U.S., N. Mexico 47–56 mm. (1·85–2·2 in.)
This quite large Actiid remained undescribed until fairly recently. During the last few decades, J. G. Franclemont, L. Martin and other N. American entomologists before them, have discovered the wealth of species inhabiting the mountains of Arizona and New Mexico: *A. thaumasta* is one of these. The caterpillar has several tufts of yellow and brown hair, like those of *Halisidota* 'Tussock' caterpillars; it feeds on oak *(Quercus)*. **368**t

**Apoda** Haworth (LIMACODIDAE)
A genus of moths that is widespread over Europe, temperate Asia and W. Africa with a few species in North America. The biology of a number of species has been studied in some detail.

**avellana** Linnaeus FESTOON MOTH
Europe including Britain 26–29 mm. (1·02–1·14 in.)
The forewings are brownish yellow, or brownish yellow with some black scales. A dark line runs across the wings from the front margin, with another near the outer margin. The hindwings are blackish. This is one of the few species of the mainly tropical and subtropical family to reach Europe. The caterpillar is green, rather slug-like, with 3 rows of shiny warts and yellow lines on the back, edged with red, and a yellow line along each side. It feeds on oak *(Quercus)*. The moth is widely distributed over Europe in June and July, occurring in many places and occasionally is common. In England it is found mainly in the south.

**Apodemia** Felder (NEMEOBIIDAE) METALMARKS
Mexico is the centre of distribution of this small genus, which ranges northwards to S. and W. U.S. and is also represented by 3 species in Brazil. The species are deceptively similar to small fritillary butterflies.

**nais** Edwards NAIS METALMARK
U.S., Mexico 25–33 mm. (0·98–1·3 in.)
This species occurs southwards from the Rockies of central Colorado. Specimens vary to some extent in the ground colour of the wings. The only recorded foodplant of the caterpillar is wild plum *(Prunus)*. **268**v

**phyciodoides** Barnes and Benjamin LOST METALMARK
S.W. U.S. 12–15 mm. (0·47–0·59 in.)
The Chiricahua Mountains of Arizona are the home of this small metalmark butterfly. It was dsecribed in 1924 from several specimens but has not been captured since, despite the facts that these mountains have been a favourite collecting place for lepidopterists and the American Museum of Natural History, New York City, maintains a field station there. This is not necessarily an extinct species, but is certainly a 'lost' species at present.

**apollinaris** see **Pseudaphelia**
**apollinus** see **Archon**
**apollo** see **Hypochrysops** and **Parnassius**
APOLLO see **Pseudaphelia apollinaris**
APOLLO BUTTERFLIES see **Parnassius**
APOLLO, FALSE see **Archon apollinus**
**apollonia** see **Lypropteryx**

**Apomyelois** Heinrich (PYRALIDAE)
This genus of Pyralid moth was described in 1956 for one N. American species. The wing venation separates it from an allied genus, *Myelois.*

**bistriatella** Hulst
Europe including Britain, U.S. 19–22 mm. (0·75–0·87 in.)
This moth was described from North American specimens in 1887. It is a grey-brown moth with white transverse lines on the forewings and smokey coloured hindwings. The caterpillar was unknown when the moth was originally described. In 1963 it was found that a species described in Britain in 1915 under the name *A. neophanes* was in fact a subspecies of the N. American *A. bistriatella*. In Britain the caterpillars feed on a fungus *Daldinia concentrica* which grows on Birch *(Betula)*. The caterpillar of the American subspecies has not yet been discovered but will probably be a fungus feeder. The British subspecies is illustrated. **56**f

**Apoprogones** Hampson (APOPROGENIDAE)
There is a single known species in this genus of moths.

**hesperistis** Hampson
South Africa 33–40 mm. (1·3–1·57 in.)
Nothing is known about the life history of this apparently very rare species. Its scientific name refers to the Hesperiid-like antennae. The end of the abdomen is missing from the illustrated specimen. **272**f

**Apoprogonidae**
There is a single genus, *Apoprogenes,* and single species in this family. It is found only in South Africa. There is some doubt at present as to its place in the classification, but it is often associated with other members of the superfamily Geometroidea. See *Apoprogenes.*

**Aporia** Hübner (PIERIDAE)
A large genus of butterflies found throughout Europe and temperate Asia with a few species reaching Formosa and into N. India.

**agathon** Gray GREAT BLACK VEIN
India, Pakistan, Burma 80–90 mm. (3·15–3·54 in.)
A black and white butterfly with the veins broadly lined with black and with white showing as small marks below the outer margins of the wing. It is common over much of its range where its floating flight in open woodlands makes it very conspicuous. The caterpillar feeds on species of *Berberis* (Berberidaceae). Some of the subspecies have less black along the veins.

**crataegi** Linnaeus BLACK VEINED WHITE
Europe including Britain, temperate Asia including Japan, N. Africa 62–78 mm. (2·44–3·07 in.)
There are several subspecies over its range, the Japanese being distinct from the European one. Widely distributed in Europe and Asia and sometimes a pest in orchards. The caterpillar feeds on hawthorn *(Crataegus)* and *Prunus* (Rosaceae). In Britain this butterfly is no longer a resident, although during the last century and the early part of this one it was relatively widespread. Finally it disappeared, and now causes excitement amongst entomologists when it occasionally reappears in Britain. **125**

**appendiculata** see **Atychia**

**Appias** Hübner (PIERIDAE)
From India through S.E. Asia there are a number of species of this genus of butterflies which are common. Several species are known from Australia where they are found in Queensland. There are also a few species in Africa. The males generally have very pointed forewings giving a characteristic shape to the butterfly.

**drusilla** Cramer PURE WHITE
C. and S. America and into the W. Indies 56 mm. (2·2 in.)
The almost pure white male has just a trace of black at the apex. It differs from the female which is yellowish at the base, turning whitish further out with a broad brown band all round the margins. Generally found on the edge of forests or in forest rides, it is commonly seen drinking at damp patches of soil. This butterfly is found from as far south as Brazil and ranges widely further north.

**epaphia** Cramer DIVERSE WHITE
Africa 48–54 mm. (1·89–2·13 in.)
The male has a narrow black edge to the white wing; the females have a large black border, often with the cell in the forewing all black. The underside of the hindwing is yellowish. This butterfly is widespread from tropical Africa, south to Natal and Mozambique. The caterpillar feeds on species of *Capparis* (Capparidaceae). At times the butterfly is common and appears in large numbers.

**nero** Fabricius ORANGE ALBATROSS
N. India, Pakistan, S.E. Asia, the Philippines 70–75 mm. (2·76–2·95 in.)
A widespread butterfly from Sikkim to Java and Sulawesi. The males are common at water in forest clearings, often in large numbers, but the females are generally found in the upper canopy of the forests. The male is orange with black veins in the forewing and yellowish in the hindwing, with a streak of yellow on the inside edge of the hindwings. The female is similar but has a black border round the wings and a band of black on the forewing. The caterpillar feeds on species of Capparidaceae. Many subspecies have been named in different parts of the range of this species. It is probably unique in being the only all-orange coloured butterfly. **119**r

**pandione** Geyer: syn. *A. lalage* SPOT PUFFIN
India, Pakistan, Burma, Thailand, Indonesia 55–80 mm. (2·17–3·15 in.)
This is a white species, with a white-spotted, black margin to the apex of the wing and a black spot at the apex of the cell of the forewing. The hindwings are white with black spots round the edge. The underside is greyish white with black way lines. The female has more black on the forewing, with a white streak below the front margin and a white patch behind; the rest is black. The hindwings have the apical third of the the wing black. In the dry season form there is even less black on the wings. This species is found com-

monly in N. India from sea-level to 1830 m. (6000 ft), it is less common further south. It flies in forest and open country from February to May.

**paulina** Cramer: syn. *A. albina* COMMON ALBATROSS
India, Pakistan, Sri Lanka, Burma, Indonesia, New Guinea, Australia 60–75 mm. (2·36–2·95 in.)
An all white butterfly on the upperside, with pale yellow underside to the hindwing in the male. The female has some black marks on the upperside of the wing. This is a fast flying butterfly with rather pointed wings (like *A. nero,* plate **119**r) sometimes migrating in thousands across country. It generally flies in the mornings and the late afternoon avoiding the heat of the day. The caterpillar feeds on species of Capparidaceae. Subspecies are found in different parts of the range of this butterfly, the Australian one being *A. p. ega* Boisduval. This occurs from Queensland down the east coast to New South Wales, where the foodplant of the caterpillar is said to be species of *Drypetes* (Euphorbiaceae).

**phaola** Doubleday CONGO WHITE
Africa 55–60 mm. (2·17–2·36 in.)
This species is found in the forests of W. Africa, Zaire, Uganda and W. Tanzania. It is a white butterfly with a black apex to the forewing and black spots round the hindwing. The female is greyish white, with a black apex and yellowish or whitish marginal spots. The life history is not known.

APPLE, PAINTED see **Orgyia anartoides**
APPLE
CODLIN see **Cydia pomonella**
LEAF MINER see **Lyonetia clerckella**
SKIN WORM see **Argyrotaenia citrana**
APPLE-WORM, YELLOW-NECKED see **Datana ministra**
**aprica** see **Acontia**
APRICOT see **Phoebus argante**
APRICOT, YELLOW see **Phoebus philea**

**Apsarasa** Moore (NOCTUIDAE)
This small genus of 4 strikingly coloured black, yellow and white species of moth, is found in tropical Asia, from India and the Andaman Islands to New Guinea.

**radians** Walker
Sikkim, India, Sumatra, Sulawesi (Celebes) 44–48 mm. (1·73–1·89 in.)
Nothing is known about the biology of this distinctively patterned species. The dramatic colour pattern may be a warning signal of distastefulness to possible predators. It could prove to be cryptic if the moths were to rest on a spiny bush or tree where the conspicuous, black 'spinose' marking would effectively disguise it. **391**k

AQUATIC MOTHS see **Acentria**
**aquilo** see **Agriades**
ARAB, LARGE SALMON see **Colotis fausta**

**Arachnis** Geyer (ARCTIIDAE)
This is a genus of about 15 N. American and C. American moths related to *Ecpantheria*. Many species are strikingly patterned in red, or yellow, and black. About 10 species are represented in N. America; many of them found also in Mexico.

**zuni** Neumogen ZUNI TIGER-MOTH
S.W. U.S., Mexico 48–54 mm. (1·89–2·13 in.)
The bright yellow ground colour of the forewings is a distinctive feature of this rather rare species. The male hindwing is usually chiefly white above, with a yellow area along the front and inner edges of the wing and a few greyish brown distal spots; the female hindwing is yellow, with 3 broad transverse brown bands and a brown outer marginal line. Its range extends from Arizona and New Mexico southwards into Mexico. **369**d

**Araschnia** Hübner (NYMPHALIDAE)
MAP BUTTERFLIES
A small genus of butterflies with species in Europe and temperate Asia with a few reaching into N. India. Most are strongly patterned and are popularly known as Map butterflies.

**levana** Linnaeus MAP BUTTERFLY
W. Europe, temperate Asia including Japan 32–40 mm. (1·26–1·57 in.)
The upperside colour is generally a dark brown with a white band across the middle and thin reddish lines round the margin of the wings with some white near the base of the forewing. The underside is strongly patterned with thin lines and some broader bands. The caterpillar feeds on nettles *(Urtica)* where it lives gregariously. The butterfly is out in May and June with a second generation in August and September; it flies in woodlands and on the edges of them. The Japanese specimens are distinct and are placed in a separate subspecies from the European ones. **163**b

**arcas** see **Parides**
**arcesilaus** see **Faunis**
ARCHDUKE see **Euthalia dirtea**
ARCHDUKE, DARK see **Euthalia khasiana**
ARCHES,
BLACK see **Lymantria monacha**
BUFF see **Habrosyne pyritoides**
KENT BLACK see **Nola albula**
**archidona** see **Coenophlebia**
**archippus** see **Limenitis**

**Archips** Hübner (TORTRICIDAE) BELL MOTHS
A worldwide genus of moths with many species. When at rest they usually hold the wings slightly flattened giving the bell-shaped appearance to the insects from which they get their popular name.

**argyrospila** Walker FRUIT TREE LEAF ROLLER
U.S., Canada 20 mm. (0·79 in.)
Forewings mottled with a mixture of yellow and brown, with 2 creamy spots on the forewing margin. The hindwings are grey-brown. The caterpillar feeds on leaves and fruit which it covers with webbing. It feeds on apples causing rough scars on the fruit but it will also feed on most deciduous fruit trees. The Fruit Tree Leaf Roller moth is out in June and is common in the U.S.

**cerasivorana** Fitch UGLY-NEST CATERPILLAR
U.S. 18–25 mm. (0·71–0·98 in.)
The moth has dull orange forewings with purple iridescence, the hindwings are orange. The name comes from the habits of the caterpillars, which are defoliators of Choke Cherry *(Prunus)*, in spinning huge communal webs on trees and hedges. These may extend over 30 m. (100 ft) along hedgerows. The moth is widely distributed in the eastern States where the huge nests of the caterpillars are a common and remarkable sight.

**oporana** Linnaeus: syn. *A. piceana*
Europe including Britain, temperate Asia 20–25 mm. (0·79–0·98 in.)
One of the species of Tortricid moths whose caterpillars feed on pine *(Pinus)*. The moth is local, in Britain confined to a few S. counties and is out in July. In some areas it can be a nuisance to the forester growing the soft wood pines commercially but is generally not one of the more serious pine pests. **16**

**podana** Scopoli FRUIT TORTRIX
Europe through temperate Asia to Japan 19–26 mm. (0·75–1·02 in.)
The eggs are laid in a compact mass on apple trees *(Malus)*, but it will also lay eggs on gooseberry and blackcurrant *(Ribes)*, marrow *(Cucurbita)*, elder *(Sambucus)* and has been recorded on citrus fruit. The caterpillars damage flower buds and fruitlets and early stages feed on the leaves which they draw together with silk. Later they feed on buds and will damage the fruit itself. The moth, at rest, resembles a dead leaf and has the typical shape of the Tortricid or Bell moths. Serious financial losses to fruit growers are caused by this moth wherever it occurs and it is a major pest.

**Archon** Hübner (PAPILIONIDAE)
There is only one known species in this primitive genus of butterflies. It is classified in the same subfamily of Papilionidae as the Apollo butterflies *(Parnassius)*.

**apollinus** Herbst FALSE APOLLO
S.E. Europe, W. Turkey, the Middle East, S.W. Russia 40–48 mm. (1·57–1·89 in.)
As its common name suggests, this butterfly is similar in appearance to many species of *Apollo* in which the colours red, black, and white or yellow, predominate. There is considerable variation in the ground colour of the wings, which may be more or less heavily speckled with black or red than in the illustrated specimen. The adult emerges from the chrysalis early in the spring, flying at elevations of up to 1520 m. (5000 ft), invariably in rocky areas. The red and white spotted, black caterpillar feeds on a species of *Aristolochia.* **94**b

**Archonias** Hübner (PIERIDAE)
A small genus of butterflies found in C. and S. America.

**bellona** Cramer
C. & S. America 55 mm. (2·17 in.)
The female is illustrated. The male is black with a lemon-yellow patch on the forewing. The underside of the hindwing of the male has red lines along the veins. Nothing is known of their biology. **119**c

**arcius** see **Rhetus**

**Arctia** Schrank (ARCTIIDAE) TIGER-MOTHS
About 12 species are placed in this genus of typical tiger-moths. They are found in N. Africa, Europe including Britain, and temperate Asia to Japan, and in Canada and the U.S.

**caja** Linnaeus GARDEN TIGER-MOTH
Europe, temperate Asia including Japan, N. America 50–70 mm. (1·97–2·76 in.)
This common and widely distributed moth has many colour varieties which are highly prized by collectors. It has three types of display, sometimes accompanied by the production of sound, all of which warn potential predators of its unpalatability. The poisonous substances found in adults include acetycholines which they are apparently able to produce within their tissues without the need for foodplants in which precursors of these chemicals are present. However, other poisons, including alkaloids, are produced from plant sources such as *Senecio* species. The caterpillar is a large, hairy creature popularly known as a Woolly Bear. Its hairs may cause a skin irritation in some people. **374**

**flavia** Fuessly YELLOW TIGER-MOTH
Europe, temperate W. Asia 50–64 mm. (1·97–2·52 in.)
This is an extremely rare species, found only at high altitudes, and regarded as a prize by collectors. Adults are capable of ejecting noxious secretions from their thoracic glands up to a distance of 200 mm. (7·87 in.). The caterpillar, which overwinters twice before changing into a chrysalis, feeds at night on the foliage of *Cotoneaster vulgaris* and other alpine plants, and rests during the day in rock crevices. It is dark brown with white-tipped, greenish yellow hairs. **369**p

**villica** Linnaeus CREAM-SPOT TIGER-MOTH
N. Africa, Europe including Britain, W. Asia 50–65 mm. (1·97–2·56 in.)
Although not as variable as its relative *A. caja*, this too is a very variable species. The cream spots of the forewing may nearly obliterate the black, and, conversely, the black areas may almost entirely obscure the cream areas. Moths are on the wing in May and June; they are nocturnal. The caterpillar is chiefly black when fully grown, with brown hairs and red legs and head; it feeds on a variety of different low-growing plants from July until April of the following year after having hibernated at an early stage of development. **373**

ARCTIC
BLUE see **Agriades aquila**
CLOUDED YELLOW, PALE see **Colias nastes**
FRITILLARY see **Clossiana chariclea**
FRITILLARY, DINGY see **Clossiana improba**
GRAYLING see **Oeneis bore**
RINGLET see **Erebia disa**
SKIPPER see **Carterocephalus palaemon**

**Arctiidae** TIGER-MOTHS, FOOTMEN, TUSSOCKS
As accepted at present, this family of the superfamily Noctuoidea includes *Arctia* and other typically mostly orange or yellow 'tiger-moths' with brown or black markings; large groups of New World genera with reduced hindwings, often with cryptically or disruptively coloured forewings but brightly coloured hindwings and abdomen *(Bertholdia)*; and genera of broad-winged moths like *Haploa*. Also included are the footmen moths (Lithosiinae) (eg. *Lithosia)* and, by some authors, parts of the family Hypsidae (syn. Pericopidae, in part) much of which should probably be united with the Arctiidae. There are about 10,000 known species. Many caterpillars are 'woolly bears' or 'tussocks', but others are less hairy; they feed on the foliage of low-growing plants, shrubs and trees or, in the case of the Lithosiinae, on lichens, and some are pests in orchards, cocoa plantations and forests. The hairs of some species have irritant properties. Most caterpillars have a partiality for poisonous species of various families of plants, for example, ragwort (Compositae), *Laburnum* (Leguminosae), potato (Solanaceae), dog's mercury (Euphorbiaceae), foxglove (Scrophulariaceae) and nettle (Urticaceae), and are capable of extracting toxins such as cardiac glycosides and pyrrolizidine alkaloids from these plants. These toxins are passed on to the adult moth which is consequently protected against many predators (probably especially birds and bats). Some species of Arctiids seem to be able to manufacture toxins without the need for a poisonous foodplant. A correlated development in evolution has produced conspicuous 'warning' colour patterns of red, orange, yellow, black and white, which diurnal predators soon learn to associate with the unpalatability of many species. Numerous New World species have warningly coloured hindwings and abdomen but cryptically coloured forewings which cover the hindwings when the moth is at rest. These species have two stages of protective behaviour; camouflage, when immobile, as a first line of defence, followed by a display of the warningly coloured hindwings and abdomen when provoked. Sound, too, is used by Arctiids as a warning signal to predators. Experiments have shown that high frequency pulses of sound produced in flight by most nocturnal Arctiids are associated with unpalatability by bats, in the same way that colour acts as a signal to birds during the day. The sound is produced by the 'tymbal organ', a blister-like part of the thorax which is rhythmically distorted in a way comparable to that in the more familiar cicadas of the Mediterranean area and further south. At least one species, *Pyrrharctia isabella*, is probably a sound mimic; it is palatable to bats and yet produces the same type of sound as unpalatable nocturnal species of Arctiidae and apparently deceives the bats in this way. *Palustra* is a sub-aquatic genus. See *Acerbia, Amastus, Amaxia, Ammobiota, Amsacta, Apantesis, Apocrisias, Arachnis, Arctia, Bertholdia, Callimorpha, Callisthenia, Chionaema, Chlanidophora, Chrysochlorosia, Cymbalophora, Damias, Diacrisia, Diaphora, Ecpantheria, Eilema, Estigmene, Euchaetias, Eucharia, Eupseudosoma, Graphelysia, Halisidota, Holomelina, Hyperborea, Hyphantria, Leptarctia, Lithosia, Lycomorpha, Miltochrista, Nyctemera, Ocnogyna, Ormetica, Palustra, Parasemia, Parathyris, Pericallia, Phragmatobia, Platarctia, Platyprepia, Pyrrharctia, Rhipha, Rhodogastria, Setina, Spilosoma, Spiris, Tyria, Utetheisa, Viviennea.*

**Arctioblepsis** Felder (PYRALIDAE)
A moth genus in the subfamily Pyraustinae, with only one species.

**rubida** Felder
India, Pakistan, China, Indonesia (Borneo) 35–40 mm. (1·38–1·57 in.)
Not many specimens of this species are known and nothing is known of its biology. This species is unique in the subfamily Pyralinae in colour and pattern. Its exact relationship with the other genera in the subfamily is not known. There is some variation in the colour and pattern of this species and it is possible that more than one species is involved, but to determine this will need further detailed examination of the specimens. The specimen figured is from N. India. **63**

**Arctonotus** Boisduval (SPHINGIDAE)
There is a single species in this N. American genus of hawk-moths.

**lucidus** Boisduval
W. U.S. 50–60 mm. (1·97–2·36 in.)
The short 'furry' body is a distinctive feature of this species. Oregon specimens are darker than those from California, and there is some variation throughout its range in the coloration of the wings. *A. lucidus* flies very early in the year, from January in the south to March in the north. It is known to visit flowers during the day. Evening Primrose *(Oenothera)* has been recorded as a larval foodplant. **344**b

**arcturus** see **Papilio**
**arcuata** see **Agathia, Drepana** under **D. falcataria**
**arcuosaria** see **Phyle**
**ardeniae** see **Cizara**
**arduinna** see **Melitaea**
**arenella** see **Hyposmocoma**
**arethusia** see **Hamadryas**
**areuta** see **Astraeodes**
**argante** see **Phoebus**
**argaula** see **Agonoxena**

**Argema** Wallengren (SATURNIIDAE) MOON-MOTHS
The Moon-moths of Europe *(Graellsia* species) and the rest of the Northern Hemisphere *(Actias* species) are close allies of this small African and tropical Asian genus of long-tailed moths.

**mimosae** Boisduval AFRICAN MOON-MOTH
Congo Basin, Kenya, subtropical S. Africa 120–130 mm. (4·72–5·12 in.)
The green coloration of this moth, like that of other green moon-moths, rapidly fades in daylight to a yellowish white, both in living and set specimens. The caterpillars are green, with several yellow-tipped tubercles bearing yellow and black hairs. *Mimosa* strangely is not among recorded foodplants of the caterpillars which prefer foliage of the Marula tree *(Sclerocarya)*. The metallic grey cocoons are perforated with numerous small holes. **332**

**mittrei** Guérin-Méneville
Madagascar 160–180 mm. (6·3–7·09 in.)
This pale green (male) or yellow (female) species is the largest member of its genus. Its caterpillars feed on *Eugenia* and *Weinmannia*. It is similar in general appearance to the illustrated *A. mimosae* except for the immensely long tails, which, from the tip to the base of the hindwing, may be as long as 200 mm. (7·87 in.).

**argenissa** see **Polystichtis**
ARGENT and SABLE see **Rheumaptera hastata**
**argentana** see **Eana**
**argentata** see **Halisidota**
**argentea** see **Argyritis**
**argenteus** see **Argyrophorus**
**argentifera** see **Nephele**
**argentina** see **Spatalia**
**argia** see **Nepheronia**
**argiades** see under **Everes**
**argillacea** see **Anomis**

**Argina** Hübner (HYPSIDAE)
About 7 species of brightly coloured, chiefly orange or yellow, and black moths belong in this genus. It is represented in Africa, tropical Asia and Australia. Some of the species are very common. Included in the caterpillars' foodplants are species of *Crotalaria* from which many species of moths are able to extract and store substances toxic to vertebrates.

**cribraria** Clerck
India, China, Madagascar, S.E. Asia to Australia 36–42 mm. (1·42–1·65 in.)
The wings of this species are either yellow or orange above (palest in the female), with numerous pale-edged, black spots on the forewing and a few larger black spots, chiefly marginal, on the hindwing. The abdomen is banded with yellow or orange, alternating with black. This is a common woodland species in much of its range. The caterpillar is dark grey with short hairs over most of the body and long hairs at the rear end.

**argiolus** see **Celastrina**

**Argopteron** Watson (HESPERIIDAE)
This is a small genus of skipper butterflies found in S. America. The species in it are remarkable for the metallic scales on the underside of the wing.

**aureipennis** Blanchard GOLD SKIPPER
Chile 34–36 mm. (1·34–1·42 in.)
The upperside of this butterfly is brown with 3 large orange-gold spots in the forewing. There is no indication there of the remarkable colour on the underside. The related species, *A. puelmae*, is smaller and has a similar underside but the upperside has yellow hindwings and the spots on the forewings are larger. *A. puelmae* is also found in S. America. **80**c

**argus** see **Plebejus**
ARGUS,
ALPINE see **Albulina orbitulus**
BLUE see **Junonia orithya**
BROWN see **Aricia agestis**
MEADOW see **Junonia villida**
SCOTCH see **Erebia aethiops**

**Argynnis** Fabricius (NYMPHALIDAE)
A large genus of butterflies, mostly European and Asian with a few extending into the Oriental region. They are popularly known as fritillaries and are generally powerful fliers, speeding up and down woodland glades or over rough ground.

**paphia** Linnaeus SILVER-WASHED FRITILLARY
Europe including Britain, N. Africa, temperate Asia including Japan 54–70 mm. (2·13–2·76 in.)
The butterfly lives in woodland clearings and flies from the end of June through August. The caterpillar feeds on violets *(Viola)*. Several subspecies of *A. paphia* have been described, one in Japan. There is a colour form found only in female specimens, (form *valesina)*, which has a pale grey with green suffusion on the upperside of the wing, the underside is typical of *paphia*. **194**

**Argyresthia** Hübner (YPONOMEUTIDAE)
A large genus of over 100 species of micro-moths. The genus was first described in 1826 and species are found in most parts of the world but the majority occur in Europe and N. America. A few have been described from New Guinea and Africa but the study of the genus is far from complete and the true relationship of many of those in the genus is not known. Many species of *Argyresthia* rest, as do others in the subfamily Argyresthiinae, with the head downwards and the body raised up at an angle. Generally they are rather narrow-winged species.

**ephippella** see **A. pruniella**

**goedartella** Linnaeus
Europe including Britain, N. America 11–13 mm. (0·43–0·51 in.)
The forewings are a shining white, often suffused with a golden colour and usually with 3 bands of coppery-gold across them. The hindwings are greyish. The caterpillar feeds on the shoots of birch *(Betula)* and alder *(Alnus)*. There is a possibility that the American specimens of this species are not correctly identified and that they may be a distinct species.

**pruniella** Clerck: syn. *A. ephippella*
Europe including Britain 10–12 mm. (0·39–0·47 in.)
The whitish caterpillars live in the shoots of Sour Cherry *(Prunus cerasus)*. The forewings are narrow, reddish brown with some yellow streaks and a thick white streak forming a complicated pattern. The hindwings are grey. Both sexes are similarly coloured. This species is common over much of its range and the moth is out in July. **41**c

**Argyritis** Hübner (NOCTUIDAE)
There is a single species in this genus of Old World moths. The green and silver forewings are highly distinctive.

**argentea** Hufnagel
Europe E. to the Balkans and S.W. Asia 34–42 mm. (1·34–1·65 in.)
This is one of the prettiest Noctuid moths found in Europe. Its caterpillar is green, with a white line along the back and brownish red tubercles. The food-plant is wormwood *(Artemisia)*. **391**q

**Argyrogrammana** Strand (NEMEOBIIDAE)
There are about 10 species in this genus of C. American and tropical S. American butterflies. The colour pattern of many species differs radically between the sexes. The name *Argyrogramma* Stichel, which was originally applied to this genus, has been used earlier in zoological literature and was replaced in 1932 by the present name.

**saphirina** Staudinger
N.W. S. America 26–30 mm. (1·02–1·18 in.)
The male of this rare species is a brilliant iridescent blue, with 4 black bands across the forewing and 3 incomplete black bands on the hindwing; the outer margins of the wings are also black except for a narrow metallic pale bluish green line which runs parallel to the outer margin. The remarkably different females are light orange in ground colour and have 5 dark brown bands across both wings. The under surface of both male and female is similar in colour pattern to the upper surface of the female.

**Argyrophorus** Blanchard (NYMPHALIDAE)
There are only 2 known species in this genus of S. American butterflies. They are characterized by the remarkable silver scales on the upperside of the wing. The genus belongs to the subfamily Satyrinae but is strikingly different in colour from the brown butterflies which are more typical of this subfamily.

**argenteus** Blanchard SILVER BUTTERFLY
Chile 45–55 mm. (1·77–2·17 in.)
The underside of the silver butterfly is strongly patterned and quite different from the upperside. The female also has brown round the margins of the wing. A smaller related species has the same silver coloured forewings but the hindwings are black. **174**n

**argyrospila** see **Archips**

**Argyrotaenia** Stephens (TORTRICIDAE)
Most of the moths in this genus are found in N. America but 2 species are found in Europe. There are many species of economic importance, particularly those attacking pines. The caterpillars make cylindrical tubes out of clusters of pine needles. Other species feed on oak *(Quercus)*, hickory *(Carya)* and other trees.

**citrana** Fernald ORANGE TORTRIX or APPLE SKIN WORM
U.S., Canada 11–19 mm. (0·43–0·75 in.)
The moth is variable in colour but is generally brown or orange-brown with black diagonal bands across the forewings. These form a V mark when the wings are folded across the body. The hindwings are white. The light green caterpillar feeds on the tips of twigs and on leaves, blossom and buds of citrus, apricot *(Prunus)*, pears *(Pyrus)*, apples *(Malus)*, and many other fruits. It probably has a wider range of host plants than almost any other N. American moth. In the northern parts of its range it can also be a pest in greenhouses.

**Aricia** R.L. (LYCAENIDAE)
The 12 or so species of this genus of butterflies are found in Europe, N. Africa, the Middle East, Kashmir, and temperate Asia as far as Siberia. Males are blue above; females are at least partly brown. The identity of 'R.L.' is unknown.

**agestis** Denis & Schiffermüller BROWN ARGUS
Europe, temperate Asia to the Pacific 23–28 mm. (0·91–1·1 in.)
Limestone or chalk regions are preferred by this brown 'Blue' butterfly. The caterpillars which are often attended by ants, feed on rock rose *(Helianthemum)* and species of Geraniaceae. The first reference in British literature to this species is by Petiver in the year 1717. (See also under *Freyeria trochylus*). **252**

**aristeus** see **Graphium**
**aristodemus** see **Papilio**
**aristolochiae** see **Pachliopta**
**arion** see **Maculinea**
**arizonae** see **Gnophaela**
**arja** see **Polyura**
**armida** see **Automeris**
**armigera** see **Heliothis**
ARMYWORM see **Mythimna unipuncta**
ARMYWORM MOTHS see **Spodoptera**
**aroensis** see **Milionia**

**Arrhenophanes** Walsingham (ARRHENOPHANIDAE)
This genus of micro-moths is found from Brazil to Honduras.

**perspicilla** Stoll
Brazil, Peru, Guyanas, Honduras 38–68 mm. (1·5–2·68 in.)
The females are much larger than the males and have pectinate antennae. The biology is unknown. There is some variation in the wing pattern but relatively few specimens have ever been collected. The study of this species, being a member of a small and little known family, would be scientifically rewarding. **11**a

**Arrhenophanidae**
A small family of micro-moths found in C. and S. America. It is probably related to the Tineidae but the exact systematic position is uncertain. See *Arrhenophanes*.

**artemis** see **Actias**
**arthemis** see **Limenitis**
**arundinata** see **Scoparia**

**Arycanda** Walter (GEOMETIDAE) CURRANT or MAGPIE MOTHS
This genus is related to *Abraxas*. It contains a large number of species in Indonesia and New Guinea. The biology is unknown.

**emolliens** Walker
Indonesia; Sulawezi (Celebes), Sarawak 60–65 mm. (2·36–2·56 in.)
The underside is similarly patterned to the upperside. There are several related species but without the large blotchy pattern. **296**r

**vinaceostriga** Prout
New Guinea 50–55 mm. (1·97–2·17 in.)
The underside is a plain smokey black with black spots on each wing contrasting with the heavily patterned upperside. **296**t

**Arzama** see under **Bellura**

**Ascalapha** Hübner (NOCTUIDAE)
One species is currently placed in this genus. The generic names *Erebus* and *Otosema* have been applied at times to this genus.

**odorata** Linnaeus BLACK WITCH
S. U.S., C. and tropical S. America, Hawaii 80–150 mm. (3·15–5·91 in.)
Occasional stragglers of this imposing species reach Canada but as a North American breeding species it is commonest in Florida and the Gulf States. Its caterpillar feeds on the legume genera *Pithecellobium* and *Cassia*. (Species of *Cassia* are the source of senna, a laxative drug.) *A. odorata* was introduced to Hawaii and is now virtually a pest there. Cardinals *(Richmondena)*, mynah birds *(Gracula)* and sparrows *(Passer)* are recorded as predators of *A. odorata* in Hawaii. **392**a

**ascanius** see **Parides**
ASCANIUS SWALLOWTAIL see **Parides ascanius**

**Ascia** Linnaeus (PIERIDAE)
This genus of butterflies is restricted to the New World where species occur from the U.S., through C. America and the West Indies into S. America.

**josephina** Latreille GIANT WHITE
C. America, S. U.S.A. 62–72 mm. (2·44–2·83 in.)
This is a very chalky white butterfly with a black spot at the apex of the cell in the forewing and a small mark at the apex of the cell in the hindwing. Although common in Central America, in N. America it is only found in S. Texas and occasionally in Kansas.

**monuste** Linnaeus GREAT SOUTHERN WHITE or CABBAGE WHITE 45–56 mm. (1·77–2·2 in.)
C. & S. America, S. U.S. and into the W. Indies
This species sometimes appears in large swarms off the coast of Florida. These migrants, like many other migrating species, have a very purposeful flight not wandering about as is more usual for a non-migrating butterfly. There are related species in C. America which may stray into the U.S. The caterpillars of the Great Southern White feed on cultivated cabbages *(Brassica)*, and various wild Cruciferae.

**asclepia** see **Symmachia**
**ascolius** see **Papilio**
**asella** see **Heterogenea**

**Asota** Hübner (HYPSIDAE)
Currently containing about 40 species, this genus is represented in India, Sri Lanka, China and Japan, through S.E. Asia to New Guinea, the Solomons and Australia. The colour pattern is composed of yellow, brown and black. The caterpillars of some species are known to feed on *Ficus*.

**iodamia** Herrich-Schäffer
N. and N.E. Australia 47–64 mm. (1·85–2·52 in.)
The forewings of this species are yellowish brown with a pale yellow basal area enclosing 6 black spots. The hindwings are orange-yellow with a single broken transverse black band. The caterpillar feeds on *Ficus*.

**aspasia** see **Cepora** and **Danaus**
ASPEN BLOTCH MINER see **Phyllonorycter tremuloidiella**
**assimilis** see **Correbidia** and **Hestinalis**
ASSMANN'S FRITILLARY see **Mellicta britomartis**
ASSYRIAN, ROYAL see **Terinos terpander**
**astarte** see **Callicore**

**Asterocampa** Röeber (NYMPHALIDAE)
HACKBERRY BUTTERFLIES
These butterflies are commonly seen flying round Hackberry trees *(Celtis* sp, Ulmaceae) resting on the leaves, then flying off fast with erratic flight to alight on mud puddles or, not infrequently, on people. The females are larger than the males. This genus, which is restricted to America, is closely related to *Doxocopa*.

**clyton** Boisduval TAWNY EMPEROR
U.S. 48–65 mm. (1·89–2·56 in.)
This butterfly is widespread from New England westwards to Michigan, Iowa, Nebraska and S. to Texas and the Gulf States. The caterpillar feeds on Hackberry *(Celtis)*. The adult butterfly is out in June. There are several subspecies of this butterfly in different parts of the U.S. The female is much larger than the male and paler coloured. **216**p

**Asterope** Hübner: syn. *Crenis* (NYMPHALIDAE)
Mostly brown or fawn coloured species with a few, larger, violet ones, with orange undersides. The flight of these butterflies is rapid but brief and when they settle on a treetrunk, they are well camouflaged. They are attracted to over-ripe fruit, damp mud and even to human perspiration. In the tropics, they occur in large swarms. The species in the genus are widespread over Africa S. of Sahara.

**rosa** Hewitson LILAC NYMPH
South Africa, Mozambique, E. & C. Africa 62–74 mm. (2·44–2·91 in.)
This species is common in the bush or savannah areas and sometimes in forest clearings. It is large, violet-coloured with a few dark spots near the outer wing margin in the male, and the female with dark blotches in the middle of the forewing. This contrasts sharply with the underside colour. Although common, little seems to be known of its life history. At times the

butterfly migrates in large swarms. Both wing surfaces are shown. **217**b, c

**asterope** see **Ypthima**

**Asthenidia** Westwood (OXYTENIDAE)
There are about 12 species in this genus of moths. Most are white in ground colour, marked with yellowish brown transverse bands, and with a short hindwing tail. Their general appearance is very similar to that of many species of the white Uraniidae (Microniinae).

**lactucina** Cramer
Tropical S. America from Venezuela to Paraguay 41–55 mm. (1·61–2·17 in.)
This moth is typical of its genus in colour pattern and wing-shape; some species have the brown transverse lines less well marked, others have darker and straighter lines. The caterpillars of *A. lactucina* are green, with a pair of eyespots near the front and a dorsal, Sphingid-like horn at the rear end; they feed on *Ourouparia* (Rubiaceae). **318**b

**astraea** see **Heliconius**

**Astraeodes** Schatz (NEMEOBIIDAE)
There is a single known species in this genus of butterflies.

**areuta** Doubleday
Brazil 34–38 mm. (1·34–1·5 in.)
The wings of *A. areuta* are bright yellow, crossed by 2 brownish yellow lines and an outer marginal row of small metallic gold spots. Nothing is known about the life history of this species.

**atala** see **Eumaeus**
ATALA see **Eumaeus atala**
**atalanta** see **Vanessa**
**ataxus** see **Chrysozephyrus**
**atergatis** see **Lycorea**

**Aterica** Boisduval (NYMPHALIDAE)
A small genus of African butterflies found in Madagascar and throughout the mainland south of Sahara.

**galene** Brown FOREST GLADE NYMPH
Africa 55–80 mm. (2·17–3·15 in.)
A brownish black butterfly with 2 rows of creamy-yellow spots on the forewing and a large oval creamy-yellow patch on the hindwing. The female is similar but the spots are often larger and whiter. This common species is found throughout the forests of tropical Africa, south to Malawi, Mozambique and E. Rhodesia. It generally flies close to the ground and stays in shady areas. The adults are about in all months of the year.

**athalia** see **Mellicta**
**athene** see **Weymeria**

**Athletes** Karsch (SATURNIIDAE)
This is a genus of 3 large tropical African species which have large eyespots on the hindwing and a short tail. The caterpillars are green, with lustrous metallic markings.

**steindachneri** Rebel STEINDACHNER'S EMPEROR
Congo Basin, E. Africa, subtropical S. Africa 130–155 mm. (5·12–6·1 in.)
This species was also known as *A. gigas*, a specific name which reflects its impressive size. The caterpillar is green and blue with metallic yellow spines. It feeds on *Julbernardia*. **326**

**atlantica** see **Plebicula**
**atlas** see **Attacus**
ATLAS MOTHS see **Attacus**
ATLAS, AFRICAN see **Epiphora vacuna**
ATLAS BLUE see **Plebicula atlantica**

**Atlides** Hübner (LYCAENIDAE)
This is a small genus of hairstreak butterflies, mostly tropical American in distribution but represented by a single species, *A. halesus*, in the U.S.

**halesus** Cramer GREAT PURPLE HAIRSTREAK
S.W. U.S., Mexico 27–38 mm. (1·06–1·5 in.)
This large hairstreak is one of the most brilliantly coloured N. American butterflies. Its caterpillar feeds on the foliage of mistletoe *(Phoradendron)*, Mesquite *(Prosopis glandulosa)* and Desert Ironwood *(Olneya tesota)*. Adult females (not illustrated) have a much broader marginal band on both wings than in the male, and have a violet and green iridescence. **263**c

**atomaria** see **Ematurga**
**atrata** see **Odezia**
**atricincta** see **Napata**
**atripennis** see **Dahana**
**atropos** see **Acherontia**

**Attacus** Linnaeus (SATURNIIDAE) ATLAS MOTHS
Some of the largest moths in the world belong to this genus. The few species, found only in E. and S.E. Asia, are now protected by law in some countries. Allied genera of large moths are found in Africa *(Epiphora)* and S. America *(Rothschildia)*.

**atlas** Linnaeus ATLAS MOTH
Sri Lanka and India to China, Malaysia and Indonesia 160–300 mm. (6·3–11·81 in.)
Females of this species with wingspans of over 300 mm. (11·81 in.) have been collected. The only species of Lepidoptera which rival it in wing length, though not in area are the S. American Noctuid *Thysania agrippina* and the females of some *Ornithoptera* butterflies. The caterpillars, which are chiefly green speckled with brown, and have 6 tubercles on each segment, may be up to 100 mm. (3·94 in.) in length and over 20 mm. (0·79 in,) in diameter. They feed on the leaves of a number of different trees and shrubs. Numerous subspecies of *A. atlas* have been described but there is doubt as to whether they are correctly associated together as a single species. The illustrated specimen from Borneo is a very small example of its species. **327**

**Atteva** Walker (YPONOMEUTIDAE)
A large genus with many colourful species. The caterpillars, where known, live together in groups, each group living in a web they have spun, rather similar to species of *Yponomeuta*. The caterpillars of *Atteva* live in these webs on species of Simarubaceae. Most of the species of *Atteva* known are from Indonesia, New Guinea and Australia but a few have been described from C. and S. America. The relationships of these to those from the Far East is still a matter of some speculation and a study of the New World species of *Atteva* is needed.

**mathewi** Butler
New Guinea, Solomon Islands 30–33 mm. (1·18–1·3 in.)
There are several very similar species. The moths are conspicuously coloured and may possibly be distasteful to birds or other predators. Nothing is known of their biology. **37**b

**megalastra** Meyrick
Australia, New Guinea, Indonesia, India 28–30 mm. (1·1–1·18 in.)
This moth is found over a large area where it is common in most parts. The specimen illustrated is from Australia. When the species has been studied in detail it will probably be found that there are several distinct subspecies in different parts of its range. The white spot and conspicuous orange hindwings make this a colourful moth. Nothing is known about the life history. **37**e

**monerythra** Meyrick
Galapagos Islands 26 mm. (1·02 in.)
There are 2 rather similar species described from the Galapagos Islands. The specimens which were used for the original description of the species were collected by members of the *St George* Expedition in 1924. The *St George* was a barquentine-rigged sailing vessel of 841 tons (with auxiliary steam power) which left Dartmouth, England in 1924 with a staff of 8 scientists to spend a year studing natural history in the Pacific area. Amongst the many species new to science they discovered was this small moth. Nothing is known of their biology. A research station to study the strange fauna of the Galapagos Islands, long associated with the great 19th-century naturalist, Charles Darwin, has been set up under the enlightened guidance of the Government of Ecuador, to whom the Archipelago of Colon (Galapagos) belong. **37**a

**punctella** Cramer: syn. *A. pustulella*, *A. aurea*
AILANTHUS WEBWORM
N. S. & C. America (Panama, Venezuela Paraguay, Brazil, Trinidad, Guyana) 24 mm. (0·94 in.)
The undersides of the wings are a smokey grey colour with none of the brilliant colours of the uppersides. Several other closely allied species look like this but differ in the amount of orange on the wings. The adult moths have been observed feeding during the day on a number of different species of flowers, but are generally crepuscular. The caterpillars live gregariously in a common web. They are longitudinally striped with orange, black and white and feed on *Ailanthus*. **37**d

**atthis** see **Heliconius**

**Atychia** Latreille (GLYPHIPTERIGIDAE)
This genus of micro-moth was at one stage regarded as sufficiently distinct to be placed in its own family (Atychiidae) but is now placed in the Glyphipterigidae though still regarded by some experts as a subfamily. Most of the species are similar to the one illustrated.

**appendiculata** Esper
Europe, N. Africa 16–20 mm. (0·63–0·79 in.)
This species is quite common in S. Europe where the adults fly from May to July. The caterpillar lives amongst the roots of grasses. **34**j

**atymnus** see **Loxura**

**Audre** Hemming (NEMEOBIIDAE)
About 30 C. and S. American species are placed in this genus of butterflies. The majority of these were at one time placed in *Hamearis* which now includes only a single European species. Some species are found as far south as the Andes of Argentina and Chile and in S. subtropical grassland regions.

**colchis** Felder
Brazil 37–45 mm. (1·46–1·77 in.)
This is a somewhat fritillary-like species, dark brown in ground colour, with numerous white spots on both wings and a single row of orange spots towards and parallel to the outer margin of the wings.

**auge** see **Cosmosoma**

**Aulacodes** Guenée (PYRALIDAE)
A genus of Pyralid moths in the subfamily Nymphulinae many of whose species have caterpillars which feed on aquatic plants. The genus is a large one, with many in the Indo-Australasian region and a few in America. Most are attractively patterned with a delicate, porcelain look.

**splendens** West
Philippines 35–45 mm. (1·38–1·77 in.)
Although known only from the one locality there are many other relatively similar related species which have the same type of hindwing pattern and broadly similar forewing. They all have the same delicate 'porcelain' look. Nothing is known of the biology of this species. **56**aa

**Aulocera** see **Satyrus**
**aurantia** see **Tharsanthes**
**aurantiaca** see **Diacrisia** and under **Holomelina**
**aurantialis** see **Daulia**

**Aurea** Evans (LYCAENIDAE)
Four species are placed in this genus of butterflies. It ranges from Malaya to Borneo.

**aurea** Hewitson
Malaya, Sumatra, Borneo 32–35 mm. (1·26–1·38 in.)
The male of this species is strikingly different in coloration from the golden female. It is a bright violet-blue above, broadly edged with dark brown on

both wings. The undersurface of both sexes is brown and cryptically patterned. Nothing is known about the life history of this beautiful butterfly. **240**a, c

**aurea** see **Atteva punctella**
**aureipennis** see **Argopteron**
**aurelia** see **Diaethria**
**aurella** see **Nepticula**
**aureus** see **Teinopalpus**
**auricrinella** see **Micropterix**
**aurifer** see under **Hypochrysops ignita** and **H. piceata**
**aurifluella** see **Ethmia**
**aurinia** see **Euphydryas**
**aurivitta** see **Cydosia**
**aurora** see **Alcides, Allancastria, Cithaerias** and **Miniophyllodes**
**aurota** see **Anaphaeis**
**ausonia** see **Euchloe**
AUSTRALIAN
ATLAS see **Coscinocera hercules**
CROW see **Euploea core**
GULL see **Cepora perimale**
VAGRANT see **Vagrans egista**
**australica** see **Momophila**
**australis** see **Cerura** and **Colias**
**autobrocha** see **Anticrates**
**autocrator** see **Parnassius**

**Autographa** Hübner (NOCTUIDAE)
This is a very small genus of moths found in Europe, temperate Asia and North America. The best known species is the extremely common *A. gamma*, previously placed in the now restricted genus *Plusia*.

**gamma** Linnaeus SILVER-Y MOTH
Temperate Europe including Britain, Asia, N. Africa 35–40 mm. (1·38–1·57 in.)
This is one of the best known and commonest migratory species in the Old World. In Britain, early summer migrants produce offspring with such efficiency that by late summer *A. gamma* is often very common. Variation in the ground colour of the wings is common; melanics are also known. Its caterpillar will feed on almost any low-growing plants, including food-crops on which it sometimes reaches pest proportions. **387**

**Automeris** Hübner (SATURNIIDAE)
This is a large genus of over 100 species, 4 of which occur in N. America, including one which reaches Canada, but which is essentially tropical American in range. Almost all the species have a conspicuous eyespot on the hindwing. The caterpillars are armed with numerous, highly irritant spines. Caterpillars of 2 S. American species, *A. armida* Fabricius and *A. erischton* Boisduval, are pests of Cocoa (*Theobroma cacao*). The identity of many species has been revealed only recently by Claude Lemaire (Paris).

**io** Fabricius IO MOTH
Mexico, U.S., Canada 50–80 mm. (1·97–3·15 in.)
This variable species is one of the best known of N. American moths. As in other similarly patterned species, the hindwing eyespots are thought to deter birds and other potential enemies which mistake eyespots for the eyes of a large and dangerous adversary. The female (not illustrated) forewing is mostly reddish to purplish brown. The handsome caterpillars are green, pink and white; they are at first gregarious but later solitary feeders on a vast number of broad-leaved trees and herbaceous plants including cotton (*Gossypium*) and corn (*Zea*). **331**m

**iris** Walker
C. America and S. U.S. 60–74 mm. (2·36–2·91 in.)
This is a rather larger moth than *A. io* and has a conspicuous dark straight transverse line close and nearly parallel to the outer margin of the forewing.

**Automolis** Hübner (THYRETIDAE)
Many species of this African genus of moths have been better known for many years under the later and consequently invalid generic name *Metarctia* Walker. Over 100 other species placed in *Automolis* do not belong there and are being transferred to various genera of Arctiidae by one of the present authors (A.W.).

**meteus** Stoll
South Africa 35–40 mm. (1·38–1·57 in.)
This is typical of its genus in wingshape and other general features. Nothing is known about the life history of this moth. **357**m

AUTUMN LEAF see **Doleschallia bisaltide**
AUTUMN-LEAF VAGRANT see **Eronia leda**
**autumnalis** see **Alcathoe**
**avalona** see **Strymon**
**avellana** see **Apoda**
**avellaneda** see **Phoebus**
**avis** see under **Callophrys rubi**
AWL, PEACOCK see **Allora doleschalli**

**Axia** Hübner (AXIIDAE)
This is a group of 4 species of moths found in N. Africa, S. Europe and some of the Mediterranean islands, eastwards through Turkey to S.E. Russia. All except the Armenian *A. olga* have one or more metallic white markings on the forewing.

**margarita** Hübner
N. Africa, S. Europe 23–30 mm. (0·91–1·18 in.)
This is an apparently rare species. The illustrated specimen of the subspecies *A. margarita atlasica* was captured in the Atlas mountains of Morocco. The caterpillar of *A. margarita* feeds possibly on *Arundo*, a genus of grasses (Gramineae). **272**n

**Axiidae**
This is a family of 2 genera, *Axia* and *Epicmelia* found in N. Africa and from S. Europe to S.E. Russia. It is a member of the superfamily Geometroidea and is sometimes combined with the family Thyatiridae. See *Axia*, *Epicmelia*.

**Axiocerses** Hübner (LYCAENIDAE)
This is a small genus of orange and brown, or red and brown butterflies. They are found in tropical and subtropical Africa, including the Sudan and Ethiopia.

**bambana** Grose-Smith SCARLET BUTTERFLY
E. Africa, South Africa 30–35 mm. (1·18–1·38 in.)
This is a common butterfly in scrubland and in open woodland areas. The upper surface of the female is orange-red with black at the base and outer margins of the wings, and a few black central spots. Males have larger areas of black above as a result of the coalescence of the wing-markings. The under surface is orange-grey, beautifully marked with black-edged metallic gold spots.

AZALEA LEAFMINER see **Caloptilia azaleella**
**azaleella** see **Caloptilia**
AZURE see **Ogyris**
AZURE,
DARK PURPLE see **Ogyris abrota**
SPRING see **Celastrinia argiolus**
SYDNEY see **Ogyris ianthis**
AZURE HAIRSTREAK see **Iolaus coeculus**

**Baeotus** Hemming (NYMPHALIDAE)
A small S. American genus of butterflies with less than 6 species included at present.

**baeotus** Doubleday & Hewitson
S. & C. America 85 mm. (3·35 in.)
The underside is in striking contrast to the upperside, illustrated, being all white with dark spots and eyespots on the hindwing surrounded by yellow. Probably the colour contrast helps in camouflage. As the butterfly settles, closing its wings, it presents a contrasting colour and pattern to a watching predator and may sufficiently confuse potential enemies that they actually lose sight of it once it has landed. **209**k

BAGWORM MOTHS see **Thyridopteryx**
**baldus** see **Ypthima**
BALKAN MARBLED WHITE see **Melanargia larissa**
**balluca** see **Plusia**
**bambana** see **Axiocerses**
BAMBOO
PAGE see **Metamorpha stelenes**
PAGE, SCARCE see **Philaethria dido**
TREE BROWN see **Lethe europa**
BANDED, PURPLE see **Limenitis arthemis**
BANDED
EMPEROR see **Cinabra hyperbius**
GOLD TIP see **Colotis eris**
GRAYLING, GREAT see **Brintesia circe**
KING SHOEMAKER see **Prepona meander**
NYMPH, BLUE see **Cynandra opis**
WOOLLY BEAR see **Pyrrharctia isabella**

**Banisia** Walker (THYRIDIDAE)
A genus of 30 or more moths, usually of a rather reddish brown colour and frequently with a translucent patch in the forewing. Often they are patterned to look like dead leaves. The genus is known from C. and S. America, Africa, India, Indonesia and Australia and is found in all warmer regions of the world.

**myrsusalis** Walker
America, Africa, India, Indonesia 22–25 mm. (0·87–0·98 in.)
This moth is abundant throughout the tropics and has been separated into several subspecies in different parts of the world. It has a red-brown forewing speckled with black, with 3 or more small round translucent spots on the forewing, all close together. There is generally a reddish or orange-red patch on the front margin of the forewing. The underside has a purplish tinge. The caterpillar feeds on the leaves of Sapodilla (*Manilkara zapota*). This is the tree from which an ingredient of chewing gum is obtained. The moth is so widespread that it seems possible that it was carried round the world by man since we know that the Sapodilla host was transported widely throughout the tropics and planted for its edible fruit.

BANK-NOTE BLUE see **Evenus regalis**
**banksiae** see **Danima**

**Baorisa** Moore (NOCTUIDAE)
There is a single colourful species of moth in this tropical Asian genus.

**hieroglyphica** Moore
N. India, Sikkim, S.E. Asia to the Philippines 40–50 mm. (1·57–1·97 in.)
This is one of the most beautifully marked species of the family Noctuidae. Nothing appears to be known about its biology. **391**c

BAR, TREBLE see **Aplocera plagiata**
**barberae** see under **Brephidium**
**barnsi** see **Sabalia**
BARON see **Euthalia aconthea**
BARONET see **Symphaedra nais**

**Baronia** Salvin (PAPILIONIDAE)
Only one species of this unusual butterfly genus is known to exist. It is considered to be a primitive genus in an evolutionary sense.

**brevicornis** Salvin
Mexico 45–60 mm. (1·77–2·36 in.)
This rare species is unknown outside Mexico, where it has been captured only in the mountains of Sierra Madre del Sur at an altitude of about 1370 m. (4500 ft). The hairy head and short antennae of the adult are characteristic. Almost completely brown female specimens are known with only a few pale yellow spots on the forewings. **94**a

BARRED BLUES see **Spindasis**
BARRED SAILOR, BROKEN see **Neptis trigonophora**
BARRED-TIP YELLOW see **Napata walkeri**
**barrei** see **Anaphe**
**basalis** see **Leptoclanis** and **Milionia**
**basimacula** see **Cyanopepla**

**Bassaris** Hübner (NYMPHALIDAE)
A genus of butterflies related to *Vanessa* and *Cynthia* found only in Australia, New Zealand and a few Pacific islands. They are rapid and strong fliers.

**itea** Fabricius
Australia, Tasmania, Norfolk Island, New Zealand, Loyalty Islands, Rapa Island 46–61 mm. (1·81–2·4 in.)

This is dark brown, almost black on the forewing with a large yellow area over the middle of the forewing and whitish or yellowish spots below the apex of the forewing. The hindwing is brown with four small blue spots near the margin of the wing. The general shape is broadly similar to the species of *Vanessa*. The caterpillar feeds on a species of nettle, *Urtica incisa*.

**Batesia** Felder (NYMPHALIDAE)
There is only one species of butterfly in this genus, found in S. America. The generic name commemorates the great 19th century explorer naturalist, H. W. Bates, whose observations and collections, made during expeditions up the Amazon still form one of the sources of information on the butterflies of that area.

**hypochlora** Felder
S. America 78–95 mm. (3·07–3·74 in.)
Although this species is common in Peru, nothing seems to be known of its biology. Variable in colour it is generally rather dark coloured, mostly blue with black edges with a rich pinky red patch on each forewing. The underside is a striking contrast, the yellow hindwings having a narrow, even band of black just below the outer margin. The forewings on the undersides are paler grey-black and the patch on the wing is pink rather than red. **216**a

BATH WHITE see **Pontia daplidice**
BATH WHITE, SMALL see **Pontia chloridice**
**bathyllus** see **Thorybes**
**batis** see **Thyatira**
**battoides** see **Philotes**

**Battus** Scopoli (PAPILIONIDAE)
The 14 species of this genus of butterflies are restricted to the New World, 3 being found in N. America. *Battus* belongs to the tribe Troidini whose members are generally distasteful to predators and are frequently mimicked by similarly patterned but palatable species of other groups of Papilionidae and other families. Their caterpillars, which bear several fleshy tubercles, normally feed on the foliage of species of *Aristolochia* and related genera of the generally poisonous family Aristolochiaceae.

**belus** Cramer
Tropical S. & C. America, including Mexico 70–90 mm. (2·76–3·54 in.)
The male of this butterfly is similar to the illustrated *B. laodamas* except that the yellow markings on the hindwing are reduced to a single elongate patch at the front edge of the wing. Rare examples have an orange patch above on the forewing and iridescent ultramarine blue hindwings. The female of this species has a black abdomen, but is otherwise similar to the male.

**laodamas** Felder
N., C. & tropical S. America 70–98 mm. (2·76–3·86 in.)
There is some variation in the extent of the greenish yellow markings on the upper surface of the hindwings of this species, and in some specimens there is a row of faintly marked yellowish white spots on the forewing, parallel to its outer margin and continuous with the yellow band of the hindwing. The under surface of the wings is greenish brown, with a short row of white spots at the posterior end of the outer margin of the forewing and a row of red, crescentic markings close to the outer margin of the hindwing. **112**g

**lycidas** Cramer LYCIDAS SWALLOWTAIL
C. & tropical S. America 94–98 mm. (3·7–3·86 in.)
The female of this mostly iridescent green and black species differs from the male in the black not green abdomen. The male (not illustrated) is similar to *B. belus* and *B. laodamas* (illustrated) from which it differs in the reduction of the yellow band to a patch at the front of the hindwing, with small yellow spots posterior to this in a few specimens, and the presence of a yellow band along the inner margins of the hindwing; the ground colour of the wings is a similar but paler and more greenish colour.

**philenor** Linnaeus PIPE-VINE or BLUE SWALLOWTAIL 75–114 mm. (2·95–4·49 in.)
S.E. Canada to Costa Rica
The caterpillar of this well-known species feeds on the foliage of the pipe-vine *(Aristolochia)* but also on knotweed *(Polygonum)* and wild ginger *(Asarum*, a genus of Aristolochiaceae). Female adults (not illustrated) lack the green iridescence of the male and have a row of white marginal spots on the forewing in addition to the hindwing row. The undersurface of the hindwing of both sexes has a row of large orange spots corresponding in position to the white spots of the upper surface but extending as far as the front margin of the wing. *B. philenor* is the distasteful model for *Papilio polyxenes, P. troilus,* the females of 2 species of Nymphalidae *(Speyeria* and *Limenitis)* and possibly the male of the Saturniid moth *Callosamia promethea*. **112**h

**polydamas** Linnaeus POLYDAMAS SWALLOWTAIL, BLACK PAGE or GOLD RIM 68–90 mm. (2·68–3·54 in.)
S.E. U.S., C. & S. America to N. Argentina
Other than *B. philenor,* this is the only species of its genus which regularly breeds in N. America. It differs from *B. philenor* (illustrated) in the absence of a tail and the presence of yellowish green, not white, marginal spots on the hindwing. There are several geographical forms of *B. polydamas,* especially in the islands of the Antilles. The caterpillars feed on several species of *Aristolochia*. Records of this species from California are considered to be doubtful.

**streckerianus** Honrath
Peru 60–86 mm. (2·36–3·39 in.)
This is a relative of the much commoner *B. polydamas*. The greenish yellow-grey central band on the hindwing varies greatly in intensity between specimens but is rarely absent. The undersurface of the forewing is brown, with a scattering of greenish yellow scales basally; the hindwing is brown with a greyish yellow central band (terminating in red at the inner margin of the wing) and a yellow outer band edged proximally with a band of violet-white. Little seems to be known about the biology of this butterfly. **112**f

**zetes** Westwood
Haiti 68–75 mm. (2·68–2·95 in.)
This is a fairly close relative of *B. philenor* which it resembles in wingshape, and basic pattern of the wings. It is known only from a few female specimens which are dark brown above with a broad band of orange on the hindwing and a row of orange spots along the outer margin of the forewing.

**baueri** see **Annaphila**
**bazochii** see **Strymon**
B. D. BUTTERFLY see **Callicore astarte**
BEAK BUTTERFLY see **Libythea geffroyi**
BEAK BUTTERFLY, EUROPEAN see **Libythea celtis**
BEAN BLUE see **Lampides boeticus**
BEAN MOTH see **Etiella behrii**
BEAN MOTH,
  CAROB see **Ectomyelois ceratoniae**
  JUMPING see **Cydia saltitans**
BEAN-BORER MOTHS see **Etiella**
BEAUTIFUL
  ANTINEPHELE see **Antinephele maculifera**
  CHINA-MARK see **Parapoynx stagnata**
  MONARCH see **Danaus formosa**
  SWIFT MOTH see under **Hepialus fusconebulosa**
  UTETHEISA see **Utetheisa ornatrix**
  WOODLING see **Egira pulchella**
BEAUTY,
  BRINDLED see **Lycia hirtaria**
  BORDERED see **Epione repandaria**
  CAMBERWELL see **Nymphalis antiopa**
  DARK BORDERED see **Epione paralellaria**
  LILAC see **Salamis cacta**
  MOTTLED see **Alcis repandata**
  MOUNTAIN see **Meneris tulbaghia**
  OAK see **Biston strataria**
  PAINTED see **Cynthia virginiensis**
  PINE see **Panolis flammea**
  WILLOW see **Peribatodes rhomboidaria**

**Bebearia** Hemming: syn. *Euryphene* (NYMPHALIDAE)
A colourful, large genus of butterflies found in the forest of tropical Africa, with a few species further south in Rhodesia and Zambia and into South Africa.

**senegalensis** Herrich-Schäffer SPECTRE BUTTERFLY
Africa 50–75 mm. (1·97–2·95 in.)
This species is found from W. Africa to Kenya, south to the Cape. Several subspecies have been described in different parts of Africa. The male is reddish with an orange band below the apex of the forewing and orange dots along the margin of fore- and hindwings. The female is different, being similar in colour and pattern to the African Monarch *(Danaus chrysippus)* but the wings of the Spectre are much broader. This species is common near species of *Phoenix* and other palms (Palmae), the host plants of the caterpillars. The butterfly settles on the ground or on low bushes, with its wings spread out flat.

BEDSTRAW
  HAWK-MOTH see **Hyles gallii**
  HAWKLET see **Microsphinx pumilum**
BEE
  BUTTERFLY see **Chorinea faunus**
  HAWK-MOTHS see **Hemaris**
  MOTHS see **Aphomia** and **Melittia**
  TYGER HAWK see under **Acherontia atropos**
BEET
  ARMYWORM see **Spodoptera exigua**
  WEBWORM see **Margaritia sticticalis**
BEETLE MIMIC, YELLOW-BANDED see **Correbidia assimilis**
**behrii** see **Etiella**
**belae** see **Didasys**
**belemia** see **Euchloe**

**Belenoptera** Herrich-Schäffer (THYRIDIDAE)
A small genus of S. American moths. All are very leaf-like in appearance and although no information is available on their biology, it is reasonable to assume that this is excellent camouflage in their natural surroundings.

**cancellata** Warren
S. America 50 mm. (1·97 in.)
As can be seen from the illustration, this leaf-like moth has an unusual pattern. The few specimens collected were found in Peru. **51**b

**belia** see **Anthocharis**
**belisama** see **Delias**
BELL-MOTHS see **Acleris, Archips** and **Tortrix**
**bella** see **Adela**
BELLA see **Utetheisa ornatrix**
**belladona** see **Delias** and **Hamadryas**
**bellargus** see **Lysandra**
**bellatrix** see **Amphicallia**
**bellona** see **Archonias**

**Bellura** Walker (NOCTUIDAE)
The eggs of these semi-aquatic moths are laid on the foodplant well above water level and are covered with hair-like scales from the abdominal tuft of the female. The first stage caterpillar is a leaf-miner, above water level, but in later stages it tunnels within the stalks of the foodplant which are mostly under water. The caterpillars of some species have special adaptations to the spiracles (breathing holes) as a protection against partial submergence. The species of this small genus have been placed at times in the closely related genus *Arzama*.

**diffusa** Grote BROWN-TAILED DIVER MOTH
Canada, N. U.S. 35–45 mm. (1·38–1·77 in.)
The forewings of this species are reddish brown, with 2 broad, transverse bands of pale yellowish brown; the hindwings are pale yellowish brown. The caterpillars live inside the stems of yellow waterlily leaves *(Nuphar)*, which are usually below water level.

BELTED CLEARWING, RED see **Conopia myopaeformis**

**Bematistes** Hemming: syn. *Planema* (NYMPHALIDAE)
This is one of the genera of distasteful butterflies in the subfamily Acraeinae. They are confined to the forests of tropical Africa, particularly on the west coast and in Zaire and include species which are models for a number of more palatable species which mimic them.

**aganice** Hewitson WANDERER
South Africa, E. Africa N. to Ethiopia and S. Sudan 55–82 mm. (2·17–3·23 in.)
This is a forest species occurring up to 1520 m. (5000 ft) with a number of distinct subspecies described from different localities. It is the model for *Pseudacraea eurytus*. The female is larger than the male and has yellow-white or white where the male is orange. The rest of the wings are generally a more grey-brown and the areas of the white or yellow are larger in the female than the corresponding orange-brown areas of the male. The caterpillar feeds on *Passiflora* and *Adenia* (Passifloraceae). The illustration is of the male. **163**d

**Bembecia** Hübner (SESIIDAE)
These are mostly European and temperate Asiatic clearwing micro-moths usually resembling wasps or related insects which generally protects them from predators.

**scopigera** Scopoli: syn. *ichneumoniformis* SIX-BELTED CLEARWING
Europe including Britain, temperate Asia, Turkey, N. Africa 17–20 mm. (0·67–0·79 in.)
A common species which gets its name from the number of rings of yellow on the abdomen of the female. The male has 7 yellow belts. The caterpillar feeds in the roots of Bird's-foot trefoil *(Lotus)* or Kidney Vetch *(Anthyllis vulneraria)*. The moth is out in July and August and is to be found in chalk pits or chalk slopes, often in the early morning or evening. **29**

**Bena** Billberg (NOCTUIDAE)
There are 2 known species in this European and N. African genus, *B. prasinana* and *B. africana* Warren.

**prasinana** Linnaeus SCARCE SILVER LINES
Europe, including Britain, Turkey 32–43 mm. (1·26–1·69 in.)
There has been considerable confusion in the past about the name of this species. It was known for well over a century as *fagana* (see *Pseudoips*) a name which is now currently applied to the Green Silver Lines moth, previously known for years as *prasinana*. Reference to the original specimens in the 200-year-old Linnaean collection enabled entomologists to correctly associate moths and their names in this instance, but where no original specimens exist and the species has not been illustrated or described accurately by its author, it is sometimes impossible to determine which name to apply to a particular species. The adult Scarce Silver Lines is similar to the Green Silver Lines but is readily distinguished by the presence of only 2 transverse, nearly parallel lines on the forewing. Its caterpillar feeds chiefly on oak *(Quercus)* foliage. **403**

**bennetii** see **Agdistis**
BENT-WING SWIFT MOTH see **Zelotypia stacyi**
**berbera** see **Amphipyra**
**berenice** see **Zingha zingha**
BERGER'S CLOUDED YELLOW see **Colias australis**

**Bertholdia** Schaus (ARCTIIDAE)
This is a particularly homogeneous group of just over 20 species, most of which are confined to tropical America, but one species, the illustrated *B. trigona*, occurs in the U.S.

**trigona** Grote
S.W. U.S., Mexico 30–39 mm. (1·18–1·54 in.)
*B. trigona* is found in Colorado, Arizona, New Mexico and N. Mexico. The large pale area on the forewing is transparent. There seems to be nothing known about the biology of this species. **368**l

**betulae** see **Thecla**

**Bhutanitis** Atkinson (PAPILIONIDAE)
There are 4 species in this genus of butterflies which extends from N. India along the slopes of the Himalayas to Burma and S. China. It is probably most closely related to *Luehdorfia*, another genus of the tribe Zerynthiini, to which the S. European Southern Festoon and Spanish Festoon *(Parnalius* species) belong.

**lidderdalei** Atkinson
Kashmir, N. India, Nepal, Bhutan, Burma, S. China 100–115 mm. (3·94–4·53 in.)
This is distinguishable from the illustrated *B. thaidina* chiefly by the broader scarlet patch, and more strongly toothed margin of the hindwing. It flies at altitudes between 1520–2740 m. (5000–9000 ft), in forested regions of the Himalayas. At rest, these butterflies are difficult to locate as the colourful upper surface of the hindwing is covered by the forewings. Many insects, moths and grasshoppers in particular, combine general cryptic or camouflage with 'flash coloration' – bright colourful patches which can be exposed intermittently. Flash coloration enables an insect, which is conspicuous in flight, to merge with its surroundings when at rest and the flash colour is suddenly concealed. Similarly, when disturbed by an enemy, a cryptically patterned insect suddenly becomes brightly coloured which may temporarily disconcert an enemy and allow enough time for escape.

**thaidina** Blanchard
Himalayan China, Tibet 70–88 mm. (2·76–3·46 in.)
*B. thaidana* is typical of its genus in the shape and colour pattern of the wings. The female is similar to the illustrated male. Little is known about the life history of this species, but there is probably only one brood per year. Adults are on the wing in June and July. **96**b

BI-COLOUR COMMODORE see **Limenitis zayla**
**bianor** see **Papilio**

**Biblis** Fabricius (NYMPHALIDAE)
A genus of butterflies found in America. At present only one species is known in it, but this has several subspecies in different parts of C. and S. America.

**hyperia** Cramer: syn. *B. biblis* RED RIM
S. & C. America, S. U.S. 50–65 mm. (1·97–2·56 in.)
This is a butterfly of damp shady places where it can be seen flying slowly in the shadows. Its popular name is very apt and this type of pattern is seen in other species, for example, in the genus of moths *Dysschema*, except that species of *Dysschema* have an extra white line on the forewing. Presumably the species of these two genera fly together, otherwise the advantages of mimicry would not operate. It is not clear if the Red rim is the distasteful model or the palatable mimic, or even if both are distasteful. If the last is the case (Müllerian mimicry) it is assumed that predators quickly learn to avoid this type of pattern and both species therefore benefit. **217**g

**biblis** see **Cethosia**
**bibulus** see **Lachnocnema**
**bicolor** see **Trosia**
**bicoloria** see **Leucodonta**
**bifasciata** see **Aglaope**
BIG
- GREASY BUTTERFLY see **Cressida cressida**
- POPLAR SPHINX see **Pachysphinx modesta**
- ROOT-BORER see **Melittia gloriosa**

**bilineata** see **Camptogramma**
**bilunaria** see **Selenia dentaria**
**bimaculella** see **Glyphipteryx lathamella**

**Bindhara** Moore (LYCAENIDAE)
There is a single known species in this genus.

**phocides** Fabricius PLANE BUTTERFLY
India through S.E. Asia to New Guinea N. Australia and the Solomons 38–42 mm. (1·5–1·65 in.)
This distinctive long-tailed butterfly is a rain-forest species. It has a rapid and skipper-like flight. The female (not figured) differs from the male chiefly in the white and black tail and the absence of blue on the hindwing. The male is pale brown and brownish yellow beneath, the female white; both sexes have a similar but not identical pattern of dark brown markings. The caterpillars feed inside the fruits of a creeper *(Salacia)* and when fully grown tunnel into bark before transformation into a chrysalis. **244**bb

**biopes** see **Evagora**
**biplaga** see **Earias**
**bipunctella** see **Ethmia**
BIRD'S WING see **Dyptergia scabriuscula**
BIRD-DROPPING MOTH see **Acontia aprice**
BIRDWINGS see **Ornithoptera** and **Troides**
**bisaltide** see **Doleschallia**
BISCUIT BUTTERFLY see **Anartia jatrophae**
**bisignata** see **Semiothisa**
**bisselliella** see **Tineola**

**Biston** Leach (GEOMETRIDAE)
Most of the moths in this genus are found in Europe and Asia. The species range from Europe, including Britain, through to Japan where related species occur. A few species occur in the U.S.

**betularia** Linnaeus PEPPER-AND-SALT MOTH or PEPPERED MOTH
Europe including Britain, W. and C. Asia, including Japan 45–52 mm. (1·77–2·05 in.)
This moth was made famous by studies done in connection with the colouring of the wings. The moth is variable in colour but there are two extremes, a light colour form and a black variety called *B. b. carbonaria* which was first found in Manchester (England) in the 1850s. This black form soon became common there and spread to other parts of Britain. Basically it is believed that in smoky areas where the posts and vegetation on which the moths rest become blackened, the black varieties has less chance of being seen by predators than the white and therefore has a higher chance of survival. This phenomenon is called 'industrial melanism'. Similarly in areas where the environment is less polluted, then the lighter coloured moths are less conspicuous on a lighter background. The Peppered moth has the melanic form in many industrial areas of Europe. **302, 303**

**strataria** Hufnagel OAK BEAUTY
Europe including Britain 46–52 mm. (1·81–2·05 in.)
A widespread and fairly common moth over most of its range, this species when at rest on tree trunks or fences is very hard to see. The pattern of the moth, and its colour is very similar to the lichens found in those situations, and as can be seen in the illustration, it has an excellent camouflage. The caterpillar, which feeds on oak *(Quercus)*, birch *(Betula)*, elm *(Ulmus)* and some other trees, is very twig-like, generally a shade of brown, if disturbed, stays absolutely straight and still. The moth is out in March and April and flies after dusk. **280**

**bistriatella** see **Apomyelois**
**bjerkandrella** see **Tebenna**
BLACK
- ARCHES see **Lymantria monacha**
- ARCHES, KENT see **Nola albula**
- BORDERED CHARAXES see **Charaxes pollux**
- CROW BUTTERFLY, DOUBLE-BRANDED see **Euploea coreta**
- CROW BUTTERFLY, STRIPED see **Euploea alcathoe**
- CUTWORM see **Agrotis ipsilon**
- HAIRSTREAK see **Strymonidia pruni**
- NYMPH, SPECKLED see **Cremna actoris**
- PAGE BUTTERFLY see **Battus polydamus**
- PRINCE see **Ludia goniata**
- RAJAH see **Charaxes fabius**
- SATYR see **Satyrus actea**
- SWALLOWTAIL see **Papilio polyxenes**
- SWORDTAIL see **Graphium colonna**
- UNDERLEAF see **Alesa amesis**
- VEIN BUTTERFLY, GREAT see **Aporia agathon**
- VEINED WHITE see **Aporia crataegi**
- WITCH see **Ascalapha odorata**

BLACK AND WHITE
- TIGER BUTTERFLY see **Danaus affinis**
- TIT BUTTERFLY see **Hypolycaena danis**

BLACK-BACKED BLUE see **Mithras hemon**
BLACK-CHEQUERED BLUE see **Menander superba**
BLACK-VEINED TIGER BUTTERFLY see **Danaus melanippus**

**Blastobasidae**
A family of micro-moths in which the species are very small and are worldwide in distribution. They typically have a hairy fringe on the hind margin of the forewing and on both sides of the slender pointed hind-

wing. This is a character they share with others in the Gelechoid group to which they belong. See *Holocera*.

BLEU, COCOA MORT see **Caligo teucer**
BLOODSUCKING MOTH see **Calpe eustrigata**
BLOSSOM PLUSIA, PEACH see **Eosphoropteryx thyatiroides**
BLOTCH MINER, ASPEN see **Phyltonorycter tremuloidiella**
BLUE BUTTERFLIES see **Agriades, Agrodiaetus, Amblypodia, Brephidium, Celastrina, Cupido, Danis, Evenus, Everes, Glaucopsyche, Hemiargus, Icaricia, Iolana, Jalmenus, Lampides, Lepidochrysops, Leptotes, Lycaenopsis, Lysandra, Maculinea, Menander, Mithras, Myrina, Narathura, Philotes, Plebejus, Plebicula, Polyommatus, Pseudolycaena, Spindasis, Syntaracus, Tajuria, Vacciniina, Virachola, Zintha, Zizeeria, Zizula**
BLUE
ADMIRAL see **Vanessa canace**
APOLLO, COMMON see **Parnassius hardwickei**
ARGUS see **Junonia orithya**
BANDED NYMPH see **Cynandra opis**
CHARAXES, LARGE see **Charaxes bohemani**
CROW BUTTERFLY, STRIPED see **Euploea mulciber**
DOCTOR see **Rhetus periander**
DUKE see **Euthalia darga**
GEM see **Poritia erycinoides**
GRECIAN see **Heliconius wallacei**
GRECIAN, SMALL see **Heliconius sara**
METALMARK see under **Lasaia sessilis**
MIME, GREAT see **Papilio paradoxa**
MONARCH see **Danaus limniace**
MORPHO see **Morpho menelaus** and **M. rhetenor**
MOUNTAIN BUTTERFLY see **Papilio ulysses**
NIGHT see **Cepheuptychia cephus**
NIGHT, IRIDESCENT see **Magneuptychia junia**
NYMPH, BRIGHT see **Paiwarria venulius**
OAK LEAF see **Kallima horsfieldi**
PANSY see **Junonia orithya**
PEACOCK see **Papilio arcturus**
SALAMIS see **Salamis temora**
SWALLOWTAIL see **Battus philenor**
SWALLOWTAIL, GIANT see **Papilio zalmoxis**
THAROPS see **Menander menander**
THEOPE, ROYAL see **Theope excelsa**
TIGER see **Danaus limniace**
TRANSPARENT BUTTERFLY see **Ithomia drymo**
TRIANGLE see **Graphium sarpedon**
BLUE-BANDED
EGGFLY see **Hypolimnas alimena**
SWALLOWTAIL, NARROW see **Papilio nireus**
BLUE-BRANDED KING CROW BUTTERFLY see **Euploea leucostictos**
BLUE-SPOT COMMODORE see **Junonia westermanni**
BLUE-STRIPED
NETTLE GRUB see **Parasa lepida**
PALMFLY see **Elymnias patna**
BLUEBOTTLE, COMMON see **Graphium sarpedon**
**blumei** see **Papilio**
**bobana** see **Eucosma**
**boeticus** see **Lampides**
**boetifica** see **Eubergia**
BOG
COPPER see **Lycaena epixanthe**
ELPHIN see under **Lycaena epixanthe**
FRITILLARY see **Proclossiana eunomia**
BOGONG MOTH see **Agrotis infusa**
**bohemani** see **Charaxes**
**boisduvali** see **Miletus**
**boisduvalii** see **Pseudacraea**
**bolina** see **Hypolimnas**
BOLLWORM MOTHS see **Earias, Heliothis** and **Platyedra**

**Boloria** Moore (NYMPHALIDAE)
This genus of butterflies has species in Europe and temperate Asia with a few species in N. America. In Asia some species are found as far south as Tibet. The genus includes butterflies popularly known as Fritillaries.

**pales** Denis & Schiffermüller SHEPHERD'S FRITILLARY BUTTERFLY 34–40 mm. (1·34–1·57 in.)
Alpine areas from Europe through Asia
Similar to the other species of *Boloria* in pattern but with the markings in the forewing darker and as spots, not lines as in some of the other species. The adult flies in July and August in mountainous areas over cranberry *(Vaccinium)* from the tree line to over 2440 m. (8000 ft). The caterpillar feeds on violets, especially *Viola calcarata*. Several subspecies have been described and there are also several closely allied species, so specimens have to be studied carefully to ensure correct identification.

**Bombycidae**
Only a few small genera belong to this Asiatic family of moths. The forewing is variously hooked at the apex. *Bombyx mori*, the Chinese Silk-moth, is the best known species. See *Bombyx*.

**Bombyx** Linnaeus (BOMBYCIDAE)
There are about 6 species in this genus of moths. They occur in India, China, Formosa, Korea and Japan.

**mandarina** Moore
China, Formosa, Korea, Japan 34–47 mm. (1·34–1·85 in.)
This species is currently accepted as the wild progenitor of the solely domestic *B. mori*. The closely related *B. huttoni* Moore, which occurs in N. India is sometimes regarded as a subspecies of *B. mandarina*. **317g**

**mori** Linnaeus, THE COMMON SILK-MOTH, SILKWORM 40–60 mm. (1·57–2·36 in.)
There are no known wild populations of this most famous of silk-moths. The origin of all the commercially bred silk-moths is thought to be China where the literature on silk production dates back several hundred years. The Empress Si-Ling-Chi was one of the earliest and most distinguished breeders of silk-moth in the 18th century B.C. In Europe, the Greek philosopher Aristotle was the first writer to give an account of silk production. The techniques of silk manufacture were kept a closely guarded secret in China for over two thousand years and the punishment for revealing its secrets or exporting the moth was death. However, in A.D. 555 two monks in the service of the Roman emperor Justinian in Constantinople were eventually able to smuggle out some eggs inside their holy bamboo staffs and the first breeding unit was established in the West. James I of England tried to establish a silk breeding industry in Britain but was unsuccessful. The variably coloured caterpillar is swollen near the head; it is fed normally on black or white mulberry *(Morus)*, but it will also feed on osage-orange *(Maclura)* and lettuce *(Lactuca)*. **317e**

**bonasia** see **Acraea**
**bonplandi** see **Perisama**
**bonplandii** see **Elzunia**
**bootes** see **Papilio**
BORDERED
BEAUTY MOTH see **Epione repandaria**
BEAUTY MOTH, DARK see **Epione paralellaria**
CHARAXES, BLACK see **Charaxes pollux**
FRITILLARY, PEARL see **Clossiana euphrosyne**
GREY see **Selidosema brunnearia**
STRAW, SCARCE see **Heliothis armigera**
WHITE see **Bupalis piniaria**
**bore** see **Oeneis**
**borealis** see under **Calephelis muticum**
**boreata** see **Operophtera fagata**
BORER, PINK see **Sesamia calamistis**
**bosniackii** see under **Luehdorfia**
**bowesi** see **Leioptilus**

**Bracharoa** Hampson (LYMANTRIIDAE)
These moths are found in Ethiopia, tropical Africa and South Africa. Eight species are known. Males are yellow and brown in colour, females of most species are wingless or nearly so.

**dregei** Herrich-Schäffer
S. Africa (male) 15–22 mm. (0·59–0·87 in.)
The female of this species never leaves the cocoon. It is fertilized within the cocoon, lays its eggs and dies there; it is usually ignominiously eaten by its offspring. The male wings are yellowish brown, with one or more small black spots on the middle of the forewing and a dark brown outer band on the hindwing.

**brachyura** see **Eudaemonia**

**Brahmaea** Walker (BRAHMAEIDAE)
There are less than 10 described species of this genus. Their range extends from S. Europe to Japan, China and S.E. Asia as far east as Sulawesi (Celebes).

**europaea** see **Acanthobrahmaea**

**wallichii** Gray
N. India, Sikkim, Nepal, China, Formosa, Japan 90–155 mm. (3·54–6·1 in.)
The illustrated specimen is an example of the Japanese subspecies *B. wallichii japonica*. Its remarkable caterpillar bears a pair of spiral spines on the second and third segments of the thorax and 3 shorter but similarly shaped processes at the rear-end – at least during most stages of its life. **340d**

**Brahmaeidae**
This is a small family of about 20 species of moths. They are similar in size and wingshape to many species of Saturniidae and considered to be a primitive member of the same superfamily, Bombycoidea. It is an entirely Old World group, found in Europe, Africa and in temperate and tropical Asia. The colour pattern of many species is highly intricate; most have an eyespot marking at the apex of the forewing. The European representative of this family was described as recently as 1963 (see *Acanthobrahmaea*). See *Acanthobrahmaea, Brahmaea, Calliprogonus, Dactyloceras.*

BRANCH-BORER, RED see **Zeuzera coffeae**
**branickiaria** see **Ophthalmophora**
BRASS, BURNISHED see **Plusia chrysitis**
**brassicae** see **Mamestra** and **Pieris**

**Brassolinae** (NYMPHALIDAE) OWL BUTTERFLIES
A subfamily of the Nymphalidae, formerly considered as a distinct family of butterflies. The Brassolids are confined to tropical America. They have many similarities to the Amathusiinae with which they were formerly associated, although the latter are an Old World group. Brassolids have prominent eyespots, often very large, on the underside, and the caterpillars differ from the Nymphaline Nymphalids and are more like Satyrine caterpillars. Nearly all Brassolids are large or very large and there is only one genus where all the species are small. The Brassolids tend to fly towards dusk and only a few species are out in the midday sun. The butterflies are readily attracted to ripe fruit where they feed freely. They are popularly known as Owl butterflies from the large eyespots on the underside of the wings.

**brassolis** see **Liphyra**
**braziliensis** see **Cynthia**

**Brenthis** Hübner (NYMPHALIDAE)
A small genus of butterflies with species in Europe, temperate Asia to Japan. Popularly known as Fritillaries, they are fast and powerful fliers.

**daphne** Denis & Schiffermüller MARBLED FRITILLARY 42–52 mm. (1·65–2·05 in.)
S.W. Europe, temperate Asia including Japan
The upperside pattern is typical of a fritillary and is reddish brown with black marks. The underside is paler, with a row of black spots on the hindwing and a yellowish band nearer the base. The butterfly is out in June and early July in lower parts of valleys where it can often be seen on the flowers of bramble *(Rubus)*. The Japanese specimens differ slightly from the European and are considered a distinct subspecies. The caterpillar feeds on violets *(Viola)* and brambles.

**ino** Rottemburg LESSER MARBLED FRITILLARY
C. & N. Europe, temperate Asia including Japan 34–40 mm. (1·34–1·57 in.)
Typical fritillary in appearance, it is smaller than *B. daphne* which it closely resembles. It can be separated

from the latter by the continuous black marginal line and by some other structural characters. Several distinct subspecies have been described, including 3 in Japan. The caterpillar feeds on Meadow Sweet *(Filipendula ulmaria)*, Great Burnet *(Sanguisorba officinalis)* and raspberry *(Rubus)*. The butterfly is out in June and July in marshy meadows and is locally common.

**Brephidium** Scudder (LYCAENIDAE) PYGMY OR DWARF BLUE
There are 3 known species in this genus: *B. barberae* Trimen (Africa), *B. pseudofea* and *B. exilis*.

**exilis** Boisduval
U.S. S. to Guatemala 14–20 mm. (0·55–0·79 in.)
This minute species is amongst the smallest butterflies in the world. Its range in the United States is western, from Oregon and Nebraska southwards. The female (not illustrated) is brown above, with white fringes and a variable concentration of blue scales at the base of the wings. **260**b

**pseudofea** Morrison EASTERN PYGMY BLUE
S.E. U.S., the Antilles 15–20 mm. (0·59–0·79 in.)
Coastal salt-marshes are preferred by this species. It is similar in colour pattern to the illustrated *B. exilis*, but there is no blue coloration on the upper surface, and the wings are brown, not white, beneath at their base. The caterpillar feeds on glasswort *(Salicornia)*. There are apparently at least 3 broods per year.

**brephos** see **Leucidia**
**breteaudeaui** see **Diacrisia**
**brevicornis** see **Baronia**
**brevifasciata** see **Sagaropsis**
BRIGHT
BLUE NYMPH see **Paiwarria venulius**
OAK BLUE see **Narathura madytus**
**brigitta** see **Eurema**
**brillians** see **Pampa**
BRILLIANT FLASH see **Rapala sphinx**
BRIMSTONE BUTTERFLIES see **Gonepteryx**
BRIMSTONE MOTHS see **Opisthagraptis**
BRINDLE, SCARCE see **Apamea lateritia**
BRINDLED BEAUTY see **Lycia hirtaria**

**Brintesia** Fruhstorfer (NYMPHALIDAE)
A very large genus of butterflies belonging to the subfamily Satyrinae. They are found in Europe and Asia with a few species extending into India. The caterpillars feed mainly on species of grasses.

**circe** Fabricius GREAT BANDED GRAYLING
W. Europe, Turkey, 66–72 mm.
Iran to the Himalayas (2·6–2·83 in.)
The butterfly is out in June and July generally in open woodlands up to 1520 m. (5000 ft). It is common and widely distributed over most of Europe. The underside is particularly well camouflaged and when at rest on the ground or trunks of trees the butterfly is almost invisible. The caterpillar feeds on species of *Bromus*, *Lolium* and other grasses (Gramineae). **165**

**briseis** see **Chazara**
**britanniodactyla** see **Capperia**
**britomartis** see **Mellicta**
BROAD-BORDERED GRASS-YELLOW see **Eurema brigitta**
BROKEN BARRED SAILOR see **Neptis trigonophora**
BRONZE COPPER see **Lycaena thoe**
BRONZE UNDERWING see **Catocala cara**
BROOKE'S BIRDWING, RAJAH see **Troides brookiana**
**brookiana** see **Troides**
BROOM-TIP MOTH see **Chesias rufata**
BROWN BUTTERFLIES see **Chazara, Lethe, Maniola, Malanitis, Pseudonympha** and **Satyrus**
BROWN
ARGUS see **Aricia agestis**
CHINA-MARK see **Nymphula nympheata**
FRITILLARY, HIGH see **Fabriciana adippe**
HAIRSTREAK see **Thecla betulae**
HOODED OWLET MOTH see **Cucullia convexipennis**
HOUSE MOTH see **Hofmannophila pseudospretella**
PLAYBOY see **Deudorix antalus**
BROWN-TAIL MOTHS see **Euproctis**
BROWN-TAILED DIVER see **Bellura diffusa**
BROWN-VEINED WHITE see **Anaphaeis aurota**
**brumata** see **Operophtera**
**brunnearia** see **Selidosema**
**brunneata** see **Semiothisa**
**bryaxis** see **Mesenopsis**
**bucephala** see **Phalera**
BUCK MOTHS see **Hemileuca**
BUCKEYE see **Junonia lavinia**

**Buckleria** Tutt (PTEROPHORIDAE) PLUME MOTHS
A small genus of plume-moths whose caterpillars feed on Sundew *(Drosera)*. In all probability the species in this genus should be transferred to *Trichoptilus*, a related genus, and the name *Buckleria* not used, but further research on this is needed.

**paludum** Zeller
Europe including Britain 12–13 mm. (0·47–0·51 in.)
The caterpillar of this tiny plume-moth feeds on the leaves of the insectivorous plant Sundew *(Drosera)*. While feeding they keep out of the way of the insect-catching hairs on the leaves. The moth flies at dusk and is a reddish or greyish brown colour with the forewings, which are slightly patterned, deeply divided into 2 lobes. The hindwings are divided into 3 lobes. There is a related species in America which has a similar biology, feeding on the leaves of *Drosera*. It is a nice quirk of evolution to have produced an insect which can feed on an insectivorous plant.

BUDWORM MOTHS see **Choristoneura** and **Heliothis**
BUFF BUTTERFLY, SPOTTED see **Pentilia tropicalis**
BUFF MOTH, CLOUDED see **Diacrisia sannio**
BUFF
ARCHES see **Habrosyne pyritoides**
ERMINE see **Spilosoma luteum**
BUFF-TIP MOTH see **Phalera bucephala**
BULRUSH WAINSCOT see **Nonagria typhae**

**Bunaea** Hübner (SATURNIIDAE)
This is a small genus of mostly large, orange, brown and yellow moths found in Madagascar, Ethiopia and Africa south of the Sahara, which typically have a transparent patch on the forewing and an eyespot marking on the hindwing.

**alcinoe** Stoll COMMON EMPEROR
Africa S. of the Sahara, 100–160 mm.
Madagascar (3·94–6·3 in.)
This species has become adapted to very different climatic conditions throughout its extensive range. Equally comprehensive is the array of foodplants eaten by the caterpillar which consumes the foliage of several different plant families with apparently equal relish, including Ulmaceae *(Celtis)*, Euphorbiaceae *(Croton)* and Combretaceae *(Terminalia)*. **333**d

**Bunaeopsis** Bouvier (SATURNIDAE)
This is a moderately sized genus of moths distributed in tropical and S. Africa, with rounded wings and usually with the forewing eyespot as large as that on the hindwing. The caterpillars, which are usually armed with short spines, feed on grasses, reeds and other grass-like plants *(Graminae)*.

**jacksoni** Jordan JACKSON'S EMPEROR
Africa S. of the Sahara, except the 100–124 mm.
Congo Basin and S. Africa (3·94–4·88 in.)
This richly coloured moth has been confused in the past with the similar *B. arabella* from which it differs most noticeably in the red ground-colour of the forewing. The caterpillar has been recorded from an unidentified reed-like grass. **340**f

**buoliana** see **Rhyacionia**

**Bupalis** Leach (GEOMETRIDAE)
An European and Asian genus of moths with several pest species, some of which are important and cause serious defoliation of trees.

**piniaria** Linnaeus BORDERED WHITE OR PINE LOOPER 32–35 mm. (1·26–1·38 in.)
Europe including Britain, Asia
The caterpillar is greenish, marked with white or yellow lines edged with black. It feeds on pine *(Pinus)* and larch *(Larix)* and at times can be destructive to plantations. The moth is out in May or June, and later in the N. parts of Europe. There are 2 forms, one is the small Scandinavian race (which occurs in Britain). The other is the larger S. form which also occurs in S. Britain. Where these 2 forms occur together hybrids between them also occur. **312**

**Burlacena** Walker: syn. *Sesiomorpha* (GLYPHIPTERIGIDAE)
The species in this small genus of micro-moths are all mimics of wasps or bees. They have yellow and black coloured bodies. Wasp mimics occur in many families in the Lepidoptera, the best known being in the family Sesiidae.

**vacua** Walker
Indonesia 15–22 mm. (0·59–0·87 in.)
This species has been collected on the island of Sulawesi (Celebes). Although no information is available on its biology, it is possible that the mimicry – as in some other wasp imitators – extends to the buzzing sound made by flying wasps but this has not been verified. **34**b

BURMESE LASCAR see **Pantoporia hordonia**
BURNET MOTHS see **Zygaena**
BURNET COMPANION see **Euclidia glyphica**
BURNISHED BRASS MOTH see **Plusia chrysitis**
BURREN GREEN see **Calamia tridens**
BURROWING WEBWORMS see **Acrolophus**
BUSH-BROWN BUTTERFLIES see **Mycalesis**

**Busseola** Thurau (NOCTUIDAE)
This is a solely African genus. One species, *B. fusca*, is an important pest of cereal crops.

**fusca** Fuller MAIZE STALK-BORER
Africa S. of the Sahara 26–34 mm.
(chiefly in wet savannah regions) (1·02–1·34 in.)
Few species approach *B. fusca* in destructiveness as a pest of maize, or corn, *(Zea)* and *Sorghum* in Africa. There is variation in the coloration of this species, especially in some Ghana localities where the typical reddish brown coloration of the forewing is replaced by black. Insecticidal control measures are employed against *B. fusca*, usually with considerable success, as the caterpillars spend their early stages on the outside of the host plants where chemicals can reach them.

**c-album** see **Polygonia**
**c-nigrum** see **Amathes**
CABBAGE BUTTERFLY see **Pieris rapae**
CABBAGE LOOPER see **Trichoplusia ni**
CABBAGE MOTH see **Mamestra brassicae**
CABBAGE WEBWORMS see **Hellula**
CABBAGE WHITES see **Ascia, Pieris**
CACAO MOTH see **Ephestia elutella**
**cacica** see **Amauta**

**Cacoecimorpha** Obratzov (TORTRICIDAE)
This generic name was proposed in 1954 for one moth which was considered rather distinct from other species in the genus in which it was formerly placed *(Archips)*. It is known from Europe and N. Africa only.

**pronubana** Hübner CARNATION TORTRIX
Europe including Britain, N. Africa 21 mm. (0·83 in.)
Although the forewings are a rather dull greyish brown, the hindwings of this species are orange, bordered with black, making it conspicuous when it spreads its wings. At rest the colourful hindwings are hidden by the more camouflaged forewings. The caterpillar, as the popular name implies, feeds on carnations *(Dianthus)*, at times being a pest of these and causing concern to horticulturalists.

**cacta** see **Salamis**

**Cactoblastis** Ragonot (PYRALIDAE) CACTUS MOTHS
This S. American genus has proved to be valuable to man for the biological control of Prickly-pear cactus *(Opuntia)*.

**cactorum** Berg CACTUS MOTH
Argentina, Uruguay, Paraguay, 23–40 mm.
S. Brazil, Australia (introduced) (0·91–1·57 in.)
This species has been studied in detail in America where it is widespread in S. America. Its cactus-feeding habits have been put to good use in biological control. After Prickly-pear cactus had been introduced into Australia and had spread rapidly over areas of Queensland, many thousands of acres of grazing land were lost. Although many attempts were made to destroy the cactus, none was successful until this moth was introduced to Australia and released in large numbers. They established themselves and spread rapidly killing most of the cacti, which were finally destroyed by a combination of caterpillar damage and a rot which followed this. This enabled most of the grazing areas to be reclaimed and is one of the most successful examples of biological control. It has been suggested, with good reason, that this small moth deserves a place in the Queensland coat of arms. **56**e

CACTUS MOTHS see **Cactoblastis**

**Cadarema** Moore (PYRALIDAE)
Only one species is at present known in this genus of Pyralid moths. It was transferred from a large group of species, collectively known as *Margaronia,* a generic name which is not now used.

**sinuata** Fabricius
Africa 38 mm. (1·5 in.)
This species is common over most of Africa and south of Sahara, but accounts of its biology do not seem to have been published. It is a common moth in collections from Africa and is easily recognized. **63**e

CADDIS WATER-VENEER, FALSE see **Acentria nivea**
**Cadra** see under **Ephestia**
**caerulescens** see **Allostylus**
**caeruleocephalus** see **Dilopa**
**cagnagella** see **Yponomeuta**
**caja** see **Arctia**

**Calamia** Hübner (NOCTUIDAE)
There are few species in this genus of temperate Old World and S. African moths.

**tridens** Hufnagel BURREN GREEN
Europe including Ireland, 35–45 mm.
to the S. Causasus (1·38–1·77 in.)
The Irish subspecies of *C. tridens* was discovered as late as 1949 in the Burren area of County Clare. Adults fly by day. The caterpillars feed on grasses (Gramineae), either inside or outside the stems. **391**d

**calamistis** see **Sesamia**
**calcareus** see **Polyptychus**

**Calephelis** Grote & Robinson (NEMEOBIIDAE)
METALMARKS
About 50 species are currently placed in this N., C. and S. American genus. The term metalmark refers to the metallic markings on the under surfaces of the wings.

**muticum** McAlpine SWAMP METALMARK
U.S. (central N. states) 22–25 mm. (0·87–0·98 in.)
This is a species of swamps, bogs and wet meadows where there is a supply of the foodplant, the swamp thistle *(Cirsium).* In general features, it resembles a very large *C. virginiensis* (illustrated) or an extremely small fritillary. Until its description in the 1930s, *C. muticum* was confused with another similarly patterned species the Northern Metalmark *(C. borealis* McAlpine). The long white hairs of the caterpillar probably have a protective function during hibernation.

**virginiensis** Guérin LITTLE METALMARK
U.S. (Texas and Florida to 18–20 mm.
Virginia and Ohio) (0·71–0·79 in.)
Grassy fields and wet meadows provide the home for this minute metalmark. It is uncommon in the northern part of its range but often common in the south. Very little is known about its life history. This is the smallest species of N. American metalmark butterflies. **260**d

CALICO BUTTERFLIES see **Hamadryas**
**calida** see **Hyles**
CALIFORNIA SILKMOTH see **Hyalophora euryalus**
**californiae** see **Leptarctia**

**Caligo** Hübner (NYMPHALIDAE)
OWL BUTTERFLIES
This is a large genus of butterflies in the subfamily Brassolinae. They are mostly very large broad-winged species with a strong pattern and prominent eyespots on the underside, hence their popular name, 'Owl' butterflies. Mostly they fly at dusk or early morning in woods and forests.

**idomeneus** Linnaeus OWL
S. America 140 mm. (5·51 in.)
This huge butterfly is found as far south as Argentina as well as into the more N. parts of S. America. The owl-like eyes on the underside are conspicuous when it is at rest and are regarded as a means of frightening birds or small mammals who might be potential predators. The butterfly is often common and occasionally the caterpillar becomes a pest in banana *(Musa)* plantations. The butterfly is out towards dusk and again in the early morning. **155**

**prometheus** Kollar
S. America 150 mm. (5·9 in.)
This species is found in the forests of Colombia and Ecuador. Nothing seems to be known of its life history. The males and females are similarly coloured although the females usually have some black spots near the apex of the forewing. The underside of the wings are brown and strongly patterned with lines and with an enormous eyespot on each hindwing. Both wing surfaces are illustrated. **152, 153**

**teucer** Linnaeus COCOA MORT BLEU
C. and S. America, 98–115 mm.
into the West Indies (3·86–4·53 in.)
The caterpillar of this butterfly feeds on all species of bananas (Musaceae). The male butterfly is a dark metallic blue on the middle of the hindwings, the rest of the wing is greyish or black with a yellow band from the front almost to the hind margin of the forewing. The females are larger than the males, with 2 yellow bands. The general shape is similar to **152**. The butterfly generally avoids the bright sunshine and is about in the late afternoon and at dusk. The underside of the wing has large 'Owl eyes' patterns which are conspicuous when at rest. The adults are frequently found feeding on over-ripe fruit. This butterfly has several subspecies over its range from Costa Rica to the Guyanas and Ecuador.

**callanga** see under **Leptotes webbianus**

**Callicore** Hübner: syn. *Catagramma*
(NYMPHALIDAE)
A large genus of butterflies with many species in S. America and some in C. America and the West Indian islands. Many of the life histories are unknown but a few have been described. The caterpillars are long and slender with horns sticking out from the head, some of these horns have rosettes of smaller spines on them. Most of the butterflies have metallic colours and finely patterned undersides. These intricately patterned undersides always attract attention and fascinate entomologists as well—not many years ago an entire book was written about the underside patterns of these butterflies and the theory of how they might have evolved. The butterflies in this genus are graceful fliers. See also under *Lucillella camissa.*

**astarte** Cramer 'B.D.' BUTTERFLY
S. America 45–55 mm. (1·77–2·16 in.)
The upperside of this species is shown; the underside is broadly similar to *C. cynosura* but has red on the forewings with some black and a yellow line below the apex. The biology is unknown. This butterfly is widespread throughout Brazil, Venezuela, Ecuador, Colombia, the Guyanas and into Trinidad. The female is brown instead of red under the forewing. The name 'B.D.' is derived from the underside hindwing pattern which resembles these letters. **217**e

**cynosura** Doubleday
S. America 55–60 mm. (2·17–2·36 in.)
The underside of this species is illustrated. The upperside is broadly similar to *C. astarte* but has broader orange-red bands on the forewing. The underside differs in having orange, not red, on the forewing. **217**l

**maimuna** Hewitson EIGHTY-EIGHT BUTTERFLY
S. America, W. Indies 48–56 mm. (1·89–2·2 in.)
The female butterfly has a large reddish brown patch on the forewing surrounded by blackish brown, with a small yellow patch near the apex. The hindwings are blackish brown. The most striking feature of this species is the conspicuous '88' in blue and black on the underside. The males are red in the inner part of the forewing, otherwise entirely black with a pale red band at the apex. This butterfly is found from Brazil to Peru and Colombia, often being common in parts of its range. **229**u

**mionina** Hewitson
S. America 42–48 mm. (1·65–1·89 in.)
This species is known from Colombia. It differs from the related species mainly in the pattern of the underside of the wing. The figure shows the contrasting pattern. Little seems to be known about its life history. **229**t

**sorona** Godart
S. America 50–60 mm. (1·97–2·36 in.)
This species lives near rivers or streams and flies around scrublands and in forested areas of Brazil, Bolivia and Paraguay. There are several subspecies of this brightly coloured species. **215, 217**h

**callidice** see **Pontia**

**Callidula** Hübner (CALLIDULIDAE)
There are probably about 50 species in this genus of day-flying moths. Their range includes Sikkim and Himalayan India, eastwards through S.E. Asia to the Solomons. Several new species have yet to be described and named. One species, *C. scotti,* has been described from Cape York, Australia.

**lunigera** Butler
Bismarck Archipelago 23–28 mm. (0·91–1·1 in.)
*C. lunigera* is typical in colour of most species of its genus. Some specimens have a yellow marginal band on the under surface of the hindwing and a trace of this band on the upper surface. Nothing is known about the life history of this species. **272**d

**Callidulidae**
This is a small family of probably less than 100 species of day-flying moths. They are generally considered to be members of the superfamily Geometroidea and to be fairly closely related to the Pterothysanidae. The Callidulidae are combined by some authors with the latter in the superfamily Calliduloidea. The flight of these moths is butterfly-like, the resting position of some species with the wings held over the back is similar to that of most butterflies and the antennae are clubbed. Most species are dark brown, with orange markings. They are found in India and Sri Lanka, China, E. U.S.S.R., Japan and E. to the Philippines, New Guinea and the Solomons. The species *Callidula scotti,* from Australia, is at present only doubtfully identified as a Callidulid. See *Callidula, Pterodecta.*

**calliglauca** see **Neopseustis**
**callima** see **Milionia**

**Callimorpha** Latreille (ARCTIIDAE) TIGER-MOTHS
This is a genus of about 13 species of moths related to the N. American *Haploa* and other genera, united by some authors under the subfamily name Callimorphinae. *Callimorpha* is represented in Europe and temperate Asia to Japan.

**dominula** Linnaeus SCARLET TIGER-MOTH
Europe including Britain, 45–52 mm.
temperate W. Asia (1·77–2·05 in.)
This is a highly variable moth of which both black and yellow forms are known; the normal colour of the hindwings, covered by the forewing in the illustrated specimen, is scarlet-red, with black marginal mark-

ings. *C. dominula* is often common in June and July in damp or marshy areas and along river banks. The caterpillar is black, with pale blue tubercles and 3 white-dotted, longitudinal yellow lines; it feeds chiefly on nettle *(Urtica)*, but also on groundsel *(Senecio)*, comfrey *(Symphytum)* and other plants in July and August and then overwinters to complete its growth the following year in April and May. **378**

**Callionima** Lucas (SPHINGIDAE)
One member of this genus of hawk-moths, *C. parce*, occurs in the S. U.S.; the remaining species are solely tropical American in range. Most adult moths are orange or brown, with silvery markings on the forewing.

**parce** Fabricius
Tropical C. & S. America, S. U.S. 65–80 mm. (2·56–3·15 in.)
In the U.S. *C. parce* occurs in S. Florida and Texas and in S.E. Arizona. Nothing is known of the life history. There is appreciable variation in the colour pattern of the upper surface of the wings which may be well marked or lack dark brown markings completely. **344**t

**calliope** see **Stalachtis**

**Callioratis** Felder (GEOMETRIDAE)
There are a few species in this genus of moths, all with similar striking patterns of orange and grey-black lines.

**milliaria** Hampson
South Africa, Natal 55–60 mm. (2·17–2·36 in.)
The underside and upperside are similar. Nothing is known of its biology, but possibly these bright colours are a warning to predators that it is distasteful. As this species is remarkably similar to a distasteful Hypsid, *Amphicallia* (figure **380**), it is possible that some mimetic association is involved. Some field studies on its biology would be useful to assess this. **296**s

**Callipia** Guenée (GEOMETRIDAE)
This genus of moths is found in S. America. There are several related species, some with more red on the hindwing.

**paradisea** Thierry-Meig
S. America 50 mm. (1·97 in.)
Nothing is known of the biology of this rather delicately coloured species which has been found only in Peru. **296**u

**Calliprogonos** Mell & Hering (BRAHMAEIDAE)
Only one species is known. Its colour pattern differs radically from that of other moths of this family.

**miraculosa** Mell & Hering
China 55–65 mm. (2·17–2·56 in.)
The three known specimens of this rare, unusually marked species were captured at an altitude of about 1520 m. (5000 ft) on the forested slopes of Tapai-shan (Tapai Mount) in the Chinese province of Shensi. The eyespot markings, present at the apex of the forewing in other species of Brahmaeidae, are absent in *C. miraculosa;* however, their function may be performed by the markings at the base of the forewing of this species which, when the wings are in the resting position, would resemble a pair of closely-set eyes. **340**e

**Callisthenia** Hampson (ARCTIIDAE)
About 7 species are placed in this genus of C. American and tropical S. American moths of the subfamily Lithosiinae. The species are minute compared with most other Arctiids; they have mostly orange or yellow hindwings and longitudinally striated forewings crossed by a yellow bar.

**plicata** Butler
Tropical S. America 12–17 mm. (0·47–0·67 in.)
The colour pattern of *C. plicata* is typical of its genus. There is an area of scent scales in a fold on the middle of each forewing of the male. **367**j

**Callithea** Boisduval (NYMPHALIDAE)
A S. American genus with many iridescent species.

**markii** Hewitson
Brazil 58–60 mm. (2·28–2·36 in.)
The underside of the forewing is grey, with a yellow base and a few black spots near the apex. The hindwing has yellow on the underside at the base and inner edge of the wing with rows of black spots round the edge as in *C. salvinii*. **229**n

**salvinii** Staudinger
S. America 55 mm. (2·17 in.)
The upper and undersides of this species are figured. The butterfly is found in Peru, but nothing is known of its life history. **229**l, m

**Callithomia** Bates (NYMPHALIDAE)
This small genus of butterflies is a member of the subfamily Ithomiinae. It is represented in C. America and N. S. America.

**hezia** Hewitson
S. Mexico to Colombia and N.W. Venezuela 66–72 mm. (2·6–2·83 in.)
The forewings of this attractive species are black or dark brown, with some reddish brown at the base, one or two yellow spots in the middle of the wing, and a double row of yellow spots near the outer margin. The hindwings are light reddish brown, with a black marginal band and sometimes a few white marginal spots. Seven subspecies have been described, each differing in the size, shape and number of the yellow spots.

**Callophrys** Billberg (LYCAENIDAE)
This is a large and widespread genus of butterflies found in Europe, Asia, and in N., C. and S. America. It contains several superficially distinctive subgenera which were long thought to represent separate genera.

**avis** see under **C. rubi**

**crethona** Hewitson JAMAICAN HAIRSTREAK
Jamaica 29 mm. (1·14 in.)
The upper side of the male *C. crethona* is a beautiful iridescent blue; the female is bluish grey. The under surface of both sexes is a rich green, with brown and white markings on the hindwing. This cryptic coloration of the under surface of the wings ensures that once the butterfly has alighted and folded its wings together, it becomes almost invisible when resting amongst foliage. Nothing is known about the life history of this rare species.

**rubi** Linnaeus GREEN HAIRSTREAK
Europe including Britain, N. Africa, temperate Asia to the Pacific 25–30 mm. (0·98–1·18 in.)
This is a common species of heathland and open country where the caterpillar feeds chiefly on heather *(Calluna)*, gorse *(Ulex)*, broom *(Cytisus)*, bramble *(Rubus)* and *Vaccinium*. Northern specimens are darker than others, while those from N. Africa are redder. In Britain this is the smallest species of hairstreak. Both sexes are chiefly dark brown above and light green beneath, differing from the similar Chapman's Green Hairstreak, *C. avis*, in the green front of the head and the absence of a white transverse line on the under surface of the wings. **256**

**callopisma** see **Lactura**

**Callosamia** Packard (SATURNIDAE)
The 3 known species of this genus are found only in E. N. America.

**promethea** Drury PROMETHEA MOTH
Canada, U.S. 75–115 mm. (2·95–4·53 in.)
The range of this well-known moth extends from Ontario and Quebec provinces to S.E. U.S. The dark coloured, day-flying male is thought to be a mimic, when in flight, of the Pipe-vine Swallowtail butterfly *Battus philenor*. The female, in contrast, is nocturnal. Foodplants include a great number of deciduous trees and shrubs, including orchard trees. **329**

**Calodesma** Hübner (HYPSIDAE)
About 11 species of moths are placed in this genus. Its range includes tropical S. America and C. America, including Mexico. Some species, for example *C. melanochroia* and *C. uraneides*, are members of mimetic partnerships with Ctenuchids, Nemeobiid butterflies and species of other families.

**maculifrons** Walker
Central America including Mexico 39–45 mm. (1·53–1·77 in.)
The ground colour of the forewings of this species is normally orange, the hindwings either orange or yellow. It is similar in general features to the Hypsid species *Mesenochroa guatemalteca*. Nothing is known about the life history. **376**e

**melanochroia** Boisduval
C. America including Mexico 45–55 mm. (1·77–2·17 in.)
This species together with another Hypsid *(Hypocrita euploeodes)* is commonly quoted as the distasteful model for the Nemeobiid butterfly species *Esthemopsis clonia*. The similarity between the 3 species is certainly impressive. Its life history is unknown. See also under *Esthemopsis clonia*. **381**h

**Caloptilia** Hübner (GRACILLARIIDAE)
A large worldwide genus of micro-moths with many species. Both the fore- and hindwings are very long and narrow. The caterpillars which generally lack legs on the last segment, are mostly leaf-miners, the entire life cycle, egg to adult, taking place between the upper and lower surface of one leaf. When the moths sit they have a characteristic pose, with the head end held upwards and the front and mid legs clearly visible.

**acidula** Meyrick
India, Pakistan 6 mm. (0·24 in.)
This minute moth is green, with 3 narrow white bands across the slender forewings. The hindwings, like those of most species in the genus, are reduced to slender strips, surrounded by long scales and looking like feathers. When at rest the moth sits with its head upwards, the wings curved upwards at the apex, and characteristically waves its antennae about. The tiny caterpillar is a leaf-miner and is found in the 'leaflets' of *Phyllanthus* (Euphorbiaceae). The caterpillar causes the leaflet to curl and when ready to pupate, it ties this leaflet together to make a tiny cone-shaped cell in which to pupate. These 'cones' can be seen on the plants which have been attacked.

**azaleella** Brants AZALEA LEAF-MINER
Europe including Britain temperate Asia including Japan, N. America 9 mm. (0·35 in.)
This tiny moth is a pest on azaleas *(Rhododendron)* in greenhouses or, in the warmer parts, out of doors. It has been imported accidentally from Japan. The forewing is elongate, with basal and apical fringes mainly black, and with a yellow blotch along the front margin of the forewing. The hindwings are grey with a long fringe. The caterpillar mines into the azalea leaves, later emerging and curling the tops of the leaves back with silk. In bad attacks it will defoliate whole bushes.

**elongella** Linnaeus
Europe including Britain, N. America 14–16 mm. (0·55–0·63 in.)
This moth has very varied wing patterns and is often best recognized from its host plants, alder *(Alnus)* and birch *(Betula)*. The forewings are generally red or reddish brown, usually unmarked. There may be some yellow marks and often black dots on the margins of the wing. The hindwings are greyish. The moth is common in Europe and N.E. N. America. The caterpillars roll the leaves of the host plants into conical nests, after an early stage of mining them which marks them with a blotch.

**Calpe** Treitschke (NOCTUIDAE)
This is chiefly an E. and S.E. Asian group of about 50 species. It is, however, represented by a few species in the New World (including a species in Canada and N. U.S.), Africa, and a single species in S. Europe. The shape of the rear or inner margin of the forewing is distinctive. An Asian species *C. eustrigata*, is a bloodsucker in the adult stage; other species are fruit-piercing moths and may cause damage in orchards, and a few are eye-frequenters *(C. minuticornis)*.

**eustrigata** Hampson VAMPIRE or BLOODSUCKING MOTH 35–44 mm.
S.E. Asia (Malaya, India, Sri Lanka) (1·38–1·73 in.)
Although this is a fruit-piercing species like many of its allies in *Calpe*, Dr H. Bänziger, a Swiss entomologist, has discovered recently that it is capable of puncturing the skin of mammals, especially cattle and deer (species of Bovidae and Cervidae), and subsequently sucking blood from them through its proboscis. It is, therefore, a possible carrier of diseases in the same way that mosquitoes transmit malaria. During the skin-puncturing process, the moth oscillates its head from side-to-side, rapidly moving alternate sides of the divided proboscis and bringing into action its rasping spines and eversible barbs. **399**h

**minuticornis** Guenée EYE-FREQUENTING CALPE
S.E. Asia, Japan to Australia 35–45 mm. (1·38–1·77 in.)
As well as being a typical fruit-piercing moth, the adult *C. minuticornis* is also one of several species of moth, mostly Noctuidae and Pyralidae, which suck lachrymal secretions ('tears') from the eyes of humans and other mammals and are potential carriers of the trachoma virus, and may be responsible for other eye diseases such as ophthalmia and keratoconjunctivitis. This moth is almost indistinguishable externally from the illustrated *C. eustrigata*.

**thalictri** Borkhausen
S. Europe, C. Asia to Japan 35–45 mm. (1·38–1·77 in.)
This is the only species of its genus to be found in Europe. The adult moth, which flies in June and July, is similar in general pattern to *C. eustrigata* (illustrated) but has much more pronounced lobes at the middle and outer angle of the rear margin of the forewing. The yellowish green and black caterpillar feeds on the foliage of Meadow Rue *(Thalictrum)*.

**calthella** see **Micropterix**
**camadeva** see **Stichophthalma**
CAMBERWELL BEAUTY see **Nymphalis antiopa**
CAMBRIC WAVE see **Venusia cambrica**
**cambria** see **Venusia**
CAMBRIDGE BLUE see **Pseudolycaena marsyas**
CAMBRIDGE VAGRANT see **Nepheronia thalassina**
**camilla** see **Ladoga**
**camillus** see **Marpesia**
**camissa** see **Lucillella**

**Campaea** Lamarck (GEOMETRIDAE)
This genus of moths is found in Europe and temperate Asia. The majority of species in the genus are green, with various patterns.

**margaritata** Linnaeus LIGHT EMERALD
Europe including Britain 36–48 mm. (1·42–1·89 in.)
The moth is out in July and is common in woodlands where the caterpillar feeds on oak *(Quercus)*, birch *(Betula)*, elm *(Ulmus)* and other trees. In N. America a related, similar looking species (*C. perlata* Guenée) is called the Light Emerald. **300**

**perlata** see under **C. margaritata**

**Camptogramma** Stephens (GEOMETRIDAE)
An European and Asian genus of moths. They are very typical of many Geometrid moths in appearance.

**bilineata** Linnaeus YELLOW-SHELL MOTH
Europe including Britain Asia 23–28 mm. (0·91–1·1 in.)
The moth is common in hedgerows and meadows throughout the summer over most of its range. The caterpillar feeds on various herbaceous plants, including chickweed (Caryophyllaceae), Dock (Polygonaceae) and many other low herbaceous plants. It is a bluish green colour, with darker lines along the back, the centre line darker than the outer. Several subspecies of this moth have been recognized. **276**

**Camptoloma** Felder (HYPSIDAE)
A single species of moth is placed in this genus.

**interiorata** Walker
Japan, China 32–41 mm. (1·26–1·61 in.)
The subspecies *C. i. binotatum* Butler, described from N.E. India, in fact probably represents a separate Indian species. The cryptically patterned, hairy caterpillars are gregarious on the leaves of chestnut *(Castanea)* trees, and spin communal nests on the bark. **376**l

**Campylotes** Westwood (ZYGAENIDAE)
Moths of this genus are found throughout Sikkim, Burma, Tibet, China and Formosa. There are only about 15 species in the genus but they are widely distributed. They are generally brightly coloured moths, with red and black patterns, most probably warning colours to deter predators from eating them. The genus belongs to the subfamily Chalcosiinae.

**desgodinsi** Oberthür
Tibet, N. India, Borneo, S. China 55–70 mm. (2·17–2·76 in.)
A subspecies occurs in S. China which is much paler in colour than the one illustrated. The males and females are similar in colour, both being so vivid that as in the case of the others in the genus, they probably want to advertize, rather than conceal themselves because they are distasteful. However no information is available on this or on the life history of the species. **45**f

**histronicus** Westwood
N. India, Assam, W. China 60–80 mm. (2·36–3·15 in.)
There are species related to this colourful moth from N. India which look similar and there are also a number of colour forms, some of which have been named. The males and females are alike in colour and pattern. The thorax has bright yellow anterior edges (actually on the tegulae) which are very conspicuous. Altogether a very remarkable looking moth.

**canace** see **Vanessa**

**Canaea** Walker (THYRIDIDAE)
Large brown moths with a pattern of round translucent patches on a brown background. Although the biology is not known, it is possible that the pattern, which makes the moths look like dead leaves, is for camouflage. The genus contains 10 species which are found in Indonesia, Solomon Islands, New Guinea and Australia. Recently specimens have been found as far N. as the Ryuku Islands.

**rusticata** Whalley
Indonesia, W. Irian, Papua, New Guinea, Solomon Islands 36–46 mm. (1·42–1·81 in.)
The thorax and wings are reddish brown, both fore- and hindwings having a pattern of white almost circular patches. Three subspecies are known, one is in W. Irian (Indonesia), the second is eastwards in the Papuan part of New Guinea, while a paler subspecies occurs on the Solomon Islands. The moths are quite common in collections from these areas but nothing is known of their biology or of the host plants of the caterpillars. Some observations on the living insects would be very interesting.

CANARIES' TIGER-MOTH see **Diacrisia rufescens**
CANARY BLUE see **Leptotes webbianus**
CANARY ISLAND BRIMSTONE see **Gonepteryx cleopatra**
**cancellata** see **Belenoptera**
**candidata** see **Hermathena**
CANKER WORM, FALL see **Alsophila pometaria**
**canthus** see **Satyrodes**
**caparea** see **Semomesia**
CAPER WHITE see **Anaphaeis java**

**Capillamentum** Pinhey (NOCTUIDAE)
There is a single known species in this genus of moths.

**nigrofasciatum** Pinhey
Ethiopa, E. Africa 43–52 mm. (1·69–2·05 in.)
This is one of the most colourful species of Noctuidae. A peculiar feature of the head and thorax is the covering of long, raised white scales on the dorsal surface. These scales have a long, narrow stem and a 2- to 4-pronged apex. Nothing is known about the life history of this moth. **399**b

**Capperia** Tutt (PTEROPHORIDAE)
A genus of plume-moths with species mostly in Europe, temperate Asia and Japan, but with a few species described in N. America.

**britanniodactyla** Gregson
Europe including Britain 18–22 mm. (0·71–0·87 in.)
There are several rather similar looking species in the same and related genera and careful examination is needed to identify them correctly. The caterpillar feeds on the underside of the leaves of Wood Sage, *Teucrium scorodonia*. It overwinters and then in the spring eats the young shoots, causing them to tip over as though wilting. The caterpillar is yellowish green, with lines and raised brown dots, each with a group of white hairs. The moth flies at dusk in June and July at at the edge of woods where the food plant grows. Frequently there are 2 generations a year. **74**

**cara** see **Catocala**
**carcharota** see **Dialectica**
**cardamines** see **Anthocharis**
CARDINAL see **Pandoriana pandora**
**cardui** see **Cynthia**
**carduidactyla** see **Platyptilia**

**Caria** Hübner (NEMEOBIIDAE)
About 10 species are placed in this tropical S. American genus of butterflies.

**mantinea** Felder
Tropical S. America 20–31 mm. (0·79–1·22 in.)
This is one of the most elegantly marked species of Nemeobiidae. The dark brown under surface of the wings lacks the iridescent green and gold of the upper surface and has a reddish orange patch at the base of the forewing. Its life history is not known. **268**j

**carinenta** see **Libythea**
CARNATION TORTRIX see **Cacoecimorpha pronubana**
CARNATION WORM, SOUTH AFRICAN see **Epichoristodes acerbella**
**carnella** see **Oncacera semirubella**
CAROB BEAN MOTH see **Ectomyelois ceratoniae**
CAROLINA SPHINX see **Manduca sexta**
CARONI FLAMBEAU see **Dryadula phaetusa**

**Carpella** Walker (GEOMETRIDAE)
A small genus of moths with some 12 species described from S. America.

**districta** Walker
S. America 54–62 mm. (2·13–2·44 in.)
This species is found in Peru and Bolivia. The females are larger than the males but their patterns are similar. The lines on the wings are unusually straight and look as though they were drawn with a ruler. **272**v

CARPENTER WORM see **Prionxystus robiniae**
CARPENTER WORM MOTHS see **Cossidae**
CARPENTERS see **Xyleutes**
CARPET MOTHS see **Cleora, Colostygia, Perizoma, Thera, Trichophaga, Xanthorhoe**

**Carposina** Herrich-Schäffer (CARPOSINIDAE)
A worldwide genus of micro-moths with many species in Australia. Some of the species have caterpillars which tunnel in *Eucalyptus;* others live in fruit or galls. The moths are usually night-flyers and their patterns are for concealment on tree trunks where they rest during the day. Some of the darker or patterned ones are almost invisible on fire-blackened tree trunks.

**adreptella** Walker
New Zealand 14 mm. (0·55 in.)
This small grey-brown moth has a typical camouflage pattern with white hindwings which are hidden when at rest. The biology is not known.

**Carposinidae**
A family of micro-moths found all over the world. While they tend to have rather pointed forewings, the hindwings are generally broad. The caterpillars feed in various places, generally concealed in bark, galls or in fruit. See *Carposina, Heterogymna, Meridarchis.*

**Carriola saturnioides** see under **Macrauzata**
CARROTWORM see **Papilio polyxenes**

**Carterocephalus** Lederer (HESPERIIDAE)
This genus of Skipper butterflies is found throughout Europe and temperate Asia with species extending into N. America.

**palaemon** Pallas ARCTIC or CHEQUERED SKIPPER
Europe including Britain 28 mm. (1·11 in.)
temperate Asia including Japan, N. America
In America the 'Checkered Skipper' is the popular name for another species, *Pyrgus communis*. *C. paleamon* is widely distributed throughout the more northerly parts of Europe but local or absent from Denmark, Italy and Spain. In N. America it ranges from central and S. Canada S. to Connecticut and New York, N. Michigan, Minnesota and Pennsylvania. The adults are out in June and July and there is generally only one brood a year. The caterpillar, which is cream coloured with a faint dorsal stripe, or narrower yellowish stripes along the sides, feeds on grasses, especially *Bromus*. The butterfly is usually found in light woodlands or woodland glades. The flight is slow and rather weak compared with the rapid flight of other Skippers. Several subspecies have been described, including a distinct one in Japan. **81**

**Carthaea** Walker (CARTHAEIDAE)
There is a single species in this genus of moths. It was placed originally by its author in the Noctuidae in spite of the fact that he was aware of its resemblance to the family Saturniid as evidenced by the name he chose for the single included species.

**saturnioides** Walker
S.W. Australia 75–84 mm. (2·95–3·31 in.)
The eyespots on the hindwing of this species are displayed when the moth is disturbed, in much the same way as in many species of Saturniidae (see *Automeris*). The orange and brown caterpillar has eyespots like those of many Sphingid caterpillars and a comparable dorsal horn at the rear end, except in the final stage. It feeds on the foliage of *Dryandra*. **317**f

**Carthaeidae**
This solely Australian family was erected in 1966 by I. F. B. Common of Canberra to accommodate a single species of moth distantly related to Saturniidae species but with several primitive features. It is placed in the superfamily Bombycoidea. See *Carthaea*.

**caryae** see **Halisidota**
**carye** see **Cynthia**
**casanella** see **Deuterotinea**

**Casbia** Walker (GEOMETRIDAE)
An Australian and New Guinean genus of moths with some 30-40 species at present known. Little is known of their biology.

**fasciata** Warren
New Guinea 16–22 mm. (0·63–0·87 in.)
This species is at present known only from Fergusson Island, a small island in the D'Entrecasteaux group which lies off the eastern tip of New Guinea. Few specimens have been collected and little is known about the biology of the species. **294**m

CASE MOTH, WHITE-TIPPED CLOVER see **Coleophora frischella**
CASEBEARER MOTHS see **Acrobasis, Coleophora**
CASE-BEARING CLOTHES MOTH see **Tinea pellionella**
CASE-MAKING CLOTHES MOTH see **Tinea pellionella**
**cassius** see **Leptotes**
CASSIUS BLUE see **Leptotes cassius**
**casta** see **Eucharia, Psyche**

**Castalius** Hübner (LYCAENIDAE)
This is a widespread genus of butterflies found in tropical Africa, Sri Lanka, India, the Malay Archipelago to New Guinea and the Bismarck Archipelago. The species are mostly white, with black markings, but some have a trace of blue and purple. The caterpillars feed on species of Rhamnaceae.

**rosimon** Fabricius COMMON PIERROT
Sri Lanka, India, Burma, Thailand, 24–32 mm.
the Malay Archipelago E. to (0·94–1·26 in.)
the Lesser Sundas
*C. rosimon* is commonest in open woodlands, but is also found in drier localities. In the Himalayas, it occurs up to 2130 m. (7000 ft). Its flight is weak and close to the ground. The male (not illustrated) is white above, with some pale blue scales at the base of the forewing and has narrower dark brown outer margins to the wings. The caterpillar feeds on a species of *Zizyphus*. **263**m

**roxus** Godart STRAIGHT PIERROT
Burma and Indonesia to Sulawesi 26–30 mm.
(Celebes) and the Philippines (1·02–1·18 in.)
This is a butterfly of lowland forests where it often congregates in large numbers on muddy patches in forest clearings and along the banks of streams. The upper surface of the wings is dark brown, with the white transverse band (narrower in the female) on the hindwing continuous with a similar but incomplete white band on the forewing. The under surface is white with a black oblique basal band, a few black central spots, and a lunulate marginal band – a pattern which is basically similar to the upper surface pattern of its close relative *C. rosimon* (illustrated). Little seems to be known about the life history, but the caterpillar possibly feeds on *Zizyphus*, the foodplant of other species of *Castalius*.

**castaneela** see **Lyonetia**
**castelnaui** see **Laringa**

**Castnia** Fabricius (CASTNIIDAE)
This Central American and tropical S. American genus of day-flying moths, together with other Castniids, were at one time grouped with the butterflies they resemble in antennal shape, wingshape and coloration.

**licus** Fabricius GIANT SUGARCANE-BORER
Tropical S. & C. 64–100 mm.
America (2·52–3·94 in.)
The caterpillar of this gaudy moth bores into the stems of sugar-cane and was first noticed as a pest at the beginning of this century in Guyana. Unlike some of its relatives this species turns into a chrysalis inside the stem of its food-plant. **45**s

**Castniidae**
This is a family of just over 30 genera and nearly 200 described species of moths, mostly tropical C. and S. American in distribution but also represented in tropical Asia, Australia and Madagascar. Many are brightly coloured, some are probably mimics of unpalatable butterflies of the families Nymphalidae (subfamilies Heliconiinae, Ithomiinae and Acraeinae) and Papilionidae. They fly by day and are frequent visitors to flowers and tree blossoms, especially before noon. The forewings of many species are cryptically coloured, the hindwings often brightly coloured and covered by the forewings when the moth is at rest (an example of 'flash-coloration'). The caterpillars of many species feed inside the stems of plants or are root-feeders and are therefore able to avoid dessication in dry climates; others feed well above the ground on the roots of epiphytic bromeliads (Bromeliaceae) and orchids (Orchidaceae). The affinities of this family probably lie with the 'microlepidoptera'. Some of their primitive characters are shared with Tortricidae, Sessiidae and Cossidae. It is possible that the Castniids may be related to the ancestors of the butterflies. See *Amauta, Castnia, Cyanostola, Enicospila, Gazera, Neocastnia, Synemon, Riechia, Tascina, Zegara*.

**castrensis** see **Malacosoma**

**Cataclysta** Hübner (PYRALIDAE)
A large genus of Pyralid moths, mostly world-wide, whose caterpillars feed on water-plants. The moths are beautifully marked, often with a group of silvery blue spots around the hindwings.

**lemnata** Linnaeus SMALL CHINA-MARK
Europe including Britain 17–21 mm. (0·67–0·83 in.)
The pale green caterpillar lives in a portable oval case made of leaf fragments on duckweed *(Lemna)* in ponds and ditches and feeds on the floating leaves. Only the early stage caterpillar lives under the water, the later stage crawls around with its case on the surface of duckweed. **66**

**Catacroptera** Karsch (NYMPHALIDAE)
This genus of African butterflies has only one species known at present, although this has been separated into a number of distinct subspecies and various colour forms have been described.

**cloanthe** Cramer PIRATE BUTTERFLY
South to E. Africa 56–67 mm. (2·2–2·64 in.)
This orange butterfly, with dark brown bars and spots on the forewing and a row of blue eyespots on the hindwing, is common over much of its range, generally in grassy hollows and swampy localities. The hindwing has a slight tail and is angled in the middle with tufts of coarse scales on the margin. The upperside has a fresh violet bloom when newly emerged. In the wet season the underside is pale brown, while the dry season forms have dark brown undersides. The caterpillar feeds on *Gomphocarpus* (Asclepiadaceae) and *Justicia* (Acanthaceae).

**Catarhoe** Herbulot (GEOMETRIDAE)
A widely distributed genus of moths with species in N. America, Europe and temperate Asia. Many of the species are common.

**rubidata** Denis & Schiffermüller FLAME MOTH
Europe including Britain, 23–28 mm.
temperate Asia (0·91–1·1 in.)
The name refers to the reddish colour on the wings. The moth is common in Europe where it flies in June. The caterpillar feeds on the leaves of *Galium* (Rubiaceae). **277**

**Catasticta** Butler (PIERIDAE)
This is a large S. American genus of butterflies. Many of the species are dark coloured and heavily patterned, often with some red on the wing.

**uricoecheae** Felder
S. America 46 mm. (1·81 in.)
The underside is a dazzling zig-zag pattern of yellow and black forming a very disruptive camouflage when the insect is at rest. This species has been collected in Colombia. **119**b

CATERPILLAR, SALTMARSH see **Estigmene acrea**
CATERPILLAR, UGLY NEST see **Archips cerasivorana**

**Catocala** Schrank CATOCALA or UNDERWING MOTHS
Over 200 colourful species belong to this genus. Most are northern temperate in distribution, but some occur in C. America, India, Formosa and the mainland of S.E. Asia. They are a speciality of some collectors, especially in North America, where 100 or so species of *Catocala* are known. The red, orange, yellow, blue or black hindwings are either hidden or partly hidden by the cryptically coloured forewings when the moth is at rest. *Catocala* caterpillars feed on the leaves of various trees and shrubs; they are capable of jumping quite strongly.

**cara** Guenée BRONZE UNDERWING
Canada, U.S. to Florida 70–88 mm. (2·75–3·46 in.)
The ground-colour of the forewings of this handsome moth is a mixture of yellowish green and reddish brown; the 2 irregularly shaped, transverse lines are black. The hindwings are pale red or reddish orange, with a black base, a black central band and a black marginal band; the extreme outer margin is pale yellow. The caterpillar is black or dark grey with orange markings; it feeds chiefly on willow *(Salix)* but also on poplar *(Populus)*.

**fraxini** Linnaeus CLIFDEN NONPAREIL
Europe including Britain, temperate 75–95 mm.
Asia to E. Russia and Japan (2·95–3·74 in.)
This is the largest species of Noctuid found in Britain where it breeds in the counties of Kent and Norfolk. Its caterpillars feed from May to July on the leaves of poplar and aspen *(Populus)*. They are strange twig-

like creatures, about 75 mm. (2·95 in.) in length and chiefly brown or grey. *C. fraxini* is one of many species which apparently have extended their range northwards during this century. **392**e

**nupta** Linnaeus RED UNDERWING
Europe including Britain, temperate Asia to E. Russia and Japan 70–80 mm. (2·76–3·15 in.)
The hindwing ground-colour varies between specimens, and can be a darker red than in the illustration, more orange in colour, or, very rarely, blue. Although there is no lack in Britain of the larval foodplants, poplars *(Populus)* and willow *(Salix)*, it apparently does not breed there and occurs only as a migrant. **394**

**relicta** Walker WHITE UNDERWING
Canada, U.S. 75–80 mm. (2·95–3·15 in.)
This is a widespread species in Canada and N. U.S. but also extends S. to Arizona. The amount of white on the forewing varies considerably, individually and geographically, and some specimens are dark grey, with indistinct markings. The hindwings are black, with a white transverse band parallel to the outer margin and a white marginal band. Recent experiments by Sargent show that adults have an innate response to their background and actively seek a tree-trunk which matches their colour pattern. Caterpillars of this species feed on the leaves of poplar *(Populus)* and willow *(Salix)*. **392**b

**Catocalopsis** Rindge (GEOMETRIDAE)
Although a genus of Geometrid moths, its sole species is very similar to species in a Noctuid genus *Catocala* and different from the more typical, delicate-looking Geometrids to which it is related. It is characterized by the very long last segment on the palp.

**medinae** Bartlett-Calvert
Chile 50–60 mm. (1·97–2·36 in.)
The original specimen (holotype) of this species appears to have been lost but recent work has indicated the correct identity of this species. The moth flies in the first 3 months of the year. Nothing is known of the early stages of its life history. **272**s

**Catonephele** Hübner (NYMPHALIDAE)
A genus of brightly coloured S. American butterflies with a few species in C. America.

**acontius** Linnaeus ORANGE-BANDED SHOEMAKER
S. America, W. Indies 58 mm. (2·28 in.)
This species is found from the Guianas to Colombia and Paraguay, and in the W. Indies. It is a dark, almost black butterfly with a broad band of reddish brown on the forewing. This reaches from the middle of the hind margin and just over half way towards the apex, with a similar broad band in the middle of the hindwing. The female has brown wings with yellowish bands in both wings and purplish spots near the apex of the forewing. **229**g, h

**numilia** Cramer GRECIAN SHOEMAKER
C. and S. America, W. Indies 70–75 mm. (2·76–2·95 in.)
This species reaches as far south as Paraguay and Uruguay where it is found mainly in forests. The adults are commonly found feeding on rotting fruit. The female is larger with yellow on the wing and without the blue of the male. It is suggested that the female pattern mimics the non-palatable Heliconid butterflies. The caterpillar feeds on *Alchornea* (Euphorbiaceae) and *Citharexylum* (Verbenaceae). The male and female are very different in this species. **229**a, b

**Catopsilia** Hübner (PIERIDAE)
This is a large genus of broad-winged butterflies most species of which are variable in pattern with many named forms. Often the males and females have different patterns (sexual dimorphism). The species are found on all continents and are particularly numerous in the tropics of India, Pakistan and S.E. Asia through to Australia. Many species have been described from Africa and one reaches into S. Europe and temperate Asia. The foodplants of the caterpillars vary but they are often species of Leguminosae. S. American species referred to as *Catopsilia* should be placed in the genus *Phoebus*.

**florella** Fabricius AFRICAN MIGRANT
Africa, Canary Islands, Egypt, India, China 50–65 mm. (1·97–2·56 in.)
This widespread species is all white in the male with a faint greenish tinge and a black spot on the forewing. The female is a yellowish colour. The caterpillars feed on *Cassia* (Leguminosae). It is common over the whole of Africa south of Sahara, and recently was recorded on Gran Canary and Tenerife. Several forms have been described over its wide range. This common and rapid flying butterfly migrates over most of the continent of Africa. **120**

**pomona** Fabricius LEMON MIGRANT
India, Pakistan, S.E. Asia, Madagascar, Mauritius, Australia 65–70 mm. (2·56–2·75 in.)
Several subspecies of this species have been described in parts of its wide range. In Australia a subspecies occurs which is found mostly in the northern and western region where the caterpillar feeds on species of *Cassia* (Leguminosae). This butterfly has been reported migrating in thousands in Upper Perak (Malaysia) and in many other parts of its range it appears periodically in large numbers, apparently on a purposeful migration. Nothing is known of the causes of these sudden fluctuations and movements of this butterfly. **119**n

**Catoptria** Hübner (PYRALIDAE) GRASS-VENEER MOTHS
Mostly European and Asian in distribution, the species of this genus of Pyralid moths are all relatively similar in appearance, with the palps sticking out in front of the head like a beak. They sit in a characteristic 'crambid' way on grass stems. The caterpillars feed on various grasses.

**pinella** Linnaeus PEARL GRASS-VENEER
Europe including Britain, Asia including Japan 18–24 mm. (0·71–0·94 in.)
Several subspecies of this have been described. The caterpillar feeds on the stalks of Tufted Hair Grass *(Deschampsia cespitosa)*, Bog Cotton *(Eriophorum vaginatum)* and other grasses. It lives in silken tubes in the dense tussocks. **53**

CATSEYE, DARK see **Zipaetis scylax**
CATTLE HEART BUTTERFLIES see **Parides**
**catullus** see **Phaliosora**
**caudipennis** see **Rhodogastria**
**cautella** see **Ephestia**
**cavella** see **Phyllonorycter**
CEANOTHUS SILK-MOTH see **Hyalophora euryalus**

**Cechena** Rothschild (SPHINGIDAE)
A genus of about 6 species of moths found in E. Asia.

**mirabilis** Butler
N. India 80–90 mm. (3·15–3·54 in.)
The pink and green adults of the *C. mirabilis* are among the most beautifully coloured species of hawk-moth. There is a green form and a chiefly brown form of the caterpillar; its horn is yellow, tipped with black, and there is an eyespot marking on each side a short distance behind the head. The foodplant is vine *(Vitis)*. **343**

**Cecidothyris** Aurivillius (THYRIDIDAE)
This genus is known only from Africa where it is distinguished from all the others in the family by having only 2 segments to the labial palps under the front of the head (all others in the family have 3 segments). The moths are reddish brown, often with brown rings (reticulations) on the fore- and hindwings. The species whose life histories are known, are borers in the twigs of species of *Terminalia* (Combretaceae).

**parobifera** Whalley
Nigeria, Zaire, Kenya 28–31 mm. (1·1–1·22 in.)
Typical of the genus, this is a reddish brown moth with a pattern made up of prominent round white spots on both fore- and hindwings. Nothing is known of the biology of this species.

**pexa** Hampson
Africa 21–45 mm. (0·83–1·77 in.)
This reddish brown moth has a prominent pattern of dark brown across the base of the fore- and hindwings, and on closer examination the pattern of the whole wing is seen to be made up of reddish brown rings. These are not as conspicuous as in some related species. Two subspecies have been recognized; one is from South Africa to Rhodesia, the other extends from Rhodesia through Uganda, Zaire and Kenya to the Sudan, then W. to the W. African countries. The moth is common over all its range. The caterpillars feed in the twigs of *Terminalia* trees (Combretaceae) which are widespread throughout Africa. A gall is produced by the plant in response to the caterpillars' presence and the infected twigs show swellings inside which the caterpillars feed.

**cecilia** see **Pyronia**
**cecropia** see **Hyalophora**
CECROPIA see **Hyalophora cecropia**
**Celama** see under **Nola**

**Celastrina** Tutt (LYCAENIDAE)
The 30 or 40 species in this genus of 'blues' are chiefly tropical Asian in distribution, but there are also representatives in Europe.

**argiolus** Linnaeus HOLLY BLUE or SPRING AZURE
N. America, C. America as far S. as Panama, Europe, N. Africa, temperate Asia to Japan 20–32 mm. (0·79–1·26 in.)
This is often a fairly common species and is sometimes relatively abundant in urban London. There are even some records from the gardens of Buckingham Palace. In N. America this is one of the first butterflies to appear in the spring and is sometimes seen before the snow has finally disappeared. The Holly Blue is typically a species of open woodland and heaths in Europe, where the caterpillar feeds chiefly on holly *(Ilex)*, but also on buckthorn *(Rhamnus)*, bramble *(Rubus)*, gorse *(Ulex)*, broom *(Cytisus)* and ivy *(Hedera)*. In N. America the foodplants include dogwood *(Cornus)*, blueberry *(Vaccinium)*, and New Jersey Tea *(Ceanothus)*. **249**

**Celerio** see **Hyles**
**celerio** see **Hippotion**
CELERY WEBWORMS see **Nomophila**
CELERYWORM see **Papilio polyxenes**
**celtis** see **Libythea**
**centaureae** see **Pyrgus**
**centaureata** see **Eupithecia**
**centuriella** see **Gesneria**

**Cepheuptychia** Forster (NYMPHALIDAE)
This genus of butterflies has only one species. It is related to *Euptychia* and is in the subfamily Satyrinae.

**cephus** Fabricius BLUE NIGHT BUTTERFLY
S. America, W. Indies 35 mm. (1·38 in.)
The iridescent blue fore- and hindwings edged with black makes the male a very conspicuous Satyrine butterfly. The females are brown with a thin blue line round the outer margins of the wings. The popular name refers to the colour, rather than the habit, of the butterfly which is a day flying species. It occurs from S. Brazil to Surinam, Colombia and is also known on Trinidad. **174**e, f

**cephus** see **Acraea, Cepheuptychia**

**Cepora** Billberg (PIERIDAE)
A widespread genus from India through S.E. Asia with one species in Australia.

**aspasia** Stoll
Philippines 45–55 mm. (1·77–2·17 in.)
This beautiful species of *Cepora* has very fresh colours in the male, while the female is more soberly coloured. She is mostly brown with less white on the forewing and only patches of yellow on the hindwing. **119**d

**nerissa** Fabricius COMMON GULL
India, Pakistan, Sri Lanka, Burma 40–65 mm. (1·57–2·56 in.)
The forewings are white, with the veins black, generally darker round the cell, and with a black margin. The hindwing is similar, with slightly less black. The underside is dark yellow on the hindwing, with the

veins outlined in dark green, and the forewing is similar to the upperside but with less black. This is the wet season form. In the dry season form, the black is reduced and the hindwings are pale yellow, relatively unmarked. The butterfly is very common throughout its range up to 1220 m. (4000 ft) in the N. and up to 2440 m. (8000 ft) in the S. The caterpillar feeds on *Capparis* (Capparidaceae). The butterfly is out in most months of the year and is a strong flier, fond of feeding at flowers.

**perimale** Donovan AUSTRALIAN GULL
Indonesia, Sulawesi (Celebes), New Guinea, New Caledonia, Fiji, New Hebrides and Australia — 45–55 mm. (1·77–2·17 in.)
Several subspecies of this white butterfly have been described. The wings have a black margin on front and hind, sometimes with white patches in this. The underside forewing has a paler margin with a brownish patch at the apex and the underside hindwing is either brown or grey-brown, with or without a darker margin. The caterpillar is green with yellow dots and brown hairs and feeds on species of *Capparis* (Capparidaceae). This common species is very variable in the colour pattern of the underside and these differences are used to separate the subspecies.

**Cerace** Walker (TORTRICIDAE)
This genus was placed in a family of its own some years ago (Ceracidae) but its species are now regarded a specialized Tortricids. The species are large multicoloured moths with wing venation like typical Tortricid moths but otherwise they are distinct. Some have been bred from *Cinchona* species in Java. These are trees from the bark of which quinine is obtained and the caterpillars can cause serious damage to them. One species has been bred from oaks *(Quercus)* in the Himalayas but in the main their biology is not well known. The species in the genus are widespread in tropical Asia, N. to Japan.

**xanthocosma** Diakonoff
Japan — 35–50 mm. (1·38–1·97 in.)
The male is illustrated. The female is larger, pale-coloured but has more yellow on the fore- and hindwings. The pattern is very striking so this, and a similar related species from India, are easily recognized. The pattern varies, as does the colour to some extent. The specimen illustrated was used in the original description of the species and is a paratype. **14g**

**Cerapteryx** Curtis (NOCTUIDAE)
About 6 species are placed in this genus of moths. Its range includes Europe and Asia eastwards to China and Siberia.

**graminis** Linnaeus ANTLER MOTH
Europe including Britain, temperate Asia to Siberia — 28–33 mm. (1·1–1·3 in.)
The brown and yellowish brown striped caterpillar is a pest of grass roots. In Britain it is commonest in N. England and in Scotland where very high concentrations of the caterpillars sometimes occur. Adults fly between 8 a.m. and noon, but also at night. Sallow *(Salix)* catkins often attract day-flying specimens. The antler-like markings on the forewings are a characteristic of this species. **388**

**Cercophana** Felder (CERCOPHANIDAE)
There are 3 species in this essentially Chilean genus of moths. Males have an outwardly directed tail to the hindwing.

**venusta** Walker
Chile — 58–72 mm. (2·28–2·83 in.)
The ground-colour of the wings of this species can be yellow, as in the illustrated male, pink or brownish red. The female hindwings lack a tail but are lobed at the anal angle. The black, orange and blue caterpillar feeds on the foliage of *Maytenus* (Celastraceae) and *Aristotelia* (Elaeocarpaceae). The closely woven cocoon is firmly attached to the side of a twig. **318f**

**cerasivorana** see **Archips**
**Ceratinia** see under **Eueides dianassa, Hypothyris**
**ceratoniae** see **Ectomyelois**

**Cercophanidae**
This is a small family of about 10 species of moths found in Chile and N. along the Andes as far as Colombia. It belongs in the superfamily Bombycoidea where it is probably most closely related to the family Oxytenidae. It was at one time placed as a subfamily of the Oxytenidae. See *Cercophana.*

**cerealella** see **Sitotroga**
**ceres** see **Lycorea**
**cerisyi** see **Allancastria**
CERULEAN, METALLIC see **Jamides alecto**

**Cerura** Schrank (NOTODONTIDAE)
There are about 40 or 50 species in this almost cosmopolitan genus of moths. The caterpillars of most resemble the strange *C. vinula* caterpillar described below.

**australis** Scott
Australia — 54–73 mm. (2·13–2·87 in.)
This is a species of the rain-forests of N. Australia. Its wings are a brilliant lustrous white, marked with numerous black spots on the forewing and an outer marginal row of black spots on the hindwing. Its caterpillar feeds on the foliage of *Scalopia brownii.* The cocoon, like that of the Palaearctic *C. vinula,* is a hard oval structure placed on the side of the trunk or a branch of the foodplant.

**scitiscripta** Walker
U.S., Mexico — 35–45 mm. (1·38–1·77 in.)
This species is similar in general appearance to the European *C. vinula,* but the lines crossing the forewing are darker, less dentate and more strongly marked. Its caterpillar is similar in shape to that of *C. vinula,* but is bluish white, light green and dark brown in colour. The foodplants are willow *(Salix)*, poplar *(Populus)*, and wild cherry *(Prunus)*.

**vinula** Linnaeus PUSS-MOTH
Europe including Britain, N. Africa, temperate Asia to Japan, Siberia — 65–80 mm. (2·56–3·15 in.)
*C. vinula* is remarkable for the number of protective devices it employs. The adults are cryptically coloured when at rest and the striped pattern of the forewings blends with the surrounding bark. The well-known 'puss-moth' caterpillar assumes a grotesque attitude when disturbed, with its head withdrawn into the thorax and its front and rear end elevated. At the rear end of the caterpillar there is a pair of filamentous, eversible red processes which are spun around and thought to deter parasitic Ichneumonid wasps and parasitic flies. The caterpillar is also capable of ejecting an irritant jet of formic acid from its thoracic glands if persistently provoked. The extremely tough cocoon is firmly glued to the surface of a branch and as it contains particles of wood is extremely difficult to distinguish from its surroundings. The chrysalis is provided with a cutting edge and is also able to secrete a caustic potash solution which enables it to penetrate a thin part of this almost impregnable cocoon and allow the eventual escape of the adult moth. The caterpillar feeds on various trees including willows *(Salix)* and poplars *(Populus)*. **356**

**cesonia** see **Colias**

**Cethosia** Fabricius (NYMPHALIDAE)
A large genus of butterflies found from India through S.E. Asia to New Guinea.

**biblis** Drury RED LACEWING
N. India, Pakistan, China, Burma, Malaysia, Indonesia, the Philippines — 80–90 mm. (3·15–3·54 in.)
This butterfly flies for most of the year and the name is derived from the lacewing pattern of the underside in reds and oranges. The general colour is deep reddish brown with black margins and a white 'V' or 'U' between the apex of the veins on the wing margin. There is a larger black area on the apex of the forewing with similar markings. The outer margins of the fore- and hindwings are toothed. The butterfly is found along forest borders and in forests. The caterpillar feeds on species of passion flower *(Passiflora)*. In Indonesia it goes as far south as Sulawesi (Celebes).

**myrina** Felder
Indonesia — 80–110 mm. (3·15–4·33 in.)
This butterfly has been collected only on the island of Sulawesi (Celebes). The females are larger than the males but lack the blue near the wing base and the general purplish suffusion. Little seems to be known about their biology, although it is said to be strongly scented when alive and for a short while after death. **185m**

**Chaerocina** Rothschild & Jordan (SPHINGIDAE)
The few species of this genus of hawk-moths are restricted to Africa.

**dohertyi** Rothschild & Jordan
E. Africa — 80–100 mm. (3·15–3·94 in.)
This species was described originally from material captured between 1980 and 2740 m. (6500–9000 ft) on the Kikuyu escarpment of E. Africa. More recent material, also from high altitudes. has led to the description by R. H. Carcasson of a new subspecies from Zambia and neighbouring states. *C. dohertyi,* named after its collector, is somewhat like an enormous *Hyles calida* (illustrated) or *H. gallii.* It has greenish brown forewings, with a single, oblique dark brown line, and pink hindwings with a black base and a black band near the outer margin.

CHALKHILL BLUE see **Lysandra coridon**
**chalybacma** see **Epicephala**

**Chamaelimnas** Felder (NEMEOBIIDAE)
About 10 species belong in this Central and tropical S. American genus of butterflies. They are mimics, together with species of the moth genus *Cyllopoda* (Geometridae), of day-flying Dioptid and Ctenuchid moths.

**joviana** Schaus
Bolivia, Peru — 34–36 mm. (1·34–1·42 in.)
Both fore- and hindwings are black above, each with a bright red band extending from the base to the outer margin of the wing, that on the forewing slightly angled in the middle. *C. joviana* is a mimic of the Dioptid *Josia fulva* which is also mimicked by one or more species of the Nemeobiid genus *Mesenopsis.* Nothing is known about the life history of this species.

**splendens** Grose-Smith
Bolivia — 25–32 mm. (0·98–1·26 in.)
There are several similarly coloured species of Ctenuchidae and some Hypsidae which fly with *C. splendens* and are probably its models in a mimetic association of species. They all have black or dark brown wings crossed by a single oblique yellow or orange band and at least partly iridescent blue hindwings.

**chaon** see **Amaxia**
CHAPMAN'S GREEN HAIRSTREAK see under **Callophrys rubi**
CHARACTER MOTH, SETACEOUS HEBREW see **Amathes c-nigrum**
CHARACTER MOTH, CHINESE see **Cilix glaucata**

**Charagia** Walker (HEPIALIDAE)
A genus of some 20 species in Australia and New Guinea. These are in the group popularly known as micro-moths.

**daphnandra** Dodd
Australia — 55–100 mm. (2·17–3·94 in.)
This green species is a typical Hepialid shape. There are larger species in the same genus but this is one of the most brightly coloured. The large caterpillar bores into the stems of trees, killing young ones and often seriously damaging the larger ones. **7b**

**mirabilis** Rothschild
Australia — 120–170 mm. (4·72–6·69 in.)
This is a colourful green species with distinctly different males and females. The males are pale lime-green and are smaller than the females and have white hindwings. The female's forewings are a darker green with grey-brown lines across and an intricate circular pattern; the hindwings are yellowish brown. This moth is the same general shape as *C. daphnandra.* It is known only from N. Australia.

**scotti** Scott
Australia 70–90 mm. (2·76–3·54 in.)
This is a common species, one of the many with green forewings. The original specimens were collected in Australia in the 19th century and the species itself was described a hundred years ago.

**Charaxes** Ochsenheimer: syn. *Eriboea* (NYMPHALIDAE)
The butterflies of this genus are particularly common in Africa, although occurring further east in the tropics with one species, *C. jasius* in the Mediterranean region. They are well patterned and colourful, often with tails on the hindwing. The forewings are broad and rather triangular. They feed on rotten fruit or tree sap, the males regularly visiting patches of damp mud, possibly to drink or in search of minerals. The genus belongs to the subfamily Charaxinae. Generally they are very powerful fliers. The caterpillars have a pair of retractile forked processes behind the head.

**ameliae** Doumet
W. Africa 88 mm. (3·46 in.)
This is found from the W. African forests to Malawi and Uganda. The male, illustrated, has the electric blue replaced in the female by white, with the rest of the wings brown. In both sexes the underside is a pale greenish colour with a prominent dark pattern, and, in the female, a white band across the hindwings and part of the forewings. The chrysalis of this butterfly is described as resembling a gooseberry. **237**a

**bohemani** Felder LARGE BLUE CHARAXES
E. and South Africa, Zaire 74–105 mm. (2·91–4·13 in.)
This bright blue *Charaxes* is readily recognized. The female has a large white patch across the forewing and slightly less blue than the male. The undersides of both are a dull grey with a mauvish tinge. There are 2 related, but smaller, blue species but these have reddish brown undersides. The caterpillars feed on *Afzelia* (Leguminosae) and *Sorghum* (Gramineae). The butterfly is generally found in bush areas and not in forests. **237**b

**eupale** Drury GREEN CHARAXES
W. Africa to Kenya 54–60 mm. (2·13–2·36 in.)
The underside of the wing of this species is dark green. The butterflies are common in forest areas feeding on rotting fruit or any decaying matter. They are easily attracted to bait. Females are larger than males but otherwise similar. There are subspecies with similar patterns found in Zaire. *C. eupale* is found from Sierra Leone and Guinea to Kenya, south to Tanzania, Angola and Zaire. **237**c

**fabius** Fabricius BLACK RAJAH
Sri Lanka, India, Pakistan, Burma 70–80 mm. (2·76–3·15 in.)
This is a typical *Charaxes* shape with 2 tails. The wings are brownish black with yellow spots down the middle of the forewing, continued as a yellow line on the hindwing. There are a series of small yellow spots along the margin. The underside is pale grey-brown with darker marks and a white band over the hindwing and part of the forewing. The margin of the wing has yellow patches edged with black. The females have longer tails than the males but are otherwise similar. The caterpillar feeds on *Tamarindus* (Leguminosae).

**jasius** Linnaeus TWO-TAILED PASHA
Mediterranean coasts, Ethiopia, equatorial Africa, including Sudan and Uganda 76–85 mm. (2·99–3·35 in.)
Several subspecies of this butterfly have been described including *C. j. epijasius* Reiche and *C. j. saturnus* Butler. These latter 2 subspecies have at times been regarded as distinct species from *C. jasius*. The butterfly is found from the coast of Greece, including the larger islands in the Mediterranean. The bright green caterpillar has 2 rounded eyespots on the 6th and 8th segments and a fine yellow line laterally and with 4 reddish horns on the head. It feeds on *Arbutus* (Ericaceae) in the Mediterranean area and the adult is on the wing in May and June with a second brood in August-September. The females are generally larger than the males. **237**e

**opinatus** Heron
Africa 55–65 mm. (2·17–2·56 in.)
A brown *Charaxes* with a reddish brown margin to the hindwing and the 2 tails similarly coloured. There is a row of blue spots inside, round the hindwing margin. The female is similar but has a broad white band through the hindwing and part of the forewing with some white spots near the apex of the cell. This is a rare mountain *Charaxes* known only from the forests on the Ruwenzori and Kigezi highlands in Uganda.

**pollux** Cramer BLACK BORDERED CHARAXES
Tropical Africa S. to Mozambique 76–88 mm. (2·99–3·46 in.)
This is a widespread species found across much of central and E. Africa. It is common in Uganda, where the specimen figured was photographed. The caterpillars feed on *Bersama* (Melianthaceae) and *Sorindeia* (Anacardiaceae). **227**

**polyxena** Cramer TAWNY RAJAH
India, Pakistan, Sri Lanka, Burma 90–115 mm. (3·54–4·53 in.)
The female of this beautiful *Charaxes* has a dark purple-black border to the orange-brown forewing and a broad white band from the front almost to the back of the forewing. The hindwing has only a trace of this band but the margin has black patches with white spots. The male is darker and has a white band across the fore- and hindwings and a blue tinge to the outer edge of it. The hindwing margins have an orange border. The underside is strongly patterned. The female has one tail, the male two. The caterpillar, which is dark green with 4 red spiny horns on the head and red spots on the abdomen, feeds on species of Anonaceae, Meliaceae and Leguminosae. This is a jungle species, with the butterfly flying fast round the treetops. Several subspecies have been described.

**smaragdalis** Butler: syn. *C. butleri*
W. Africa 80 mm. (3·15 in.)
This species is common from Sierra Leone to Ghana. The male is generally a rather rich dark blue, the female usually a lighter colour. There is considerable variation in the pattern on the wings, particularly on the margin of the hindwings and many subspecies and forms have been described. The underside of the fore- and hindwings in both sexes is blue with a strong pattern. The female has white on the forewing underside.

CHARAXES see **Charaxes, Hadrodontes varanes**
**chariclea** see **Clossiana**
**charitonius** see **Heliconius**
**charltonius** see **Parnassius**
**charonda** see **Sasakia**

**Chazara** Moore (NYMPHALIDAE)
This genus is one of the 'Browns' (Satyrinae), closely related to and sometimes considered a subgenus of, *Satyrus*.

**briseis** Linnaeus HERMIT BUTTERFLY
N. Africa, Spain, S. France to W. Asia, Iran 42–60 mm. (1·65–2·36 in.)
A grey-brown butterfly with creamy white band on the forewing interrupted by the veins and with two black, white centred, eyespots. The hindwing is similar, with a broad creamy white band, but this is not so interrupted and there are no eyespots. The underside forewing has more creamy white than the upper but the hindwing is a grey-brown with faint black marks and indistinct middle lighter band. Although this is quite rare in S. Europe it is quite common in N. Africa in dry stony places from lowlands to 1830 m. (6000 ft). The caterpillars feed on various grasses and the adults are about from June onwards.

CHECKERED SKIPPER see **Pyrgus communis**
CHECKERED WHITE see **Pontia callidice**
CHECKERSPOT, SILVERY see **Chlorosyne nycteis**
CHEQUERED SKIPPER see **Carterocephalus palaemon**
CHEQUERED SWALLOWTAIL see **Papilio demoleus**

**Cheritra** Moore (LYCAENIDAE)
Only a few species are known. They are all tropical Asian in distribution.

**freja** Fabricius COMMON IMPERIAL
Sri Lanka and India, E. to Borneo 38–42 mm. (1·5–1·65 in.)
This long-tailed species is a fairly common forest butterfly. Both sexes are brown above, the male shot with purple. The amount of white on the hindwing varies between the subspecies. Wet forests are the home of this species in India, but drier country is the habitat in Sri Lanka. The long white tails of the hindwing are conspicuous in flight. Its caterpillars have been reported from various plants, including species of the family Leguminosae.

CHERRY CASEBEARER see **Coleophora pruniella**

**Chersonesia** Distant (NYMPHALIDAE)
This is a small genus of butterflies found in S.E. Asia. Some of the species have been bred from *Ficus* (Moraceae).

**risa** Doubleday COMMON MAPLET
India, Pakistan, Burma 40–45 mm. (1·57–1·77 in.)
This is a reddish brown butterfly with almost parallel lines running from the forewing continuing across the hindwing. Usually there are 8 or 9 of these dark lines across each wing. The butterflies tend to settle on the underside of leaves where they are less conspicuous. This is generally a lowland species but nothing seems to be known of its biology.

**cherubina** see **Doxocopa**

**Chesias** Treitschke (GEOMETRIDAE)
A genus of moths with species in Europe, Asia and N. Africa. The caterpillars are typical of the family.

**rufata** Fabricius BROOM-TIP
Europe including Britain, Turkey 31–33 mm. (1·22–1·3 in.)
Several subspecies of this moth have been described, 2 of which occur in Britain. The dark marks on the forewing gave rise to the older English name of 'Chevron' for this species (but this is now used for a different species). The caterpillar is greenish blue above, pale on the underside. It has a dark line on the back and 2 white lines on the sides. It feeds on broom (*Cytisus*). The moth is out in May and July but its time of appearance varies. **287**

CHESTNUT TIGER see **Danaus sita**

**Chetone** Boisduval (HYPSIDAE)
About 11 species are placed in this C. and tropical American genus of day-flying moths. Most species are apparently mimetic partners of various species of Heliconiinae, Ithomiinae and other butterflies and moths. Nothing seems to be known about the life history of these moths.

**angulosa** Walker
C. America including S. Mexico, tropical S. America 65–90 mm. (2·56–3·54 in.)
Numerous butterflies are associated with the day-flying *C. angulosa* in a well known mimetic association of species (see *Melinaea lilis*) which includes several species of Ithomiinae, 1 of Danainae, 2 of Heliconiinae, 2 of Nymphalinae and 2 of Pieridae. In the so-called subspecies *C. angulosa felderi* (possibly a different species) the markings in the outer half of the forewings are white and partly transparent. **376**p

**histrio** Felder
N.W. South America, Amazon Basin 70–100 mm. (2·76–3·94 in.)
One form of the subspecies *C. h. hydra* is a mimetic partner of various species of *Heliconius* (Heliconiinae), *Papilio bachus* (Papilionidae), various species of Ithomiinae (eg. in *Mechanitis* and *Napeogenes*) and the Castniid species *Duboisvalia pellonia* Druce. *C. h. histrio* (illustrated) is part of another mimetic association which includes a species of *Lycorea* (Danainae). **376**n

**phyleis** Druce
N.W. South America, Amazon Basin 65–75 mm. (2·56–2·95 in.)
One of the most impressive examples of mimicry between Hypsids and species of the butterfly subfamily

Heliconiinae is that which exists between *C. phyleis* and various forms of *Heliconius* species. One unillustrated form of *C. phyleis* is a much paler orange colour at the base of the forewing than the illustrated specimen. The life history of this species is unknown. **381**c

CHEVRON see **Chesias rufata**
CHIEF BUTTERFLY see **Amauris echeria**
**chilo** see **Acraea**
**chimaera** see **Ornithoptera**
CHIMNEY-SWEEPER see **Odezia atrata**
CHINA MARK MOTHS see **Parapoynx, Nymphula**
CHINA-MARK, SMALL see **Cataclysta lemnata**
CHINESE CHARACTER see **Cilix glaucata**
CHINESE OAK SILK-MOTH see **Antheraea pernyi**

**Chionaema** Herrich-Schäffer (ARCTIIDAE)
There are about 200 species in this genus of moths of the subfamily Lithosiinae. It is found in Africa, India, China, Japan and tropical S.E. Asia to New Guinea and Australia. The colour pattern of most species is similar to that of the illustrated *C. moupinensis*. *C. coccinea* has been reported as a pest of tea *(Thea)* in N.W. India.

**moupinensis** Leech
Central and Himalayan China 38–47 mm. (1·5–1·85 in.)
This rather unusually coloured species of *Chionaema* was at one time regarded as an aberration of the Chinese *C. phaedra* Leech which is similar in pattern but has the orange areas replaced by reddish orange. The recorded localities for this species range from 920–1680 m. (3000–5500 ft). The caterpillar is unknown. **364**n

**scintillans** Rothschild
Bismarck Archipelago (New Ireland) 25–34 mm. (0·98–1·34 in.)
This is one of a group of about 20 species, all brightly coloured and atypical of their genus. Its latin name refers to the iridescent green and gold patches on each side of the red patch on the forewing. Very few specimens of this species exist in collections. Nothing is known about the life history. **364**l

**chiridota** see **Lacosoma**
**chiron** see **Marpesia**

**Chlanidophora** Berg (ARCTIIDAE)
Two species are placed in this genus of moths; *C. patagiata* (illustrated) and *C. culleni* Brethes. Both are found in temperate S. America.

**patagiata** Berg
Argentina (Patagonia) 35–40 mm. (1·38–1·57 in.)
This is apparently a solely temperate S. American species with no close relatives in the tropics of S. America. It is unlikely to be confused with *C. culleni*, also an Argentinian species, which is roughly similar in pattern, but has hardly a trace of red coloration. Nothing seems to have been published about the life history of *C. patagiata*. **369**f

**Chliara** Walker (NOTODONTIDAE)
There are about 10 species in this genus of C. American and tropical S. American moths. Most species have metallic gold markings on the forewings.

**croesus** Cramer
C. America, tropical S. America 45–54 mm. (1·77–2·13 in.)
The presence of numerous conspicuous, metallic gold markings on the forewings is probably the reason why Cramer chose for this species the name of the reputedly fabulously wealthy, 6th century King Croesus of Lydia. Nothing is known about the life history. **363**

**chloridice** see **Pontia**

**Chloroclystis** Hübner (GEOMETRIDAE)
PUG MOTHS
A widespread genus of relatively small Geometrid moths with very many species in Europe, Asia and Africa, but only one species is at present known from N. America.

**rectangulata** Linnaeus GREEN PUG
Europe including Britain, recently N. America. 17–21 mm. (0·67–0·83 in.)
This is a common moth throughout most of Britain and the rest of Europe. It is very variable in pattern and there are many named varieties. Generally the fore- and hindwings are green with thin blackish lines with similar patterns on both wings. In some specimens there are few traces of pattern and the wings are grey-brown. The adult moth is out in June and July. The caterpillar, which is pale yellow-green with a reddish or dark line along the back, feeds on the flowers of apples *(Malus)* and pears *(Pyrus)* in April and May and can often be very destructive. In 1972 the first specimens of this species were collected in Nova Scotia. It is thought to have been accidentally introduced.

**Chlorostrymon** Clench (LYCAENIDAE)
This is a genus of essentially tropical American, tailed hairstreak butterflies, 3 species of which reach S. U.S.

**maesites** Herrich-Schäffer MAESITES HAIRSTREAK
U.S. (S. Florida) the Greater Antilles (including the Bahamas) 20–23 mm. (0·79–0·91 in.)
The wings of this species are a brilliantly iridescent purplish blue above in the male, less bright in the female which has a broad, black marginal band on both wings. Like many other species of its family, *C. maesites* is attracted to the flowers of Spanish Needles *(Bidens)*. There are apparently two broods each year. A closely related species, *C. telea* Herrich-Schäffer, occurs from Texas S. into S. America. The life history of neither species is known.

**Chlorosyne** Butler (NYMPHALIDAE)
These are fritillary butterflies from America which were formerly associated with European fritillaries in the genus *Melitaea*. They are not now regarded as so closely related and are placed in a separate genus.

**lacinia** see under **C. saundersii**

**nycteis** Doubleday SILVERY CHECKERSPOT
Canada, U.S. 38–42 mm. (1·5–1·65 in.)
This butterfly is common in meadows and roadsides from S. Canada through the Maritime Provinces W. to Manitoba, S. to New Jersey and in N. Carolina, Ohio, Indiana, Illinois, Missouri and Kansas. The caterpillar, which is velvet black with a dull orange stripe on the back and purple streaks on the side, is covered in white spots, each with a black hairy spine. It feeds on *Helianthus*, *Aster* and *Actinomeris* (Compositae). The butterfly is out in early June, with one brood in the N., 2 in the S. The black pattern on the upperside leaves a broad orange-brown area across the middle of the fore- and hindwings; the underside hindwing is very white with a satiny lustre. **163**e

**saundersii** Doubleday LITTLE SOLDIER
Mexico to Paraguay, W. Indies 45 mm. (1·77 in.)
This butterfly is orange-brown on the upperside with wide black margins round both forewings and a black patch at the base. The hindwings are mostly orange with a black border. Two rows of white and reddish brown spots run along the margins of the forewing. The female is larger than the male. The butterfly is a relatively slow flier, alternately flapping and gliding in the sunshine. It is often abundant with considerable variation in pattern between specimens from different localities. Now regarded by some authorities as a subspecies of *C. lacinia* Geyer.

CHOCOLATE PANSY see **Junonia iphita**
CHOCOLATE TIGER see **Danaus melaneus**

**Choerapais** Jordan (AGARISTIDAE)
There is a single known species in this genus of moths.

**jucunda** Jordan
Angola 35–52 mm. (1·38–2·05 in.)
The female of this disinctively patterned species (not illustrated) has a dark brown, crescent-shaped mark on the middle of the hindwing and an orange-spotted, black band along the outer margin of this wing; the forewing is similar in colour pattern to the male. Nothing is known about the life history. The author of *C. jucunda*, Karl Jordan, was one of the foremost entomologists of the first half of this century. **399**l

**Choreutis** Hübner (GLYPHIPTERIGIDAE)
Mostly these micro-moths are found in Europe or N. America with a few species in Africa and India. They are very colourful with iridescent scales on the wings which gleam when the light catches them. The N. American species often have tiny golden patches on the wings. The biology of most known species is unknown. Some European species have been bred from Compositae and Labiatae.

**dolosana** Herrich-Schäffer
Europe, Algeria 10 mm. (0·39 in.)
The hindwings are black, metallic-looking but without patches. The caterpillar mines in the leaves of *Aristolochia* (Aristolochiaceae). When fully grown it pupates in a small circular flattened cocoon. **34**k

**Chorinea** Gray: syn. *Zeonia* (NEMEOBIIDAE)
This is a small C. and S. American genus of about 7 species. Some authors have likened these butterflies, with their partly transparent wings, to bees or wasps which predators generally avoid. This resemblance therefore may have some protective value to the butterflies.

**faunus** Fabricius BEE BUTTERFLY
C. America, tropical S. America 28–32 mm. (1·1–1·26 in.)
The fast wing-beat and transparent wings of this butterfly lend it a distinctly bee-like appearance. As in other species of *Chorinea*, the red element of the wing pattern may be replaced by orange, a deep yellow, or, rarely, by violet. The life history is not known. **268**f

**sylphina** Bates
Bolivia, Ecuador, Peru 34–40 mm. (1·34–1·57 in.)
This is similar to the illustrated *C. faunus* except for the broader, transverse band and marginal bands, and the more extensive area of red on the anal area of the hindwing. *C. heliconides* Swainson from Brazil closely resembles *C. sylphina* and *C. faunus* in colour pattern except for the broken red band on the hindwing. Nothing is known about its life history.

**Choristoneura** Lederer (TORTRICIDAE)
Species of this genus of micro-moth range over the whole of the N. Hemisphere, with many species in N. America.

**fumiferana** Clemens SPRUCE BUD WORM
U.S., Canada 20–25 mm. (0·79–0·98 in.)
The forewings are mottled grey with brown or reddish-brown, the hindwings are grey. The brown caterpillar has paler sides, two rows of white dots on the back and a V-mark on the posterior part of the thorax. In the early stages it mines into conifer leaves; later it attacks the young buds, cones or twigs. The needles are cut off and a conspicuous web is formed with the caterpillars living in tubes. The caterpillars feed on Balsam Fir *(Abies balsamea)*, spruce *(Picea)* and on several other conifers. It is regarded as the third most destructive insect in the U.S. after the Cotton Boll Weevil and the Corn Earworm. It is widespread in Canada where it is also a defoliator of conifers. There are 2 forms, the Western and the Eastern form, now recognized which differ in colour.

**Chorodnodes** Warren (GEOMETRIDAE)
Only one species has been described in this genus. The moth is known only from W. Africa.

**rothi** Warren
Africa 42–52 mm. (1·65–2·05 in.)
This species has been collected in Nigeria, Guinea, Sierra Leone and Zaire and also on the island of Sao Thôme in the Gulf of Guinea. The species was named after the collector, Dr Roth, who collected thousands of specimens in many parts of Africa and whose name is perpetuated in the many new species he discovered. **294**x

**Chresmarcha** Meyrick (TORTRICIDAE)
A genus of micro-moths with only 3 species, all from

New Guinea. They are rather similar, silvery white and delicate looking. Nothing is known of their biology.

**delphica** Meyrick
New Guinea 18–20 mm. (0·71–0·79 in.)
The related species from New Guinea are similar to this but have a grey hindwing. The females generally have some grey on the apex of the hindwings (the specimen illustrated is a male). This species was collected at altitudes of 1520 m. (5000 ft) and over by a famous collector, A. S. Meek, who collected insects for Lord Rothschild in the early part of this century. **11**f

CHRISTMAS BUTTERFLY see **Papilio demodocus**
**christophi** see **Mirina**

**Chrysamma** Karsch (LIMACODIDAE)
This genus of moths has 3 species in it, all found in Africa. Nothing is known of their biology.

**purpuripulcra** Karsch
Rhodesia, Kenya, Angola, Sudan, Ghana, Nigeria 28–38 mm. (1·1–1·5 in.)
Related African species have white wings. This species is very constant in pattern over its range, with the females larger than the males. In spite of being such a widespread species nothing appears to be published about its biology. **54**f

**chrysias** see **Eretmocera**
**chrysippus** see **Danaus**

**Chrysiridia** Hübner (URANIIDAE)
There are 2 species in this genus of day-flying moths. One is African, the other Madagascan; both are brilliantly iridescent in colour. Its closest relative is probably the similarly coloured and patterned S. American genus *Urania.*

**croesus** Gerstaecker
E. Africa, Zanzibar 50–68 mm. (1·97–2·68 in.)
Very similar in colour pattern to the illustrated *C. ripheus,* the chief differences in pattern are the more finely striate appearance of the forewing distal to the broad green transverse band, and the striate, rather than spotted pattern of the outer part of the hindwing.

**riphearia** Drury
Madagascar 80–103 mm. (3·15–5·06 in.)
This species outshines many of the spectacular *Papilio* butterflies which it resembles in the shape of its wings. It has been called the most beautiful moth in the world by several authors and found favour with Victorian jewellers who used the iridescent parts of its wings in costume jewellery together with kingfisher feathers and the wings of *Morpho* butterflies. Less use is now made of *C. ripheus* wings, but *Morpho* wings are still widely used commercially. *C. ripheus* flies throughout the year, but is commonest from May to July; it often visits flowering trees. Its caterpillar is yellow and black, with several long clubbed hairs; it feeds on the foliage of *Omphalea,* a poisonous genus of Euphorbiaceae. **314**e

**chrysitis** see **Plusia**

**Chrysocale** Walker (CTENUCHIDAE)
There are 11 described species in this genus of tropical S. American moths. The wings of most species are iridescent green, blue or bronze, with a few pale, translucent markings. Nothing is known about their life history.

**gigas** Rothschild
N.W. S. America 55–65 mm. (2·17–2·56 in.)
The forewings of *C. gigas* are a beautiful unmarked iridescent green; the hindwings are brown, with an iridescent green patch near the middle of the outer margin.

**ignita** Herrich-Schäffer
N. S. America 46–51 mm. (1·81–2·01 in.)
This is probably the most richly coloured species of a colourful genus. The pale markings on the wings are transparent and there is a red sheen on the forewing when viewed at certain angles. **364**c

**Chrysochlorosia** Hampson (ARCTIIDAE)
The 5 or so species of the Central American and tropical S. American moths have brilliant, iridescent, green or blue forewings and are amongst the most colourful members of the subfamily Lithosiinae.

**splendida** Druce
Ecuador 26–32 mm. (1·02–1·26 in.)
This is a very rare species in collections. The under surface of the wings is more greenish than the beautiful turquoise upper surface. Nothing is known about the life history of this species or the significance of the wing colour, which matches that of some S. American Zygaenidae. **382**f

**Chrysoclista** see **Glyphipteryx**
**chrysophanes** see **Conopia**

**Chrysopoloma** Aurivillius (CHRYSOPOLOMIDAE)
Little is known of the biology of the species in this genus of African moths.

**similis** Aurivillius
South Africa 42–50 mm. (1·65–1·97 in.)
The female has simple antennae (not pectinate) and pale grey-brown forewings and is generally larger than the male. The male illustrated has paler hindwings than many specimens, some of which have very dark brown hindwings. The biology is unknown. **48**b

**Chrysopolomidae**
A small family of moths, related to the Zygaenidae and Limacodidae. They are known only from Africa. See *Chrysopoloma.*

**chrysorrhoa** see **Scirpophaga**
**chrysorrhoea** see **Euproctis**

**Chrysoteuchia** Hübner (PYRALIDAE)
CRAMBIDS or GRASS MOTHS
The Pyralid moths in this genus, about 30–40 species, occur in large numbers in pasture lands where at times the caterpillars cause severe damage to the grass and reduce the amount of grazing land. Many also feed on winter and spring sown barley or wheat. The adult moths mostly fly at dusk but are easily disturbed during the day when they dash off in a zig-zag flight. The caterpillars feed at or just below the soil surface.

**culmella** Linnaeus: syn. *C. hortuella*
GARDEN GRASS VENEER 18–22 mm. (0·71–0·87 in.)
Europe including Britain, Asia including Japan
This species has been separated into several subspecies in different parts of Europe. The moths are a typical crambid shape (see under Crambidae). The caterpillar, apart from causing trouble to farmers, can also be a nuisance to golfers by ruining the greens. It is a translucent, yellowish white creature, often with a green tinge, with a shining yellow-brown head, thorax and tip of the abdomen. This species is common and widespread throughout Europe and Asia but does not occur in N. America. There is a very similar species there which was formerly believed to be *C. culmella* but it is now known to be a distinct species. **58**

**chrysotheme** see **Colias**

**Chrysozephyrus** Shirozu Yamamoto (LYCAENIDAE)
About 30 species are included in this genus of butterflies. They are found in Afghanistan and N. India to S. China, Formosa and Japan.

**ataxus** Doubleday WONDERFUL HAIRSTREAK
N. India, China, Russia, Japan 40–46 mm. (1·57–1·81 in.)
The male of this rare species is iridescent green and black above; the female (not illustrated) is brown, with a large blue, central patch and usually 2 preapical, orange markings. The under surface of both sexes is mostly white, with brown and orange markings. *C. ataxus* has been described frequently as the most beautiful species of hairstreak butterflies. It is normally restricted to hilly and mountainous country at elevations of 1670 m. (5500 ft) and above. **244**s

**syla** Kollar SILVER HAIRSTREAK
Afghanistan, N. India, Sikkim, Bhutan 42–45 mm. (1·65–1·77 in.)
The male of *C. syla* is an iridescent, golden green above, with a slight violet sheen and brown borders. The female forewing is pale blue above, with an oblique, white bar in the middle and black borders; its hindwing is brown, with pale, greenish blue rays near the outer margin. The under surface of both sexes is an unusual silver colour, with brown markings, and a black spot and an area of orange at the base of the hindwing tail. Most captures of this butterfly have been made between 1830–3350 m. (6000–11,000 ft) in the Himalayas. Its caterpillar is known to feed on oak *(Quercus).*

**Cicinnus** Blanchard (MIMALLONIDAE)
This is a rather heterogeneous group of about 60 almost exclusively Central and S. American moths, but with a single N. American and Mexican species, *C. melsheimeri.*

**despecta** Walker
S. Brazil, Argentina 34–60 mm. (1·34–2·36 in.)
The female of this fairly common species differs from the illustrated male in the absence of transparent patches on the wings and a greater concentration of black speckles. Some specimens of both sexes are marked with pink on the upper surface of the abdomen and at the inner margin of the hindwing immediately distal to the transverse band. The caterpillar is without processes or conspicuous hairs and is tapered towards the head. The curious, tough cocoon is firmly secured at its base to a twig; it is open at both ends, the upper opening protected by a hood-like projection of the cocoon. **318**k

**melsheimeri** Harris MELSHEIMER'S SACK BEARER
Canada (S. Ontario), N.E. U.S. S. to Mexico 32–48 mm. (1·26–1·89 in.)
This is the only species of its genus found in N. America. It is similar in wingshape to the illustrated S. American *C. despecta,* but has less acutely pointed forewings, a less oblique, transverse line and lacks transparent patches in the middle of the forewing. The ground-colour of the wings is a pinkish grey. The caterpillar feeds on oak *(Quercus).*

**Cilix** Leach (DREPANIDAE)
*Cilix* is found in Europe and temperate Asia, including India, E. to E. U.S.S.R. and Japan. There are 6 known species, all of them small, white and similar in colour pattern and with atypical (for Drepanids) rounded wings. The most romantically named member is *C. tatsienluica,* described from the Himalayan village of Ta-tsien-lu in the Chinese province of Szechwan. *C. asiatica* is a minor pest of non-citrus fruit trees in Israel.

**glaucata** Scopoli CHINESE CHARACTER
Europe, N. Africa, Cyprus 18–24 mm. (0·71–0·94 in.)
This is one of several species of moths which, when at rest, resemble bird-droppings and are probably well camouflaged in natural conditions. The older name for this species, 'Goose-egg' probably refers to the shape of the greyish brown markings on the middle of the forewing. **270**

**Cinabra** Sonthonnax (SATURNIIDAE)
The moths of this African genus are very similar to those of the related *Automeris,* of the New World.

**hyperbius** Westwood BANDED EMPEROR
Kenya to Transvaal 75–120 mm. (2·95–4·72 in.)
The forewing of this moth can be either yellow, orange or dull red; the hindwings yellow, orange or pink. Females have one or more transverse bands of purplish grey on the forewing. Both sexes have a conspicuous black-edged eyespot on the middle of the hindwing. The bright green caterpillar bears several short black spines and is marked with red, orange and white. It feeds on species of *Brachystegia, Julbernardia* and *Protea.*

**cinctaria** see **Cleora**
**cinerealis** see **Neadeloides**
**cingalensis** see **Timyra**

CINGALESE BUSH-BROWN see **Mycalesis rama**
**cingulatus** see **Agrius**
CINNABAR see **Tyria jacobaeae**
**cinxia** see **Melitaea**
**cippus** see **Tajuria**
**cipris** see **Phoebus**
CIRCE see **Hestinalis nama**
**circe** see **Brintesia**

**Cithaerias** Hübner: syn. *Callitaera* (NYMPHALIDAE)
These delicate looking butterflies are found in C. and S. America. They have thin bodies and transparent, thinly scaled and rather rounded wings. They are found in the tropics where they can be seen flying slowly in the early morning. When they land their gauzy wings make their actual shapes difficult to see. The genus is in the subfamily Satyrinae.

**aurora** Felder
Brazil, Colombia, Ecuador 50–60 mm. (1·97–2·36 in.)
This delicate butterfly is almost invisible with its transparent ghost-like wings. It is found in forests, generally flying slowly, low down, in the early morning or late evening. Nothing is known of its biology. The female has less pink on the hindwing than the male. **159**

**esmerelda** Doubleday
Brazil, Peru 52 mm. (2·05 in.)
This species, like most others in the genus, lives in tropical forests in S. America. The colour of the patch on the hindwings varies from blue to purple but the completely transparent wings make this one of the more unusual looking butterflies, especially in the Satyrinae where the majority of species are brown in colour. In a famous passage about this butterfly, the 19th century explorer naturalist, H. W. Bates, after seeing this butterfly in the Amazon forests, wrote 'It has one spot only of opaque colouring on its wings, which is of a violet and rose hue; this is the only part visible when the insect is flying low over dead leaves in the gloomy shades where alone it is found, and it then looks like a wandering petal of a flower'. **174d**

**Citheronia** Hübner (SATURNIIDAE)
This is a genus of just over 20 species of conspicuously coloured moths. Three species are found in the U.S.; one of them, *C. splendens*, is chiefly known from specimens caught in Peña Blanca, a favourite haunt of moth collectors in S. Arizona. The remaining species are C. and S. American in distribution.

**regalis** Fabricius ROYAL WALNUT or REGAL MOTH, HICKORY HORNED DEVIL
E. U.S. 100–130 mm. (3·94–5·12 in.)
The huge multicoloured horned caterpillar of this species, known as the Hickory Horned Devil, is even more impressive than the moth. It has been reported as a pest of cotton *(Gossypium)* in the S. U.S. earlier this century, but its chief foodplants are walnut *(Juglans)*, hickory and pecan *(Carya)*, sumach *(Rhus)* and other trees. The adult is similar to the illustrated *C. splendens* in general features of the pattern but is a much more reddish moth with yellowish white wing markings, and a mainly brownish orange hindwing and abdomen.

**splendens** Druce
U.S., Mexico 100–130 mm. (3·94–5·12 in.)
*C. splendens* is known only from Mexico and from S. Arizona. The illustrated specimen is a male of the subspecies *C. s. sinaloensis* captured in Peña Blanca, Arizona. The flight period of this species seems to be restricted to July and August. Its caterpillars have been recorded from *Schinus*, wild cotton *(Gossypium)* and walnut *(Juglans)*. **333e**

**citrana** see **Argyrotaenia**
**citri** see **Prays**
**citrodesma** see **Metachanda**
CITRUS SWALLOWTAILS see **Papilio demodocus, P. thoas**

**Cizara** Walker (SPHINGIDAE)
There are 3 species in this genus of hawk-moths. Its range extends from India to Australia.

**ardeniae** Lewin
N. Australia 52–68 mm. (2·05–2·68 in.)
The adult moth is similar in general plan to *Rethera komarovi* (illustrated), but has a much darker green forewing, with white-edged, green areas and chiefly dark brown hindwings unlike the latter. The caterpillar feeds on *Grevillea, Cissus* and *Embothrium*.

**clarissa** see **Terinos**
**clarus** see **Epargyreus**
**clathrata** see **Semiothisa**
**claudia** see **Agrias**
CLEARWING MOTHS see **Alcathoe, Bembecia, Conopia, Hemaris, Parathrene, Sesia, Synanthedon**
**cleobaea** see **Eueides**
**cleopatra** see **Gonepteryx**
CLEOPATRA see **Gonepteryx cleopatra**

**Cleora** Curtis (GEOMETRIDAE)
A large genus of moths found in the tropics of America, Africa and Asia with species in Europe and N. America. In Africa some of the species have caterpillars which defoliate introduced trees in softwood plantations.

**cinctaria** Denis & Schiffermüller RINGED CARPET
Europe, temperate Asia including Japan 30–35 mm. (1·18–1·38 in.)
The caterpillar of this moth feeds on heaths *(Erica)* and on birch *(Betula)* and a number of other trees. It is a light green caterpillar with several paler lines on the back. The moth is out in May and June and is locally common in Europe. There are several subspecies, one occurring in Japan while two different subspecies are known from Britain. **285**

**clerckella** see **Lyonetia**
CLIFDEN NONPAREIL see **Catocala fraxini**
CLIPPER see **Parthenos sylvia**
CLOAK, MOURNING see **Nymphalis antiopa**
**cloanthe** see **Catacroptera**
**cloanthus** see **Graphium**
**clodius** see **Parnassius**
**clonia** see **Esthemopsis**
**clorana** see **Earias**
**clorinde** see **Anteos**
CLOSEWINGS see **Pediasia**

**Clossiana** Reuss (NYMPHALIDAE)
A large genus of fritillary butterflies with many species in Europe and temperate Asia and some in N. America.

**chariclea** Schneider ARCTIC FRITILLARY
Arctic Europe, Greenland, N. America 32–36 mm. (1·26–1·42 in.)
This species is circumpolar in distribution. It is very similar to *C. titania* and has been confused with it, although it is smaller, with more pointed forewings and narrower dark markings above. The Arctic Fritillary lives up to its name, having been collected further N. than any other species (81° 41′ N.). The food-plant of the caterpillar is not known. The adults fly at the end of June to July or later if the northern weather is bad.

**euphrosyne** Linnaeus PEARL BORDERED FRITILLARY 36–46 mm. (1·42–1·81 in.)
Europe including Britain, temperate Asia
This butterfly is similar the *C. selene* but is less heavily marked on the upperside. The adults fly in April-May with a second generation in July and August. The caterpillars feed on violets *(Viola)* and are locally common in parts of their range, particularly in woodlands.

**freija** Thunberg FREJYA'S FRITILLARY
N. Europe, N. Asia including Japan, N. America 36–44 mm. (1·42–1·73 in.)
In N. America this species ranges from Alaska to Labrador, S. into Ontario and in the Rockies at or above the timberline to Colorado. Several subspecies are known including a distinct one in Japan. The caterpillar feeds on Cloudberry *(Rubus chamaemorus)* and Bog Bilberry *(Vaccinium uliginosum)* in Europe. In N. America it is recorded on Bearberry *(Arctostaphylos uva-ursi)* and on *Rhododendron aureum* in Japan. The butterfly, which is out in May or June, flies over moors, mountain heaths and tundra. This butterfly must not be confused with *Melitaea arduinna*, Freyer's Fritillary.

**frigga** Thunberg SAGA or FRIGGA'S FRITILLARY
N. Europe, N. Asia, N. America 40–46 mm. (1·57–1·81 in.)
Several subspecies of this species, which is circumpolar in distribution, have been described. In N. America it is widespread in Canada and extends along the Rockies S. to Colorado. The butterfly is out in June and July and flies over sphagnum bogs and moors. The caterpillar feeds on Cloudberry *(Rubus chamaemorus)*. **163g**

**improba** Butler DUSKY-WINGED or DINGY ARCTIC FRITILLARY 30–34 mm. (1·18–1·34 in.)
N. Europe, N. America, not recorded yet from N. Asia
The butterfly is out in July on dry mountain slopes from 370–1070 m. (1200–3500 ft). The foodplant of the caterpillar is unknown. This species is smaller and darker than the other arctic fritillaries with rather indistinct markings on the upperside.

**polaris** Boisduval POLAR FRITILLARY
N. Europe, N. Asia, N. America 36–40 mm. (1·42–1·57 in.)
This, with *C. chariclea*, is one of the true Arctic butterflies although its range extends further south than *C. chariclea*. The butterfly is out in June or July in dry tundra regions. The foodplant of the caterpillar is not known, but is believed to be *Dryas octopetala* (Rosaceae). This butterfly is not common in Europe, but occurs in Greenland and widely in the Canadian Arctic.

**selene** Denis & Schiffermüller SILVER-BORDERED or SMALL PEARL-BORDERED FRITILLARY
Europe including Britain, temperate Asia, N. America 36–42 mm. (1·42–1·65 in.)
This is common over most of its range where the European subspecies lives in woodland clearings and by streams near woods. The American subspecies, (there are several) are common in marshy meadows and brushy swamps. The greenish brown mottled caterpillar feeds on violets *(Viola)*. **195**

**titania** Esper PURPLE LESSER or TITANIA'S FRITILLARY 42–46 mm. (1·65–1·81 in.)
W. Europe, temperate Asia, N. America
This species is found from Canada down the Rockies to New Mexico. Several subspecies have been described. The caterpillar feeds on *Viola* and *Polygonum*. In Europe it tends to fly on woodland margins or clearings at 910-1520 m. (3000-5000 ft), whereas in America it occurs in boggy meadows. The butterfly is out in June-July. The underside of the hindwing is brightly marked with purplish brown, yellow and black. **201**

CLOTHES MOTH, FALSE see **Hofmannophila pseudospretella**
CLOTHES MOTHS see **Tinea, Tineola, Endrosis, Trichaphaga**
CLOUDED
BUFF see **Diacrisia sannio**
FLAT see **Tagiades flesus**
MAGPIE see **Abraxas sylvate**
MOTHER-OF-PEARL see **Salamis anacardii**
SULPHUR see **Colias philodice**
YELLOWS see **Colias**
CLOUDY WING, SOUTHERN see **Thorybes bathyllus**
CLOVER
CASE MOTH see **Coleophora frischella**
HAYWORM see **Hypsopygia costalis**
WEBWORM see **Hypsopygia costalis**
CLUBTAIL, COMMON see **Parides coon**
**clymena** see **Diaethria**
**clytemnestra** see **Hypna**
**clytia** see **Papilio**
**clyton** see **Asterocampa**

**Cnephasia** Curtis (TORTRICIDAE)
A large micro-moth genus, mainly found in the Eurasian land mass, with one species known in N. Africa.

**interjectana** Haworth: syn. *C. virgaureana*
Europe including Britain, Asia, N. America 15–17 mm. (0·59–0·67 in.)
The forewings are greyish or brownish, finely speckled with white and with irregular marks. The hindwings are pale grey and unpatterned. The grey-green caterpillar has small black spots and black thorax. It feeds on plantain *(Plantago)*, *Lathyrus* (Leguminosae) and germander *(Teucrium)*. The moth flies in June and July and is widespread in Europe but little seems to be known of its N. American distribution.

**coccidivora** see **Laetilia**
**coccinata** see **Cymothoe**
**Cochlididae** see **Limacodidae**

**Cochylidae** (PHALONIIDAE)
Related to the Tortricidae, the moths in this family are abundant in N. America and in Europe. Species also occur in S. America and in Japan, India and Indonesia, but only one species is known at present in Australia. See *Aethes, Agapeta, Cochylis, Commophila.*

**Cochylis** Treitschke (COCHYLIDAE)
A large genus of micro-moths found in Europe, N. and S. America and in the Oriental region.

**roseana** Haworth
Europe including Britain, Turkey, Iran 10–14 mm. (0·39–0·55 in.)
The light yellow colour at the front of the forewings becomes rose-pink posteriorly. There is a dark brown central band on the light grey hindwings. The caterpillar feeds on seeds of *Dipsacus* (Dipsaceae) and the moth, which is out in June and August, is common throughout its range.

**cocles** see **Cyrestis**
COCOA MORT BLEU see **Caligo teucer**
COCOA MOTH see **Acrocercops cramerella**

**Cocytia** Boisduval (AGARISTIDAE)
There is at present only one species in this genus of strange, transparent-winged moths. However, at least one of the several subspecies which have been described probably merits specific status.

**durvillii** Boisduval
New Guinea and the associated islands of Aru, Amboim, the Bismarck Archipelago 62–75 mm. (2·44–2·95 in.)
The long antennae are one of the many unusual features of this generally bizarre wasp- or bee-like species. The various subspecies of *C. durvillii* differ from each other most noticeably in the size of the orange patch at the base of the forewing. Nothing seems to be known about the life history of this species. **399**r

**Cocytius** Hübner (SPHINGIDAE)
Five species of this genus of hawk-moths are known, all of them essentially tropical Central and S. American in distribution. Two of them stray into N. America: *C. antaeus,* a regular visitor, and *C. duponchel,* recorded only once, from Texas.

**antaeus** Drury GIANT SPHINX
S. U.S., Central and tropical S. America 140–160 mm. (5·51–6·3 in.)
There are many records from Florida and Texas of this large tropical species. The forewings are yellowish and greenish brown with mostly dark brown and black markings; the hindwings are without scales and transparent in the central area, with a dark brown outer band and orange-yellow at the base of the wings. Its caterpillars are green with a pink line along the back, 5 faint oblique stripes and a pink and grey horn. They feed on the leaves of custard apple *(Annona)*.

CODLIN MOTH, APPLE see **Cydia pomonella**
**codrus** see **Graphium**
**coeculus** see **Iolaus**

**Coeliades** Hübner (HESPERIIDAE) SKIPPERS
Large robust skipper butterflies with broad, short tails on the hindwings, found mainly in Africa.

**forestan** Cramer STRIPED POLICEMAN
Tropical to S. Africa 44–50 mm. (1·73–1·97 in.)
The forewings are a dark greyish brown, and the hindwings have long orange scales at the base and in the middle; the fringe of the hindwing by the 'tail' is orange. Below the hindwing is a broad white band. The antennae have the hooked and slightly swollen end typical of skipper butterflies. This Skipper is found in forests and in the bush, patrolling backwards and forwards, a habit which gave it its popular name. The caterpillar feeds on a very wide range of host families *Combretum* (Combretaceae), *Geranium* (Geraniaceae), *Phaseolus* and *Robinia* (Leguminosae), and *Solanum* (Solanaceae). **80**b

**Coenobasis** Felder (LIMACODIDAE)
An African genus of moths with 7 or 8 species plus one described from S. Sinai.

**amoena** Felder
South Africa 25–35 mm. (0·98–1·38 in.)
Related species occur in the Ruwenzori mountains in W. Uganda, but generally these lack the stripes across the green wings. The underside is a pale yellow-brown, quite unmarked. The caterpillars feed on species of *Acacia*. When they pupate they form a large round cocoon which hardens like a small nut and is generally attached to the stem of the host plant. A small circular patch of the cocoon wall is thinner than the rest and this is where the moth emerges, leaving the pupal case sticking out of the nut-like cocoon. **54**d

**Coenonympha** Hübner (NYMPHALIDAE)
This is a genus of butterflies with many species found in Europe, temperate Asia and a few species in N. America. The genus is in the subfamily Satyrinae and are popularly known as heath butterflies from the type of area in which they are commonly found.

**leander** Esper RUSSIAN HEATH
Hungary, Bulgaria, Romania, Yugoslavia, S. U.S.S.R., Turkey, Iran 32–34 mm. (1·26–1·34 in.)
The butterfly is out in May and June in rough grassy areas from lowlands to several thousand feet. The foodplant of the caterpillar is not known. The original specimens (including the holotype) were collected in the U.S.S.R. in 1784, and the vernacular name is derived from this. **175**

**pamphilus** Linnaeus SMALL HEATH
Europe including Britain, N. Africa, Turkey, Lebanon, Iraq, W. Asia 28–32 mm. (1·1–1·26 in.)
A widespread species with a number of subspecies from different parts of Europe. The butterfly is out in April or May and has several broods each year. It flies in meadows from sea-level to 1830 m. (6000 ft). The caterpillar feeds on various grasses, including *Poa* and *Nardus*. It is a clear green with darker stripes along the back. At times, particularly later in the year in the second or later generation, this can be a very common butterfly found in large numbers in rough pastures and heathlands. **169, 172**

**tullia** Müller LARGE HEATH or RINGLET
Europe including Britain, temperate Asia, N. America 36–40 mm. (1·42–1·57 in.)
The underside is illustrated; on the upperside the hindwings are less grey than in the picture. In N. America several subspecies, popularly known as Ringlets are found. They occur from California, through the western mountains to the eastern States. More subspecies are found across the rest of the range. The butterfly is out in June or early July in bogs, peat mosses and rough meadows amongst Cotton Grass *(Eriophorum)* which is the foodplant of the caterpillar. The caterpillar is green with white dots and white lines along the side. **171**

**Coenophlebia** Felder (NYMPHALIDAE)
This S. American genus of butterflies has only one known species. It is fairly widespread and has been separated into 3 subspecies.

**archidona** Hewitson
S. America 95–120 mm. (3·74–4·72 in.)
The remarkable elongated wing tips of this butterfly are probably amongst the most exaggerated of the many modifications of this part found in butterfly families. The underside of the wing is a very dark leaf-like brown with 'leaf veins' and silvery patches of 'fungus' on the leaf, providing excellent camouflage for the butterfly when it is at rest. **236**g

**coeruleonitens** see **Milionia**
**coffeae** see **Zeuzera**
**colchis** see **Audre**

**Coleophora** Hübner (COLEOPHORIDAE) CASEBEARERS
The caterpillars of the micro-moths in this genus are usually leaf-miners in their early stage. As they get larger they make cases for themselves which are characteristically shaped and can often be used to recognize species groups. The caterpillars feed on leaves or on seeds. The genus is worldwide with many species. Coleophorid moths have long narrow pointed wings with very long fringes.

**frischella** Linnaeus WHITE-TIPPED CLOVER CASE
Europe including Britain, U.S. 17–18 mm. (0·67–0·71 in.)
This small coleophorid moth was first found in the U.S. in 1966. The caterpillars live in cylindrical cases 7–8 mm. (0·28–0·32 in.) long, of hollowed out seeds from the foodplants. These are species of clover, including *Melilotus* (Leguminosae). The adult moth is a metallic green colour. It is believed to have been accidentally introduced into America probably with imported clover seeds, and is a minor pest which could become serious.

**malivorella** Riley PISTOL CASEBEARER
U.S., Canada 12–14 mm. (0·47–0·55 in.)
This moth gets its name from the shape of the case carried round by the caterpillar. The case is black and white, fibrous, about 8–9 mm. (0·32–0·35 in.) long with a distinct bend at the open end, giving it the shape of a pistol. The caterpillar carries this case with it as it moves about feeding on buds, young fruit and leaves of apple *(Malus)*. The case projects at an angle to the leaf when the caterpillar walks about. The moth is out in early July in the Atlantic States of America.

**pruniella** Clemens CHERRY CASEBEARER
U.S. 12·5–13 mm. (0·49–0·51 in.)
The fore- and hindwings of this moth are a uniform dark smokey grey-brown. The moth has been reared from *Prunus serotina* (Rosaceae) in Pennsylvania and Ohio in June. The case, which the whitish caterpillar carries about, is a double one, made of a small curved silken case inserted into a larger one cut from the edge of the leaf. It is described as resembling a pistol butt protruding from a holster.

**virgatella** Zeller
Europe 12–15 mm. (0·47–0·59 in.)
A small moth widespread in S. Europe where the young caterpillars mine in the leaves of sage *(Salvia)*. The case carried by the older caterpillar is curved and covered with small pieces of leaf, often those which had been mined by the caterpillar when it was young. **44**g

**Coleophoridae** CASEBEARERS
These are a worldwide family of very small micro-moths. They are characterized by the caterpillars' habit of making a small case which they carry around. At times they can be pests on flowerheads such as clover *(Melilotus)*, causing loss of seed. The family is commonest in the Northern Hemisphere. There are over 100 species in Britain and many more in the U.S. They are narrow winged insects usually rather inconspicuous. The number of genera in the family is in dispute. Formerly there were less than a dozen known, but some recent work has separated one genus *(Coleophora)* into many new ones. The validity of these genera is still actively under debate. See *Coleophoria.*

**coleoxantha** see **Lactura**

**Colias** Fabricius (PIERIDAE) YELLOW or SULPHUR
This genus is widespread in the N. Hemisphere with many species in Europe, Asia and N. America. A few species are found in Africa, India and S. America. Generally the males and females are different colours

(sexual dimorphism). The caterpillars are often found on species of Leguminosae, but there are still many species whose life histories are unknown.

**alexandra** Edwards ALEXANDRA SULPHUR
U.S., Canada 48–60 mm. (1·89–2·36 in.)
Males are pale canary yellow with black edges to the wings; the females are paler, generally without the black border. As in most sulphurs, there is a small black spot in the middle of the wing. This butterfly is found in W. N. America, S. British Colombia and Manitoba, south to California, Arizona and New Mexico. The caterpillar is yellow-green with white lines on each side, each having an orange-red line through it. It feeds on species of *Astragalus* and other Leguminosae. Several different populations with quite distinctive local colours are known.

**australis** Verity BERGER'S CLOUDED YELLOW
S. and C. Europe, occasionally reaching Britain, N. Africa 42–54 mm. (1·65–2·13 in.)
This butterfly, like many other Clouded Yellows, is known for extensive migrations over Europe. On rare occasions specimens are taken in Britain, to the delight of the local collectors. The caterpillar feeds on various species of Vetch including *Hippocrepis*, *Coronilla* (Leguminosae). Externally this species is similar to *C. hyale* with which it has often been confused. *C. australis* is generally brighter yellow on the upperside than *C. hyale* and has less black near the apex of the forewings.

**cesonia** Stoll DOGFACE BUTTERFLY
U.S., C. and S. America 45–60 mm. (1·77–2·36 in.)
This species is recognized by the pointed apex to the forewing and the 'dog's-head' shape in the black border of the yellow forewing. The hindwings are yellow with a thin black edge. The caterpillar feeds on *Amorpha* and *Trifolium* (Leguminosae). The butterfly does not reach Canada and is rare on the northerly part of its range. The adult flies in early spring and has several generations each year.

**chrysotheme** Esper LESSER CLOUDED YELLOW
E. Europe to the Altai Mountains 40–48 mm. (1·57–1·89 in.)
Generally this butterfly is smaller than the others, but with more distinctive characters such as the yellow streaks along the veins which run through the black area on the wing margins. The most westerly recorded specimens have been collected near Vienna. The foodplant of the caterpillar is *Vicia hirsuta* (Leguminosae). (See also *C. eurytheme*). **119k**

**crocea** Geoffroy CLOUDED YELLOW
S. and C. Europe, including Britain, N. Africa, W. Asia 46–54 mm. (1·81–2·13 in.)
This widespread butterfly is a common migrant, sometimes moving in vast numbers. In N. Europe the butterfly does not survive the winter and is replaced each spring by immigrants from further south. Generally there are two broods each year, the caterpillars feeding on vetches *(Vicia)* and other Leguminosae. There is a white female form of this species whose genetics have been investigated. A small percentage of the wild population have the genes which produce the white population—an example of polymorphism. The yellow male, although variable, occurs only in one form. The migration of these butterflies into N. Europe and Britain always arouses the interest of entomologists and in some years they arrive in vast numbers which are talked about years afterwards. Spectacular invasions, documented in entomological magazines in Europe (and Britain) occurred in 1877 and 1941 when the butterflies arrived in clouds. The reason many more survive in one year to produce this phenomenon is not known. **121**

**erate** Esper EASTERN PALE CLOUDED YELLOW
E. Europe, temperate Asia to Japan, N. India and Formosa, Africa 46–52 mm. (1·81–2·05 in.)
This species is similar to *C. hyale* with bright lemon-yellow uppersides. It flies in May and June with a second brood later in the year. The foodplant of the caterpillar is not known. Several subspecies have been described in different parts of its range. In Africa it is known from Somalia and Ethiopia.

**eurytheme** Boisduval ALFALFA or ORANGE SULPHUR
S. Canada, U.S., Mexico 41–60 mm. (1·61–2·36 in.)
This butterfly is widespread and at times abundant over much of its range, except at the N. or S. extremes. It is very variable but can always be recognized by the orange patch on the hindwing, which separates it from the closely allied *C. philodice*. The caterpillar feeds on alfalfa *(Medicago)*, clover *(Trifolium)* and other Leguminosae. There are 3 or more broods in a year, the first adults appearing in April, although in some years earlier if they have hibernated. At times the caterpillar can be a pest on alfalfa. This is closely allied to the European *C. chrysotheme* and is regarded by many as a subspecies of it. However there are varying differences which have been suggested and for the present they are kept as separate species.

**hecla** Lefebre NORTHERN CLOUDED YELLOW
Arctic Europe, N. America and Greenland, Circumpolar 40–46 mm. (1·57–1·81 in.)
The butterfly is out in late June or July and is found in rough grassy areas from sea-level up to 910 m. (3000 ft). The caterpillar feeds on *Astragalus* (Leguminosae). Several subspecies of this butterfly are recognized from different localities. The upperside of the male is orange-yellow with a reddish reflection, in general pattern similar to figure **121**; the female has dark veins and is greyish yellow with spots on the margin between the veins.

**hyale** Linnaeus PALE CLOUDED YELLOW
Europe including Britain, W. temperate Asia 42–50 mm. (1·65–1·97 in.)
This butterfly is a migrant and repopulates the more northerly parts of its range each year. Some years they are more numerous and travel further northwards. Unfortunately the closely allied *C. australis*, which is very similar, is often confused with it so the exact distribution of both species, which fly together, still has to be worked out. The differences are so slight that a detailed study of the specimens is needed (see, for example, Riley and Higgins *Field Guide* 1970). The caterpillar of *C. hyale* feeds on lucerne and various vetches (Leguminosae). **119h**

**nastes** Zetterstedt PALE ARCTIC CLOUDED YELLOW
Circumpolar 44–48 mm. (1·73–1·89 in.)
This species occurs in Arctic Europe, Greenland and Labrador. The European subspecies differs slightly in colour from the N. American one. The caterpillar feeds on *Astragalus* (Leguminosae). The upperside of this butterfly is a pale yellow-green with a very small yellow spot on the forewing and none on the hindwing – characteristics found in the other related species, but they are usually darker. It generally lacks the grey suffusion found in the allied Mountain Clouded Yellow, *C. phicomone*.

**palaeno** Linnaeus MOORLAND CLOUDED or PALAENO SULPHUR YELLOW
Europe, Asia including Japan, N. America 50–54 mm. (1·97–2·13 in.)
This widespread species flies in June and July in N. Europe. The Japanese specimens are considered as a distinct subspecies. The caterpillar feeds on *Vaccinium* (Ericaceae). The nominate subspecies is found in Finland and Sweden, while a distinct subspecies occurs in Belgium, N.E. France and Germany. In Canada it is found in Manitoba and Labrador, W. it is widespread.

**phicomone** Esper MOUNTAIN CLOUDED YELLOW
Europe 40–50 mm. (1·57–1·97 in.)
This butterfly, which is greenish yellow in the male and greenish white in the female, is similar to *C. nastes* in appearance, but contrasts in distribution with that widespread species. *C. phicomone* is known from the main mountains of Europe, but has not been recorded in the Jura, Apennine or S. Carpathians. The caterpillar, which is dark green with white stripes on the side and yellow spots below, feeds on *Vicia* (Leguminosae).

**philodice** Godart COMMON or CLOUDED SULPHUR, MUD PUDDLE
N. America, Guatemala 33–48 mm. (1·3–1·89 in.)
This is one of the commoner butterflies in the U.S. where it is also known as the Mud Puddle butterfly. It can be seen in swarms at times over clover *(Melilotus)* fields or feeding on damp mud. The caterpillar feeds on various leguminous plants including *Trifolium*, *Vicia*, *Medicago*. The males are generally a lemon colour with black margins to the wings. Traces of yellow veins can usually be seen in this black near the apex of the forewing. There is generally a black spot at the apex of the cell in the forewing and a yellow-orange spot in the middle of the hindwing. The underside is paler without the black margin and with a larger spot in the centre of the hindwing, generally with a white centre.

**collenettei** see **Cozola**

**Colobura** Billberg: syn. *Gynaecia* (NYMPHALIDAE)
A genus of butterflies found in C. and S. America with only one species which is also found on most West Indian Islands.

**dirce** Linnaeus ZEBRA BUTTERFLY
C. & S. America, West Indies 55–75 mm. (2·17–2·95 in.)
This species derives its popular name from the remarkable stripes on the underside of the wing. There are alternate rows of brown and white stripes making a complicated geometrical pattern. The upperside, which is mostly brown, has a yellow band across the forewing; the hindwing has a short tail with small blue spots. These spots are clearer and larger on the underside. The caterpillar, which feeds under a small protective tent of silk, eats the leaves of *Cecropia pellata* (Asclepiadaceae). The adult is a fast flier, feeding on succulent fruits. Males and females are similar in pattern. **226**

**Coloradia** Blake (SATURNIIDAE)
The 4 species of this genus of moths are restricted to pine forests in the mountains of Mexico and W. U.S.

**pandora** Blake PANDORA MOTH
W. U.S. to Mexico 76–102 mm. (2·99–4·02 in.)
The caterpillars of this species were considered delicacies by the Mono Indians of California. They are sometimes pests of Ponderosa, Lodgepole and Jeffrey pines *(Pinus)* in the U.S. The forewings of the adult are chiefly grey; the hindwings a yellowish pink, with grey medial and marginal transverse bands. Most other species of *Coloradia* differ from this species in only small details of the coloration and pattern.

**colonna** see **Graphium**
**colorata** see **Darna**

**Colostygia** Hübner (GEOMETRIDAE)
Moths in this genus are mainly found in Europe and Asia and in N. America.

**pectinataria** Knoch GREEN CARPET
Europe including Britain, Asia 24–28 mm. (0·94–1·1 in.)
The moth illustrated is a female; the male is similar. The olive brown caterpillar is marked with a reddish V on the dark central line on the back. It feeds on bedstraw *(Galium)* but is also recorded on other plants. The moth flies in June, at dusk. It is common over most of its range. **274**

**Colotis** Hübner: syn. *Teracolus* (PIERIDAE)
A very large genus of butterflies, generally characterized by the coloured apical patch on the forewing, which may be present in both sexes. It has many species in Africa with a few in India and reaching into Europe. Many species have well developed seasonal forms.

**achine** see **C. antevippe**

**antevippe** Boisduval: syn. *C. achine* RED TIP
Africa 34–50 mm. (1·34–1·97 in.)
The male, in the dry season, is white with a large bright red patch at the apex of the forewing whereas in the wet season form the red is enclosed by a black line and black rays extend outward from the base of the margin. In between there are varying amounts of black on the wing. The female generally has more

brown or black than the male, a large black patch on the base of the wing and an oblique bar on the hindwing. The tip of the wing in this butterfly varies and may be yellow, red or orange. Thus with seasonal differences, and differences between the sexes, this is very variable from the basic white butterfly with a red tip to the wing. It is common over most parts of Africa, from the Cape to the equator, but generally absent in forest regions. The caterpillar feeds on species of *Capparis* (Capparidaceae).

**danae** Fabricius CRIMSON TIP
Iran, Pakistan, India, Sri Lanka, Africa 45 mm. (1·77 in.)
The name aptly describes this red-tipped butterfly. The female has more black scales, giving a greyer look and the wing tip markings are divided and less vivid than those of male. Subspecies of this occur in South Africa, and there is a distinct one across from Iran to Sri Lanka. The underside of this species varies between the wet and dry season forms. The butterflies are out in hot sunshine and are widespread in the lowland areas. They are often found congregating on the foodplants where they roost for the night. The caterpillar feeds on *Cadaba*, *Capparis* and *Maerua* (Capparidaceae). **119q**

**eris** Klug BANDED GOLD TIP
Tropical and E. Africa to South Africa 48–51 mm. (1·89–2·01 in.)
The male is white, with brownish gold-tipped forewings and a broad black band along the outer margin which also runs along the hind margin of the forewing, leaving a roughly oval patch of white. The hindwings have a blackish brown band along the front edge and some black scales at the apex of the vein; the rest of the wing is white, with a few black scales at the base. The female has a blacker apex to the forewing with white or yellow dots. This butterfly is found in the hot dry areas and is a rapid flier. Several seasonal forms of *C. eris* have been described. Caterpillars feed on *Capparis* (Capparidaceae). **119l**

**evagore** Lucas DESERT ORANGE TIP
Africa, S. Arabia 30–36 mm. (1·18–1·42 in.)
This butterfly has a large number of both seasonal and local colour forms. There are several broods each year and in Tunisia, Algeria and Morocco the subspecies *C. e. nouna* occurs. *C. evagore* is a small orange-tip butterfly which is otherwise broadly similar to *C. cardamines* but more delicate, and with much less orange on the forewing. It reaches into S. Africa, where there is a distinct subspecies which also feeds on species of Capparidaceae.

**evippe** Linnaeus SMOKY ORANGE TIP
Africa 30–45 mm. (1·18–1·77 in.)
This is a common and widespread species throughout Africa S. of Sahara, in grasslands and open bush country. It can be seen in all months of the year. The caterpillar feeds on species of Capparidaceae. The male dry season form is white with a red-orange tip to the wing. This is surrounded by a black band and is broken up by black lines. The hindwings are white, with black spots on the margins. The dry season female has a large black patch along the hind margin of the forewing and the front margin of the hindwing. The hindwing also has a black margin and a large patch of black in the middle. The black and orange-red tip is larger than in the male. The wet season male is similar to the dry season female but with smaller black bands on the wing, and less prominent margin of black round the hindwing.

**fausta** Olivier LARGE SALMON ARAB
India, Pakistan, Sri Lanka, Syria, Iran, Israel, Afghanistan 40–50 mm. (1·57–1·97 in.)
The male is a deep orange-pink colour with a series of black-edged pinkish spots at the apex of the forewing and a black spot over the apex of the cell. The hindwings are the same colour, with just a few black spots along the hind margin. There are several forms of this species in which the females vary in colour either being similar to the male, or white, sometimes with traces of the salmon pink. This species is generally a faster flier than the others in the genus and is found in drier parts, rarely in forests.

**halimede** Klug ORANGE PATCH WHITE
Africa 45–55 mm. (1·77–2·16 in.)
This is a white butterfly with striking orange yellow bands across the hind part of the forewing and the anterior margin of the hindwing. The apex of the forewing is greyish and the hindwing has a few grey spots below the margin. The female is generally greyer and has smaller orange-yellow patches and more spots in the forewing and along the margin of the hindwing. This is a common and widespread species from W. Africa to the Sudan, Ethiopia and Somalia and S. through Kenya to Tanzania.

**ione** Godart PURPLE TIP
E. to South and S.W. Africa 48–62 mm. (1·89–2·44 in.)
The males have white forewings with a row of blue or violet apical spots surrounded by black. The hindwings are white, with veins showing blackish. The females are similar but have the purple spots replaced by an orange band. This species is, however, very variable in the female which has many forms (polymorphic) mainly linked with the wet or dry season in which the adults fly. The species is common over its range but generally does not occur in forests.

**regina** Trimen QUEEN PURPLE TIP
South and C. Africa 60 mm. (2·36 in.)
Locally common over its range, this butterfly is out in most months of the year. The male is white with purple tipped forewings edged with black, and the white hindwings have the veins marked with black. The female has rows of small reddish violet spots on the dark tip of the forewing. The butterfly ranges from Uganda, Tanzania and Kenya, south through Malawi to South Africa. Generally found in bush country. The foodplant is not known but believed to be species of Capparidaceae.

**zoe** Grandi PURPLE TIP
Madagascar 35–50 mm. (1·38–1·97 in.)
This species is confined to Madagascar where it is fairly common. The females have much less purple on the forewing and more brown on the edge of the wings. The underside is pale lemon-yellow without the purple colour but with a conspicuous black spot on the underside of the forewing. **119m**


COLOUR SERGEANT see **Parathyma nefte**
**comariana** see **Acleris**
**comma** see **Spiramiopsis**
COMMA BUTTERFLIES see **Polygonia**
COMMA, FALSE see **Nymphalis vau-album**
COMMANDER see **Limenitis procris**
COMMODORES see **Junonia, Limenitis, Precis**
COMMON
ALBATROSS see **Appias paulina**
ARMYWORM see **Spodoptera exempta**
BIRDWING see **Troides helena**
BLUE see **Polyommatus icarus, Syntaracus telicanus**
BLUE APOLLO see **Parnassius hardwickei**
BLUEBOTTLE see **Graphium sarpedon**
BUSH-BROWN see **Mycalesis janardana**
CLEARWING see **Hemaris thysbe**
CLOTHES MOTH see **Tineola bisselliella**
CLUBTAIL see **Parides coon**
DUFFER see **Discophora sondaica**
EGGFLY see **Hypolimnas bolina**
EMPEROR see **Bunaea alcinoe**
FAUN see **Faunis arcesilaus**
FIVERING see **Ypthima baldus**
GLIDER see **Neptis sappho**
GRASS-BLUE see **Zizina otis**
GUAVA BLUE see **Virachola isocrates**
GULL see **Cepora nerissa**
HEATH see **Ematurga atomaria**
HOPPER see **Platylesches moritili**
IMPERIAL see **Cheritra freja**
IMPERIAL BLUE see **Jalmenus evagoras**
INDIAN CROW see **Euploea core**
JEZEBEL see **Delias eucharis**
MAP see **Cyrestis thyodamus**
MAPLET see **Chersonesia risa**
MIME see **Papilio clytia**
MORMON see **Papilio polytes**
OAK BLUE see **Narathura micale**
PALMFLY see **Elymnias hypermnestra**
PIERROT see **Castalius rosimon**
POSY see **Drupadia ravindra**
RED FLASH see **Rapala jarbus**
ROSE see **Pachliopta aristolochiae, P. hector**
SAILOR see **Neptis sappho**
SATYR see **Satyrus swaha**
SILK-MOTH see **Bombyx mori**
SOOTY-WING see **Pholisora catullus**
SULPHUR see **Colias philodice**
TIT see **Hypolycaena phorbus**
WALL see **Lasiommata schakra**
WHITE see **Pontia callidice**
WHITE, AFRICAN see **Anaphaeis creona**
WINDMILL see **Parides philoxenus**


**Commophila** Hübner (COCHYLIDAE)
This monobasic genus of micro-moths is known only from Europe.

**aeneana** Hübner
Europe, including Britain 12–16 mm. (0·47–0·63 in.)
The moth is only known in a few localities in Britain, but is more widespread in W. Europe, where it is found in France, Belgium, N. Italy, Switzerland and Germany. The caterpillar feeds on the roots of ragwort *(Senecio)*. The moth is a typical example of the family Cochylidae. **14**b

**communis** see **Pyrgus**

**Comocritis** Meyrick (YPONOMEUTIDAE)
A large genus with several similar looking species found in N. India, Indonesia, Formosa and south to New Guinea. Nothing is known of their biology and up to the present the moths have been little studied.

**olympia** Meyrick
Burma, India 30–40 mm. (1·18–1·57 in.)
The white edges to the striped pattern of the wings are conspicuous. The hindwings are grey and the undersides of the wings are grey-brown with white edges. This is one of the larger species in the genus. A related species from Formosa has all white forewings with few black spots in each wing. The foodplant of the caterpillar is not known. **37**c


**complana** see **Eilema**
**complanella** see **Tischeria ekebladella**


**Composia** Hübner (HYPSIDAE)
Two colourful species are placed in this genus of moths.

**credula** Fabricius
Antilles, N. S. America 42–57 mm. (1·65–2·24 in.)
*C. credula* is one of relatively few species of moths and butterflies which is essentially Antillean in distribution. It is also represented in N. S. America, but few specimens have been captured there so far. Its caterpillar is covered with reddish brown hairs at the front and rear, but with black hairs in the middle. This species is possibly mimicked in Guyana by the Nemeobiid butterfly *Amarynthis meneria* Cramer, which is similarly coloured but has a red band across the wings unlike *C. credula*. **376**b

**fidelissima** Herrich-Schäffer
U.S., Central and tropical S. America 50–65 mm. (1·97–2·56 in.)
*C. fidelissima* differs from the illustrated *C. credula* in the absence of white spots on the abdomen, at the base of the forewing and on the hindwing except at the outer margin. The ground-colour of the wings is iridescent blue or greenish blue and black over most of the hindwing and in two areas at the base and end of the forewing cell.

**Compsoctena** Zeller (COMPSOCTENIDAE)
A small genus of micro-moths related to the Psychid moths. The species in the genus are mostly rather dull coloured and are found in Europe, India, Pakistan, temperate Asia and S. Africa. The caterpillars are stem-borers of plants or live in the soil in tubes, feeding on roots.

**primella** Zeller
South Africa 12–20 mm. (0·47–0·79 in.)
This is a widespread species of grey-brown moths with black hindwings. **11**e

**Compsoctenidae**
A small family of moths which was only recognized as a distinct group in 1970. They were formerly considered as Psychid moths (q.v.) but differ in certain ways. The species in the genus are mostly similar to the one illustrated but little detail of their biology is available. See *Compsoctena*.

COMPTON TORTOISESHELL
see **Nymphalis vau-album**
**comyntas** see **Everes**
CONE MOTHS, PINE see **Eucosma**
**confusa** see **Thyridia**
CONGO WHITE see **Appias phaola**

**Coniodes** Hulst (GEOMETRIDAE)
An American genus of moths with typical 'looper' or 'earth measurer' caterpillars, see Geometridae entry.

**plumogeraria** Hulst WALNUT SPANWORM
U.S. 38 mm. (1·5 in.)
The male has silvery grey forewings with 4 heavy brown crossbands and hindwings with a black spot near the middle. The shape is typical of other Geometrid moths illustrated. The female, a brownish grey insect, is wingless and runs when disturbed. The black caterpillar has white patches. It is a pest on oak *(Quercus)*, walnut *(Juglans)* and other trees in S. California.

**Conopia** Hübner (SESIIDAE) CLEARWINGS
A large worldwide micro-moth genus, with more than 60 species, all of them resembling bees or wasps. When they emerge their wings are covered in scales, many of which they lose on the first flight, hence the popular name clearwings. The caterpillars are generally internal feeders, boring into stems.

**chrysophanes** Meyrick
Australia, New Britain 22 mm. (0·87 in.)
This small wasp-like moth is common in Queensland. The caterpillar tunnels in the branches of trees. In Australia it has been recorded in fig trees *(Ficus)*; in New Britain it has been found boring the stem of cocoa *(Theobroma)*. The actual identity of the latter specimens needs checking as clearwing moths are notoriously difficult to identify. The abdomen varies in colour; often there is a lot of orange colour on it. **27**b

**myopaeformis** Borkhausen
RED BELTED CLEARWING
Europe including Britain 17–21 mm. (0·67–0·83 in.)
The caterpillar feeds on the bark of fruit trees including apples *(Malus)* and pears *(Pyrus)*. It takes nearly 2 years to become adult, the moths appearing in June. The chrysalis can be seen protruding from the trunks of trees when the moth has emerged. The insect is widespread and can even be found in fruit trees growing in large cities. The moth is a typical clearwing (see figure **29**) with a black body and prominent belt on the abdomen towards the base of the thorax. This belt is generally red in both sexes, but on occasions specimens may be found with orange or even yellow belts. The fore- and hindwings are transparent with black edges and the slender black veins showing clearly across the wing. There is a small black band across the forewing just below the apex of the wing.

**consociella** see **Acrobasis**

**Consul** Hübner: syn. *Protogonius* (NYMPHALIDAE)
From Mexico to Brazil large numbers of butterflies were described in this genus but most of these are now considered as members of one very variable species. However, further investigation of the biology and morphology of this is needed. Several of the forms or subspecies have been reared on *Piper* (Piperaceae) while another was reared on *Mespilus* (Rosaceae). When at rest the butterflies look like long-stalked dry leaves; while in flight they resemble a distasteful species of *Heliconius*.

**hippona** Fabricius
S. America 70–92 mm. (2·76–3·62 in.)
All over S. America from Venezuela to S. Brazil W. to Peru this butterfly is relatively common, but very variable in pattern and colour. **236**c

CONTINENT, SIX see **Hypolimnas misippus**
**contraria** see **Ochrogaster**
**convecta** see **Mythimna**
**convexipennis** see **Cucullia**
**convolvuli** see **Agrius**
CONVOLVULUS HAWK-MOTH see **Agrius convolvuli**
**conwagana** see **Psendargyrotoza**
COOLIE see **Anartia amathea**
**coon** see **Parides**

**Copidryas** Grote (AGARISTIDAE)
This is a small genus of moths found in N. America S. to Argentina.

**gloveri** Grote & Robinson GLOVER'S PURSLANE
U.S., Mexico 40–45 mm. (1·57–1·77 in.)
The caterpillar of *C. gloveri* is known to feed only on purslane *(Portulaca)* foliage. A few days prior to changing into a chrysalis, it forms a lined, tubular burrow in the soil at the foot of the foodplant. It is often common in the mid-W. and S.W. U.S. **399**p

COPPER see **Lycaena, Palaeochrysophanus, Poecilmitis**
COPPER JEWEL see **Hypochrysops apelles**
COPPER UNDERWING see **Amphipyra**
COPPERY SWALLOWTAIL
see **Graphium latreillianus**

**Copromorpha** Meyrick (COPROMORPHIDAE)
Most species of this genus of micro-moths are found in the Indo-Australian region.

**tetracha** Meyrick
Solomon Islands 24–40 mm. (0·94–1·57 in.)
The males are much smaller than the females, and paler in colour. Nothing is known of their biology. **47**d

**Copromorphidae**
A family whose affinities are uncertain but which has been associated with the Carposinidae and Alucitidae in a major grouping, Copromorphoidea. Certain features suggest relationships with the Gelechiidae but the proboscis is without scales which are present on the proboscis of species of Gelechiidae. See *Copromorpha*.

**cordigera** see **Anarta**
**core** see **Euploea**
**corena** see **Ithomeis, Napeogenes**
**coresia** see **Marpesia, Stibochiona**
**coreta** see **Euploea**
**corethrus** see **Euryades**

**Coreura** Walker (CTENUCHIDAE)
The members of this genus have unusually broad wings. There are about 10 species, distributed in Central and tropical S. America.

**fida** Hübner
Tropical S. America 40–45 mm. (1·57–1·77 in.)
The wings of this colourful species are black, with a red or orange oblique band across the forewing, a red or orange outer marginal band on the hindwing, and a slight blue sheen at the base of the hindwing.

**coridon** see **Lysandra**
CORN BORERS see **Ostrinia**
CORN EARWORM see **Heliothis**
CORN MOTH see **Nemapogon**
CORNELIAN BUTTERFLIES see **Deudorix**
**cornifoliella** see **Antispila**
**coronata** see **Thecla**

**Coronidia** Westwood (SEMATURIDAE)
Four species are placed in this genus of moths. The wings are brown with a central oblique white band on the forewing and an iridescent blue and violet band on the hindwing. Their range includes Mexico, C. and S. America as far S. as Paraguay. The species of *Homidiana* Strand have been placed, at times, in *Coronidia*; they mostly have orange, red or yellow patches on the hindwing. **314**d

**orithea** Stoll
Mexico to Paraguay 42–55 mm. (1·65–2·17 in.)
Males differ from the illustrated female in the regularly pectinate antennae which are not swollen near the tip, and in the shape of the blue and violet marking, which is narrower and does not reach the outer margin of the hindwing. **314**d

**Correbidia** Hampson (CTENUCHIDAE)
Species of this genus are generally considered to be mimics of similarly patterned, distasteful Lycid beetles such as the N. American *Lycomorpha*. Those *Correbidia* species which have been the subject of experiments have proved to be palatable to frogs, lizards, mantids (carnivorous insects related to grasshoppers) and spiders but were rejected by an oriole (Icteridae).

**assimilis** Rothschild
YELLOW-BANDED BEETLE MIMIC
Tropical America 21–28 mm. (0·83–1·1 in.)
There is a very close resemblance between this species and certain Lycid beetles, not only in appearance but in the movements of the antennae and the raising and lowering of the forewings when walking. **367**h

**corsicum** see **Ocnogyna**

**Corymica** Walker (GEOMETRIDAE)
A genus of moths with species in Borneo, Hainan, Java, Malaya and Sulawesi (Celebes), New Guinea to India, and in Japan.

**vesicularia** Walker
Malaya 20–25 mm. (0·79–0·98 in.)
This bright yellow species has many similar related species ranging from New Guinea to India. **294**j

**Coscinocera** Butler (SATURNIIDAE)
There is a single known species in this genus of moths.

**hercules** Miskin AUSTRALIAN ATLAS
N. Australia, New Guinea and associated islands of the Louisiade and Bismarck Archipelagos 165–210 mm. (6·5–8·27 in.)
This huge rain-forest species is the Papuan and Australian equivalent of the related Atlas moths *(Attacus)*. It is similar in general appearance to *Attacus atlas*, except for the lone hindwing tails of the male (up to 170 mm., 6·69 in., from the base of the wing to the tip of the tail) and the short, broad hindwing tail (or lobe) of the female. Its caterpillars feed on the foliage of various trees.

**Cosmopterix** Hübner (MOMPHIDAE)
A large genus of micro-moths with species all over the world. Most of them are small moths with rather narrow pointed forewings, and very slender hindwings. Some have brilliant markings on the wings. The adults of many species tend to sit with the head downwards.

**scribaiella** Hübner
Europe, Asia 9–10 mm. (0·35–0·39 in.)
This species was first described from Austria, where the moth was collected near Vienna. It is common in S. and Central Europe but is not found in Britain, although similar looking species occur there. The caterpillars mine in the leaves of reeds *(Phragmites)*. In Japan there is a subspecies where the specimens differ slightly from the European ones. **44**b

**Cosmosoma** Hübner (CTENUCHIDAE)
This is a basically tropical genus of over 80 species found in Central and S. America. The range of one species extends N. to S. Florida (U.S.).

**auge** Linnaeus
Central America including Mexico, tropical S. America, S. U.S. 30–38 mm. (1·18–1·5 in.)
In the U.S. this beautiful red and black wasp-mimic is found only in S. Florida, where the caterpillar has been recorded from *Mikania scandens*, a species of the family Compositae. The final stage caterpillar is densely covered with black and white hairs, there are 2 pairs of lateral hair-tufts. **364**h

**Cossidae** GOAT OF CARPENTER WORM MOTHS
These range from very small to very large moths, up to 240 mm. (9·45 in.) wingspan. The caterpillars of these moths are generally wood-borers. This family contains many species which are of economic importance. The caterpillar of one species *(Cossus cossus)* smells like a goat, hence one of the popular names. The females, where known, deposit large numbers of eggs – over 15,000 eggs from one female have been found. In Australia some of the large grubs are eaten by the aborigines. The family is worldwide and is particularly well developed in Australia where there are many species. See *Allostylus, Cossus, Prionxystus, Xyleutes, Zeuzera.*

**Cossus** Fabricius (COSSIDAE)
GOAT MOTH OF CARPENTER WORMS
Generally large moths with rather long, narrow forewings. The adults do not usually have a proboscis and thus cannot feed. They are all species with rather stout bodies. Usually the caterpillars bore into trees.

**cossus** Linnaeus GOAT MOTH
Europe including Britain, W. and Central Asia, N. Africa 70–95 mm. (2·76–3·74 in.)
The name is derived from the caterpillars which are said to smell like goats. The large caterpillars bore into the wood of willow *(Salix)*, elm *(Ulmus)*, ash *(Fraxinus)* and other trees. They take 3 or 4 years to reach maturity by which time they are up to 14 mm. (0·55 in.) thick. They cause considerable damage to the trees. The 'goat' smell remains on the tree long after the caterpillar has crawled out to look for a place to pupate. The moths generally emerge in June and July. **25**

COSTA, TAWNY see **Telchinia violae**
**costalis** see **Hypsopygia**
COSTER BUTTERFLY, YELLOW see **Acraea vesta**
COTTON BOLLWORMS see **Heliothis, Earias**
COTTON LEAFWORMS see **Anomis, Spodoptera**
COTTON SQUARE BORER see **Strymon melinus**
COURTIER, WESTERN see **Sephisa dichroa**
COXCOMB PROMINENT see **Ptilodon capucina**

**Cozola** Walker (LYMANTRIIDAE)
The 5 species of this unusually coloured genus are restricted to the Moluccas and Sulawesi (Celebes). The females of most species are apparently mimics of *Euploea* species.

**collenettei** Nieuwenhuis
Sulawesi (Celebes) 35–55 mm. (1·38–2·17 in.)
The male is consistently smaller and less regularly patterned with white than the illustrated female which resembles and presumably mimics certain *Euploea* butterflies (Danainae). However, nothing is known about the habits or the life history of this moth. **357r**

**crabronis** see **Pseudosphex**
CRACKER BUTTERFLIES see **Hamadryas**
**Crambidia allegheniensis** see **Eilema complana**
**Crambinae** see **Chrysoteuchia, Chilo, Crambus, Pediasia, Pyralidae**

**Crambus** Fabricius (PYRALIDAE) GRASS MOTHS
This is a large genus of Pyralid moths in the subfamily Crambinae, with many species all over the world. As their popular name implies, they are the small moths commonly disturbed when one is walking through grass. At rest, the moths sit on the grass stem with the wings rolled round the body, looking like a slender wedge-shaped object along the grass stem. They fly during the day and at dusk and are very active. The larvae feed on grasses and can be very destructive in pasture lands and cause trouble when they attack the greens on golf courses. Many are important agricultural pests.

**mutabilis** Clemens STRIPED SOD-WEB WORM MOTH
Canada, U.S. 25 mm. (0·98 in.)
The forewings are dull grey with white, grey and black streaks from base of the wing through the middle towards the outer margin; the hindwings are pale brown. The caterpillar feeds on blue grass, timothy grass and other common pasture grasses. The moth, which flies in August, is widespread in the U.S. and Canada.

**occidentalis** Grote
U.S. 23–25 mm. (0·91–0·98 in.)
This species is common in California. There are several related species in N. America with a rather similar external pattern. The caterpillar is a grass-feeder as are the others in the genus. **56u**

**pascuellus** Linnaeus
Europe including Britain 21–24 mm. (0·83–0·94 in.)
Rather similar to the American one above in external appearance, this species is common in grasslands throughout its range. The rapid zig-zag flight through the grass when disturbed, and sitting with the wings rolled round the body are typically Crambine habits. The moth is characteristic of damp or marshy areas where it appears in June or July.

**perlella** Scopoli
Europe including Britain, temperate Asia including Japan, N. America 20–27 mm. (0·79–1·06 in.)
This species is widespread and common over much of its range. Several subspecies have been described including one in Japan. The caterpillars feed on various species of grass. The forewings vary in colour from a pearly white and almost unmarked, to very strongly marked, with brown lines from the base of the wing to the outer margin, mostly along the veins. **57**

**cramerella** see **Acrocercops**
CRANBERRY BLUE see **Vacciniina optilete**
**crassipes** see **Parides**
**crassus** see **Eurytides**
**crataegi** see **Aporia**
CREAM WAVE see **Cyclophora floslactata**
CREAM-BANDED SWALLOWTAIL see **Papilio leucotaenia**
CREAM-BORDERED GREEN PEA MOTH see **Earias clorana**
CREAM-SPOT TIGER-MOTH see **Arctia villica**
**credula** see **Composia**

**Cremna** Doubleday (NEMEOBIIDAE)
There are 3 species in this tropical S. American genus of butterflies.

**actoris** Cramer SPECKLED BLACK NYMPH
N. S. America 32–35 mm. (1·26–1·38 in.)
*C. actoris* flies apparently only at dusk, always returning to rest on the same leaf or other suitable perch. The female (not illustrated) has bluish white markings on the wings. **263s**

**thasus** Stoll RED-BANDED ZEBRA
N. S. America 20–29 mm. (0·79–1·14 in.)
This is a rare butterfly in collections. Although its flight is slow, the disruptive pattern of the wings makes it very difficult to detect, especially when at rest under a leaf. Both sexes are brown, with numerous parallel transverse bands of orange (male) or white (female). There is some evidence of an association between ants and the caterpillars of this butterfly as in many species of the family Lycaenidae.

**creona** see **Anaphaeis**
CRESCENT, PEARL see **Phyciodes tharos**
**cresphontes** see **Papilio**

**Cressida** Swainson (PAPILIONIDAE)
There is a single known species in this genus. *Cressida* is probably most closely allied to the S. American genus *Euryades*; both are members of the tribe Troidini or *Aristolochia*-feeding swallowtails.

**cressida** Fabricius BIG GREASY BUTTERFLY
Australia, New Guinea 66–75 mm. (2·6–2·95 in.)
The first specimen of this species was captured at Cooktown, Australia, during one of the historic voyages of the British explorer and navigator Captain James Cook. The colour pattern of the male hindwing is similar to that of another Australian species *Pachliopta polydorus*. The female wings are much paler and less well-marked than the male, partly the result of the rapid loss of scales which produces a greasy effect. Its caterpillars feed on a native species of *Aristolochia* and on the introduced Dutchman's Pipe *(Aristolochia elegans*; they remain inactive for long periods then suddenly recommence feeding. **111a**

**crethona** see **Callophrys**
**cribraria** see **Argina**
**cribrella** see **Myelois**
**cricosoma** see **Symmachia**
CRIMSON SPECKLED see **Utetheisa pulchella**
CRIMSON TIP see **Colotis danae**
CRINGLED FLANNEL see **Megalopyge crispata**
**crispata** see **Megalopyge**
**cristana** see **Acleris**

**Crocallis** Treitschke (GEOMETRIDAE)
This genus of moths has species in Europe and temperate Asia with a few in N. Africa and the New World.

**elinguaria** Linnaeus SCALLOPED OAK
Europe including Britain, Asia 34–37 mm. (1·34–1·46 in.)
The caterpillar varies in colour from pale to dark grey, tinged with purple, with diamond-shaped marks along the back. It feeds on most trees and bushes in the spring. The moth flies in July and August and is common across most of Europe except in the north. **297**

**crocea** see **Colias**
**crocicapitella** see **Monopis**

**Croesia** Hübner (TORTRICIDAE)
This genus of micro-moths is found in Europe, including Britain, temperate Asia to Japan and in N. America. Recently one species was described from Madagascar. Some of the species in the genus are widespread, occurring in both Europe and N. America.

**forsskaleana** Linnaeus
Europe including Britain, U.S. 11–14 mm. (0·43–0·55 in.)
This species is common over most of its range. The moth is out in July and August, earlier in S. Europe. The pale yellow caterpillar feeds on the leaves of maple *(Acer campestre)*. There is some variation in the pattern of the species but generally it is similar to the one figured. In N. America this species has been found in Connecticut, New Jersey and New York. It flies from July to August in the U.S. **23**

**croesus** see **Chliara, Chrysiridia, Ornithoptera**
CROW BUTTERFLIES see **Euploea**
CRUISER see **Vindula erota**

**Cryptophasa** Lewin (XYLORYCTIDAE)
A genus of micro-moths with many species in Australia and New Guinea. They are large, for so-called 'micro-lepidoptera' and look rather like fat-bodied Noctuid moths. They are generally grey-white or white and brown coloured. Those that have been reared have been bred from *Acacia* (Leguminosae) or from *Grevillea* (Proteaceae).

**nephrosema** Turner
Australia 32–48 mm. (1·26–1·89 in.)
This grey-white and brown species, with a touch of a reddish brown on the forewings and yellow hindwings is a striking looking moth. Related species have a more distinct brown border to the wing. The biology of this species is not recorded but like other species in the genus, the caterpillar probably tunnels into the stems or feeds on the bark of species of *Acacia*. **42c**

**Ctenucha** Kirby (CTENUCHIDAE)
There are about 50 species in this mainly nocturnal New World genus, most of which are restricted to tropical S. and Central America. Grasses (Gramineae) form the foodplants of several of the species. The family name, Ctenuchidae, is derived from the name of this genus.

**virginica** Charpentier VIRGINIAN CTENUCHA
E. Canada, N.E. U.S. 40–50 mm. (1·57–1·97 in.)
This is both a nocturnal and a day-flying species. Grasses are the chief, but not exclusive, foodplant of the caterpillar. The conspicuous blue abdomen is

hidden by the hindwings when the moth is at rest. **364a**

**Ctenuchidae**
Many of the 3000 or so species of this family of moths are mimics of various genera of wasps and Lycid beetles. Others are brightly coloured like their close relatives in the family Arctiidae and are probably mostly distasteful to predators. Many are day-fliers. The family is chiefly tropical in range but is represented by a few species in N. America and a single species in Europe, *Syntomis phegea*. See *Amata, Chrysocale, Coreura, Correbidia, Cosmosoma, Ctenucha, Cyanopepla, Dahana, Dasysphinx, Didasys, Euchromia, Gymnelia, Horama, Macrocneme, Napata, Phaeosphecia, Pompilopsis, Pseudosphex, Sphecosoma, Syntomeida.*

**cubana** see under **Dahana**

**Cucullia** Schrank (NOCTUIDAE)
SHARK OF HOODED OWLET MOTHS
About 200 species are placed in this genus of moths. Most are Northern Temperate in range (especially the Old World) but the genus is also represented in S. America, Africa and tropical Asia. The caterpillars of many species are colourful and feed in exposed positions on low-growing plants. The chrysalis overwinters. Adult moths are frequently attracted to flowers.

**convexipennis** Grote & Robinson
BROWN HOODED OWLET 40–48 mm.
Canada, U.S. (1·57–1·89 in.)
*C. convexipennis* ranges from Saskatchewan to Nova Scotia and southwards through the U.S. to California and Tennessee. The beautiful, striped, mostly red and black caterpillar is a general feeder on herbaceous plants. It is frequently found in gardens feeding on the foliage of golden rod *(Solidago)* and asters *(Aster)*.

**verbasci** Linnaeus MULLEIN MOTH
Europe including Britain, 42–45 mm.
W. Asia (1·65–1·77 in.)
The common name of this species refers to the caterpillar's chief foodplant, mullein *(Verbascum)*, on which it feeds in June and July. It will also feed on the foliage of figwort *(Scrophularia)*. The caterpillar, which is a particularly colourful creature, is greenish white in ground-colour, banded with yellow and dotted with black. The moths fly in late April and May. Some specimens are much more brownish than in the illustration. **399f**

**cucurbitae** see **Melittia**
**Culapa** see under **Mycalesis**
**culleni** see under **Chlanidophora**
**culmella** see **Chrysoteuchia**
**cunea** see **Hyphantria**
**cunina** see **Phiala**

**Cupido** Schrank (LYCAENIDAE)
This is a large genus of about 175 species of butterflies. It is represented in most parts of the World.

**minimus** Fuessly LITTLE BLUE
Europe including Britain, 18–28 mm.
temperate Asia to the Pacific (0·71–1·1 in.)
*C. minimus* is often common on grassland, especially on chalk or limestone hills up to 2440 m. (8000 ft). Males are dark brown above, with numerous silvery blue scales on both pairs of wings. Females lack these blue scales. The under surface of both sexes is light grey, marked with white-edged, blue scales. The caterpillar feeds on the flowers and seeds of several species of Leguminosae. **247**

**cupido** see **Helicopis**
**cuprina** see **Dysphania**

**Curetis** Hübner (LYCAENIDAE)
SUNBEAM BUTTERFLIES
This is a widely distributed genus found in Sri Lanka, India, Malay Archipelago, New Guinea and the Solomons. The males of the genus are red above, the females white or orange. The underside is a lustrous white in both sexes. All the species are tailless. Over 40 species have been described.

**thetis** Drury INDIAN SUNBEAM
Sri Lanka, India and Burma to 40–48 mm.
New Guinea (1·57–1·89 in.)
The flight of this species is powerful and is put to advantage during pursuit of any butterfly intruding into its territory. Low elevation scrubland is the home of *C. thetis*. The female (not illustrated) is brown above with a large yellow or white patch on each wing. The caterpillar has two whip-like, eversible processes at the rear which can be whirled about rapidly when provoked and may afford protection against parasitic wasps. **244k**

**curiosa** see **Cyclosia**
**curius** see **Lamproptera**
CURRANT BORER see **Synanthedon salmachus**
CURRANT CLEARWING see **Synanthedon salmachus**
CURRANT MOTHS see **Arycanda**
CURRANT MOTH, DRIED see **Ephestia cautella**
CURRENT MOTHS see **Abraxas**
**cuspidae** see **Euclidia**
**cuspidella** see **Scythris**
CUTWORM see **Agrotis ipsilon, Amanthes c-nigrum, Peridroma saucia**
**cyanea** see **Polystichtis**
**cyaneus** see **Parthenos**
**cyaniris** see **Myscelia**

**Cyanocrates** Meyrick (XYLORYCTIDAE)
A genus of micro-moths with only 2 species, both from W. Africa. The shape and colour of the wings are different from others in the family. The species are said to resemble distasteful species in another family, from which it is assumed they derive some measure of protection from predators.

**grandis** Druce
W. Africa 60 mm. (2·36 in.)
This species is a mimic of an Agaristid moth, many of which are known to be distasteful to predators and are thus avoided. *C. grandis* therefore is protected by this mimetic resemblance. However nothing is known of its biology and the mimetic resemblance needs further investigation. **42e**

**Cyanopepla** Clemens (CTENUCHIDAE)
There are over 30 Central and S. American species in this genus of moths. Most of them have a brilliantly iridescent blue or green head, thorax and abdomen; their wings are dark brown in ground-colour, marked with red, orange, yellow, green or blue, or a combination of colours.

**basimacula** Hampson
Ecuador 36–38 mm. (1·42–1·5 in.)
The brown forewing of this species has a single crimson spot at its base. The hindwing is a brilliant, iridescent blue with dark brown margins, contrasting with the equally dazzling green of the abdomen.

**pretiosa** Burmeister
Argentina, Ecuador 34–38 mm. (1·34–1·5 in.)
There are several other species with variations on this colour pattern; for example, *C. phoenicia* Hampson from Peru, which has a continuous red band on the forewing and a red patch on the hindwing. **364f**

**Cyanostola** Houlbert (CASTNIIDAE)
There are 3 species in this tropical Central and S. American genus of moths. The hindwings are basally iridescent blue and purple.

**diva** Butler
Tropical Central and 65–95 mm.
S. America (2·56–3·74 in.)
The superbly coloured, iridescent basal two-thirds of the hindwings vary in colour from gold to blue or purple depending on the angle at which they are viewed. **45j**

**hoppi** Hering
Colombia 70–75 mm. (2·76–2·95 in.)
*C. hoppi* is similar in coloration to *C. diva*, but has a broader orange band at the outer margin of the hindwing and a strongly pointed apex to the forewing.

**cyara** see **Phylaria**
**cybele** see **Speyeria**

**Cyclidia** Guenée (CYCLIDIIDAE)
About 10 species of moths are placed in this genus. Except for *C. orciferaria*, which is dark purplish brown, all the species are white in ground-colour, with grey, brown or black markings. The genus is represented in India, Sikkim, Burma, China, Japan and through S.E. Asia to Borneo.

**dictyaria** Swinhoe
Sri Lanka, S. India 56–70 mm. (2·2–2·76 in.)
This is one of the most distinctive species of its genus. It is probably a derivative of the Chinese and N. Indian *C. substigmaria* Hübner but has evolved this striking black and white pattern after having become isolated geographically. There are many other examples in the fauna of Sri Lanka and adjacent S. India, at both species and subspecies level, which illustrate this zoogeographical phenomenon. **272p**

**Cyclidiidae**
This small family was erected in 1962 by the Japanese lepidopterist Hiroshi Inoue to accommodate a group of moths previously placed in the Drepanidae. Most of the species are large, broad-winged moths, chiefly grey and white in colour and restricted to S.E. and E. Asia. There are 2 genera: *Cyclidia* and *Mimozethes*. Cyclidiids belong to the superfamily Geometroidea; their closest allies are probably the Thyatirids. See *Cyclidia*

**Cyclophora** Hübner (GEOMETRIDAE)
A widespread genus of moths with many species found in Europe, N. Africa and with some species in N. America. They are mainly rather delicately marked moths, mostly nocturnal.

**floslactata** Haworth CREAM WAVE MOTH
Europe including Britain, 26–30 mm.
Asia (1·02–1·18 in.)
This is a white geometrid with pale lines across the wings. It is common in suitable localities over its whole range. Generally during May and June it will be found in woods. The moth flies at dusk, or night, and is attracted to light. The caterpillar feeds on woodruff *(Asperula)* and bedstraw *(Galium)*. It is a yellowish grey or brown with irregular darker marks and a pale line along the back. There are a number of related species, popularly known to collectors as 'Waves' which are variously patterned, generally with more dark marks on the wings than the Cream Wave.

**Cyclosia** Hübner (ZYGAENIDAE)
A large genus of moths with over 100 species. They occur in India, Indonesia, Thailand and the Philippines. Most are striking looking, large moths with iridescent colours, popular with collectors. Little is known of their biology or life history. See also under *Euploea alcathoe*.

**curiosa** Swinhoe
Java 44–55 mm. (1·73–2·17 in.)
There are several related, similar species in Borneo and India. The related species have more white on the wing while *C. curiosa* is usually lemon-yellow. The underside is patterned like the upperside which is unusual in moths. Some of the specimens of *C. curiosa* are much darker with less yellow on the wing. **45h**

**midamia** Herrich Schäffer
Java, Borneo, India 60–80 mm. (2·36–3·15 in.)
There are several subspecies of this beautiful moth. They differ mainly in the amount of blue on the wings. The underside has more white on than the upperside and has less blue around the edges. The females generally have less blue than the males. The life history is unknown. **45b**

**pieridoides** see under **Graphium delesserti**

**Cyclotorna** Meyrick (CYCLOTORNIDAE)
A small Australian genus of rather stout-bodied moths, greasy looking and soberly coloured. They are remarkable for their life history, in which they are associated with ants and leaf-hoppers. The adults have the mouth

parts reduced and do not feed. The whole family consists of only the one genus, all probably with similar life histories.

**monocentra** Meyrick
Australia 22–30 mm. (0·87–1·18 in.)
The caterpillars of this moth were first discovered by an Australian entomologist, who found them in the nests of ants, feeding on the ant grubs. He was unable to find any early stages there and after a careful search discovered that the young caterpillars, after hatching, attach themselves to leaf-hoppers or other bugs (Hemiptera). The ants are in attendance on the leaf-hoppers from whom they obtain a sweet secretion. The caterpillar, which was on the leaf-hopper, spins a shelter beneath the bark of the tree in which it moults to become a brightly coloured caterpillar, attractive to ants who carry it to their nest. There they feed on a secretion produced by the caterpillar which in turn eats the ant grubs. Thus the caterpillar gets its food in the ants' nest and is protected by the ants which keep off other predators. It seems however that the ants are the losers in this situation, for although they eat some secretion from the caterpillar, their offspring are eaten by the caterpillar. **54**b

**Cyclotornidae**
This family of moths is confined to Australia. They are generally stout-bodied, grey species which do not feed in the adult stage. They have a remarkable life history, being associated with ants and leaf-hoppers. The caterpillar starts off as an external parasite on a leaf-hopper (Hemiptera-Homoptera) and then later is predatory on ant larvae. See *Cyclotorna*.

**Cydia** Hübner (TORTRICIDAE)
Worldwide. This genus contains many species whose caterpillars feed on various fruits, particularly apples *(Malus)*, peaches *(Prunus)*, etc., and are serious pests in many countries. The genus belongs to the subfamily Oleuthreutinae.

**egregiana** Felder
New Guinea, Solomon Islands, Moluccas 18–20 mm. (0·71–0·79 in.)
This species is variable in pattern and the specimen illustrated, which is from the Solomon Islands, differs from 'true' *C. egregiana* in having a black streak across the hindwings. It is possible that the specimen illustrated may represent an undescribed species. Nothing is known of the biology of this species. **14**h

**molesta** Busck ORIENTAL FRUIT MOTH
Worldwide 10–15 mm. (0·39–0·59 in.)
The forewings are dark grey with chocolate-brown markings and black dots, the hindwings blackish grey with white fringe. The caterpillar is whitish grey or purplish. It spins a web on leaves or twigs. The young caterpillar bores into twigs causing leaf fall while the older ones enter the fruit–this makes the fruit susceptible to rot. It will feed on apple *(Malus)*, apricot *(Prunus)*, quince *(Cydonia)*, and is probably the worst single pest of peaches. It can cause serious damage in warm climates where there may be two or three generations a year and 80% of a crop may be ruined. The moth is believed to have originated in Japan from where it has been accidentally introduced into various countries. Some of the earliest records of it in the U.S. are in the 1920s.

**nigricana** Fabricius PEA MOTH
Europe, and introduced into the U.S. and Canada 12–14 mm. (0·47–0·55 in.)
The moth's satiny forewings are mainly brown with a black line close to and parallel with the hind margin. There are some black and white marks along the front margins. The moths are out in Europe from June to August and are a typical Tortricid (Bell moth) shape. The eggs are laid on pea plants *(Pisum sativum)* usually on the sepals around young pods. The caterpillars bore through the pods and into the developing peas, generally with only two caterpillars in any one pod. When fully grown they bite their way out of the pod and make silk cocoons on the ground where they hibernate until the following spring. The pea moth is a serious pest locally in England, parts of Europe, and in several places in the U.S.

**ninana** Dyar
N. America 17–20 mm. (0·67–0·79 in.)
This species is known from Arizona and it is thought that it arrived in the U.S. from the tropics. Its biology and country of origin are not known. **14**f

**pomonella** Linnaeus APPLE CODLIN MOTH
Worldwide 15–22 mm. (0·59–0·87 in.)
This is one of the more notorious pests of fruit, particularly apples *(Malus)*. The caterpillars which live inside the fruit are popularly called 'maggots'. The feed on the pips of the apples and the flesh around the core. When fully fed the caterpillar leaves the apple to find somewhere to pupate. This habit has given rise to the occurrence of Codlin moth caterpillars in many different foods including sugar and flour. When found in unusual situations it means that at some stage the product was stored near apples – from which the caterpillars emerged. In N. America, where the moths are widespread, from California to Canada, the caterpillars are recorded in pear, walnut *(Juglans)* and quince and they are likely to attack these fruits wherever they are grown throughout the world. The popular name often given is 'Codling' moth. **14**e, **19**

**saltitans** Westwood JUMPING BEAN MOTH
C. America 19–22 mm. (0·75–0·87 in.)
Jumping beans are a popular toy all over the world. The 'beans' are the seeds of a Mexican spurge (Euphorbiaceae) which contain a small caterpillar. The moth lays its eggs on the developing fruit of the spurge and, when it hatches, the young caterpillar burrows into the seed. The seed contents are eaten and the caterpillar lines the seed with a layer of silk which is produced from silk glands below the head. When the pods containing these seeds are ripe they burst, flinging out the seeds. The seeds which contain caterpillars are the 'jumping beans'. This jumping is done by the caterpillar contracting, causing its body to catapult against the inside of the bean and making the bean jump. Jumping takes place when the bean is exposed to warmth and sunlight and is believed to be the caterpillar's attempts to move out of direct sunlight. Before the caterpillar turns into a pupa it cuts a circular window in the side of the seed which it keeps closed with silken threads. When the moth emerges from the pupa it pushes through the window. The bean is left with the pupal case sticking out. Needless to say, once the caterpillar pupates, the bean no longer jumps. **14**d

**Cydosia** Westwood (NOCTUIDAE)
The 10 or so species in this genus of moths are found in N. Central and tropical S. America.

**aurivitta** Grote & Robinson
U.S. 21–25 mm. (0·83–0·98 in.)
This is a solely S. U.S. moth. It is similar in general appearance to the illustrated tropical American species, *C. nobilitella*. The wings are dark brown with a greenish blue sheen, and dull orange markings on the forewings. Little seems to be known about the life history of this species.

**nobilitella** Cramer
N. S. America, the Antilles including the Bahamas 24–32 mm. (0·94–1·26 in.)
The colour pattern of this small but pretty moth is typical of several species of its genus. Nothing is known about the biology of this moth. **390**b

**Cymbalophora** Rambur (ARCTIIDAE)
This genus of about 6 species is found in N. Africa, Europe and W. Asia. The colour pattern of the wings is similar to that in many species of the related N. American genus, *Apantesis*.

**pudica** Esper FESTOONED TIGER-MOTH
N. Africa, Europe 32–38 mm. (1·26–1·5 in.)
The male of this species is unusual in having a much larger tymbal organ (sound-producing equipment) than the female and has been heard to produce a loud, crackling sound during flight at dusk. It is not known what function or functions these large male tymbals perform. The tymbals of most Arctiids, which do not differ in size between the sexes, operate at night as warning signals to bats of their unpalatability. The caterpillar of *C. pudica* is the overwintering stage; it is pale grey, with a yellowish white, lateral line and black tubercles bearing tufts of yellow hair. It feeds on grass. **369**m

**cymela** see **Euptychia elegans**

**Cymothoe** Hübner (NYMPHALIDAE)
The male and female butterflies are very different in pattern showing strong sexual dimorphism. Often there are several female forms with one male (polymorphism). They are mostly woodland butterflies in Africa, rare in the south and east and absent from Madagascar. The genus is not known from outside Africa.

**coccinata** Hewitson
W. Africa to Uganda, Zaire 55–60 mm. (2·17–2·36 in.)
The male is illustrated. The female is variable in colour, generally brown with an orange base to the fore- and hindwings and prominent large black spindle-shaped marks below the margin of the wing. Some forms have females which are mostly white through the middle of the wing. The underside of both sexes is pale, the female being paler than the male. **216**l

**herminia** Grose-Smith
Africa 60–65 mm. (2·36–2·56 in.)
This is a widespread species in E. and C. Africa. The specimen photographed is the subspecies *C. h. johnstoni* Butler from Uganda. There is considerable variation in colour in this species. **225**

**lucasi** Doumet
W. Africa 70–85 mm. (2·76–3·35 in.)
The female is strongly cross patterned with white and is quite different from the male illustrated. This is a richly coloured species with a fiery orange above and a dull yellowish brown on the underside with a dark line through the middle of both pairs of wings. The female is larger than the male. **231**a

**sangaris** Godart BLOOD RED CYMOTHOE
W. Africa to W. Uganda, W. Kenya and W. Tanzania 60–70 mm. (2·36–2·76 in.)
This is a forest species, where the adult can be found in most months of the year. It is similar to *C. coccinata* but possibly even brighter red than that species. The female is variable in colour, usually grey brown with orange brown at the base of the wings. The underside of the wing is dark brown with a red median line. The amount of black on the hindwing varies, some specimens being prominently marked. The life history of this species is not known. **216**g

**cyna** see **Zizula**

**Cynandra** Schatz (NYMPHALIDAE)
A genus of butterflies with only 2 included species, both from Africa.

**opis** Drury BLUE BANDED NYMPH
Africa 45–55 mm. (1·77–2·17 in.)
A blackish butterfly with brilliant iridescent blue lines across the fore- and hindwings of the male, roughly parallel to the wing margin and with some whitish spots near the apex of the forewing. The female is larger, dark brown with whitish yellow markings. The blue iridescence of the male is very conspicuous in the field. The butterfly is on the wing all the year and flies in the forest areas of W. Africa, Zaire and W. Uganda. It tends to fly in the undergrowth of the forests.

**cynorta** see **Papilio**
**cynosbatella** see **Epiblema**
**cynosura** see **Callicore**

**Cynthia** Fabricius (NYMPHALIDAE)
PAINTED LADY BUTTERFLIES
This is a worldwide genus of butterflies, many of which are migrants. Recent work on the species in the genus has shown that there are 7 American species, one Australian and one other species which is worldwide. One of the American species has recently become established on Hawaii and the Azores. The generic name for the species formerly used was *Vanessa* but the recent work indicated that the Painted Lady butter-

flies were not as closely related to the Vanessid butterflies.

**anabella** Field WEST COAST LADY
Canada, U.S., Guatemala, Mexico 34–54 mm. (1·34–2·13 in.)
This species can be separated from *C. cardui* by the lack of the reddish tints on the wing band and in having the bar just beyond the cell on the upperside of the forewing brownish instead of white. It has 4 blue-centred eyespots on the hindwing. The caterpillar feeds on species of Malvaceae and a few other plant families. In Canada this species has been found from British Columbia to Alberta; in the U.S. from California to Washington, Utah, Wyoming, Colorado, New Mexico and Texas.

**braziliensis** Moore
Venezuela, Colombia, Ecuador, Peru, Bolivia, Brazil, Paraguay, Uruguay, Argentina 40–60 mm. (1·57–2·36 in.)
Typical, in general appearance, of species of *Cynthia*, this species is separated from the others in having, on the upperside of the forewing, a white band round the inner side of the black at the end of the cell. There are other differences in its internal morphology. The caterpillar feeds on various species of Compositae.

**cardui** Linnaeus PAINTED LADY
Almost worldwide 44–80 mm. (1·73–3·15 in.)
This is probably one of the best known butterflies in the world. It is widespread on most continents with the exception of S. America, where it is rare, and Australia. where it has not yet been recorded. The only other place it has not reached is New Zealand. The spectacular migrations of this butterfly, often with many thousands of individuals, have been well documented in popular literature for many years. The caterpillar feeds on species of thistle (Compositae). The Painted Lady cannot survive the winter in N. Europe and Britain and each year these areas are repopulated by insects which have survived further south. **200**

**carye** Hübner
S. America, Easter Island, Tuamotu Archipelago 30–55 mm. (1·18–2·17 in.)
This species can be separated from the more widespread *C. virginiensis* by the bar on the forewing being a tawny colour instead of white, and 4 blue-centred eyespots on the upperside of the hindwing. The caterpillar feeds on species of Malvaceae, Compositae, and Geraniaceae. It is interesting that it has been found on the Tuamotu Archipelago which is over halfway from Chile to New Zealand. In S. America this species has been found in Colombia, Ecuador, Bolivia, Chile and Brazil.

**virginiensis** Drury: syn. *C. huntera*
PAINTED BEAUTY or HUNTER'S BUTTERFLY
N., S. America 38–70 mm. (1·5–2·76 in.)
This widespread American species ranges from Canada, through the U.S. and C. America, to Colombia. In S. America several subspecies have been named. It has also been found on several W. Indies Islands and has now become established on Hawaii and the Azores, the Canary Islands and Madeira. It is thought to have migrated to these islands although it might have been accidentally introduced.

**cynthia** see **Samia**
CYNTHIA MOTH see **Samia cynthia**

**Cyphura** Warren (URANIIDAE)
About 20 species are placed in this genus of moths. It is found in New Guinea and associated islands, and the Solomons. This is one of several genera of the subfamily Microniinae, characterized by their flimsy, Geometrid-like build, and the white ground-colour of the wings. *Cyphura* species have a short, outwardly directed tail on the hindwing.

**pardata** Warren
New Guinea 31–42 mm. (1·22–1·65 in.)
This is one of the more brightly coloured species of its genus and subfamily. Nothing has been published about the life history of this species. **317**b

**cypris** see **Morpho**

**Cyrestis** Boisduval (NYMPHALIDAE) MAP-WING BUTTERFLIES
These butterflies look like pieces of paper blown by the wind when they are in flight. The genus is known from India and S.E. Asia as far as New Guinea. Formerly the African species of Map butterflies were included in this genus, but they are now in *Marpesia*.

**cocles** Fabricius MARBLED MAP
India, Pakistan, Burma 50–60 mm. (1·97–2·36 in.)
The butterfly is pale green with faint lines across the wing from front to back. It has a short tail but the wings are much more rounded than those of *C. thyodamus*. The biology of this species is not known.

**nivea** Zinken-Sommer
Burma, Malaysia, Indonesia 40–50 mm. (1·57–1·97 in.)
There are many species with a similar type of pattern, often with a brown ground-colour, but with similar lines along the wings. Nothing is known of the biology of this species. It is common over its range, with some variation in colour and pattern and has been separated into subspecies in different parts. **229**e

**thyodamus** Boisduval COMMON MAP
N. India, Pakistan, Burma, S. China, Taiwan, Japan 60–70 mm. (2·36–2·76 in.)
This butterfly is generally found in hilly jungles and is broadly similar to *C. nivea*. The map-like markings make this very easily recognized; the flight too is characteristic with jerky irregular movements interspersed with a gliding action. The caterpillar feeds on species of *Ficus* (Moraceae). There is seasonal variation in the pattern as well as geographical variation over its wide range with several described subspecies.

**cyrnus** see **Graphium**
**Cyrtogone** see **Micragone**
**cytherea** see **Imbrasia**
**czekanowskii** see **Hyperborea**

**Dabasa** Moore (PAPILIONIDAE)
Two species are known. Both are restricted to S.E. Asia. *Dabasa* belongs to the same group of butterflies as *Graphium*. The early stages are unknown.

**payeni** Boisduval
Sikkim, Bhutan, N.E. India, Burma, Indochina to Borneo 75–120 mm. (2·95–4·72 in.)
This is a butterfly of medium elevations. The forewing of the female is even more finely pointed, apically, than that of the male (illustrated). There is a row of silvery lunules across the under surface of the hindwing in both sexes. Nothing is known about the caterpillar. **105**h

**Dahana** Grote (CTENUCHIDAE)
Two species are placed in this New World genus of moths: the illustrated *D. atripennis* and the smaller *D. cubana* Schaus from Cuba.

**atripennis** Grote BLACK-WINGED DAHANA
U.S. (Florida and Alabama) 28–38 mm. (1·1–1·5 in.)
The forewings of this distinctive N. American species are dark greenish brown, with a conspicuous yellow streak or triangle at the anal angle. The hindwings are a uniform iridescent blue and black, a colour which is repeated at the sides of the otherwise orange-yellow abdomen.

**Dactyloceras** Mell (BRAHMAEIDAE)
There are probably under 10 species in this genus of tropical African moths. The colour pattern of the various species is similarly intricate in each species, but differs in detail. The caterpillars bear numerous spiny tubercles.

**widenmanni** Karsch
E. Africa 105–112 mm. (4·13–4·41 in.)
Very few specimens of this species exist in collections and nothing is known about its life history. **340**g

**daedalus** see **Hamanumidia**
**daeta** see **Hyalyris**
DAGGER MOTHS see **Acronicta**
DAGGER-WING, RUDDY see **Marpesia petreus**

**Dalcera** Herrich-Schäffer (DALCERIDAE)
A genus of moths with species in South America. The species are characterized by tufts of scales on the antennae. The moths have rather rounded wings.

**abrasa** Herrich-Schäffer
S. America 38 mm. (1·5 in.)
This is a reddish yellow species with rounded wings. There are several similar, related species from different parts of S. America. The caterpillars are not well known but related species are described as creeping over leaves like snails. This species has been found in Colombia, Venezuela and the Guianas. **48**c

**Dalceridae**
A small family of moths related to the Limacodidae. The species in the family are found mainly in S. America and a few reach into the S. U.S. The moths are mostly white or yellow, usually with little pattern. See *Dalcera*.

**daltonae** see **Antistathmoptera**
**damajanti** see **Laxita**
**damaris** see **Orinoma**

**Damias** Boisduval (ARCTIIDAE)
The 45 or so species of this genus of moths are found in India and China, through S.E. Asia to the Solomons, and Australia. *Damias* is a member of the subfamily Lithosiinae. Most of the species are orange, black and yellow in coloration or orange, black and white.

**esthla** Prout
Goodenough Island 20–24 mm. (0·79–0·94 in.)
Nothing is known about this apparently rare species. It is typical in colour pattern of its genus. **364**m

**damone** see **Anthocharis**
**danae** see **Colotis**
DANAID EGGFLY see **Hypolimnas misippus**

**Danainae** (NYMPHALIDAE)
These are predominantly tropical African and S.E. Asian butterflies, but there are also about a dozen New World species. The caterpillars feed chiefly on plants of the families Asclepiadaceae and Apocynaceae which contain poisonous substances. The adult butterflies, which store heart-poisons synthesized from foodplant poisons eaten by the caterpillar, are largely unpalatable to vertebrate predators and advertise this fact by flaunting their bright colours and conspicuous patterns, both at rest and during their normally slow unhurried flight. Like other distasteful species of butterflies and moths, they are extremely tough and resilient in the adult stage, a characteristic which will allow a bird to take a trial peck without effecting serious damage to the butterfly. Many of the species look very much like one another and will be able to benefit from a relative's unpleasant experience in that a bird once having bitten a distasteful species will invariably be shy of attempting to eat any butterfly that looks like that species. The genera recognized at present are *Amauris*, *Danaus*, *Egialea*, *Diogas*, *Idea*, *Ideopsis*, *Ituna*, *Lycorea* and *Euploea*. The almost cosmopolitan *Danaus* is the most widespread, and the Old World *Euploea* the largest genus in number of species. See *Amauris*, *Danaus*, *Diogas*, *Euploea*, *Idea*, *Ideopsis*, *Ituna*, *Lycorea*.

**Danaus** Kluk (NYMPHALIDAE) TIGER BUTTERFLIES
This genus of the subfamily Danainae is found in N., C. and S. America, Africa, S.E. Asia, Australia and New Zealand. It is best represented in the Old World. Like most genera of Danainae its species are chiefly tropical in distribution. They are typically either orange, white or blue, with brown markings. The caterpillars of most species feed on members of the families Asclepiadaceae or Apocynaceae, and as adults, are generally distasteful to birds and other predators. Often flying together with species of *Danaus* are similarly patterned, palatable species of Papilionidae and the Nymphalid subfamily Nymphalinae, which are able to deceive predators by resembling the noxious species of *Danaus*. Open country, where flowers abound, are

the favourite haunts of *Danaus* species which are seldom seen in forested areas.

**affinis** Fabricius BLACK AND WHITE TIGER
S.E. Asia to Australia and the Solomons 48–60 mm. (1·89–2·36 in.)
This is a dark brown species, with typical 'tiger' markings of white in most subspecies, or orange and white in others. Numerous subspecies have been described, some of them restricted in range to small islands. The caterpillar is blue, yellow and white and has 3 pairs of fleshy red-based, black processes. In Australia it feeds on a species of *Cynanchum* (Asclepiadaceae), a trailing plant associated with reeds in brackish areas.

**aglea** Cramer
Sri Lanka, India to Formosa and Burma 60–70 mm. (2·36–2·76 in.)
The colour pattern of this species is typical of many members of its genus; there is a basal ray-like pattern and a row of marginal spots. In this species the ground-colour is dark brown or black and the markings translucent white. It is common in N.E. India where it occurs up to 1830 m. (6000 ft) in the Himalayas, and in Sri Lanka is one of the commonest species of butterfly. The caterpillar is yellow with white spots and a black head; it feeds on a species of Asclepiadaceae. Black and golden dots decorate the otherwise green chrysalis. *D. aglea* is mimicked by *Elymnias nesaea*.

**aspasia** Fabricius
Burma to Borneo and the Philippines 64–80 mm. (2·52–3·15 in.)
The forewing of this rather beautifully marked small species of *Danaus* is dark brown, with broad, ray-like bluish white markings at the base of the wings and a row of bluish white spots at the outer margin; the basal half of the hindwing is lemon yellow, with brown veins, the outer half is dark brown with bluish white marginal spots. In some subspecies there is more than the usual trace of lemon-yellow at the base of the forewing and most of the ray-like markings are yellow, not white.

**chrysippus** Linnaeus PLAIN TIGER, AFRICAN MONARCH, LESSER WANDERER, or GOLDEN DANIID
Canary Islands, Africa, Middle East, Japan and S.E. Asia to New Guinea, Australia and Fiji 66–82 mm. (2·6–3·23 in.)
The female colour forms of this species are mimicked in various parts of its range by the females of *Hypolimnas misippus*, *Elymnias hypermnestra*, by several other species of butterflies and by moths of the genus *Phaegorista* (Agaristidae) and others. *H. misippus* is probably the best known mimic; it has 2 distinct colour forms, each of which mimics a corresponding colour form of *D. chrysippus*. Apart from the Monarch *(D. plexippus)* this is the only species of its genus to have reached Europe as a migrant; in Africa it is one of the commonest butterflies. Up to 12 generations per year have been recorded for the tropical elements of *D. chrysippus*, whereas there are only 1–2 generations in the temperate parts of its range. The caterpillar is a beautiful yellow, black and pale blue creature; the normal foodplants include species of *Calotropsis* and other genera of Asclepiadaceae, but it has also been recorded from species of Rosaceae and Scrophulariaceae. **127**b

**eresimus** Cramer ERESIMUS BUTTERFLY
U.S. to Panama and the Caribbean islands 65–75 mm. (2·56–2·95 in.)
This species breeds in the S. U.S. (Texas and Florida) but is an essentially C. American and Antillean butterfly. It is similar in general appearance to *D. gilippus* (illustrated), but has well marked, black veins on the hindwing.

**erippus** see **Diogas erippus**

**formosa** Godman BEAUTIFUL MONARCH
Tropical Africa 70–94 mm. (2·76–3·7 in.)
This species is found both in forested areas, and in gardens where it is attracted to flowers. It is mimicked by a form of *Papilio rex* (illustrated) which is much larger than *D. formosa* but closely similar in colour pattern. Its caterpillar feeds on species of Asclepiadaceae.

**gilippus** Cramer QUEEN BUTTERFLY
S. U.S., C. America including Mexico, tropical S. America 70–75 mm. (2·76–2·95 in.)
In N. America, specimens have been captured as far north as Nebraska, but *D. gilippus* is normally only a Gulf State resident. Unlike the Monarch *(D. plexippus)*, this delicately marked species is non-migratory, a feature which has allowed the evolution of several subspecies. Jamaica, for example, has its own subspecies, the Jamaica Queen *(D. gilippus jamaicensis)*. In Florida, the Queen is more common than the Monarch and becomes the model for the Viceroy, *Limenitis archippus*, which has a darker brown coloration in this part of its range, more like that of the Queen than the paler Monarch it mimics in the north. The courtship behaviour of this species has been studied recently in some detail. The male has a pair of specialised hair-pencils which disperse aphrodisiac scents prior to mating. Some male scents have been shown to be highly specific in their action and probably provide a mechanism whereby wasteful mating between different species is inhibited, especially in those day-flying species belonging to Müllerian assemblages of externally similar species in which visual recognition of the male by a female is difficult. The caterpillars of *D. gilippus* feed mainly on species of Asclepiadaceae, but also feed occasionally on the foliage of other plants. Adults derived from non-*Asclepia* feeding caterpillars are not unpalatable to birds and are, rather anomalously, Batesian mimics of their own species. **127**a

**hamata** McLeay
Sri Lanka, India, most of S.E. Asia to the S.W. Pacific islands 72–95 mm. (2·83–3·74 in.)
This species has one of the most extensive ranges of S.E. Asian Danaine species and is a common butterfly in parts of its range. Most of its numerous subspecies are black or dark brown above, with white or greenish white streaks and spots; there is considerable variation in the shape and size of these markings. The name *D. melissa* Cramer has been applied incorrectly to this species in the past. (See also under *Graphium megarus*.)

**juventa** Cramer
Malaya to New Guinea and the Solomons 60–70 mm. (2·36–2·76 in.)
There is considerable geographical variation in the colour pattern of this widespread and often common S.E. Asian species, both in the adult and the caterpillar. The wings are dark brown, with the usual *Danaus* pattern of basal rays and marginal spots which can be yellowish, bluish or greenish white in colour and can vary both in size and in the degree of transparency.

**limniace** Cramer BLUE TIGER or BLUE MONARCH
Sri Lanka, India, Burma, Thailand, Borneo, Sulawesi (Celebes), Africa S. of the Sahara 90–100 mm. (3·54–3·94 in.)
The strong flight of this butterfly is more powerful than in many other species of its subfamily. It is very common both in low-lying areas and in the hills. Like many other Danaines, it is mimicked by supposedly palatable species of other groups. The mimics include a form of *Papilio clytia* in India, and *Graphium leonidas* in Africa. **127**d

**melaneus** Cramer CHOCOLATE TIGER
Sikkim, N. India, Burma, S. China to Malaya & Java 85–95 mm. (3·35–3·74 in.)
This is a generally common species in the hills of N.E. India where it occurs up to 2740 m. (9000 ft). The wings can be either black or dark brown, with pale bluish white markings corresponding on the forewing to those of *D. sita* (illustrated). It is a model for the butterfly *Graphium xenocles*, a species of Papilionidae.

**melanippus** Cramer WHITE or BLACK-VEINED TIGER
India & Burma to the Philippines & Sulawesi (Celebes) 80–95 mm. (3·15–3·74 in.)
This butterfly is similar in pattern to the Monarch, *D. plexippus*, but differs in the ground colour of the hindwing which is normally white, not orange. Its several subspecies differ from each other in details of the coloration and pattern. In one subspecies the ground colour of the hindwing is reddish brown, somewhat similar to that of a dark Monarch.

**melissa** see under **D. hamata**

**plexippus** Linnaeus MONARCH, MILKWEED or WANDERER BUTTERFLY
N., C. and S. America, Canary Islands, Indonesia, New Guinea, Australia, New Zealand 75–100 mm. (2·95–3·94 in.)
This is one of the most successful species of butterfly. Originally a New World species, its migratory habits, amongst other factors, produced dramatic extensions of its range during the last century. In N. America successive broods move northwards as far as Canada during the summer, the progeny eventually returning southwards towards California, Florida and Mexico where many overwinter in communal roosts, some of which become tourist attractions. Experimental evidence shows that heart-poisons stored by the caterpillar from its milkweed *(Asclepia)* diet and passed on to the adult butterfly, are, not surprisingly, noxious to birds which learn, even after only one unpleasant experience, to associate the Monarch's colour pattern with unpalatability. The best known mimic of this species is probably the female of the N. American Viceroy butterfly *(Limenitis archippus archippus)*, a member of the subfamily Nymphalinae whose species are usually quite unlike those of the subfamily to which *Danaus* belongs (Danainae). Fairly recent work has shown that some specimens of *D. plexippus* contain no cardiac glycosides (heart poisons), probably because their caterpillars feed on foodplants other than the normal milkweeds, and were therefore edible to birds. These edible specimens are mimics of their own species (automimics). Other automimics will probably be discovered and may explain some of the anomalous mimetic relationships in the Heliconiinae and other groups. **128** (communal winter roost), **129**

**pumila** Boisduval
New Hebrides, New Caledonia, the Loyalty islands 35–49 mm. (1·38–1·93 in.)
This is one of the smallest species of its genus. It is an unusual, transparent green in colour. Relatively few specimens are represented in most national collections. Nothing is known about its life history. **127**g

**schenckii** Koch
The Lesser Sundas, New Guinea to the Solomons 62–70 mm. (2·44–2·76 in.)
The wings of the male, especially the forewings, are less whitish and opaque than those of the illustrated female of this beautiful species. **127**e

**sita** Kollar: syn. *D. tytia* Gray CHESTNUT TIGER
N. India, Sikkim, Bhutan, Nepal, Burma, China, S. Japan, Malaya, Sumatra 72–110 mm. (2·83–4·33 in.)
This beautifully marked species flies in forested areas, most commonly between 910 and 2440 m. (3000–8000 ft), but has been caught as high as 3050 m. (10,000 ft). It is mimicked by the Tawny Mime *(Papilio agestor)*, the Great Zebra *(Graphium zenocles)* and the Circe *(Aldania nama)*. Its caterpillar has been recorded from a species of *Marsdenia* (Asclepiadaceae). **127**f

**danava** see **Limenitis**
DANIID, GOLDEN see **Danaus chrysippus**
**danilovi** see **Stamnodes**

**Danima** Walker (NOTODONTIDAE)
A single species of moth is placed in this genus.

**banksiae** Lewin BANKSIA MOTH
Australia 56–73 mm. (2·2–2·87 in.)
This is one of the most colourful species of the family Notodontidae. Females differ from the illustrated male in the brown, not white hindwings. The conspicuously spotted caterpillar has a Sphingid-like horn at the rear end and is capable of everting an x-shaped, flesh-coloured process from beneath the front segments when suddenly disturbed. The function of this process is doubtful; it may be used to disconcert predators or to ward off parasitic wasps and flies. The foodplants of the caterpillar include *Banksia* and related genera. **357**e

**Danis** Fabricius (LYCAENIDAE)
This is a tropical genus of butterflies found in the Philippines, Moluccas, New Guinea, Australia and the Solomons. Males are typically blue and white above, with black borders, and with white and green markings on the under surface.

**danis** Cramer LARGE GREEN-BANDED BLUE
Moluccas to New Guinea and N.E. Australia 38–40 mm. (1·5–1·57 in.)
The upper surface of the male and the under surface of the female are illustrated. The male under surface is similar to that of the female; the female upper surface is dark brown, with the transverse white band continued nearly to the front edge of the forewing. The caterpillars have been recorded from red ash *(Alphitonia)*. **242**b, d

**danis** see **Hypolycaena**
**daos** see **Ideopsis**
**daphalis** see **Pyrausta**

**Daphnis** Hübner (SPHINGIDAE)
There are under 10 species in this almost cosmopolitan genus of hawk-moth.

**nerii** Linnaeus OLEANDER HAWK-MOTH
Africa, India, Mediterranean Europe 80–120 mm. (3·15–4·72 in.)
The caterpillar of this beautiful moth feeds on Lesser Periwinkle *(Vinca minor)*, and on Oleander *(Nerium)* a genus of plants poisonous to vertebrates. It does not store toxins from *Nerium* leaves, however, and the adult is consequently not chemically protected from predators and is cryptically, not warningly coloured. *D. nerii* is recorded in Britain as a migrant from time to time, and very rarely in Japan. **344**a

**daphnandra** see **Charagia**
**daphne** see **Brenthis**

**Daphoenura** Butler (NOCTUIDAE)
There is only one species in this genus of moths.

**fasciata** Butler
Madagascar 55–60 mm. (2·17–2·36 in.)
Both pairs of wings of this very distinctive species are deep yellowish orange above. The forewings have 4 black transverse bands, a single black central marking, and are black at the base. The hindwings are black marginally and at their base. The head and thorax are red above; the abdomen black with a red posterior zone. **399**c

**daplidice** see **Pontia**
DAPPLED WHITE BUTTERFLIES see **Euchloe**
**dardanus** see **Papilio**
DARK
ARCHDUKE see **Euthalia khasiana**
BORDERED BEAUTY see **Epione paralellaria**
CATSEYE see **Zipaetis scylax**
DAGGER see under **Acronicta psi**
GRASS-BLUE see **Zizeeria knysna**
GREEN FRITILLARY see **Mesoacidalia aglaja**
PURPLE AZURE see **Ogyris abrota**
SWORD-GRASS MOTH see **Agrotis ipsilon**
TUSSOCK MOTH see **Dasychira fascelina**
YELLOW UNDERWING, SMALL see **Anarta cordigera**
DARK-BARRED TWIN-SPOT CARPET MOTH see **Xanthorhoe ferrugata**
DARK-BRANDED BUSH-BROWN see **Mycalesis mineus**

**Darna** Walker (HYPSIDAE)
There are at present 18 species in this exclusively tropical S. American genus of moths. One group of brilliantly coloured species which includes *D. colorata*, is a model for species of *Lucillella*, a genus of Nemeobiid butterflies. Another group is mimetically associated with Dioptidae.

**colorata** Walker
Colombia, Peru, Ecuador 32–42 mm. (1·26–1·65 in.)
*D. colorata* may be mimicked in Ecuador by the Nemeobiid butterfly *Lucillella camissa*. The male, as in others of its group, bears a patch of scent scales on the part of the hindwing overlapped by the forewing. **381**k

**rubriplaga** Warren
Ecuador 35–43 mm. (1·38–1·69 in.)
This is similar to *D. colorata*, but the forewing patch is slightly smaller and is a much more reddish orange in colour in most specimens, and the iridescent blue of the forewings is replaced by a rich violet.

**Dasychira** Hübner (LYMANTRIIDAE)
There are probably over 500 species in this almost cosmopolitan genus of moths. The only major region of the world without representatives is tropical America. The caterpillars have a characteristic 'toothbrush' of 4 dorsal tufts on the middle of the back and a further tuft on the eighth segment of the abdomen. *D. grotei* and others are pests of teak *(Tectona)* in India, and 4 species attack tea *(Thea)* in N. India.

**fascelina** Linnaeus DARK TUSSOCK
Europe including Britain, through temperate Asia to W. Siberia 40–50 mm. (1·57–1·97 in.)
There is some variation in the ground-colour of the forewing but both wings are usually darker in colour than in the closely related and similarly patterned *D. pudibunda*. The handsome black caterpillar bears tufts of yellow, grey, black and white hairs; it feeds on hawthorn *(Crataegus)*, willow *(Salix)*, broom *(Cytisus)* and ling heather *(Calluna)*.

**pudibunda** Linnaeus PALE TUSSOCK
Europe including Britain, through temperate Asia to Japan 42–64 mm. (1·65–2·52 in.)
The particularly beautiful caterpillar of this species is green or greenish yellow, with black intersegmental bands, 4 yellow dorsal tufts, and a long, red pencil of hairs near the end of the abdomen. It feeds from July to September on the foliage of numerous deciduous trees, and on hops *(Humulus)* where before the days of insecticidal sprays it was the well-known 'Hop-dog' in the hop fields of Kent, S.E. England. **360**

**Dasysphinx** Felder (CTENUCHIDAE)
There are 17 described species in this genus of moths, but there are several others already in collections yet to be described. They are restricted to C. America and tropical S. America. Most species have transparent wings with dark edges.

**torquata** Druce
Brazil 41–50 mm. (1·61–1·97 in.)
This is one of the more colourful species of its genus. The orange-red end to the abdomen produces a distinctly bumblebee-like appearance. There is no possibility of mimicry between *D. torquata* and bumblebees, however, as the latter are not found in S. America. **364**d

**Datana** Walker (NOTODONTIDAE)
HAND-MAID MOTHS
The few species of *Datana* are essentially N. American in distribution but the genus is also represented in C. America. The forewing of most species bears a series of transverse lines. Some are minor pests.

**ministra** Drury YELLOW-NECKED APPLE-WORM, YELLOW-NECKED CATERPILLAR
Canada, U.S. 38–48 mm. (1·5–1·89 in.)
The head and front of the thorax of this species are a rich chocolate brown above; the forewings yellowish or reddish brown, with darker brown transverse lines; the hindwings are yellowish brown. Several other N. American species are similar in pattern to *D. ministra* but differ in details of the pattern. The yellow-striped caterpillars of this widespread N. American species are at first gregarious on one leaf, their heads pointing towards the edge. The foliage of numerous fruit trees, blueberry *(Vaccinium)* and several deciduous forest trees are eaten by the caterpillars which are sometimes pests, especially in apple orchards *(Malus)*.

**Daulia** Walker (PYRALIDAE)
Pyralid moths of the Nymphulinae subfamily many of whose caterpillars are aquatic. This is a small genus with species from India to Australia and one in S. America.

**aurantialis** Hampson
Malaysia N. India, Burma 15–16 mm. (0·59–0·63 in.)
The delicate silver lines on the wings are characteristic of species in the genus. The moths have a very delicate appearance. Nothing is known of the biology of this species. **63**d

DEATH'S-HEAD HAWK-MOTH see **Acherontia atropos**

**Decachorda** Aurivillius (SATURNIIDAE)
There are about 10 species of moth in this tropical African genus. They are atypical of many species of Saturniidae in that the forewings are rounded and there are no eyespot markings or transparent patches on the wings. The caterpillar feeds on various grasses.

**rosea** Aurivillius
Subtropical E. Africa, S. Africa 34–46 mm. (1·34–1·81 in.)
Some males of this round-winged species are uniformly red above, others are orange-yellow. Females are usually brownish orange. The grass-feeding caterpillar is very hairy and similar in appearance to the Woollybear caterpillars of *Arctia caja* (Arctiidae).

**decolor** see **Ectomyelois**
**decora** see **Amnosia, Ecpantheria**
**defectalis** see **Megalorhipida**
**defoliaria** see **Erannis**
**degeerella** see **Nemophora**
**deidamia** see **Morpho**

**Deilephila** Laspeyres (SPHINGIDAE)
The 5 species of this hawk-moth genus are found in Europe, eastwards to Japan and as far south as N. India.

**elpenor** Linnaeus ELEPHANT HAWK-MOTH
Europe, through temperate Asia to Japan 58–65 mm. (2·28–2·56 in.)
The caterpillar of *D. elpenor* tapers towards the head in a distinctly trunk-like way, while the 2 pairs of eyespots on the thorax accentuate this long-nosed appearance; there is a green form and a brown form. It feeds on the leaves of willowherb *(Epilobium)*, bedstraw *(Galium)*, *Fuschia* and other plants. The adults are often seen sucking nectar from honeysuckle *(Lonicera)* blooms at dusk. **348**

**deiphile** see **Prepona**
**dejanira** see **Hypocrita**
**delesserti** see **Graphium**

**Delias** Hübner (PIERIDAE)
A large genus of butterflies with species from India and Tibet to Australia. There are many species in New Guinea. The caterpillars live in a communal silk web spun amongst the leaves and branches of the food-plants which are generally species of mistletoe *(Loranthaceae)*.

**aglaia** Linnaeus RED-BASE JEZEBEL
India, Pakistan, Burma, Indonesia Formosa, Malaysia 70–90 mm. (2·76–3·54 in.)
The flight is slow and, as with the others of the genus, the colour is a warning one. The insect is believed to be distasteful and to be avoided by birds or other potential predators. The forewing has black veins with grey-black between white; the hindwings have greyish outer marks and a yellow middle part with white spots edged with black along the margin. The underside forewing is darker than the upperside while the hindwing has a dark red base, black veins with grey-yellow between and a black border. This species is widespread and often common. Many subspecies have been described.

**aganippe** Donovan WOOD WHITE
Australia 60–65 mm. (2·36–2·56 in.)
This species is common throughout S. Australia, occasionally reaching further north. The females are similar to the males but have more black on the wing margin. The underside has many red spots around the margin of the hindwing and at the base, and yellow spots near the apex of the forewing and along the inner margin of the costa. The caterpillar feeds on *Amyema cambegei* and other species of Loranthaceae.

**belisama** Cramer
Indonesia 60–80 mm. (2·36–3·15 in.)
This species is common in Java. The striking underside pattern, with a red margin to the yellow hindwing, contrasts with the upperside which is less colourful.

In the male the upper side is white, with a large sooty black patch covering the apex of the forewing and a narrow black margin to the hindwing. The females tend to have the white areas in the male replaced by an orange-yellow colour. There is considerable variation in the pattern and amount of black on the wing and a number of subspecies have been described. **127k**

**belladona** Fabricius HILL JEZEBEL
India, Pakistan, Burma, S. China, Indonesia 70–90 mm. (2·76–3·54 in.)
This is a common species between 610–3050 m. (2000–10,000 ft). It is attracted to *Buddleia* and other flowers. The butterflies are about from April and in many areas there are 3 broods. Their flight is generally slow and they usually occur gregariously, but fly rapidly if disturbed. This is a black butterfly with variable grey-white or creamy coloured, usually rounded, patches near the apex of the cell. There are more grey-white patches in the hindwing, with the veins clearly darker. The hindwing also has the conspicuous yellow patch at the front near the base and another patch of yellow at the posterior tip of the hindwing.

**descombesi** Boisduval RED-SPOT JEZEBEL
India, Pakistan, Burma, S.E. Asia, Indonesia 65–90 mm. (2·56–3·54 in.)
This is a butterfly of warm valleys, common in the foothills of the Himalayas up to 1520 m. (5000 ft). It flies from March to December and is commonly seen feeding on flowers. The forewings are black with grey patches, particularly in a row below the outer margin. The hindwing has a pale grey-white centre and a black margin. Below, the forewing is similar, with paler edges to the veins; the hindwing is deep yellow with a bright red, black-edged, patch along the front edge. The margin is black, with grey patches between the veins on the margin. A number of subspecies have been described.

**eucharis** Drury COMMON JEZEBEL
India, Pakistan, Sri Lanka, Burma 66–83 mm. (2·6–3·27 in.)
The butterfly is greyish white above with black veins and a black line down the wing near the margin. The hindwing is similar, but paler. Along the hindwing margin are patches of pinky-red, edged by the submarginal line. The underside forewing is like the upper, but the hindwing is yellow, with black veins and large patches of red along the wing margin. The butterfly is very common over most of its range and is found almost everywhere there are trees, even in the middle of towns, where it can often be seen flying slowly about. The caterpillars feed on species of Loranthaceae.

**ninus** Wallengren MALAYAN JEZEBEL
Malaysia, Indonesia 60–75 mm. (2·36–2·95 in.)
This is one of the commonest Malaysian species of *Delias* and can be found in every hill station even above 910 m. (3000 ft). The forewing veins are black, rather broadly so, giving the appearance of a blackish rather than a white butterfly. The hindwing is yellow with a red basal patch, edged with bluish grey. On the underside the yellow extends in large oval patches between the veins on the hindwing. There is considerable variation in the pattern, both of the upper and underside of the wing and a number of subspecies have been described. Some of them have white rather than yellow on the upperside of the hindwing. The butterflies are strong fliers and regularly feed at flowers, particularly in the earlier part of the morning. There are several related species with a broadly similar appearance to *D. ninus* and identification of these needs care.

**delius** see **Antanartia**
**delphica** see **Chresmarcha**
**delphis** see **Polyura**
DELTOID MOTHS see **Hypena**
**demetrius** see **Papilio**
**democles** see **Nirachola**
**demodocus** see **Papilio**
**demoleus** see **Papilio**
**demophoon** see **Prepona**

**Dendrolimus** Germar (LASIOCAMPIDAE)
This is a small genus of robust moths, related to *Lasiocampa*, found in Europe and C. Asia.

**pini** Linnaeus PINE LAPPET MOTH
Europe, C. Asia 52–80 mm. (2·05–3·15 in.)
This is a common and sometimes destructive species in the pine woods of Europe. It is widely distributed except in Britain, where a single specimen was reported in 1909. The moths are highly variable in coloration, especially in the ground-colour of the forewing which can be greyish white or one of many shades of brown or grey and almost black in some specimens. The caterpillars are a serious pest of pines, especially Scots Pine (*Pinus sylvestris*). **318**n

**dentaria** see **Selenia**
**dentella** see **Ypsolopha**

**Dercas** Doubleday (PIERIDAE)
A small genus of butterflies found in India and S. China with species in Burma, Malaysia and into Indonesia.

**verhuelli** Hoeven TAILED SULPHUR
India, Pakistan, Burma, S. China, Hainan, Thailand 60–70 mm. (2·36–2·76 in.)
A rather square-looking, pale yellow butterfly with black tips to the forewings, which have wavy edges, and a slight tail to the hindwing. A small orange double spot and a thin line across the wing make up the rest of the pattern. The males are common on flowers and can be seen drinking at damp patches of sand.

DESERT ORANGE TIP see **Colotis evagore**
**descombesi** see **Delias**
**desgodinsi** see **Campylotes**
**despecta** see **Cicinnus**
**desumptana** see **Hypertropha**

**Deudorix** Hewitson (LYCAENIDAE)
CORNELIAN BUTTERFLIES
The range of this small genus extends from India and Formosa through S.E. Asia to Australia, the Solomons and Samoa. There are both red and blue species. The males are generally more colourful than the females and are frequently found on tree-clad hilltops in company with other butterflies which are similarly addicted to heights. *Deudorix* caterpillars feed inside various fruits.

**antalus** Hopffer BROWN PLAYBOY
Africa 25–30 mm. (0·98–1·18 in.)
The wings of this species are an unusual light bluish brown (almost white) above, with a pinkish violet iridescence. It occurs throughout Africa in scrubland and savannah, often on the crests of hills. The foodplants of the caterpillar include *Crotalaria* and *Acacia* (Leguminosae).

**dinomenes** Grose-Smith
Tropical E. Africa 20–27 mm. (0·79–1·06 in.)
The male of this rare, short-tailed butterfly is coppery orange above with a black apex to the forewing, a small brown sex-brand near the front edge of the hindwing and a black spot near the base of the tail. The female wings are brown above with a weak, violet iridescence and with paler central areas on each wing.

**epijarbas** Moore CORNELIAN
Sri Lanka, India, Formosa and S.E. Asia to Samoa 34–44 mm. (1·34–1·73 in.)
The caterpillars feed on seeds inside fruits of various trees. The seeds of pomegranate (*Punica*) form one of the food sources in Malaya and India; in the Himalayas the seeds – 'conkers' of horse chestnut (*Aesculus*) may be attacked, and in Australia tulipwood (*Harpullia*) seeds. The forewings of the males are brownish orange above, with a broad, dark brown front and outer marginal band; the hindwings are similarly brownish orange, with brown at the base and along the inner margin, and with an orange-bordered black spot near the base of the single hindwing tail. Females are dull brown above with a hindwing spot, as in the male, and with a trace of orange on the middle of the forewing in some specimens.

**hypargyria** Elwes SCARCE CORNELIAN
N.E. India, Burma, Malaysia 40–44 mm. (1·57–1·73 in.)
This species is similar to *D. epijarbas*, but the male is a more yellowish orange above and the female has a white patch on the hindwing. It is extremely rare in collections as its popular name indicates.

**lorisana** Hewitson
E. Africa 19–30 mm. (0·75–1·18 in.)
The caterpillars of *D. lorisana* are sometimes a pest of coffee (*Coffea*) plantations in E. Africa. The adult butterfly has orange-red hindwings and black forewings which bear a triangular orange-red marking in most male specimens.

**Deuterotinea** Rebel (DEUTEROTINEIDAE)
A small genus of micro-moths with species in N. Africa and Europe.

**casanella** Eversmann
Europe 22 mm. (0·87 in.)
This is a rather grey coloured species with narrow fore- and hindwings. The outer margin of the forewing has a white border, the hindwings are unmarked greyish white. The adults have been found flying in birch woods.

**Deuterotineidae**
A small and rather obscure family of micro-moths of uncertain affinities. The species were formerly included in the Tineidae, but currently are considered nearer the Psychidae.

DEVIL, HICKORY HORNED see under **Citheronia regalis**
DEW MOTH see **Setina irrorella**
DEWITZ'S PRINCELING see **Vegetia dewitzi**
**dewitzi** see **Vegetia**
**dexithea** see **Hypolimnas**

**Diacrisia** Hübner (ARCTIIDAE)
This is a large Old World genus which currently includes many species that should be transferred to other genera – especially to *Spilosoma*. The caterpillars of *D. aurantiaca* Holland, *D. maculosa* Cramer and *D. rattrayi* Rothschild are pests of cocoa (*Theobroma*) trees in W. and C. Africa.

**breteaudeaui** Oberthür
China (Tibet), Sikkim 24–37 mm. (0·94–1·46 in.)
Most specimens of this brilliantly coloured moth have been captured between 3960 and 4880 m. (13,000–16,000 ft) in the Himalayas. Females are invariably smaller than the males (illustrated) but are probably dimorphic for size: some have very short, stunted wings measuring about 24 mm. (0·95 in.) from tip to tip, others are about 32 mm. (1·26 in.) in wingspan. **368**c

**metelkana** Lederer SWAMP TIGER-MOTH
C. Europe to Japan 35–48 mm. (1·38–1·89 in.)
This is a species of marshy country. The caterpillar feeds on *Iris* and other marshland plants and is able to swim to safety across the surface of water if it falls from its foodplant. The adult is similar in some features to the allied *D. sannio*; the forewings are usually yellow with a few black specks in the male, and pale orange with diffuse orange-red markings in the female; the hindwings are orange-red, with a single black medial marking, and 4 or more black, outer marginal spots. *D. metelkana* is placed by some authors in the genus *Rhyparioides* Butler. **368**d

**purpurata** Linnaeus
Europe, temperate Asia to China and Japan 43–60 mm. (1·69–2·36 in.)
This species is common throughout its range, a result, partly, of the wide variety of plants accepted by the caterpillar which will feed on the foliage of many low-growing plants (for example *Galium*) as well as shrubs and trees (such as *Salix* and *Prunus*). The caterpillar overwinters and then continues to feed until May. The moths emerge in June and July of the same year. There is a considerable amount of variation in the colour pattern of the wings: the ground-colour of the hindwing is essentially red, but is sometimes yellow, and both the

forewing and hindwing spots can be heavily marked – more so than in the illustration – or nearly absent. **369**q

**rufescens** Brullé CANARIES' TIGER-MOTH
Canary Islands 48–60 mm. (1·89–2·36 in.)
*D. rufescens* is known only from the islands of Tenerife and Gomera in the Canaries group. The forewings are yellow brown or reddish brown, with 4 large black markings along the front edge of the wing, and usually with a scattering of black scales. The hindwings are pinkish brown, with a trace of a blackish brown transverse band towards the outer margin. Its caterpillars feed on *Rumex* and *Kleinia* foliage.

**sannio** Linnaeus CLOUDED BUFF
Europe including Britain, through temperate Asia to Japan 35–43 mm. (1·38–1·69 in.)
The male and female of *D. sannio* are so different in colour pattern that Linnaeus described each sex as a different species. Females differ from the illustrated male in the brownish red or yellow ground-colour, the absence of a dark marking on the forewing and the more extensive black markings on the hindwing. The moths emerge in June or July; males fly during the day in the sunshine but females seldom fly before the evening. *D. sannio* occurs in wooded areas and on heaths. The hairy brown caterpillar overwinters at an early stage in its growth – it feeds on dandelion *(Taraxacum)*, dock *(Rumex)* and other low-growing plants. **371**

**Diachrysia** see **Plusia**
DIADEM BUTTERFLIES see **Hypolimnas**

**Diaethria** Billberg (NYMPHALIDAE)
This butterfly genus is related to *Catagramma* and has species with beautiful metallic colours on the upperside and a striking pattern on the hindwings. There are many species in C. and tropical S. America and in the West Indies.

**aurelia** Guenée '89' BUTTERFLY
Trinidad 38–42 mm. (1·5–1·65 in.)
This is one of the common species in Trinidad. The underside hindwing pattern forms the '89' from which the butterfly gets its name. The sexes are similar in this species. The caterpillar feeds on *Trema micrantha* (Ulmaceae). The upperside is black with a narrow green line below the apex of the forewing and a broad green band in the middle of the wing, both greens are bright and iridescent. **217**n

**clymena** Cramer FIGURE OF EIGHT or JEWEL BUTTERFLY 44 mm. (1·73 in.)
C. & S. America, rare visitor S. U.S.
There are several similar looking species with a strongly patterned underside, all of which are rather difficult to separate. Generally they are found in forests on mountain slopes. The butterflies are readily attracted to rotting fruit. The foodplant of the caterpillar is not known for certain although it is suggested as species of *Trema* (Ulmaceae). The underside and upperside are different but there is little difference in pattern or colour between the males and females. The species is found as far south as Paraguay, north to Guatemala. This beautiful jewel-like butterfly is common round Rio de Janeiro where it is a great favourite with children who collect it. Many subspecies have been described in different parts of its range. **221, 229**s

**Dialectica** Walsingham (GRACILLARIIDAE)
A small genus of micro-moths with species mostly in Africa, although the limits of the genus are not well known. The moths in the genus are very small and their caterpillars pass their entire life between the upper and lower surface of one leaf.

**carcharota** Meyrick
South Africa, Rhodesia 9 mm. (0·35 in.)
The slender forewings are grey-brown along the costa with an irregular white band along the hind margin. The hindwings are grey-brown. The head and thorax of the moth are white. The caterpillar makes an irregular shaped blotch mine in the leaves of *Lithospermum* (Boraginaceae). The moth is very similar to the European *D. scalariella* Zeller.

DIAMOND BACK MOTH see **Plutella xylostella**
**dianassa** see **Eueides**

**Diaphania** Hübner (PYRALIDAE)
One of the many genera which have been separated from the genus incorrectly known as *Margaronia*. There are a number of species with rather similar appearance from the Old and New Worlds but at present the generic name *Diaphania* is restricted to the New World species.

**superalis** Guenée: syn. *translucidalis*
S. America 28–55 mm. (1·1–2·17 in.)
This species, or ones like it, are widespread in S. America. The distinctive pattern is also found in some Old World species. *D. superalis* has been found in Brazil, Bolivia, Ecuador, Peru and the Guyanas. Information on the life history of this and allied species would help in sorting out the problem of complex groups of similar-looking species. **63**k

**Diaphora** Stephens (ARCTIIDAE)
This genus was recently re-established to accommodate a single species, *D. mendica*.

**mendica** Clerck MUSLIN MOTH
Europe including Britain, W. & C. Asia 30–40 mm. (1·18–1·57 in.)
As in many other spotted species there is considerable individual variation in the size and number of the spots. In male specimens the ground-colour of the wings is either dull yellow, or grey, whereas females are usually white. Its hairy brown caterpillar feeds in July on numerous low-growing plants, shrubs and, less commonly, on the foliage of trees. The species overwinters as a chrysalis. **372**

**Diatraea** Guilding (PYRALIDAE)
A large American genus of Pyralid moths with probably over 100 rather similar looking species, many of them pests of crops. Usually they have simple patterns and triangular forewings with straight outer edges to them. The genus is in the subfamily Crambinae.

**saccharalis** Guilding SUGAR-CANE BORER
C. and S. America, Florida, West Indies 26 mm. (1·02 in.)
The straw coloured forewings are marked with black dots which form a V when the wings are closed. The young caterpillars feed on the leaves while the older ones bore into the stems of sugar-cane *(Saccharum)* and rice *(Oryza)*. This moth is often a major pest of sugar and rice crops and attacks by it can cause heavy losses. In the W. Indies attempts are being made to control it by introduction of a parasite which attacks the caterpillar. The moth is believed to have been accidentally introduced to America in the middle of the 19th century.

**Dichocrocis** Lederer (PYRALIDAE)
A large and very mixed genus of Pyralid moths, found in the warmer parts of the world, except the Americas. The genus has not recently been studied and many of the species in it, including the one figured, could be transferred to other genera.

**zebralis** Moore
India 26 mm. (1·02 in.)
This species is found in N. India. The antennae of the specimen figured have been broken, normally they are longer. The biology of this species is not known. **56**b

**Dichogama** Lederer (PYRALIDAE)
A genus of large, robust Pyralid moths, more like Noctuids in appearance than Pyralids. The forewings are rather square-tipped and the abdomen is long. The caterpillars live together in silken nests on leaves of species of *Capparis* in America and the W. Indies.

**redtenbacheri** Lederer
C. America to Florida 30–35 mm. (1·18–1·38 in.)
This is a pearly white moth with prominent brown wavy lines on the wings. It is widespread in tropical America and Florida. The species is variable in pattern and there are rather similar looking specimens, perhaps of another species, from Peru.

**dichroa** see **Sephisa**
**dictyaria** see **Cyclidia**

**Didasys** Grote (CTENUCHIDAE)
There is a single known species in this moth genus.

**belae** Grote DOUBLE-TUFTED WASP-MOTH
U.S. (Florida, and rarely in other S. states) 21–25 mm. (0·83–0·98 in.)
The two long, anal scent-pencils are present only in the male. Yellow replaces orange in the wing-markings in some specimens, and in a minority of examples all red coloration is replaced by yellow except at the anal angle of the abdomen. This species is presumably a wasp mimic like many other members of its family. **367**g

**didius** see **Morpho**
**dido** see **Philaethria**
**didyma** see **Melitaea**
**dietzi** see **Setiostoma**
**diffusa** see **Bellura**

**Digama** Moore (HYPSIDAE)
The range of these chiefly yellow, or yellow and brown moths includes much of Africa south of the Sahara and India through S.E. Asia to Australia.

**marmorea** Butler
Australia, New Guinea 29–33 mm. (1·14–1·3 in.)
This is a Lithosiid-like species with cryptically patterned dark brown and brownish white forewings, and yellow hindwings marked with brown at the apex. It is common in N. Australia. The caterpillar feeds on species of *Carissa*, a genus of the poisonous plant family Apocynaceae.

**Diloba** Boisduval (NOTODONTIDAE)
A single species is currently placed in this genus of moths.

**caeruleocephala** Linnaeus FIGURE OF EIGHT MOTH
Europe including Britain, W. Asia 35–42 mm. (1·38–1·65 in.)
The middle of the otherwise chiefly brown forewings of this common species bears 2 greyish white markings, the innermost of which (sometimes both) is often in the shape of a figure eight. It flies in the autumn, as late as November. The yellow and blue caterpillar feeds often at the tips of shoots and branches in conspicuous positions but is apparently avoided by birds. The foodplants include hawthorn *(Crataegus)*, sloe *(Prunus)* and other wild and cultivated trees of the family Rosaceae.

**dilutata** see **Epirrita**
DINGY
ANTINEPHELE see under **Antinephele maculifera**
ARCTIC FRITILLARY see **Clossiana improba**
SKIPPER see **Erynnis tages**
SWALLOWTAIL see **Papilio anactus**
**dinizi** see under **Hebomoia**
**dinomenes** see **Deudorix**
**diocletiana** see **Euploea**

**Diogas** d'Almeida (NYMPHALIDAE)
*Diogas* was erected for a single species, *D. erippus*, which was formerly classed as a member of the genus *Danaus*. It belongs to the Nymphalid butterfly subfamily Danaiinae.

**erippus** Cramer
C. America to Argentina (Patagonia) 80–100 mm. (3·15–3·94 in.)
This species is similar in pattern and colour to the illustrated *Danaus plexippus*, but differs chiefly in the absence of a dark band along the inner margin of the forewing and the presence of an orange area in the middle of the apical part of the forewing.

**Dione** Hübner (NYMPHALIDAE)
There are 4 species in this genus of the butterfly subfamily Heliconiinae. Each of them has silvery markings on the undersurface of the wings as in species of *Agraulis* (q.v.).

**juno** Cramer SCARCE SILVER-SPOTTED FLAMBEAU
C. and tropical S. America 63–74 mm. (2·48–2·91 in.)
In pattern, this species is very similar to *Dryas iulia* (illustrated) with which it flies, and to *Eueides vibilia*.

The female is more reddish in ground-colour than the male. The gregarious caterpillars feed on several species of passion flower (Passifloraceae). Adults of the various subspecies of *D. juno* differ in their colour preferences when feeding – the subspecies *D. juno juno* prefers red flowers; subspecies *D. juno suffumata* is attracted to both red and blue flowers.

**lucina** Felder
C. America 62–72 mm. (2·44–2·83 in.)
This species has been confused in the past with *Agraulis vanillae*. It is, however, easily distinguished from the latter, and more closely resembles *D. juno* and *D. moneta* in colour pattern. It differs from all 3 species in the absence of silvery spots on the basal half of the under surface of the hindwing. It is most frequently seen in forest clearings and along the margins of forests.

**moneta** Hübner
C. America, N.W. S. America 54–68 mm. (2·13–2·68 in.)
This species flies both in open country and in forested areas. At night it roosts on the tips of grasses in company with several others of its species. Red, white and yellow flowers are visited for nectar. The upper surface (not illustrated) of this butterfly is similar to that of *D. juno* and of the illustrated *Dryas iulia*, but the forewing invariably lacks a dark, apical patch and pre-apical band, and the wing veins are more heavily marked. The solitary caterpillar feeds on species of *Passiflora*.

**Dioptidae**
This is a small family of moths, closely related to the Notodontidae and a member of the superfamily Noctuoidea. It is solely C. American and tropical S. American in distribution. The majority of the species have bright yellow or orange markings. Most of the species of one large genus, *Josia*, have a similar, distinctive colour pattern, a common feature in groups of species which are distasteful to predators and advertise this quality with the same visual signals. Several species of *Josia* and other Dioptid genera form mimetic partnerships with species of Ctenuchidae, Arctiidae, Hypsidae, Geometridae and the butterfly families Nemeobiidae and Nymphalidae (subfamily Ithomiinae). See *Dioptis, Josia, Scea.*

**Dioptis** Hübner (DIOPTIDAE)
About 30 species belong in this genus of C. American and tropical S. American moths. The colour pattern of most of these partly transparent moths is repeated in species of Hypsidae and in the butterfly groups Ithomiinae and Nemeobiidae (q.v.).

**egla** Druce
Ecuador and the upper Amazon Basin of Brazil 33–38 mm. (1·3–1·5 in.)
There are extremely good matches for the colour pattern of this species in the 3 family groups mentioned under the genus (see for example the illustration of the Ithomiine *Napeogenes corena*). It is highly probable that functional mimicry is involved here but there is no experimental evidence as yet. **362**a

**diores** see **Thaumantis**

**Diphthera** Hübner (NOCTUIDAE)
This genus of moths was previously known as *Noropsis*. There is a single species.

**festiva** Fabricius: syn. *D. hieroglyphica*
Tropical S. and C. America, S.E. U.S. 44–50 mm. (1·73–1·97 in.)
This curiously marked moth is chiefly tropical in distribution, but extends as far north as Florida. Pecan *(Carya)* and coconut palms *(Cocos)* have been tentatively recorded as foodplants of the caterpillar. **383**

**Diplosaridae** see under **Aphthonetus**
**dirce** see **Colobura**
**dirtea** see **Euthalia**
**disa** see **Erebia**
DISA ALPINE see **Erebia disa**

**Discophora** Boisduval (NYMPHALIDAE)
This is one of the butterfly genera in the subfamily Amathusiinae, whose species are mostly large broad-winged insects which fly at dusk in forest areas in S.E. Asia. Although many species were described in this genus it is now realized that the species are very variable and many of those previously regarded as species are now considered forms or at the most subspecies.

**sondaica** Boisduval COMMON DUFFER
India, Pakistan, Burma, Tibet, S. China, Hainan, Thailand, Malaysia, Philippines 80–90 mm. (3·15–3·54 in.)
This butterfly has wet and dry season forms which were originally described as separate species. The forewing is pointed and dark brown with 3 rows of small bluish spots. These spots are more conspicuous in the female The hindwing of the male has a large central black spot on the uppersides. This butterfly is common over most of its range wherever bamboo, the foodplant of the caterpillar, grows. It is widely distributed and many forms and subspecies have been described.

**discigera** see **Plutodes**

**Dismorphia** Hübner (PIERIDAE)
This genus of butterflies is in the subfamily Dismorphinae. The species are found from the S. U.S., throughout C. America and the West Indies to S. Brazil. A large number of the species mimic species of distasteful Ithomiid, Danaid or Heliconiid butterflies. This presumably gives the species of *Dismorphia* a measure of protection. In some species of *Dismorphia* the males are white but the females are strongly patterned Helicoid or Ithomiid mimics, totally different in appearance from the males. These mimics fly with the butterflies they are imitating thus Ithomiid-like mimics are shade-lovers, while some of the more brightly coloured species of *Dismorphia* fly in the sunshine. Apart from these different forms, there is also seasonal variation in many of the species.

**amphione** Cramer TIGER PIERID
C. & S. America, W. Indies 44 mm. (1·73 in.)
The female is illustrated. The male has a large silver white patch on the front part of the hindwing and less orange on the forewing. They mimic the distasteful Danaid butterflies. They are typical Danaid mimics and are always found in their company. Several subspecies and forms of *D. amphione* have been described. **119**t

**astyocha** see under **Heliconius nattereri**
**crise** see under **Gazera linus**

**melia** Godart
C. & S. America 54–62 mm. (2·13–2·44 in.)
The male is yellowish, the female orange-brown. These are mimics of female Ithomiine butterflies, which are distasteful. This species is common in Brazil. **154**

DISMORPHIA, WHITE see **Enantia licina**
**dispar** see **Lycaena**, **Lymantria**

**Disphragis** Hübner (NOTODONTIDAE)
This is a chiefly S. American genus but with a few representations in N. America. There are probably over 100 species. Many of the included species have been placed incorrectly under *Heterocampa* Doubleday, a later name.

**guttivitta** Walker SADDLED or MAPLE PROMINENT MOTH
S.E. & E. Canada, U.S. 34–42 mm. (1·34–1·65 in.)
*D. guttivitta* is rare in S. U.S. but common and occasionally a pest in the N. of its range where caterpillars feed on maple *(Acer)*, especially sugar-maple, beech *(Fagus)*, apple *(Malus)* and many other broad-leaved trees and shrubs. The unusual first stage caterpillars are armed with long spines and branched processes. The forewings of the adults are cryptically patterned with green and brown. Part of the forewing pattern is repeated on the front edge of the otherwise unpatterned hindwing; this part of the hindwing is not covered by the forewings when the moth is at rest.

**dispula** see **Anamologa**
**disstria** see **Malacosoma**
**distincta** see **Gonodonta**
**districta** see **Carpella**

**diva** see **Cyanostola**
DIVER, BROWN-TAILED see **Bellura diffusa**
DIVERSE WHITE see **Appias epaphia**
**divitiosa** see **Tortyra**

**Dixeia** Talbot (PIERIDAE)
A genus of butterflies with species in Africa, south of Sahara ranging from Ethiopia to South Africa and Kenya through to W. Africa. Some of the species are common in parts of their range. One species has been described from Madagascar.

**doxo** Godart AFRICAN SMALL WHITE
Africa 40–50 mm. (1·57–1·97 in.)
This species is widely distributed throughout Africa S. of Sahara. It flies in most months of the year. The caterpillar feeds on species of Capparidaceae. The species is very like the European Small White, *Pieris rapae*, but lacks the spots on the wings, having only a black tip, and black extending down the outer margin of the forewing. The female has a larger black margin, and usually 2 black spots on the wing.

DJATI see **Hyblaea pura**
**doctissima** see **Trisophista**
DOCTOR, BLUE see **Rhetus periander**
**dodecella** see **Exoteleia**

**Dodona** Hewitson (NEMEOBIIDAE)
There are about 20 species in this genus of Asian butterflies. They range from India and Sri Lanka to S. China and Indonesia. Little is known about the early stages, but the caterpillar of one species is known to feed on species of the grass family Gramineae. Most species are restricted to mountainous regions above 910 m. (3000 ft). Adults tend to jump from leaf to leaf like species of *Abisara*. The colour pattern of the under surface is similar to that of many hairstreak butterflies (*Thecla* and its allies) of the family Lycaenidae.

**adonira** Hewitson
Bhutan, Sikkim, Nepal, N. Burma 28–34 mm. (1·1–1·34 in.)
This is a rare species found most frequently between 2130 and 2740 m. (7000–9000 ft) in the Himalayas. Its life history is not known. The forewing is triangular in shape and brownish red in colour above, with a black basal and marginal band and 3 black transverse medial bands. The hindwing is similarly marked and tapers to a bilobed 'tail'. The under surface of the wings are similarly striped.

DOG BUTTERFLY, ORANGE see **Papilio anchisiades**
DOGFACE BUTTERFLY see **Colias cesonia**
**dohertyi** see **Alucita, Chaerocina, Himantopterus**
**dolabraria** see **Plagodis**
**doleschalli** see **Allora**

**Doleschallia** Felder (NYMPHALIDAE)
A genus of butterflies with species from India, Pakistan and Sri Lanka to the Philippines, and through Indonesia south to New Guinea and Australia. A few species are found on the Solomon Islands, New Caledonia, New Hebrides and Fiji. These butterflies are found in rain-forests. One characteristic of this genus is the remarkable leaf-like underside pattern. This provides an excellent camouflage when the butterflies are at rest.

**bisaltide** Cramer AUTUMN LEAF or LEAFWING BUTTERFLY
India, Pakistan, Sri Lanka, Philippines, Japan, Thailand, Indonesia, Solomon Islands, New Caledonia, Loyalty Islands, New Britain, New Hebrides, Rennell Islands 60–70 mm. (2·36–2·76 in.)
This is a very widespread and common species which has been separated into many subspecies over its range. It has apparently not yet been seen on New Guinea, but will probably be found there. The bright orange coloured wings have black margins and the hindwing is produced into a short tail. On the underside it has an orange and grey pattern with a prominent middle line and eyespots along the margin. The caterpillar feeds on *Pseuderanthemum* (Acanthaceae).

**dolli** see **Sphinx**
**dolon** see **Polyura**
**dolosana** see **Choreutis**
**dolus** see **Agrodiaetus**
**dominica** see **Hemiargus**
**dominula** see **Callimorpha**
**dorcas** see under **Lycaena helloides**
**dorilas** see **Syrmatia**
**doris** see **Heliconius**
DORIS BUTTERFLY see **Heliconius doris**
**dorothea** see **Sphaerelictis**
**dorsalis** see under **Lasiocampidae**
DOTTED BORDER,
TRIMENS see **Mylothris trimenia**
TWIN see **Mylothris poppea**
DOUBLE-BRANDED BLACK CROW BUTTERFLY see **Euploea coreta**
DOUBLE-TUFTED WASP-MOTH see **Didasys belae**
**doubledayi** see **Neurosigma**
DOUGLAS FIR TUSSOCK MOTH see **Hemerocampa pseudotsugata**

**Douglasia** Stainton (DOUGLASIIDAE)
A genus of micro-moths found in Europe and Asia with less than a dozen species. Only one species is known in Britain. The forewings of all the species are long and narrow, coming to a point. The caterpillars of this genus are often found on species of borage (Boraginaceae).

**anchusella** Berander
Denmark 9 mm. (0·35 in.)
This moth was recognized and first described in 1936. The caterpillar had been found mining leaves of *Anchusa officinalis* (Boraginaceae). This plant is quite widespread in Europe and occurs in Britain, so it is probable that these small moths are also more common but they are relatively little studied. **41**d

**Douglasiidae**
A small family of micro-moths worldwide in distribution but with only one species known from Australia. The moths are very small and their caterpillars generally tunnel in stems or flower heads. The hindwings are very pointed and narrow with very reduced venation. See *Douglasia.*

**doxo** see **Dixeia**

**Doxocopa** Hübner: syn. *Chlorippe* (NYMPHALIDAE)
These butterflies are found in C. and S. America and the West Indies. They are popularly known as 'reflecting butterflies' from their magnificent blue reflection. While most species have this, a few have a green reflection, while in others these iridescent colours may be absent. Many species are common, the males, with their bright colours, being very conspicuous. The caterpillar feeds on species of *Celtis* (Ulmaceae) although in many cases the life histories are unknown.

**agathina** Cramer
S. America 60–70 mm. (2·36–2·76 in.)
The specimen figured is the subspecies *D. a. vacuna* Godart from Brazil. The purple iridescence of the male is absent in the female and the orange band on the forewing is complete in the female. The underside in both sexes is paler brown. This is a fast flying butterfly widely distributed with several subspecies recognized. **233**

**cherubina** Felder
S. America 60–70 mm. (2·36–2·76 in.)
This forest butterfly is often found feeding on rotting fruit or any decaying matter. For the rest of the time it flies high in the tree tops. The male is brown with a prominent green band in the middle of each wing, running from the costa backwards. The caterpillar feeds on species of *Celtis* (Ulmaceae). The butterfly is widespread in the Upper Amazon of Brazil, Bolivia, Peru, Ecuador, Colombia and Venezuela.

**lavinia** Butler
S. America 55–60 mm. (2·17–2·36 in.)
The female is larger than the male and lacks the intense blue colour in the middle of the wings. Instead she has a whitish band in the middle of the hindwing and a pale yellow-brown one on the middle of the forewing. The underside of *D. lavinia* is pale grey-white, relatively unmarked on the hindwings but with some black and white spots on the yellow-orange band on the wing. The apex of the wing is grey-white. The female underside is similar but browner on the hindwing. This butterfly has been found in Bolivia and Peru. **216**h

**Dracaenura** Meyrick (PYRALIDAE)
A genus of Pyralid moths in the subfamily Nymphulinae. While many genera in this subfamily have species with aquatic caterpillars, information on *Dracaenura* is not available. The genus is a small one with species from all over the world, many of them probably unrelated. It is in need of study.

**stenosoma** Felder
Fiji 35–50 mm. (1·38–1·97 in.)
This is a curiously shaped Pyralid moth with elongate forewings and rather rounded hindwings. The antennae are very long. At present the species is known only from Fiji and does not appear to have any immediate close relatives on nearby islands. **63**g

DRAGON BUTTERFLY, GREEN see **Alesa prema**
DRAGONTAIL, GREEN see **Lamproptera meges**
**dravidarum** see **Papilio**
**dregei** see **Bracharoa**

**Drepana** Schrank (DREPANIDAE)
About 10 species are placed in this genus of hook-tip moths. Two species, *D. curvatula* Borkhauser and *D. falcataria*, are European and temperate Asian in range; *D. arcuata* is N. American; the remainder are Indian and Chinese.

**arcuata** see under **D. falcataria**

**falcataria** Linnaeus PEBBLE HOOK-TIP
Europe including Britain, temperate C. and W. Asia 30–40 mm. (1·18–1·57 in.)
The illustration shows a specimen of the subspecies *D. f. scotica*; this differs from the European *D. f. falcataria* in the nearly white, not yellowish brown ground-colour of the wings. The green and brown caterpillars feed on birch *(Betula)*, alder *(Alnus)* and sometimes on the foliage of shrubs. The moths usually emerge in April and May, mate and lay eggs to produce a second brood of adults in July and August, but in Alpine regions of Europe there is only one emergence in June and July. *D. falcataria* is fairly closely matched in colour pattern by the N. American species *D. arcuata* Walker. These 2 species are indicators, with many other such 'trans-atlantic pairs', of the existence in past ages of a once continuous forest belt which extended across most of the northern temperate and sub-arctic regions of the world – the 'Arcto-tertiary Geoflora'. **272**r

**Drepanidae** HOOK-TIP MOTHS
This is a chiefly Old World tropical and subtropical family of about 800 species of moths. It is best represented in S.E. Asia, but with a few temperate European and American and Australian species and a rather larger representation in Africa and Madagascar and in temperate Asia. No species are known from C. America or S. America. Most are sombrely coloured moths, usually with a hooked, pointed apex to the forewing. The closest relatives of the Drepanidae in the superfamily Geometroidea are the families Cyclidiidae, Thyatiridae and Epiplemidae. The caterpillars are unique in that they lack the terminal pair of false-legs and taper to a point posteriorly; they feed almost exclusively on the foliage of broad-leaved trees and shrubs. *Epicampoptera* is a pest of coffee plants *(Coffea)* in much of tropical Africa, and *Gonoreta* is a minor pest of tea *(Thea)* in Kenya. See *Cilix, Drepana, Epicampoptera, Macrauzata, Oreta, Tridrepana.*

**Drepanoptera** see under **Epiphora**
DRIED CURRANT MOTH see **Ephestia cautella**
DRINKER MOTH see **Philudoria potatoria**
**drucei** see **Anaxita**

**Drupadia** Moore (LYCAENIDAE)
About 16 species are currently placed in this genus of butterflies. They are found in India eastwards to the Philippines. Most are brown, or brown and orange above, with some blue on the hindwing which has 3 tails.

**ravindra** Horsfield COMMON POSY
Burma to Indonesia and Borneo 23–34 mm. (0·91–1·34 in.)
This is a forest species, often common in Malaya. Like many other Lycaenids it often rests upside down with partly closed wings in which position the hindwing tails look like antennae and the spots at their base resemble eyes. This is thought to be a protective device against predators which often first attack the head of their potential prey. An attack on the apparent head, the actual tail of the butterfly, may allow the prey to escape unharmed except for a damaged tail. In fact, it is common to find what appears to be damage caused by birds' beaks on this area of the hindwing in this species and other tailed Lycaenidae. The caterpillars' foodplants include a species of *Eugenia* (Myrtaceae), *Derris* and *Albizzia* (Leguminosae). Ants are normally found together with the caterpillars but the nature of this association is unknown. **263**k

**drusilla** see **Appias**
DRY-LEAF BUTTERFLY see **Precise tugela**
**Dryad** see **Minois dryas**

**Dryadula** Michener (NYMPHALIDAE)
A single species is placed in this genus of the butterfly subfamily Heliconiinae.

**phaetusa** Linnaeus CARONI FLAMBEAU
C. & tropical S. America 73–76 mm. (2·87–2·99 in.)
The female (not illustrated) is a paler and more brownish red colour than the male. Both sexes appear to prefer open grassland, marshes or scrubland areas. Doubtful records of this species exist from U.S. (Florida), but as *D. phaetusa* is known to occur in Mexico it is probable that specimens will continue to be recorded in N. America from time to time. White, red and yellow flowers appear to be particularly attractive to this species. Its caterpillars feed on species of *Passiflora.* **142**k

**Dryas** Hübner (NYMPHALIDAE)
There is a single known species in this genus of the subfamily Heliconiinae.

**iulia** Fabricius FLAMBEAU
N., Central and tropical S. America 70–95 mm. (2·76–3·74 in.)
This seems to be a generally common species. It frequently visits garden flowers and damp patches, often in considerable numbers. Its caterpillars are armed with long branching spines and feed on species of passion flower *(Passifloraceae)*; they are sometimes cannibalistic. Caterpillars of many species will eat one another when in crowded conditions; some species tend to such villainy more often than others. In N. America this is not normally a breeding species except in S. Florida and S. Texas. Numerous subspecies are recognized; the illustrated *D. iulia delila*, from Jamaica, has lost the black oblique, pre-apical band on the forewing which is present in the other subspecies. **138**

**dryas** see **Minois**
**drymo** see **Ithomia**
**dryope** see **Eurytela**
**dryopterata** see **Epiplema**
**dubitalis** see **Scoparia arundinata**
DUCHESS, GRAND see **Euthalia patala**

**Dudgeonea** Hampson (DUDGEONEIDAE)
Brightly patterned moths related to the Cossidae, but with a more striking pattern than most Cossidae. The species are found in New Guinea and Australia, with one species recorded from Africa, although this may not have been correctly placed in this family.

**actinias** Turner
Australia 26–28 mm. (1·02–1·1 in.)
A brightly coloured species, similar in appearance to a related one from New Guinea. The caterpillars have been bred from *Canthium* (Rubiaceae). **8**b

**Dudgeoneidae**
A small family of micro-moths related to the Cossidae. The species are found in Africa, India, Pakistan and Australia. The family is still of uncertain status and is included by many in the Cossidae although by some specialists it is considered to be related to the Sesiidae. See *Dudgeonea.*

DUFFER, COMMON see **Discophora sondaica**
DUKE BUTTERFLIES see **Euthalia**
DUKE OF BURGUNDY FRITILLARY see **Hamearis lucina**
**dumi** see **Lemonia**
**duponchel** see under **Cocytius**
**duponchelii** see **Euryades**
**duprei** see **Salamis anacardii**
**durga** see **Euthalia**
**durranti** see **Mictopsichia**
**durvillii** see **Cocytia**
DUSKY
HEDGE BLUE see **Lycaenopsis vardhana**
SWORDTAIL see **Graphium polistratus**
DUSKY-WINGED FRITILLARY see **Clossiana improba**
DUSKYWING, JUVENAL'S see **Erynnis juvenalis**
DWARF BLUE BUTTERFLIES see **Brephidium**

**Dynamine** Hübner (NYMPHALIDAE)
A genus of butterflies found in C. and S. America with some 50 known species. Some of these have been reared from caterpillars feeding on species of *Dalechampia* (Euphorbiaceae).

**mylitta** Cramer LARGE DYNAMINE BUTTERFLY
S. and C. America, West Indies 36–45 mm. (1·42–1·77 in.)
The pattern of this species is very variable. The underside of the male is rather like the upperside of the female, but paler. This species reaches Peru and Brazil and north to Honduras and Guatemala. The upper wing surfaces of both sexes are shown. **217**k, p

DYNAMINE BUTTERFLY, LARGE see **Dynamine mylitta**

**Dynastor** Doubleday (NYMPHALIDAE)
This is a small genus of butterflies and is in the subfamily Brassolinae. The species are confined to C. and S. America. The underside of the wings have a very leaf-like pattern and the butterfly must be well camouflaged at rest.

**napoleon** Doubleday, Hewitson
S. America 125–160 mm. (4·92–6·3 in.)
Very little is known about this forest-living species which is found only in Brazil. It is said to fly at dusk. The sexes are similarly patterned but the females are much larger than the males. **150, 151**

**Dypterygia** Stephens (NOCTUIDAE)
This is a fairly large, widespread genus found in temperate and tropical zones in both the Old and New Worlds.

**scabriuscula** Linnaeus BIRD'S WING MOTH
Europe including Britain, temperate Asia, N. America 30–38 mm. (1·18–1·5 in.)
The common name of this widespread species of the Northern Hemisphere refers to the highly distinctive brownish white, wing-shaped markings on the forewing. The caterpillar feeds on dock, sorrel *(Rumex)*, *Polygonum* and other species of the family Polygonaceae; it is chiefly reddish brown in colour, with a whitish line along its back, whitish lateral patches and spotted with black and yellow. **384**

**dysmenia** see under **Theorema**
**dysmephila** see **Zophopetes**

**Dysphania** Hübner (GEOMETRIDAE)
This is a large group with a distinctive wing pattern. The antennae are pectinate (feathery) in both sexes. They are mostly day-flying moths, some of them are mimetic, imitating supposedly distasteful species for their own protection. The caterpillars are green, looking like hawk-moth caterpillars with a horn on the tail end. They sit extended on the foodplant and do not look a bit like a normal Geometrid caterpillar which is generally rather twig-like in appearance. The moths are avoided by birds and often fly with butterflies, such as Papilios and Pierids, which are also distasteful to birds.

**cuprina** Felder
Indonesia, Java, Sumatra, Philippines, New Guinea, India, Pakistan 70 mm. (2·76 in.)
This is a widespread moth with several similar looking related species. Little is known about its biology although specimens are quite common in collections from the Far East. **272**c

**Dysschema** Hübner: syn. *Pericopis* (HYPSIDAE)
About 50 species are placed in this genus of essentially C. American and tropical S. American moths. At least 2 species have been recorded from S. U.S. Over 200 names have been applied to the various subspecies, male and female forms and colour forms. Some of the currently accepted species may prove to have 2 names, one for the female and another for the male, both of which are treated at present as valid species names. A few species are mimetic partners of Heliconiine, Papilionid and Ithomiine butterflies.

**mariamne** Geyer
S. U.S., C. America 60–78 mm. (2·36–3·07 in.)
This is one of the few species of its genus represented in N. America. The female differs from the illustrated male in the yellow or orange ground-colour of the upper surface of the hindwing. The males of C. American specimens have rather more transparent hindwings than in northern specimens. **376**a

**tricolora** Sulzer
C. & tropical S. America 50–65 mm. (1·97–2·56 in.)
The remarkable females of *D. tricolora* are probably mimetic partners of species of the butterfly genus *Parides*. Males are 'typical' members of *Dysschema*, and have dark-edged, transparent greyish white wings. **376**c

**Eacles** Hübner (SATURNIIDAE)
This is a genus of about 20 species of large moths, 2 of which occur in the U.S. and the remainder in C. and tropical S. America.

**imperialis** see under **E. oslari**

**oslari** Rothschild
N. America, Mexico 100–130 mm. (3·94–5·12 in.)
There are several colour forms of this handsome moth, ranging from yellow to pinkish or purplish brown. The moth can often be attracted in large numbers to ultraviolet light, especially if a female is present, but it is becoming scarce in and near large urban areas. Close relatives of this species are found in C. and S. America as far south as Argentina. *E. imperialis*, a species found in S.E. U.S., is almost identical in colour pattern to *E. oslari*. The illustrated female of *E. oslari* was captured in S. Arizona at Peña Blanca. **333**f

**Eana** Billberg (TORTRICIDAE)
A European genus of micro-moths, similar to the other bell-moths (Tortrix) with a few species extending into N. America. The life histories are similar to species of *Eulia.*

**argentana** Clerck
Europe including Britain, Asia, India, Japan, N. America 20–25 mm. (0·79–0·98 in.)
The forewings are silvery white, shading to brown near the margins; the hindwings are yellowish-white. The moth is out in July in the U.S. and in Europe. In Britain the moth is local but it is more widespread in the rest of Europe. In appearance it is the typical shape of a tortricid moth. Although widespread, little is known of its biology, or whether it has spread of its own accord to N. America from Europe or vice versa. It feeds on a wide variety of plants and is recorded on spruce *(Picea)*, grasses *(Gramineae)* and moss. There is a related species in N. America, *E. subargentana* Obraztsov which is very similar in appearance but is smaller in size.

**subargentana** see under **E. argentana**

**Earias** Hübner (NOCTUIDAE)
About 50 species are included in this economically important genus. It has an extensive distribution over much of the Old World. The caterpillars feed on species of the family Malvaceae, including cotton *(Gossypium)*.

**biplaga** Walker SPINY BOLLWORM or STEM-TIP BORER 18–23 mm. (0·71–0·91 in.)
Africa, south of the Sahara, Madagascar
The forewings of this species are either green or orange, and there is some variation in the distinctness of the brown central band. It is similar in general appearance to the illustrated *E. insulana,* but usually has a darker forewing fringe, and straighter transverse lines when these are present. The caterpillar is a serious pest of the shoots and flower-buds of cocoa *(Theobroma)* and cotton *(Gossypium)*.

**clorana** Linnaeus
CREAM-BORDERED GREEN PEA MOTH 20–23 mm.
Europe including Britain, W. Asia (0·79–0·91 in.)
The forewings of this moth are green, without markings except for a greenish white area along the basal half of the front margin (much like those of the Tortricid *Tortrix viridana* in colour and shape); the hindwings are almost entirely white. The caterpillar feeds in July and August on willow *(Salix)*, on which it spins together terminal leaves to form a shelter. *E. clorana* is most common in damp and marshy areas.

**insulana** Boisduval EGYPTIAN BOLLWORM or COTTON SPOTTED BOLLWORM
S. Europe, W. and S. Asia, Africa, Australia 15–18 mm. (0·59–0·71 in.)
The caterpillar of this pest-species feeds inside the bolls and buds of cotton *(Gossypium)* and other species of Malvaceae and in the buds of carob *(Ceratonia)* and *Hibiscus*. **390**g

EARLY
HAIRSTREAK see **Erora laeta**
THORN see **Selenia dentaria**
EARED COMMODORE see **Precis tugela**
EARTH-MEASURERS see under **Geometridae**
EARWORM, CORN see **Heliothis zea**
EASTERN
FESTOON see **Allancastria cerisyi**
GRAPELEAF SKELETONISER see **Pampa americana**
PALE CLOUDED YELLOW see **Colias erate**
TAILED BLUE see **Everes comyntas**
TENT see **Malacosoma americana**
ORANGE-TIP see **Anthocharis damone**
PYGMY BLUE see **Brephidium pseudofea**

**Echenais** Hübner (NEMEOBIIDAE)
There are about 30 mostly small species in this genus of C. and tropical S. American butterflies. The posterior part of the male hindwing of many species is white or a very pale colour.

**alector** Butler WHITE-TAILED ERYCID
Tropical S. America 20–30 mm. (0·79–1·18 in.)
This is a common species throughout much of its range. Its female (not illustrated) has a white, oblique, preapical band on the forewing and the whole of the hindwing patterned with brown and white markings. **268**s

**echeria** see **Amauris**
**echerius** see **Abisara**
**echo** see **Taygetis**
**eclipsis** see **Gonepteryx**

**Ecpantheria** Hübner (ARCTIIDAE)
This is a large genus of typically white moths, with black markings, and a posteriorly produced hindwing especially in the male. The centre of distribution is in tropical regions of the New World, but there is one quite common species in N. America. The caterpillar of *E. muzina* Oberthür is a defoliator of cocoa *(Theobroma)* trees in Trinidad. *E. oslari* Rothschild, which occurs in Texas, lacks hindwing 'tails' and may be wrongly placed in this genus.

**decora** Walker
Mexico, Cuba, Haiti 45–60 mm. (1·77–2·36 in.)
*E. decora* is typical in colour pattern of most species of its genus. The lobed, inner part of the hindwing bears specialized scent-distributing scales. The hindwing of the female (not illustrated) lacks scent scales and is less strongly lobed; it is patterned with markings similar to those of the forewing. **369**g

**muzina** see under **Ecpantheria**
**oslari** see under **Ecpantheria**

**scribonia** Stoll GREAT LEOPARD MOTH
S.E. Canada, U.S., Mexico 58–80 mm. (2·28–3·15 in.)
*E. scribonia* is similar in general features of the colour pattern to the illustrated *E. decora,* but the wing-markings are typically much darker – the lines are thicker, and obliterate the white centres at the costa of the forewing and elsewhere on the wing in some specimens – and the abdomen is mostly black above. Its caterpillars feed on low-growing plants but also on cherry *(Prunus)*, maple *(Acer)*, tangerine *(Citrus)* and *Bougainvillea.*

**Ectomyelois** Heinrich (PYRALIDAE)
This generic name was first used in 1956 for a group of American and European Pyralid moths which formerly were part of a larger grouping in the genus *Myelois.*

**ceratoniae** Zeller CAROB BEAN MOTH
Cosmopolitan 16–26 mm. (0·63–1·02 in.)
This is a pest, in the caterpillar stage, of castor-oil seed, dried fruits and nuts. It has probably been widely transported by man and probably reached even oceanic islands like Hawaii in this way. The caterpillar feeds inside the nuts or beans and little evidence of the damage can be seen from outside. The moth has rather narrow grey forewings, speckled with white, with 2 black lines across the wings. The hindwings are whitish.

**decolor** Zeller
Cuba, C. America, Bahamas, Jamaica, Venezuela, Guyana, Brazil 19–30 mm. (0·75–1·18 in.)
This moth is similar to the preceding one externally. The caterpillars also feed on fruit and nuts. There is considerable variation in the size of specimens.

**Edule** Hübner (GEOMETRIDAE)
A genus of moths with species in C. and S. America. Frequently they are rather delicate looking moths with distinctive wing shape and coloration.

**ficulnea** Druce
Peru, Ecuador 27–35 mm. (1·06–1·38 in.)
There are several species with rather similar patterns to the one illustrated; many have delicate lines along the wing veins. The patches on the wing of *E. ficulnea* vary in different specimens from red to orange and the amount of colour on the hindwings is also variable. **294**p

**edwardsi** see **Euproctis**
**efformata** see under **Aplocera plagiata**
**egaensis** see **Mechanitis**
**egea** see **Polygonia**
EGGAR, OAK see **Lasiocampa quercus**
EGGFLY BUTTERFLIES see **Hypolimnas**
**egina** see **Acraea**

**Egira** Duponchel (NOCTUIDAE)
Species of this genus have been known for some time under the name *Xylomiges* Guenée. There are about 20 species, found in N., C. and S. America and in the temperate regions of the Old World.

**pulchella** Smith BEAUTIFUL WOODLING
W. Canada 32–40 mm. (1·26–1·57 in.)
British Columbia is the only recorded area for this attractive species. The forewings are patterned with pale yellow, brown and brownish orange, with a black marginal area and a central white spot; the hindwings are brownish white.

**egista** see **Vagrans**
**egla** see **Dioptis**

**egregiana** see **Cydia**
EGYPTIAN
BOLLWORM see **Earias insulana**
COTTON LEAFWORM see **Spodoptera littoralis**
**eichhorni** see **Milionia**
EIGHT-SPOTTED FORESTER see **Alypia octomaculata**
EIGHTY-EIGHT BUTTERFLY see **Callicore maimuna**
'89' BUTTERFLY see **Diaethria aurelia**

**Eilema** Hübner (ARCTIIDAE) FOOTMEN MOTHS
This is a member of the subfamily Lithosiinae. It is a large, almost cosmopolitan genus of moths. The species of *Eilema* are typically mostly dull brown in colour and often impossible to identify without examining internal genitalic structures. The caterpillar's hairs, in some species, are known to have irritant properties.

**complana** Linnaeus SCARCE FOOTMAN
Europe, temperate Asia to Siberia, E. U.S. 30–35 mm. (1·18–1·38 in.)
It seems likely that this is one of many moths introduced accidentally into N. America, where it was re-described in 1903 as *Crambidia allegheniensis.* It occurs in woods, where the caterpillars probably feed on tree lichens, and in open country where moss, bramble *(Rubus)* and various legumes are among the food-plants. **364**k

**ekebladella** see **Tischeria**
**ekthlipsis** see **Nymphula**

**Elachista** Treitschke (ELACHISTIDAE)
There are many species in this genus and they are to be found in all parts of the world, with many species in Europe. The forewings tend to be slender and pointed and the species are all very similar externally. The caterpillars generally make mines in the leaves of grasses *(Gramineae)* or sedges *(Cyperaceae)*.

**regificella** Sircom
Europe including Britain 9 mm. (0·35 in.)
The female moths have white tips to the antennae which make them easy to distinguish from the males. The caterpillars feed on the leaves of woodrush *(Luzula)*. **44**l

**Elachistidae**
A family of micro-moths found all over the world. The caterpillars are generally leaf- or stem-miners in grasses (Gramineae) or sedges (Cyperaceae). The adult moths have a characteristic type of hindwing venation. They are mostly rather small moths. See *Elachista.*

**electra** see **Hemileuca**
**elegans** see **Loxotoma, Milionia**
ELEGANT ACRAEA see **Acraea egina**
ELEPHANT HAWK-MOTH see **Deilephila elpenor**
**eleus** see **Euphaedra**
**eleusina** see **Euploea**
ELFIN, BOG see under **Lycaena epixanthe**
**elima** see **Spindasis**

**Elimniopsis** Fruhstorfer (NYMPHALIDAE)
This is a small genus of butterflies in the subfamily Satyrinae. The species are found in Africa and are widespread but their life history is unknown.

**lise** Hemming: syn *E. phega*
W. Africa 70 mm. (2·76 in.)
This species is widespread in W. Africa from Liberia to Gabon. The upper and undersides of the wing are similarly coloured and patterned. **174**c

**elinguaria** see **Crocallis**
**ello** see **Erinnyis**
**elongella** see **Caloptilia**
**elpenor** see **Deilephila**
**elsa** see **Sagenosoma**
**elutella** see **Ephestia**

**Elymnias** Hübner (NYMPHALIDAE) PALMFLIES
This genus of Satyrine butterflies has many species from India to Australia. Unlike most other Satyrids, these are often brightly coloured and some appear to mimic Danaid butterflies.

**agondas** Boisduval PALMFLY
New Guinea, Australia, Kai and Aru Islands 70–90 mm. (2·76–3·54 in.)
The velvet black male has a band of cream-coloured marks below the outer wing margin. The hindwing is similarly patterned, but with a broader cream band with small spots towards the back of the hindwing and a faint bluish tinge round them. The underside is similar, but the hindwing band is more yellowish. The eyespots, surrounded by orange, are more conspicuous with white centres. The female has eyespots visible on the upper and lower surfaces and broad white patches on the hindwing, with large white patches on the forewing. The butterflies, as their name implies, are associated with palms (Palmae), on which the caterpillars feed, and are shade-loving, rain-forest insects.

**hypermnestra** Linnaeus COMMON PALMFLY
Formosa, China, India, Malaysia, Indonesia 60–80 mm. (2·36–3·15 in.)
This is a bluish black butterfly with a series of pale spots along the hindwing margin. The underside is richly marked brown with a paler shade at the apex of the forewing and a white spot on the margin in the hindwing. The butterfly is a weak flier, fluttering around in the shade and generally avoiding bright sunshine. The caterpillar feeds on Coconut palms *(Cocos)* and the species is usually common where this occurs. Several subspecies have been described in different parts of its range. Some of the subspecies are quite good mimics of distasteful Danainae butterflies.

**malelus** see under **Euploea mulciber**

**nesaea** Linnaeus TIGER PALMFLY
India, Pakistan, Burma 75–85 mm. (2·95–3·35 in.)
This species is a mimic of the Glassy Tiger *(Danaus aglea)*. It has a short tail on the hindwing and is pale greenish blue with the veins broadly marked with brown. The undersides are heavily streaked with fine reddish brown lines. It flies generally in the lower lying jungles where the caterpillars feed on palms (Palmae). It is a slow flier, avoiding the sunshine and keeping to the deeper shades of the jungle.

**patna** Hewitson BLUE-STRIPED PALMFLY
India, Pakistan, Burma, Hainan, Malaysia 80–100 mm. (3·15–3·94 in.)
Very little is known about this species which appears to mimic one of the distasteful Danainae, *Euploea harrisi* with which it flies. The butterfly is brown, with blue marks across the forewings, less conspicuous on the hindwing. The hindwing has a row of small bluish white spots.

**Elzunia** Bryk (NYMPHALIDAE)
About 6 species of butterflies are placed in this C. American and tropical S. American genus of the subfamily Ithomiinae. The species have broad, *Danaus*-like wings and are nearly black in ground-colour.

**bonplandii** Guérin-Méneville
N.W. S. America 73–92 mm. (2·87–3·62 in.)
This relatively common butterfly is one of the largest species of the subfamily Ithomiinae and the largest species in its genus. The white spots of some specimens are replaced by yellow spots, but *E. bonplandii* is otherwise similar to the generally smaller *E. pavonii* on the upper surface. The under surface of the hindwings has only one row of brown spots compared with the 2 rows in *E. pavonii.*

**pavonii** Butler
C. America, N.W. S. America 60–74 mm. (2·36–2·91 in.)
In Ecuador, *E. pavonii* is a mimetic partner of another distasteful species *Heliconius atthis.* The under surface of the wings is paler than the illustrated upper surface and has a brownish orange band parallel to the outer margin of the wings. This is the smallest species of its genus. **131**b

**Ematurga** Lederer (GEOMETRIDAE)
A genus of moths with species across the Northern

Hemisphere. The caterpillars, which are typical Geometrid 'loopers' feed on species of Leguminosae and Ericaceae.

**atomaria** Linnaeus COMMON HEATH
Europe including Britain, Asia 24–29 mm. (0·94–1·14 in.)
This is a very variable species with several named aberrations and at least 2 subspecies in Britain. The caterpillar, which feeds on *Calluna, Erica* (Ericaceae), *Lotus* and *Trifolium* (Leguminosae), is equally variable in colour, all shades of grey to brown, sometimes marked on the back. The moth, which flies in May and June, is common and widely distributed in Europe where heathlands occur. **299**

EMERALD,
LARGE see **Geometra papilionaria**
LIGHT see **Campaea margaritata**

**Emmelia** Hübner (NOCTUIDAE)
The 13 or so species of this group of moths are chiefly temperate Old World in distribution; but the genus is almost cosmopolitan.

**trabealis** Scopoli SPOTTED SULPHUR
Europe including England, temperate Asia to Japan 20 mm. (0·79 in.)
The Breckland region of Norfolk is the British home of this otherwise more generally distributed, colourful species. Its caterpillars feed on Lesser Bindweed *(Convolvulus)*; there are two broods each year. The moths fly in June, July and August. **389**d

**Emmelina** Tutt (PTEROPHORIDAE)
This genus of plume moths has only one species. This is widespread in the N. Hemisphere and many varieties have been described.

**monodactyla** Linnaeus
Europe including Britain, temperate Asia, N. America 21–26 mm. (0·83–1·02 in.)
This is a common plume-moth, occurring in a variety of habitats but is commonest in hedgerows. The moth is out in July and again in September, hibernating until the following spring. It is active at dusk and is commonly attracted to light. It is widespread in the U.S. and ranges from Mexico into Canada. There is some dispute as to whether all the specimens from this enormous range are the same species and a more critical examination is needed. The caterpillar feeds on various plants but has a preference for bindweed *(Convolvulus)*. **76**

**emolliens** see **Arycanda**
EMPEROR BUTTERFLIES see **Apatura, Asterocampa, Helcyra, Polyura**
EMPEROR MOTHS see **Aglia, Athletes, Bunaea, Bunaeopsis, Cinabra, Eochroa, Imbrasia, Saturnia, Urota, Ubaena**
EMPEROR SWALLOWTAIL see **Papilio ophidicephalus**

**Enantia** Hübner (PIERIDAE)
A small genus of butterflies with 2 known species, both from S. America. They are related to species in the genus *Dismorphia*.

**licina** Cramer WHITE DISMORPHIA
S. America, W. Indies 48 mm. (1·89 in.)
The male has a black tip to the white wings, with a small white patch in it. The hindwings are noticeably broader than the forewings. The apex of the hindwing protrudes beyond the outer margin of the forewing. There is a small black line on the edge of the under part of the hindwing. The females have a narrower hindwing and the black on the forewing extends right along the margin. The butterflies are out in the sunshine and frequently visit flowers. Several subspecies have been described in different parts of its range including one in Trinidad. **119**s

**endochus** see **Graphium**
**endocina** see **Theope**

**Endromidae**
There are 2 genera in this family of moths: *Endromis* and *Mirina* Staudinger, but the latter is certainly wrongly placed here. The family is possibly fairly closely related to the Bombycidae and Lasiocampidae and is a member of the superfamily Bombycoidea. See *Endromis, Mirina*.

**Endromis** Ochsenheimer (ENDROMIDAE)
There is a single known species of moth in this genus.

**versicolora** Linnaeus KENTISH GLORY
Europe including Britain 52–70 mm. (2·05–2·76 in.)
The males of this richly coloured species are normally day-fliers whereas the females fly only at night. In Britain, it is best represented in Scotland, especially in areas where the main larval foodplant, birch *(Betula)*, is plentiful. The caterpillars, which are gregarious at first, are remarkably like hawk-moth caterpillars in shape and colour pattern, with a row of oblique stripes along each side. However, these stripes run from the front downwards to the rear of each segment, unlike those of, for example, the Privet Hawk-moth *(Sphinx ligustri)* which run from the rear of each segment downwards to the front edge. In Britain, the chrysalis sometimes overwinters twice before the moth emerges. **320**

**Endrosa** see **Setina**

**Endrosis** Hübner (OECOPHORIDAE)
One species of this genus of micro-moths is found only in Africa. The other species is worldwide and described below. The species in the genus are related to *Hofmannophila*, another detritus-feeding genus.

**sarcitrella** Linnaeus WHITE-SHOULDERED CLOTHES OR HOUSE MOTH 14–20 mm. (0·55–0·79 in.)
Worldwide
Aptly described as a 'domestic tramp', this moth is usually closely associated with human habitations where the caterpillar feeds on almost any organic detritus, food waste or dry refuse. The moth is very distinctive with grey-white mottled wings but with a conspicuous white head and thorax. They have even been found in storehouses built on remote subantarctic islands. It is rarely a pest in the sense that it obviously destroys large quantities of food or clothing, but tends to live on accumulations of debris between floor boards or under cupboards. **35**

**endymion** see **Helicopis**

**Enicospila** Houlbert (CASTNIIDAE)
One species of moth is placed in this New World genus.

**marcus** Jordan
Tropical S. American 75 mm. (2·95 in.)
Both pairs of wings of this extremely rare and unusually patterned moth are iridescent. The forewings are green, brown and dull violet; the hindwings brownish orange, brown and dull violet. The highly conspicuous, cream spot on the forewing is comparable with a similarly placed marking in females of the unpalatable butterfly genus *Parides*, some of which also have orange areas on the hindwing which are partly matched by the orange ingredient of the iridescence of *E. marcus* hindwings. The wings of *Parides* are broader, but the resemblance between them and this species may be sufficiently good to endow an advantage to *marcus* where predator pressures are high and suitable *Parides* models are available for this mimicry to be effective. Other possible models are species of *Heliconius* such as *H. eanes* Hewitson. **45**q

**Ennomos** Treitschke (GEOMETRIDAE)
This genus of moths is found in temperate Asia including Japan, and Europe.

**autumnaria** Werneburg LARGE THORN MOTH
Europe including Britain, Asia including Japan 42–48 mm. (1·65–1·89 in.)
The caterpillar feeds on birch *(Betula)*, hawthorn *(Crataegus)* and sloe *(Prunus)* and the moth is fairly widespread across Europe, but it is rare in Britain. The specimen illustrated is a female. The male is slightly smaller and a darker yellow-brown. The Japanese specimens differ from the European ones and are a distinct subspecies. **298**

**Enodia** Hübner (NYMPHALIDAE)
This is a small genus of butterflies in the subfamily Satyrinae. The species are characterized, amongst other features, by the presence of fine hairs over the surface of the eyes. They are found in C. and N. America.

**portlandia** Fabricius PEARLY EYE BUTTERFLY
Canada, U.S. 40–50 mm. (1·57–1·97 in.)
Several subspecies of this wide ranging woodland butterfly have been described in N. America. It ranges south to the Gulf States and westwards to the eastern Great Plains. The males, as with a number of other species of butterflies, adopt territories and chase off other males of their species. The adults are out in June with one brood in the north, two in the south. The caterpillar feeds on species of grass. **174**p

**Eochroa** Felder (SATURNIIDAE)
Brilliant colours and relatively long antennae characterize the only known species of this genus of moths.

**trimeni** Felder ROSEATE EMPEROR
Known only from N.W. Cape Province, South Africa 58–71 mm. (2·28–2·8 in.)
The black and yellow caterpillar of this colourful species feeds on *Melianthus* (Melianthaceae). **338**

**Eosphoropteryx** Dyar (NOCTUIDAE)
This is a genus of the subfamily Plusiinae (Plusia moths) which, like many others of its subfamily, is characterized by metallic markings on the wings. It contains a single known species.

**thyatiroides** Guenée PEACH BLOSSOM PLUSIA
Canada, N. U.S. E. of the Rockies 35–38 mm. (1·38–1·5 in.)
The forewings are grey above, with a faint gold and green iridescence towards the inner margin, 2 metallic silver markings in the middle, a pink basal area and a pinkish grey outer marginal band. The hindwings are a lustrous pale yellowish brown. The caterpillars of this beautifully marked moth feed on *Thalictrum*, a genus of the buttercup family, Ranunculaceae. It hibernates when young and commences feeding again in the spring. **391**m

**epaphia** see **Appias**
**epaphus** see **Metamorpha**

**Epargyreus** Hübner (HESPERIIDAE)
This is a genus of skipper butterflies found in America. About 15 species are known.

**clarus** Cramer: syn. *E. tityrus*
SILVER SPOTTED SKIPPER 45–50 mm.
Canada, U.S., C. & S. America (1·77–1·97 in.)
A large, dark skipper with yellow-orange patches on forewing and dark hindwing which is slightly produced into a short blunt tail. The most conspicuous feature, visible even in flight, is a brilliant white patch on the underside of the hindwing. It is a fast and powerful flier common in gardens, fields and along roadsides. A widespread species which has one brood in the north and 2 or more in the south. The caterpillar feeds on *Robinia, Wisteria, Acacia, Gleditsia* and other Leguminosae. It makes a nest of leaves on its foodplant and is yellow-green with a brown head which has 2 eye-like red spots.

**Epermenia** Hübner (EPERMENIIDAE)
Epermenids are micro-moths, most easily recognizable by the scaled tufts on the hind margin of the forewing. The genus is worldwide in distribution but little is known about its biology. The known species have caterpillars which mine leaves at first, then feed externally on the leaves, spinning pieces of the leaves together. The pupal case stays inside the cocoon when the moths emerge. In many other families the pupal case is pushed partly out of the cocoon by the emerging moth.

**pontificella** Hübner
C. & S. Europe 14–17 mm. (0·55–0·67 in.)
The caterpillar is said to feed on *Thesium* (Santalaceae). **41**a

**Epermeniidae**
A family of micro-moths with species on most continents. They are usually very small moths generally with stiff bristles on the hind tibiae and with scale tufts along the inner margin of the forewing. The caterpillars mine in leaves in their early stages, then externally feed on the leaves, living in a small web. See *Epermenia*.

**epeus** see **Spalgis**
**ephemeraeformis** see **Thyridopteryx**

**Ephestia** Guenée (PYRALIDAE)
FLOUR or MEAL MOTHS
A genus with many species which directly affect man and are of considerable economic importance. The species, particularly those which are pests of stored foods, are worldwide. The caterpillars feed on flour, meal, wheat, nuts, beans, currants and dried fruit and many other food products, such as bread, cakes, chocolates which they spoil in vast quantities by covering it with silk and rendering it unfit for human consumption – although probably not dangerous, the food is obviously unpleasant. The moths can multiply rapidly under warm conditions in, for example, warehouses, holds of ships, and shops. Apart from the vast losses caused directly by their damage, the presence of an infestation in a consignment of, for example, currants can cause the consignment to be refused entry at ports – with great loss of revenue or large insurance claims. The adults are all rather similar looking dull coloured insects, but for all their small size and inconspicuous colours they play a significant part in restricting the food available for man. The genus is frequently divided into several subgenera including *Anagasta* and *Cadra*. These have also been regarded as distinct genera.

**cautella** Walker DRIED CURRANT MOTH, FIG or ALMOND MOTH
Cosmopolitan 14–20 mm. (0·55–0·79 in.)
This species is mainly a pest of dried fruit and nuts, although it will attack other stored products. It is frequently introduced to the more temperate regions where it becomes a pest in warehouses. Generally the moths appear in May and June but in artificial heat they may have more generations a year.

**elutella** Hübner CACAO or TOBACCO MOTH
Cosmopolitan 14–19 mm. (0·55–0·75 in.)
One of the major pests of stored food, the caterpillars of this species have been recorded on a wide range of stored foods, including cereals, tobacco, dried fruit, nuts, seeds, sugar, and it is frequently found by shopkeepers or housewives in some recently purchased food. The caterpillars can eat their way through many packing materials – paper, cardboard, polythene – and their presence can be detected by the silk webbing which ties together the flour or sugar or other food they are on. The small grey moths, flying round flour mills or food factories, often get trapped in the dough or baked in the cakes, thus they are frequently the source of complaints or legal actions for 'foreign bodies' in manufactured food. **56**r

**kuehniella** Zeller MEDITERRANEAN FLOUR MOTH
Cosmopolitan 20–23 mm. (0·79–0·91 in.)
Probably one of the commonest pests of stored food in many parts of the world and it is by no means as restricted as its name might suggest. The caterpillars are commoner on flour and wheat, but may occur on many other foods. They are particularly troublesome in bakery and food stores, where larvae get into the bread and cakes either in the dough or in the final cooked product making it unattractive and may cause the firm selling it to be liable for legal prosecution. **56**e, s

**Ephestris** Hübner (HYPSIDAE)
A single species of moth is placed in this genus.

**melaxantha** Hübner
Brazil 62–73 mm. (2·44–2·87 in.)
This unusually patterned Hypsid is like an enormous Dioptid, with wings and body conspicuously striped with yellow and black. **376**j

**ephippella** see **Argyresthia pruniella**

**Epia** Hübner (APATELODIDAE)
This is a genus of about 10 C. American and tropical S. American moths. They are apparently cryptically coloured, with a combination of chiefly brown and green, or brown and yellow markings.

**muscosa** Butler
Mexico, through C. America to Peru 35–54 mm. (1·38–2·13 in.)
Females differ markedly from the illustrated male in the paler, more yellowish or orange-brown ground-colour of the wings and the reduction in the amount of green, or greenish yellow, at the outer margin of the forewing. The triangular patch of green at the front margin of the forewing is poorly marked or absent in the female. **318**g

**Epiblema** Hübner (TORTRICIDAE)
A large genus of micro-moths, mostly in the temperate parts of the Northern Hemisphere. The species in it are separated into several distinct groupings within the genus (subgenera). They are similar in general appearance to other tortricid bell-moths.

**cynosbatella** Linnaeus
Europe including Britain, W. and C. Asia 14–20 mm. (0·55–0·79 in.)
The forewings are whitish with a faint grey suffusion near the apex, while their base is brown mixed with grey, contrasting sharply with the white part of the wing. The hindwings are grey, with the moth a typical Tortricid shape. The adult moth is out in June and July and is common over most of its range. The reddish brown caterpillar feeds on the leaves of rose or brambles (Rosaceae).

**foenella** Linnaeus
Europe including Britain, temperate Asia 17–25 mm. (0·67–0·98 in.)
This moth, although widespread, is rarely common. The yellowish caterpillar feeds in the stem and roots of *Artemesia vulgaris* (Compositae). The moth is out in July and flies at night and is attracted to light. **17**

**Epicampoptera** Bryk (DREPANIDAE)
There are currently 19 species in this genus of hook-tip moths. All the species are confined to Africa south of the Sahara or to Madagascar. The caterpillars of some species are pests of coffee *(Coffea)*. The closest relative of this genus is probably the Indo-Australian genus *Cyclura* Warren whose species have similarly shaped wings.

**marantica** Tams
Tropical Africa 33–43 mm. (1·3–1·69 in.)
*E. marantica* is typical in colour pattern of most species of its genus. The wings are dull reddish or yellowish brown, or dark greyish brown, with a weakly marked transverse band. Both male and female have a short process on the outer margin of the wings, but these are better developed in the male which has an almost tailed hindwing. The caterpillar of *E. marantica* is a common pest of coffee foliage.

**Epicausis** Butler (NOCTUIDAE)
There is only one species in this Madagascan genus.

**smithi** Mabille
Madagascar 55–65 mm. (2·17–2·56 in.)
Both sexes bear a highly conspicuous, massive tuft of bright red hair-scales at the posterior end of the abdomen. The colours red, black and yellow are a common combination in species which are distasteful to predators and have developed a warning colour pattern. It is a reasonably safe prediction that *E. smithi* will prove to store toxic chemicals and to be rejected by birds as a source of food. **392**g

**Epicephala** Meyrick (GRACILLARIIDAE)
A large genus of micro-moths with species in Australia, India and Africa.

**chalybacma** Meyrick
India, Pakistan, Sri Lanka 8 mm. (0·31 in.)
This moth is green with darker lines along the slender, rather pointed forewing, and a black, white-edged, patch near the tip of the wing. There is a long fringe on the hind margin of the forewing. The forewing in many specimens is often mottled with brown. The hindwings are slender, pointed, with a long fringe. The caterpillar feeds on flowers, species of *Caesalpina* (Leguminosae). When it is fully grown it gnaws its way out of the flower and lowers itself on a long thread. As soon as it touches a leaf or other flat surface it starts spinning a cocoon which is unique in having the upper-side covered with minute droplets. These are produced by the caterpillar who fixes them onto the back of the cocoon where they look a mass of glistening globules. The function of these is obscure. It has been suggested that they are imitation parasite cocoons so that when the parasitic wasp, searching for a host, sees them, it refrains from laying any eggs in the caterpillar assuming that it has already been attacked. However there is no experimental evidence for this.

**Epichoristodes** Diakonoff (TORTRICIDAE)
A small genus of micro-moths found in Africa and Madagascar.

**acerbella** Walker: syn. *E. ionephela* SOUTH AFRICAN CARNATION WORM
Europe including Britain, Africa 18–22 mm. (0·71–0·87 in.)
This moth has been imported into Europe on carnation *(Dianthus)* cuttings and other plants from South Africa. In Europe it is mainly a greenhouse pest. The moth has reddish brown forewings scattered with brown spots; the hindwings are grey. The eggs are laid in clusters of about 25 on the plant, near the base or apex of the leaves. The yellowish caterpillar has a brown head and eats the leaf edges. After 4–8 weeks it is fully grown and green, with pale dorsal and lateral stripes. After pupating, the moth emerges in 9–26 days. There may be 5 generations a year in South Africa. The caterpillars kill the plants when they bore into the shoots.

**Epicmelia** Korb (AXIIDAE)
Only one species is known. This is one of the 2 genera of the moth family Axiidae.

**theresiae** Korb
Turkey, Syria 37–40 mm. (1·46–1·57 in.)
The forewing of this species may be more yellowish than in the illustrated specimen and, rarely, greenish brown in ground colour. The hindwing varies between specimens from pale yellow to dark orange. Its caterpillar is thought to feed on the foliage of *Astragulus* (Leguminosae). **272**h

**Epicoma** Hübner (NOTODONTIDAE)
This is a small, endemic Australian genus of the subfamily Thaumetopoeinae (Processionary moths). The species are unusual in that they lack a proboscis. Females have a dense mass of loose scales at the end of the abdomen which are used to cover and protect the eggs. The caterpillars are not invariably gregarious or processionary.

**melanospila** Wallengren
E. Australia 35–42 mm. (1·38–1·65 in.)
Adult males differ from the illustrated female in the extremely long pectinations (branches) of the antennae. There is some individual variation in pattern, especially in the hindwing which may be uniformly dark brown except for the yellow marginal band. The caterpillars of *E. melanospila* feed on *Eucalyptus*. **357**h

**Epicopeia** Westwood (EPICOPEIDAE)
This is the only genus of the family Epicopeidae. It occurs in N. India, Sikkim, Bhutan, China and Japan, southwards to the mountains of Sumatra. The moths fly by day, feed at flowers, and are thought to be mimics of *Parides* butterflies.

**hainesi** Holland
C. China, Japan 50–70 mm. (1·97–2·76 in.)
In general plan, *E. hainesi* is similar to the much larger *E. polydora* (illustrated) in coloration, but the hindwing tail is relatively larger, there is no red spot at the apex of the tail and the red markings are represented by a single row of lunules. In most specimens, the outer part of the hindwing is much darker in ground-colour than the rest of the wing. There are 2 broods per year. The caterpillar is green, and covered with waxy granules or threads; it feeds on *Lindera* (Lauraceae).

**mencia** Moore
C. China 60–77 mm. (2·36–3·03 in.)
*E. mencia* is similar to *E. polydora* in general coloration and wingshape, but has a larger tail to the hindwing, no white patch, and a double row of red spots or lunules at the outer margin of the hindwing. There is a pair of red patches at the front of the thorax. It is a fairly good mimic in wingshape and coloration of the Papilionid butterflies *Parides mencius* and *P. alcinous* and their relatives. The caterpillar, which is black and covered with a bloom of waxy granules, feeds on elm *(Ulmus)*. Before pupating it spins several leaves together as a protective shelter for the chrysalis which is also covered with a waxy deposit.

**polydora** Westwood
N. India, Tibet, N. Burma 80–114 mm. (3·15–4·49 in.)
This species apparently forms a mimetic association with species of the butterfly genus *Parides* such as *P. philoxenus*. Some specimens are without markings on the hindwing, except for the red spots at the apex of the tail. The caterpillar is covered with strings of waxy secretion and has been compared with a colony of scale-insects (Coccidae) in appearance. The chrysalis is encased in a flimsy cocoon which is attached to the branch by a silken thread. **314f**

**Epicopeidae**
This is a small but fascinating family of Asian moths. Its affinities are uncertain and it has been associated at times with both the Uraniidae and Chalcosiidae, but is usually placed next to the former in the superfamily Geometroidea. Many species are closely similar in wingshape and colour pattern to species of the generally distasteful butterfly genus *Parides*. As adult Epicopeids are capable of producing strong-smelling yellow fluid from their thoracic glands, it is probable that they are as unpalatable to their enemies as the species of *Parides* and that their colour pattern is as 'warning' in character as that of the butterflies. A common warning pattern is an economic method of advertising unpalatability as the potential predator has only one pattern to recognize. The caterpillars of this family secrete waxy deposits, sometimes in the forms of threads, which are presumably protective in function. See *Epicopeia*.

**epijarbus** see **Deudorix**
**epilais** see **Syntomeida**
**epimenis** see **Psychomorpha**
EPIMENIS, GRAPE-VINE see **Psychomorpha epimenis**

**Epinotia** Hübner (TORTRICIDAE)
This is a large genus of micro-moths with many species in Europe and temperate Asia. The exact limits of the genus is not known; further studies are needed to define it.

**stroemiana** Fabricius: syn. *E. similana*
Europe including Britain, temperate Asia, N. America 17–21 mm. (0·67–0·83 in.)
There is some doubt as to whether the American specimens are this species, they are rather yellower than the European ones. The caterpillar feeds on birch *(Betula)* and the moth, which is common over most of its range, is out in September. **15**

**Epione** Duponchel (GEOMETRIDAE)
A genus of moths found in Europe and temperate Asia. The caterpillars are typical of Geometrid moths.

**paralellaria** Denis & Schiffermüller
DARK BORDERED BEAUTY
Europe including Britain 23–28 mm. (0·91–1·1 in.)
The moth is out in July and August. The difference between this species and the following one are given under *E. repandaria*. The 2 species are closely related. The caterpillar feeds on birch *(Betula)*, aspen *(Populus)*, willow *(Salix)* and other trees and shrubs. The moth is found in N. and C. Europe but is not common in Britain. **295**

**repandaria** Hufnagel BORDERED BEAUTY
Europe including Britain, Asia 25–29 mm. (0·98–1·14 in.)
The moth, which flies in July, is common in W. Europe and S. Britain. The caterpillar is brownish and has black marks in more or less diamond shapes along the back, and a yellow line along the side. It feeds on willow, sallow *(Salix)* and alder *(Alnus)*. The orange-yellow colour of the moth is similar to the related *E. paralellaria*, the Dark Bordered Beauty, but the marginal dark areas have more or less wavy edges on the latter, whereas in *E. repandaria* they are straighter on the forewing. In *E. paralellaria* the females are distinct in colour and pattern from the male.

**Epipagis** Hübner (PYRALIDAE)
A mainly tropical genus of Pyralid moths found on most continents. Many species have been described in the genus but the characters of it are not well known.

**pictalis** Swinhoe
N. India, Pakistan, Malaya 22–25 mm. (0·87–0·98 in.)
This species is fairly typical of a wide range of Pyralid moths. Nothing is known of its biology but the species is usually well represented in collections. **56ff**

**Epiphile** Doubleday (NYMPHALIDAE)
A small genus of butterflies with some 20 species in C. and S. America.

**hubneri** Hewitson
Brazil 48 mm. (1·89 in.)
This species is variable in pattern. The underside of the forewings are pale yellowish; the hindwings are brown. There are several related species with broadly similar patterns. The female is browner on the hindwing and has less iridescence than the male. **229f**

**Epiphora** Wallengren (SATURNIIDAE)
This is a large genus of moths restricted to Africa, south of the Sahara. Both fore- and hindwings bear a round or crescent-shaped, transparent patch, as in the related S. American *Rothschildia* and the Asian *Attacus*. Some species are uncharacteristic of their family in having a functional proboscis. The nominal genus *Falcipennae* Pinhey (a replacement name for *Drepanoptera* Rothschild) is currently regarded as a subgenus of *Epiphora*.

**vacuna** Westwood AFRICAN ATLAS
Tropical & subtropical Africa 100–180 mm. (3·94–7·09 in.)
There is some dispute amongst collectors concerning the validity of the several subspecies of *E. vacuna*. The W. African *E. v. plotzi*, for example, is variously regarded as such or as a distinct species. The females of all the subspecies differ from the males in the less strongly hooked apex to the forewing. In general appearance, *E. vacuna* is similar to the illustrated *Attacus atlas*.

**epiphron** see **Erebia**

**Epiplema** Herrich-Schäffer (EPIPLEMIDAE)
This is a large genus of small, mostly cryptically patterned moths found in Africa, temperate America (U.S. and Canada) and tropical America, and in temperate and tropical Asia eastwards to Australia and the Solomons. The caterpillar of one species has been reported feeding on the foliage of coffee *(Coffea)* in Madagascar.

**dryopterata** Grote
Canada, N. U.S. 16–19 mm. (0·63–0·75 in.)
This is a small brown species, its wings crossed by 2 nearly parallel dark brown lines. It flies from May to June and again in August. The black-spotted, greenish white caterpillar feeds on *Viburnum prunifolium*.

**himala** Butler
Sikkim, N. India, N. Burma, Himalayan China 20–27 mm. (0·79–1·06 in.)
This is possibly the most colourful species of its genus. There is some doubt as to whether the Chinese specimens represent the same species as those from India. Specimens of *E. himala* have been taken at elevations of up to 2440 m. (8000 ft) in the Himalayas. **272e**

**irrorata** Moore
India, Sri Lanka 12–14 mm. (0·47–0·55 in.)
This is one of the smallest species of its genus. The wings are grey, marked with bands of yellowish brown and black and speckles of dark brown. The forewing has 3 short processes at the outer margin; the hindwing 2 rather larger processes (almost tails). Its life history is unknown.

**Epiplemidae**
The family Uraniidae is generally considered to be the nearest relative of this mainly tropical, African, Asiatic, Australian and American family of moths. It belongs in the superfamily Geometroidea. There are probably about 500 species of Epiplemids in the world. most are small, fragile, Geometrid-like moths, usually with emarginate outer margins to the wings. At rest, the wings are folded in various ways. The hindwing is usually folded along the side of the abdomen, while the forewing can be either flat or rolled. The caterpillars are gregarious web-makers in their early stages. See *Epiplema*.

**Epipomponia** Dyar (EPIPYROPIDAE)
A small genus of moths whose caterpillars feed externally on leaf-hoppers or cicadas. The species in the genus are known from S. America and Japan.

**nawi** Dyar
Japan 25 mm. (0·98 in.)
This small moth, whose caterpillar is an ectoparasite on another insect, generally a species of cicada, is known only from Japan. **54c**

**Epipyropidae**
A small family of moths at present placed near the Zygaenidae and Cyclotornidae, whose caterpillars live as ectoparasites on various species of bugs (especially cicada, leaf-hopper and related groups). There is some dispute as to whether they feed on the waxy secretaions on the outside of the host insect or actually feed on the insect itself (like a flea on a mammal). Certainly in some cases they eventually kill their hosts. This unusual habit of feeding externally on leaf-hoppers is shared by the caterpillar of the Cyclotornidae. See *Epipomponia*, *Heteropsyche*.

**Epirrita** Hübner: syn. *Oporinia* (GEOMETRIDAE)
A genus of moths with species in Europe and temperate Asia and a few in N. America. The caterpillars are typical of most Geometrid moths.

**dilutata** Denis & Schiffermüller NOVEMBER MOTH
N. & C. Europe including Britain, W. & C. Asia 25–36 mm. (0·98–1·42 in.)
Records of this moth from N. America have proved to be of a similar looking, related species. The caterpillar is green, often marked with red in various lines or spots along the back. It feeds on elm *(Ulmus)*, oak *(Quercus)*, birch *(Betula)* and fruit trees including apples *(Malus)*, pears *(Pyrus)* and plums *(Prunus)*. The moth is out in October and November in woodlands and is common over much of its range. It is readily attracted to light.

**Episcea** Warren (HYPSIDAE)
Two species of tropical S. American moths are placed in this genus: *E. sancta* and *E. extravagans* Warren.

**sancta** Warren
Brazil 15–24 mm. (0·59–0·94 in.)
This is very similar in colour pattern to the Dioptid *Scea auriflamma* and is probably a member of a mimetic complex involving Nemeobiid butterflies and day-flying moths.

**epixanthe** see **Lycaena**
**equitella** see **Glyphipterix**

**Erannis** Hübner (GEOMETRIDAE)
A genus of moths whose caterpillars generally feed on foliage of trees. The species are often abundant and their caterpillars can be very destructive, defoliating large areas of forest and plantation. They are found mostly in Europe, temperate Asia and N. America.

**defoliaria** Clerck MOTTLED UMBER
Europe including Britain, N., C., & W. Asia 35–40 mm. (1·38–1·57 in.)
This moth can be a serious pest to the forester. The caterpillars feed on birch *(Betula)*, oak *(Quercus)* and

many forest trees. At times they occur in vast numbers and can defoliate a forest, turning early summer green to winter brown. The growth of the trees so attacked may be affected for years. Their devastations at times are so serious that control measures have been attempted by spraying from the air with insecticides. The moth flies late in the year, from October to December and is one of the species which is attracted to light in the early part of the winter. The female is wingless. **294**d, **308**

**Erasmia** Hope (ZYGAENIDAE)
There are about 6 species in this genus of moths from China, Formosa, Hainan and in Okinawa. The caterpillars are velvet black, coloured with red and yellow spots. These probably make them conspicuous and it is possible that they are distasteful to birds. They often occur on wild coffee *(Coffea)* where the cocoon is spun upon the leaf when they pupate.

**pulchella** Hope
India, Formosa, S. China, Buru 70–85 mm. (2·76–3·35 in.)
There is considerable variation amongst specimens of this species in the amount of iridescent metallic green-blue on the wings. The angle of the illumination completely alters the colour of this moth from green to purple. Several subspecies have been described from different areas. Nothing appears to be known of its biology although it is such a widespread and relatively common species. **45**e

**sanguiflua** Drury
India, Assam 80–110 mm. (3·15–4·33 in.)
This is strikingly different from the others in the genus. There is some variation in colour in specimens from different localities. The males have generally more black colour on the wing. The underside is without the brown colour on the wing fringes. Nothing is known of its biology. **45**g

**erate** see **Colias**

**Erateina** Doubleday (GEOMETRIDAE)
A large genus of American moths which often have the hindwings extended into long tails. There is a complete series which can be seen in the moths of this genus from species with rounded hindwings, to those with increasingly elongate tails to the hindwing. Some of the more extreme forms are illustrated. Quite a number of the species in the genus have rather similar patterns with variation in the size of the patches being the main external difference between the species. In the more extreme tailed forms some of the species approach the Zygaenid genus *Himantopterus* which has the entire hindwing reduced to a long thin process. Although some of the *Erateina* have extreme elongation of the wing, generally the base of the hindwing is still relatively broad.

**julia** Doubleday
Venezuela 35 mm. (1·38 in.)
The very long tails on the orange hindwings are characteristic of this species. Nothing is recorded of its life history. The underside has more pattern with thin markings. **294**s

**leptocircata** Guenée
Colombia 26–35 mm. (1·02–1·38 in.)
The hindwings are very elongate in this species. The tails are larger than in the similar *E. julia* and orange-yellow, this colour patch being larger than in *E. julia*. The underside has pronounced white stripes on the hindwing and 2 bands across the forewing; only one of these is visible on the upperside.

**lineata** Saunders
S. America 35 mm. (1·38 in.)
The hindwings are chestnut red and slightly elongate, much darker in colour than the related species. The underside is strongly patterned with white lines. Nothing is known of the biology of this species. **294**q

**erato** see **Heliconius**

**Erebia** Dalman (NYMPHALIDAE)
A huge genus, with several hundred species of butterflies described in it, mostly from Europe and Asia, but with one New Zealand species and a few N. American. The genus is in the subfamily Satyrinae.

**aethiops** Esper SCOTCH ARGUS
Europe including Britain, Turkey, Ural Mountains and Caucasus to W. Asia 44–52 mm. (1·73–2·05 in.)
This butterfly is out in August and September in hilly country – usually near coniferous trees. It is found at sea-level in some areas (eg round Ostend) but more frequently in the hills. The caterpillar feeds on grasses, including *Poa, Dactylis*. There is some variation in size and colour of this species, several forms have been named. **181**

**disa** Thunberg ARCTIC RINGLET or DISA ALPINE
Arctic Europe, N. America 46–50 mm. (1·81–1·97 in.)
The butterfly is found in bogs and wet moorland in the far north. The European and American subspecies are distinct. The foodplant of the caterpillar is not known. There are a number of similar species generally dark brown with black, usually white-centred, rings on the wings, particularly the forewings.

**epiphron** Knoch MOUNTAIN RINGLET
Europe including Britain 34–36 mm. (1·34–1·42 in.)
Several subspecies have been described of this mountain butterfly in different parts of Europe. It is found in Britain over 610 m. (2000 ft) and in Europe generally over 1220 m. (4000 ft). The butterfly is out in July and is usually found in high moorland or on mountain slopes. The caterpillar, which is green with stripes along the side, feeds on various grasses, including *Deschampsia caespitosa* and *Nardus stricta*. There is some variation in the pattern of the adults, particularly in the eyespots on the wings. **173**

**meolans** de Prunner PIEDMONT RINGLET
Europe 38–42 mm. (1·5–1·65 in.)
This species, which is typical *Erebia*, has several subspecies, all from different parts of Europe. The butterflies are out in June and July and generally fly over stony slopes on mountains at 1520–1830 m. (5000–6000 ft). The caterpillar feeds on species of grass.

**rossii** Curtis ROSS'S ALPINE BUTTERFLY
N. America, Asia 40 mm. (1·57 in.)
This species flies in the arctic and subarctic regions of N. America and N. Asia. Nothing is known of its biology. The butterfly is a brownish black with eyespots ringed with reddish brown. The American and Asian butterflies are considered as distinct subspecies of *E. rossii*. **174**g

**Erebomorpha** Walker (GEOMETRIDAE)
The only species in this genus of moths ranges from India to China.

**fulguritia** Walker
N. India, Pakistan, China 70–80 mm. (2·76–3·15 in.)
Several subspecies of this species have been described. The specimen illustrated is from India. **272**q

**eresimus** see **Danaus**
ERESIMUS BUTTERFLY see **Danaus eresimus**

**Eretmocera** Zeller (HELIODINIDAE)
Species in this genus of micro-moth occur in Europe, W. and S. Africa and the Far East. It contains a large number of species but very little is known about them. Most have been collected in small numbers and some are known only from single examples. Many are brilliantly coloured. Their narrow wings have long fringes.

**chrysias** Meyrick
Africa 13–16 mm. (0·51–0·63 in.)
This species found from Zaire to Natal and in Uganda. The red abdomen and golden spots on the wings make it a very beautiful insect, but, like the others in the genus, being small, it is seldom noticed except by lepidopterists and many of them would probably miss it in its natural surroundings where its colours blend with the surroundings. **41**b

**fuscipennis** Zeller
W. Africa 14 mm. (0·55 in.)
This is a bright yellow and red species. There are many related species with slightly differing colours including orange and yellow ones with golden wing spots. **41**g

**impactella** Walker
India, Burma, Sri Lanka, Singapore, Formosa 12–13 mm. (0·47–0·51 in.)
This moth is a typical *Eretmocera* with black wings and 4 small yellowish patches, a curved hind margin to the forewing and 2 black bands on a yellow-orange abdomen. The caterpillar feeds on species of *Amaranthus* and spins a web, pulling the heads of the flowers together. It is sometimes a minor pest.

**laetissima** Zeller
W. Africa 12 mm. (0·47 in.)
This is a widespread species which at present is believed to extend from W. Africa into S. Africa although there is some dispute on the identity of the specimens from part of this range. Little is known about the biology of this species. **41**e

**Erinnyis** Hübner (SPHINGIDAE)
This is a chiefly tropical New World genus of hawk-moths, but many species stray at times into N. America and 8 species probably breed there from time to time.

**ello** Linnaeus
N., C. & tropical S. America 44–100 mm. (1·73–3·94 in.)
This is a common species in most of its extensive range, including N. America where it is a breeding species in the S. and a regular stray as far N. as New York State and Michigan. The caterpillar is a pest of poinsettia *(Euphorbia)* in S. Texas and feeds also on guava *(Psidium)* in tropical America. **344**h

**Eriocrania** Zeller (ERIOCRANIIDAE)
This small genus of micro-moths is found in Europe including Britain, and N. America. It is one of the more primitive groups of moths. The caterpillars mine in leaves, or occasionally in seeds, of pines *(Pinus)*. Generally the moths are day-fliers. One of the American species makes blotch-mines in oaks *(Quercus)*. Although rare over much of the range, they may be quite common in certain restricted localities.

**sparmannella** Bosc
C. & S. Europe, Britain 9–11 mm. (0·35–0·43 in.)
The whitish caterpillar of this moth forms blotchy patches when it mines in the leaves of birch *(Betula)*. The moths are out in June and July. **1**g

**Eriocraniidae**
A small family of primitive, mostly small moths which usually fly in the day time. The caterpillars of these moths are leaf-miners and are often legless. Some of the family have small tongues (proboscis) but the mouthparts are regarded as more primitive than most other Lepidoptera (except the Micropterygidae). One of the genera in this family *(Agathiphaga)* from Queensland has recently been placed in a separate family *(Agathiphagidae)*. See *Eriocrania*.

**eriphia** see **Pinacopteryx**
**erippus** see **Diogas**
**eris** see **Colotis**
**erischton** see **Automeris**
ERMINE MOTHS see **Spilosoma**

**Erocha** Walker (AGARISTIDAE)
This is a small genus of about 10 species of C. American and tropical S. American moths.

**leucodisca** Hampson
Ecuador, Peru 42–55 mm. (1·65–2·17 in.)
The unusual pattern of the forewings, which cover the hindwings when at rest, is presumably cryptic, but observations on the choice of resting place in the field are needed before this can be confirmed. Specimens have been caught up to about 610 m. (2000 ft) in elevation. **399**j

**Eronia** Hübner (PIERIDAE)
A small genus of butterflies with species throughout Africa extending northwards into the United Arab Republic.

**leda** Boisduval ORANGE AND LEMON or AUTUMN-LEAF VAGRANT BUTTERFLY
Tropical and C. Africa to S. Africa 56–64 mm. (2·2–2·52 in.)

The male is a bright sulphur yellow with a crimson tip to the forewing, edged on the outer margin with brown. The head and antennae are reddish. The female is paler, with a brown apex with black spots on the forewing and a few black spots round the hindwing. The dry season form is more speckled on the underside. This fast flying species is found in the thorn bush country in the E. Cape, through Mozambique up to the tropics.

**Erora** Scudder (LYCAENIDAE)
There are 3 species in this genus of butterflies. It is found in N. America, Mexico and C. America.

**laeta** Edwards EARLY HAIRSTREAK
E. Canada, U.S. 20–24 mm. (0·79–0·94 in.)
This is typically a butterfly of forested areas wherever there are beech trees *(Fagus)* which are the caterpillars' foodplants. It is not a common species and is regarded as something of a prize by collectors.

**erota** see **Vindula**
ERYCID, WHITE see **Hermathena candidata**
**ericinoides** see **Poritia**

**Erynnis** Schrank (HESPERIIDAE)
A genus of skipper butterflies found throughout Europe and temperate Asia and in N. America. In N. America they are popularly known as Dusky-wings.

**horatius** see under **E. juvenalis**

**juvenalis** Fabricius JUVENAL'S DUSKY-WING
S. Canada, U.S. west to the Rocky Mountains 30–42 mm. (1·18–1·65 in.)
This is a common, rather large, Dusky-wing. It has clear white spots near the apex of the forewing and 2 white spots just below the outer forewing margin. There is a closely allied species, *E. horatius* Scudder & Burgess, which has a similar appearance and can only be reliably separated from *E. juvenalis* on structures which need examination under a microscope.

**tages** Linnaeus DINGY SKIPPER
Europe including Britain, temperate Asia 26–28 mm. (1·02–1·1 in.)
There are several similar-looking related species in Europe. The butterfly is out in May or June, often with a second brood later in the more southerly parts of its range. It flies in meadows, particularly in limestone areas. The caterpillar feeds on *Lotus corniculatus* (Leguminosae), *Eryngium* and other umbellifers. Sunshine is needed to get this butterfly on the wing, otherwise it sits on grass or dead seed heads, with its wings over its back, rather like a noctuid moth, when it is difficult to detect. **85**

**Eryphanis** Boisduval (NYMPHALIDAE)
This genus of butterflies is in the subfamily Brassolinae. The species are known from C. and S. America.

**polyxena** Meerburg PURPLE MORT-BLEU
S. & C. America, W. Indies 82–102 mm. (3·32–4·02 in.)
Several subspecies of this large blue butterfly have been described. The black edges round the midnight blue on the wings makes it a striking species when it opens its wings. At the base of the hindwing in the male there is an oval patch of yellowish scales on a silvery area. The female lacks this and has less blue on the wings. The caterpillar feeds on bamboo *(Gramineae)*. The butterflies are darker coloured on the underside. The adult flies in the late afternoon and at dusk and has a rapid flight, interspersed with a slower, dipping flight as it moves round forest edges or glades.

**erythropsalis** see **Hypsidia**
**esmerelda** see **Cithaerias**
ESSEX SKIPPER see **Thymelicus lineola**

**Esthemopsis** Felder (NEMEOBIIDAE)
There are about 15 species in this C. and tropical S. American genus of butterflies. They mimic moths of the genera *Hypocrita* and *Calodesma* and day-flying moths of the families Hypsidae, Ctenuchidae and Arctiidae. The species of *Uraneis* are similarly mimetic.

**clonia** Felder
Mexico to N.W. S. America 35–42 mm. (1·38–1·65 in.)
This is probably a mimic of *Hypocrita euploeodes* and *Calodesma melanchroia* (Hypsidae), both of which occur together with *E. clonia* in C. America. The female has a less strong blue iridescence on the wings than the illustrated male. **266**d

**sericina** Bates
Tropical S. America 36–42 mm. (1·42–1·65 in.)
*E. sericina* is a mimic of several day-flying Ctenuchid and Arctiid species which fly together with it. One of the 'best' models is *Euagra splendida* Butler, an iridescent, blue and black species of Ctenuchidae which has transparent patches in the wings.

**thyatira** Hewitson
Tropical S. America male 30–34 mm. (1·18–1·34 in.) female 36–48 mm. (1·42–1·89 in.)
The range of wing-measurements indicates the unusually large difference in size as well as colour pattern (see illustrations) between males and females. **268**k, l

**esthla** see **Damias**

**Estigmene** Hübner (ARCTIIDAE)
About 50 species are at present placed in this genus of moths. Most are tropical Old World in range and are certainly not closely related to the type-species of the genus, *E. acrea*.

**acrea** Drury SALTMARSH CATERPILLAR
Canada, U.S., C. America 46–64 mm. (1·81–2·52 in.)
The female hindwing of *E. acrea* is white in much of its range in N. America, but in S.W. U.S., and further south, females have yellow wings like the male (illustrated). Some males have greyish hindwings, especially in the north of its range. The grey or brown 'woollybear' caterpillar is by no means restricted to saltmarshes, as the popular name might imply, and feeds on a variety of low-growing plants including garden plants, water-melon *(Citrullus)*, corn *(Zea)* and at times on the foliage of fruit trees; it is gregarious at first, under leaves. Masses of caterpillars will sometimes move together to a new feeding area. **369**j

**Eterusia** Hope (ZYGAENIDAE)
The caterpillars of moths in this genus feed on a variety of plants, including ones as different as tea *(Thea)* bushes and roses *(Rosa)*. At times they are minor pests in tea plantations. Little is known of the biology of the species in the genus.

**rajah** Moore
India 36 mm. (1·42 in.)
Several related species, one of them larger than this, are found in India. The bright green forewings and yellow hindwings are a contrast to the illustrated species. The abdomen is an electric blue iridescent colour above.

**repleta** Walker
India, Hainan 75–80 mm. (2·95–3·15 in.)
This strikingly coloured insect should be compared with the related species *E. rajah*. **45**a

**ethillus** see **Heliconius**

**Ethmia** Hübner (ETHMIIDAE)
A worldwide genus of micro-moths with numerous species, many with caterpillars feeding on species of Boraginaceae. All the known caterpillars are brightly coloured. The species often have very striking metallic patterns. At one time the genus was included in the Yponomeutidae.

**aurifluella** Hübner
S. Europe 16–29 mm. (0·63–1·14 in.)
The caterpillar of this species has been reared on *Thalictrum* (Ranunculaceae) and *Anchusa* (Boraginaceae). **39**e

**bipunctella** Fabricius
Europe including Britain, Canary Islands, Turkey, Canada 24 mm. (0·94 in.)
The moths are out in May and August and are common in the warmer parts. The caterpillar feeds on *Echium* (Boraginaceae) and is found among the flowers and leaves in July and September. It is bright orange and black. **39**f

**hilarella** Walker
Sri Lanka, S. India, Formosa 28–35 mm. (1·1–1·38 in.)
There are several similar species with the grey spotted forewings and yellow hindwings. They are slightly different in pattern but generally have to be examined closely under a microscope to separate the species. **39**a

**lineatonotella** Moore
India, Formosa 38–45 mm. (1·5–1·77 in.)
This species has been collected in the mountains of N. India. Nothing is known of its biology. **39**c

**Ethmiidae**
A family of micro-moths related to the Gelechiidae with many species in the tropics. It is generally separated from other families by different venation characteristics in the hindwing. The forewings often have brightly coloured or metallic spots. The life histories of a few species are known; generally the caterpillars feed in webbing they spin on the leaves of species of Boraginaceae. See *Ethmia*.

**Ethopia** Walker (PYRALIDAE)
This genus of Pyralid moths has only one species. It is unusual in the subfamily Galleriinae in being brightly coloured as the majority of species in this subfamily are rather sombre coloured.

**roseilinea** Walker
New Guinea 75–110 mm. (2·95–4·33 in.)
This species is found only in New Guinea and some of the neighbouring islands. Although there is no English name for it, the scientific name is particularly descriptive. Nothing is known of the life history of this moth, not even if it flies in the day or at night. **63**r

**Etiella** Zeller (PYRALIDAE) BEAN-BORER MOTHS
This genus occurs throughout the tropics and subtropics, with most species in the Australian-New Guinea area. One species is cosmopolitan and is a serious pest of beans in many countries. Most of the species are rather similar and there is much confusion in the literature on the identity of the species. The genus belongs to the subfamily Phycitinae.

**behrii** Zeller BEAN MOTH
Australia, New Guinea 20–26 mm. (0·79–1·02 in.)
This species is similar to *E. zinckenella* and has been much confused with that species in the past. The two appear to have similar habits although a detailed study of this species is needed. Although relatively restricted in distribution there is some evidence that this species is spreading and could be as potentially serious a pest as the more widespread one.

**zinckenella** Treitschke LIMA-BEAN POD-BORER
Cosmopolitan between 50° N. and 50° S. 20–30 mm. (0·79–1·18 in.)
This widespread species causes serious loss of crops of Lima beans *(Phaseolus)* and other legumes in many countries. In the U.S. much work on the biology has been done in an effort to understand the reason for its spread and to control the insect by insecticide or biological control. The caterpillar feeds inside the bean-pod on the developing beans. While it is a problem in the U.S., it can be even more serious in less developed countries where the beans form an important source of protein in the diet. The moth is a strong flyer and migrations of many hundreds of them have been recorded. This undoubtedly helps in its widespread dispersal, although there is some evidence for accidental introduction by man. **56**p

**Euagra splendida** see under **Esthemopsis sericina**

**Eubergia** Bouvier (SATURNIIDAE)
This is a small genus of conspicuously patterned, tropical American moths. Its closest relative is possibly *Hylesia* with which it is united as a subgenus by some authors. The striped colour pattern is highly diagnostic; no other Saturniids have one comparable.

**boetifica** Druce
Peru, S. Brazil, Paraguay, Argentina 34–50 mm. (1·34–1·97 in.)
This striking species is typical of its genus in colour pattern. **331**j

**Eublemma** Hübner (Noctuidae)
About 50 species are currently placed in this genus of small moths. Most of the species are Old World in distribution (from Europe to Australia and Africa) but there are also representatives in N. America.

**ostrina** Hübner PURPLE MARBLED MOTH
N. Africa, Egypt, Madeira, Canary Islands, Turkey, S.E. Asia, S. Europe 18–28 mm. (0·71–1·1 in.)
This is a rare migrant in N. Europe and Britain, but not uncommon elsewhere. It is highly variable in coloration which may lack purple entirely, and in the colour pattern which may be almost completely absent. The caterpillar feeds on the shoots of Carline Thistle *(Carlina)* and *Helichrysum*. **390**f

**eucalypti** see **Xyleutes**

**Euchaetias** Lyman (Arctiidae)
Canada, U.S., C. America
The caterpillars of this small, mostly N. American genus of moths feed on species of poisonous Asclepiadaceae or Apocynaceae and probably pass on distasteful qualities of these plants to the adult moth. Many species are white, with red or yellow on the head, legs and abdomen; others have mostly grey or brown wings.

**murina** Stretch MOUSE-COLOURED EUCHAETIAS
S. U.S., Mexico 23–31 mm. (0·91–1·22 in.)
This small moth is one of the commonest species of Arctiidae in S. U.S. Its forewings are greyish brown, the hindwings are the same colour, or mostly white, and the abdomen reddish orange. The caterpillar feeds on milkweeds (Asclepiadaceae).

EUCHAETIAS, MOUSE-COLOURED see **Euchaetias murina**

**Eucharia** Hübner (Arctiidae)
There is only one known species in this genus of moths.

**casta** Esper
S. C. Europe 30–32 mm. (1·18–1·26 in.)
This is a fairly rare species which flies during May. The forewings of *E. casta* are brown with two irregularly shaped, transverse pink bands; the hindwings are pink except for brown outer marginal markings. Its caterpillars feed on bedstraws *(Galium)* and *Asperula*. **368**a

**eucharis** see **Delias**

**Euchlaena** Hübner (Geometridae)
A N. American genus of moths with some 60–100 species known. This genus is widespread in the U.S. The caterpillars are generally twig-like in appearance.

**irraria** Barnes & McDunnough
U.S., Canada 38–45 mm. (1·5–1·77 in.)
The specimen illustrated has darker wing margins then many others of this species, which is very variable in colour. The caterpillar feeds on oak *(Quercus)*. It is rare in Canada but reasonably common in the U.S. **294**c

**Euchloe** Hübner (Pieridae)
This genus of butterflies has species throughout Europe, temperate Asia, N. America with a few species in Africa.

**ausonia** Hübner DAPPLED WHITE
Europe, N. Africa, Asia, N. America, from Alaska to Arizona 40–48 mm. (1·57–1·89 in.)
Several subspecies have been described of this widespread butterfly. The hindwings have a green pattern on the underside which shows as a grey shadow through the upperside of the wing. The apex of the white forewings have black tips, rather similar to *Pieris napi*. There are several related species which look similar and a careful study is needed to recognize the differences. The adults fly generally in mountain valleys and meadows over 1520 m. (5000 ft). The caterpillars feed on *Sisymbrium*, *Barbarea* and other species of Cruciferae.

**belemia** Esper GREEN-STRIPED WHITE
S. Europe, Canary Islands, N. Africa, Iran to Baluchistan 36–44 mm. (1·42–1·73 in.)
This species is similar to *E. falloui* but can be separated by the presence on the underside of the forewing of a white line in the middle of the black spot – in *E. falloui* the spot is all black. The hindwings have the green arranged in bands across the wing. A subspecies has been described from mountains in the Canary Islands. The foodplant of the caterpillar is not known; it is probably a species of Cruciferae.

**falloui** Allard SCARCE GREEN-STRIPED WHITE
N. Africa 36–38 mm. (1·42–1·5 in.)
The undersides of the hindwings have the greenish pattern typical of the genus, but it is arranged in distinct bands across the wings rather than along the veins. There is a black spot on the forewing and the underside has greenish lines. The foodplant is not known. The butterfly ranges from Egypt and Libya south to the Tibesti.

**olympia** Edwards OLYMPIA BUTTERFLY
U.S. 38 mm. (1·5 in.)
This is common in the Middle West of the U.S. where it can be seen in early spring in open woodlands or meadows. The butterfly is white, with a green marbling colour on the underside of the hindwing which shows through on the upperside. There are a few dark marks towards the apex of the forewing. The green caterpillar feeds on *Arabis lyrata*, *Sisymbrium officinale* and other Cruciferae.

**tages** Hübner PORTUGUESE DAPPLED WHITE
S.W. Europe, N. Africa 30–44 mm. (1·18–1·73 in.)
This is similar to *E. ausonia* in pattern but the hindwings are evenly rounded, whereas in *E. ausonia* there is a blunt angle opposite the anterior vein in the hindwing. The foodplants include *Iberis* (Cruciferae). Two subspecies have been described, one in S. France, the other in Spain and N. Africa.

**Euchloron** Boisduval (Sphingidae)
There is one known species in this genus of hawk-moths.

**megaera** Linnaeus VERDANT HAWK-MOTH
Africa, S. of the Sahara 72–120 mm. (2·83–4·72 in.)
Most specimens of this striking species resemble the illustration which is of subspecies *E. m. serrai* from the island of São Thomé, but others have more orange in the markings and may even have orange forewings. The caterpillar, which feeds on grape-vine *(Vitis)* and Virginia Creeper *(Parthenocissus)* (both Vitaceae), has a pair of eyespot markings near the front which may well deter small birds from making an attack. **344**g

**euchroia** see **Podotricha**

**Euchromia** Hübner (Ctenuchidae)
There are approximately 50 species in this genus of moths. Most of them are restricted to S.E. Asia, but 8 occur only in tropical Africa. The wings are dark brown or black, with white, yellow or red markings. Even more colourful is the abdomen, which is ringed with a variety of colours, some of which can be iridescent.

**guineensis** Fabricius
W. Africa, from Nigeria to Sierra Leone 42–52 mm. (1·65–2·05 in.)
This is somewhat similar to *E. lethe* (illustrated) but lacks the transverse white band towards the posterior end of the abdomen, and has deeper yellow areas on the wings.

**lethe** Fabricius
W. Africa and the Congo Basin 43–52 mm. (1·69–2·05 in.)
This is a species frequently discovered amongst consignments of African bananas in the docks and warehouses in Britain, although it has not been recorded in the literature as a pest of banana palms *(Musa)*. **364**b

**madagascarensis** Boisduval
Madagascar, the Comores Islands 42–50 mm. (1·65–1·97 in.)
The colour pattern of this species is similar to that of *E. guineensis* but the abdomen is only weakly iridescent blue, the thorax is marked with red, and basal markings on both fore- and hindwings are tinged with red proximally.

**euchromozona** see **Milionia**

**Euclidia** Ochsenheimer (Noctuidae)
The 10 or so species of this European and African genus have a geometrical wing pattern, no doubt the reason for Ochsenheimer's use of the name *Euclid*, the ancient Greek mathematician.

**cuspidea** Hübner
Canada S. to Arizona and Georgia 30–40 mm. (1·18–1·57 in.)
The colour pattern of this moth is similar to that of its close European relative *E. glyphica* (illustrated). Its caterpillar has comparable foodplant preferences but will also eat grasses.

**glyphica** Linnaeus BURNET COMPANION
Europe including Britain, temperate Asia to Japan 26–32 mm. (1·02–1·26 in.)
*E. glyphica* flies in May and June, often in company during the day with Burnet moths (Zygaenidae) whose caterpillars have similar foodplant needs – clovers and trefoils *(Trifolium* and *Lotus)*, amongst other legumes. **385**

**Eucosma** Hübner (Tortricidae)
PINE CONE MOTHS
This is a worldwide genus of micro-moths with many hundreds of species described. In N. America alone there are more than 150 described species. The caterpillars are mostly stem- and root-borers although some have specialized in the green branch tips and cones of pine trees *(Pinus)*. In the early part of this century, controversy arose over the naming of species in this genus. One entomologist described many species and called them *bana*, *wandana*, *landana*, and so on. Many purists who liked to derive names from Latin or Greek words were upset, and an attempt was made to change them. This failed because the earliest proposed name must be used.

**bobana** Kearfott PINE CONE MOTH
S.W. U.S. 17–22 mm. (0·67–0·87 in.)
This species has reddish-brown forewings, each with a clearly marked squarish patch of a darker colour Often these marks are narrowly bordered with white. The females are generally larger than the males and usually a slightly darker colour. The moth is widespread from C. Texas to the edge of the Mojave Desert in California and N. to Utah and C. Nevada. The caterpillar feeds on pine trees *(Pinus)*. They burrow into the scales on the cone and both these and developing seeds are eaten.

**ponderosa** Powell
U.S. 16–26 mm. (0·63–1·02 in.)
This species is similar in appearance to the preceding one and was first recognized as a distinct species in 1968. It is generally larger than *E. bobana* with which it had previously been confused and has orange forewing markings, often without the black scaling. It can only be reliably distinguished from *E. bobana* by its internal morphology. The moth was described from species reared from cones of *Pinus ponderosa* and is at present known only from the Blue Mountains in Oregon southwards through the eastern slopes of the Cascades and Sierra Nevada into the mountains of S. California.

**Eudaemonia** Hübner (Saturniidae)
There are 5 tropical African species in this genus. The extraordinarily long tails of these moths are up to five times as long as the remainder of the hindwing.

**brachyura** Drury
W. Africa, the Congo Basin 35–54 mm. (1·38–2·13 in.)
Little seems to be known about this species, and there

is no certainty, for example, concerning the function of the long tails. One theory suggests that predators may be more attracted to waving tails than to the rest of the moth and that an attack is in this way directed away from more vital parts. **333**c

**eudamippus** see **Polyura**

**Eudaphnaeura** Viette (NOCTUIDAE)
The 3 species of this genus of moths are restricted in distribution to Madagascar.

**splendens** Viette
Madagascar 45–57 mm. (1·77–2·24 in.)
It seems highly probable that this vividly coloured, red-tailed, orange, black and grey species will prove to be warningly coloured and to harbour poisonous substances in its tissues and consequently be distasteful to predators. **399**d

**Eudonia** Billberg: syn. *Witlesia, Scoparia* (PYRALIDAE)
Very large genus of similar looking black and white or grey and white Pyralid moths. They usually have some darker markings on the wings. The early stages, where known, tunnel in moss. A species described from Hawaii has a caterpillar which bores into Club Moss *(Lycopodium)*. In Hawaii there are at least 100 species of *Eudonia*. The islands have been particularly favourable for evolution of the group. There are also many species in New Zealand, one in Fiji and species on most of the continents.

**franclemonti** Munroe
U.S. (Arizona) 15–16 mm. (0·59–0·63 in.)
This small grey *Eudonia* was described in 1972 from specimens collected by a well-known American entomologist Dr J. G. Franclemont. He collected the specimens at West Fork, Flagstaff (U.S.) in 1964. The species is smaller than most of the other N. American *Eudonia*.

**mercurella** Linnaeus: syn. *E. mercurea*
Europe including Britain 15–18 mm. (0·59–0·71 in.)
This is a variable species, with some interesting local forms. One of these, *E.m. portlandica*, is found in Britain and has the central area of the wing darker with white edges. The moth is common over most of its range but is rather inconspicuous, the adult particularly blending in colour with lichen-covered rocks. It flies in June and July, generally at night, although easily disturbed from its resting places on trees, rocks or walls during the day. The caterpillar feeds in galleries which it makes in thick clumps of moss growing on trees or other sites. It is a pale yellowish colour, with a faint brown line along the back and a dark head and prothorax. It pupates under the moss in a silken cocoon. There are many similar looking species which are difficult to distinguish from one another without close study. **64**

**Eueides** Hübner (NYMPHALIDAE)
This genus of the butterfly subfamily Heliconiinae was united for a time with *Heliconius*, from which it differs in the shorter, more strongly clubbed antennae, but it is currently regarded as a separate genus. Many of its species closely resemble, and presumably mimic species of *Heliconius* and *Lycorea* (Danainae). There are about 9 species which are chiefly tropical S. American in distribution and found mainly in open country. Only one species, *E. cleobaea*, has been recorded with any frequency from the U.S.

**aliphera** Godart SMALL FLAMBEAU
C. & tropical S. America 48–50 mm. (1·89–1·97 in.)
In both wing shape and coloration, this butterfly resembles species of *Dryas, Dione* and *Dryadula* but is most like *Dryas iulia*, and is apparently a member of a mimetic complex.

**cleobaea** Geyer
U.S., C. America, the Antilles 75–90 mm. (2·95–3·54 in.)
At times, this has been a fairly common species in the southernmost part of Texas in the area just north of the Rio Grande. Hurricanes repeatedly alter the pattern of distribution of species in this area, and the Texan population of *E. cleobaea* is now becoming established again after the last climatic catastrophe at the end of the 1960s. Its colour pattern is similar to that of several species of *Lycorea* (Danainae) and species of Ithomiinae.

**dianassa** Hübner
Brazil 60–65 mm. (2·36–2·56 in.)
This butterfly is a member of a mimetic complex which includes *Heliconius narcea* and species of the Ithomiine genera *Melinaea, Mechanitis* and *Ceratinia*. It is similar in colour pattern to the illustrated *E. isabella*, but differs in details, especially in the presence of white apical spots on the forewing.

**heliconioides** Felder
Colombia, Ecuador 50–60 mm. (1·97–2·36 in.)
Several species of the genus *Heliconius* have forms which are very similar in colour pattern to that of *E. heliconioides*. There are basal rays of orange on both wings, larger on the hindwing, and a yellow patch towards the apex of the forewing.

**isabella** Cramer
C. & tropical S. America 54–72 mm. (2·13–2·83 in.)
The various forms of this usually common butterfly are members of mimetic partnerships with species of *Heliconius*. *E. isabella* is a forest species with a preference for hilltops and an addiction for white and yellow flowers as a source of nectar. The solitary caterpillars feed on species of *Passiflora*. The illustrated specimen is an example of *E. isabella hippolinus*. **142**e

**lineata** Salvin & Godman
C. & N. S. America 55–64 mm. (2·17–2·52 in.)
This is very similar to the male of *E. vibilia* (illustrated) but lacks the incomplete black band on the basal two-thirds of the forewing. It is a deeper yellow brown in colour than *E. pavana*. Its life history is unknown.

**lybia** Fabricius
C. & N.W. S. America 58–62 mm. (2·28–2·44 in.)
This is one of the most distinctively marked species of *Eueides*. The illustrated specimen is of the subspecies *E. lybia olympia* Fabricius, which has orange not white markings on the forewing. **140**

**pavana** Ménétriés
Brazil 64–72 mm. (2·52–2·83 in.)
This is a forest species, preferring elevations from 609–1520 m. (2000–5000 ft). White and yellow flowers are usually chosen as a source of nectar by these butterflies. The caterpillars feed on species of *Passiflora*. There are 2 female colour forms, one like the male (similar to *Dione juno*), the other pale yellow. *E. pavana*, together with the female of *E. vibilia* and species of *Actinote* (Acraeinae) are partners in a mimetic complex of species.

**ricini** see **Heliconius ricini**

**vibilia** Godart
C. & tropical S. America 58–62 mm. (2·28–2·44 in.)
This butterfly exhibits a remarkable, sex-linked dual mimicry. Males are mimetic partners of other species of *Eueides* and of *Dione juno*. Females, in contrast, are members of a mimetic association with species of *Actinote* (Acraeinae). The caterpillars of *E. vibilia* are gregarious.

**zorcaon** Reakirt
C. America including Mexico 69–94 mm. (2·72–3·7 in.)
This species is a member of a well known mimetic association of species described under *Melinaea lilis* It differs little from *E. cleobaea* and species of the Danaiine genus *Lycorea* and some species of Ithomiinae.

**Eugraphe** Hübner (NOCTUIDAE)
There are about 10 species in this genus of moths. It is found in Europe including Britain, through temperate Asia to China and Japan.

**subrosea** Stephens ROSY MARSH MOTH
Europe, temperate Asia to E. Russia, China 40–45 mm. (1·57–1·77 in.)
This species has had an unusual history in Britain where it was first found in Yaxley Fen, Huntingdonshire, in 1837, seen again in 1846 and was presumed to be extinct in about 1851. It was rediscovered, however, in Wales a few years ago. The forewings are reddish brown or grey above with 2 transverse lines and some dark brown scales in the cell. The hindwings are yellowish brown.

**eulema** see **Ithomeis**

**Eulia** Hübner (TORTRICIDAE)
A genus of micro-moths which are typical of those in the N. Hemisphere, with species from Europe, Japan and N. America. They are related to the bell-moths (Tortrix) and the caterpillars usually roll the tips of leaves together or live in silken tubes.

**ministrana** Linnaeus
Europe including Britain, N. America 17–25 mm. (0·67–0·98 in.)
The forewings are pale yellow with greyish scales. They have indistinct marks of orange, brown and red and the outerwing margins are reddish; the hindwings are grey. This small moth is widely distributed throughout Europe and the E. coast of N. America. The caterpillar feeds on many kinds of trees and shrubs. It overwinters and the moth finally emerges in the spring. There are generally 2 broods a year. The moth is a typical bell-moth shape.

**eulimene** see **Pantoporia**

**Euliphyra** Holland (LYCAENIDAE)
This genus, found in Africa, is a close relative of the Australian and Papuan 'moth-butterfly' genus *Liphyra*.

**mirifica** Holland
Tropical W. Africa 35–52 mm. (1·38–2·05 in.)
The chiefly brown female resembles a skipper (Hesperiidae) in general appearance. Males are white with brown marginal bands to both fore- and hindwings. Its strange caterpillars live inside the nests of tailor-ants and are fed with regurgitated food and cared for by the ants as if they were ant-grubs. Caterpillars have been observed to extract food from inside the mouth of returning ants. Some chemical mimicry on the part of the Lycaenid caterpillars possibly misleads the ants into acting in this benevolent way, as there is no 'honey-gland' in this species of Lycaenid, which in other species provides the ants with food.

**Eulithis** Hübner: syn. *Lygris, Euphia* (GEOMETRIDAE)
A genus of moths with species mostly in Europe and temperate Asia.

**pyraliata** Denis & Schiffermüller
BARRED STRAW MOTH 34–38 mm. (1·34–1·5 in.)
Europe including Britain, temperate Asia
This is a widespread and common species over much of its range. There are a number of named colour varieties. The name is derived from its straw-like colour, with generally one or two black bars on the wings. The caterpillar feeds on species of Rubiaceae, including goosegrass or cleavers *(Galium aparine)* and species of bedstraw such as *Galium mollugo*. It is reddish yellow with many lines along it. It has several other recorded host plants, including *Hypericum* (Hypericaceae). **284**

EULYPA see **Rheumaptera**

**Eumaeus** Hübner (LYCAENIDAE)
There are under 10 species in this genus of butterflies which ranges from Texas to Brazil. All the species are partly iridescent green or blue above, and have at least the under surface of the abdomen orange and an orange patch on the inner margin of each hindwing. The caterpillars feed on cycads *(Cycadaceae)*, a primitive group of plants and a most unusual choice of foodplant. Earlier authors had considerable difficulty deciding where *Eumaeus* belonged, and at times classified it in the Nemeobiidae.

**atala** Poey ATALA BUTTERFLY
U.S. (Florida), the Greater Antilles 38–42 mm. (1·5–1·65 in.)
Although once common in S. Florida this beautiful tropical butterfly is now extremely rare on the mainland as a breeding species. The under surface of the

wings is dark brown with numerous silvery spots, and a single orange spot on the hindwing. The foodplant, Coontie *(Zamea)*, was also at one time the food of the Seminole Indians of Florida. The caterpillars of a common N. American Tiger-moth, *Seirarctia echo* Smith have been reported as occasional contemporary competitors for the same foodplant. **243**a, c

**Eumelea** Duncan (GEOMETRIDAE)
A widespread genus of moths from the Bismarck Archipelago, a group of islands off the coast of New Guinea, through Indonesia and as far N. as Formosa.

**fumicosta** Warren
Solomon Islands 45–54 mm. (1·77–2·13 in.)
The moth illustrated is the subspecies which is common in the Solomon Islands. Other subspecies occur on some of the other islands in the group. The underside of the wings of the moth are a deeper purple than the upperside. Nothing is known of the biology of the species. The subspecies *E. fumicosta fumicosta* is illustrated. **296**v

**eumenia** see **Theorema**
**Eumenis** see **Hipparchia**
**eumolphus** see **Narathura**

**Eumorphia** Hübner (SPHINGIDAE)
Most species of this genus of hawk-moth are solely tropical American, but the range of some includes the S. parts of the U.S. Many are large and have a generally uniform colour pattern. The caterpillars feed on species of dogbane (Apocynaceae), evening primrose (Onagraceae) and grape (Vitaceae).

**pandorus** Hübner PANDORUS SPHINX
Canada & U.S., from Nova Scotia to Texas & Florida 95–115 mm. (3·74–4·53 in.)
This species is a minor pest of Virginia creeper *(Parthenocissus)* and grape-vine *(Vitis)* in the U.S. The adult moth is chiefly green, with a simplified jig-saw like pattern of various shades of green on the forewings, but with 2 areas of black on the hindwing.

**Eunica** Hübner (NYMPHALIDAE) PURPLE WINGS
The butterflies in this large genus are known only from S. America with a few species reaching into C. America and the West Indies. The species are very variable in pattern in different areas, many of the same species having been given different names in these areas. Most of the species are found in the Amazon region and N. Peru with only a few further S. The name *Evonyme* Hübner has often been used for species in this genus but the older name *Eunica* is now used. Few species have been reared, at least one species has caterpillars recorded as feeding on *Sebastiania* (Euphorbiaceae).

**alcmena** Doubleday & Hewitson
S. & C. America 70–80 mm. (2·76–3·15 in.)
This species has been found from Mexico all over C. America and S. to Ecuador and Peru. Generally it is found in mountain forests, where the butterfly is especially active on sunny days and is readily attracted to rotting fruit. The life history is unknown. The male is a velvet black with a deep blue band curving just below the forewing margin from the body towards the apex. The hindwing is similar with a deep blue band along the margin separated into squares by the black veins. The female is brown, with white bands on the forewing and without the blue iridescence. **209**e

**eurota** Cramer
S. America 52–55 mm. (2·05–2·17 in.)
The female is brown with a white patch on the forewing. The underside is brown. Many forms have been named of this wide ranging species which is found from S. Brazil to Surinam and Colombia.

**sophonisba** Cramer
Brazil 50–55 mm. (1·97–2·17 in.)
The female has a broad white band below the apex of the forewing and a blackish blue, darker margin; the underside is similar to the one figured. The upper wing surfaces of both sexes are illustrated. **217**m, q

**tatila** Herrich-Schäffer FLORIDA PURPLE WING
C. & S. America, West Indies, U.S. 40–50 mm. (1·57–1·97 in.)
This species reaches into Florida where it is found in the S. part. It is a brownish black butterfly with a strong purple sheen and 7 white spots in the forewing. The butterfly can be found in the dense shade of tropical hardwood forests. Its underside pattern renders it nearly invisible when it lands on a tree trunk.

**eunomia** see **Proclossiana**
**eupale** see **Charaxes**

**Euphaedra** Hübner (NYMPHALIDAE)
This genus is related to *Euryphene,* differing in the orange colour of the palps of the species. They are generally large, powerful butterflies usually with some iridescence in the male. The females are larger than the males. Some species are very variable, while others are similar over a wide area. The caterpillar has a row of finely branched spines which spread out horizontally. Most of the species are found in W. Africa with a few in other parts but the genus is not known outside Africa. It is a large genus with more than 100 known species, many of which are shade-loving forest species.

**eleus** Drury ORANGE FORESTER
W. Africa to Uganda, S. to Angola 70–90 mm. (2·76–3·54 in.)
The female is reddish brown, but the pattern is broadly similar to the male. There is some variation in pattern and subspecies have been described in different regions. The biology is unknown. This butterfly has a rather Danaid-like appearance and it is possible that it is a mimic of one of the unpalatable Danaine butterflies. The underside is paler, orange-brown with the white bar on the forewing and white spots on the hindwing but generally is less well marked compared with the upperside. **228, 231**l

**eusemoides** Smith & Kirby
W. Africa to Uganda 64–74 mm. (2·52–2·91 in.)
The pattern varies in this butterfly and several subspecies have been described. The underside of the hindwing and base of the forewing are a mauve colour with black spots near the base of the wing. The hindwing veins, which are not clear on the upperside, are conspicuous below and lined with black. Nothing seems to be known of the biology of this species. **237**f

**francina** Godart
Africa 68–85 mm. (2·68–3·35 in.)
This species has a greenish male, while the female is larger and more bluish in colour. There is considerable variation in the colour, which may be purple instead of blue. The underside is olive green or an orange-green with large black spots near the base of both wings. The butterfly is widespread in W. Africa. **237**j

**medon** Linnaeus
Africa 55–70 mm. (2·16–2·75 in.)
This species is widespread in W. Africa where it occurs from Sierra Leone to Zaire and Angola and across Africa to Uganda. There are several subspecies described in different parts of the range. The specimen illustrated was photographed in Uganda. Nothing is known of its biology. **206**

**neophron** Hopffer GOLD-BANDED FORESTER
Africa 60–75 mm. (2·36–2·95 in.)
The forewing has a purple base, with black dots in the cell, the apical half consists of a broad orange band at an angle, edged with black on each side and a small orange tip to the wing. The hindwings are purple with a darker suffusion round the edge. The female is similar but larger. This species is common in the coastal forests from E. Kenya, through Tanzania to Mozambique and is found further S. in Rhodesia and Natal. It is a fast flier, but keeps low down, settling with the wings spread out flat. The caterpillar feeds on species of *Deinbollia* (Sapindaceae). **237**h

**spatiosa** Mabille
Africa 85–111 mm. (3·35–4·37 in.)
This species is found in the forests of W. Africa, the Congo, W. Uganda and W. Tanzania. It flies low down near the ground, and when settling, spreads its wings out flat. It is commonly attracted to rotting fruit. The fore- and hindwings are a deep olive brown with a small white mark at the apex of the forewing and an irregular yellow brown band from the front margin just before the apex towards, but not reaching, the outer margin of the forewing. The underside is a pale green on fore- and hindwings. The caterpillar feeds on species of *Phialodiscus* and *Paullinia* (Sapindaceae). **231**j

**themis** Hübner
W. Africa 80 mm. (3·15 in.)
There are many colour forms of this butterfly, which is found from Sierra Leone to Zaire. Its biology is not known. **237**g

**Euphia** see **Eulithis**
**euphorbiae** see **Hyles**
**euphrosyne** see **Clossiana**

**Euphydryas** Scudder (NYMPHALIDAE)
A large butterfly genus with species in the whole of the N. Hemisphere.

**aurinia** Rottemburg MARSH FRITILLARY
Europe including Britain, temperate Asia 34–40 mm. (1·34–1·57 in.)
This species is found in bogs, meadows and lake margins, from lowlands up to 1520 m. (5000 ft). The adults fly in May and June. The caterpillar is black with white dots and rows of short black spines, it feeds mainly on plantain *(Plantago)* or scabious *(Scabiosa)*. The dark spots in each segment on the upperside of the hindwing near the margin are characteristic of this species. **189** (Now placed in **Eurodryas**).

**Eupithecia** Curtis (GEOMETRIDAE)
A huge genus of moths which is worldwide. All the caterpillars were believed to be plant feeders until a predatory one was discovered in Hawaii in 1972. The caterpillars are typical of Geometridae with the 'looper' habit. The species discovered in Hawaii is unique in the lepidoptera in that the caterpillar feeds on flies. These it catches when they land near the caterpillar which bends with surprising speed using its forelegs to catch the fly. The caterpillars of all the other species in the genus are known feeders on various plants. (See also under *Geometridae.*)

**centaureata** Denis & Schiffermüller
LIME-SPECK PUG MOTH
Europe including Britain, temperate Asia 19–22 mm. (0·75–0·87 in.)
The popular name, as can be seen from the illustration, is most appropriate. The moth at rest on a leaf is said to resemble a bird dropping in this way disguising itself and avoiding predators. The slender, green or brownish caterpillar has a line along the back and may have a series of up to 5 3-toothed marks in reddish brown along the back. It feeds on the flowers and seeds of many species of Compositae and Umbelliferae. **288**

**linariata** Denis & Schiffermüller TOADFLAX PUG
Europe including Britain, W. Asia 17–20 mm. (0·67–0·79 in.)
This is very similar externally to the Foxglove Pug *(E. pulchellata* Stephens). Their names are derived from the plants on which the caterpillar usually feeds. The 2 can only be distinguished externally with practice. The caterpillar of the Toadflax Pug feeds on *Linaria* (Scrophulariaceae), eating the flowers. It will also eat the flowers of snapdragon *(Antirrhinum)*. The moth is abundant over most of its range. Several colour varieties have been named. The caterpillar of the Foxglove Pug feeds on *Digitalis*. Both moths fly in May and June.

**pulchellata** see under **E. linariata**

**venosata** Fabricius NETTLE PUG
Europe including Britain 22–25 mm. (0·87–0·98 in.)
Early names for this moth included 'The Pretty Widow Moth'. There are several subspecies of this moth in Europe, 3 of them occurring in Britain. The caterpillar is a fat, greyish brown with pale green below and 3 dark lines along the back. It feeds on Catchfly *(Silene)*. The moth is common in May and June flying at night. **289**

**Euplagia** Hübner (ARCTIIDAE)
A single species of moth is placed in this genus.

**quadripunctaria** Poda JERSEY TIGER-MOTH
Europe including Britain 55–62 mm.
W. Asia (2·17–2·44 in.)
This is a rare species in Britain, where it is found normally only in S. Devon and the Channel Islands, but in continental Europe it can be quite common. Adult moths migrate in their thousands to mate in the 'Valley of the Butterflies' at Petaloudes on the island of Rhodes during June and July. This is probably the only moth to have become a tourist attraction. There is considerable individual variation in the colour pattern of the wings, especially of the hindwings, which are usually red in colour, but can be any shade from red through orange to yellow. The caterpillars feed on dandelion *(Taraxacum)*, white dead-nettle *(Lamium)*, ground ivy *(Glechoma)*, elm *(Ulmus)* and *Rubus*, amongst many other plants. **379**

**Euplocia** Hübner (HYPSIDAE)
A single species of moth is placed in this genus.

**membliaria** Stoll
India and Burma to the Philippines 69–78 mm.
and Sulawesi (Celebes) (2·72–3·07 in.)
The possibly unique structure of the male scent organ on the front edge of the forewing is characteristic of this species (see illustration). In most specimens the front lip of this scent organ is folded over the scent scales and it is not certain whether the moth is able to open and close this structure at will. Some of the so-called subspecies of *E. membliaria* should probably be treated as distinct species. Specimens of the Sulawesi (Celebes) subspecies, for example, not only differ in wing shape and colour pattern, but also in the shape and colour of the male scent organ. Females of this species have a broad, orange bar at the base of the forewing and a black-marked, greyish white area between the orange bar and the front margin of the wing. **381**d

**Euploea** Fabricius (NYMPHALIDAE) CROW BUTTERFLIES
This is the largest genus of the subfamily Danainae. It is tropical Asian and Australian in distribution, extending W. across the Indian Ocean to Mauritius and the Seychelles. Its caterpillars like those of *Danaus* feed chiefly on species of Asclepiadaceae and Apocynaceae which contain chemicals poisonous to vertebrate predators; the caterpillars are able to store and pass on to the adult butterflies this chemical protection. Several species, however, will feed on *Ficus* (Moraceae) as an alternative source of food. Many of these butterflies are 'crow-like' in coloration, with a dark brown or black and blue iridescence and white markings. Several are mimicked by supposedly palatable species of *Papilio* and some Nymphaline genera. The close similarity between some species of the moth family Zygaenidae and *Euploea* species is different in character, as both partners in this association are unpalatable to predators. *E. leucostictos* feeds on an unusual foodplant and may be a mimic of other *Euploea* species. These butterflies are usually confined to lowland forests, but some Indian species, such as *E. core* are found as high as 2130 m. (7000 ft) in the valleys of the Himalayas. The males of many species produce a strong smelling secretion from their abdominal scent organs which is presumed to have an aphrodisiac function prior to mating. There is a marked difference in colour pattern between the sexes of various species.

**alcathoe** Godart STRIPED BLACK CROW
S. Burma, Malaya 75–85 mm.
to Borneo (2·95–3·35 in.)
Probable mimics of this species include a rare form of the Blue Mime *(Papilio paradoxa)*, another form of which mimics *Euploea mulciber*, and a species of the moth genus *Cyclosia* (Zygaenidae). The male forewing of this species is dark brown above with a weak blue iridescence; the hindwing is brown with 2 rows of white spots near the outer margin. The female is brown above, with an additional, marginal row of white spots on the forewing and with the spots of the inner row on the hindwing elongated to form a series of parallel dashes.

**core** Cramer COMMON INDIAN CROW, AUSTRALIAN CROW, OR OLEANDER BUTTERFLY
Sri Lanka, India, through S.E. Asia to 85–95 mm.
New Guinea & Australia (3·35–3·74 in.)
This is the commonest species of *Euploea* in India and is a frequent visitor to gardens where it is attracted to flowers. *E. core* is mimicked by both sexes of the Common Mime *(Papilio clytia)*, the Malabar Raven *(Papilio dravidarum)*, the female of *Hypolimnas bolina* (Nymphalinae) and others. Oleander (Apocynaceae) is the chief food of the caterpillar but it also feeds on species of Asclepiadaceae, Moraceae and Urticaceae. In Australia it is a minor pest of introduced Oleander and other ornamental shrubs and trees. The illustrated specimen is a male of the subspecies *E. core kalaona* Fruhstorfer. **130**f

**coreta** Godart DOUBLE-BRANDED BLACK CROW
Sri Lanka, S. India 85–95 mm. (3·35–3·74 in.)
The 'double brand' refers to the 2 pairs of scent-scale patches on the upper surface of the male forewing. The only other markings on the weakly iridescent, blue and dark brown wings are 2 rows of white spots along the outer margins. The species of Papilionidae and other families which mimic the similarly patterned *E. core* (illustrated) are also considered to be mimics of *E. coreta*. Several other species of *Euploea* differ from *E. coreta* in only small details of the colour pattern; one of these, *E. sylvester* Fabricius, a widespread tropical Asian and Australian species, also has a pair of sex-brands on each forewing and is called the Two-brand Crow by Australian collectors. *E. coreta* is classed as a subspecies of *E. sylvester* in some collections.

**diocletiana** Fabricius MAGPIE CROW
Sikkim, N. India, E. through the Malay 80–90 mm.
Archipelago to Sulawesi (Celebes) (3·15–3·54 in.)
This is a less uniformly coloured species than many of its relatives. It is mimicked by both sexes of a form of *Papilio paradoxa*, by female forms of *Euripus halitherses* and *Idrusia nyctelius* (Nymphalinae). The abdominal scent brushes of the male disseminate a strong but pleasant, vanilla-like scent, which is doubtless seductive to the female *E. diocletiana*. There is considerable variation in colour pattern between the various subspecies of this butterfly. Females differ from the illustrated male in the white, not blue marginal spots on the wings. **130**h

**eleusina** Cramer
Java, Sulawesi (Celebes), 58–65 mm.
Sumbawa, Lombok (2·28–2·56 in.)
The male of this species is dark brown above with a blue iridescence. Both pairs of wings are marked with pale blue – chiefly at the outer margin of the forewings, which bear a row of large, elongate patches with white centres, and at the outer margin of the hindwing where there is a row of similar markings without white centres. The female is chiefly brown above, with smaller blue markings. *Mniszechi*, sometimes listed as a species of *Euploea*, is currently regarded as a subspecies of *E. eleusina*.

**harrisi** Felder
India, Sikkim, Burma, 70–90 mm.
Indo-China, Malaya (2·76–3·54 in.)
This is a typical crow butterfly, with a strong, blue or violet iridescence on the large dark brown wings. It is highly variable geographically and individually. Some forms have light blue patches near the apex of the forewing; others have a row of white spots along the outer margin of the hindwing. One of its forms is mimicked by *Elymnias patna*.

**leucostictos** Gmelin BLUE-BRANDED KING CROW
Burma, Thailand, Malay Archipelago 62–85 mm.
to New Guinea, Solomons, Fiji (2·44–3·35 in.)
There are some anomalous features about this species. Its caterpillars feed on the foliage of species of the non-toxic genus *Ficus*, not on the more usual, toxic species of Asclepiadaceae or Apocynaceae. It is, however, conspicuously patterned in black, red and yellow colours which are normally associated with toxic properties in warningly coloured insects. The adult butterfly, unlike many other species of *Euploea*, has no exact mimics in other groups of butterflies – evidence that it is possibly not distasteful to predators, and that the caterpillar is likely to prove similarly palatable. The conspicuous coloration of both adults and caterpillars may function as general mimetic resemblance of the coloration exhibited by unpalatable species of *Euploea*. **130**d

**mniszechi** see under **E. eleusina**

**modesta** Butler
Indo-China & Burma to 65–80 mm.
Malaya & Sumatra (2·56–3·15 in.)
The wings of this common and typical species of *Euploea* are iridescent brown and dark violet above, with a double row of white spots along the outer margin of the hindwing and usually with a few white spots on the forewing.

**mulciber** Cramer STRIPED BLUE CROW
India, S. China, Malay Archipelago to 90–100 mm.
the Philippines & Sulawesi (Celebes) (3·54–3·94 in.)
This is a forest species, most often seen in clearings and along roadsides. The iridescent, dark blue and black male (not illustrated) is spotted above with pale blue, in contrast to the slightly paler female which is normally marked with numerous bluish white streaks or stripes across the hindwing, as in many species of the related genus *Danaus*. Oleander (Apocynaceae) and species of Aristolochiaceae have been recorded as foodplants of the caterpillar, but non-toxic *Ficus* is also eaten. Both sexes of *E. mulciber* are mimicked, sex for sex, by *Papilio paradoxa* and *Elymnias malelus* (Satyrinae). Numerous subspecies have been described. **130**e

**sylvester** see under **E. coreta**

**tobleri** Semper
Philippines 80–90 mm. (3·15–3·54 in.)
This is a rare, mountain species, apparently confined to the island of Luzon. It is highly variable in colour pattern, which like that of several other species of *Euploea* is dark brown with a single incomplete row of white markings on the hindwing – an outer row of spots, and an inner row of parallel dashes, with a broad oblique white band towards the apex of the forewing, and with much of the hindwing white, crossed by brown veins.

**euploeodes** see **Esthemopsis clonia**

**Euproctis** Hübner (LYMANTRIIDAE)
There are well over 500 species in this genus of moths. They are essentially Old World in distribution and are found in Europe, Asia, Africa and Australia. The coloration of most species is a combination of one or more of the colours orange, white, yellow or brown. The caterpillars of several species are pests of forest and orchard trees, and several cocoa *(Theobroma)* pests have been recorded from Africa, Sri Lanka, New Guinea and the New Hebrides islands. In N. America, the introduced *E. chrysorrhoea* is a pest species.

**chrysorrhoea** Linnaeus BROWN-TAIL
N. Africa, Canary Islands, Europe 32–45 mm.
including Britain, Turkey, (1·26–1·77 in.)
E. Canada, N.E. U.S.
Since its introduction into the U.S., this has become a serious pest of orchard and forest trees, even of conifers in some areas. Caterpillars overwinter in communal webs and complete their growth the following summer, building one or more new webs as they develop. The irritant barbed hairs of the caterpillar are incorporated into the structure of the cocoon and are swept-up by the tail-tuft of the emerging adult female who deposits them later as a protective covering for her eggs. This tail-tuft is usually dark brown in colour, but is yellowish brown in some specimens which in the absence of wing-markings may be difficult to distinguish from the Yellow-tail, *E. similis*. **357**p

**edwardsi** Newman MISTLETOE BROWN-TAIL
Australia 30–53 mm. (1·18–2·09 in.)
The gregarious caterpillars of *E. edwardsi* feed on mistletoes (Loranthaceae) parasitic on species of *Eucalyptus*; their hairs can cause severe urticaria or nettle rash. The forewings of the adult are yellowish brown; the hindwings similar in colour distally but a darker brown basally. The abdomen is dark brown above, with a brownish yellow tip.

**similis** Fuessly YELLOW-TAIL or GOLD-TAIL (U.S.)
U.S., Europe including Britain, 28–45 mm.
temperate Asia to Japan (1·1–1·77 in.)
This differs from the illustrated Brown-tail (*E. chrysorrhoea*) in that there is usually a brown patch at the posterior angle of the forewing of the male and that the abdominal tuft (tail) is golden (rarely brownish yellow). The Yellow-tail, unlike the Brown-tail, is only occasionally reported from the N. Atlantic states of the U.S., but it is not uncommon in much of the Old World. The foodplants of the caterpillar include the foliage of oak (*Quercus*), apple (*Malus*) and many other orchard and forest trees, but the commonest food-source is hawthorn (*Crataegus*). **358**

**Euproctoides** Bethune-Baker (LYMANTRIIDAE)
Six tropical African species belong in this genus of moths. They are either orange and yellow or red and yellow in colour. Females are fully winged.

**acrisia** Plötz
Tropical Africa 29–43 mm. (1·14–1·69 in.)
This is one of the more colourful species of its family. It is apparently fairly common in parts of its range. **357**n

**Eupseudosoma** Grote (ARCTIIDAE)
Nine species are included in this genus of mainly tropical New World moths. The wings of most species are white, the abdomen white below but bright scarlet above. One species, *E. involuta*, occurs in the U.S.

**involuta** Sepp SNOWY EUPSEUDOSOMA
S. U.S., C. & tropical 38–45 mm.
S. America (1·5–1·77 in.)
This is a typical red and white *Eupseudosoma* species. Its caterpillars feed on the foliage of guavas (*Psidium*) and *Eugenia* (genera of Myrtaceae). Adult moths have been shown to be distasteful to predators. **368**g

EUPSEUDOSOMA, SNOWY
see **Eupseudosoma involuta**

**Eupterotidae**
This is a small family of moths related to emperor moths (Saturniidae) and a member of the superfamily Bombycoidea. The caterpillars are covered with dense, irritant (at least in some species) hairs; some are known to be gregarious and live in communal webs. Some species cause damage to pine (*Pinus*), *Eucalyptus* and other trees in Australia. See *Acrojana*, *Phiala*.

**Euptychia** Hübner (NYMPHALIDAE)
The species in this genus of butterflies are found mainly in S. America. They are small to medium in size, brown or grey with prominent eyespots on the wing. Species have been described from many places, although the majority are American. Little is known of the early stages of the butterflies in the genus. Most of the species are found in tropical parts but a few reach S. Brazil, Argentina, and N. in the S. U.S. They are often common in parts of their range, even in towns. The genus belongs to the subfamily Satyrinae. (See also under *Magneuptychia*.)

**eurytus** Fabricius: syn. *E. cymela*
LITTLE WOOD SATYR
S. Canada, W. to Nebraska, S. to the 44 mm.
Gulf States and Texas (1·73 in.)
This is found in open deciduous woods and wet meadows surrounded by woods and is common over most of its range. The pale, greenish brown, finely hairy caterpillar has thin stripes along the body with whitish spots. It feeds on grasses (*Gramineae*). The butterfly is out in mid May with one brood in the N. and probably 2 in the S. part of its range. It is a brown butterfly with a lighter fringe and 2 black eyespots edged with a narrow orange line on the forewing and 2–3 eyespots in the hindwing, each eyespot with a small blue centre. The underside is paler grey brown with 2 lines across the fore- and hindwings. The conspicuous eyespots have broader orange borders than on the upperside.

**Eurema** Hübner: syn. *Terias* (PIERIDAE)
A widespread genus of butterflies with many species in the tropics. The species are generally yellow with black margins to the wings, and have a characteristic erratic flight, close to the ground. Various foodplants have been recorded for the caterpillars including *Desmodium* and *Aeschynomene* and other Leguminosae, and Euphorbiaceae. Most of the males of species in the genus have a distinctive patch of scales (sex-brand) on the wing.

**brigitta** Cramer BROAD-BORDERED GRASS-YELLOW or NO-BRAND GRASS-YELLOW 40–50 mm.
Africa, India, Pakistan, Sri Lanka, (1·57–1·97 in.)
China, Indonesia, New Guinea, Australia
This butterfly is common in open bush areas and gardens and in Africa is found roughly from 5 °N. to South Africa. It occurs in large numbers at puddles and wet mud and can be found in all months of the year. The male is bright yellow with a broad black border on the wing, thickest near the apex, which continues round the outer edge of the hindwing. The rest of the hindwing is bright yellow. The caterpillar feeds on species of *Cassia* (Leguminosae). The female butterfly is a paler yellow than the male, with some black scales scattered over the wings. There are seasonal forms of this butterfly, a heavily marked wet season form and an almost unmarked dry season one. There are many described subspecies over the extensive range of *E. brigitta*. In the Australian subspecies it has been found that the caterpillar is very specific in its feeding habits and will only feed on *Cassia mimusoides* (Leguminosae).

**gratiosa** Doubleday JAGGED-EDGED YELLOW
C. & S. America & into the W. Indies 38 mm. (1·5 in.)
This small yellow gets its popular name from the irregular inner edge to the broad black borders round the yellow wings. The hindwing has a yellow patch near the outer front margin; the rest of the wing is white, with the black border. The irregular inner edge of the black margin is not so clear in the female. The butterfly is common over parts of its range from Panama to Venezuela and Colombia.

**herla** Macleay SPOTLESS GRASS YELLOW
India, Pakistan, Burma, Sri Lanka, 30–45 mm.
Indonesia to Australia (1·18–1·77 in.)
A very common bright yellow butterfly with a black apex to the forewing, the hindwing having a narrow border of black. On the underside of the forewing there is a black spot over the apex of the cell. The female has broader black borders. It occurs in dry and wet season forms with differences in the amount of black on the wings. This butterfly is common over most of its range, flying in most months of the year. It is rather a delicate looking insect, as it flutters around the grass and low bushes, mostly during mid morning and mid afternoon, often congregating in large numbers on damp patches. The caterpillar feeds on species of Leguminosae. Recent work has suggested that *E. herla* is restricted to Australia and that the name *E. laeta* Boisduval should be applied to the more widespread species.

**laeta** see under **E. herla**

**proterpia** Fabricius LITTLE JAUNE
Mexico & C. America 42 mm.
to Venezuela (1·65 in.)
The orange wings bordered with black, with thicker black on the front edge of the forewing and black lines and veins are characteristic of this species. The flight is slow over roadsides or across grasslands. In some years this species can be quite common. In spite of this, and that it has recently been bred, little has been published of its biology.

**salome** Felder: syn. *E. damaris* SALOME SULPHUR
C. & S. America, possibly into the 35–45 mm.
S. U.S. (1·38–1·77 in.)
This is a lemon yellow butterfly with black outer wing margins, broader on the forewings. The hindwings are slightly pointed and the base of the wing near the body is black. Underneath the hindwings are orange-yellow but the black margin of the upperside is absent, with an orange-brown edge instead. This butterfly is common in Mexico but it is still disputed as to whether it really does reach into the U.S.

**venusta** Boisduval LITTLE YELLOWIE
S. America & into the 30–35 mm.
West Indies (1·18–1·38 in.)
This butterfly is common, often abudant, from Colombia, Venezuela, Guyanas, Brazil and the West Indies. The male has a bright yellow forewing, with a black margin along the side and front, broad at the apex. The hindwings are pale yellow with a narrow black border. The females are generally paler on the forewing, often with the black on the hindwing absent or reduced to small spots. The caterpillar feeds on *Mimosa* (Leguminosae). The butterfly is out in most months, usually in fairly open habitats.

**euridice** see **Satyrodes canthus**
**eurimedia** see **Aeria**
**Euripus halitherses** see under **Euploea diocletiana**
**euriteus** see **Euselaisia**
**europa** see **Lethe**
**europaea** see **Acanthobrahmaea**
EUROPEAN
BEAK see **Libythea celtis**
CORN-BORER see **Ostrinia nubilalis**
GRAIN MOTH see **Nemapogon granella**
FRUIT-PIERCING MOTH see **Scoliopteryx libatrix**
PINE SHOOT MOTH see **Rhyacionia buoliana**
SKIPPER see **Thymelicus lineola**
**eurota** see **Eunica**

**Eurrhypara** Hübner (PYRALIDAE)
This genus of Pyralid moths is in the subfamily Nymphulinae although the caterpillar, unlike many in the genus, lives on land plants. Only 2 species have been described in this genus, one European, the other from Japan.

**hortulata** Linnaeus: syn. *E. urticata*
SMALL MAGPIE MOTH 30–32 mm. (1·18–1·26 in.)
Europe including Britain, temperate Asia
The popular name reflects a likeness to the larger Magpie moth (Geometridae). The moth is common in the summer months, and often abundant in low herbage. It is attracted to light and easily disturbed during the day when it flies off fairly slowly. The caterpillars feed on nettles (Urticaceae) and occasionally mint and other Labiatae. They leave the host plant to pupate, often wandering far and frequently turning up in houses; they have even been found in large numbers at the tops of tall tower blocks. Generally the infestation can be traced to a nearby nettle bed. The caterpillars are harmless, if unpleasant, in the house. **71**

**Euryades** Felder (PAPILIONIDAE)
There are 2 known species, *E. corethrus* and *E. duponchelii*, both of which are restricted to temperate S. America. Their closest relative is probably the Australian *Cressida*. Their caterpillars feed on species of Aristolochiaceae as do other genera of Troidini such as *Parides* and *Battus*.

**corethrus** Boisduval
Argentina, Paraguay 70–85 mm. (2·76–3·35 in.)
The colour pattern of the male is similar to that of *E. duponchelii* but the ground-colour is not such a dark brown and the yellow markings are a less clear yellow. The most obvious difference between the species is the lack of a hindwing tail in *E. corethrus*. Females resemble those of *duponchelii*, but are tailless and have only marginal pink spots on the hindwing. Nothing is known about the life history of this species.

**duponchelii** Westwood
Argentina 65–84 mm. (2·56–3·31 in.)
The brown and pale yellow females (not illustrated) of this species contrast with the warningly coloured males; they lack the yellow band on the wings except for the 2 spots at the front edge of the forewing, and the red spots of the male hindwing are replaced by smaller pink spots. **97**b

**euryalus** see **Hyalophora**
**Euryphene** see under **Euphaedra**
**eurypylus** see **Graphium**

**Eurytela** Boisduval (NYMPHALIDAE)
A genus of butterflies found over most of Africa and Madagascar with less than a dozen species at present described in it.

**dryope** Cramer GOLDEN PIPER
Tropical to South Africa 50–62 mm. (1·97–2·44 in.)
The brown wings have an orange marginal band, broader in the hindwing, and reddish brown undersides. Both fore- and hindwing margins are slightly crenulate, often with black marks along them. This species is a typical Nymphalid shape. The caterpillar feeds on *Ricinus* and *Tragia* (Euphorbiaceae). Some years the moths appear in large numbers, in others they are rare.

**eurytheme** see **Colias**

**Eurytides** Hübner (PAPILIONIDAE)
This is a genus of about 50 species, found in Canada, S. to Argentina. One group of 15 species (see *E. pausanias*) mimics poisonous species of Heliconiinae or of the Papilionid tribe Troidini.

**crassus** Cramer
Tropical S. America 82–110 mm. (3·23–4·33 in.)
The yellow streak on the forewings of this distinctively marked butterfly varies in size and is absent in some specimens. The life history of this species is not known. **97**d

**harmodius** Doubleday
Tropical S. America 70–90 mm. (2·76–3·54 in.)
This is a butterfly of the Andean slopes, especially in Bolivia, Ecuador and Peru where it mimics species of the genus *Parides* (see *P. aeneas*). The wings are dark brown above, with a green iridescence. There is a white patch on the forewing and a transverse row of crimson spots on the hindwing. Some males have a row of white lunulate markings between the crimson spots and the outer margin of the hindwing. Females sometimes lack a white patch on the forewing; the hindwings are either like those of the male or have a row of white or yellow patches. The caterpillar is not known.

**marcellus** Cramer ZEBRA SWALLOWTAIL
E. Canada, U.S. 50–70 mm. (1·97–2·76 in.)
This distinctively striped species is unlike any other N. American swallowtail. It is a breeding species in the S., but a non-breeding migrant in the N. of its range. There is a high degree of seasonal variation in colour, pattern and size in this butterfly. The caterpillar feeds chiefly on the foliage of Papaw *(Asimina)*; it is green, with yellow and black transverse bands. **110**

**marchandi** Boisduval
Mexico to W. Ecuador 70–90 mm. (2·76–3·54 in.)
In parts of C. America this is a fairly common species up to elevations of 760 m. (2500 ft) in rain-forests. Females are seldom seen but males will visit flowers in forest clearings and margins and sometimes assemble on wet mud or sand with butterflies of other species. The caterpillar is unknown. **97**f

**pausanias** Hewitson
Tropical S. America & C. America 70–90 mm. (2·76–3·54 in.)
In flight, both sexes of this species resemble the butterfly *Heliconius wallacei* to some extent, but the female is the truly mimetic sex. Other species related to *E. pausanias* in this subdivision of the genus are known to be mimics. **97**e

**protesilaus** Linnaeus SWORDTAIL
C. & tropical S. America 75–90 mm. (2·95–3·54 in.)
A 'magnolia-like' tree has been recorded as a foodplant of the caterpillar, but little is otherwise known about this species. The apical part of the forewing is transparent or partly so in some specimens. **97**a

**telesilaus** Felder SOUTHERN WHITE PAGE
C. & tropical S. America 65–75 mm. (2·56–2·95 in.)
This elegant butterfly is another rain-forest resident. Like the illustrated *E. protesilaus* and other swordtail species, it is capable of powerful flight and is consequently extremely difficult to capture unless it is feeding or drinking. The female is similar in colour pattern to the male.

**thyastes** Drury
Tropical S. America 85–105 mm. (3·35–4·13 in.)
This is one of several species which are similar in colour pattern to the illustrated *E. marchandi*. It differs from the latter in the yellow, or orange-yellow markings on the wings and the presence of a lunulate, scarlet marking just posterior to the middle of the inner margin of the hindwing. This is a rain-forest species and is seldom seen except in forest clearings and along river banks where males assemble on wet mud or visit flowers.

**xynias** Hewitson
Bolivia, Peru 70–90 mm. (2·76–3·54 in.)
The general appearance of this rare butterfly is much like that of South American species of the genus *Parides* which it apparently mimics. There is a greenish yellow patch in the middle of the rear edge of the forewing and a red or reddish orange, ovate patch or pair of patches near the inner margin of the short-tailed hindwing. The ground colour of both wings is greenish black. The female is not known; neither is the caterpillar.

**eurytus** see **Euptychia, Pseudacraea**

**Euschemon** Doubleday (HESPERIIDAE)
This is a peculiar Australian skipper butterfly genus. It is interesting in that it shows certain characteristics intermediate between butterflies and moths. It has a wing coupling mechanism by means of a frenulum and retinaculum which is similar to that found in most moths. This structure is present in the male only. At one time it was believed to be a day-flying moth.

**rafflesia** Macleay REGENT SKIPPER
Australia 48–60 mm. (1·89–2·36 in.)
This species is found in Queensland and New South Wales where the adults fly in the early months of the year. The caterpillars feed on species of *Wilkiea* (Monimiaceae). Two subspecies of *E. rafflesia* have been described, both from Australia. This is a striking black and yellow skipper with a rapid flight in bright sunshine. It is generally found on the edges of rain-forests, feeding freely on flowers. **79**a

**Euselasia** Hübner (NEMEOBIIDAE)
There are over 100 species in this C. and tropical S. American genus of butterflies.

**euriteus** Cramer
N. S. America 21–34 min. (0·83–1·34 in.)
The female (not illustrated) lacks a blue iridescence on the wings; its forewing is brown above, its hindwing brown with an orange patch at its rear angle. Both wing surfaces of the male are illustrated. **263**p, q

**issoria** Hewitson
Tropical S. America 27–40 mm. (1·06–1·57 in.)
The illustrated under surface of the male exhibits almost every colour of the rainbow, when viewed from different angles. The upper surface of the male is a much more sombre dark brown in colour with a slight blue sheen. The female is patterned with brown above and below and lacks any iridescence. **263**r

**eusemoides** see **Euphaedra**
**eustrigata** see **Calpe**
**euterpe** see **Stalachtis**
**euterpinus** see **Papilio**

**Euthalia** Hübner: syn. *Adolias* (NYMPHALIDAE)
A huge genus of butterflies, with many species from India through Indonesia to New Guinea. They tend to live in the jungle areas where they may be seen along the paths and glades. The caterpillars have rows of long feathery spines along the side, giving them a curious appearance. The pupae are suspended from the leaves of the foodplants and often have brilliant metallic spots on them. The genus is in the subfamily Satyrinae.

**aconthea** Cramer: syn. *garuda* BARON
India, Sri Lanka, Pakistan, China, Burma, Thailand, Indonesia (2·17–2·36 in.)
The male is dark brown with darker markings and an almost complete row of white spots across the middle of the wing. The hindwing has a line of black spots below the margin and the base of the wing is darker than the outer part. The underside is a similar dark brown. The female has the outer part of the wing a paler brown. This is a widely distributed species and is found everywhere that mangos *(Mangifera)* or cashew nut *(Anacardium)* trees, the foodplants of the caterpillars, are grown. The butterfly is a strong flier, often basking on the ground with the wings spread out. It feeds on sap and rotting fruit. The caterpillar is very well camouflaged and difficult to see as it rests on the midrib of a mango leaf. The body has a pale dorsal line which matches the midrib of the leaf while the spines sticking out from the body of the caterpillar resemble the veins.

**adonia** Cramer
Indonesia 50–70 mm. (1·97–2·76 in.)
The male is smaller than the female and there are several described subspecies. Nothing is known of the life history of this species which is found in Java, Sumatra and Borneo. **216**d, f

**dirtea** Fabricius ARCHDUKE
Malaysia, Burma, India, Pakistan, Indonesia, Hainan 80–105 mm. (3·15–4·13 in.)
The male is dark brown with a blue border to the hindwing. The female is also dark brown but profusely spotted with yellow. The underside is spotted as above and the hindwing is a pale greyish green, without the blue of the male. The males are particularly attracted to rotting pineapples *(Ananas)* on which they feed.

**durga** Moore BLUE DUKE
N. India, Indonesia 95 mm. (3·74 in.)
This beautiful butterfly is found in Sikkim and the Naga Hills in N. India, but is now becoming rare. It flies with a powerful sailing flight at 1220–1520 m. (4000–5000 ft) in the foothills of the Himalayas and is attracted to rotting fruit or sap oozing from trees. **230**e

**evelina** Stoll REDSPOT DUKE
India, Pakistan, Sri Lanka, Burma, Hainan, Philippines, Indonesia 80–115 mm. (3·15–4·53 in.)
This is a dark green butterfly, with a white line near the apex of the forewing and a red spot near the base. The hindwing is similar, with a less conspicuous red spot. The underside has a broad, irregular, translucent band which narrows towards the hindwing and there are some darkish marks round the hindwing margin. This butterfly is found in thick jungle. It has a powerful flight but settles on the ground on damp patches or on sap flowing from trees. The caterpillar feeds on species of Ebenaceae. There are many descriptions of this green butterfly disappearing in the green of the forest and its appearance and disappearance are well expresssed by 'a green thought in a green shade'.

**khasiana** Swinhoe DARK ARCHDUKE
India, Pakistan, N. Burma 80–105 mm. (3·15–4·13 in.)
Two subspecies of this have been described in N. India. The male is black above with hardly any trace of pattern except for a slight blue border towards the hind part of the margin of the forewing. The hindwing has a narrow purple marginal band with a row of black spots on it. The underside is a rich reddish brown. The female has a row of bluish white spots across the wing. This species is now regarded by some authorities as a subspecies of *E. dirtea*.

**patala** Koller GRAND DUCHESS
India, Pakistan, Burma 80–120 mm. (3·15–4·72 in.)
The female has the patches in the forewing white rather than the yellowish colour of the male. This species is single brooded in the Himalayas and appear from June to August. It is common in oak *(Quercus)* forests between 1520 and 2290 m. (5000–7500 ft). The adult feeds on over-ripe fruit and sap from trees, the young on *Quercus leucotricophora* (Fagaceae). **230**a

**Euthyatira** Smith (THYATIRIDAE)
About 4 species belong in this genus of moths. It is possibly closely related to the Old World species of *Thyatira*.

**pudens** Guenée
Canada, N. U.S. 40–50 mm. (1·57–1·97 in.)
The normal form of this species is similar to the European *Thyatira batis* but the wings are narrower and the

arrangement of the pink markings differs, especially in the absence of the pink spot at the anal angle of the forewing. Specimens of both sexes occur without any pink spots on the forewing or with the pink markings much reduced. The caterpillars feed on species of dogwood *(Cornus)* inside a rolled leaf.

**Euxanthe** Hübner (NYMPHALIDAE)
This butterfly genus is in the subfamily Charaxinae. Most are found in well-wooded areas of Africa, S. of Sahara.

**tiberius** Grose-Smith
E. Africa 85–105 mm. (3·35–4·13 in.)
The female is illustrated; the male has a pale yellowish patch in the forewing and all black hindwings. It is similar to *E. trajanus* but the hindwings are black in the middle with a few spots near the apex. The median band of the forewing is narrower and greenish. The biology is not known. **231**c

**trajanus** Ward
W. Africa 90 mm. (3·54 in.)
From Nigeria to Angola this butterfly occurs in well wooded areas. Nothing is known of its biology. It is broadly similar to *E. tiberius* with large white spots near the margin. There is a reddish brown streak near the base of the forewing. The hindwings are yellowish white with dark margins and 2 rows of white spots along the margin. There is less yellowish white in the male hindwing. The veins on the underside are prominently lined with black.

**wakefieldii** Ward FOREST QUEEN
E. Africa to S. Rhodesia, Mozambique & Natal 80–104 mm. (3·15–4·09 in.)
A forest species, with a pattern to match the light and dark of the habitat. The butterfly has very rounded margins to the forewing and is black patterned with streaks and spots of white, rather like *Graphium leonidas*. The hindwing has a large white basal area and a row of smaller white spots below the margin. The forewing spots look blue in some light and white in others. The female, which is larger than the male, has all the markings white, not blue. The caterpillar feeds on *Deinbollia* (Sapindaceae). The adults generally fly high round the trees in shady forests and can be found all the year. They are readily attracted to fermenting fruit.

**Evagora** Clemens (GELECHIIDAE)
A genus of micro-moths whose caterpillars mine in pine *(Pinus)* needles. The species in the genus are found in N. America and are related to the European species of *Recurvaria* Haworth.

**biopes** Freeman
Canada 13–14 mm. (0·51–0·55 in.)
This small moth with rather grey-brown, narrow wings has paler markings across the wing. The hindwing, which has a long white fringe, is grey. The caterpillar feeds in the leaves of the Lodgepole Pine *(Pinus contorta)*. At present this moth is known only from one locality in Canada. The caterpillar makes its mine in the pine needles, pupating in the mine usually near the base of a needle.

**evagoras** see **Jalmenus**
**evagore** see **Colotis**
**evelina** see **Euthalia**
EVENING BROWN see **Melanitis leda**

**Evenus** Hübner (LYCAENIDAE)
A genus of butterflies, large for Lycaenids, brilliantly iridescent blue or green, found in C. and tropical S. America. About 10 species are known.

**regalis** Cramer BANK-NOTE BLUE
C. America, including Mexico, tropical S. America 38–48 mm. (1·5–1·89 in.)
This is among the most colourful species of the family Lycaenidae. The male is an especially brilliant, iridescent turquoise, bordered with black; both sexes have 2 tails to the hindwing and a red spot near the base of the tails. The under surface (illustrated) is even more colourful. Males are in the habit of resting in groups on a tree during the middle of the day. **244**f

**Everes** Hübner (LYCAENIDAE)
This small genus of butterflies occurs in the New World, Europe, Asia and Australia. One species, the Short-tailed blue *(E. argiades* Pallas*)*, is found in Britain as a rare migrant. The hindwing of each included species has a single, slender tail.

**amyntula** see under **E. comyntas**

**argiades** see under **Everes**

**comyntas** Godart EASTERN TAILED BLUE
S. Canada, U.S. to C. America 20–25 mm. (0·79–0·98 in.)
This is one of the commonest butterflies in the U.S. It flies close to the ground and is a frequent visitor to flowers and wet mud. Males are purplish blue above, with dark brown or black margins to the wings and usually 2 black-centred, orange spots near the base of the slender hindwing tail. Females are mainly brown above. Various species of Leguminosae are eaten by the caterpillar which prefers the flowers and buds of the foodplant. *E. comyntas* is replaced by *E. amyntula* in the N.W. of the U.S.

**lacturnus** Godart
India, China, S.E. Asia to N. & E. Australia, the Solomons 18–20 mm. (0·71–0·79 in.)
This is a small purplish blue (male) or dark brown (female) butterfly with a row of small spots along the outer margin of the hindwing. It is a species of grassland areas throughout its extensive range. This is the only species of *Everes* represented in Australia. Its caterpillars feed on species of Leguminosae.

**eversmanni** see **Parnassius**
**Evetria** see **Rhyacionia**
**evippe** see **Colotis**
**Evonyme** see under **Eunica**
**excellens** see **Hilarographa**
**excelsa** see **Theope**
**exempta** see **Spodoptera**
**exigua** see **Spodoptera**
**exilis** see **Brephidium**

**Exoteleia** Wallengren (GELECHIIDAE)
A small genus of micro-moths with species in N. America and Europe. Their hindwings are the typical Gelechiid shape (trapezoidal).

**dodecella** Linnaeus
Europe including Britain, Canada 11–13 mm. (0·43–0·51 in.)
The reddish brown forewings have white marks and darker bands across them. There are white scales on the legs. The hindwings are greyish with a long fringe. There are several similar looking related species. The caterpillar mines in the needles of pines *(Pinus)* in the early stages but in buds at a later stage and can be destructive in pine plantations. At present this European species has been recorded only from Ontario in N. America.

**extravagans** see under **Episcea**
**exultans** see **Milionia**

**Exyra** Grote (NOCTUIDAE)
*Exyra* caterpillars live inside the pitchers or urns of insectivorous pitcher-plant leaves and are apparently immune to its secretions which are lethal to most insects. They close the mouth of the pitcher with a silken web and hibernate inside it. Adult moths also shelter inside the pitchers. Four species are currently recognized; they are N. American in distribution.

**rolandiana** Grote PITCHER-PLANT MOTH
Canada, E. U.S. 16–26 mm. (0·63–1·02 in.)
The caterpillar of this species lives inside the pitchers of *Sarrecenia purpurea*. The adults are dull purplish red on the head, thorax and forewings, with yellowish orange markings in the middle of the wing. The hindwings and abdomen are dark brown above.

EYE BUTTERFLY, PEARLY see **Enodia portlandia**
EYE-FREQUENTING MOTHS see **Lobocraspis griseifusa, Calpe minuticornis**
EYED
BLUE BUTTERFLIES see **Hemiargus**
BROWN see **Satyrodes canthus**
HAWK-MOTHS see **Smerinthus**

**fabius** see **Charaxes**
**fabriciana** see **Anthophila**

**Fabriciana** Reuss (NYMPHALIDAE)
This genus of fritillary butterflies has many species which are variable in pattern and many subspecies and forms have been described. They are found in Europe across temperate Asia to Japan. The species in the genus are closely related to the Argynnid fritillary butterflies.

**adippe** Denis & Schiffermüller HIGH BROWN FRITILLARY
Europe including Britain, N. Africa, temperate Asia including Japan 50–62 mm. (1·97–2·44 in.)
This butterfly is found in woodland clearings and meadows where the caterpillar feeds on violets *(Viola)*. The adults are out in June and July. There is considerable variation in the numbers of silver spots on the underside of the hindwing and several forms, with different numbers of spots have been named. As well as these forms, there are a number of distinct subspecies of *F. adippe* described, including one from Japan. The photograph is of a female. **193**

**niobe** Linnaeus NIOBE FRITILLARY
Europe, temperate Asia, Turkey to Iran 46–60 mm. (1·81–2·36 in.)
Similar to *F. adippe* but with small yellow spots below the median vein near the base of the cell which has a minute black centre and is one of the characters used to separate the 2 species. The butterfly is out in June and July and flies in meadow lands and sub-alpine pastures. The caterpillars feed on violets *(Viola)*. A number of forms have been described based on variations in the silver spots on the underside of the hindwings.

**fagana** see **Pseudoips**
**fagata** see **Operophtera**
**fagi** see **Hipparchia, Stauropus**
**falcataria** see **Drepana**
FALCATE ORANGE-TIP see **Paramidea genutia**
**falciferella** see **Ypsolopha**
**Falcipennae** see under **Epiphora**
**falkensteinii** see **Uranothauma**
FALL
ARMYWORM see **Spodoptera frugiperda**
CANKER WORM see **Alsophila pometaria**
WEBWORM see **Hyphantria cunea**
**falloui** see **Euchloe**
FALSE
ACRAEA, TRIMEN'S see **Pseudacraea boisduvalii**
APOLLO see **Archon apollinus**
CADDIS WATER-VENEER see **Acentria nivea**
CLOTHES MOTH see **Hofmannophila pseudospretella**
COMMA see **Nymphalis vau-album**
FRITILLARY see **Pseudargynnis hegemone**
WANDERER see **Pseudacraea eurytus**
**fasciata** see **Casbia**
**farinalis** see **Pyralis**
**fascelina** see **Dasychira**
**fasciata** see **Daphoenura**
**fasciata** see **Strophidia**
FAUN, COMMON see **Faunis arcesilaus**

**Faunis** Hübner (NYMPHALIDAE)
Mostly Indian or S.E. Asian butterflies with a few species in temperate Asia. This genus is in the subfamily Amathusiinae.

**arcesilaus** Fabricius COMMON FAUN
India, Burma, Malaysia, S.E. Asia 65–75 mm. (2·56–2·95 in.)
This is rather a plain brown butterfly; the underside is a deep brown with dark longitudinal stripes and a series of spots along the margins of both wings. It is a large butterfly which is on the wing from November to February and is found in dense forests where it is attracted to species of Zingiberaceae. The caterpillar has been found feeding on banana *(Musa)*. In India it is common in the lower bamboo jungles.

**faunus** see **Chorinea**
**fausta** see **Colotis**
FEATHER-WING MOTHS see **Alucita**
FEATHERED FOOTMAN see **Spiris striata**
**felderi** see **Pterodecta**
**felinaria** see **Percnia**
**fenestra** see **Hyalurga**
**fenestrata** see **Hecatesia**

**Feniseca** Grote (LYCAENIDAE)
There is a single known species in this genus.

**tarquinius** Fabricius HARVESTER
Canada, E. U.S. 34 mm. (1·34 in.)
The strange feeding habits of this orange and black species were unknown for nearly a century after its discovery. It is usually found only in marshy areas where the adult is frequently seen on alder *(Alnus)* sucking aphid honeydew. The caterpillar is carnivorous and preys on woolly aphids (allies of greenflies) amongst which it spins a loosely knit web which becomes covered with the remains of aphids. The chrysalis resembles a miniature monkey's face in shape. **244**d

FENTON'S WOOD WHITE see **Leptidea morsei**
FERENTIA see under **Hamadryas feronia**
**feronia** see **Hamadryas**
**ferrugata** see **Xanthorhoe**
**festiva** see **Ammobiota, Diphthera**
FESTOON BUTTERFLIES see **Allancastria, Parnalius**
FESTOON MOTH see **Apoda avellana**
FESTOONED TIGER-MOTH see **Cymbalophora pudica**
**festucae** see **Plusia**
**ficulnea** see **Edule**
**fida** see **Coreura**
**fidelissima** see **Composia**
FIERY JEWEL see **Hypochrysops ignita**
FIG MOTH see **Ephestia cautella**
FIGHTER, PALM TREE NIGHT see **Zophopetes dysmephila**
FIGTREE BLUE see **Myrina silenus**
FIGURE OF EIGHT BUTTERFLY see **Diaethria clymena**
FIGURE OF EIGHT MOTH see **Diloba caeruleocephala**
FIGURE OF EIGHTY see **Tethea ocularis**
FILBERT WORM see **Melissopus latiferreanus**
**filipendulae** see **Zygaena**

**Filodes** Guenée (PYRALIDAE)
A small genus with species in Africa, India, Malaysia, Indonesia to Australia. Little is known of their biology but recent interest has been aroused by the discovery that some species are attracted to the eyes of animals where they feed on the liquids round the eye. Thus there is the possibility that they may transmit eye diseases. Several species of Pyralid are known to do this, as are species of Noctuidae and Geometridae. All the species in the genus have very long antennae.

**fulvidorsalis** Hübner
India, Pakistan, Malaysia, Africa 28–38 mm. (1·1–1·5 in.)
This is believed to be a very widespread species but further study will probably show that several distinct species are involved. This species is one which has been studied feeding on the eye secretions of cattle, water buffalo, pigs, horses and elephants. They are therefore potential vectors of eye diseases and are at present being studied from this aspect. **56**g

FIVE-SPOT BURNET see **Zygaena trifoli**
FIVEBAR SWORDTAIL see **Graphium antiphates** and **G. aristeus**
FIVERING, COMMON see **Ypthima baldus**
FLAMBEAU BUTTERFLIES see **Agraulis, Dione, Dryadula, Dryas, Eueides, Marpesia**
FLAME MOTH see **Catarhoe rubidata**
**flammea** see **Panolis**
**flammealis** see **Stericta**
**flammula** see **Polystichtis**
FLANNEL MOTHS see **Megalopyge**
FLASH BUTTERFLIES see **Rapala**
FLAT BUTTERFLIES see **Tagiades**
**flava** see **Tridrepana**
**flavago** see **Xanthia**
**flavia** see **Arctia**
**flaviplaga** see **Orybina**
**flesus** see **Tagiades**
**floralis** see **Phyllodes**
**florella** see **Catopsilia, Syngamia**
FLORIDA PURPLE-WING see **Eunica tatila**
FLOUR MOTHS see **Ephestia**
**floslactata** see **Cyclophora**
FLUFFY TIT see **Zeltus amasa**
**fluctuata** see **Xanthorhoe**
**fluonia** see **Hypothyris**
**foenella** see **Epiblema**
**follicula** see **Hepialodes**
FOOTMEN see **Eilema, Lithosia, Miltochrista, Spiris**
FOREST
BLUE see **Mithras lisus**
GLADE NYMPH see **Aterica galene**
QUEEN see **Euxanthe wakefieldii**
TENT CATERPILLAR see **Malacosoma disstria**
**forestan** see **Coeliades**
FORESTER MOTHS see **Adscita, Alypia, Euphaedra**
**formosa** see **Danaus, Pempelia**
**formosissima** see **Ancycluris**
**fornax** see **Hamadryas**
**forsskaleana** see **Croesia**
FOUR-SPOTTED FOOTMAN see **Lithosia quadra**
FOURBAR SWORDTAIL see **Protographium leosthenes**
FOX MOTH see **Macrothylacia rubi**
FOXGLOVE PUG see under **Eupithecia linariata**
**francina** see **Euphaedra**
**franclemonti** see **Eudonia**
**fraxini** see **Catocala**
**freija** see **Clossiana**
**freja** see **Cheritra**
FREJYA'S FRITILLARY see **Clossiana freija**
FREYER'S FRITILLARY see **Melitaea arduinna**

**Freyeria** Courvoisier (LYCAENIDAE)
This is a genus of about 6 tailless species of butterflies found in Europe, Africa and the Indo-Australian region.

**trochylus** Freyer GRASS JEWEL
S.E. Europe, tropical & subtropical Africa, Asia 8–13 mm. (0·31–0·51 in.)
This is the smallest species of European butterfly and one of the smallest in the world. It is similar in the colour pattern of the upper surface to the much larger Brown Argus, *Aricia agestis*, of Europe. The forewings are marked with black spots beneath, and there are orange, black and metallic marginal spots under the hindwing. Its caterpillars feed on the flowers of species of legumes (Leguminosae) and are attended by ants. **260**a

FRIAR see **Amauris niavius**
**frigga** see **Clossiana**
FRIGGA'S FRITILLARY see **Clossiana frigga**
**frigidana** see **Nycteola**
**frischella** see **Coleophora**

**Fritillaries**
Popular name for rather similarly patterned butterflies in several genera, see *Agraulis, Argynnis, Boloria, Brenthis, Chlorosyne, Clossiana, Euphydryas, Fabriciana, Hamearis, Issoria, Melitaea, Mellicta, Mesoacidalia, Proclossiana* and *Speyeria*. There is still some dispute on the status of the generic names to be used for these butterflies. In this book the ones mentioned are used in a generic sense although some authors consider this group of butterflies too finely separated and use these names as subgenera, uniting them together into one genus, *Argynnis* Fabricius.

FRITILLARY, FALSE see **Pseudargynnis hegemone**
FROG FOOT MOTH see **Antistathmoptera daltonae**
**frugiperda** see **Spodoptera**
FRUIT MOTH, ORIENTAL see **Cydia molesta**
FRUIT
TORTRIX see **Archips podana**
TREE LEAF-ROLLER see **Archips argyrospila**
FRUIT-PIERCER, TROPICAL see **Othreis fullonia**
FRUIT-PIERCING MOTH, EUROPEAN see **Scoliopteryx libatrix**
**fugax** see **Rhodinia**
**fulguritia** see **Erebomorpha**
**fuliginosa** see **Phragmatobia**
**fulleborni** see **Papilio**
**fulleborniana** see **Ubaena**
**fullonia** see **Othreis**
**fulvidorsalis** see **Filodes**
**fumicosta** see **Eumelea**
**fumiferana** see **Choristoneura**
**funebris** see **Anania**

**Furcula** Lamarck (NOTODONTIDAE)
This is an almost cosmopolitan genus of about 30 species. The wing pattern is similar in type to that of *Dicranura* (Puss-moths), but many of the species have a brilliant silvery white ground-colour.

**rivera** Schaus
Tropical S. America 30–42 mm. (1·18–1·65 in.)
At least some females differ from the illustrated male in the greyish brown, not white hindwing, and there is some variation between specimens in details of the intricate pattern of black lines on the wings. Nothing is known about the biology of this species. **357**b

**furnacalis** see **Ostrinia**
FURRY BLUE see **Agrodiaetus dolus**
**furvivestia** see **Ratarda**
**fusca** see **Busseola**
**fuciformis** see **Hemaris**
**fuscinervis** see **Himantopterus**
**fuscipennis** see **Eretmocera**
**fusconebulosa** see **Hepialus**
**gaika** see **Zizula hylax**
**galathea** see **Melanargia**
**galanthis** see **Siderone**
**galene** see **Aterica**

**Galleria** Fabricius (PYRALIDAE) WAX MOTHS or HONEYCOMB MOTHS
This is a genus of one species which is cosmopolitan. The moth is typical of the rather sombre coloured Galleriinae to which it belongs. They can prove troublesome to bee-keepers, particularly if the standards of hygiene are not high. The moth has been used in recent years as a laboratory animal for research work as it is easy to rear in large numbers.

**melonella** Linnaeus GREATER WAX MOTH
Europe including Britain, N. America, Africa, Australia 20–40 mm. (0·79–1·57 in.)
The moth is easily recognized by its characteristic shape (see illustration). The caterpillar, which feeds on honeycombs, moves rapidly if disturbed. The honeycombs become riddled with their silk-lined tunnels. When it pupates it spins a strong, papery, cocoon. Generally a large number of caterpillars pupate together and these tough cocoons survive long after the moths have emerged. **56**a

**gallii** see **Hyles**

**Galona** Karsch (NOTODONTIDAE)
Only one species is known.

**serena** Karsch
Tropical Africa 40–66 mm. (1·57–2·6 in.)
This is a fairly common, brightly coloured moth in parts of its extensive range across Africa. The forewing of the female differs from that of the illustrated male in the restriction of the oblique pink streak to the basal half of the wing; the hindwing is white basally and dark brown distally. **357**k

**gamma** see **Autographa**

**Gangara** Moore (HESPERIIDAE)
Large skipper butterflies, often unmarked on the upperside, although some have a few large spots. They are separated from the other skippers by their blood-red eyes. The caterpillars feed on palms (Palmae), bamboos (Gramineae) and plantains *(Plantago)*.

**thyrsis** Fabricius GIANT REDEYE
India, Pakistan, Sri Lanka, Burma, Indonesia — 70–76 mm. (2·76–2·99 in.)
One of the largest skipper butterflies. This is generally found near palm trees in hilly jungle areas. It tends to fly at dusk, and immediately after dawn for a short time. The wings are dark chocolate brown with a large square, semi-transparent yellow spot over the cell with 2 smaller squarish spots near the outer margin of the wing and 2 or 3 small ones below the apex.

GARDEN
CARPET see **Xanthorhoe fluctuata**
GRASS VENEER see **Chrysoteuchia culmella**
TIGER-MOTH see **Arctia caja**
WEBWORM see **Loxostege rantalis**
WHITE see **Pieris rapae**
**garuda** see **Euthalia aconthea**

**Gastropacha** Ochsenheimer (LASIOCAMPIDAE)
This is a small genus of moths, typically with a crenulate outer margin to the wings. They are found in Europe, Africa and in Asia as far east as Japan and Java. The caterpillars have a series of processes or lappets along each side.

**quercifolia** Linnaeus LAPPET MOTH
Europe, Asia to Siberia, China, Japan — 40–78 mm. (1·57–3·07 in.)
The unusual resting position of this moth produces a shape much like that of a bunch of dead leaves which must be highly effective as a method of camouflage. It is on the wing in June or July. Its large, flattened, hairy dark grey caterpillar feeds on the foliage of a variety of trees, including sloe *(Prunus)*, apple *(Malus)*, sallow *(Salix)* and hawthorn *(Crataegus)*. It overwinters after feeding for a few weeks at the end of the summer and completes its development the following spring. The caterpillars will eat laurel *(Laurus)* leaves which are rich in hydrogen cyanide, apparently without ill-effects, just as species of Zygaenidae are able to eat the similarly poisonous leaves of vetches *(Vicia)*. **321**

GATEKEEPER BUTTERFLIES see **Pyronia**
**gaura** see **Ideopsis**

**Gazera** Herrich-Schäffer (CASTNIIDAE)
This is a genus of about 5 species, all of them mimics of Ithomiine butterflies. The wings are black or dark brown with large transparent areas – the latter a result of the placement of the scales on their edge, at 90° to the surface of the wing.

**linus** Cramer
Tropical S. America. — 70–100 mm. (2·76–3·94 in.)
This moth closely resembles the Ithomiine butterfly species *Thyridia thermisto* and *T. confusa*. *T. confusa* (and therefore probably also *G. linus*) is known to be a member of a mimetic complex which includes *Ituna phanerete* (Danainae) and *Dismorphia crise* (Pieridae). **45**m

**geffroyi** see **Libythea**

**Gelechia** Hübner (GELECHIIDAE)
This is a huge genus of micro-moths, species of which genus found on all continents. Many are rather sombre in colour, brown or grey brown with dark marks, but some are strikingly patterned. The shape of the rather pointed hindwing is characteristic of the genus.

**ophiaula** Meyrick
S. America — 16 mm. (0·63 in.)
This species has been collected in Argentine, Paraguay and Brazil and may well be widespread in S. America. Nothing is known of its biology. **43**c

**rhombella** Denis & Schiffermüller
Europe including Britain — 13–14 mm. (0·51–0·55 in.)
The species is local in England but is widespread through Central Europe. The caterpillar feeds on apple *(Malus)* trees where it ties the leaves together with its silk. Several related species feed on different plants and the identity of the host plant is often a guide to the identity of the species of moth found on it. **43**g

**Gelechiidae**
A large family, most species with a characteristically shaped hindwing (trapezoidal) with a pointed apex. The caterpillars normally feed amongst leaves or shoots or seedheads which they spin together with their silk. Less frequently they mine leaves or construct portable cases. The family includes species of great economic importance, whose ravages of crops annually cause losses of many hundreds of thousands of pounds sterling all over the world. See *Anarsia, Evagora, Exoteleia, Gelechia, Phthorimaea, Physoptilia, Platyedra, Sitotroga.*

GEM BUTTERFLIES see **Poritia**
**gemmifera** see **Gymnelia**
**genoveva** see **Ogyris**
**genutia** see **Paramidea**
**geoffrella** see **Alabonia**

**Geometra** Linnaeus (GEOMETRIDAE)
Species of moths in this genus are found in Europe and temperate Asia.

**papilionaria** Linnaeus LARGE EMERALD
Europe including Britain, temperate Asia including Japan — 45–56 mm. (1·77–2·2 in.)
This green moth is common throughout Europe and Asia. The caterpillar is green and well camouflaged on the leaves of birch *(Betula)*, hazel *(Corylus)* and other trees and shrubs. The moth flies in June and July and is frequently attracted to light. Its colour presumably forms a camouflage when it is at rest during the day. **273**

**Geometridae**
A huge family of moths with many genera. There is a wide variety of shape and colour but the 'typical' geometrid, if such a thing exists, has rather rounded wings, a complicated pattern of lines across the wings and usually a rather slow, fluttery flight. However, there are probably almost as many exceptions to these statements as species that fit them. Geometrid moths are found all over the world; many are important agricultural, horticultural or forestry pests.

One of the most characteristic features of the Geometridae is their caterpillar. While most moths have caterpillars with 3 pairs of legs on the thorax, 4 pairs along the middle of the body and a pair at the end (claspers) Geometrid caterpillars lack some or all of the middle 4 pairs. As a result they have a very characteristic means of locomotion, drawing the hind part of the body up to the front and 'looping' the middle. This is stretched out as the front part moves forward. The hind part then 'loops' up. The popular name of the caterpillars reflects this, Looper caterpillars, Earth-measurers, Inch worms amongst them. All the species were thought to be plant feeders in the caterpillar stage but in 1971 a species was discovered in Hawaii which feeds on flies, these are caught as they settle nearby. Subsequently several species on Hawaii currently in the one genus *Eupithecia* have been found to do this. This discovery really stirred up Entomologists, for although some other caterpillars are predatory, feeding on greenfly or scale-insects *(Coccina)*, catching active insects like flies had never previously been known. Another feature of many Geometrid caterpillars is the habit of freezing when threatened and sticking straight out from the plant. Since the caterpillars are frequently brown, they look just like a stick. The family is divided into a number of distinct subfamilies, 6 of these divisions are commonly recognized. The moths at rest sit with their wings raised above the bodies in the manner of butterflies in many species, although other modes are adopted.
See *Abraxas, Agathia, Alcis, Alsophila, Amnemopoyche, Angerona, Anisozyga, Aplocera, Arycauda, Biston, Bupalis, Callioratis, Callipia, Campaea, Camptogramma, Carpella, Casbia, Catarhoe, Catocalopsis, Chesias, Chloroclystis, Chorodnodes, Cleora, Colostygia, Coniodes, Corymica, Crocallis, Cyclophora, Dysphania, Edule, Ematurga, Ennomos, Epione, Epirrita, Erannis, Erateina, Erebomorpha, Euchleena, Eulithis, Eumelea, Eupithecia, Geometra, Gonora, Idaea, Iotophora, Isturgia, Lycia, Milionia, Odezia, Operophtera, Opthalinophora, Opisthograptis, Osteosema, Ourapteryx, Percnia, Peribatodes, Perizoma, Phyle, Pityeja, Plagodes, Plutodes, Pseudopanthera, Rheumaptera, Rhodometra, Rhodophthilius, Selenia, Selidosema, Semiothisa, Stamnodes, Thalaina, Thera, Venusa, Xanthabraxas, Xanthorhoe.*

**germana** see **Amata**

**Gesneria** Hübner (PYRALIDAE)
In the Pyralid moth subfamily (Scopariinae) to which they belong species of *Gesneria* are relatively large. There are not many species in the genus which is characteristic of northern mountains of Europe, Asia and N. America.

**centuriella** Denis & Schiffermüller
N. America, Europe, Greenland — 14–20 mm. (0·55–0·79 in.)
This moth, with one subspecies in C. Europe, has several subspecies in N. America. The moth is smokey grey with dark markings. The American subspecies occurs in the Appalachian mountains and as far south as Arizona. The subspecies differ in colour and pattern. The life history of this common moth is not known.

GHOST SWIFT MOTH see **Hepialus humuli**
GIANT
BLUE see **Lepidochrysops gigantea**
BLUE SWALLOWTAIL see **Papilio zalmoxis**
REDEYE see **Gangara thyrsis**
SPHINX see **Cocytius antaeus**
SUGAR-CANE-BORER see **Castnia licus**
SULPHUR BUTTERFLIES see **Phoebus**
SWALLOWTAIL see **Papilio cresphontes**
SWALLOWTAIL, AFRICAN see **Papilio antimachus**
WHITE see **Ascia josephina**
**gigantalis** see **Ramphidium**
**gigantea** see **Lepidochrysops**
**gigas** see **Chrysocale**
**gilippus** see **Danaus**
GLANVILLE FRITILLARY see **Melitaea cinxia**
GLASSWING see **Acraea andromache**
**glaucata** see **Cilix**
**glaucippe** see **Hebomoia**

**Glaucopsyche** Scudder (LYCAENIDAE)
There are about 11 species in this genus of blue butterflies. They are found in North America, Europe, N. Africa, and temperate Asia as far as Japan.

**alexis** Poda GREEN-UNDERSIDE BLUE
Europe through C. Asia to the Pacific — 24–32 mm. (0·94–1·26 in.)
*G. alexis* is found near woodlands up to 1220 m. (4000 ft). The origin of the common name of this species is the unusual green area on the under surface at the base of the hindwing (illustrated). Males are blue above with black outer margins; females are brown above, usually with some blue scales at the base of the wings. The caterpillars feed on various species of Leguminosae. **263**n

**lygdamus** Doubleday SILVERY BLUE
Canada, U.S. — 23–31 mm. (0·91–1·22 in.)
The lustrous, silvery blue of the male upper side explains the popular name for this butterfly. It flies in both open woods and fields where it is a frequent visitor to flowers and wet patches. The caterpillars, which are attended by ants, feed on Everlasting Peas *(Lathyrus)* and possibly a vetch (Leguminosae).

**glaucus** see **Papilio**
**glenwoodi** see **Tolype**
GLIDERS see **Neptis**
**gloriosa** see **Melittia**
GLORY BUTTERFLY, JUNGLE see **Thaumantis diores**
GLORY MOTH, KENTISH see **Endromis versicolora**
GLOVER'S PURSLANE see **Copidryas gloveri**
**gloveri** see **Copidryas**
**glycine** see **Phalaenoides**
**glyphica** see **Euclidia**

**Glyphipterigidae**
A large family of micro-moths found over most of the world. They are generally small day-flying species which often have bright metallic or bright orange

colours on the wings. The caterpillar feeds amongst seeds or tunnels into shoots. See *Anthophila, Atychia, Burlacena, Choreutis, Glyphipterix, Hilarographa, Imma, Mictopsichia, Tebenna, Tortyra.*

**Glyphipterix** Hübner (GLYPHIPTERIGIDAE)
A very large genus of micro-moths with over 200 species. The genus is worldwide. There are many species in Australia and New Zealand and some in N. and S. America. They are mostly day-fliers, particularly active in bright sunshine.

**equitella** Scopoli
Europe including Britain 9-10 mm. (0·35-0·39 in.)
The caterpillars live in the shoots of a stonecrop *(Sedum acre)*. The species is common over most of its range. **34**h

**Glyphipteryx** Curtis: syn. *Chrysoclista* (MOMPHIDAE)
The genus of micro-moths is mostly European and Asian, with species in Japan, and must not be confused with the similar *Glyphipterix* which contain species in another family. The caterpillars feed in bark or on the fruit of trees and shrubs. The two names with only one letter difference has lead to some confusion in the past.

**lathamella** Fletcher: syn. *G. bimaculella*
Europe including Britain 10–15 mm. (0·39–0·59 in.)
A local European species whose caterpillar feeds on White Willow *(Salix alba)*. The moths are to be found in June and July. **34**c

**lineella** Clerck
Europe including Britain, Turkey, U.S. 10–13 mm. (0·39–0·51 in.)
Smaller than *G. lathamella*, which it otherwise resembles, with orange and yellow forewing with black base and fine silvery streaks on the wing. The caterpillar feeds in the bark of lime *(Tilia)*. The moths are out in July and August. One single specimen of this species was recently collected in New Jersey in the U.S., possibly accidentally introduced. It is difficult to imagine how so small a moth would otherwise cross the Atlantic. **34**g

**gnoma** see **Pheosia**

**Gnophaela** Walker (HYPSIDAE)
U.S., Mexico, C. and tropical S. America is the range of the 4 or so species of this genus of day-flying moths. The caterpillar of at least one species, *G. latipennis* (W. U.S.), feed on species of Boraginaceae, a family generally considered to be toxic to vertebrates. *G. latipennis* may well be able to extract these toxins and achieve some form of chemical protection from predators.

**arizonae** French
S.W. U.S., Mexico 45–50 mm. (1·77–1·97 in.)
This is a slow-flying, typically high elevation species, most active in the sunshine when numbers can be seen visiting flowers in meadowland. The illustrated male was captured at about 2590m. (8500 ft) in the mountains of N. Arizona by A. Watson while collecting with Ron Wielgus of Phoenix. **376**h

**latipennis** see under **Gnophaela**

GOAT MOTHS see **Cossus**
GOAT-WEED BUTTERFLY see **Anaea andria**
**goedartella** see **Argyresthia**
**goeldii** see **Pachypodistis**
GOLD, PURPLE-BORDERED see **Idaea muricata**
GOLD
RIM BUTTERFLY see **Battus polydamus**
SPOT see **Plusia festucae**
TIP, BANDED see **Colotis eris**
GOLD-BANDED FORESTER see **Euphaedra neophron**
GOLD-DROP HELICOPIS see **Helicopis cupido**
GOLD-FRINGE WEBWORM see **Hypsopygia costalis**
GOLD-SPOTTED SYLPH see **Metisella metis**
GOLD-TAIL see **Euproctis similis**
GOLDEN
BIRDWING see **Troides aeacus**
COPPER see **Lycaena thetis**
DANIID see **Danaus chrysippus**
LONG HORN see **Adela reamurella**
PIPER see **Eurytela dryope**
**goliath** see **Ornithoptera**
GOLIATH BIRDWING see **Ornithoptera goliath**
GONATRYX, YELLOW-SPOTTED see **Anteos chlorinde**

**Gonepteryx** Leach (PIERIDAE) BRIMSTONE BUTTERFLIES
The species in this genus are found mainly in Europe and temperate Asia with a few species in the Orient. These bright yellow butterflies are conspicuous whereever they occur. In N. America there are related genera with similar species. One of the species described in this genus was the centre of an intriguing fraud which may have involved the great naturalist, Carl Linnaeus.

**cleopatra** Linnaeus CLEOPATRA BUTTERFLY
N. Africa, Europe 50–60 mm. (1·97–2·36 in.)
This species is similar in shape to *G. rhamni* but has orange-red on the forewing surrounded by yellow. The caterpillar feeds on *Rhamnus* (Rhamnaceae). A number of subspecies of *G. cleopatra* have been described from different parts of its range. Several of these have popular names including *G. c. cleobule* Hübner, the Canary Island Brimstone. This subspecies is out in February-March over rough ground on the islands of Tenerife and Gomera, another similar subspecies occurs on La Palma.

**eclipsis** Linnaeus PILTDOWN BUTTERFLY
N. America 56–67 mm. (2·2–2·64 in.)
Unfortunately there is no such species! Nevertheless in 1763 the great Swedish naturalist, Linnaeus, described an insect under this name, mentioning that it came from N. America. Several subsequent authors listed and discussed this 'species' but it was only when the specimens in the Linnean collection were examined that the insect was found to be a Brimstone butterfly with hand-painted wings. The specimens, which are beautifully painted, are still in existence. Strangely enough Linnaeus in his description refers to a figure in a book by James Petiver. Many of Petiver's figures in his book are rather inaccurate, but not the one of *G. eclipsis*. This matches exactly the Linnean specimens. The question is, who was playing the joke? Was it one of Linnaeus's students, or could it have been someone making some money out of James Petiver by selling him the specimens? The English name suggested above is after another celebrated fake, the Piltdown skull, which fooled anthropologists for many years.

**rhamni** Linnaeus BRIMSTONE
Europe including Britain, N. Africa, Asia including Japan 56–67 mm. (2·2–2·64 in.)
This species ranges widely over Europe and N. Africa. In Japan a distinct subspecies has been described. The butterfly hibernates in the adult stage through the winter, but any mild winter day will tempt it out. It is also one of the first butterflies to be tempted out by some spring sunshine. The caterpillar feeds on buckthorn *(Rhamnus)*. The Brimstone is one of the contenders for the name 'butterfly', it is suggested that this 'butter-coloured-fly' as it could be described, was contracted to butterfly. **118**

**goniata** see **Ludia**

**Gonodonta** Hübner (NOCTUIDAE)
There are about 40 species in this New World genus of moths. The shape of the inner margin of the forewing is characteristic (see illustration of *G. distincta*). Adults are fruit-piercers and are minor pests in citrus orchards. The caterpillars are semi-loopers.

**distincta** Todd
N.W. S. America 34–38 mm. (1·34–1·5 in.)
The wing shape and colour pattern of this species are typical of its genus, although the inner margin of the forewing is more extremely emarginate than in most other species. The foodplant is unknown. **391**l

**unica** Neumoegen
Florida, Cuba 32–40 mm. (1·26–1·57 in.)
This is somewhat similar in appearance to the illustrated *G. distincta*, but the abdomen is orange. The caterpillar feeds on pond-apple and sugar-apple *(Annona)*.

**Gonora** Walker (GEOMETRIDAE)
A genus of S. American moths with about half a dozen known species.

**hyelosioides** Walker
S. America 50–55 mm. (1·97–2·17 in.)
This species has been found in Ecuador, Colombia, Brazil and Peru. There are several related and rather similar species with transparent wings. The males and females are similar. Nothing is known of their life history. **296**w

**Goodia** Holland (SATURNIIDAE)
This is a genus of about 10 species of chiefly brown and orange moths found in tropical and S. Africa. The caterpillars bear tubercles and long branched hairs like those of *Ludia* species.

**kuntzei** Dewitz LUNAR PRINCE
E. Kenya to Transvaal 50–60 mm. (1·97–2·36 in.)
Parthenogenesis is known to occur in this remarkable species – virgin females lay unfertilized eggs which successfully hatch and produce normal caterpillars. Four generations have been bred without any males being produced, although males are known to exist. The wing coloration varies greatly, from yellowish white to red or greenish brown. *Acacia, Brachystegia* and *Julbernardia* are recorded as foodplants of the caterpillar. **331**l

GOOSE-EGG MOTH see under **Cilix glaucata**
**gossypiella** see **Platyedra**
**goughi** see **Peridroma**
**gracilis** see **Hepialus**

**Gracillariidae** LEAF BLOTCH MINERS
Very small micro-moths. The family is worldwide, containing many species whose caterpillars are leaf miners or roll the leaves into conical chambers in which they live. Generally all have narrow rather pointed forewings and very slender pointed hindwings. See *Acrocercops, Caloptilia, Dalechea, Epicephala, Phyllonorycter, Polysoma.*

**Graellsia** Grote (SATURNIIDAE)
There is one species in this genus of moths.

**isabellae** Graëlls
C. Spain, the French Alps 63–85 mm. (2·48–3·35 in.)
This elegantly marked moth was known only from Spain for many years. It is the only tailed member of its family to be found in Europe. The brown and white marked, green caterpillar feeds on pine *(Pinus)*. The strongly pectinate antennae of the male are well shown in the illustration. There is a single brood per year which overwinters as a chrysalis. **337**b

GRAIN MOTHS see **Nemapogon, Sitotroga**
**graminis** see **Cerapteryx**

**Grammodes** Guenée (NOCTUIDAE)
This is a group of about 30 species of moths, found in Europe, Africa, temperate and tropical Asia eastwards to Australia.

**stolida** Fabricius
Africa, Madagascar, India, S.W. Asia, S.E. Europe 24–36 mm. (0·94–1·42 in.)
*G. stolida* has an unusually wide climatic tolerance – from the Mediterranean through tropical Africa and into India. In colour pattern, the adult moth is typical of several species of *Grammodes*. Its caterpillar feeds on the foliage of numerous shrubs and trees, including *Rubus, Quercus* and *Zizyphus* (the genus to which *Z. lotus*, the Lotus of antiquity belongs). The illustrated specimen was caught in S.W. Turkey. **386**

**Grammodora** Aurivillius (LASIOCAMPIDAE)
There is a single species in this African genus of moths.

**nigrolineata** Aurivillius
E. Africa S. to South Africa 38–68 mm. (1·5–2·68 in.)
Few species of Lasiocampidae are as strikingly coloured as this species. Females differ from the illustrated male in mainly dark brown hindwing. The caterpillars are clothed in a mixture of white hairs and

a few brown hairs and are variously marked with brown, yellow and brownish red. The tightly woven cocoon is attached to the side of a twig and usually at least partly concealed in a rolled leaf or a few leaves joined together with silk by the caterpillar. **318**m

GRAND DUCHESS see **Euthalia patala**
**grandipennis** see **Scythris**
**grandis** see **Cyanocrates, Milionia**
**granella** see **Nemapogon**
GRAPE PLUME MOTH see **Pterophorus periscelidactyla**
GRAPE-VINE EPIMENIS see **Psychomorpha epimenis**
GRAPELEAF SKELETONISER,
EASTERN see **Pampa americana**
WESTERN see **Pampa brillians**

**Graphelysia** Hampson (ARCTIIDAE)
A single species is placed in this genus.

**strigillata** Rothschild
Peru 42–46 mm. (1·65–1·81 in.)
The significance of the unusual, striate pattern of the forewing of this species is not known. It is doubtful if it can be considered cryptic in function, as the head and thorax are partly red and clearly visible to avian predators. **368**e

**Graphium** Scopoli SWORDTAIL and SWALLOWTAIL BUTTERFLIES (PAPILIONIDAE)
This is a large genus of about 130 species. They are essentially tropical African and Asian in distribution, but there are 10 species in Europe and temperate Asia. The closest relatives of this genus are *Iphiclides*, *Eurytides* and *Protographium*. Many species have sword-like hindwing tails. Those caterpillars which are known have 3 pairs of spines on the part of the body immediately behind the head.

**agamemnon** Linnaeus TAILED JAY or GREEN-SPOTTED TRIANGLE BUTTERFLY
N. India, E. China, the Malay Archipelago, E. to New Guinea, Australia and the Solomons 85–100 mm. (3·35–3·94 in.)
This beautiful, dappled, green and black butterfly is found mainly in open country. It is a difficult species to capture except when engrossed in feeding on the blooms of *Lantana* and other flowering trees and shrubs along the margins of rain-forests. The caterpillar's usual food plants are species of Annonaceae, including the cultivated custard apple *(Annona)*, and sour-sop *(Annona)* – an introduced plant which is native to the Caribbean. The illustrated butterfly is a specimen of the subspecies *G. agamemnon agamemnon*. **99**c

**androcles** Boisduval
Sulawesi (Celebes), Sula Island 80–88 mm. (3·15–3·46 in.)
In general appearance *G. androcles* is similar to the widespread *G. antiphates* (illustrated) but it differs from the latter in its larger size, the two complete black lines at the base of the forewing and in the presence of two incomplete lines at the base of the hindwing. There is a relatively high number of endemic moths and butterflies in the island of Sulawesi, which to a large extent retained its water barriers during the glacial stages of the Pleistocene period of geological history. The geographically associated islands of Borneo, Sumatra and Java were at times joined by land connections with each other and the mainland of S.E. Asia which allowed the free interchange of insects, other animals and plants. Faunal interchange with Sulawesi was limited or prevented by the sea and evolution was able to follow separate paths.

**antiphates** Cramer FIVEBAR SWORDTAIL
Sri Lanka, India, S. China, Malaya to the Lesser Sundas 58–70 mm. (2·28–2·76 in.)
This species was described originally from a Chinese specimen, but is commoner in the south-east of its range where it flies at low elevations in rain-forest areas. The caterpillar feeds on the foliage of Annonaceae species. The popular name for this butterfly is based on the 5 incomplete black bands on the forewing. **99**e

**aristeus** Cramer FIVEBAR SWORDTAIL
N. India to the Philippines, New Guinea and Australia 45–52 mm. (1·77–2·05 in.)
This widespread species is named after the 5 incomplete bars on the forewing. Other butterflies have a similar pattern: another 'Fivebar', *G. antiphates*, and the Australian 'Fourbar', *Protographium leosthenes*. Its caterpillar probably feeds on species of Annonceae.

**cloanthus** Westwood
Kashmir, N. India, Nepal to S. China Japan and Sumatra 50–75 mm. (1·97–2·95 in.)
Several subspecies of this elegantly marked species have been described. They differ little from each other in colour pattern, but in the Sumatran subspecies the translucent greenish blue markings of the upper surface are replaced by a more opaque green colour. The caterpillar of this species is green, with 2 longitudinal yellow lines above and greenish blue beneath; it feeds on species of *Machilus*, a genus of Lauraceae. **99**a

**codrus** Cramer
Sulawesi (Celebes) and the Philippines to New Guinea and the Solomons 70–95 mm. (2·76–3·74 in.)
There is some variation between the subspecies of *G. codrus* in the colour of the forewing band, which can be lime yellow as illustrated, a deeper yellow, or bluish green, and in one subspecies the band is interrupted in the posterior half of the wing. Females are similar to the males but lack the green iridescence on the forewing. The adults are tree-canopy dwellers, but are attracted to ground level by rotting fruit, flowers and damp mud or sand. Nothing is known of the life history of this species. **97**g

**colonna** Ward MAMBA or BLACK SWORDTAIL
Kenya, Tanzania to Mozambique and South Africa 65–86 mm. (2·56–3·39 in.)
This butterfly is found in the neighbourhood of coastal forests and in dense scrubland where it is seldom seen far from ground level. The foodplants of the caterpillar are species of *Artabotrys* (Annonaceae). The irregularly shaped green chrysalis is particularly well camouflaged against its background. **99**f

**cyrnus** Boisduval
Madagascar 65–80 mm. (2·56–3·15 in.)
This is a rather rare species, at least in collections. The female (not illustrated) differs little in colour and pattern from the male. The under surface is brown, with pink at the base of the wings and pale yellow markings with several black spots between the veins of the hindwing. Madagascar contains many species of moths and butterflies not found elsewhere. Most of these are closely related to African species but some are without such close affinities and appear to have more characters in common with S. Asian species, probably as a result of extinction in Africa of forms ancestral to present-day Madagascan species. **99**d

**delesserti** Guérin-Méneville
Malaya to Borneo and the Philippines 64–90 mm. (2·52–3·54 in.)
This is one of a group of *Graphium* species which mimic species of unpalatable Danainae. The ground-colour of the wings is white or greenish white, with the veins marked in black and a yellow spot at the anal angle of the hindwing. The colour pattern of the female is nearly identical to that of *Ideopsis daos*, *I. gaura* (Danainae) and *Cyclosia pieridoides* (Zygaenidae). The male is similar in pattern, but the yellow spot on the hindwing is larger and there is more black apically on the forewing and towards the outer margin of the hindwing. Males frequently congregate on wet mud at the sides of forest streams. Both sexes fly in a leisurely manner like species of Danainae.

**endochus** Boisduval
Madagascar 60–70 mm. (2·36–2·76 in.)
Very few specimens of this species exist in collections. It is similar in pattern to the closely related African species *G. pylades*, but is mostly white above, without white spots on the marginal black borders and with an orange spot at the rear angle of the hindwing in some specimens. There is little difference in colour pattern between the sexes. The caterpillar is unknown.

**eurypylus** Felder PALE GREEN TRIANGLE BUTTERFLY
Sri Lanka and India to China, Japan, and Malaya to New Guinea and N. Australia 55–68 mm. (2·17–2·68 in.)
There are 13 described subspecies in this very widely distributed species. Only *G. sarpedon* has a wider distribution in this genus. The caterpillars of *G. eurypylus* are sometimes a minor pest of the cultivated custard apple *(Annona)*. Other recorded foodplants are species of Sapindaceae, Lauraceae and wild species of Annonaceae. The female (not illustrated) is generally slightly paler in coloration than the male. **99**b

**idaeoides** Hewitson
The Philippines 100–130 mm. (3·94–5·12 in.)
This huge and atypical species is a remarkably accurate mimic of *Idea leuconoe* (see the illustration of the similar *I. idea*). There is little difference between the two sexes of this butterfly, both of which are greyish white ornamented with large black spots.

**illyris** Hewitson YELLOW-BANDED SWORDTAIL
W. Africa and the Congo Basin 70–90 mm. (2·76–3·54 in.)
This rain-forest species is typical in wing shape of other swordtails, the hindwings of which are produced into a characteristic blade-like process. It derives its common name from the continuous curved pale yellow band which crosses the dark brown upper surface of both wings. There are pale yellow spots on the hindwing between the yellow band and the outer margin, and a yellow spot at the apex of the hindwing tail. The under surface of the wings is a paler brown and the transverse band is white; the hindwing has additional dull red and dark brown markings.

**latreillianus** Godart COPPERY SWALLOWTAIL
W. Africa from Sierra Leone to Angola and the Congo Basin 65–90 mm. (2·56–3·54 in.)
This is a beautifully marked butterfly above, with unusual markings on the under surface. The wings are dark brown above, with a transverse row of large lime-green spots on the forewing and a ray-like pattern of parallel green dashes on the hindwing. Below, the wings are pinkish brown with green spots on the forewing, but with several black spots on the hindwing which produce a deceptively *Acraea*-like appearance.

**leonidas** Fabricius VEINED SWALLOWTAIL
Africa S. of the Sahara 70–80 mm. (2·76–3·15 in.)
Except in mountainous and desert regions, this tailless species occurs in most of Africa, but is most common in scrubland, savanna and along the edges of forests. Some forms have the green markings replaced by white and are mimics of *Danaus limniace* and of *Amauris echeria* which they resemble both in colour pattern, especially when the wings are closed, and the characteristic gliding flight. Several specimens will roost together overnight, with their wings hanging downwards, in the manner of Danaine species. The caterpillars feed on various species of Annonaceae, including the cultivated custard apple *(Annona)*.

**macleayanus** Leach MACLEAY'S SWALLOWTAIL
Australia and associated islands 46–52 mm. (1·81–2·05 in.)
The under surface of this species is almost as colourful as the upper surface – the black areas are mostly replaced by light brown and the pale blue of the hindwing is replaced by green. Males of this species closely guard their chosen tract of territory and will attempt to challenge other males of the same species and species of a similar size. There are 4 subspecies, the most northerly of which flies at higher elevations than the others. The caterpillar feeds on several native species of the families Monimiaceae and Winteraceae and on introduced Camphor laurel *(Cinnamomum)*. **97**c

**megarus** Westwood SPOTTED ZEBRA BUTTERFLY
N.E. India, Indo-China, Burma, Malaya to Borneo 65–90 mm. (2·56–3·54 in.)
Both sexes of this species are brown, with *Danaus*-like markings, and appear to be mimics of *Danaus hamata*. This is a close relative of *G. delesserti*, another mimetic species. Males of *G. megarus* congregate on wet patches on river banks together with many other species of

butterflies; females are almost invariably confined to the forests.

**nomius** Esper SPOT SWORDTAIL
Sri Lanka, India, Sikkim, Burma, Thailand 75–90 mm. (2·95–3·54 in.)
Forested lowland or hilly country is the home of this attractive and sometimes common species. It is somewhat similar to *G. aristeus* in appearance, but has mostly rounded, not lunulate white spots along the outer margin of the forewing, and two well marked oblique bands at the base of the hindwing. The caterpillar is highly variable in colour: it can be black, banded with white above and green below, or black in the middle and yellow at either end, or entirely green. The foodplants are species of *Saccopetalum* and *Polyalthia* (Annonaceae).

**polistratus** Grose-Smith DUSKY SWORDTAIL
E. Africa including Tanzania and Malawi 60–70 mm. (2·36–2·76 in.)
This butterfly is similar to *G. colonna*, but has broader stripes and spots, and lacks the crimson spots on the hindwing. The under surface of the wings is mostly brown and white, but the hindwing has a central red band. It is a rather uncommon species and is confined mostly to coastal forests.

**pylades** Fabricius ANGOLA WHITE LADY SWALLOWTAIL
Africa S. of the Sahara 60–80 mm. (2·36–3·15 in.)
This is one of a group of dark brown and white, relatively small, tailless species found in Africa. There is a large white area in the basal half of each wing and several white spots near the outer margins of the wings. The under surface of the forewing differs from the upper surface in being brownish red basally at the front edge of the wing and light yellowish brown apically. The caterpillar feeds on species of *Annona* and *Sphedamnocarpus*.

**ridleyanus** White ACRAEA SWALLOWTAIL
Tropical W. Africa, the Congo Basin, E. Africa, sub-tropical South Africa 64–85 mm. (2·52–3·35 in.)
This scarlet and black, tailless butterfly is an inhabitant of the rain-forests and is widely distributed in suitable regions in the west, as far as Sierra Leone, in much of the immense forests of the Congo and in isolated areas of forest in Uganda. Throughout most of its range it closely resembles species of the genus *Acraea* with which the males frequently assemble while drinking on patches of wet mud. Not only is their shape and colour *Acraea*-like, but also their manner of flight, so that the total impression received by a predatory bird must be nearly completely deceptive. Females (not illustrated) are rarely captured; they differ from the males in the almost straight outer margin of the forewing. **105**c

**sarpedon** Linnaeus BLUE TRIANGLE or COMMON BLUEBOTTLE BUTTERFLY
Sri Lanka, India, China, Japan, Malay Archipelago, to New Guinea, Australia and the Solomons 80–90 mm. (3·15–3·54 in.)
This is a species of forests, open country, and of gardens where *Lantana* and *Buddleia* blossoms attract the adults. The caterpillars feed on a great variety of species of at least 5 different families of plants, including the myrtles *(Myrtaceae)*, laurels *(Lauraceae)* and the 'chewing-gum' family, *Sapotacaea*, This adaptability is probably one of the reasons why this species has successfully established itself over the whole of S.E. Asia. In Australia, *G. sarpedon* caterpillars are a minor pest of Camphor laurel *(Cinnamomum)*. **108**

**weiskei** Ribbe PURPLE SPOTTED SWALLOWTAIL
Ceram to New Guinea and the Goodenough Islands 75–80 mm. (2·95–3·15 in.)
This exquisitely coloured species is a butterfly of the forested Owen Stanley Mountains of New Guinea and high elevations in associated islands. It normally flies at tree-canopy level, and is most likely to be seen early in the morning when feeding on flowers. Its caterpillar is unknown. The female (not illustrated) is brown, with the blue markings of the male replaced by bluish green. *G. weiskei* is closely related to the Australian species *G. macleayanum*. **115**

GRASS MOTHS see **Agriphila, Chrysoteuchia, Crambus, Pediasa**
GRASS JEWEL see **Freyeria trochylus**
GRASS-BLUE BUTTERFLIES see **Zizeeria, Zizina**
GRASS VENEER, PEARL see **Catoptria pinella**
GRASS-YELLOW BUTTERFLIES see **Eurema**
**gratiosa** see **Eurema**
GRAYLINGS see **Brintesia, Hipparchia, Oeneis**
GREASY BUTTERFLY, BIG see **Cressida cressida**
GREAT
BANDED GRAYLING see **Brintesia circe**
BLACK VEIN BUTTERFLY see **Aporia agathon**
BLUE MIME see **Papilio paradoxa**
EGGFLY see **Hypolimnas bolina**
MORMON see **Papilio memnon**
LEOPARD see **Ecpantheria scribonia**
NAWAB see **Polyura eudamippus**
ORANGE-TIP see **Hebomoia glaucippe**
PEACOCK see **Saturnia pyri**
PURPLE HAIRSTREAK see **Atlides halesus**
SATYR see **Satyrus padma**
SOUTHERN WHITE see **Ascia monuste**
SPANGLED FRITILLARY see **Speyeria cybele**
GREATER WAX MOTH see **Galleria melonella**
GRECIAN,
BLUE see **Heliconius wallacei**
SMALL BLUE see **Heliconius sara**
GRECIAN SHOEMAKER see **Catonephele numilia**
GREEN, BURREN see **Calamia tridens**
GREEN
CARPET MOTH see **Colostygia pectinataria**
CHARAXES see **Charaxes eupale**
DRAGON see **Alesa prema**
DRAGONTAIL see **Lamproptera meges**
FRITILLARY, DARK see **Mesoacidalia aglaja**
HAIRSTREAK see **Callophrys rubi**
HAIRSTREAK, CHAPMAN'S see under **Callophrys rubi**
OAK BLUE see **Narathura eumolphus**
OAK TORTRIX see **Tortrix viridana**
PEA MOTH, CREAM-BORDERED see **Earias clorana**
PUG MOTH see **Chloroclystis rectangulata**
SAPPHIRE see **Heliophorus androcles**
SILVER LINES see **Pseudoips fagana**
SPHINX see **Tinostoma smaragditis**
SWEET-OIL BUTTERFLY see **Aeria eurimedia**
TRIANGLE, PALE see **Graphium eurypylus**
GREEN-BANDED BLUE, LARGE see **Danis danis**
GREEN-PATCH SWALLOWTAIL see **Papilio phorcas**
GREEN-SPOTTED TRIANGLE see **Graphium agamemnon**
GREEN-STRIPED
WHITE see **Euchloe belemia**
WHITE, SCARCE see **Euchloe falloui**
GREEN-UNDERSIDE BLUE see **Glaucopsyche alexis**
GREEN-VEINED WHITE see **Pieris napi**

**Greta** Hemming (NYMPHALIDAE)
This is a C. American and tropical S. American genus of the subfamily Ithomiinae. There are about 24 species.

**quinta** Staudinger
Tropical S. America 65–73 mm. (2·56–2·87 in.)
There are very few specimens of this almost colourless, transparent-winged butterfly in collections, but the chances of discovering and then capturing the nearly invisible *G. quinta* in dense forest must be remote. Its life history is not known. **131**p

GREY, BORDERED see **Selidosema brunnearia**
GREY
CRACKER see under **Hamadryas feronia**
DAGGER MOTH see **Acronicta psi**
DAGGER MOTH, AMERICAN see **Acronicta interrupta**
HAIRSTREAK see **Strymon melinus**
**griseana** see **Antaeotricha**
**griseifusa** see **Lobocraspis**
**grisella** see **Achroia**
GRIZZLED SKIPPERS see **Pyrgus**
**grossulariata** see **Abraxas**
GROUND LACKEY see **Malacosoma castrensis**
GRUB, BLUE-STRIPED NETTLE see **Parasa lepida**
**guatemalteca** see **Mesenochroa**
GUAVA BLUE, COMMON see **Virachola isocrates**
**guessfeldti** see **Acanthosphinx**
GUINEA FOWL BUTTERFLY see **Hamanumidia daedalus**
**guineensis** see **Euchromia**
GULF FRITILLARY see **Agraulis vanillae**
GULL BUTTERFLIES see **Cepora**
**gumia** see **Synploca**
**gundlachi** see **Xylophanes**
**guttata** see **Platyprepia**
**guttivitta** see **Disphragis**

**Gymnautocera** Guérin (ZYGAENIDAE)
There are about a dozen species in this genus of moths from India, Thailand, Sulawesi (Celebes) and the Philippines. They are large species with metallic green and blue colourings. Generally their bodies are red and black or orange and black. These are probably warning colours to tell predators that they are distasteful. Some of them are mimics of the poisonous *Aristolochia*-feeding *Papilio* butterflies.

**rhodope** Cramer
W. China 65–80 mm. (2·56–3·15 in.)
The specimens of this moth vary from blue to a greenish colour. The unusual wing shape differs from most of the others in the family, the fore- and hindwings forming a colourful 'X' when the wings are set, although in life the position in which the wings are held is slightly different. These are several related species with a similar wing shape from India and Sumatra. Some of these have green instead of blue wings. The antennae have a feathery appearance (pectinate). **45**d

**Gymnelia** Walker (CTENUCHIDAE)
Over 50 C. and tropical S. American species have been described, and more await description. The wings of *Gymnelia* species are typically transparent, yellowish, and with dark margins and a dark apex to the forewing. The abdomen is transversely banded with yellow and black, and bears iridescent green or blue spots.

**gemmifera** Walker
Colombia, Venezuela 38–47 mm. (1·5–1·85 in.)
This is one of the largest species of its genus and has a particularly broad abdomen, ornamented with brilliant iridescent green. Very few specimens are known. **364**e

**Gynaecia** see **Colobura**
GYPSY MOTH see **Lymantria dispar**

**Habrosyne** Hübner (THYATIRIDAE)
About 17 species are placed in this widespread genus of mostly rather elegantly marked moths. It is represented in Canada and the U.S., from Alaska to Arizona, in Europe and much of Asia as far E. as Japan and Java.

**pyritoides** Hufnagel BUFF ARCHES
C. Europe and Asia to China, Japan 35–40 mm. (1·38–1·57 in.)
This is seldom a particularly common species. The beautifully marked wings are probably cryptic in effect when the moth is at rest amongst foliage, especially if there are dead leaves about. The moths emerge in June and July. Its caterpillars feed in July and August on bramble *(Rubus)*, hawthorn *(Crataegus)* and hazel *(Corylus)*, chiefly in wooded areas. **269**

**scripta** Gosse
Canada, N. U.S. 30–35 mm. (1·18–1·38 in.)
This is generally similar to the illustrated European *H. pyritoides* in colour pattern but differs in several details. Its caterpillars feed on various species of the family Rosaceae, such as blackberry and thimbleberry *(Rubus)*. **272**j

HACKBERRY BUTTERFLIES see **Asterocampa**
**Hadena lateritia** see **Apamea lateritia**
HADENA, RED-WINGED see **Apamea lateritia**

**Hadrodontes** Stoneham (NYMPHALIDAE)
A small genus of African butterflies very similar to *Charaxes* to which they are related.

**varanes** Cramer PEARL CHARAXES
South Africa to tropical Africa 78–100 mm. (3·07–3·94 in.)
This is a common species over much of its range and can be recognized, even in flight, by the broad white area at the base of the wings and the white body. The rest of the wings are reddish brown with darker marks. The underside is variably patterned, but always closely resembling a dead leaf, even to the 'mid rib'. This butterfly is found in forests or thick bush, where, when it settles, its underside camouflage makes it 'disappear'. The caterpillar feeds on *Allophyllus* and other species of Sapindaceae and Anacardiaceae.

**haemagrapha** see **Mazuca**

**Haetera** Fabricius (NYMPHALIDAE)
This genus of butterflies is in the subfamily Satyrinae. There are only 2 species at present known in this C. American genus. They are delicate transparent-winged butterflies unlike the typical Satyrids to which they are related.

**macleannania** Bates
Costa Rica, Paraguay 74 mm. (2·91 in.)
The female is illustrated; the male lacks the red on the hindwings and is smaller. The wings in both sexes are absolutely transparent. Many subspecies and forms of this have been described but little seems to be known about its biology. **174**a

HAG see **Phobetron pithecium**
**hahneli** see **Parides**
**hainesi** see **Epicopeia**
HAIRSTREAK BUTTERFLIES see **Atlides, Callophrys, Chlorostrymon, Chrysozephyrus, Erora, Iolaus, Panthiades, Quercusia, Strymon, Strymonidia, Thecla**
**halesus** see **Atlides**
**halia** see **Lycorea**
**halimede** see **Colotis**

**Halisidota** Hübner (ARCTIIDAE)
This is a large, exclusively New World genus represented by about a dozen species (some of them pests) in N. America and over 100 C. or S. American species, with probably many more yet to be described or discovered in the tropics. Adults typically feign death, when disturbed at rest, and drop to the ground where they remain motionless for some time and probably escape further investigation by a possible predator. The caterpillars have conspicuous tufts of hair, often of a contrasting colour.

**argentata** Packard SILVER-SPOTTED TIGER-MOTH
Canada, U.S. and probably N. Mexico 38–52 mm. (1·5–2·05 in.)
In Canada *H. argentata* is restricted almost entirely to coastal S. W. British Colombia, but is elsewhere generally distributed in coniferous forests. It is similar in appearance to the illustrated *H. caryae*, but the ground-colour of the forewings is much darker and the markings are white. The caterpillar, which bears tufts of brown or black irritant hairs, is gregarious in a web at first, but after hibernation it feeds singly on one of several coniferous trees, including Douglas Fir *(Pseudotsuga)* other firs *(Abies)* and various pines *(Pinus)*.

**caryae** Harris HICKORY TUSSOCK
E. Canada, U.S., C. America 36–50 mm. (1·42–1·97 in.)
The caterpillar of *H. caryae* is a common pest of hickory *(Carya)* in the U.S., Black Walnut *(Juglans)* in Canada, and of many other broad-leaved forest trees. It is covered with pale grey hairs and has a black tuft, or tussock, of hairs on the first and seventh segments of the abdomen and a white tuft on the eighth segment. **368**s

**harrisii** Walsh SYCAMORE TUSSOCK
U.S. 35–50 mm. (1·38–1·97 in.)
The exact distribution of this species is unknown as adults are extremely difficult to distinguish from *H. tessellaris* except by comparison of the moth genitalia. The caterpillars are much easier to identify; they are nearly white, with orange and white hair-tufts, and feed, apparently exclusively, on American Sycamore *(Platanus)*.

**schausi** Rothschild SOUTHERN TUSSOCK
S. U.S.A. to Peru 38–56 mm. (1·52–2·24 in.)
This species seems fairly common in much of its range. Moths have been caught in most months of the year, from sea-level up to 2200 m. (10400 ft). The caterpillar is hairy, with conspicuous tufts of hairs and is known to feed on hackberry (*Celtis*) foliage and on a species of *Cestrum*, a member of the nightshade family. This is one of many species whose adults are attracted to dried *Eupatorium* plants. **368**r.

**Hamadryas** Hübner (NYMPHALIDAE) CRACKER or CALICO BUTTERFLIES
This is an American genus of butterflies which has many species in S. America, a few reaching C. America and the W. Indian islands and just into the S. U.S. The butterflies of this genus generally rest head downwards on tree trunks with their wings spread out. A mechanism on the wing enables them to make clicking noises in flight, hence one of their popular names of 'Crackers'. There are descriptions of larvae of butterflies in this genus but little information about their foodplant.

**arethusia** Cramer QUEEN CRACKER
Mexico, America, W. Indies 65 mm. (2·56 in.)
Velvet black with iridescent blue spots on the wing in the male. The female has more blue spots and a whitish band across the centre of the forewing. When they fly they make a crackling noise with the wings, particularly when several are flying together. The undersides are camouflaged and when they settle on a tree they are very hard to see. **209**b

**belladona** Bates
Brazil 63–66 mm. (2·48–2·6 in.)
The hindwing on the underside is patterned rather like the Monarch butterfly, although they are not related. The orange-brown on the hindwing extends nearly to the margin, with black veins and a few blue spots on the black wing margin. The forewing is dark, with white spots and a trace of orange-brown near the base of the wing.

**feronia** Linnaeus CRACKER
S. America and into the W. Indies 55–64 mm. (2·17–2·52 in.)
The mottled bluish grey with white patches and black eyespots with white centres gives an interesting pattern. Males and females are similar externally. The species can be separated from the rather similar Grey Cracker *(H. ferentina* Godart) by the lack of the reddish rings round the eyespots on the hindwings. **209**a

**fornax** Hübner YELLOW SKIRTED CALICO
C. America, as strays into the S. U.S. 40–45 mm. (1·57–1·77 in.)
The general colour is bluish grey with zig-zag cross lines of black and greyish brown. The hindwing has large white patches. Nothing seems to be known of its biology.

**velutina** Bates
S. America 68–75 mm. (2·68–2·95 in.)
The underside is dark blue with red spots around the posterior margin of the hindwing and at the base of the anterior margin of the hindwing. The females are slightly larger and have a white bar across the forewing below the apex. The markings on the wing in this species are generally smaller than in related species. **209**j

**hamana** see **Agapeta**

**Hamanumidia** Hübner (NYMPHALIDAE)
This is a genus of African butterflies with only one included species. There are a number of forms described of this species which is widespread of S. Sahara.

**daedalus** Fabricius GUINEA FOWL BUTTERFLY
South Africa, N. to the tropical region 50–64 mm. (1·97–2·52 in.)
The leaden grey with rows of white dots and rather rounded forewing give this a very different appearance from most African butterflies. Its popular name was given because of its similarity to the Guinea Fowl's plumage of grey and white speckles. The underside is orange-brown with rows of grey and white dots. There are several seasonal forms of this species. This is a common bush veldt species which, when at rest, spreads the wings out flat. The caterpillar, which is covered with long feathery spines, feeds on species of *Combretum* (Combretaceae).

**hamata** see **Danaus**

**Hamearis** Hübner (NEMEOBIIDAE)
There is a single known species, *H. lucina*. The name *Hamearis* was incorrectly applied, between 1875 and 1934, to a group of S. American species for which Hemming, a British lepidopterist and biographer of Hübner, erected the genus *Audre* in 1934.

**lucina** Linnaeus DUKE OF BURGUNDY FRITILLARY
Europe including S. England, as far E. as the Balkans 28–33 mm. (1·1–1·3 in.)
This butterfly is the only European representative of an essentially tropical family. It is a species of sunlit woodland clearings, and also of hilly scrubland in S. and C. Europe. The caterpillar feeds on cowslips and primroses *(Primula)*. There are 2 broods in continental Europe; only one in England, where the adults fly in May and June. It hibernates as a chrysalis. **267**

HAND-MAID MOTHS see **Datana**
HANDKERCHIEF see **Phyciodes ianthe**
**Haploa** see under **Callimorpha**
**haquinus** see **Taxila**
**hardwickei** see **Parnassius**
HARLEQUIN see **Taxila haquinus**
**harmodius** see **Eurytides**
**harmonia** see **Tithorea**

**Harpeptila** Diakonoff (YPONOMEUTIDAE)
A small genus of micro-moths described in 1967 in the subfamily Plutellinae. It has the general shape of *Ypsolopha* but has different venation in the wings. Nothing is known of its biology. This is one of the many genera which have been recognized in recent years as more intensive studies are made of collections. Many of the genera which are newly described are based on specimens collected perhaps 100 or more years ago, but which have been little studied or overlooked because of their similarity to commoner species.

**prasina** Diakonoff
Philippines 24 mm. (0·94 in.)
The forewings are glossy white and densely marked with olive, yellow and dark brown near the apex. They are rather pointed. This moth is known only from a single specimen which was described in 1967. Nothing is known of its life history.

**harrisi** see **Euploea, Halisidota**

**Harrisimemna** Grote (NOCTUIDAE)
Only two species are known; one N. American, the other Japanese.

**trisignata** Walker HARRIS'S THREE-SPOT
Canada, U.S. 30–38 mm. (1·18–1·5 in.)
*H. trisignata* is found east of the Rockies southwards from Canada to Texas. Although cryptic in its natural surroundings, the colour pattern is decidedly beautiful. Its caterpillars feed chiefly on winterberry *(Ilex)* and lilac *(Syringa)* foliage; when disturbed they raise the front part of the body, and sway from side to side with a quivering motion. The chrysalis is formed in a tunnel inside dead wood, the entrance to its retreat having been sealed with silk by the caterpillar. **391**p

HARRIS'S THREE-SPOT MOTH see **Harrisimemna trisignata**
HARVESTER see **Feniseca tarquinius**
**hastata** see **Rheumaptera**
**hastiana** see **Acleris**
HAWK-MOTH see **Acherontia, Agrius, Antinephele, Chaerocina, Cizara, Daphnis, Deilephila, Euchloron, Hemaris, Hippotion, Hyles, Laothoe, Leptoclanis, Macroglossum,**

**Mimas, Nephele, Polyptychus, Smerinthus, Sphinx, Xenosphingia**
HAWKLET, BEDSTRAW see **Microsphinx pumilum**
HEATH BUTTERFLIES see **Coenonympha**
HEATH FRITILLARY see **Mellicta athalia**
HEATH, LATTICED see **Semiothisa clathrata**
HEATH MOTH, COMMON see **Ematurga atomaria**

**Hebomoia** Hübner (PIERIDAE)
A genus of butterflies with 2 species from India, S.E. Asia to Japan and one described in 1936 from Brazil (*H. dinizi* de Toledo Piza). One of the species, *H. glaucippe* which is very widespread in the Orient, has nearly 50 different names applied to it. It has a large number of subspecies and forms.

**glaucippe** Linnaeus GREAT ORANGE-TIP
China, India, Pakistan, Malaysia, Japan 70–100 mm. (2·76–3·94 in.)
This is the largest Pierid butterfly in Asia. The upper surface of the wings are white, the apex of the forewing has a large orange triangle, outlined in black. The female has a row of spots along the wing margin. Underneath the wings the apex of the forewing is a brown mottled colour while the entire hindwing is mottled brown giving a very leaf-like appearance. This species, which generally has 2 generations per year, is common in most of its range, occasionally visiting hilly country. The wet season form is generally larger than the dry season form. The caterpillar feeds on species of *Capparis* (Capparidaceae). The female butterfly tends to stick to the forest areas while the males are more conspicuous over open country as well. Because of their more secretive habits the females tend to be rarer in collections. There are several subspecies of *H. glaucippe* in different parts of its range, in Japan the subspecies is *H. g. liukiuensis* Fruhstorfer.

**hebrus** see **Menander**

**Hecatesia** Boisduval (AGARISTIDAE)
WHISTLING MOTHS
This is a small genus of Australia moths, the males of which are capable of producing a Cicada-like sound from their wings, see *H. fenestrata*. Other Agaristids can produce sounds by using different methods (see *Platagarista* and *Musurgina*).

**fenestrata** Boisduval WHISTLING MOTH
E. Australia 24–29 mm. (0·94–1·14 in.)
The day-flying males of this common species produce a loud whistling noise. The sound is produced during flight from a specialized, ribbed, concave area on the forewing which is distorted when the forewings meet at the top of each wing-beat. Cicadas, and the native Australian 'wobbleboard' produce sound in much the same way, by distortion of a convex surface. **402**d

**hecla** see **Colias**
**hector** see **Pachliopta**
HEDGE BLUE BUTTERFLIES see **Lycaenopsis**

**Hednota** Meyrick (PYRALIDAE) WEBWORM MOTHS
These Pyralid moths, typical in appearance and shape of the subfamily Crambinae, to which they belong, are serious pests in pastures in Australia. The damage they do is similar to that of species of *Crambus* and *Pediasia*. At present the genus contains mainly species from Australia.

**panteucha** Meyrick WEBWORM MOTH
Australia 21–27 mm. (0·83–1·06 in.)
This is a serious pest in pastures and cereal crops. The name is derived from the habits of the caterpillars which use a silken webbing to line their burrows in the soil and the tubes of silk they construct among the food plants. The moths have a life cycle, from egg to adult, which takes one year. Although there are several species which are pests, *H. panteucha* is the predominant one in Western Australia. The adult moth is a yellow-brown with dark marks between the veins, and a white line near the margin of the forewing. The hindwings are unmarked and whitish. The two long palps, which all the Crambine species have, protrude straight out in front of the head like a beak.

**Helcyra** Felder (NYMPHALIDAE)
A genus of butterflies with only 2 included species. Both of these have a number of named forms and subspecies in different localities in their wide range from India, S.E. Asia and China.

**hemina** Hewitson WHITE EMPEROR
India, Pakistan, Burma, Indonesia, Formosa, China 65–75 mm. (2·56–2·95 in.)
A silvery white butterfly with a large black apex to the forewing with an irregular inner edge. There are 2 black spots, one near the front margin, the other near the hind margin. The hindwing is silvery white with a row of irregular black spots below the margin and a continuous, slightly irregular, line between these spots and the margin. The underneath is silvery colour. The specimens from China come from the W. and are regarded as a distinct subspecies.

HELEN, RED see **Papilio helenus**
**helena** see **Troides**
**helenus** see **Papilio**
**heliconides** see **Anthomyza** and under **Chorinea sylphina**

**Heliconiinae** (NYMPHALIDAE)
The species of this subfamily of Nymphalidae are restricted to C. America, S. America and the U.S. It is largely a tropical group, however, and very few species venture into subtropical and temperate U.S. The caterpillars are clothed in long, branching spines; nearly all of them feed on species of poisonous Passifloraceae (passion flowers). The adult butterflies are distasteful to predators, especially birds, and advertise their unpalatability with a multitude of bright colours and striking patterns. Numerous species are mimicked by quite edible species of other groups of butterflies and moths, and there are many mimetic associations between Heliconiine species and equally unpalatable species of Danainae, Ithomiinae, Acraeinae and the day-flying moth group Pericopidae (a synonym of Hypsidae, in part). Many species of Heliconiinae have several different colour forms, each of which may exactly copy species of other genera or even species of its own genus occurring in the same locality. *Heliconius erato*, for example, has no fewer than 30 different colour forms and is popular with geneticists and other evolutionary biologists as an experimental insect. See *Agraulis, Dione, Dryadula, Dryas, Eueides, Heliconius, Philaethria, Podotricha*.

**heliconioides** see **Eueides**

**Heliconisa** Walker (SATURNIIDAE)
A single species of moth is placed in this genus.

**pagenstecheri** Geyer
S. Brazil, Argentina 65–85 mm. (2·56–3·35 in.)
This is a species of grassland areas where the caterpillars' foodplants include giant grasses of the genus *Paspalum*. Fully grown caterpillars are brown and are protected by numerous branched spines; they feed at night and rest during the day at the base of the grass. Adult males, and possibly females, are day-fliers, and apparently difficult to capture. In the 1920s a Minister of Agriculture at Buenos Aires described the chasing of *H. pagenstecheri* on horseback as 'the most exciting sport imaginable', and to be compared only to pig-sticking or polo. **337**d

**Heliconius** Kluk (NYMPHALIDAE)
A few species of this well known genus of the subfamily Heliconiinae have been recorded from S.E. Texas, and one species is a resident in Florida, but the genus is essentially C. American and tropical S. American in distribution. Several species have many distinctively different colour forms which are matched in other species of *Heliconius*, by species of Ithomiinae and Danainae and by a few moths of the family Hypsidae. These associations of unpalatable species are mimicked in turn by supposedly edible Pieridae, Nymphalinae and Castniidae species. The caterpillars of *Heliconius* feed almost exclusively on the foliage of passion flowers (Passifloraceae). See also under *Eueides*.

**anderida** Hewitson
C. & N.W. S. America 85–94 mm. (3·35–3·7 in.)
The illustrated specimen of the subspecies *H. zuleika* is similar in pattern to several species of the subfamily Ithomiinae with which it is mimetically associated. Recent work shows that this butterfly is probably a form of *H. ethillus*. **142**g

**antiochus** Linnaeus
Tropical S. America 65–74 mm. (2·56–2·91 in.)
The wings of this butterfly are black, with 2 oblique white or yellow bands across the forewing. The yellow basal streak on the forewing and the broad yellow band at the front edge of the hindwing are either present or absent.

**aristionus** Hewitson
Tropical S. America 70–85 mm. (2·76–3·35 in.)
This species is very similar in pattern to *Tithorea harmonia* (illustrated) which it mimics, together with several other species. The orange band at the posterior margin of the hindwing is, however, replaced by a ray-like row of orange streaks in some forms of *H. aristionus*.

**astraea** Staudinger
Brazil 80–85 mm. (3·15–3·35 in.)
There are few specimens in European collections of this orange, yellow and black species. It is quite closely related to the more common *H. doris*. The illustrated specimen is an example of the subspecies *H. astraea rondonia*; it is feeding from a flower of *Passiflora glandulosa*, a plant whose leaves provide food for the caterpillar. Red flowers are particularly attractive to the adult butterfly, which will investigate almost any red object. Some authorities consider this butterfly to be a form of *H. egerius*. **141**

**atthis** Doubleday
N.W. S. America 68–76 mm. (2·68–2·99 in.)
This species is a mimetic partner of *Elzunia pavonii*, a species of the subfamily Ithomiinae. Nothing appears to be known about its life history.

**charitonius** Linnaeus ZEBRA
U.S., C. & N.W. S. America 76–86 mm. (2·99–3·39 in.)
The range of this mainly tropical species extends northwards as far as S. Carolina but it is common in N. America only in Florida and Texas. It is the only species to breed regularly in large numbers in the U.S., and elsewhere in its range it is often the commonest species of its genus. Like many other species of *Heliconius*, it flies in and near dense forests. Unlike most other *Heliconius* species, there are no close copies of its colour pattern in the many forms of other species of its genus. The illustrated specimen from Jamaica represents the subspecies *H. charitonius simulator*. **139**

**chrysonymus** see under **H. ricini**

**doris** Linnaeus DORIS BUTTERFLY
C. & tropical S. America 80–90 mm. (3·15–3·54 in.)
This colourful and variable species is a butterfly of open country. There are 3 main colour forms; one with orange and black hindwings, another with green and black hindwings, and a third (not illustrated) with blue and black hindwings. The flight of *H. doris* is not particularly strong and it is an easy species to capture. Several other species of *Heliconius* have similar colour forms to those of *H. doris* which makes identification of the various species extremely difficult. **142**c, f

**eanes** see under **Enicospila marcus**

**erato** Linnaeus SMALL POSTMAN
C. & tropical S. America 55–82 mm. (2·17–3·23 in.)
This is an unusually variable species, both geographically and sometimes also individually. The illustrations show some of these colour varieties. Almost every colour form of *H. erato* is matched in the similarly distributed *H. melpomene*. Like several other species of *Heliconius*, it rests at night in communal roosts, illustration **136**, often together with other species. Marked individuals of both sexes have been found to return to exactly the same roost each night, although at times it may be migratory (for example at Rancho Grande, Venezuela). The life expectation of Trinidad specimens averages about 50 days, a longevity which may

be correlated with their unpalatability. The solitary, notoriously cannibalistic caterpillar feeds on several species of *Passiflora* and *Tetrastylis* (Passifloraceae). See also under *Heliconiinae*. Two specimens are illustrated to give an idea of the variability of the species. **135, 136, 145**

**ethillus** Godart RARE TIGER
C. & tropical S. America 60–82 mm. (2·36–3·23 in.)
There are very many known colour forms of this species, one of which is figured; this form may occur with much less yellow in the pattern. They are distinctly Ithomiine in pattern and are probably a mimetic partner of certain Ithomiines. It appears to prefer forests at medium elevations and is uncommon in the lowlands. See also *H. anderida*. **142**a

**hermathena** Hewitson
Tropical S. America 57–85 mm. (2·24–3·35 in.)
This butterfly is apparently restricted to pockets of dried shrub in the Amazon Basin, an unusual habitat for a species of *Heliconius* which is an essentially forest-dwelling genus. **137**

**hortense** see under **H. ricini**

**melpomene** Linnaeus POSTMAN
C. & tropical S. America 62–84 mm. (2·44–3·31 in.)
Many forms of this species are hardly distinguishable from the corresponding forms of its more common relative, *H. erato*. Red flowers, as a source of nectar, are particularly attractive to this butterfly. The solitary caterpillar has been recorded from several species of passion flowers (Passifloraceae). *H. melpomene*. One of the illustrated specimens represents the subspecies *H. melpomene cytherea*, **147**. The other is a rather complicated laboratory cross between a specimen of *H. melpomene thelxiope* and the results of a cross between *H. melpomene melpomene* and *H. melpomene thelxiope*. **146**. Interbreeding of this sort between subspecies and species of *Heliconius* occurs regularly in natural conditions. **146, 147**

**metharme** Erichson
C. & tropical S. America 64–80 mm. (2·52–3·15 in.)
The wings of this rather unusually marked species are black, with a yellow apical bar and a yellow, or yellow and white, central patch on the forewing, and with a row of parallel, blue and white streaks at the outer margin of the hindwing. The strikingly coloured under surface has orange, ray-like markings on the basal half of the hindwing, and a narrow, yellow streak at the base of the forewing, in addition to the markings in common with those on the upper surface.

**nattereri** Felder
Brazil 70–84 mm. (2·76–3·31 in.)
This rare butterfly is confined to areas of virgin rainforest in E. Brazil but is sadly in the process of extinction as these forests are felled. It prefers steep hillslopes and is particularly attracted by blue, magenta and red flowers. It is the only known species of *Heliconius* to exhibit sexual dimorphism – the males are black and yellow, the females black, yellow and orange; both sexes having red dots at the base of the wings. The female is a partner in a mimetic complex which includes several species of *Heliconius* and Ithomiine species of the genera *Mechanitis*, *Hypothyris*, and *Napeogenes*. Mimics of this group include the Pierid *Dismorphia astyocha* and a colour form of the Nymphaline species *Phyciodes lansdorfi*. The solitary caterpillars feed exclusively on *Tetrastylis ovalis* (Passifloraceae). **132**

**numatus** Cramer
C. & tropical S. America 80–100 mm. (3·15–3·94 in.)
Together with several other species (see *Melinaea lilis*), *H. numatus* forms an association of mimetic butterflies, all of which advertise their noxious qualities to potential predators with almost identical colour patterns. This is one of the largest species of its genus.

**ricini** Linnaeus
Amazon Basin & N. S. America 54–70 mm. (2·13–2·76 in.)
*H. ricini* closely matches forms of the much larger species *H. hortense* and *H. clysonymus* in colour pattern except for the presence of a yellow, apical band on the forewing in addition to the yellow, central band. All 3 species fly together in a typical mimetic partnership. The caterpillar feeds on *Passiflora laurifolia*. This species was placed in the genus *Eueides* until fairly recently. **142**b

**sapho** Drury
C. & N.W. S. America
The specimen shown in the illustration of this species is an example of *H. sapho* form *primulans*. In other forms, the 2 pale yellow bars on the forewing are replaced by a single white bar, and the yellow, marginal band on the hindwing by a white band. In a few specimens the hindwing is unmarked. **144**

**sara** Fabricius SMALL BLUE GRECIAN
C. & tropical S. America 48–56 mm. (1·89–2·2 in.)
In colour pattern, this butterfly closely resembles the blue form of *H. doris* and certain other species of *Heliconius*, but unlike *H. doris* it tends to be more rigidly restricted to the undergrowth of forests. The adults are known to roost overnight in groups of up to 40 individuals. The gregarious caterpillars of *H. sara* feed on species of *Passiflora* or *Tetrastylis* (Passifloraceae).

**telesiphe** Doubleday
N.W. S. America 64–88 mm. (2·52–3·46 in.)
This is a species of the high Andes; it normally flies above 1370 m. (4500 ft) and is the only species of *Heliconius* found above elevations of 2440 m. (8000 ft). The subspecies *H. telesiphe sotericus* flies together with the similarly patterned *Podotricha telesiphe*; it has normally rounded wings, however, like other species of *Heliconius* and lacks the indentations present in *P. telesiphe*. The wing markings of one form of this species are bluish white, not orange.

**vulcanus** Butler
W. Ecuador, W. Colombia 62–75 mm. (2·44–2·95 in.)
Few specimens of this butterfly exist in collections. The bluish white, marginal band on the hindwing is absent in some specimens. It is treated by some authorities as a form of *H. melpomene*. **142**h

**wallacei** Reakirt BLUE GRECIAN
Tropical S. America 57–82 mm. (2·24–3·23 in.)
The resemblance between this species and the Papilionid species *Eurytides pausianus*, at least in flight, suggests that the latter is a mimic of the presumably distasteful *H. wallacei*. Other Heliconiines similar to this species include certain colour forms of *H. sara* and *H. doris*. See also under *Eurytides pausanias*. **142**d

**Heliocopis** Fabricius (NEMEOBIIDAE)
There are currently only 3 species in this genus. Specimens of *Helicopsis* were amongst the first butterflies to attract the attention of the famous naturalists Bates and Wallace during their travels in S. America. Conspicuous, metallic silver or gold spots decorate the under surfaces of the wings of these butterflies.

**acis** Fabricius
N. S. America 40–50 mm. (1·57–1·97 in.)
The male (not illustrated) differs from the female in the brown base to the forewing. Caterpillars of *H. acis* are mostly white, covered chiefly with white hairs, but with a tuft of red hairs behind the head. **264**

**cupido** Linnaeus GOLD-DROP HELICOPIS
Trinidad, Venezuela, the Guyanas and Brazil 34–38 mm. (1·34–1·5 in.)
This is a remarkable, many-tailed butterfly which, like the rarer *H. endymion*, is apparently restricted to swamps where the caterpillar feeds on Wild Tania (*Montrichardia*). It differs above from *H. endymion* (illustrated) chiefly in the more extensive yellowish white area on the forewing which nearly reaches the outer margin. In *H. cupido* form *crotica*, the wings are mostly yellowish white above, with only marginal markings.

**endymion** Cramer SIX-TAILED HELICOPSIS
N. S. America, Trinidad 33–38 mm. (1·3–1·5 in.)
As the popular name of this rare species suggests, no fewer than 6 tails adorn the hindwing of this butterfly. The female (not illustrated) lacks the brown patch at the base of the forewing. *H. endymion* is found in swampy, coastal districts, where it forms loosely knit colonies. **263**t

**Helicoverpa** see **Heliothis**
HELICOPIS see **Helicopis**

**Heliodinidae**
This is a family of micro-moths with few genera or species. The species themselves are mostly very small. There is no real agreement on the species in the family, one of the characters often used is the position at rest of the hind legs. These are raised above the back, although this habit is often found in species of *Stathmopoda* (Stathmopodidae). See *Eretmocera*.

**Heliophorus** Geyer (LYCAENIDAE) SAPPHIRE BUTTERFLIES
The males of this genus are amongst the most brilliantly coloured species of Lycaenidae; females are largely brown. The 13 or so species are restricted to S.E. Asia.

**androcles** Doubleday GREEN SAPPHIRE
Kashmir, N. India, Burma 30–35 mm. (1·18–1·38 in.)
The male of this species is a brilliant, iridescent green and blue. The female (not illustrated) is brown above, with an orange bar on the forewing and a lunulate submarginal band on the hindwing. The under surface of both sexes is a contrasting yellow, marked with brown lines on the forewing and a brown line and an orange marginal band on the hindwing. *H. androcles* occurs between elevations of 1220–3660 m. (4000–12,000 ft) in the Himalayas. **263**g

**tamu** Kollar
Himalayan China 25–30 mm. (0·98–1·18 in.)
The upper surface coloration contrasts sharply with the illustrated under surface. The male of *H. tamu* is brown above, with an extensive iridescent bluish green, basal area on the forewing and a lunulate, scarlet band at the outer margin of the hindwing. The hindwing of the female is similar to that of the male, but the forewing is dark brown, with a short, oblique orange bar. **263**h

**helios** see **Hypermnestra**

**Heliothis** Hübner (NOCTUIDAE)
The caterpillars of this large, economically important genus feed on the flowers and fruits of their foodplants and cause immense damage to many crops. They occur extensively between 40°N. and 40°S. in both the Old and New Worlds. Many of the species included here are placed in the genus *Helicoverpa* by Dr D. Hardwick, the Canadian entomologist, and others. Control is difficult because insecticides have little effect on the caterpillars when they are feeding inside fruits.

**armigera** Hübner SCARCE BORDERED STRAW or OLD WORLD BOLLWORM
Africa, Europe, Asia, Australia 32–39 mm. (1·26–1·54 in.)
In England this is a chiefly migrant species, but possibly a resident in S. Devon. In Africa and Australia it is a notorious pest of cotton *(Gossypium)* and a lesser pest of corn cobs *(Zea)*. It is also a well-known pest of tomato *(Lycopersicon)*; the caterpillar feeds inside the fruit of the tomato and is occasionally canned with its food-source. Other foodplants include lucerne *(Medicago)* and linseed *(Linum)*; and in New Guinea cocoa *(Theobroma)* is sometimes attacked. A further odious characteristic of the caterpillars is their habit of eating one another when in crowded conditions.

**virescens** Fabricius TOBACCO BUDWORM
C. & S. America, U.S., Canada, Hawaii 28–36 mm. (1·1–1·42 in.)
In N. America this is a breeding species normally only in the S. and in California, although it occurs as far N. as New York State and Ontario. The forewings of *H. virescens* are dull green and are crossed by 3 nearly parallel, pale edged, greenish brown lines; the hindwings are yellowish white, often with reddish brown towards the outer margin. The caterpillar is a serious pest of cotton *(Gossypium)* and other crops in the S. U.S. Early broods seem to prefer tobacco *(Nicotiana)*; the later broods attack cotton, tomato *(Lycopersicon)* and peas of various genera (Leguminosae). Wild

tobacco, also a foodplant, may be the overwintering refuge in S. Texas. Female sex-scent traps have been most effective in attracting and exterminating males when the traps have been painted a fluorescent green.

**zea** Boddie CORN EARWORM, COTTON BOLLWORM or TOMATO FRUITWORM 35–42 mm.
N., C. & S. America, Hawaii (1·38–1·65 in.)
Apart from corn *(Zea)* and cotton *(Gossypium)*, the caterpillar of this infamous species will also attack tomatoes *(Lycopersicon)*, tobacco *(Nicotiana)*, *Sorghum*, melons *(Citrullus)* and other crops, and has had several other common names from time to time as a result. It has been recorded from Hawaii in recent years. The Aztecs of ancient Mexico found infested corn-cobs to their liking and apparently ate both corn and caterpillar with equal pleasure. **390**k

**Heliozelidae** SHIELD BEARERS
Small day-flying micro-moths with scales giving a metallic look to the wings. The caterpillars are leaf-miners, usually forming a blotch-mine in the leaf. In contrast to the neat spirals of some leaf-miners visible on the outside leaf, the leaf-mine of most Heliozelids appears as an irregular shaped patch. See *Antispila*.

**helliodes** see **Lycaena**

**Hellula** Guenée (PYRALIDAE)
These are rather broad-winged, pale coloured Pyralid moths with a brownish colour and pattern, often with a conspicuous rounded eyespot on the forewing. The genus is worldwide but was recently divided into 2 main groups, one exclusively American, the other worldwide with some American species.

**rogatalis** Hulst AMERICAN CABBAGE WEBWORM
U.S., Canada 13–17 mm. (0·51–0·67 in.)
This is a widespread pest of cabbages and other Cruciferae from N. Carolina to Florida, from Texas, S. California N. to Nova Scotia. There are two closely related and similar species which have previously been confused. Both are an olive-brown colour with a rather kidney shaped mark on the forewing.

**undalis** Fabricius CABBAGE WEBWORM
Europe including Britain, Africa, India, Hawaii 16 mm. (0·63 in.)
This widespread pest of cabbages *(Brassica oleracea)* and related plants is similar in appearance to the preceding species. The small caterpillars mine the leaves and stems and feed on the surface of the leaf as well. The caterpillar pupates in a loose cocoon with particles of soil mixed in with it. Although once considered worldwide in distribution, the specimen found in N. America and recorded under this name, is now known to have been *H. rogatalis*.

**Hemaris** Dalman (SPHINGIDAE) BEE HAWK-MOTHS
These are moths of the N. Hemisphere and are found throughout much of N. America, Europe, Asia and Mediterranean N. Africa. All the species resemble bumble-bees *(Bombus)* especially when in flight. The transparent areas of the wings are at first covered with scales, but these are lost during the first flight.

**fuciformis** Linnaeus
Europe including Britain, N. & C. Asia, N. Africa 38–43 mm. (1·5–1·69 in.)
This is a day-flying species which visits the blooms of honeysuckle *(Lonicera)*, *Rhododendron* and other plants in search of nectar, most commonly in the morning between ten and noon. It can be seen in May and June. Honeysuckle is a common foodplant of the caterpillar. It is pale green above, darker green along the sides, and brown beneath. The dark brown chrysalis overwinters. **341**

**thysbe** Fabricius HUMMING-BIRD or COMMON CLEARWING HAWK-MOTH
Canada, U.S. 45–62 mm. (1·77–2·44 in.)
*H. thysbe* is distributed over much of Canada, and in the U.S. from Alaska to Texas and Florida but excluding the S.W. Recorded foodplants of the caterpillar include cherry and plum *(Prunus)*, hawthorn *(Crataegus)*, honeysuckle *(Lonicera)* and snowberry *(Symphoricarpos)*. The adults fly by day and visit the blooms of numerous different plants. The head, thorax and front part of the abdomen of the adult are green; the rest of the abdomen is black or black with terminal area of green. The transparent areas of the wings are bordered with reddish brown except for the base of the forewing which is green.

**Hemerocampa** Dyar (LYMANTRIIDAE)
The females of this genus are either wingless or virtually so. The caterpillars of most species are general feeders and often reach pest proportions. About 18 species are known; they are found in N., C. and S. America.

**leucostigma** Smith WHITE-MARKED TUSSOCK
Canada E. of the Rockies, U.S. (male) 28–36 mm. (1·1–1·42 in.)
The female of *H. leucostigma* is a greyish white, wingless creature, which seldom lives long after depositing her eggs on the recently vacated cocoon. Males are rather similar to those of *Orgyia antiqua* (illustrated) in pattern but less reddish in colour. The brightly coloured, solitary caterpillar feeds on the foliage of most deciduous trees, and many coniferous trees, and is occasionally a pest of apple *(Malus)*, plum *(Prunus)* and other trees, both broad-leaved and coniferous. Balsam Fir *(Abies)* is the commonest foodplant in E. Canada, among over 60 other trees.

**pseudotsugata** McDunnough DOUGLAS FIR TUSSOCK
N. America (Rocky Mountains, from British Colombia to Colorado and Nevada) (male) 23–26 mm. (0·91–1·02 in.)
The caterpillar of this species is a major pest of Douglas Fir *(Pseudotsuga)* and other firs *(Abies)*. Other coniferous trees are attacked, but less frequently. The adult male is similar in colour pattern to the illustrated *Orgyia antiqua* but the forewings are less reddish in colour and have a better marked pattern. The female is wingless.

**Hemiargus** Hübner (LYCAENIDAE) EYED BLUE BUTTERFLIES
This genus is distributed throughout most of tropical C. and S. America and in the U.S. as far north as Nebraska. The species have 2 or 3 silvery eyespots on the under surface of the hindwings.

**dominica** Möschler
Jamaica, the Antilles 21 mm. (0·83 in.)
This small blue is not well represented in collections – possibly the result of collecting techniques rather than the degree of rarity in nature. The adult butterfly often feeds on flowers with other blues but is very easily disturbed and probably escapes the collectors net more frequently than other species.

**isola** Reakirt REAKIRT'S BLUE
W. U.S., S. to Costa Rica 19–27 mm. (0·75–1·06 in.)
This is a generally common butterfly of open spaces. Details of its life history are uncertain, but the caterpillars feed on mesquite *(Prosopis)* in Texas, and probably on many other species of Leguminosae. Males are purplish blue above. Females have a black spot near the base of the tail.

**thomasi** Clench MIAMI BLUE
U.S. (Florida and associated islands), the Bahamas 20–28 mm. (0·79–1·1 in.)
Although described only recently, this is a generally common butterfly in Florida where it is a frequent visitor to flowers, especially Spanish Needles *(Bidens)*. It is a typical 'blue' butterfly, with a brown outer margin, broader in the spring brood, and a few marginal spots on the hindwing.

**Hemileuca** Walker (SATURNIIDAE)
There are about 25 species in this N. American and Mexican genus of moths. Some species are day-fliers. The caterpillars are at first gregarious, later solitary; they have irritant spines which produce painful reactions in sensitive individuals.

**electra** Wright
U.S. (California. Arizona) 43–55 mm. (1·69–2·17 in.)
This colourful species flies by day, and is on the wing during September and November. Its caterpillar feeds on buckwheat *(Eriogonum)*. The illustrated moth is a specimen of subspecies *H. electra clio*. Males of *H. electra electra* are partly transparent and differ in details of the colour pattern. **331**a

**hera** Harris SAGEBRUSH SHEEP-MOTH
W. U.S. 54–65 mm. (2·13–2·56 in.)
This strikingly patterned species inhabits the mountains of W. U.S. from near the Canadian border southwards. The illustrated specimen, which was caught in Reno, Nevada, belongs to the subspecies *H. hera hera*. Some specimens have much broader black markings than in the illustration, others have these markings greatly reduced. Little is yet known about the life history, but sagebrush *(Artemisia)* is possibly the main foodplant. The moth flies in July and August. A specimen of *H. hera* had the distinction of being illustrated in Audubon's famous *Birds of America*. **331**d

**maia** Drury BUCK MOTH
Canada, U.S. 50–65 mm. (1·97–2·56 in.)
The common name of this species possibly refers to the fact that the moths fly at the time of year when deer hunting is permitted. *H. maia* occurs from Nova Scotia to Florida and westwards to New Mexico and Colorado. The ground-colour of the wings is a slightly transparent, dark brown, nearly black, with a single yellowish white band crossing both fore- and hindwings. The black abdomen terminates in a conspicuous brownish orange tail-tuft in the male. The gregarious dark brown and yellow caterpillars feed on oak *(Quercus)* and willow *(Salix)* foliage.

**tricolor** Packard TRICOLOR BUCK
U.S. (Arizona), New Mexico 40–65 mm. (1·57–2·56 in.)
*H. tricolor* has an unusually early flight period, from February to April. Females and most males have narrower white bands on the forewing than in the illustrated specimen. Its caterpillar feeds on palo verde *(Cercidium)*. **331**e

**hemina** see **Helcyra**
**hemionata** see **Xanthabraxes**
**hemizonalis** see **Jocara**
**hemon** see **Mithras**

**Henotesia** Butler (NYMPHALIDAE)
These small butterflies are in the subfamily Satyrinae. They are found in Africa and Madagascar where they favour the savannah bushveldt in warmer parts, especially near wet areas. There are probably over 50 species in the genus.

**simonsii** Butler
Africa 35–40 mm. (1·38–1·57 in.)
Dry and wet season forms of this species are known; it ranges from Malawi to Transvaal. The caterpillar feeds on *Panicum* (Gramineae). The butterfly is a pale sandy yellow colour, tinged with reddish brown at the wing margin, with 2 small spots on the hindwing and 2 larger spots on the forewing, the front one often being surrounded by an orange-brown ring. This is the dry season form; the wet season form is a dark brown. The butterfly can be found in all months of the year, generally in shady areas of long grass, particularly near streams.

**Hepialidae** SWIFT MOTHS
A worldwide family of moths with many species in Australia including some of the largest ones. In spite of the Hepialids being classified as micro-moths, some Australian species have wingspans of 180 mm. (7·09 in.). The family contains moths which are regarded as rather primitive and which have a number of basic differences from other lepidoptera. In most moths the fore- and hindwing are linked by a bristle (frenulum) but in Hepialids this is effected by a simple lobe (jugum) from the forewing overlapping the hindwing. Another difference is in the venation of the wings; most moths have differences between the veins in the fore- and hindwings but in Hepialids the venation of the fore- and hindwings is similar. The cater-

pillars tunnel into wood or feed on roots in the ground. In wood the tunnels may be up to 50 cm. (20 in.) long and cause serious damage to the trees attacked. The root feeders are often problems for the farmers. See *Abantiades, Charagia, Hepialus, Leto, Zelotypia.*

**Hepialodes** Guenée (THYRIDIDAE)
A genus of one, odd-looking species which is found in S. America. This is one of the Thyridid moths whose general appearance is leaf-like. They are believed to live in forests but no observations have been published on their biology.

**follicula** Guenée
S. America 70–85 mm. (2·76–3·35 in.)
This species is known from Brazil but no other information is available about it. As a leaf-mimic, it certainly is as well adapted as the more famous Orange Oak or Indian Leaf butterfly *(Kallima)*. **51**a

**Hepialus** Fabricius (HEPIALIDAE) SWIFT MOTHS
This is a large worldwide genus. Many of the caterpillars feed on roots of plants and are likely to be pests. The related genera are particularly well-developed in Australia. The genus is in the group known as micro-moths.

**fusconebulosa** De Geer MAP-WINGED SWIFT
Europe including Britain 30–48 mm. (1·18–1·89 in.)
This has had several popular names in British entomological literature including the 'Beautiful Swift' and the 'Northern Swift'. The caterpillar which is a greyish white with brown marks and pale raised dots feeds on the underground parts of bracken *(Pteridium)*. The moth is out in June and July and flies at dusk. It is common throughout its range. **10**

**gracilis** Grote
U.S., Canada 30–40 mm. (1·18–1·57 in.)
The wing shape is typical of the genus with the fore- and hindwings virtually similar in their venation. The brownish grey, mottled forewing has the characteristic jugum (lobe) which interlocks with the hindwing and enables them to beat together. There are several colour varieties, one of the Canadian ones being smaller and darker. The species flies by day and at dusk. The caterpillars are believed to bore into ferns.

**humuli** Linnaeus GHOST SWIFT
Europe including Britain, W. Asia 44–64 mm. (1·73–2·52 in.)
The Ghost Swift or Ghost Moth gets its name from the male which is all white with rather shiny white wings. The females have dull yellow wings with some orange-brown markings. The Ghost Swift is common in meadow lands where the caterpillars feed on the roots of grasses (Gramineae) and hops *(Humulus)*. The males generally fly to and fro a few feet above the ground, looking like puppets on strings. Their flight begins at dusk and with the constant appearance and disappearance of the white, they attract the females. This is the reverse of the usual process in moths where males generally assemble to the females. The moth flies in June and July. In the Shetland Islands there is a subspecies with a non-white male. It has been suggested that as the June nights are not dark there, there is no need for the dazzling white colour to attract the females.

**hera** see **Hemileuca**
HERALD see **Scoliopteryx libatrix**
**hercules** see **Coscinocera**

**Herdonia** Walker (THYRIDIDAE)
A small genus of Thyridid moths with species in S. America and the Orient. It is probable, when the species are more closely examined, that the Oriental species will be placed in a different genus from the American species. At present the genus contains species which look broadly similar in external appearance.

**osacesalis** Walker
New Guinea, Indonesia, Burma, Malaysia 35–50 mm. (1·38–1·97 in.)
This remarkable looking species is probably a complex of several related species but for the present all those with the similar pattern are placed under the one specific name. Nothing is known of the biology of the species or the significance of the odd pattern. **56**x

**heritsia** see **Phylaria**
**herla** see **Eurema**

**Hermathena** Hewitson (NEMEOBIIDAE)
There is one species in this butterfly genus.

**candidata** Hewitson WHITE ERYCID
Tropical Central & S. America 40–42 mm. (1·57–1·65 in.)
This unusually coloured and rare species has an equally unusual slow flight. Its total resemblance to a species of Pieridae (Whites and Sulphurs) is very close. **263**cc

**hermathena** see **Heliconius**
**herminia** see **Cymothoe**
HERMIT BUTTERFLY see **Chazara briseis**
**heroldella** see **Swammerdamia**
**Herpaenia** see **Pinacopteryx**

**Hesperia** Fabricius (HESPERIIDAE)
A genus of skipper butterflies found throughout the N. hemisphere.

**leonardus** Harris LEONARDUS SKIPPER
Canada, U.S. 28–36 mm. (1·1–1·42 in.)
This is a common species in open field and damp meadows in August and September. The orange-brown upperside has darker brown on the wing margins with only a small patch of orange-brown on the hindwing. The underside has red on the hindwing with cream coloured spots. The caterpillar feeds on various grasses (Gramineae). Subspecies have been described from Ohio; the nominate subspecies is found from S. Maine and Ontario to Florida, Missouri and Kansas.

**Hesperiidae** SKIPPERS, AWL OF DUSKY-WINGS
A worldwide group, easily recognized on most continents by their large head, thorax and abdomen with relatively short wings. The antennae are often bent backwards at the apex giving a slightly hooked appearance. They fly fast, in a darting manner. The name skippers alludes to their flight. The caterpillars are mostly external feeders on grass (Gramineae). *Abantis, Alloea, Argopteron, Carterocephalus, Epargyreus, Erynnis, Euschemon, Gangara, Leucochitonea, Metisella, Mimoniathes, Muschampia, Ochlodes, Oreisplanus, Pholisera, Platysches, Pyrgus, Spialia, Tagiades, Thorybes, Thymelicus, Urbanus, Zophopetes.*

**hesperistis** see **Apoprogenes**
**hesperus** see **Papilio**

**Hestinalis** Bryk: syn. *Hestina, Diagora* (NYMPHALIDAE) SIREN BUTTERFLIES
This genus of butterflies, formerly known as *Hestina* was given a new name because *Hestina* had been used earlier for another genus of animals. Under the rules covering the names of animals (the International Code of Zoological Nomenclature), the newer name, when discovered to be the same as an earlier one in the same or any other group of animals must be replaced. This is a small genus with African and some temperate Asian species.

**assimilis** Linnaeus
China, Korea 70–100 mm. (2·76–3·94 in.)
There is some variation in the thickness of the black veins on the wings, which in turn affects the amount of yellow-white showing. The underside is similar to the upperside. The specimen illustrated is a male; the females generally have more yellow between the vein and are larger. Some related species have no red spots in the hindwing and much thinner black lines along the veins. Others have the yellow replaced by white between the veins. It is altogether a fairly variable species about which little seems to be known. **236**a

**nama** Doubleday CIRCE
India, Pakistan, Burma 95–105 mm. (3·74–4·13 in.)
This butterfly is a good mimic of the Chestnut Tiger *(Danaus tytia)* but the pale markings between the dark veins are thinner. Both sexes mimic the *Danaus* and the female has adopted the same flight and may be seen soaring about in open spaces just like the Chestnut Tiger. The males feed on overripe fruit and visit damp sand. The butterflies are about for most of the year in parts of their range in the Himalayas.

**persimilis** Westwood SIREN
India 65–75 mm. (2·56–2·95 in.)
This butterfly is creamy white in between black, heavily marked, veins. It is found in the Himalayas up to about 2290 m. (7500 ft). This species is regarded as possibly a mimic of a blue *Danaus*. The caterpillar feeds on species of *Celtis* (Ulmaceae).

**Heterogenea** Knoch (LIMACODIDAE)
This genus of moths is found in Europe and temperate Asia. There is one species in N. America. The caterpillars are generally rather slug-like.

**asella** Denis & Schiffermüller TRIANGLE MOTH
Europe including Britain, Asia into Japan 15–20 mm. (0·59–0·79 in.)
This is one of the only 2 species of the family Limacodidae which occurs in Britain. It is a rather drab brown or blackish brown moth with rather triangular shaped forewings (hence its name). The caterpillar is rather like a woodlouse *(Oniscus)*, green with a reddish band on the back that broadens in the middle of the back. It feeds on the leaves of beech *(Fagus)* or oak *(Quercus)* and is recorded on other trees as well. It is not common anywhere over its range and is very local in S. and C. Britain. It is reported as having been introduced into N. America although this may be a misidentification of a similar looking local species.

**Heterogymna** Meyrick (CARPOSINIDAE)
Species of this genus of micro-moths are found in Australia, India and a few in the islands of the E. Indies. There are possibly less than 20 species known at present, mostly similar in colour.

**pardalota** Meyrick
N. India 25 mm. (0·98 in.)
This is a silvery white moth with conspicuous black spots on the upperside of the wings. The undersides are plain smoky grey. There are related, similar looking species from India, similarly patterned but with more black on the forewing. There is no information on the life history of these moths. **47**a

**Heterogynidae**
A family of micro-moths found in Europe and N. Africa. There is still no real measure of agreement as to their exact relationships with other families in the Lepidoptera. At present they are regarded as part of the Zygaenoid complex. See *Heterogynis.*

**Heterogynis** Rambur (HETEROGYNIDAE)
A small genus of moths, currently placed in the major grouping Zygaenoidea, although formerly considered related to the Psychidae. There are few species in the genus, mostly from Europe and N. Africa.

**pennella** Linnaeus
Spain, S. France 21 mm. (0·83 in.)
The caterpillar is short, wood-louse *(Oniscus)* shaped, pale yellow with 2 dorsal bands. It feeds on *Genista* (Leguminosae). The moth is out in July in S. Europe, where it is common. **54**a

**Heteropsyche** Perkins (EPIPYROPIDAE)
A small genus of moths found in Australia whose species have a peculiar life history. The caterpillars, instead of being plant feeders, feed externally on other insects. They can be found on Jassid and Fulgorid leaf-hoppers (Hemiptera-Homoptera) where they attach themselves externally to the leaf-hopper, feeding on the wax which covers these insects. The eggs are laid in groups on the foodplant of the leaf-hoppers. The newly hatched caterpillar searches for a leaf-hopper, to which it attaches itself. The effect on the host varies and in some cases the caterpillar eventually kills the leaf-hopper. There is still some dispute on the exact mode of feeding of the caterpillar. It has been suggested that it may actually feed on the leaf-hopper rather than only on the wax which surrounds it.

**melanochroma** Perkins
Australia, New Guinea 10 mm. (0·39 in.)
This is a small black moth whose caterpillar feeds externally on species of leaf-hoppers. There is some evidence that this species may be parthenogenic. Sometimes more than one caterpillar may attach itself to one host leaf-hopper but only one will complete its development and pupate.

**hewitsonii** see **Nessaea**
**hexadactyla** see **Alucita**
**hezia** see **Callithomia**

**Hibrildes** Druce (Pterothysanidae)
This is a small genus of 2 species. They are either white and grey; white, grey and orange; or grey and orange. *H. norax* is a mimic of an Acraeine butterfly.

**norax** Druce
Malawi, Zambia and Mozambique 50–65 mm. (1·97–2·56 in.)
*H. norax* form *neavi* (illustrated) is a good copy of the butterfly *Acraea anemosa*, and *H. norax* form *crawshayi* an equally good copy of *Acraea natalica* (Nymphalidae, Acraeinae). The resemblance of the moth mimics to the species of *Acraea* is particularly good when the moths are at rest with their wings placed together over the back. **272**b

HICKORY HORNED DEVIL see under **Citheronia regalis**
HICKORY TUSSOCK see **Halisidota caryae**
**hiemalis** see **Neomyrina**
**hieroglyphica** see **Baorisa, Diphthera festiva**

**Hieroxestidae**
A small family of micro-moths, related to the Lyonetidae. Formerly they were included in this subfamily. (See also under *Oinophila.*)

**hierta** see **Junonia**
HIGH BROWN FRITILLARY see **Fabriciana adippe**
**hilarata** see **Parasa**
**hilarella** see **Ethmia**

**Hilarographa** Zeller (Glyphipterigidae)
This is a worldwide genus of micro-moths. Most of the species are found in S. America with one or two in N. America. One species is found in Japan, others occur in North India and Sri Lanka to New Guinea. Species of this genus have also been described from Madagascar. There is some doubt as to the accuracy of some of these descriptions, whether in fact the species described really is in the genus *Hilarographa*. Thus the real extent of this genus will not be known until there is further research on it.

**excellens** Pagenstecher
New Guinea, Bismarck Archipelago 15 mm. (0·59 in.)
Several other related species occur in New Guinea. All have the conspicuous metallic wing colours, with strong iridescence and general orange colour to the wings. Nothing is known at present of their biology. **34**f

HILL JEZEBEL see **Delias belladona**
HILL-SIDE BROWN, SMALL see **Pseudonympha narycia**
**himala** see **Epiplema**
HIMALAYAN JESTER see **Symbrenthia hypselis**
**Himantopteridae** see under **Himantopterus**

**Himantopterus** Wesmael (Zygaenidae)
A small genus of moths from Burma and N. India characterized by the extremely long and narrow hindwings. The caterpillars are said to live in association with termites (white ants, *Isoptera*) in their nests. The moth resembles an insect in another Order *(Neuroptera)* although the significance of this is not known. When disturbed the moths fall to the ground and pretend to be dead to avoid detection. They are weak fliers with their long hindwings trailing behind. This genus is considered as forming a distinct family, Himantopteridae, by some specialists.

**dohertyi** Elwes
India, Malaya 20–25 mm. (0·79–0·98 in.)
Each hindwing is as long as the total forewing span in this curiously shaped moth. There are many others with the hindwings modified into long tails. The specimen illustrated is subspecies *H. d. elwesi* Jordan. Many of the other related species are rather similar. **46**c

**fuscinervis** Westwood
Java, Sumatra 30–35 mm. (1·18–1·38 in.)
In this moth each hindwing is larger than the full wingspan. Nothing is known of the life history of this species. The sexes are similar in external appearance, unlike some moths where the males and females are very different (dimorphic). **46**b

**hintza** see **Zintha**
HINTZA BLUE see **Zintha hintza**

**Hipparchia** Fabricius: syn. *Eumenis* (Nymphalidae)
This is a large genus of butterflies with species in Europe and temperate Asia. The species are often very variable in pattern and many varieties have been named. The genus is in the subfamily Satyrinae.

**fagi** Scopoli WOODLAND GRAYLING
Europe 66–76 mm. (2·6–2·99 in.)
The butterfly, which is on the wing in July-August flies amongst bushes and trees in areas from sea-level to 910 m. (3000 ft). The caterpillars feed on various grasses, including *Holcus*. There are a number of broadly similar, related species throughout Europe and Asia, which are often difficult to separate from one another without careful examination. The butterfly is a grey brown with an eyespot near the apex of the wing in a yellowish-white band. The hindwing is similar but the white band below the margin is broad. The underside has clearer eyespots on the yellowish-brown forewing band and a strongly banded hindwing with a whitish area across it. Generally it resembles the Grayling, *H. semele.*

**semele** Linnaeus GRAYLING
Europe including Britain, S. Russia 42–50 mm. (1·65–1·97 in.)
This is a common and widespread species of open heaths and rough hillside areas. The brown upperside has 2 black eyespots, with white centres, one below the other on the yellow-brown or orange-brown outer part of the forewing. The hindwings, as with the forewings, are dark brown at the base with a single eyespot on the orange patches just below the wing margin. The underside forewing has mostly yellow or orange-brown and the hindwing has a white, zig-zag band of variable thickness in the middle of the grey-brown wing. This is a variable species, with several subspecies including some in Britain. The butterfly is out in July and August in heathlands etc. where the caterpillar feeds on species of grass (Gramineae). When the butterfly settles on the ground, amongst the rocks, the patterned hindwing is an excellent camouflage amongst the stones, the underside of the forewing is hidden. The butterfly (as some other species do) will often lean towards the direction of the sun, thus reducing its shadow and making it even less conspicuous.

**hippona** see **Consul**
**hippothoe** see **Palaeochrysophanus**

**Hippotion** Hübner (Sphingidae)
There are between 30 and 40 species in this genus of hawk-moths. Their range includes S. Europe, Africa, Asia and Australia.

**celerio** Linnaeus SILVER-STRIPED HAWK-MOTH
S. Europe, Africa, S. Asia Malay Archipelago to Australia 72–80 mm. (2·83–3·15 in.)
This species is often fairly common in most of its range. Migrants occasionally reach beyond the geographical limits listed above, especially in Europe where Britain is included in the list of northerly countries regularly visited by *H. celerio* although the yearly count is sometimes only one or two. Cultivated grape vines *(Vitis)* are sometimes attacked by the caterpillar, but more common foodplants are bedstraws *(Galium)*, *Fuschia* and Virginia creeper *(Parthenocissus)*. It is an occasional minor pest of cotton *(Gossypium)* in E. Africa. The illustrated specimen was captured in the Melanesian Admiralty Islands in the Bismarck Archipelago. **344**q

**hirtaria** see **Lycia**
**histrio** see **Chetone**
**histronaria** see **Pityeja**
**histronicus** see **Campylotes**
**hodeei** see **Amauta**
**hodeva** see **Hyantis**

**Hofmannophila** Spuler (Oecophoridae)
HOUSE MOTHS
One of the most persistent pests generally associated with man, the species of this genus of micro-moths are probably the most widespread pests in houses, warehouses and shops in the temperate region. The caterpillars feed on any organic dust under floorboards, skirting-boards or in any crevices. In the N. Hemisphere the caterpillars are common in old birds nests, but it is in the home that they cause most problems. The caterpillars will feed on almost any animal or plant product, particularly favouring items with some wool in them. Of the newer man-made articles damaged by these caterpillars, nylon, polythene and polystyrene are commonly attacked. While the caterpillar is not known to derive nourishment from these, its damage, and the debris it leaves makes it very unpopular.

**pseudospretella** Stainton BROWN HOUSE or FALSE CLOTHES MOTH
Worldwide 19–23 mm. (0·75–0·90 in.)
Frequently found in houses, where the caterpillars may damage soft furnishings or man-made fibre carpets. It regularly gets into food in larders. It is a difficult species to eradicate from a house and re-infestation is common since the moth is readily attracted to lights in houses in summer. This species has caused many problems in the home, where it will eat through polythene to get at the clothes wrapped up 'to keep the moth out' and has been found damaging goods as far removed as toilet paper and top hats! It is not as common in food products as some species but at some time or other has been found in most manufactured goods, or damaging soft furnishings. **36**

**holantha** see **Zacorisa**
HOLLY BLUE see **Celastrina argiolus**

**Holocera** Clemens (Blastobasidae)
These micro-moths, of which there are some 60 species, are mostly American, occurring in both N. and S. America; there is one species in India. They are separated from other genera by differences in the hindwing venation. Some have caterpillars which feed on plants while those of other species are predatory on scale insects (Coccidae). The caterpillars eat these soft-bodied insects which themselves are plant feeders, and often pests in their own right. The caterpillars of many Blastobasids are useful as biological control agents.

**iceryaeella** Riley
Australia, introduced into N. America 12–16 mm. (0·47–0·63 in.)
This species was introduced into N: America to be used for the Biological control of scale insects (Coccidae). This means that instead of spraying plants with insecticide, the caterpillars of this species will feed on the scale insects without harming anything else. **44**f

**Holocerina** Pinhey (Saturniidae)
The forewing apex of these moths is strongly hooked, especially in the males and the wing has an irregularly shaped, transparent patch. The caterpillars are armed with tubercles and irritant hairs. Six species are known, from tropical and S. Africa.

**angulata** Aurivillus
Tropical Africa 45–75 mm. (1·77–2·95 in.)
The males (not figured) of this species are much smaller, with strongly hooked forewings and a more yellowish coloration. The illustrated female was caught in Malawi. Nothing is known about the life history of this species. **331**f

**rhodesiensis** Janse
Malawi, Rhodesia 40–70 mm. (1·57–2·76 in.)
This is the most distinctively shaped and coloured species of *Holocerina*. The wings may vary in colour from violet to red in both sexes; in the female the outer margins are markedly crenulate. The gregarious caterpillars feed on Cabbage-tree foliage *(Cussonia)*.

**Holomelina** Herrich-Schäffer (ARCTIIDAE)
These are small moths, mostly orange, yellow and red species, found in Canada, the U.S. and C. America. Some of them, *H. aurantiaca* and allies, are very much alike and difficult to identify. There are about 35 species.

**lamae** Freeman
Canada (Nova Scotia) 18–22 mm. (0·71–0·87 in.)
This is a species of Nova Scotian peat-bogs. It is more easily identifiable from its colour-pattern than some of its close relatives. Little is known of its life history, but the moth flies in July and August. **382**b

**ostenta** Edwards
U.S., Mexico 28–38 mm. (1·1–1·5 in.)
This brightly coloured moth ranges from Colorado S. to Mexico. Its forewings are yellowish brown, with a dull red costal margin, the hindwings are red basally and in an irregularly shaped zone at the front of the wing, and dark brown or black over the remainder of the wing.

**homerus** see **Papilio**

**Homidiana** Strand (SEMATURIDAE)
There are about 20 species in this C. and tropical S. American genus of moths. The ground-colour of their wings is brown above; most species have a red, orange or yellow band on the hindwing.

**tangens** Strand
Ecuador 55–60 mm. (2·17–2·36 in.)
This species is rather similar in wing shape and general features of the colour pattern to *Coronidia orithea* except for the replacement of the blue and violet patch on the hindwing by an almost equally conspicuous red area.

HONEY HOPPER see **Platylesches moritili**
HONEY MOTHS see **Achroia**
HONEYCOMB MOTHS see **Galleria**
HOOK-TIP MOTHS see **Drepanidae**
HOOKED TIP, TOOTH-STREAK see **Ypsolopha dentella**
HOP HYPENA see **Hypena humuli**
HOP LOOPER see **Hypena humuli**
HOP VINE THECLA see **Strymon melinus**
HOPPER BUTTERFLIES see **Platylesches**
**hoppi** see **Cyanostola**

**Horama** Hübner (CTENUCHIDAE)
There are about 10 described species in this tropical American genus. Most of the species are dull brown in colour and difficult to identify on superficial characters. They are being studied currently at the U.S. National Museum of Natural History, Washington, D.C., by R. E. Dietz.

**oedippus** Boisduval
Mexico, Guatemala 28–40 mm. (1·1–1·57 in.)
This is typical of several other species of *Horama* all of which apparently mimic species of wasps. **367**e

**horatius** see under **Erynnis horatius**
**hordonia** see **Pantoporia**
HORNED DEVIL, HICKORY see **Citheronia regalis**
HORNET CLEARWING MOTH see **Sesia apiformis**
HORNWORM MOTHS see **Agrius, Manduca**
**horsfieldi** see **Kallima**
**hortulata** see **Eurrhypara**
HOUSE MOTHS see **Hofmannophila**
**hubneri** see **Epiphile**
**humeralis** see **Uzucha**
HUMMING-BIRD HAWK-MOTHS see **Hemaris, Macroglossum**
**humuli** see **Hepialus, Hypena**
HUNGARIAN GLIDER see **Neptis rivularis**
HUNTER'S BUTTERFLY see **Cynthia virginiensis**

**huttoni** see under **Bombyx mandarina**
**hyale** see **Colias**

**Hyalophora** Duncan (SATURNIIDAE)
Each of the 4 species of this genus of large moths is confined to N. America except for minor incursions into Mexico. One of the species, *H. colombia*, has a solely N. range and is unknown in the S. U.S., its caterpillar feeds on larch *(Larix)*. The caterpillars of each species have brightly coloured tubercles; those on the second and third abdominal segments are greatly enlarged.

**cecropia** Linnaeus CECROPIA or ROBIN MOTH
S. Canada, U.S., Mexico 120–150 mm. (4·72–5·91 in.)
This is probably the largest species of Lepidoptera found in the U.S. and Canada, but is nearly matched in size by *Antheraea polyphemus*. In colour pattern it resembles the illustrated *H. euryalus* but is generally darker in colour. Its range extends as far S. as N. Florida and S. Texas, and a few specimens have been captured in Mexico. The flight period is between March and June. The caterpillar is pale green, with red and yellow tubercles; its foodplants include Manitoba Maple *(Acer)*, wild cherry and plum *(Prunus)*, apple *(Malus)* and willow *(Salix)*. **339**

**euryalus** Boisduval CEANOTHUS or CALIFORNIA SILKMOTH
W. Canada, U.S., N.W. Mexico 80–114 mm. (3·15–4·49 in.)
*H. euryalus* is found along the Pacific coast from British Columbia to Baja California. Most specimens are characterized by a more pinkish coloration than in other species of the genus. Its caterpillar feeds on the foliage of *Ceanothus*, maple *(Acer)*, *Ribes*, willow *(Salix)* and many others; it differs from other species of *Hyalophora* in having solely yellow dorsal tubercles. **328**

**Hyalurga** Hübner (HYPSIDAE)
About 18 species are placed in this genus of C. American and tropical S. American moths. Some are mimetic partners of similarly patterned species of the butterfly subfamily Ithomiinae.

**fenestra** Linnaeus
C. America and tropical S. America 49–78 mm. (1·93–3·07 in.)
Some specimens of this species are paler in colour than the illustrated specimens but there is otherwise little significant variation between specimens or between the sexes. Many species of the butterfly subfamily Ithomiinae are similar in general appearance to *H. fenestra*. **381**e

**osiba** Druce
Amazon Basin and N.W. S. America 32–50 mm. (1·26–1·97 in.)
This is a member of a large mimetic complex of butterflies and moths, which includes the Ithomiine *Napeogenes corena*. The illustrated moth is an example of the subspecies *H. osiba batesi*. The big difference in size between males and females of *H. osiba* suggests that they may be members of different mimetic complexes. **381**g

**Hyalyris** Boisduval (NYMPHALIDAE)
There are about 12 species in this tropical S. American genus of the subfamily Ithomiinae. Some species are black or brown, with orange markings, and are similar in pattern to species of the Ithomiine genera *Mechanitis* and *Melinaea*; others are mostly transparent. One authority has stated that most of the included species probably belong in the genus *Hypothyris*.

**daeta** Boisduval
Amazon Basin and N. S. America 50–60 mm. (1·97–2·36 in.)
In E. Brazil this is part of a mimetic association of species listed under *Napeogenes xanthone*. It is one of several species of its genus which have a transparent band across each wing; this is yellowish on the forewing of *H. daeta* and brownish on the hindwing. Its caterpillars feed on species of Solanaceae.

**Hyantis** Hewitson (NYMPHALIDAE)
This genus is in the subfamily Amathusiinae. *Hyantis* is a small genus with 2–3 species which are found in New Guinea. Each species has a large number of described subspecies or forms.

**hodeva** Hewitson
New Guinea 75–85 mm. (2·95–3·35 in.)
There are many subspecies and aberrations named from different parts of its range in New Guinea. The underside is mostly white with a large black, yellow-edged, eyespot below the apex of the forewing and 2 large ones on each hindwing; one near the front margin of the wing, the other nearer the middle. There is considerable variation in the extent of the black on the forewing. **160**

**Hyblaea** Fabricius (HYBLAEIDAE)
Stout-bodied moths of uncertain affinities formerly thought to be related to Noctuid moths, now believed to be closer to the Pyralidae. Most species are found in the tropics or sub-tropics but little is known of their biology.

**pura** Cramer TEAK or DJATI MOTH
Worldwide, in the tropics 38 mm. (1·5 in.)
This wide ranging species has been recorded on most continents through the tropics. It is possible that this may be separated into several species when they are examined more critically. It is broadly similar to *H. sanguinea* but with dark forewings and yellow patches on the black hindwings. The caterpillar is a serious pest of teak *(Tectona)* in Java where it defoliates the trees. In Jamaica it has been recorded feeding on *Catalpa longissima* (Bignoniaceae). The caterpillar is dark purplish grey with short silky hairs, olive green below and with white lines dorsally and laterally.

**sanguinea** Gaede
Fiji 40 mm. (1·57 in.)
The specimen illustrated is typical of the genus. Many of the species have colour patterns on the hindwings and dark coloured forewings. Nothing is known of the biology of this species. **56**m

**Hyblaeidae**
The systematic position of these moths has been in dispute for years. Originally considered part of the Noctuidae (macro-moths) they are now regarded as related to the Pyralidae. They have certain features in common with these. One of the species is regarded as a pest of teak *(Tectona)* which is defoliated by its caterpillars. See *Hyblaea*.

**hyceta** see **Pierella**
**Hydrocampa** see **Nymphula**
**hydrographus** see **Abantiades**
**hyelosioides** see **Gonora**
**hylax** see **Zizula**

**Hygrochora** Hübner: syn. *Apatelodes* (APATELODIDAE)
This is a genus of about 70 species of essentially tropical American moths. Two species are found in N. America. The valid generic name, *Hygrochora*, has priority over the more generally known but later name *Apatelodes*.

**torrefacta** Smith
Canada and E. U.S. including Florida and Texas 35–40 mm. (1·38–1·57 in.)
*H. torrefacta* has 2 broods per year in the S. of its range, only one in the N. Moths of the second brood are generally darker than those of the first. The caterpillar is yellow or grey when fully grown. It is densely hairy with tufts at the front and rear, much like the caterpillars of the Noctuid genus *Acronicta*. The foodplants include numerous broad-leaved trees and shrubs. **318**h

**hypargyria** see **Deudorix**

**Hyles** Hübner: syn. *Celerio* (SPHINGIDAE)
The species of this genus of hawk-moths were known formerly under *Celerio*, a generic name unavailable according to the international committee which controls the application of zoological names. *Hyles* is an almost cosmopolitan genus; there are few species, but a multitude of names which have been applied to geographical units, seasonal and colour forms, and to individual specimens of some of the species, especially

to *H. euphorbiae* which has over 200 names associated with it.

**calida** Butler
Hawaii 58–70 mm. (2·28–2·76 in.)
Both subspecies of *H. calida* are confined to the islands of Hawaii; *H. calida calida* (illustrated) is found on Kauai, Oahu and Molokai, *H. calida hawaiensis* Rothschild & Jordon only on Hawaii. The European and Asiatic *H. euphorbiae* and the N. hemisphere species *H. gallii* resemble *H. calida* most closely in colour pattern but are less richly coloured and lack the white or pale yellow markings on the thorax. The caterpillars feed on the foliage of various plants including *Acacia* and *Gardenia*. **344**n

**euphorbiae** Linnaeus SPURGE HAWK-MOTH
C. & S. Europe to Turkey; Canada 80–90 mm. (3·15–3·54 in.)
An attempt has been made recently to control *Euphorbia* weeds by the introduction of *H. euphorbiae* to W. Canada, and the moth is now established there. In Britain it is a rare migrant, but is fairly common in the rest of Europe. The caterpillar is a particularly conspicuous black, red and yellow creature (sometimes almost completely black) which is rejected by birds, possibly because it is able to store plant toxins in its tissues. Species of the plant family Euphorbiaceae are mostly poisonous to vertebrates, and many moth and butterfly caterpillars are known to be capable of extracting and storing plant poisons, see for example *Danaus*. The moth is similar in pattern to *H. calida*. **351**

**gallii** Rottemburg BEDSTRAW HAWK-MOTH
Canada, U.S., Europe including Britain, through temperate Asia to Japan 60–75 mm. (2·36–2·95 in.)
The colour pattern of this extremely widespread species is similar in general plan to that of *H. calida* but the pale, central band on the forewing is much narrower and paler, there are no pale yellowish white lines on top of the thorax, and the central band on the hindwing is paler, and white at the inner margin of the wing. The caterpillars apparently prefer the foliage of bedstraws *(Galium)* in Europe and willow herb *(Epilobium)* in N. America. Like many hawk-moths this is a migratory species—its status in Britain.

**lineata** Fabricius STRIPED HAWK-MOTH or WHITE-LINED SPHINX 70–80 mm. (2·76–3·15 in.)
N. and C. America, Africa, Europe, Asia
No species of hawk-moth has a wider distribution than this species. It is common in the U.S. and Canada, and a frequently encountered migrant in most of Europe from its breeding grounds along the Mediterranean and N. Africa. The moths fly at night, and by day, and often visit blue flowers. The caterpillars feed on a great variety of plants, including elm *(Ulmus)*, vine *(Vitis)*, cotton *(Gossypium)*, water-melons *(Citrullus)*, apple *(Malus)*, *Fuschia* and bedstraws *(Galium)*. Blacklight (ultraviolet) traps have been used to attract adults and control this species in the U.S. where it sometimes reaches pest proportions. Although described as 'great horrid things' by a 19th Century American collector, the caterpillars were regarded as a delicacy by Indian tribes of S.W. U.S. and formed the basis of ceremonial stews. **344**p

**vespertilio** Esper
S. Europe (except Spain) to S.W. Russia 48–58 mm. (1·89–2·28 in.)
The brownish grey forewings of this small species are almost patternless; the hindwings are dull pink, with a dark brown base and marginal band. The pale-spotted caterpillars, which have no terminal horn, feed on willow herb *(Epilobium)* and bedstraw *(Galium)*.

**Hylesia** Hübner (SATURNIIDAE)
This is a large genus of moths found in N., C. and S. America. It is, however, mainly tropical in distribution, with only 2 species represented in the U.S. The species are small and generally inconspicuous in colour pattern. Contact with the moths causes a painful irritation known in Venezuela as Caripito itch, a result of chemical secretions (as yet unidentified) in the barbed hairs of the anal tuft. The identification of the species of this genus is extremely difficult and there is much doubt about the number of species which should be recognized. The caterpillars are typically gregarious and respond to sound and touch with sudden jerky movements. The sound produced by a Chalcid wasp, a parasite of caterpillars, elicits this group-response which may have a protective function. See also under *Eubergia*.

**alinda** Druce ALINDA MOTH
C. America and S. U.S. 35–48 mm. (1·38–1·89 in.)
This is one of 2 species of *Hylesia* recorded from Arizona in S.W. U.S., but there is some doubt as to the authenticity of these N. American records for *H. alinda*. The ground-colour of the wings is dull pink, with several brown, transverse bands; the central band on the forewing is forked in the middle of the wing to form a Y.

**Hypena** Schrank (NOCTUIDAE) DELTOID or SNOUT MOTHS
This is an almost cosmopolitan genus of small, triangular winged species of moths. The front of the head bears a conspicuous tuft of scales, and the elongate palps extend forwards from near the base of the proboscis to a distance equal to the combined length of the head and thorax. The caterpillars have lost the first pair of false-legs and are semi-loopers.

**humuli** Harris HOP HYPENA, HOP LOOPER
Canada, U.S. 22–32 mm. (0·87–1·26 in.)
The destructive caterpillar of *H. humuli* is a common, but minor pest of hops *(Humulus)* in much of Canada and the U.S.; it is green, with 2 white lines along the back. The adults vary considerably in coloration and distinctness of the pattern of browns and greys. There are 2 broods each year; adults of the second brood overwinter. This is probably the commonest species of *Hypena* in N. America.

HYPENA, HOP see **Hypena humuli**
**hyperanthus** see **Aphantopus**
**hyperbius** see **Cinabra**

**Hyperborea** Grum-Grshimailo (ARCTIIDAE)
The genus was erected for one species, *H. czekanowskii*, discovered in Siberia in the 1870's.

**czekanowskii** Grum-Grshimailo
U.S.S.R. (Siberia), U.S. (Alaska) 34 mm. (1·34 in.)
Nearly 100 years after the original discoveries in Siberia of this arctic species, a specimen has been found recently on the Seward Peninsula of Alaska. It is not unusual for arctic species to be widely distributed, and there may be yet more discoveries of this sort to be made in Alaska which at its closest point on the mainland of the Seward Peninsula is just over 50 miles from the E. tip of Siberia. The forewings are brown, the wing-veins marked in white, with a broad, white costal band and other *Apantesis*-like white bands; the hindwings are white, with brown, outer marginal markings.

**hyperia** see **Biblis**

**Hypermnestra** Ménétriés (PAPILIONIDAE)
There is a single known species in this genus of butterflies. Its closest relative is the well known *Parnassius*.

**helios** Nickerl
N. Iran to Turkestan (U.S.S.R.) and Pakistan 35–50 mm. (1·38–1·97 in.)
This is a species of the steppes and semi-desert of S.W., C. Asia. There is little difference between the sexes, but there are very pale individuals of both sexes which have a greatly reduced colour pattern on the forewing and hardly a trace of the hindwing colour pattern. The adult, which has a *Parnassius*-like flight is found most commonly on hill-slopes where the caterpillars foodplant is *Zygophyllum*, a genus of succulent plants. **94**c

**hypermnestra** see **Elymnias**, **Parnalius polyxena**

**Hypertropha** Meyrick (OECOPHORIDAE)
This is a very small genus of micro-moths which has been separated from the Glyphipterigidae. There are less than half a dozen species in the family, all in Australia. They generally have metallic wing scales and most have quite bright yellow hindwings. Those which have been bred have been reared on *Eucalyptus*. The genus formerly formed part of a separate family, Hypertrophidae, this name is not now used, and the included genera are now placed in different families.

**desumptana** Walker 15 mm. (0·59 in.)
The metallic scales and yellow hindwings are conspicuous on this species. This species has been bred from *Eucalyptus maculata* where it was feeding on the leaves. The species occur in several states of Australia, including Queensland, New South Wales and Victoria. **34**l

**Hyphantria** Harris (ARCTIIDAE)
The 5 or so species of moths currently placed in this genus are probably not closely related to each other and require reclassification. They are found in Africa, Europe, temperate Asia, N. America and Mexico.

**cunea** Drury: syn. *H. textor* FALL WEBWORM
S. Canada, U.S., E. Europe, Japan 25–32 mm. (0·98–1·26 in.)
*H. cunea* is a native N. American species which has been introduced, presumably accidentally, into Hungary and Japan. *H. cunea* is white, usually with at least some greyish brown spots on the forewing and in some specimens on the hindwing; the pattern of the spots is similar to that of the illustrated *Estigmene acrea*. Its caterpillar is a highly destructive pest and feeds on the foliage of over 100 species of broad-leaved trees, including orchard species, and shrubs. Caterpillars live gregariously in large webs at the end of branches. Control measures against them have included the use of viruses. **369**e

**Hyphilaria** Hübner (NEMEOBIIDAE)
The 4 species of this genus are found only in tropical S. America.

**parthenis** Doubleday
Tropical S. America 22–28 mm. (0·87–1·1 in.)
The female differs greatly in colour pattern from the illustrated male; both wings are crossed by several alternating, parallel bands of brown and white. The life history is unknown. **263**x

**Hypna** Hübner (NYMPHALIDAE)
A small genus of butterflies in S. America which is related to *Callicore*. Many species have broadly similar patterns in these 2 genera.

**clytemnestra** Cramer
S. America 42 mm. (1·65 in.)
There are several subspecies, the one illustrated is *H. clytemnestra forbesi*. The butterfly is resident in the very dry areas in N.E. Brazil. It lives in a particularly inhospitable area and so is rare in collections. **232**

**hypochlora** see **Batesia**

**Hypochrysops** Felder (LYCAENIDAE) JEWEL BUTTERFLIES
One species of this genus occurs as far W. as Thailand, but the group is best represented in New Guinea and Australia. Many of these butterflies are exquisitely coloured. The upper surface of the wings is a brilliant green, blue, purple or orange, and there are lustrous, metallic, red, green and blue markings on the under surface. The nocturnal caterpillars are attended by ants, and can often be found in ants' nests near the caterpillars' foodplants.

**apelles** Fabricius COPPER JEWEL
Aru Island, New Guinea, E. Australia 25–28 mm. (0·98–1·1 in.)
Both sexes of this butterfly are brownish orange above, with a broad, dark brown, outer border to the forewing and a dark brown band along the front edge of the hindwing. The male wing veins are generally more heavily marked with dark brown. In coastal districts the caterpillars feed on the foliage of mangrove *(Rhizophora)*; they are attended by black ants of the genus *Crematogaster*. **244**h

**apollo** Miskin
New Guinea, N. Australia 34–36 mm. (1·34–1·42 in.)
This large, bright orange or orange-red species is confined to coastal swamps where the caterpillar feeds on

an epiphytic ant-plant *(Myrmecodia)*. The caterpillars live and feed together with ants inside the galleries of the foodplant, which itself lives on swamp trees, including swamp mahogany *(Tristania)*.

**ignita** Leach FIERY JEWEL
Australia, New Guinea 24–30 mm. (0·94–1·18 in.)
The orange and iridescent bluish green markings of the under surface of this species are similar in type to those of the illustrated *H. pythias*. The upper surface of the wings is purple, with black outer margins to both pairs of wings and an orange-yellow band along the front edge of the forewing. The caterpillars of *H. ignita* feed at night on species of *Acacia*, *Camellia*, plum *(Prunus)* and blackberry *(Rubus)*, and are attended by ants of the genus *Iridomyrmex* which guide them between their ant-nest foster home and the foodplant.

**piceata** Kerr
Australia 20–22 mm. (0·79–0·87 in.)
This small species described in 1969 is apparently restricted to S. Queensland. The male is dark purple above; the veins marked in orange and the outer margin dark brown. The under surface is similar in pattern-type to that of *H. pythias* (illustrated). Its caterpillars feed on the foliage of bull-oak *(Casuarina)* and share the attentions of the same ants of the genus *Iridomyrmex* as those which attend nearby caterpillars of a species of *Ogyris* feeding on mistletoe parasitic on the same *Casuarina* tree.

**pythias** Felder
The Moluccas, through New Guinea 27–36 mm.
to Fergusson Island (1·06–1·42 in.)
The male of this butterfly is a deep violet-blue above, with black outer margins to the wings; females are chiefly dark brown, with a light violet-blue area at the base of the forewing and a similar but smaller area at the base of the hindwing. The under surface of the wings (illustrated) is similarly brilliant in the other species of this genus. **255**

**theon** Felder
The Moluccas, Aru Island, 26–32 mm.
New Guinea, N. Australia (1·02–1·26 in.)
The male of this butterfly is a highly lustrous blue above, with black markings; the female is mostly black above, with a yellowish white patch on the forewing. Beneath (illustrated) both sexes are a brilliant iridescent green and black. Nothing is known about the life history of *H. theon*. **244j**

**Hypocrita** Hübner (HYPSIDAE)
About 26 brightly coloured species are placed in this genus of moths. Its range includes tropical S. America, Mexico and C. America.

**aletta** Stoll
C. America and 53–70 mm.
N.W. S. America (2·09–2·76 in.)
The ray-like pattern on the wings of this species is matched in some species of Heliconiinae. Its close relative *H. euploeodes*, together with another Hypsid *Calodesma melanchroia*, is mimicked by the Nemeobiid butterfly *Esthemopsis clonia*. **376**d

**dejanira** Druce
C. & N.W. S. America 53–65 mm. (2·09–2·56 in.)
This is one of the most gaudily coloured species of Hypsidae. Other species closely related to *H. dejanira* are similarly patterned but have yellow or red bands on one or both pairs of wings. There is little difference between the sexes in colour pattern. **376**g

**euploeodes** see under **Esthemopsis clonia**

**Hypoleria** Godman & Salvin (NYMPHALIDAE)
About 14 species are placed in this genus of Central and tropical S. American Ithomiinae. Their wings are largely transparent, and marked with black, orange and white.

**oreas** Weymer
Brazil 48 mm. (1·89 in.)
Several species of *Hypoleria* closely resemble *H. oreas* in colour pattern. The upper surface (not illustrated) of this species has dark brown, not orange margins to the otherwise transparent wings. **133**

**Hypolimnas** Hübner (NYMPHALIDAE) DIADEM BUTTERFLIES
These butterflies are found throughout the tropics of the Old World. The females often mimic the poisonous Danaid butterflies, differing considerably in appearance from the males. The caterpillars have finely branched spines, and similar, but longer ones on the head.

**alimena** Linnaeus BLUE-BANDED EGGFLY
Moluccas, Bismarck Archipelago, 65 mm.
New Guinea, Solomon Islands, (2·56 in.)
Kai and Aru Islands, Australia
This is found in N. Australia where the striking blue band across the dark wing of the male gives the popular name. The females lack this but have larger patches of white near the apex of the wing. There are several distinct forms of this butterfly and several subspecies have been described. The caterpillar feeds on *Pseuderanthemum* (Acanthaceae). **208k**

**bolina** Linnaeus COMMON or GREAT EGGFLY
India, Pakistan, S.E. Asia, Formosa, 70–110 mm.
New Guinea, Indonesia, (2·76–4·33 in.)
New Zealand and many W. Pacific islands
This widespread species has many described subspecies. The butterfly is common over much of its range and is often seen in gardens. The caterpillars feed on a variety of plants including species of Amaranthaceae, Malvaceae, Acanthaceae, Portulacaceae, Polygonaceae, Rutaceae and Compositae. A widespread species often has a large range of host plants and this is certainly the case with *H. bolina*. This species in India mimics *Euploea core*, one of the species probably distasteful in the subfamily Danaiinae. **208**b

**dexithea** Hewitson
Madagascar 100–120 mm. (3·94–4·72 in.)
This species is only known from the forests of N. and E. Madagascar. The biology is unknown. The underside is broadly similar to the upperside but the streaks along the hindwing margin are white instead of bluish. They are fast fliers and the males are described as very quarrelsome. Almost every species of butterfly has its devotees who believe it is the most beautiful and this one has been previously described as 'one of the most beautiful butterflies' by several authors. **208**a

**misippus** Linnaeus SIX CONTINENT BUTTERFLY, DANAID EGGFLY or DIADEM BUTTERFLY
India, Pakistan, Australia, S. America, 62–66 mm.
N. America, W. Indies (2·44–2·6 in.)
This remarkably widespread butterfly shows very little variation over its range. The caterpillar feeds on species of Portulacaceae. The butterfly is found in the N. part of S. America and in the U.S., where it has been suggested it was introduced by the slave ships 300–400 years ago. The females mimic species of Danaid butterflies and in this way are avoided by predators who have learnt that this colour and pattern means an unpalatable insect. The male is similar to *H. bolina* but the patches on the wing are white. The female is illustrated. **208**j

**pandarus** Linnaeus
Indonesia 92–120 mm. (3·62–4·72 in.)
The female is brown on the forewing with white spots in bands across the wing. The hindwings are like the male but without the purple colour, this is replaced by a yellow-brown colour. The eyespots on the hindwing are much larger in the female. **208**c

**salmacis** Drury
Sierra Leone to Uganda 90 mm. (3·54 in.)
This is very variable in pattern, with many named forms. Several distinct forms of this butterfly have been described from different parts of its range. **208**h

**hypolitus** see **Troides**

**Hypolycaena** Felder (LYCAENIDAE)
This is a small African and Indo-Australian genus of butterflies found as far E. as the Bismarck Archipelago. Males have a large, round scent-patch on the middle of the forewing. Its caterpillars feed on the flowers and foliage of several different plants.

**danis** Felder BLACK AND WHITE TIT
Moluccas to New Guinea, the Bismarck 30 mm.
Archipelago and N.E. Australia (1·18 in.)
This distinctively marked butterfly is sometimes a garden pest in Australia, where the caterpillars attack orchids (Orchidaceae). Although the orchid flowers are preferred, other parts of the plants may also be damaged. The female colour pattern (not illustrated) is similar to that of the male. **244**y

**phorbas** Fabricius COMMON TIT
New Guinea, the Bismarck Archipelago, 30–33 mm.
N. Australia (1·18–1·3 in.)
The male forewings are an iridescent greenish blue above, with a black outer and front margin and a large, round, black, central spot; the hindwings are similar in ground-colour, with a dark brown front and inner edge. Females are brownish grey, with a blue and white, central patch on the forewing and dark spots at the outer margin of the hindwing. The caterpillar feeds on a variety of plants and are attended by green tree-ants *(Oecophylla)*.

**Hyponomeuta** see **Yponomeuta**
**Hyponomeutidae** see **Yponomeutidae**

**Hyposmocoma** Butler (MOMPHIDAE)
This remarkable genus of micro-moths is found only on Hawaii, where several hundred species are known. It was formerly in a family of its own, but is now considered to be related to the genus *Cosmopterix*.

**arenella** Walsingham
Hawaii 11–14 mm. (0·43–0·55 in.)
The specimen illustrated was one of the original moths collected in 1892. Many of the related species are rather similar in appearance though nothing is known of their biology. This is a classic example of speciation that has taken place on a group of islands, the study of which would provide an interesting insight into possible mechanisms of evolution. **43**m

**Hypothyris** Hübner (NYMPHALIDAE)
About 30 species are placed in this genus of the subfamily Ithomiinae. Many were at one time placed in the genus *Ceratinia*. Its range is Central American and tropical S. American. See also *Hyalyris*.

**antea** see **Ithomia linda**

**fluonia** Hewitson
Tropical S. America 40–45 mm. (1·57–1·77 in.)
This butterfly is part of a large mimetic complex of the Nymphalid subfamilies Ithomiinae, Heliconiinae, Danaiinae and Nymphalinae and the family Pieridae, see *Tithorea harmonia*.

**oulita** Hewitson
Peru 53–60 mm. (2·09–2·36 in.)
The female of this species lacks the transparent patch in the middle of the forewing (see illustration of male) and has the basal half of the hindwing orange in colour. Nothing is known about the habits or life history of this butterfly. **131**f

**hypselis** see **Symbrenthia**

**Hypsidae**
This is a rather mixed assortment of moth genera, some of which are fairly typical Arctiids. Other sections of this family are probably no more than subfamilies of the family Arctiidae. Two of the chief groups comprise the New World genera *Chetone*, *Dysschema* and their allies (often separated as Pericopidae), and the Old World genera typified by *Asota* and *Aganais*. Several species are mimetic partners of species of Heliconiinae, Ithomiinae, Danainae, Nemeobiidae, Papilionidae, Ctenuchidae, Arctiidae and other familes, see for example *Anthomyza, Calodesma, Chetone, Darna, Dysschema* and *Hyalurga*. See *Aganais, Amphicallia, Anaxita, Anthomyza, Argina, Asota, Calodesma, Camptoloma, Chetone, Composia, Darna, Digama, Dysschema, Ephestris, Episcea, Euplocia, Gnophaela, Hyalurga, Hypocrita, Mesenochroa, Phaegorista, Phaloesia, Sagaropsis.*

**Hypsidia** Rothschild (PYRALIDAE)
This genus contains only one known species which is from Australia. It belongs to the subfamily Pyralinae which has moths of varied size and colour but with nothing else remotely resembling this species.

**erythropsalis** Rothschild
Australia 54–62 mm. (2·13–2·44 in.)
This strikingly coloured insect with the bright red pattern is known only from the tropical areas of Queensland. Nothing is known at present of its biology, or why it has such bright colours to advertise itself. It may be that it is distasteful to predators and this colour warns them off. A study of its biology and life history would be very interesting. **63**f

**Hypsopygia** Hübner (PYRALIDAE)
These small, attractive, Pyralid moths are found on many continents but the majority are European. One species is found in N. America.

**costalis** Fabricius GOLD-FRINGE, CLOVER HAYWORM or CLOVER WEBWORM
Europe including Britain, Asia, N. America 17–20 mm. (0·67–0·79 in.)
This species is found throughout the Northern hemisphere. The caterpillars feed on stored clover *(Trifolium)*, grass (Gramineae), hay and also on straws in thatches. The hay tends to be fouled with the black frass (excreta) of the caterpillars and will rot more quickly. The moth is therefore unpopular with farmers. The species is locally common in Britain and very common in the U.S. Several colour varieties have been described. **61**

**ianthe** see **Phyciodes**
**ianthis** see **Ogyris**
**iarbus** see **Lymnas**

**Icaricia** Nabakov (LYCAENIDAE)
This genus was erected to accommodate *shasta* by an author more widely known as the novelist who wrote the best-selling 'Lolita'.

**shasta** Edwards SHASTA or ALPINE BLUE
Canada, U.S. 25 mm. (0·98 in.)
This hardy little butterfly is restricted to high elevations in the Rockies and Sierras from Alberta to Colorado, and W. to California. In Colorado it is found above the tree-line. Males are mainly blue above, females brown, with orange spots along the outer margin of the hindwing. The caterpillars feed on mountain lupins *(Lupinus)*.

**icarus** see **Polyommatus**
**iceryaeella** see **Holocera**

**Idaea** Treitscke: syn. *Sterrha* (GEOMETRIDAE)
A huge genus of moths with species all over the world. All those known have typical geometrid caterpillars.

**muricata** Hufnagel PURPLE-BORDERED GOLD
Europe including Britain, temperate Asia including Japan 17–18 mm. (0·67–0·71 in.)
The moth flies in the early morning in June and July and is common in marshes, fens and boggy heaths. The caterpillar feeds on plantain *(Plantago)* and Marsh Cinquefoil *(Potentilla)*. In Japan a distinct subspecies of *I. muricata* occurs. There are several distinct colour forms of this species most of which have been given names. There is no evidence that the commoner forms are related to the geographical distribution of the species. **278**

**idaeoides** see **Graphium**

**Idea** Fabricius (NYMPHALIDAE)
TREE NYMPH BUTTERFLIES
This is a genus of nearly a dozen oriental species of butterflies belonging to the subfamily Danainae. The high ratio of wing-area to body-size, and therefore muscle, in these butterflies results in a weak, fluttery flight alternated with gliding. They are confined to densely forested regions where their strange, ghost-like appearance has given rise to many local legends. The ground colour of the wings is normally yellowish, brownish or greyish white, varying in the degree of transparency, with black or dark brown veins and spots.

**idea** Linnaeus
The Moluccas 120–140 mm. (4·72–5·51 in.)
This is one of the largest species of butterfly and is probably the best known species of its genus. It is common on some of the islands in the Moluccas group. The illustrated male is a rather small example of its species. **130**g.

**leuconoe** Eschschoff
S. China, S. Japan, Malaya, Java, Borneo, the Philippines 95–110 mm. (3·74–4·33 in.)
This is probably the most widely distributed species of its genus. In pattern it is similar to that of *I. idea* (illustrated) but is much more variable in its markings and in the ground-colour of the wings. It may be extinct in Java, where cultivation has removed nearly all of the native forests which are essential to this butterfly and hundreds of others. Amongst its mimics is the remarkable Papilionid species *Graphium idaeoides*.

**lynceus** Drury
Indochina, Malaya, Sumatra, Java, Borneo 115–165 mm. (4·53–6·5 in.)
As in other species of *Idea* the flight of this butterfly is extremely slow and gliding with very little wing movement. The colour pattern of *I. lynceus* is similar to that of the much smaller *Ideopsis daos* (illustrated) but the ground-colour is finely speckled with black.

**Ideopsis** Horsfield (NYMPHALIDAE)
There are 7 solely Oriental species in this genus of the subfamily Danainae. The males have a distinctively concave outer margin to the forewing. Their flight is slow and undulating like that of *Danaus* species. They frequently fly together with species of *Idea* along forest streams.

**daos** Boisduval
S. China, Malaya and Sumatra to Borneo and the Philippines 70–98 mm. (2·76–3·86 in.)
This is the model, with the Javanese *I. gaura*, for *Graphium delesserti* (Papilionidae), especially for its female which closely resembles *I. daos* in colour pattern and flies together with it in at least part of its range. Males (not illustrated) are slightly darker in coloration than the females and have more elongate forewings. There is disagreement amongst entomologists as to whether *I. daos* should be regarded as a subspecies of *I. gaura*. **130**b

**gaura** Horsfield SMALLER WOOD-NYMPH
Java 70–94 mm. (2·76–3·7 in.)
This is essentially a forest insect but it will also fly in more open country. It is mimicked by females of the Papilionid *Graphium delesserti* and forms a mimetic association with a day-flying Zygaenid moth *Cyclosia midamia*. It is similar in colour-pattern to the illustrated *I. daos*, especially the subspecies *I. daos costalis* from the island of Nias. *I. gaura* is replaced by *I. daos* as the model for *Graphium delesserti* outside Java.

**vitrea** Blanchard
Sulawesi (Celebes) E. to W. New Guinea 65–96 mm. (2·56–3·78 in.)
This butterfly has been described as the most beautiful species of *Ideopsis*. Males are dark brown above, with usually a broad, oblique, white band on the forewing and small, white, marginal spots; the hindwing has a white basal area, a yellow inner margin, large yellow, submarginal spots and small, yellow marginal spots. In the subspecies *I. vitrea chloris*, the forewing markings are yellow, not white, and the base of the hindwing is also yellow. Females normally have yellow markings on the forewing. The illustrated male is from Sulawesi. **130**a

**idomeneus** see **Caligo**
**Idrusia nyctelius** see under **Euploea diocletiana**
**ignita** see **Chrysocale, Hypochrysops**
**ilia** see **Apatura**
**ilione** see **Itura**
**illyris** see **Graphium**

**Imbrasia** Hübner (SATURNIIDAE)
There are many externally similar species in this large African genus of moths. They are replaced in Asia by the *Antheraea* Silk-moths. Some species have a very short tail to the hindwing, but most have no tail. The caterpillars of most species bear short, spinose tubercles. The African genus *Nudaurelia* is currently considered as a subgenus of *Imbrasia*.

**cytherea** Fabricius PINE EMPEROR
South Africa 90–125 mm. (3·54–4·92 in.)
Pine trees *(Pinus)* are often infested by the caterpillars of this common species which may sometimes reach pest proportions. The foliage of numerous other trees is also attacked, including *Cupressus, Eucalyptus, Acacia*, ebony *(Euclea)*, apple *(Malus)* and guava *(Psidium)*. The adult moth varies in coloration from pale yellow to dark brown; it has two transverse lines and an eyespot on each wing, the eyespots larger on the hindwing.

**Imma** Walker (GLYPHIPTERIGIDAE)
Well over 150 species have been described in this micro-moth genus from all over the world. Most species are found in New Guinea with a few in Africa. Generally they are brightly coloured species with some metallic iridescence on the wing, but many are rather sombre coloured.

**racemosa** Durrant
New Guinea 35 mm. (1·38 in.)
There are quite a number of species in New Guinea all rather similar to this one, with yellow and black colours on the wing. The other species have the black in thicker lines across the wing and usually have some black along the wing margins. These are useful characters for recognizing the different species. Their biology is unknown. **33**b

**saturata** Walker
Java, Malaya 35–42 mm. (1·38–1·65 in.)
Related and similar looking moths are found in China. The body and wing-base are yellow and black spotted in contrast to the wings. This may be a warning coloration to show predators that it is distasteful. Having tried one and found it unpleasant, the predator then learns to avoid similarly coloured species. There is no information on their life history. **33**a

**impactella** see **Eretmocera**
IMPERIAL BLUE, COMMON see **Jamides evagoras**
IMPERIAL, COMMON see **Cherita freja**
IMPERIAL, RED see **Suasa lisides**
IMPERIAL, WHITE see **Neomyrina hiemalis**
**imperialis** see **Teinopalpus**
IMPORTED BLUE see **Zizula cyna**
IMPORTED CABBAGEWORM see **Pieris rapae**
**improba** see **Clossiana**

**Inachis** Hübner (NYMPHALIDAE)
A genus of butterflies with a single species in it, formerly included in the genus *Nymphalis*.

**io** Linnaeus: syn. *Nymphalis io* PEACOCK
Europe, temperate Asia including Japan 54–64 mm. (2·13–2·52 in.)
The striking eyespots on this butterfly, like the 'eyes' on a peacock's tail, make this very conspicuous when the wings are open. However, when it closes its wings the butterfly practically disappears, the dark underside blending closely with its surroundings. The eyespots, which it will 'flash' by opening and closing the wings are believed to be a device for scaring off possible predators. The caterpillar, which is velvet black covered with white dots with long black spines, feeds in groups on nettles *(Urtica)*. The adult flies in July through the summer, then again in the following spring after a winter's hibernation. In Japan there is a distinct subspecies. **198**

**inachus** see **Kallima**
INCHWORMS see under **Geometridae**
**inconspicuella** see **Solenobia**

**Incurvariidae** INCURVARID MOTHS
The division of this family into subfamilies is a matter

of some dispute at present. The species are all small or very small moths many with legless, leaf-mining larvae. The forewing venation is basically similar to the larger moths and not reduced as for example in the Nepticulidae. The family contains the long-horned moths. Their popular name refers to the extremely long antennae which project in front of the head, and are many times longer than the body. The moths are usually a metallic colour. The family is worldwide. Some of the moths in the Incurvariidae have a remarkable life history, particularly in the Prodoxine moths see *Tegeticula*. See *Adela, Lampionia, Nemophora, Prodoxus, Tegeticula.*

INDIAN
CROW, COMMON see **Euploea core**
LEAF see **Kallima inachus**
MEAL MOTH see **Plodia interpunctella**
MOON-MOTH see **Actias selene**
PURPLE EMPEROR see **Apatura ambica**
RED ADMIRAL see **Vanessa indica**
SUNBEAM see **Curetis thetis**
WHITE ADMIRAL see **Limenitis trivena**
**indica** see **Vanessa**
**inferens** see **Sesamia**
**infernalis** see **Styx**
**infusa** see **Agrotis**
**innotata** see **Tryporyza**
**ino** see **Brenthis**
**inous** see **Jalmenus**
INOUS BLUE see **Jalmenus inous**
**insignis** see **Olyras**
**insulana** see **Earias**
**interiorata** see **Camptoloma**
**interjectana** see **Cnephasia**
**interpunctella** see **Plodia**
**interrogationis** see **Polygonia**
**interrupta** see **Acronicta**
**involuta** see **Eupseudosoma**
**io** see **Automeris, Inachis**
IO MOTH see **Automeris io**
**iodamia** see **Asota**

**Iolana** Bethune-Baker (LYCAENIDAE)
This is a genus of about 3 species of blue butterflies. They are found in Europe and temperate Asia E. to China.

**iolas** Ochsenheimer IOLAS BLUE
N. Africa, S. Europe, Turkey, Iran 32–38 mm. (1·26–1·5 in.)
The male of this butterfly is a lustrous violet-blue above with a thin, dark outer marginal band. The female is variable, but usually has a much broader, less dark marginal band on both wings and ill-defined marginal spots at the outer margin of the hindwing. This species is one of the largest blues found in Europe. The caterpillar feeds inside the pods of senna. *(Colutea).*

IOLAS BLUE see **Iolana iolas**

**Iolaus** Hübner (LYCAENIDAE)
This is a genus of iridescent, tree-dwelling butterflies whose caterpillars feed on mistletoes *(Loranthus)*. They are restricted to Africa S. of the Sahara.

**coeculus** Hopffer AZURE HAIRSTREAK
Tropical & S. Africa 35–40 mm. (1·38–1·57 in.)
This is a generally common butterfly of scrubland, savannah and forest margins. The wings are a particularly vivid, violet-blue, especially in the male which has narrower, darker brown margins to the wings than the female. The distinctive under surface of both sexes is pale grey, with several transverse, crimson lines across both pairs of wings. The hindwing has 2 moderately long tails.

**lalos** Druce PALE SAPPHIRE
Kenya, Tanzania, Malawi, Rhodesia 35–40 mm. (1·38–1·57 in.)
In Kenya and Tanzania this butterfly inhabits coastal scrub; in other parts of its range *I. lalos* is found in scrubland and savannah. The upper surface of both sexes of this handsome, sexually dimorphic species are illustrated. The under surface is white in both sexes, usually with a single red dash on the forewing, and with red, brown and orange marginal markings on the hindwing. **240**b, d

**pallene** Wallengren SAFFRON
Kenya, Tanzania, Mozambique, Rhodesia, South Africa 35–45 mm. (1·38–1·77 in.)
This species is remarkable for the unusual pale yellow coloration, which produces a Pierid-like appearance, especially when in flight, and possibly confers some protection on this species. Some white butterflies such as the almost cosmopolitan Small White, *Pieris rapae*, are distasteful to birds. The caterpillar of *I. pallene* feeds on mistletoe *(Loranthus).*

**iole** see **Marpesia**
**iona** see **Colotis**

**Iotaphora** Warren (GEOMETRIDAE)
A genus of moths found in temperate Asia and N. India. Only 2 species are known at present.

**admirabilis** Oberthür
China 55 mm. (2·17 in.)
The striking pattern of this moth is quite distinct from other Geometrids. Nothing is known of its life history. The only other species in the genus has a similar pattern. **294**y

**Iphiclides** Hübner (PAPILIONIDAE)
Both known species are restricted to the Old World. This genus is closely related to the widespread butterfly genus *Graphium* and to *Protographium.*

**podalirinus** Oberthür
Tibet, Himalayan China 64–84 mm. (2·52–3·31 in.)
This is similar in general colour pattern to the much more common *I. podalirius* (illustrated), but is mostly black above instead of white as a result of the much broader black markings. It is extremely rare in collections.

**podalirius** Linnaeus SCARCE SWALLOWTAIL
N. Africa, Europe and temperate Asia, E. to China 65–82 mm. (2·56–3·23 in.)
Occasional strays of this species reach England, but in much of its range it is not a particularly scarce butterfly below elevations of about 1830 m. (6000 ft). Over 100 names have been applied to various geographical and seasonal forms and individual colour-forms of this species since it was first described in the middle of the 18th century. The caterpillars will feed on the foliage of sloe *(Prunus)* and other shrubs or trees of the family the Rosaceae, including orchard trees, and on oak *(Quercus).* **93**

**iphita** see **Junonia**
**ipsilon** see **Agrotis**
IRIDESCENT BLUE NIGHT see **Magneuptychia junia**
**iris** see **Apatura, Automeris, Phlogophora**
**irraria** see **Euchlaena**
**irrorata** see **Epiplema**
**irrorella** see **Setina**
**isabella** see **Eueides, Pyrrharctia**
**isabellae** see **Graellsia, Teratoneura**
**isocrates** see **Virachola**
**isola** see **Hemiargus**

**Issoria** Hübner (NYMPHALIDAE)
A small genus of butterflies, related to the Fritillaries found mostly in Europe, temperate Asia and N. Africa. See also *Vagrans.*

**lathonia** Linnaeus QUEEN OF SPAIN FRITILLARY
Europe including Britain, N. Africa, Canary Islands, temperate Asia to the Himalayas and W. China 36–46 mm. (1·42–1·81 in.)
The underside of the hindwing has very large silver spots. There are 2–3 broods a year, the adults flying in February-March in the warmer parts, generally in meadows and rough ground up to 2290 m. (7500 ft). The caterpillar feeds on violets *(Viola* sp.). This butterfly is common in N. Africa and S. Europe, further N. it is a migrant, reaching Britain on occasions but it is resident in the S. of Sweden. **186**

**sinha** Kollar VAGRANT
India, Pakistan, Malaysia, Indonesia, W. to Samoa and Tahiti 50 mm. (1·97 in.)
This widespread butterfly is common in many open places and along forest roads. The flight is rapid and the butterfly returns frequently to its favourite spot. It is often attracted to human perspiration. The upperside is a rich orange brown with deep brownish-black bordering the front and outer wing margins. Both wings are variegated with darker patches and spots and the hindwing has a prominent, short, tail. The caterpillar feeds on *Flacourtia* (Flacourtiaceae).

**issoria** see **Euselasia**
**isthmia** see **Mechanitis**

**Isturgia** Hübner (GEOMETRIDAE)
This genus of moths has few species but they occur in Europe, Asia and N. America. The moths are generally well patterned, often rather dark coloured.

**truncataria** Warren
U.S., Canada 16–22 mm. (0·63–0·87 in.)
This moth is found as far N. as Alaska and S. to Colorado and Arizona. In the E. it goes from Labrador to New Jersey. The moth is golden brown with blue black zig-zag stripes on the forewing and 2 on the hindwing. The outer black line on the forewing has a white edge. The margins of both wings have black marks. The male has more yellow on the wing than the female and a yellowish edge to the black forewing band. The caterpillar feeds on *Arctostaphylos* (Ericaceae). The moths are out from May to July, and are day-fliers.

**iswaroides** see **Papilio**
**itea** see **Bassaris**

**Ithomeis** Bates (NEMEOBIIDAE)
There are about 10 species in this tropical C. and S. American genus of butterflies. They are mimics of Ithomiinae species and day-flying moths of the family Hypsidae.

**corena** Felder
Colombia 40–55 mm. (1·57–2·17 in.)
This is a mimic of the Ithomiinae *Napeogenes corena* and others and it closely resembles species of the Hypsid moth genus *Hypocrita,* such as *H. osiba* Druce. There is considerable individual variation in size – perhaps an adaptation related to the different sizes of the moths and butterflies it mimics. Females (not illustrated) of this species are paler in colour than the males. **268**e

**eulema** Hewitson
C. America & Colombia 45–52 mm. (1·77–2·05 in.)
Both the illustrated C. American form of this species and the Colombian form, which has orange-brown streaks on the wing, are mimics of various species of the Ithomiine genera *Napeogenes, Ithomia* and others. **268**d

**Ithomia** Hübner (NYMPHALIDAE)
About 40 species of butterfly are placed in this genus of the subfamily Ithomiinae. It is mainly tropical S. American in distribution but is also represented in C. America.

**drymo** d'Almeida BLUE TRANSPARENT BUTTERFLY
Tropical S. America 40–46 mm. (1·57–1·81 in.)
This is typical of the transparent-winged Ithomiines which are almost impossible to follow with the eye as they flit slowly among the trees of dense rain-forest. Its wings are slightly bluish in colour, with black margins, an oblique, black band across the forewing and a small white spot at the front edge of the forewing immediately distal to the oblique band.

**linda** Hewitson
Ecuador 44–54 mm. (1·73–2·13 in.)
This is similar in pattern to *I. drymo*, but has orange-tipped antennae, a nearly straight front margin to the forewing and lacks a dark bar in the middle of the forewing. It flies together with *Hypothyris antea* Hewitson and *Napeogenes glycera* (Ithomiinae) in a presumably mimetic association of species.

**Ithomiinae** (NYMPHALIDAE)
This subfamily of butterflies is essentially tropical American in distribution but is represented by a single

small genus, *Tellervo* in S.E. Asia and Australia. The Amazon Basin is the distributional centre, but several species are found in Central America and Mexico and 2 species occur in the Antilles. Most of the species are forest dwellers. They are regular visitors to flowers, often in large numbers of different species; their flight is not strong and they are not known to be migratory. Many species are members of mimetic associations with species of Heliconiinae, Danainae, Acraeinae and the moth family Hypsidae, all of which are probably distasteful to predators. These are mimicked by probably palatable species of Nymphalinae, Pieridae and the moth families Castniidae and Dioptidae. The caterpillars feed on species of the generally toxic families Apocynaceae and Solanaceae in the New World and on species of Apocynaceae and Asclepiadaceae in the Old World. Male Ithomiines have a scent-brush on the hindwings. See *Aeria, Callithomia, Elzunia, Greta, Hyalyris, Hypoleria, Hypothyris, Ithomia, Mechanitis, Melinaea, Napeogenes, Oleria, Olyras, Scada, Tellervo, Thyridia.*

**Ituna** Doubleday (NYMPHALIDAE)
The 3 species of this genus of the subfamily Danaiinae are found in tropical Central and S. America. They are large butterflies with transparent, dark-edged wings, typically commonest in mountainous country. The transparency of the wings has been achieved by a reduction in the number of scales, unlike *Thyridia* species (Ithomiinae) in which the scales have been reduced to hairs to produce the same effect. *Ituna* species are members of mimetic associations with species of Ithomiinae, Heliconiinae and the moth families Hypsidae and Castniidae.

**ilione** Cramer
Tropical S. America 72–100 mm. (2·83–3·94 in.)
This is a member of a mimetic association of transparent-winged species of several families discussed under *Thyridia confusa*. The wing shape, colour pattern and flight behaviour are all closely matched in these remarkable species. The various subspecies of *I. ilione* differ in details of the pattern, especially at the base of the forewing which can be transparent, or brown and opaque.

**phanerete** see under **Gazera linus**

**iulia** see **Dryas**

**Ixias** Hübner (PIERIDAE)
A large genus of butterflies with species over Malaysia, India and Indonesia, mostly white or orange coloured.

**marianne** Cramer WHITE ORANGE-TIP
India, Pakistan, Sri Lanka 50–55 mm. (1·97–2·17 in.)
This white butterfly has the black patch at the apex of the forewing with a large reddish-orange patch, through which the brown veins clearly show. There is a brown spot between a few of the veins, but the hindwing, although having a black border, has the rest of the wing all white. The underside is dusted with black scales. The butterfly dashes about, flying about before stopping at flowers or drinking at damp patches. It is out in the sun in the open plains and bush areas.

**reinwardti** Vollenhoven
Indonesia 45–55 mm. (1·77–2·17 in.)
This species is only known from a few of the Islands in Indonesia – Flores, Sumbawa, Alor, Sumba. The male is illustrated, the female is more brown coloured without the orange but with some pale lemon colour on the forewings. **119**f

**vollenhoferi** Wallengren
Dili, Wettar, Timor (Indonesia) 42–48 mm. (1·65–1·89 in.)
There are several forms of this butterfly, which is found in a few islands in Indonesia near New Guinea. The underside lacks the orange. **119**p

**jacksoni** see **Bunaeopsis, Kallima**
JACKSON'S EMPEROR MOTH see **Bunaeopsis jacksoni**
JACKSON'S LEAF see **Kallima jacksoni**
**jacobaeae** see **Tyria**

**Jacoona** Distant (LYCAENIDAE)
There is a single known species in this butterfly genus.

**anasuja** Felder
Malaysia, Indonesia 34–40 mm. (1·34–1·57 in.)
The patch on the front edge of the male hindwing (illustrated) is an area of scent-scales. The female wings are brown above, with a broad white band, enclosing black spots, at the base of the hindwing tail. The under surface of the wings is yellowish, with some black markings at the base of the tails. Very few specimens of this unusually shaped species have been captured. **244**x

JAGGED-EDGED YELLOW see **Eurema gratiosa**

**Jalmenus** Hübner (LYCAENIDAE)
The 9 described species of this genus are known only from Australia. The males differ from females in colour and wing-shape. The caterpillars of some species feed by day, others by night on *Acacia* where they are attended by ants. The chrysalids are sometimes attached to a communal web and are also visited by ants.

**evagoras** Donavon COMMON IMPERIAL BLUE
E. & S.E. Australia 32–35 mm. (1·26–1·38 in.)
The popular name for this butterfly applies only to the illustrated subspecies *J. evagoras evagoras*; the wings of subspecies *J. evagoras ebulus*, for example, are greenish white above. The caterpillars are gregarious and are attended by ants of the genus *Iridomyrmex*. The normal foodplants are species of *Acacia*, but the caterpillar will also eat the waxy secretions of scale-insects (Coccidae). Formation of the chrysalis takes place in webs spun by the caterpillars. **244**n

**inous** Hewitson INOUS BLUE
S.W. Australia 30–35 mm. (1·18–1·38 in.)
The male of this species is similar to the illustrated *J. evagoras evagoras* above, but has darker brown borders on the forewings and a shorter tail to the hindwing. Females are mainly brown above, with a few iridescent blue scales on the basal half of each wing. Its caterpillars, which are attended by ants, feed at night on *Acacia* foliage.

JAMAICA QUEEN see **Danaus gilippus**
JAMAICAN HAIRSTREAK see **Callophrys crethona**
**jamaicensis** see **Smerinthus**

**Jamides** Hübner (LYCAENIDAE)
This widespread but small genus of butterflies is found in Sri Lanka, India, Formosa, S.E. Asia to New Guinea and Australia, and in Fiji and Tonga. *Jamides* is typical of many groups of animals which probably first evolved on the mainland of Asia and then gradually extended their range eastwards through the islands of Malaysia and Indonesia. The wings of most species are blue or blue and white above and the hindwing has a short tail.

**alecto** Felder METALLIC CERULEAN
Sri Lanka, India, Sikkim, Burma 30–44 mm. (1·18–1·73 in.)
This is a common species in wooded, hilly country where the caterpillar feeds on the flowers and young fruit of Cardamom *(Elettaria)*, the seeds of which are used in pharmacy and oriental cookery. Females (not illustrated) are blue, with a broad, greyish brown margin to the forewing and a row of well defined, black-edged, white spots along the outer margin of the hindwing. There are some metallic scales surrounding black markings at the base of the tail on the under surface of both sexes. **263**j

**janardana** see **Mycalesis**
**jania** see **Magneuptychia**
JANSE'S HAWK-MOTH OR SPHINX see **Xenosphingia jansei**
**jansei** see **Xenosphingia**
JAPANESE OAK SILK-MOTH see **Antheraea yamamai**

**Japonica** Tutt (LYCAENIDAE)
There are 2 known species in this butterfly genus. It is found in China, Formosa, Japan and S.E. Russia.

**saepestriata** Hewitson
E. Russia, N.E. China, Japan 35–45 mm. (1·38–1·77 in.)
This unusually coloured Lycaenid is often common in parts of Japan. The under surface of the forewing is also yellow, but is heavily marked with numerous dotted black lines; the hindwing is orange with 2 black spots in the area at the base of the tail. The male is illustrated. **263**d

**japonica** see **Leuhdorfia**
**jarbus** see **Rapala**
**jasius** see **Charaxes**
**jatrophae** see **Anartia**
JAUNE, LITTLE see **Eurema proterpia**
**java** see **Anaphaeis**
JAY, TAILED see **Graphium agamemnon**
**jehana** see **Taturia**
JERSEY TIGER-MOTH see **Euplagia quadripunctaria**
JESTER, HIMALAYAN see **Symbrenthia hypselis**
JEWEL BUTTERFLIES see **Diaethria clymena, Hypochrysops**
JEWEL, GRASS see **Freyeria trochylus**
JEWELLED NAWAB see **Polyura delphis**
JEZEBEL BUTTERFLIES see **Delias**

**Jocara** Walker (PYRALIDAE)
A large genus of Pyralid moths most with a long process at the base of each antennae which are erectile. All, except 3 species described from China, are from Central and S. America. It is possible that, when the Chinese species are studied more closely, they will prove to be distinct from the true species of *Jocara*.

**hemizonalis** Hampson
S. America 24 mm. (0·94 in.)
This Pyralid moth, which has curious long processes from the base of the antennae, has been studied in the canopy of a forest in Guyana. Here at a height of 30 m. (100 ft) up in the treetops the moths were sitting on leaves of trees. When disturbed they raised this process over the head, although at rest it is normally lying along the back of the moth. Only the male has this curious device and the suggestion has been made that this is a warning device. Its sudden erection at close range might disturb a predator. However since it is present only in the male it could be connected with a courtship display mechanism of the male.

**josephina** see **Ascia**

**Josia** Hübner (DIOPTIDAE)
There are about 75 species in this Central American, including Mexico, and tropical S. American genus of moths. Most species are black or dark brown above, with a yellow or orange band extending from the base to the outer margin of each wing. These bands lie nearly parallel to the longitudinal axis of the body when the moth is at rest. This black and yellow pattern is repeated in species of several other families, including Arctiidae, Ctenuchidae, Geometridae and the butterfly family Nemeobiidae.

**lativitta** Walker
Brazil 28–33 mm. (1·1–1·3 in.)
This species is typical in colour pattern of most other species of its genus. A Nemeobiid butterfly *Mesenopsis bryaxis* is thought to be a mimic of *Josia ligata*, a more northern species almost identical in pattern to *J. lativitta*; see also *Chamaelimnas joviana*. **362**c

**joviana** see **Chamaelimnas**
**juanita** see **Prosperinus**
**jucunda** see **Choerapais**
JUDY BUTTERFLIES see **Abisara**
**jugulandis** see **Acrobasis**
**julia** see **Erateina**
JUMPING BEAN MOTH see **Cydia saltitans**
JUNGLE GLORY see **Thaumantis diores**
JUNGLE KING see **Thauria lathyi**
JUNGLE KING, TUFTED see **Thauria aliris**
JUNIPER CARPET see **Thera juniperata**
**juniperata** see **Thera**
**juno** see **Dione**

**Junonia** Hübner (NYMPHALIDAE)
Species in this butterfly genus are found on all the continents and some of the Pacific islands. The genus belong to the subfamily Satyrinae. Several species in the genus were formerly in *Precis* and there is some confusion in the literature about the species in these 2 genera.

**almana** Linnaeus PEACOCK PANSY
India, Pakistan, Burma, Sri Lanka 60–65 mm. (2·36–2·56 in.)
This species is common over much of its range. It is a light tawny coloured butterfly with black marks along the front margin of the forewing and an eyespot with a white centre near the middle of the wavy outer margin of the wing. The hindwing has a prominent eyespot, larger than the forewing one, which is surrounded by a yellow, black-edged, rim. The centre of the eyespot is black and red, usually with two lighter patches. There are three wavy black lines round the margin of the hindwing and it is produced into a short tail. The underside is leaf-like in the dry season form, but the wet season form has eyespots below and is generally not tailed. Because of the differences in the two forms they were for a long time regarded as distinct species. The leaf-like underside of the dry season form helps it to blend with the dead leaves on which it settles. The insect can be found up to 2290 m. (7500 ft) but is common in gardens at elevations below 1220 m. (4000 ft), frequently visiting flowers.

**hierta** Fabricius YELLOW PANSY
Africa, Arabia 50–60 mm. (1·97–2·36 in.)
In Africa this ranges from the Sudan to all parts south of Sahara. It is a butterfly of the open bush and is common in gardens and can be found in most months of the year. The caterpillar feeds on *Barleria*, *Justicia*, *Asystasia* (Acanthaceae). Similar looking *Junonia* species occur in India.

**iphita** Cramer CHOCOLATE PANSY
India, Pakistan, Sri Lanka, Burma 55–80 mm. (2·17–3·15 in.)
This is more sombre coloured than the others mentioned, but nevertheless is boldly patterned. It is brown above with iridescent brown bands across the wing. The eyespots on the hindwing are only faintly visible. The apex of the forewing is produced slightly making the outer wing margin concave. It is a very common species but tends to stick to the wetter well-wooded slopes in the hills up to 2740 m. (9000 ft). It settles on damp patches on the road where it can be found sunning itself, although it is commoner in shady places than the other species in the genus.

**lavinia** Cramer BUCKEYE
Tropical America N. to S. Canada 56–64 mm. (2·2–2·52 in.)
This butterfly is common and widespread in the U.S. There are a number of forms of this butterfly with much local variation. It is a fast flier and often visits flowers or mud puddles. The caterpillar is grey striped with yellow spots, and is covered with numerous short branching spines. It feeds on a variety of plants inincluding plantain *(Plantago)*, snapdragon *(Antirrhinum)*, stonecrop *(Sedum)*. The butterfly is out in early spring since it hibernates over winter in the adult stage. Some of the populations of Buckeye butterflies are migratory. **185**f

**lemonias** Linnaeus LEMON PANSY
India, Pakistan, Burma, Sri Lanka 45–60 mm. (1·77–2·36 in.)
A very common and widespread species but generally more of a forest insect than most of the others in the genus. It reaches up to 2440 m. (8000 ft) and is particularly common at flowers in gardens. The forewing has a large irregular yellow patch, edged with black all round. The apex has a few whitish spots. The hindwing has a yellow patch edged with black, but in the front of the hindwing the black edge is broad and has a large blue patch in it. The caterpillar feeds on species of Acanthaceae.

**orithya** Linnaeus BLUE PANSY or BLUE ARGUS
India, Pakistan, Burma, Africa, S. Asia, China, Philippines, Indonesia, New Guinea, Australia 40–60 mm. (1·57–2·36 in.)
This is a common species over most of its range from sea-level up to 2740 m. (9000 ft). They are found in dry regions, preferring the stony arid areas to forests. The butterflies are about in all the months of the year at the lower levels, but generally only in spring at higher elevation. They fly in the hot sunshine. This is, as it's name implies, a blue butterfly and many subspecies have been described. The caterpillar feeds on species of Acanthaceae and on *Antirrhinum* (Scrophulariaceae). **185**b

**rhadama** Boisduval
Africa, Madagascar and the Mascarene Islands 50 mm. (1·97 in.)
This is relatively local in Africa but occurs on Mauritius and Rodrigues as well as the Comoro Islands. The caterpillar feeds on *Justicia* and *Barleria* (Acanthaceae). **185**c

**sophia** Fabricius LITTLE COMMODORE
Africa 35–45 mm. (1·38–1·77 in.)
This species is common throughout the year, flying in forests and forest glades. It is the smallest African species in the genus. The caterpillar is black, with fine yellow lateral lines and short spines. Several different colour forms have been described and there are differences between the wet and dry season forms. This species ranges from Senegal in the W., through Zaire to Uganda, Kenya and Tanzania S. as far as Natal. **183**

**stygia** Aurivillius
Africa 45–55 mm. (1·77–2·17 in.)
This species is common in forests in the wetter parts of W., Central and E. Africa and extends up into the Sudan. It is a very dark brown or blackish brown butterfly with eyespots along the wing margins and some transverse reddish brown bands across the wing. The underside is paler, with a purplish tinge in the male, brown in the female. The eyespots are reduced on the underside to small dots along the wing margin; there is a light patch near the apex of the forewing. **178**

**villida** Fabricius MEADOW ARGUS
New Guinea and S.W. Pacific, Australia, Tasmania and Lord Howe Islands 42–55 mm. (1·65–2·17 in.)
This brightly coloured butterfly is common over much of its range. The greyish brown upperside has 2 large eyespots, black with white centres surrounded by red and a blue patch with red edges on the forewing. The hindwing is plainer, but has two prominent eyespots, edged with red. The margin of fore- and hindwings are marbled with lines. Underside forewing is grey with 2 orange bars and a large red eyespot, with a black and white centre. The hindwings are grey, relatively unmarked. The caterpillars feed on a wide variety of plants in various families, including Plantaginaceae, Scrophulariaceae and Compositae. **185**j

**westermanni** Westwood BLUE-SPOT COMMODORE
W. Africa 48 mm. (1·89 in.)
From Ghana to Angola this butterfly is fairly common. The male is velvet black and blue, quite distinct from the female which has blackish grey wings with black transverse streaks in the cell and some red-yellow spots. The biology of this species is unknown. **184**

**jurtina** see **Maniola**
JUVENAL'S DUSKY-WING see **Erynnis juvenalis**
**juvenalis** see **Erynnis**
**juventa** see **Danaus**

KAISER-I-HIND BUTTERFLY see **Teinopalpus imperialis**

**Kalenga** Whalley (THYRIDIDAE)
An African genus, not known elsewhere. They are lightly patterned, reddish brown moths usually about 24 mm. (0·94 in.) wingspan with paler brown round marks (reticulations). Only 3 species are known at present. Nothing is known of their biology.

**maculanota** Whalley
Africa 20–24 mm. (0·79–0·94 in.)
Reddish-brown with whitish spots over the wings, usually with a small white spot and a black spot in the middle of each wing. The caterpillar of this moth is not known and only relatively few of the moths have been collected in Zaire, Zambia and Rhodesia. The species was only recently recognized and very little is known about it. Field studies, as in so many of these little known species, would be valuable.

**Kallima** Doubleday (NYMPHALIDAE) LEAF BUTTERFLIES
These are generally found in areas of heavy rainfall and thick forests. They rarely fly far and spend most of their time settled on trees or bushes. Their flight is erratic and rapid, moving the wings quickly up and down. It has often been remarked that, while they are often pursued by birds in flight, once they land the birds are baffled by their disappearance. They are readily attracted to sap or over-ripe fruit. The butterflies in this genus are always used as examples of camouflage, probably the 'Indian Leaf-butterfly' is amongst one of the best known examples of this. Species of the genus are found in India, Africa and S.E. Asia.

**ansorgei** Rothschild ANSORGE'S LEAF BUTTERFLY
Africa 55–65 mm. (2·17–2·56 in.)
A rare species found in the forests in Congo, W. Uganda and W. Kenya where it can be found in all months of the year. The shape is typical of *Kallima* with a long tail and leaf-like wings. The upperside is a bright blue-green on the fore- and hindwings, with a broad olive brown outer margin on the forewing; the hindwing has the green surrounded by a darker olive green. There is a reddish eyespot near the tail and indistinct dark spots along the margins of the wings. The underside is a brown colour with a leaf pattern and provides a marvellous camouflage.

**inachus** Boisduval ORANGE OAK LEAF or INDIAN LEAF BUTTERFLY 92–120 mm. (3·62–4·72 in.)
India, Pakistan, Burma, S. China, Taiwan
This butterfly is always referred to in any discussion of cryptic (concealing) patterns. The underside is remarkably leaf-like even with a 'mid-rib'. The butterfly occurs in woods, often near river banks. It is commonly attracted to rotting fruit. The caterpillar feeds on *Girardinia* (Urticaceae), *Strobilanthus* (Acanthaceae) and other plants. Although the sexes are similarly patterned, the females are larger than the males. There are many forms, with slightly different patterns, which have been named. **160**d

**jacksoni** Sharpe JACKSON'S LEAF BUTTERFLY
E. Africa, Zaire 58–66 mm. (2·28–2·6 in.)
The female is paler and larger than the male. The underside of both is leaf-like, with a long 'stalk' which extends into the tails on the wing. This species is rare in collections and is often regarded as one of the most beautiful of the African butterflies. It lives in open bush and wooded savannah where it flies in all months of the year. **208**d

**rumia** Westwood AFRICAN LEAF BUTTERFLY
Africa 68–80 mm. (2·68–3·15 in.)
In the male the wings are dark brown above with a broad violet blue median band. The female is larger and the forewings have a white median band. Both sexes have the leaf-like underside. The caterpillar is reddish grey with broad black lines and darker coloured lateral streaks. The butterfly is common from the Ivory Coast to the Congo and into Uganda. The specimen figured was photographed in Uganda. The butterfly is out in all months of the year and sometimes can be found feeding on fermenting fruit. **211, 212**

**horsfieldi** Kollar: syn. *K. philarchus*
BLUE OAK LEAF BUTTERFLY
Sri Lanka, S. India 85–110 mm. (3·35–4·33 in.)
Similar shape to *K. inachus* but with indigo blue to green on the forewing and without the blue sheen of *K. inachus*. The green near the base of the forewing is edged with paler green and then a dark outer margin to the wing. The underside is just as wonderfully leaf-like as *K. inachus*. When they settle it really is impossible to distinguish them from the surrounding leaves. The caterpillar feeds on *Strobilanthes* (Acanthaceae).

**kannegieteri** see **Stibochiona coresia**
**karschina** see **Scada**
**kenedyae** see **Pseudosphex**

KENT BLACK ARCHES see **Nola albula**
KENTISH GLORY see **Endromis versicolora**
KERNEL BORER, OLIVE see **Prays oleae**
**khasiana** see **Euthalia**
KILIMANJARO SWALLOWTAIL see **Papilio sjoestedti**
KING BUTTERFLIES see **Amathusia, Thauria**
KING
CROW, BLUE-BANDED see **Euploea leucostictos**
PAGE see **Papilio thoas**
SHOEMAKER, BANDED see **Prepona meander**
SHOEMAKER, SILVER see **Prepona demophoon**
KNOT-HORN MOTHS see **Myelois**
**knysna** see **Zizeeria**
KOH-I-NOOR BUTTERFLY
see **Amathuxidia amythaon**
**kolpakofskii** see under **Acerbia**
**komarovi** see **Rethera**

**Kricogonia** Reakirt (PIERIDAE)
This is a genus with only one species which is found in the tropics of America. It occasionally strays into the southern parts of the U.S.

**lyside** Godart LIGNUM VITAE YELLOW
U.S., Central and S. America. W. Indies 38 mm. (1·5 in.)
This species is widespread, and often common over its range which extends into Venezuela north into the S. part of the U.S. The males have a large patch of yellow at the base of pale greenish white wings and the hindwings have a short black line just below the apex. The females are all over pale greenish white with only a trace of black on the apex of the forewing margins. The caterpillar feeds only on Lignum vitae *(Guaiacum officinale)* and the insect is common wherever this occurs. In Mexico, in 1973, a huge migration of this species was watched when hundreds of thousands were seen moving along a front where an estimate of the number was 10,000 butterflies per 10 m. (32 ft) of road in a solid mass of flying insects 6–8 m. (20–26 ft) high.

**kuehniella** see **Ephestia**
**kuntzei** see **Goodia**
**labdaca** see **Libythea**
**laboulbeni** see **Palustra**
LACEWING, BUTTERFLY,
RED see **Cethosia biblis**
SMALL see **Actinote pellenea**
**laceritalis** see **Parotis**

**Lachnocnema** Trimen (LYCAENIDAE)
The four species of this genus of butterflies are found in tropical Africa. They are blue or brown above, with white or orange patches. The hindwing is without a tail.

**bibulus** Fabricius
Tropical Africa 20–30 mm. (0·79–1·18 in.)
This is a common butterfly in parts of W. Africa, including urban areas. Males are entirely dark brown above; females are less dark in colour and have a large, white patch on each wing. The caterpillars are predators of plant-bugs (Heteroptera). Ants attend both the caterpillars and the plant-bugs, but the nature of the association is as yet uncertain.

**lacinia** see **Chlorosyne** under **C. saundersii**
LACKEY see **Malacosoma neustria**
LACKEY, GROUND see **Malacosoma castrensis**

**Lacosoma** Grote (MIMALLONIDAE)
Two species of this genus occur in N. America, the remainder are solely Central and S. American in distribution. Many species have scalloped outer margins to the wings.

**chiridota** Grote SCALLOPED SACK-BEARER
Canada (Ontario) to Florida and Texas 22–30 mm. (0·87–1·18 in.)
This is somewhat similar in wing-shape to that of the illustrated *L. valeria*, but the colour of the wings is a fairly uniform brown, palest and more yellowish in the female. Moths have been captured as early as April in the S. of its range and towards the end of June in the N. The caterpillar feeds on oak *(Quercus)*.

**valeria** Schaus
Venezuela 28 mm. (1·1 in.)
This is one of numerous S. American species of its genus about which virtually nothing is known except for the general characteristics of its genus, such as nocturnal flight habits and so on. **318**l

**Lacosomidae** see under **Mimallonidae**
**lactinea** see **Amsacta**
**lactucina** see **Astheridia**

**Lactura** Walker (YPONOMEUTIDAE)
A large genus of micro-moths with nearly 100 described species from New Guinea, Australia, Solomon Is., the U.S. and Central America. There is a wide variety of colours, mostly red-patterned wing and reddish hindwings. Most species are known only from a few specimens, although the reason for their apparent rarity is not known.

**callopisma** Walsingham
New Guinea 30 mm. (1·18 in.)
Very few specimens of this species have ever been found. Even large museums have only one or two specimens. The moth is a strikingly coloured insect, quite unmistakable. Although there has been fairly extensive collecting in New Guinea over the years, the island is so large and has so many almost unvisited areas that many more new and remarkable species will be found there. Probably more will then be discovered of the biology of some of the unusual looking moths like *L. callopisma* which are at present so rare in collections. **38**d

**coleoxantha** Diakonoff
Indonesia 45–65 mm. (1·77–2·56 in.)
This has more yellow colour and less red on the wings than the other species. It is known only from a few specimens from West Irian (formerly Dutch New Guinea). **38**c

**suffusa** Walker
Australia 35–40 mm. (1·38–1·57 in.)
This is a variable species and the exact identity of the species at present under *L. suffusa* Walker is not clear. The specimen illustrated is darker than many in the series of specimens examined and probably represents the distinct species, *L. obscura* Butler. However further research into this species complex is needed. The slug-like caterpillar feeds on species of *Sideroxylon* (Sapotaceae). **38**a

**lacturnus** see **Everes**

**Ladoga** Moore (NYMPHALIDAE)
A small genus of butterflies which extend from Europe through temperate Asia. The species were formerly included in the genus *Limenitis* to which *Ladoga* is closely related.

**camilla** Linnaeus WHITE ADMIRAL
Europe including Britain, temperate Asia including Japan 52–60 mm. (2·05–2·36 in.)
The caterpillar feeds on honeysuckle *(Lonicera)* and the butterfly flies in woodlands in June and July. The Japanese specimens are slightly different from the European ones and have been given a subspecific name. The butterfly is a powerful flier, frequently gliding. The upperside black and white pattern contrasts strongly with the brown and white underside. **220**

**populi** Linnaeus POPLAR ADMIRAL
C. Europe, temperate Asia including Japan 70–80 mm. (2·75–3·15 in.)
The underside is a bright orange with prominent white patches between the veins. The butterfly is out in June and July, flying in open woodlands and is often common over much of its range. Several subspecies have been described, for example the Japanese subspecies which has the white more conspicuous on the upperside than the European subspecies. The caterpillar feeds on aspen and poplars *(Populus)*.

**laertes** see **Morpho**
**laeta** see **Erova, Musurgina**

**Laetilia** Ragonot (PYRALIDAE)
Instead of the more usual plant diet, the caterpillars of species in this genus of Pyralid moths are predatory, feeding on other insects. The species in this genus are all American. The genus belongs to a subfamily of moths (Phycitinae) which have many species injurious to crops, but *Laetilia* is useful in helping to destroy insect pests.

**coccidivora** Comstock
U.S. 10–20 mm. (0·39–0·79 in.)
This is the best known species in the genus. The caterpillars eat scale insects (Coccids). There is considerable variation in size of specimens of this species. The caterpillars browse on the scale insects which are attached to the plant. Since scale insects cause damage to the plant, the caterpillars are useful in destroying them.

**laetissima** see **Eretmocera**
**laglaizei** see **Papilio**
**lalos** see **Iolaus**
**lamae** see **Holomelina**

**Lampides** Hübner (LYCAENIDAE)
A single species of butterfly is placed in this genus.

**boeticus** Linnaeus TAILED, LONG-TAILED, PEA, BEAN or LUCERNE BLUE
Europe S. of the Alps, Ascension Island, Africa, Asia, Australia, the Pacific Islands to Hawaii 24–36 mm. (0·94–1·42 in.)
A remarkably wide range of climates is tolerated by this species. It can withstand both tropical and temperate conditions and has been captured at sea level up to altitudes of 3350 m. (11,000 ft) on the slopes of Mt Everest. Migratory specimens reach the shores of England from time to time. The female (not illustrated) is mostly brown above, with some blue at the base of the wings. The caterpillar feeds on the flowers and inside the pods of numerous species of Leguminosae. It is a pest of broad beans *(Vicia)* in Hawaii and sometimes of green peas *(Pisum)* in other parts of the world. **248**

**lamprima** see **Milionia**

**Lampronia** Stephens (INCURVARIIDAE)
A genus of micro-moths with species in Europe, N. America and Africa. At present there are no clear limits for the genus which possibly contains some unrelated species.

**oehlmanniella** Treitschke
Europe including Britain 12–14 mm. (0·47–0·55 in.)
This small moth is a reddish bronze colour. It is common over most of its range. The caterpillar mines leaves in the early stages, later on it comes out of the leaf, making a small case of leaf fragments which it carries around. The moth is out in June. **5**e

**Lamproptera** Gray (PAPILIONIDAE) DRAGONTAIL BUTTERFLIES
This is a genus of strange, long-tailed species found in N.E. India, Burma, S. China, Indo-China, through the Malay Archipelago to Sulawezi (Celebes) and the Philippines. There are two known species, *L. curius* and *L. meges*. Both are said to resemble dragonflies in flight. *Lamproptera* is possibly most closely related to the Himalayan *Teinopalpus*. The caterpillars feed on species of Combretaceae.

**meges** Zinken GREEN DRAGONTAIL
Distribution as genus 35–50 mm. (1·38–1·97 in.)
This extraordinary species flies along sunlit streams and rivers in forested regions up to elevations of 1525 m. (5000 ft). Its flight is remarkably like that of a dragonfly, with rapidly beating wings; even when settled, the wings may continue to flutter like those of dragonflies. *L. meges* will hover over flowers when feeding, unlike most butterflies. Its close relative, the well named *L. curius*, differs in having a nearly white, transverse band on the wings. **94**m

**lanata** see **Megalopyge**
LANTANA PLUME MOTH see **Lantanophaga pusillidactyla**
LANTANA, SMALLER see **Strymon bazochii**

**Lantanophaga** Zimmerman (PTEROPHORIDAE)
Plume moths in this genus are not well studied with the exception of one widespread species. The forewings are divided into 2 lobes by a short cleft which does not reach the middle and with rather broad apices to these lobes. The hindwings are divided into 3 lobes. They have very long legs as do all known species of the family.

**pusillidactyla** Walker LANTANA PLUME MOTH
W. Indies, Mexico – but now widespread in the tropics and sub-tropics 10–12 mm. (0·39–0·47 in.)
This moth has been widely introduced to control the weed *Lantana* (Verbenaceae). It has been introduced into Hawaii and is widespread in India where it has been described as 'beneficial' in districts invaded by *Lantana*. The moth is reddish brown with black and white marks near the apex of the forewing. The host plant, *Lantana* can spread into areas needed for grazing and at times is a serious problem. The use of biological control, using this tiny plume-moth has many practical and financial advantages. The use of the insect is cheap, after the initial introduction, and if successful, self-perpetuating, and there is no risk to human life, as there may be with careless use of insecticides.

**Laodamia** see **Oncocera**
**laodomas** see **Battus**

**Laothoe** Fabricius (SPHINGIDAE)
There is a single species in this genus of hawk-moths.

**populi** Linnaeus POPLAR HAWK-MOTH
Europe, temperate and subarctic Asia, not Japan 70–82 mm. (2·76–3·23 in.)
This is one of the few species of moths and butterflies which are more common in Britain than elsewhere in their range. It prefers damp localities, where the caterpillar feeds from July to early October, on poplar *(Populus)*, sallows and willows *(Salix)*, alder *(Alnus)* and birch *(Betula)*. Fully grown caterpillars are green, speckled with yellow, or reddish brown and have oblique, yellow stripes on each side; the horn is tipped with red. **349**

**laothoe** see **Temenis**
LAPPET see **Gastropacha quercifolia**
LAPPET MOTH, PINE see **Dendrolimus pini**
LARGE
BLUE see **Maculinea arion**
BLUE CHARAXES see **Charaxes bohemani**
COPPER see **Lycaena dispar**
DYNAMINE see **Dynamine mylitta**
EMERALD see **Geometra papilionaria**
GREEN-BANDED BLUE see **Danis danis**
GRIZZLED SKIPPER see **Pyrgus alveus**
HEATH see **Coenonympha tullia**
NONAGRIA see **Nonagria oblonga**
OAK BLUE see **Narathura amantes**
SALMON ARAB see **Colotis fausta**
SKIPPER see **Ochlodes venatus**
SPOTTED ACRAEA see **Acraea zetes**
THORN MOTH see **Ennomos autumnaria**
TIGER see **Lycorea ceres**
TIGER-MOTH see **Pericallia matronula**
TORTOISESHELL see **Nymphalis polychloros**
VAGRANT see **Nepheronia argia**
WHITE see **Pieris brassicae**
YELLOW UNDERWING see **Noctua pronuba**
LARGER SOD WEBWORM see **Pediasia trisecta**
**larilla** see **Napeogenes**

**Laringa** Moore (NYMPHALIDAE)
A small genus of butterflies found in Indonesia and India.

**castelnaui** Felder
Indonesia: Nias Island 48–55 mm. (1·89–2·17 in.)
The females are larger than the males and brown, with white patches. Several subspecies have been described, the one illustrated is subspecies *L. c. niha* Fruhstorfer from Nias Island. Nothing appears to be known of the biology of this species. **208**f

**larissa** see **Melanagria**

**Lasaia** Bates (NEMEOBIIDAE)
There are 11 species in this chiefly C. American and tropical S. American genus of butterflies. Two species occur in subtropical S. Texas. They fly in scrubland and thinly wooded country, especially where there are streams. Flowers are visited regularly, at least by the males.

**maria** Clench
Mexico, Guatemala 29–33 mm. (1·14–1·3 in.)
Although common in the mountains of C. and S. Mexico, and known since 1911, *L. maria* had been misidentified with other species of *Lasaia*, mostly *L. sula*, until 1972 when its identity as an undescribed species was revealed. The male is a dull, slightly iridescent blue above, with black lines and some white and reddish brown markings. Females are reddish or greyish brown above, with a blue sheen which is especially noticeable in fresh specimens. Most captures of this species have been made at elevations of between 600–1830 m. (2000–6000 ft).

**scotina** see under **L. sessilis**

**sessilis** Schaus
Mexico, C. America 34–40 mm. (1·34–1·57 in.)
This weakly iridescent, dark bronze-coloured butterfly is at present known only from the foothills of the Sierra Madre, from Mexico to Guatemala. *L. sessilis* had been misidentified frequently prior to the revision by Clench in 1972. The species known as the Blue Metalmark, and referred to as *L. sessilis* by many authors, is the species *L. sula*. The only species of *Lasaia* with similar coloration to that of *L. sessilis* is the much smaller *L. scotina*. **268**n

**sula** see under **L. maria** and **L. sessilis**

LASCAR, BURMESE see **Pantoporia hordonia**

**Lasiocampa** Schrank (LASIOCAMPIDAE)
EGGAR MOTHS
This is a moderately sized genus of round-winged moths found chiefly in temperate Europe and Asia but with a few African species.

**quercus** Linnaeus OAK EGGAR MOTH
Europe including Britain, temperate Asia to Siberia 48–75 mm. (1·89–2·95 in.)
The female (illustrated) of this species moves very little until it has been fertilized, after which it takes to the wing and lays its eggs. Males often fly during the day in the search for females. Northern specimens have a 24-month life cycle; they hibernate as a young caterpillar during the first winter and as chrysalids during the second winter. Southern examples feed more rapidly during the summer and overwinter as a chrysalis during the following winter and emerge as adults the following spring in a 12-month cycle. The foodplants include dogwood *(Cornus)*, bramble *(Rubus)*, hawthorn *(Crataegus)*, ling heather *(Calluna)* and ivy *(Hedera)*. **325**

**staudingeri** see under **Lasiocampidae**

**Lasiocampidae** TENT CATERPILLAR MOTHS
This is a fairly large family of moths found in most parts of the world but with its greatest concentration of species in the tropics. They are absent from New Zealand and there is only one record of this family from the Greater Antilles. The abdomen of most Lasiocampid species is noticeably longer than that of most other species in related families (eg. Bombycidae). Females are usually rather cumbersome creatures, incapable of sustained flight; some have greatly reduced wings, for example *Lasiocampa staudingeri* (Algeria). The caterpillars are densely hairy; some are flattened and have a series of processes along each side, which effectively break up lateral shadows and aid in concealment. A few live in communal webs or nests and all species spin a cocoon inside which the transformation into a chrysalis takes place. Limited use has been made at times by man of the silk fibres which make up these cocoons. The caterpillars of several species are pests of various forest and orchard trees; 2 species, *Leipoxais rufobrunnea* Strand (W. Africa) and *Streblote dorsalis* Walker (S.E. Asia) are pests of cocoa *(Theobroma)*. The caterpillars' hairs have irritant properties in some species. See *Dendrolimus, Gastropacha, Grammodora, Lasiocampa, Macrothyalacia, Malacosma, Philudoria*.

**Lasiommata** Westwood (NYMPHALIDAE)
This genus of butterflies is in the subfamily Satyrinae. The generic name *Pararge* Hübner is often used for the species in this genus and there is still some dispute as to the exact limits of these 2 genera and whether all the species should be grouped in the one genus.

**megera** Linnaeus WALL BUTTERFLY
Europe including Britain, Turkey, N. Africa, Syria and W. Asia 36–50 mm. (1·42–1·97 in.)
The butterfly is active in the sunshine on rough ground, gardens and edges of woodlands from March or April. There are 2 or more broods in different parts of its range. The caterpillar feeds on grasses, especially *Poa* and *Dactylis*. Several subspecies have been described. It is a common butterfly over most of its range. **176**

**schakra** Kollar COMMON WALL BUTTERFLY
India, Pakistan 55–60 mm. (2·17–2·36 in.)
A common species, found in most months of the year up to 1830 m. (6000 ft). It is common on open hillsides where it may settle on the ground or stones. The forewing is brown with a prominent eyespot of black with a white centre near the apex in a yellowish brown patch. There are several conspicuous eyespots on the hind margin of the hindwing. In general colour and appearance it is broadly similar to the Wall butterfly of Europe, *L. megera*.

**lateritia** see **Apamea**
**lathamella** see **Glyphipteryx**
**lathonia** see **Issoria**

**Lathroteles** Clarke (LATHROTELIDAE)
There is only one species in this genus of micro-moths. It is related to the Pyralid moths but differs from them in the lack of a tympanal organ. This is a structure found at the base of the abdomen of most Pyralid moths which is responsible for sound reception and which in some Pyralids is well developed, see *Ostrinia nubilalis*.

**obscura** Clarke
Rapa Islands (Tubuai or Austral Islands) 7 mm. (0·28 in.)
The forewings are black with a white band near the base of the wing and some more white near the apex of the wing, including a square white spot. Only 2 female specimens of this species are known and its life history is unknown.

**Lathrotelidae**
There is only one genus in this family of micro-moths which was proposed for an odd species from the Island of Rapa. This is in the Tubuai or Austral group S.E. of Tahiti. The family is related to the Pyralidae. See *Lathroteles*.

**lathyi** see **Thauria**
**latiferreanus** see **Melissopus**
**latipennis** see **Gnophaela**
**latistrigella** see **Lyonetia**
**lativitta** see **Josia**
**latreillianus** see **Graphium**
LATTICED HEATH see **Semiothisa clathrata**
**lautella** see **Phyllonorycter**
**lavinia** see **Doxocopa, Junonia**

**Laxita** Butler (NEMEOBIIDAE)
This is a small genus of butterflies restricted to S.E. Asia. Borneo seems to be the centre of distribution. They are related to species of *Taxila*. The ground-colour of the upper surface of many species is a purplish red, sometimes described as the colour of claret.

**damajanti** Felder
Malaya, Sumatra, Borneo 36–40 mm. (1·42–1·57 in.)
The male of this butterfly is a rich, dark red above, unmarked on the forewing, but with a broad, black outer margin to the hindwing. The female is similarly, but less richly coloured, and may have a trace of yellow, apical band on the forewing. This is the com-

monest species of *Laxita*. The under surface of both sexes is red or orange, with an iridescent black and blue pattern.

LAYMAN see **Amauris albimaculata**
LEAF BUTTERFLIES see **Kallima, Doleschallia**
LEAF BLUE see **Amblypodia anita**
LEAF BLOTCH MINERS see **Gracillariidae**
LEAF, DRY see **Precis tugela**
LEAF, YELLOW see **Phoebus trite**
LEAF-MINERS see **Caloptilia, Lyonetia**
LEAF-ROLLER HEATH, FRUIT TREE see **Archips argyrospila**
LEAF ROLLER MOTHS see **Tortricidae**
LEAFWING BUTTERFLIES see **Anaea, Doleschallia**
LEAFWORM MOTHS see **Anomis, Spodoptera**
**leander** see **Coenonympha**

**Lecithoceridae**: syn. *Timyridae*
These are small micro-moths related to the Oecophoridae. The caterpillars are believed to feed in dead leaf litter. The adult moths when at rest sit with the head downwards. Most species are found in the Oriental and Australian regions. See *Timyra*.

**leda** see **Eronia, Melanitis**

**Leguminivora** Obraztsov (TORTRICIDAE)
This genus of micro-moths is in the subfamily Olethreutinae. It is a small genus with species in India, Sri Lanka and Japan. The caterpillars, as the generic name suggests, feed on species of Leguminosae.

**tricentra** Meyrick: syn. *L. pseudonectis*
India, Pakistan 10 mm. (0·39 in.)
A reddish grey mottled moth with three small black marks near the margin of the forewing. The hindwing is the same colour but lacks the black marks and has, in the male, a conspicuous black basal area. The moth is widespread. The caterpillar tunnels in the stems of Sunn Hemp, *Crotalaria juncea* (Leguminosae) and can be a serious pest of this crop. It causes the stem to swell up and a gall to form at the axil of the leaf. It occasionally attacks the seeds of *Crotalaria*.

**leilus** see **Urania**
**Leipoxais rufobrunnea** see under **Lasiocampidae**

**Leioptilus** Wallengren (PTEROPHORIDAE)
The plume moths of this genus are typical of the family with long angular looking legs and divided fore- and hindwings. The forewings are divided into 2 lobes by a cleft which runs halfway down the wings while the hindwings have 3 lobes cleft almost to the base.

**bowesi** Whalley
Britain, France 25–29 mm. (0·98–1·14 in.)
This plume moth has pale straw-coloured forewings with a reddish-brown longitudinal streak from the cleft to the base of the forewing. There is usually a black spot at the base of the cleft in the forewing. The moth was first recognized from Britain in 1960 when it was separated from a closely allied species *L. osteodactylus* Zeller. It is known only from Kent and the adjoining counties in Britain. The caterpillar feeds on the seed heads of Golden Rod *(Solidago)*, and becomes full grown in October. Recently a French entomologist has discovered the species in S. of France. It is probably widespread but is easily overlooked or confused with the more widespread species.

**lemnata** see **Cataclysta**
LEMON MIGRANT see **Catopsilia pomona**
LEMON PANSY see **Junonia lemonias**

**Lemonia** Hübner (LEMONIIDAE)
There are about 12 species in this genus of mostly yellow and brown, medium sized moths. It is represented in temperate Asia and in Europe.

**dumi** Linnaeus
Europe to Turkey, S. Russia 44–60 mm. (1·73–2·36 in.)
This is widespread but fairly rare species in much of Europe. It is not found in Britain. Adults are on the wing in October; males fly by day, females only during the evening and at night. The antennae of the females are much less feathery than in the male (illustrated). The caterpillar is brown, covered with short hairs and has a pair of black bars on top of each segment; it feeds on the foliage of hawkweeds *(Hieracium)*, dandelion *(Taraxacum)* and other species of the family Compositae. **318**e

**sardanapalpus** Staudinger
C., E. temperate & subarctic Asia 34–42 mm. (1·34–1·65 in.)
The ground-colour of the wings of this species is similar to that of the illustrated *L. dumi*, but the markings are a much paler yellow. The hindwing pattern is also like that of *L. dumi*, but the forewing differs in the presence of 2, pale yellow, oblique bars which join the middle of the pale yellow, transverse band to the front margin of the wing. The moths fly from the end of September into October. The black and white marked, grey caterpillars are covered with black hairs which arise from small, brownish orange tubercles; they are sometimes common, unlike the adults which are rare in collections.

**lemonias** see **Junonia**

**Lemoniidae**
About 20 species are recognized currently in this Saturniid-like family of the superfamily Bombycoidea. There are 3 genera: *Lemonia, Spiramiopsis* and *Sabalia*. Some authors treat *Sabalia* as the sole genus of a separate subfamily Sabaliinae. See *Lemonia, Sabalia, Spiramiopsis*.

**leonardus** see **Hesperia**
LEONARDUS SKIPPER see **Hesperia leonardus**
**leonidas** see **Graphium**
LEOPARD MOTHS see **Zeuzera, Ecpantheria**
**leosthenes** see **Protographium**
**lepida** see **Parasa**

**Lepidochrysops** Hedicke (LYCAENIDAE)
This is a large African genus of tailed or tail-less blue butterflies averaging between 30–60 mm. (1·18–2·36 in.) wingspan.

**gigantea** Trimen GIANT BLUE
C. Africa 38–60 mm. (1·5–2·36 in.)
This is a woodland butterfly and one of the largest species of Lycaenidae. The upper surface is a violet blue, with black borders in the male and broader, white spotted borders in the female. The under surface ground-colour of both male and female is yellowish white; the markings black, edged with white.

**leprieuri** see **Ocnogyna**

**Leptarctia** Stretch (ARCTIIDAE)
There is a single known species in this genus.

**californiae** Walker
W. U.S. 28–33 mm. (1·1–1·3 in.)
S. California is the home of this small, highly variable species. The hindwings may be yellow or red, with black markings, or entirely black; the forewings are brown, usually with at least a trace of a white, transverse band across the middle of the wing.

**Leptidea** Billberg (PIERIDAE)
A large genus of butterflies found throughout Europe and temperate Asia and Japan.

**morsei** Fenton FENTON'S WOOD WHITE
Europe, Asia including Japan 42–46 mm. (1·65–1·81 in.)
Several subspecies of this butterfly have been named since the species was originally described from Japan. They have 2 or more broods a year. The caterpillar feeds on *Lathyrus* (Leguminosae) and the butterfly, which is out in June and July, flies in woodlands. Externally it is very similar to *L. sinapis* but the latter is larger. Frequently *L. morsei* has been considered a subspecies of *L. sinapis* but the more recent works separate these as 2 distinct species.

**sinapis** Linnaeus WOOD WHITE
Europe including Britain, Syria 36–48 mm. (1·42–1·89 in.)
This is mainly a woodland insect as its name implies. Its flight is weaker-looking than most white butterflies. The caterpillar feeds on various Leguminosae, including *Lotus* and *Vicia*. Many subspecies and aberrations have been described over the range. **116**

**leptocircata** see **Erateina**

**Leptoclanis** Rothschild & Jordan (SPHINGIDAE)
The 2 species of this hawk-moth genus are found in tropical Africa. The proboscis of this genus is short and poorly developed.

**basalis** Walker WOUNDED HAWK-MOTH
Rhodesia, Tanzania 72–88 mm. (2·83–3·46 in.)
The hindwing of this species has a conspicuous burgundy-red marking at its base. The forewings are either pale yellow or yellowish white, speckled with pale greenish grey. The life history is unknown.

**Leptosia** Hübner (PIERIDAE)
White butterflies with very rounded forewings and a very slow flight. They are generally rather fragile looking butterflies found mainly in Africa but with some species in the Oriental region.

**alcesta** Stoll AFRICAN WOOD WHITE
Africa 30–46 mm. (1·18–1·81 in.)
A delicate looking, white butterfly with a brown edge to the apex of the forewing, and a black spot just below it. The hindwing underside is yellowish with green markings. This butterfly is found from the tropical equatorial forests S. to Natal, E. Mozambique and N. Transvaal. The foodplant of the caterpillar is believed to be *Vernonia* (Compositae).

**nina** Fabricius PSYCHE BUTTERFLY
Burma, Pakistan, India, Sri Lanka, Indonesia, S. China, Hainan 35–50 mm. (1·38–1·97 in.)
A small white butterfly with rather rounded wings with a small black patch at the apex of the forewing and another black patch on the wing below it. The rest of the wings are unmarked. The underside hindwing is faintly streaked with green. This little butterfly is a weak flier. It flutters in and out of the undergrowth, not venturing far into the open. The caterpillar feeds on species of Capparidaceae. The butterfly is widespread in the Oriental region and is often common where it occurs.

**Leptotes** Scudder (LYCAENIDAE)
This essentially pan-tropical genus has a remarkable capacity for crossing oceans. It is represented on islands in the Indian Ocean (Socotra), the Pacific Ocean (Galapagos) and the Atlantic Ocean (Canary Islands). Two species are found in N. America.

**andicola** Godman & Salvin
Tropical S. America 40 mm. (1·57 in.)
This is a rather large violet-blue butterfly known only from between 2290 m. (7500 ft) and 3660 m. (12,000 ft) in the Andes.

**callanga** see under **L. webbianus**

**cassius** Cramer CASSIUS BLUE
Warm temperate N. America to C. & tropical S. America 22 mm. (0·87 in.)
There are several subspecies of this widespread and common New World species, including 2 which are restricted to groups of islands in the Lesser Antilles and a third unnamed subspecies in the Cayman islands. This is a butterfly of the sunshine, and flowers of all kinds, and is especially common in damp areas where there are flowering shrubs and trees. Flowers also provide food for the caterpillars which prefer species of Leguminosae. The female is white with brown markings, the male mostly brown above.

**webbianus** Brullé CANARY BLUE
Canary Islands 25 mm. (0·98 in.)
The male of this species is a rich violet-blue above; the female, in contrast, is orange. *L. webbianus* is unknown outside the Canary Islands where it occurs as high as 3050 m. (10,000 ft) on Tenerife. Its closest relative

seems to be *L. callanga* of the S. American Andes. The foodplant is not known, but the butterfly is commonly found in pine forests.

LESSER ARMYWORM see **Spodoptera exigua**
CLOUDED YELLOW see **Colias chrysotheme**
FRITILLARY, PURPLE see **Clossiana titania**
MARBLED FRITILLARY see **Brenthis ino**
PURPLE EMPEROR see **Apatura ilia**
SWALLOW PROMINENT see **Pheosia gnoma**
WANDERER see **Danaus chrysippus**
WAX MOTH see **Achroia grisella**

**Lethe** Hübner (NYMPHALIDAE)
A large butterfly genus with species in Asia, including India and in America. Several subgenera have been described which are found in different parts of the range of genus. This is a typical member of the subfamily Satyrinae.

**europa** Fabricius BAMBOO TREE BROWN
Formosa, China, India, Pakistan, Philippines, Indonesia 65–75 mm. (2·56–2·95 in.)
This richly coloured chocolate brown butterfly has a patterned underside with spots near the wing margin and some stripes across the wings. The female has a broad white subapical bar. The caterpillar feeds on Bamboo (Gramineae) and the butterfly is found round houses where these are planted. In Malaya it is common in the rainy season and is particularly active at dawn and dusk. It is one of the butterflies which are occasionally attracted to the lights of houses. **174**r

**lethe** see **Euchromia**

**Leto** Hübner (HEPIALIDAE) SWIFT MOTHS
A genus found in S. Africa in the grouping called micro-moths. This refers to their systematic position and not their size.

**venus** Stoll
S. Africa 100–160 mm. (3·94–6·3 in.)
This species is found in the Transvaal and Cape Province. The caterpillar feeds in the trunks of trees, probably feeding between the wood and the bark. **7**c

**Leucidia** Doubleday (PIERIDAE)
A genus of butterflies with only 4 species at present known, all confined to C. and S. America.

**brephos** Hübner SNOWFLAKE
S. America into the W. Indies 20–25 mm. (0·79–0·98 in.)
This small, all white butterfly, has rather rounded, almost moth-like, wings. It is found in Venezuela, Ecuador, Brazil and Colombia. Male and females are similar externally except for a small patch of scent scales under the front margin of the male forewing. The butterfly is a weak flier and tends to keep in well-shaded areas but is common over much of its range.

**leucobathra** see **Trachydora**

**Leucochitonea** Wallengren (HESPERIIDAE)
Only 3 species are known is this genus of skipper butterflies and all are confined to Africa.

**levubu** Wallengren WHITE-CLOAKED SKIPPER
S. Africa to Rhodesia and Malawi 44 mm. (1·73 in.)
This is practically an all white Skipper butterfly. The margin of the forewing has a black line with small white spots and some more black edged white spots near the apex. The white hindwings have a narrow black edge. This Skipper is found in bush country, sometimes on hilltops. Its biology is not known.

**leucodesma** see under **Phyciodes ianthe**
**leucodisca** see **Erocha**
**leucodrosime** see **Pereute**

**Leucodonta** Staudinger (NOTODONTIDAE)
There are 2 known species, *L. bicoloria* and the solely Japanese *L. nivea*.

**bicoloria** Denis & Schiffermüller
WHITE PROMINENT 32–40 mm. (1·26–1·57 in.)
Europe including Britain, temperate Asia to Japan
The orange and white colours of this species are unusual in this family of essentially cryptically coloured moths. This is a rare moth in Britain and is well established only in Ireland. Its caterpillar feeds on birch *(Betula)*. **352**

**Leucoma** Hübner (LYMANTRIIDAE)
About 50 species of white moths are placed in this widespread, Old World genus. The species previously placed under *Stilpnotia* are currently placed in *Leucoma*.

**salicis** Linnaeus WHITE SATIN or SATIN MOTH
Europe including Britain, temperate Asia to E. U.S.S.R. and Japan, E. & W. Canada, N.E. U.S. 44–55 mm. (1·73–2·17 in.)
This is an Old World species which has been introduced accidentally on both the Pacific and Atlantic coasts of N. America where its yellow-haired, black and red caterpillar is now an occasional pest during May and June of forest trees such as poplar *(Populus)* and willow *(Salix)*. The caterpillar overwinters in a silken shelter and completes its growth the following year by June or July. Moths are on the wing from June to August. **361**

**leucomelas** see **Milionia**
**leucomochla** see **Xyleutes**
**leuconoe** see under **Graphium idaeoides, Idea**
**leucostictos** see **Euploea**
**leucostigma** see **Hemerocampa**
**leucotaenia** see **Papilio**
**leucothyris** see **Olearia**
**levana** see **Araschnia**
**levubu** see **Leucochitonea**
**libatrix** see **Scoliopteryx**
**libya** see **Melanitis**

**Libythea** Fabricius (LIBYTHEIDAE)
SNOUT BUTTERFLIES
This is a worldwide genus. One of the characteristics of *Libythea* is a reduction of the front legs of the male butterfly (cf. Nymphalidae) but the female has the normal 3 pairs. The caterpillars are generally smooth, not hairy like Nymphalid caterpillars. The popular name is derived from the long palps which project in front of the head like a snout.

**carinenta** Cramer SNOUT BUTTERFLY
U.S. through C. America to Paraguay 45 mm. (1·77 in.)
Very similar to the species on the other continents. This is a dark blackish brown butterfly with three white patches in the black forewing apex and three orange marks on the rest of the forewing. The hindwings, which are paler near the inner margin, also have three small orange patches. It reaches from Texas as far S. as Buenos Aires. It is a well known migrant in America, occasionally reaching the W. Indies.

**celtis** Laicharting NETTLE-TREE or EUROPEAN BEAK BUTTERFLY 32–44 mm. (1·26–1·73 in.)
S. Europe, N. Africa, Asia including Japan, Formosa
Several subspecies of this butterfly have been described. It has a long hibernation period in Europe over the winter months before flying in the spring to lay the eggs which produce the summer brood. The caterpillar feeds on trees of *Celtis australis* (Ulmaceae). The toothing of the outside of the wing gives a striking appearance to this butterfly. **185**a

**geffroyi** Godart BEAK BUTTERFLY
Burma, Thailand, Indonesia to New Guinea, Solomon Is., Australia 48–55 mm. (1·89–2·17 in.)
This species is typically Libytheid in shape but is more brightly coloured than many. The males have a strong violet-blue colour on the upperside of the wings, sometimes with a few white spots near the apex of the wing. The females are orange brown with some white spots on the wing. This widespread species reaches into N. Queensland where two distinct subspecies have been described.

**labdaca** Westwood AFRICAN SNOUT BUTTERFLY
W. & C. Africa and Madagascar 25–50 mm. (0·98–1·97 in.)
A brownish butterfly with white spots (usually 3) near the apex of the forewing and large pale orange-brown patches on the fore- and hindwing. The forewing outer margin is sharply angled giving a blunt projecting apex. The hindwing has a sharply angled point. The underside of the forewing is like the upperside but the underside of the hindwing is purplish and less patterned than the upperside. This butterfly is common over most of its range and often migrates locally in huge swarms, sometimes in such numbers, in roads through forests, that the radiators of passing cars get clogged up with the bodies of dead butterflies and the engines overheat. Several subspecies of this butterfly have been described from different parts of Africa.

**Libytheidae** SNOUT BUTTERFLIES
A small family, with one European, one N. American species while the rest of the species are found mostly in the Old World Tropics. Their characteristic feature is the long palps which project in front of the head like a beak. Most of the species have a rather similar appearance for example, the N. American species is similar to the European, *Libythea celtis*. See *Libythea*.

**licina** see **Enantia**
**licus** see **Castnia**
**lidderdalei** see **Bhutanitis**
LIGHT EMERALD see **Campaea margaritata**
LIGNUM VITAE YELLOW see **Kricogonia lyside**
**ligustri** see **Sphinx**
LILAC
BEAUTY see **Salamis cacta**
NYMPH see **Asterope rosa**
LILAC-BORER see **Podosesia syringae**
**lilis** see **Melinaea**
LIMA-BEAN POD-BORER see **Etiella zinckenella**

**Limacodidae** SLUG-CATERPILLAR MOTHS
This family of moths was formerly known as Cochlididae and also included the Heterogeneidae and Eucleidae, names which appear in older literature. The family is regarded as part of the Zygaenoid groups. The caterpillars of Limacodids are generally slug like. There are a number of species of economic importance in the family. Species of this family are found all over the world but are commonest in the tropics. Some species have caterpillars with stinging hairs and most have very bizarre shaped caterpillars, quite unlike the more common 'tube-like' ones. See *Apoda, Chrysamma, Coenobasis, Heterogenea, Parasa, Phobetron, Sibine*.

**limatula** see **Rhodoneura**
**limbalis** see **Uresiphita**
LIME
HAWK-MOTH see **Mimas tiliae**
SWALLOWTAIL see **Papilio demoleus**
LIME-SPECK PUG see **Eupithecia centaureata**

**Limenitis** Fabricius (NYMPHALIDAE)
A huge genus of butterflies in Europe, Asia and the Orient. The caterpillars hibernate when partly grown in a rolled up tube made from the basal part of the leaf. Species included in *Limenitis* in this book may be found in *Ladoga* in some works. In N. America a number of species of *Limenitis* are known to hybridize.

**archippus** Cramer VICEROY or MIMIC
Canada, U.S., Mexico, W. Indies 65–70 mm. (2·56–2·76 in.)
This butterfly is a mimic of the Monarch *(Danaus plexippus)* which is known to be distasteful. It is assumed that the Viceroy is also avoided because of this resemblance. It is common over most of the E. States in open meadows and roadsides. Its flight differs from the Monarch, in having a faster wing beat alternating with glides on horizontal wings, the Monarch holds the wings sloping upwards. The caterpillars feed at night on *Salix*, *Populus* and a number of other trees. There are 2 broods in the N., 3 or more in the S. The butterflies are out in May or June. **216**n

**arthemis** Drury WHITE ADMIRAL or PURPLE BANDED BUTTERFLY
Canada, U.S. 70–80 mm. (2·76–3·15 in.)
This species is separated into several subspecies. It is

common in the hardwood forests and forest edges. It is often found feeding on carrion or drinking at the edge of wet mud, sometimes in large numbers. The caterpillar feeds on Black and Yellow Birch *(Betula* sp.*)* and several other trees. The adults are out in mid June. The N. subspecies, *L. arthemis arthemis,* is found from Canada into New England, New York, Pennsylvania, Michigan and Minnesota. One of the more S. subspecies, *L. arthemis astyanax* Fabricius, the Red Spotted Purple extends from the S. edge of the range of the *L. arthemis arthemis* to central Florida and Texas. It differs in the forewing pattern, in the Purple Banded, the prominent white band is present across the wing, in the Red Spotted Purple, this band is absent. The butterflies are strong fliers, gliding up and down woodland rides. **216**e

**camilla** see **Ladoga camilla**

**danava** Moore COMMODORE
India, Pakistan, Burma 80–85 mm. (3·15–3·35 in.)
This is a two-tone brown butterfly, darker towards the base and paler brown towards the wing margin. There is a greenish tinge to the hindwing margin which is lightly patterned. This butterfly is found in the hills, settling on damp patches of sand or mud. The flight is swift and sailing. The female is brownish green with two white bands across the fore- and hindwing, the inner band being broader. The biology is not recorded.

**populi** see **Ladoga populi**

**procris** Cramer COMMANDER
Sri Lanka, India, Pakistan, Burma 60–75 mm. (2·36–2·95 in.)
This species is very like the White Admiral, *Lagoda camilla*, but is more reddish brown on the upperside. The base of the underside of the wings have a greenish tinge. The caterpillar feeds on *Mussaenda frondosa* (Rubiaceae) and species of Menispermaceae. The butterfly is widespread in India and is common on the lowlands and is found in open glades and roadsides through forests.

**reducta** Staudinger SOUTHERN WHITE ADMIRAL
S. and C. Europe, W. Asia, Syria, Iran, Caucasus 46–54 mm. (1·81–2·13 in.)
Very similar to the White Admiral, *Ladoga camilla*, differing in a more black upper surface with a white mark in the cell and with the marginal black spots with a bluish tinge. The underside has a paler basal area and more red near the apex of the forewing. The female is similar. The butterfly is out in May with 2–3 broods in S. Europe, a single brood in July in Switzerland. The caterpillar feeds on Honeysuckle *(Lonicera)* and the butterfly is common over much of its range in woodlands and open bush land.

**trivena** Moore INDIAN WHITE ADMIRAL
India, Pakistan 60–75 mm. (2·36–2·95 in.)
This is very common in India and often flies in large numbers. It appears in May in the Simla Hills at 2440 m. (8000 ft) and in other parts later in the year. The butterfly is very similar to the European White Admiral, *Ladoga camilla*, but has more white on the fore- and hindwings and is generally paler below. The caterpillar feeds on Honeysuckle *(Lonicera)* which grows up to 3960 m. (13,000 ft) but the butterfly has not been found above 3350 (11,000 ft).

**weidemeyerii** Edwards WEIDEMEYER'S ADMIRAL
U.S. 65–70 mm. (2·56–2·76 in.)
This butterfly is found from E. California and Idaho to New Mexico, Colorado, Kansas and S. Dakota. It has the typical appearance of the White Admiral butterflies, *Ladoga camilla*, black with a broad white band on each wing and some white spots near the margin, but without the blue marginal colour found in allied species. The caterpillar feeds on species of *Populus* and *Salix* (Salicaceae).

**zayla** Doubleday & Hewitson
BI-COLOUR COMMODORE
India, Pakistan, Burma 80–95 mm. (3·15–3·74 in.)
This is a common butterfly over its range found in the hill forest up to 2440 m. (8000 ft) in August and September. The male is olive-brown, with a prominent yellow band down the forewing, continued by a white band on the hindwing. The margin of both wings have red lines, very irregular on the hindwing with darker brown round them.

**limniace** see **Danaus**
**linariata** see **Eupithecia**
**linda** see **Ithomia**
**lineata** see **Erateina, Eueides, Hyles**
**lineatella** see **Anarsia**
**lineatonetella** see **Ethmia**
**lineella** see **Glyphipteryx**

**Lineodes** Guenée (PYRALIDAE)
A small genus of Pyralid moths from C. and S. America. They have long, slender, forewings and long legs and consequently they are rather similar to some of the plume moths (Pterophoridae).

**peterseni** Hampson
S. America 18–20 mm. (0·71–0·79 in.)
Little is known of the relationships of these curious moths. This species is known only from Colombia and its life history is unknown. **56**z

**lineola** see **Thymelicus**
**linus** see **Gazera**

**Liphyra** Westwood (LYCAENIDAE)
This genus is represented in India, the Malay Archipelago, New Guinea and N. Australia. The 2 known species are said to be active at dawn and dusk. This characteristic together with their robust shape and powerful flight has lent the species of this genus a somewhat moth-like character. It is probably most closely related to the African genus *Euliphyra.*

**brassolis** Westwood MOTH-BUTTERFLY
India, S.E. Asia to N. Australia 60–92 mm. (2·36–3·62 in.)
The predacious caterpillars of this species live in ants' nests where they feed on ant larvae. The emerging butterfly has a curious covering of loose scales which protect it from possible ant attacks when it makes its escape from the nest. Ants become entangled in these scales if they attempt to molest the butterfly. The somewhat anomalous common name for this species results from its rather moth-like appearance. Females differ from the male in the reduction of black and dark brown apically on the forewing and on the outer part of the hindwing. **241**

**Liptena** Westwood (LYCAENIDAE)
There are over 50 species in this tropical African genus of butterflies. About half of these are white and black in colour and mimic species of Pieridae; many of the remainder are mimics of *Acraea* species. Their caterpillars are hairy and Lymantriid-like and feed on lichens and fungi like species of *Mimacraea.*

**simplicia** Möschler
Africa S. of the Sahara 25–34 mm. (0·98–1·34 in.)
The upper surface of this butterfly is white, except for a black, apical area on the forewing; beneath, the coloration is similar, but there is an additional, black, marginal band on the hindwing. Several species of *Liptena* closely resemble *L. simplicia* in colour pattern.

**liris** see **Siga**
**lise** see **Elimniopsis**
**lisides** see **Suasa**
**lisus** see **Mithras**
**Lithocolletis** see **Phyllonorycter**

**Lithosia** Fabricius (ARCTIIDAE) FOOTMEN MOTHS
This is a genus of moths of the subfamily Lithosiinae. The 20 or so species are found in Europe and temperate Asia, as far as Japan, and in Africa.

**quadra** Linnaeus FOUR-SPOTTED FOOTMAN
Europe including Britain, through temperate Asia to Japan 42–52 mm. (1·65–2·05 in.)
This is a typical Lithosiine in wing-shape, see *Eilema complana*. Only the female is 'four-spotted', it has yellow wings, with a pair of conspicuous black spots on each wing. The male has mainly grey forewings, without spots. The moths are on the wing in July and August; they start flying at dusk. The caterpillar is nearly black, with grey and black hairs, and yellow and red markings. It hibernates and recommences feeding in May on lichens, especially those growing on oaks *(Quercus)*.

**litigiosa** see **Tagiades**
LITTLE
BLUE see **Cupido minimus**
COMMODORE see **Junonia sophia**
JAUNE see **Eurema proterpia**
METALMARK see **Calephelis virgiensis**
SOLDIER see **Chlorosyne saundersii**
WOOD SATYR see **Euptychia eurytus**
YELLOWIE see **Eurema venusta**
**littoralis** see **Spodoptera**
**litura** see **Spodoptera**

**Lobocraspis** Hampson (NOCTUIDAE)
There is a single known species of moth in this genus.

**griseifusa** Hampson EYE-FREQUENTING MOTH
Burma, Thailand, Cambodia 35–38 mm. (1·38–1·5 in.)
In appearance this is an unremarkable, mostly grey, speckled moth, suffused with dull orange-yellow especially along the front edge of the forewing and marginally on the hindwing. It is, however, one of the several eye-frequenting moths found in S. Asia and in Africa and may be the cause of eye diseases in man and other mammals; see *Calpe minuticornis*.

LOBSTER MOTH see **Stauropus fagi**
LONG-HORN MOTHS see **Adela, Nemophora**
LONG-TAILED
ADMIRAL see **Antanartia schaenia**
BLUE see **Lampides boeticus**
SKIPPERS see **Urbanus**

**Lonomia** Walker (SATURNIIDAE)
This is a group of 26 sombrely coloured, medium sized Saturniids found in C. America and tropical S. America. The caterpillar of one species, *L. achelous*, has been the subject of medical research; the spines of every species are at least mildly irritating.

**achelous** Cramer
Tropical S. America 70–114 mm. (2·76–4·49 in.)
The fibrinolytic secretions produced by the spines of the caterpillar have been shown recently to produce bleeding for up to a month, from the mucous membranes of the nose and intestines of man and have produced one fatality. There are possible medical uses for a chemical having such powerful anticoagulent properties. Females differ from the illustrated male in the more greyish coloration and the slightly convex outer margin to the forewing. **337**f

LOOPER MOTHS see **Bupalis, Semiothisa, Trichoplusia** and under **Geometridae**

**Lophocorona** Common (LOPHOCORONIDAE)
Only 3 species are known in this genus, all were first described in 1973.

**pediasia** Common
Australia 11–13 mm. (0·43–0·51 in.)
The whitish forewings are heavily patterned with black near the base, leaving a white band near the apex with a few black marks on the apex of the wing. The hindwings are brownish speckled with grey. This very obscure species, recently discovered, could provide data for the continuing dialogue on the relationship of the more primitive Lepidoptera. **1**h

**Lophocoronidae**
This family of moths was first recognized and described in 1973. It is one of the more primitive families with similar fore- and hindwing venation. There is only one genus in the exclusively Australian family. It is related to the Mnesarchaeidae and the Agathiphagidae. See *Lophocorona*.

**loranthi** see **Antheraea**
**lorisana** see **Deudorix**
LOST METALMARK see **Apodemia phyciodoides**

**Loxolomia** Maassen (SATURNIIDAE)
Both species of this genus are tropical American in distribution.

**serpentina** Maassen
Brazil 110–140 mm. (4·33–5·51 in.)
This is a particularly rare species in collections. The rather beautiful, speckled brown and white wings of this species are probably effectively cryptic when the moth is at rest on bark. The pale front half of the forewing, if facing the sky (the source of light), produces an effect of roundness, and resemblance to a twig. The opposite of this is 'counter-shading', found in the caterpillars of some hawk-moths (Sphingidae), in which the surface furthest away from the light is paler in colour thus tending to nullify shadow and produce an effect of flatness. **337**e

**loxoscia** see **Mnesarcha**

**Loxostege** Hübner (PYRALIDAE)
A huge genus of Pyralid moths, which at present contains a large number of unrelated species. The American ones, which are true *Loxostege* were recently studied in detail.

**rantalis** Guenée: syn. *L. similalis* GARDEN WEBWORM
U.S., W. Indies, Mexico 17–22 mm. (0·67–0·87 in.)
The forewings are dull brownish yellow with darker indistinct lines. The hindwings are paler. The greenish caterpillar has a light mid-dorsal stripe and dark spots and tufts of hair along the side. It feeds in webbings on a wide variety of plants including Alfalfa, clover (Leguminosae) and sugar beet (Chenopodiaceae). The moth is out from April to November.

**Loxotoma** Zeller (STENOMIDAE)
Only 2 species are known in this genus of micro-moths. One is widely distributed while the other in known only from a single damaged specimen in a museum collection. The wing-shape and size of species in this genus separate it from most of the others in the family.

**elegans** Zeller
Central & S. America 34–44 mm. (1·34–1·73 in.)
This is a large species by comparison with others in the family (Stenomidae) and has a characteristic wing shape. Nothing seems to be known of its biology. The genus was studied in detail recently by an American entomologist who collected together all the known information about the morphology and relationships of this species which is widespread in Central and S. America. **42**b

**loxozona** see **Phyllonorycter**

**Loxura** Horsfield (LYCAENIDAE)
The range of this genus includes Sri Lanka, India, Burma, Malaya to the Philippines and Sumbawa Island. Two species are known, both have long, white-tipped hindwing tails. Their caterpillars feed on the foliage of monocotyledonous plants.

**atymnus** Stoll YAMFLY
Sri Lanka, India, Burma to the Philippines and Lombok 36–40 mm. (1·42–1·57 in.)
This is a fairly common butterfly at low elevations in open forest and wasteland where it normally flies some distance above the ground; but when settled it is relatively easy to capture. The male is a darker orange above than the illustrated female, has dark streaks along the outer parts of the hindwing veins and has no trace of the oblique, transverse line. The caterpillar feeds on young shoots of the Yam *(Dioscorea)* and is consequently a potential pest-species. Red ants apparently always attend the caterpillars. **244**p

**lubricipeda** see **Spilosoma**
**lucasi** see **Cymothoe**
LUCERNE BLUE see **Lampides boeticus**
**lucidus** see **Arctonotus**
**Lucilla** see under **Lucillella**

**Lucillella** Strand (NEMEOBIIDAE)
There are 3–4 species of butterflies in this N.W. S. American genus. They are possibly mimics of species of the Hypsid genus *Darna*. Hewitson's original name for this genus, *Lucilla,* has been used already by 3 earlier authors and was replaced with the present name in 1932 by Strand.

**camissa** Hewitson
Ecuador 34–38 mm. (1·34–1·5 in.)
This has been compared to closely similar species in *Callicore* a genus of Nymphalidae but it is doubtful that these will prove to be unpalatable models for *L. camissa* – one of the few known foodplants of *Callicore* species is *Trema,* a genus of Ulmaceae which is not known to be a poisonous family of plants. Other possible models for *L. camissa* are *Darna colorata* Walker and *Darna rubiplaga* Warren, both day-flying moths which occur in Ecuador and belong to the family Hypsidae, a generally toxic family. These *Darna* species are not exact copies of *L. camissa* but are fairly good matches.

**lucina** see **Dione, Hamearis**

**Ludia** Wallengren (SATURNIIDAE)
The strongly hooked male forewing of *Ludia* species bears an E-shaped transparent patch, the hindwing a large eyespot. Females differ in the weakly hooked apex to the forewing. The caterpillars, like those of *Goodia,* bear long, branched hairs. The 14 or so species are tropical or S. African in distribution.

**goniata** Rothschild BLACK PRINCE
South Africa 50–60 mm. (1·97–2·36 in.)
The range of this species is restricted to Natal and the E. part of Cape Province. It is doubtfully recorded from coastal Mozambique and Tanzania. *L. goniata* is similar in general appearance to the illustrated *L. orinoptena* except for the more strongly hooked forewing of the male. The caterpillar is unknown.

**orinoptena** Karsch
W. Africa and the Congo Basin 50–68 mm. (1·97–2·68 in.)
The colour pattern and wing-shape of this species is typical of its genus. Females differ from the illustrated male in the shape of the forewing which is only very slightly hooked at the apex. **331**c

**Luehdorfia** Crüger (PAPILIONIDAE)
There are 2 species in this genus of butterflies. It is closely related to the Mediterranean *Parnalius.* The caterpillars feed on species of the family Aristolochiaceae. The fossil species *L. bosniackii* Rebel was discovered in Tuscan deposits of the Miocene Period of geological history, dating back about 30,000,000 years.

**japonica** Leech
S.E. Russia, N.E. China, Japan, Korea 45–60 mm. (1·77–2·36 in.)
Females of this species differ little from the illustrated male in colour pattern, and individual variation is restricted chiefly to small differences in the shape of the wing markings. Its caterpillars differ from those of *L. puziloi* in the absence of yellow on the lateral tubercles of the essentially black body. **94**h

**puziloi** Erschoff
S.E. U.S.S.R., N.E. China Japan, Korea 40–50 mm. (1·57–1·97 in.)
This butterfly is similar to *L. japonica* and differs chiefly in the reduction of black at the outer margin of the hindwing. It is sometimes common in gardens, where violets *(Viola)* are often one of the attractants.

**lugubralis** see **Thyris**
LULWORTH SKIPPER see **Thymelicus acteon**
**luna** see **Actias**
LUNA MOTH see **Actias luna**
LUNAR PRINCE see **Goodia kuntzei**
**lunata** see **Zale**
**lunigera** see **Callidula**
**lunus** see **Nothus**
LURCHER BUTTERFLY see **Yoma sabina**
**luteolata** see **Opisthograptis**
**luteum** see **Spilosoma**
**lybia** see **Eueides**

**Lycaena** Fabricius (LYCAENIDAE)
This is a large, relatively homogeneous group of essentially N. temperate butterflies of both the Old and New World, but with incursions into the tropics. The species are small, robust, fast-flying, mostly coppery orange or purplish brown in colour, with several black or brown spots especially on the under surface of the wings.

**alciphron** Rottemburg PURPLE-SHOT COPPER
W. Europe, N. Africa to Turkey and Iran 30–36 mm. (1·18–1·42 in.)
The upper surface of the male of this species is a beautiful, dark spotted, iridescent purple and reddish orange; the female is dark brown above. It is found up to 910 m. (3000 ft) in the N. part of its range and up to 2740 m. (9000 ft) in the Atlas Mountains of Morocco. The caterpillars feed on sorrel *(Rumex),* usually in damp meadows. **263**e

**dispar** Haworth LARGE COPPER
Europe, temperate Asia E. to the Pacific 34–40 mm. (1·34–1·57 in.)
A now extinct subspecies of this butterfly *L. dispar dispar,* once inhabited the marshy regions of Huntingdonshire and Cambridgeshire in E. England until 1847 or 1848. The chief factor in its decline was the drainage of the marshes which was the home not only of this butterfly but of many other marshland insects, birds and plants. The present English population of *L. dispar* belongs to a Dutch subspecies *L. dispar batavus* Oberthür (illustrated), which was introduced to Britain in 1927. The caterpillars of this handsome Copper feed on the leaves of dock *(Rumex)* in marshy areas and along wet ditches. **238**

**dorcas** see under **L. helloides**

**epixanthe** Boisduval & Leconte BOG COPPER
E. Canada and the U.S. as far S. as Minnesota and New Jersey 22–26 mm. (0·87–1·02 in.)
This species, like its fellow Lycaenid the Bog Elfin *(Callophrys lanoraieensis* Sheppard) is an inhabitant of acid bogs where its caterpillar feeds on wild Cranberry *(Vaccinium).* The male is chiefly brown above, with a slight purplish iridescence and a few black spots on both wings and an orange band along the outer margin of the hindwing. The under surface of both sexes is yellow, with numerous black spots and an orange-red line close to the outer margin of the hindwing. The eggs of this species are laid singly under leaves of the foodplant where they remain until the spring.

**helloides** Boisduval PURPLISH COPPER
W. U.S., E. to Michigan 29–33 mm. (1·14–1·3 in.)
This is a butterfly of open meadows and wasteland where its caterpillar feeds chiefly on knotweed *(Polygonum).* The adult male is normally a rather unusual iridescent purple and brownish-orange above, with black spots on both wings and an orange, marginal hindwing band, but some males are much darker in colour. Females have an orange forewing and larger black spots. *Lycaena dorcas,* which has a more northerly distribution than *L. helloides,* is difficult to distinguish from the latter, but is generally larger and has several broods per year in contrast with the single-brood *L. dorcas.*

**phlaeas** Linnaeus SMALL COPPER
E. N. America and the Rockies from California to the shores of the Arctic Ocean, N. and E. Africa, Europe including Britain, temperate Asia E. to Japan 25–32 mm. (0·98–1·26 in.)
This is one of the commonest and most widespread butterflies in the N. Hemisphere. Grassy slopes from sea level to 1830 m. (6000 ft) (and higher in E. Africa) are the haunts of this species. The male is a notoriously aggressive butterfly and has been known to drive away even small birds. The male's territory covers an area of about 14 sq.m. (150 sq. ft) in which it feeds and perches while waiting for passing females. The foodplants of the caterpillar are sorrels *(Rumex)* and other genera of the family Polygonaceae. **239**

**thoe** Guérin-Méneville BRONZE COPPER
E. Canada, N. & C. U.S. 33–36 mm. (1·3–1·42 in.)
This is the largest species of its genus to be found in N.

America. It flies in wet meadows where there is a supply of the caterpillars foodplants, docks *(Rumex)* and knotweed *(Polygonum)*. The male of *L. thoe* is purplish orange above, with darker spots, and with an orange, marginal band on the hindwing. Females are light orange above.

**thetis** Klug GOLDEN COPPER
Pakistan (Baluchistan) 32–36 mm. (1·26–1·42 in.)
The female (not illustrated) of this species differs from the superb, golden copper-coloured male in the more yellowish upper surface. *L. thetis* is confined to mountainous regions towards the W. end of the Himalayas. **263**f

**Lycaenidae** BLUES, HAIRSTREAKS and COPPERS
This is a chiefly tropical and subtropical family except for the subfamily Plebiinae which is mainly N. temperate in range. It is best represented in the rainforests of S.E. Asia and Africa. Most of the species are small, often with tailed hindwings; many are a brilliantly iridescent blue, violet, green, red or orange, especially in the male. There are also a few white or yellow species. The upper surface of the wings is often very different between the sexes and almost invariably quite different from the pattern of the under surface. Most species are active only in the sunshine, but *Liphyra brassolis* flies at dusk and dawn. The caterpillars are typically small and tapered at both ends; many are associated with ants in various ways, some species living at times inside ants nests. A 'honey gland' found on the upper surface of the abdomen in most species produces a liquid which is eaten by the ants but most caterpillars seem to gain little advantage from this association. The sound-producing organs of the chrysalis have been discussed and illustrated recently by Downey and Allyn. Buds, flowers, fruits including the pods of legumes and lichens are main sources of food of most species, but some are carnivorous and feed on living aphids, scale insects and leaf-hoppers (Hemiptera). The adults of many species have an eyespot at the base of the hindwing tail or tails, which, when the butterfly is at rest with its wings over its back, produces an effect of an insect facing the opposite direction to the actual specimen. The characteristically quivering tails look like antennae and enhance this illusion. A few species are mimics of distasteful groups of butterflies see, for example *Liptena* and *Mimacraea*. See *Agriades, Agrodiaetus, Albulina, Amblypodia, Aphnaeus, Aricia, Atlides, Aurea, Axiocerses, Bindahara, Brephidium, Callophrys, Castalius, Celastrina, Cheritra, Chlorostrymon, Chrysozephyrus, Cupido, Curetis, Danis, Deudorix, Drupadia, Erora, Euliphyra, Eumaeus, Evenus, Everes, Feniseca, Freyeria, Glaucopsyche, Heliophorus, Hemiargus, Hypochrysops, Hypolycaena, Icaricia, Iolana, Iolaus, Jacoona, Jalmenus, Jamides, Japonica, Lachnocnema, Lampides, Lepidochrysops, Leptotes, Liptena, Liphyra, Loxura, Lycaena, Lycaenopsis, Lysandra, Maculinea, Megalopalpus, Miletus, Mimacraea, Mithras, Myrina, Narathura, Neomyrina, Ogyris, Paiwarria, Palaeochrysophanus, Panthiades, Pentila, Philotes, Phylaria, Plebejus, Plebicula, Poecilmitis, Polyommatus, Poritia, Pseudolycaena, Quercusia, Rapala, Rekoa, Sandia, Spalgis, Spindasis, Strymon, Strymonidia, Suasa, Syntaracus, Tajuria, Teratoneura, Thecla, Theorema, Uranothauma, Vacciniina, Virachola, Wagimo, Zintha, Zizeeria, Zizina, Zizula, Zeltus.*

**Lycaenopsis** Felder HEDGE BLUE
Males of this large genus are an unusually bright blue above, with dark borders on the forewings. The species fly in forests and their margins, less frequently in open woodland. Specimens have been recorded at elevations up to 3050 m. (10000 ft) in India. The extensive range of *Lycaenopsis* includes N. America, China, Japan, Sri Lanka, India, and E. through S.E. Asia to Australia.

**vardhana** Moore DUSKY HEDGE BLUE
N. India and N. Burma 38–44 mm. (1·5–1·73 in.)
This is one of the largest species of its genus. It is characterized by the pale greyish blue and brown upper surface of the wings of both sexes. It is most common in the vicinity of streams, at elevations of between 2440–3050 m. (8000–10,000 ft). Nothing is known about its life history.

**Lycia** Hübner (GEOMETRIDAE)
Moths in this genus are found in N. America and throughout temperate Asia and Europe. The caterpillar is typical of many geometrids.

**hirtaria** Clerck BRINDLED BEAUTY
Europe including Britain 38–50 mm. (1·5–1·97 in.)
This is a variable species and is common over much of its range, even near the centre of large cities like London. The moth is out in March and April and is to be found by day resting on tree trunks or fences, where its pattern and colour gives it good camouflage. The purplish grey caterpillar is speckled with darker marks and spotted with yellow on some of the middle segments. It feeds on lime *(Tilia)*, elm *(Ulmus)*, willow *(Salix)* and many fruit trees, occasionally being common enough locally to cause damage to apple *(Malus)* trees.

**lycidas** see **Battus**
LYCIDAS SWALLOWTAIL see **Battus lycidas**

**Lycomorpha** Harris (ARCTIIDAE)
As their name suggests, most species of this small genus closely resemble species of Lycid beetles, a group of insects known to be highly distasteful to predators. The 20 or so species fly by day and frequently visit flowers.

**pholus** Drury
U.S., Canada 23–35 mm. (0·91–1·38 in.)
This Lycid-mimicking moth is widely distributed in the U.S. and E. Canada where the caterpillars feed on lichens. In the S. subspecies, *L. pholus miniata*, the orange-yellow colour at the base of the wings is replaced by red. As expected of a Lycid mimic. *L. pholus* flies during the day and can be found in company with Lycid beetles on flowers. **367**c

**lycophron** see **Papilio**

**Lycorea** Doubleday (NYMPHALIDAE)
There are 5 species in this genus of the subfamily Danainae. They are associated in mimicry complexes with species of Heliconiinae, Ithomiinae and the moth families Hypsidae and Castniidae. Little is known of their life histories. (See also under *Eueides.)*

**atergatis** Doubleday
Tropical Central and S. America 75–88 mm. (2·95–3·46 in.)
Together with several Ithomiinae – see *Melinaea lilis* Doubleday, 2 Heliconiinae and 1 Hypsid, this species forms a mimetic association of distasteful species which exhibit an almost identical colour pattern. Other presumably palatable species, 2 Nymphalinae and 2 Pieridae, mimic the species of this complex.

**ceres** Cramer LARGE TIGER
C. America, tropical S. America, U.S. 70–83 mm. (2·76–3·27 in.)
This is a black, brown and yellow butterfly with long wings resembling those of several species of *Heliconius*, Ithomiinae and a form of the moth species *Chetone histrio* (Hypsidae). It has not yet been captured in the relatively well collected island of Jamaica, although there are suitably forested localities there. Virgin forest seems to be particularly attractive to this species. Stray specimens occasionally reach S. Texas and Florida. The form *L. ceres cinnamonea* differs from the illustrated specimen in the much darker coloration, especially of the hindwing which is basally dark brown and without markings. **130**c

**halia** Hübner
Tropical S. America 90 mm. (3·54 in.)
This butterfly is a member of a mimetic association of at least 2 species of Ithomiinae, a *Melinaea* species and a *Mechanitis* species, and one species of Heliconiinae *(Heliconius)*. The forewing of *L. halia* is similar to that of the illustrated *L. ceres;* the hindwing differs from that of *L. ceres* in the broad marginal band and the presence of a yellow central bar parallel to the front edge of the wing.

**pasinuntia** Stoll
Tropical S. America 78–94 mm. (3·07–3·7 in.)
This is another mimetic partner of various species of Ithomiinae, Heliconiinae and others. It is closest to *L. ceres* but differs chiefly in the more extensive yellow, or less commonly orange, area proximal to the pre-apical row of spots on the forewing.

**lygdamus** see **Glaucopsyche**
**Lygris** see **Eulithis**

**Lymantria** Hübner (LYMANTRIIDAE)
About 100 species are placed in this genus of moths. It is represented in both temperate and tropical zones of the Old World. The wings of most species are yellowish or pinkish white, with numerous dark transverse lines on the forewing. The caterpillars of many *Lymantria* species are forest and orchard pests. *L. ampla* Walker is a pest of cocoa *(Theobroma)*.

**dispar** Linnaeus GYPSY MOTH
E. America and Canada, Europe, Asia to Japan 36–64 mm. (1·42–2·52 in.)
The unfortunate escape in 1869 of specimens of this species, from the laboratory of a French naturalist in Massachusetts, U.S., resulted in one of the worst insect plagues known. Nearly all species of N. American deciduous and evergreen trees are attacked by the caterpillars which may cause sufficient damage to kill an infested tree. Comparable extensive damage seldom occurs in Europe; in Britain *L. dispar* became practically extinct in about 1850. Attempts to control the adults of this pest-species in America have involved the use of traps baited with a synthetic equivalent of the female Gypsy moth sex attractant scent called appropriately 'dispalure', while the bacterial insecticide *Bacillus thuringiensis* has been employed against the caterpillar. **357**s

**monacha** Linnaeus BLACK ARCHES
Europe including Britain through temperate Asia to Japan 37–52 mm. (1·46–2·05 in.)
This elegantly marked species is fairly common in much of its range, and is a minor pest in continental European orchards. Females are paler in colour than the illustrated male. In Britain it is commonest in oak woods, where it flies in July and August. The eggs are laid in crevices of the bark and remain there throughout the following winter. The mainly grey and brown, hairy caterpillars feed from April to July on oak *(Quercus)*, apple *(Malus)*, and many other broad-leaved trees, and sometimes on pine *(Pinus)*. The hairs of the caterpillar are barbed and poisonous. The unusual chrysalis is a metallic brown in colour and noticeably hairy. **357**t

**rosina** Pagenstecher
Bismarck Archipelago 47 mm. (1·85 in.)
The closest relative of *L. rosina* is possibly *L. subrosea* Swinhoe from Sri Lanka, which has a basically similar pattern. The forewings of this species are similar to those of *L. monacha* in general features of the pattern; the hindwings are a beautiful, pinkish white, with a bright pink area at the base of the inner margin of the wing. It is known only from the male.

**Lymantriidae** TUSSOCK MOTHS
This is a family of about 2000 species of average-sized moths belonging to the superfamily Noctuoidea. Their most characteristic features, compared with other Noctuoids, are probably the lack of a functional proboscis, and the consequent inability to feed as an adult, and the presence of exceptionally long antennal pectinations in the male. The caterpillars are invariably hairy and often brightly coloured; they are often easier to identify than the more uniformly coloured adults. Many caterpillars have tufts or tussocks of hairs, often in a dorsal row of 4 (a 'toothbrush'), and segments 6 and 7 of the abdomen have 2–3 eversible glands on the dorsal surface. The larval hairs of many species have irritant properties and are usually woven into the cocoon so that this too is protected. The tuft of hair-scales at the end of the female abdomen will also irritate, and they too protect another stage in the life history, the eggs, which are normally covered with these hair-scales as they are laid. Apart from their nuisance value, the caterpillars of several species are sometimes serious pests of deciduous forest and some are pests of cocoa *(Theobroma)* eg. *Euproctis* in the Old World. The females of some Lymantriids are wingless,

eg. most species of *Orgyia*, and in an extreme case, *Bracharoa*, never leave the cocoon. See *Bracharoa, Cozola, Dasychira, Euproctis, Euproctoides, Hemerocampa, Leucoma, Lymantria, Naroma, Orgyia, Periria*.

**Lymnas** Blanchard (NEMEOBIIDAE)
There are probably between 30 and 40 species in this Central American and tropical S. American genus of butterflies. Most of the species are black, with red or orange markings.

**iarbus** Fabricius UNDERLEAF BUTTERFLY
Central America and tropical S. America 37 mm. (1·46 in.)
Shaded forests are the home of this beautiful species. Both sexes are black above, with a red basal spot on each wing and oblique orange bar on the forewing. There is an orange-red band at the outer margin of the hindwing. The female is as colourful as the male but has more evenly rounded forewings.

**pixe** Boisduval
Mexico, U.S. 34–45 mm. (1·34–1·77 in.)
This is typically a Mexican species, but specimens have been recorded from Florida. The basal spots of the wings are usually red but sometimes orange. **263**aa

**xarifa** Hewitson ORANGE-TIPPED UNDERLEAF BUTTERFLY
Central America and N. S. America 34–48 mm. (1·34–1·89 in.)
This common species flies in the same localities as *L. iarbus*. The male (not illustrated) has more strongly pointed forewings and a smaller orange apical patch. Some specimens of both species lack the red spot at the base of the hindwing. **263**bb

**lynceus** see **Idea**

**Lyonetia** Hübner (LYONETIIDAE)
LYONETID MOTHS
These are beautifully marked micro-moths with wingspan from 5–10 millimetres (0·2–0·39 in.), often whitish in colour with delicate markings, particularly near the base of the wing. They are common in the N. Hemisphere. The caterpillars are all leaf-miners, making blotches or serpentine mines on leaves. The wings are slender and pointed with a large patch of scales near the base.

**castaneella** Kuroko
Japan 6–7 mm. (0·24–0·28 in.)
This little moth has shining, narrow white wings, with brown marks near the thorax and a black spot at the base. There are usually 2 small brown lines on the costa. There is a large fringe at the base of the forewing which is grey white. The caterpillar makes an irregular blotch mine in the leaves of *Castanea crenata* and *Quercus acutissima* (Fagaceae). The caterpillar can migrate from one leaf to another and 1–2 mines may be found on one leaf. The adults fly in early August, at the end of that month and again at the end of October.

**clerckella** Linnaeus APPLE LEAF-MINER MOTH
Europe including Britain, Japan, Formosa, India, Madagascar 6–9 mm. (0·24–0·35 in.)
This widespread species has shiny white forewings and a strongly patterned base to the wing of brown marks with black edges. The large fringe on the hindwing margin is greyish-white. The caterpillar makes serpentine mines on the leaves of *Prunus*, often with the mine going round the edges of the leaves. In Japan several biological races have been recognized which live on different hosts, although the adults cannot be distinguished. In Europe the pale green caterpillar makes irregular galleries in leaves of apple *(Malus)*, hawthorn *(Crataegus)*, and birch *(Betula)*.

**latistrigella** Walsingham
N. America 8–9 mm (0·31–0·35 in.)
This tiny leaf-mining species is found in the U.S. where they have been bred on species of *Rhododendron* (Ericaceae). The caterpillar makes a linear mine when it is young, widening but still linear for a few centimetres along the leaf before making a blotch, several millimetres in extent. This mine is characteristic of the species and can be seen on the leaves of *Rhododendron maximum*. **6**c

**yasudai** Kuroko
Japan 6–7 mm. (0·24–0·28 in.)
This small moth has shining white wings with a brown or reddish brown base and a black spot at the base. The fringes of the wings are long, the hindwing is white. This species was only recently discovered and is known from only one locality in Japan, although further collecting will probably show it to be more widespread. The caterpillar makes a linear mine in *Quercus acuta* (Fagaceae) and the moth is out in June.

LYONETID MOTHS see **Lyonetia** and **Lyonetiidae**

**Lyonetiidae** LYONETID MOTHS
Very small moths whose caterpillars are usually leaf-miners, living between the upper and lower surface of the leaf. They produce a mark, visible on the leaf, which is often of characteristic shape for the individual species. The family contains many species which may be unrelated and further research on it is needed. A number are of economic importance causing galls or damaging sugar cane *(Saccharum)* or bananas *(Musa)*. See *Lyonetia, Paraleucoptera*.

**Lypropteryx** Westwood (NEMEOBIIDAE)
This is a small genus of 23 Central American and tropical S. American species of butterflies.

**apollonia** Westwood
C. & tropical S. America 38–50 mm. (1·5–1·97 in.)
Females are typically as in the illustration, but may have the orange or orange-red continued along the outer margin of the forewing. The under surface of both sexes is as colourful as the upper surface; the base of both wings is black, ornamented with several brilliant purple-red spots; the rest of the wings are striped alternately with white and black in a ray pattern. The life history of this species is not known. Both sexes are illustrated. **268**t, u

**Lysandra** Hemming (LYCAENIDAE)
This is a small genus of butterflies found in Europe, W. Asia and N. Africa.

**bellargus** Rottemburg ADONIS BLUE
Europe, S.W. Russia, Turkey Iraq, Iran 30–36 mm. (1·18–1·42 in.)
The most distinctive feature of this butterfly is the intense blue of the upper surface of the male wings. Females are brown above, usually with some blue scales. It is restricted to areas of chalky soil, including parts of S. England. There are usually 2 broods per year. The caterpillars feed on Horseshoe Vetch *(Hippocrepis)* and on other species of the family Leguminosae.

**coridon** Poda CHALKHILL BLUE
Europe including Britain, Turkey 30–36 mm. (1·18–1·42 in.)
*L. coridon* is confined to hilly or mountainous country in limestone or chalky areas up to 1830 m. (6000 ft). There is a single brood per year which reaches the adult state in July and August and overwinters as a young caterpillar. Females (not illustrated) are mainly brown above, suffused with blue, extensively so in parts of the species range. The blue colour of the male wings has been described as the blue of the sky 'in the twilight of a stormy day'. The caterpillars feed on Horseshoe Vetch *(Hippocrepis)* and other species of Leguminosae. Ants often attend the caterpillars and possibly carry them to foodplants near their nests. **250**

**lyside** see **Kricogonia**

**Lyssa** Hübner: syn. *Nyctalemon* (URANIIDAE)
This is a genus of about 10 species of large brown, tailed moths, which unlike their relatives *Alcides* and *Chrysiridea* are nocturnal in habits. They are frequent visitors to flowering trees in the evening and are apparently more easily attracted by sugary bait than by a light. The species of *Lyssa* have been better known under the synonym *Nyctalemon* for many years. Its range extends from India to the Solomons.

**patroclus** Linnaeus
India, China, through S.E. Asia to the Solomons and N. Australia 110–130 mm. (4·33–5·12 in.)
There are several described subspecies of this large species, some of which may prove to represent different species. The illustrated specimen is from the Bougainville Island in the Solomons group. **314**a

**m-album** see **Panthiades**
**macculochii** see **Alypia**
**mcfarlandi** see **Sandia**
**machaon** see **Papilio**
**macleannania** see **Haetera**
MACLEAY'S SWALLOWTAIL see **Graphium macleayanum**
**macleayanum** see **Graphium**

**Macrauzata** Butler (DREPANIDAE)
This is a small genus of hook-tip moths found in India, China, Japan, the Philippines and in S.E. Asia as far E. as Sulawesi (Celebes). The colour pattern of all the species is similar to that of the illustrated *M. maxima*, which is probably a leaf-mimicking moth. The Lymantriid moth *Carriola saturnioides* Snellen has a similar, though not identical pattern of transparent patches on the wings and has a similar distribution.

**maxima** Inoue
Japan, China 40–55 mm. (1·57–2·17 in.)
This species was described quite recently by the Japanese lepidopterist Hiroshi Inoue. The illustrated moth is the type specimen of the Japanese subspecies *M. maxima maxima*. The caterpillar feeds on species of oak *(Quercus)*. **272**g

**Macrocneme** Hübner (CTENUCHIDAE)
There is a close resemblance between the species of this genus and wasps of various genera (eg. *Salius* and *Pepsis*) both in appearance and behaviour, even down to such details as long legs which are trailed when in flight. Their distribution extends from Mexico to Brazil. About 50 species are currently recognized, but a forthcoming revision by R. E. Dietz of Washington, D.C., may modify this figure.

**thyra** Möschler
Trinidad, Colombia 38–50 mm. (1·5–1·97 in.)
Flight takes place either at night or by day in this species. Most other species of *Macrocneme* differ externally from *M. thyra* in only minor details of the coloration or pattern; some are slightly darker, others have a more greenish iridescence. In laboratory tests with this species, the potential predator, a species of oriole (Icteridae), refused to touch the moth. **367**f

**Macroglossum** Scopoli (SPHINGIDAE)
HUMMING-BIRD HAWK-MOTHS
The several species of this genus are chiefly tropical Asian in distribution, but there are also representatives in temperate Asia, Africa, Europe, Australia and the islands of the Pacific.

**stellatarum** Linnaeus HUMMING-BIRD HAWK-MOTH
S. Europe, N. Africa, temperate Asia to Japan 38–50 mm. (1·5–1·97 in.)
*M. stellatarum* is probably more common in urban Britain, including London, than in the countryside. Jasmine *(Jasminium)* and many other garden flowers are visited for nectar during the day by these fearless moths. The high-pitched hum of their rapidly beating wings and the darting flight produce a remarkable resemblance to humming-birds. Bedstraws *(Galium)* are the commonest foodplants of the caterpillar. **344**d

**Macrothylacia** Rambur (LASIOCAMPIDAE)
This is a small genus of moderately sized moths found in Europe and temperate Asia.

**rubi** Linnaeus FOX MOTH
Europe, temperate Asia to E. U.S.S.R., China 45–60 mm. (1·77–2·36 in.)
Both the adult male and the caterpillar are a reddish (foxy) brown in colour. The female (illustrated) is paler and less reddish. Males fly chiefly during the afternoon and evening, in search of females; the latter are mainly active only at night. Eggs are laid in clusters on the foodplants or on a nearby fence or rock. The caterpillars, which overwinter, feed on ling heather *(Calluna)*, bramble *(Rubus)*, sallow *(Salix)* and other plants. The loosely woven cocoon is spun

close to the ground near the base of the foodplant. **323**

**maculanota** see **Kalenga**
**macularia** see **Pseudopanthera**
**maculifera** see **Antinephele**
**maculifrons** see **Calodesma**

**Maculinea** van Eecke (LYCAENIDAE)
Three species are placed in this genus of butterflies. They are found in Europe through Asia to Japan. A multitude of names have been applied to various subspecies, seasonal and other forms and to individual aberrations of the 3 species.

**arion** Linnaeus LARGE BLUE
Europe, U.S.S.R. and China 35–40 mm. (1·38–1·57 in.)
Both sexes of this butterfly are bright blue above, with white fringes to the wings. The caterpillar of the subspecies *M. arion arion* feeds at first on the flowers of thyme *(Thymus)*, but in later stages on ant grubs after having been carried by ants to their nest. It overwinters in the ants' nest, and starts feeding again on the grubs in the spring. After emergence from the chrysalis, the adult crawls through the passages of the nest to reach the outside world and a brief independent existence. In Britain, the Large Blue had a restricted range but recently became extinct.

**maculipennis** see **Plutella xylostella**
**maculosa** see **Diacrisia**
**madagascarensis** see **Euchromia**
**madytus** see **Narathura**
**maenas** see **Actias**
**maerula** see **Anteos**
MAERULA BUTTERFLY see **Anteos maerula**
**maesites** see **Chlorostrymon**
MAESITES HAIRSTREAK see **Chlorostrymon maesites**

**Magneuptychia** Forster (NYMPHALIDAE)
This is a genus of butterflies in the subfamily Satyrinea, but unlike many Satyrine genera which have brown coloured species, some of these are brightly coloured. The genus was described recently for a group of butterflies, formerly in the genus *Euptychia* from which they are now separated.

**junia** Cramer IRIDESCENT BLUE NIGHT BUTTERFLY
Brazil, Tobago 38 mm. (1·5 in.)
Most Satyrid butterflies are brown but this is one of the exceptions and is a dark greenish metallic colour on the upperside. In the female nearly all the forewing is brown with the iridescence showing only at the base of the wings. The butterflies tend to fly in the undergrowth in shaded areas. The males are similar to *Cepheuptychia cephus*.

**tricolor** Hewitson
C. & S. America 43 mm. (1·69 in.)
The purplish eyespots on the underside of the hindwing often show faintly on the upperside. This species is widespread in S. America and has been found in Peru, Brazil, Guyana, Surinam. The female has purplish transverse stripes but the underside is similar to the male. **174**l

**magnifica** see **Sindris**
**magniplaga** see **Moresa**
MAGPIE MOTHS see **Abraxas, Arycanda, Eurrhypara**
MAGPIE CROW see **Euploea**
**maha** see **Zizeeria**
**maia** see **Hemileuca**
**maimuna** see **Callicore**
MAIZE STALK-BORER see **Busseola fusca**
MALABAR RAVEN see **Papilio dravidarum**
MALACHITE BUTTERFLY see **Metamorpha stelenes**
**malachitis** see **Nacna**

**Malacosoma** Hübner (LASIOCAMPIDAE)
This is a fairly widely distributed but small genus of moths. Four species are found in Europe, N. Africa and temperate Asia and 6 in N. America and Mexico. The caterpillars of most species form silken webs in the spring. The moths are probably very short-lived.

**americana** Fabricius EASTERN TENT
S. Canada, U.S., E. of the Rockies 25–35 mm. (0·98–1·38 in.)
The relatively inconspicuous adult of this species varies from light to dark brown, dusted with white, and with 2 pale transverse lines across the forewing. This is another pest species which before the advent of modern control measures was a serious menace. It was first reported as a pest in 1646. Control by spraying is now generally effective in orchards, so that *M. americana* is better classed as a general nuisance as a defoliator than as a pest species. The caterpillars and, more particularly, their silken tents or nests are a common sight in gardens and along roadsides. They can be distinguished from the other N. American species *M. disstria*, by the pale yellowish white band on the back. The usual foodplants are species of *Prunus*, hawthorn *(Crataegus)* apple *(Malus)*, and pear *(Pyrus)* and other trees of the family Rosaceae, but in the later stages the caterpillar will attack the foliage of almost any broad-leaved tree.

**castrensis** Linnaeus GROUND LACKEY
Europe including Britain, Turkey, S.W. Russia 27–40 mm. (1·06–1·57 in.)
The wings of this generally common species can vary in the male from the normal pale yellow, through all shades of brown, to brownish red. Females are normally reddish brown in colour. *M. castrensis* is found in woods and heaths and in Britain in salt-marshes. It flies in July and August. The black, blue, orange and white caterpillars feed by day on heaths *(Erica)*, spurges *(Euphorbia)*, wild carrot *(Daucus)* and in Britain on various salt-marsh species. They rest gregariously, in webs spun close to the ground amongst the foodplants. **322**

**disstria** Hübner FOREST TENT CATERPILLAR
Canada, U.S., probably Mexico 20–37 mm. (0·79–1·46 in.)
This species was first reported as a pest in 1791. The foliage of most trees is attacked by the caterpillar of *M. disstria* but aspen *(Populus)*, blackgum and sweetgum *(Nyssa)* and oak *(Quercus)* are the commonest foodplants. In general appearance, the moth is similar to the Old World *M. castrensis*. The common name for this species is rather misleading, as the caterpillar never builds more than a platform of silk. The caterpillars cluster together when not feeding and when moulting. Pest outbreaks may last between 1–4 years before a decline in numbers occurs. During the height of an outbreak caterpillars are sometimes so numerous that trains are brought to a standstill as a result of the effective lubricating qualities of their squashed bodies on the lines.

**neustria** Linnaeus LACKEY
Europe E. to Japan 25–38 mm. (0·98–1·5 in.)
This species is similar in pattern to the illustrated *M. castrensis*, and is similarly variable in coloration. Its white, black, blue, red and yellow caterpillars hatch from overwintered eggs laid close together in a characteristic, broad band round a twig. The foodplants include many forest, hedgerow and orchard trees, which may suffer considerable damage from the ravages of the caterpillars. The large, communal webs or tents spun by the caterpillars are a common sight in some years. The timing of the various stages of the life history is similar to that of *M. castrensis*.

MALAY VISCOUNT see **Tanaecia pelea**
MALAYAN JEZEBEL see **Delias ninus**
**malivorella** see **Coleophora**
**malvae** see **Pyrgus**
MAMBA BUTTERFLY see **Graphium colonna**

**Mamestra** Ochsenheimer (NOCTUIDAE)
This is an almost cosmopolitan genus which is best represented in the N. Hemisphere. The caterpillars of most species are 'cutworms' on herbaceous plants.

**brassicae** Linnaeus CABBAGE MOTH
Europe, temperate and subtropical Asia (including India), E. to Japan 40–48 mm. (1·57–1·89 in.)
This is one of the commonest moths in Europe. It is best known from the brown or green caterpillar, which is a pest of cabbage *(Brassica)* and other species of the family Cruciferae but will also attack almost any low-growing plant and sometimes the foliage of trees. **391**h

MAMMOTH SULPHUR BUTTERFLIES see **Anteos**

**Mandarina** Leech (NYMPHALIDAE)
The single species in this genus of butterflies occurs in China. It is in the subfamily Satyrinae. Several subspecies and forms have been described on small differences in pattern.

**regalis** Leech
C. & W. China 54 mm. (2·13 in.)
The underside is black with rows of eyespots round the margin and is a typical Satyrine pattern. Species in related genera have black or purple bands on the forewing. Little seems to be known of the biology of this species. **174**h

**mandarina** see **Bombyx**

**Manduca** Hübner (SPHINGIDAE)
There are just over 40 species in this genus of New World hawk-moths, 9 of them known in N. America, of which 4 reach S. Canada. Two species, the Tobacco and Tomato Hornworms are well known pests.

**ochus** Klug
Mexico S. to N.W. S. America 95–140 mm. (3·74–5·51 in.)
Little has been published about this species and its life history is unknown. The colour pattern of the chiefly brownish orange and black wings is one of the most distinctive in its family. Specimens have been captured in C. America and in Venezuela but there appears to be no records at present from the intermediate state of Colombia. **344**u

**quinquemaculata** Haworth TOMATO HORNWORM
N.E. Canada to U.S., Hawaii 70–100 mm. (2·76–3·94 in.)
The feeding habits of the highly destructive caterpillar of this species are very similar to those of *M. sexta* (q.v.) with which it often shares the same individual plant (almost invariably a species of the family Solanaceae). It has 8 oblique, white stripes on each side and a black, terminal horn. Female sex-scent (or sex pheromone) has been used successfully in conjunction with an electric grid in traps designed to attract and destroy adult males and control this pest-species. **344**m

**rustica** Fabricius RUSTIC SPHINX
U.S., C. America, S. America, Galapagos Islands 100–150 mm. (3·94–5·91 in.)
This is a large, distinctively coloured species, especially attractive when fresh, with rich brown forewings, marked with white, yellowish brown and black. Its abdomen is greenish brown, marked laterally with black-margined, yellow patches on each side. The caterpillar feeds on numerous plants including *Bignonia* and Jasmine *(Jasminium)* and on fringe tree *(Chionanthus)*. *Petunia* blossoms frequently attract the adults.

**sexta** Linnaeus
TOBACCO HORNWORM or CAROLINA SPHINX
N. U.S. to tropical S. America 80–120 mm. (3·15–4·72 in.)
This species and *M. quinquemaculata* are the most common pests of tomato in N. America. The caterpillar attacks both the foliage and the green fruit of tomato *(Lycopersicum)* and is also a pest of peppers *(Capsicum)*, potato *(Solanum)*, tobacco *(Nicotiana)* and other species of Solanaceae. It has 7 oblique white bands on each side, and a red horn at the rear end. The moth is similar to *M. quinquemaculata* above, but is fairly easy to distinguish by the coloration of the lateral patches on the first (front) segment of the abdomen which are normally orange-yellow, not mostly white as in *M. quinquemaculata* and by the nearly black line along the outer margin of the forewing. Two successful methods of controlling the adults of this pest species have involved the use of ultra-violet lamp traps and traps baited with virgin female *M. sexta*.

**Maniola** Schrank (NYMPHALIDAE)
A large genus of butterflies with most species in Europe

and temperate Asia, a few species are known from S. America. They are typical of many of the subfamily Satyrinae being mostly brown, rather sombre coloured butterflies, but they often have intricate patterns.

**jurtina** Linnaeus MEADOW BROWN
Europe including Britain, Canary Islands, N. Africa, Turkey, Iran 44–50 mm. (1·73–1·97 in.)
Very common over most of its range in meadows and grassy places from June to September except for a short period in early August. Several subspecies have been described, including some in Britain. The caterpillar feeds on various grasses including *Poa*. The butterfly will fly in dull weather as well as in sunshine and must be amongst the commonest in meadows and pasture lands. **177**

**manni** see **Pieris**
MANROOT MOTH see **Melittia gloriosa**
**mantinea** see **Caria**
MANY-PLUMED MOTHS see **Alucita**
MAP BUTTERFLIES see **Araschnia, Cyrestis, Marpesia**
MAP-WING BUTTERFLIES see **Cyrestis**
MAP-WINGED SWIFT MOTH see **Hepialus fusconebulosa**
MAPLE PROMINENT MOTH see **Disphragis guttivitta**
MAPLET, COMMON see **Chersonesia risa**
**marantica** see **Epicampoptera**
MARBLED
FRITILLARY see **Brenthis daphne**
FRITILLARY, LESSER see **Brenthis ino**
MAP see **Cyrestis cocles**
MOTH, PURPLE see **Eublemma ostrina**
WHITE see **Melanargia galathea**
WHITE, BALKAN see **Melanargia larissa**
WHITE, WESTERN see **Melanargia occitanica**
**marcella** see **Marpesia**
**marcellus** see **Eurytides**
**marchandi** see **Eurytides**
**marcus** see **Enicospila**
**margaretta** see **Mesene**
**margarita** see **Axia**
**margaritata** see **Campaea**

**Margaritia** Stephens (PYRALIDAE)
A genus of Pyralid moths whose species are not known with certainty. Many of the species in *Loxostege* and related genera probably belong here.

**sticticalis** Linnaeus BEET WEBWORM
Europe including Britain, Asia, N. America 22–27 mm. (0·87–1·06 in.)
The forewings are dark brown or purple with white lines below the margin and a rather square white mark. The hindwings are paler. This is an inconspicuous moth when it flies at dusk. It comes readily to light. The caterpillars feed on a variety of plants including beet (Chenopodiaceae) and Alfalfa (Leguminosae). When occurring in large numbers the caterpillars, which feed under a silken webbing, will skeletonise the leaves.

**Margaronia** see **Cadarema** and see under **Diaphania, Parotis**
**marginata** see **Amsacta, Schinia**
**maria** see **Lasaia**
**mariamne** see **Dysschema**
**marianne** see **Ixias**
**markii** see **Callithea**
**marmorea** see **Digama**

**Marpesia** Hübner (NYMPHALIDAE)
Primarily an American genus, with species in the S. and C. parts of the continent and the W. Indies; there is at least one species found in Africa. A few species have been bred from caterpillars found on fig *(Ficus)*.

**camillus** Fabricius
AFRICAN PORCELAIN or AFRICAN MAP BUTTERFLY
W. Africa, Congo, Uganda, Mozambique, Rhodesia, Malawi 50–62 mm. (1·97–2·44 in.)
The wings are white with some of the veins brown with some orange or brown stripes across the wings. There is a sharply pointed tail on the hindwing with some blue and grey dots above it. This forest species settles on the underside of leaves with the wings spread flat and often is found feeding at damp mud or wet patches. The butterfly is out in most months of the year. This species has often been placed in the genus *Cyrestis* which contains the Oriental species of map-wing butterflies. It seems likely that more research is needed to establish the correct relationship of this species. **229**d

**chiron** Fabricius ROAD PAGE BUTTERFLY
C. & tropical S. America 50–62 mm. (1·97–2·44 in.)
This brown butterfly with lighter parallel lines down the wing and 3 small spots near the apex of the forewing looks a typical *Marpesia*. The 2 tails on the hindwing are unequal with the outer one being very long, and the inner one just a small protuberance. The butterfly is a fast flier and often glides with horizontal wings. Occasionally there are large migrations in parts of its range. It is also a frequent visitor to damp patches for drinking.

**coresia** Godart WAITER BUTTERFLY
C. & tropical America, rare in Trinidad 54 mm. (2·13 in.)
The black and white appearance of this species when at rest gives its popular name. It is common over much of its range. The upperside is dark brown with a slightly patterned reddish brown margin and there is one long and one short tail on each hindwing. **224**

**iole** Drury
S. America, W. Indies 48–58 mm. (1·89–2·28 in.)
The female has blackish brown outer margins with a white median band across the forewing and white spots near the apex of the wing. The underside is pale brown, crossed by transverse lines, the male underside is similar but darker. The female thus contrasts sharply in colour with the male illustrated. **216**j

**marcella** Felder
S. America, W. Indies 50–60 mm. (1·97–2·36 in.)
The male is illustrated. The female is brown with a white band along the forewing and without the blue on the hindwing. This species is common over much of its range. **229**c

**petreus** Cramer RUDDY DAGGER WING or TAILED FLAMBEAU 66–72 mm. (2·6–2·83 in.)
S. America, C. America, U.S. and West Indies
This S. American species is found occasionally in S. Florida but is rarer further N. The sexes are practically identical in pattern. In Trinidad this species turns up in the course of migratory movement. This butterfly is often common on flowers and frequently comes down to water to drink. It has a fast flight and large numbers take part in local migrations. The caterpillars feed on the leave of fig *(Ficus)* and *Mora* (Leguminosae). **216**m

MARSH
CARPET MOTH see **Perizoma sagittata**
FRITILLARY see **Euphyadryas aurina**
MOTH, ROSY see **Eugraphe subrosea**
MARSHALL'S ACRAEA MIMIC see **Mimacraea marshalli**
**marshalli** see **Mimacraea**
**marsyas** see **Pseudolycaena**
**marthesia** see **Siderone**

**Maruca** Walker (PYRALIDAE)
A small genus of Pyralid moths with one worldwide species which is very common on most continents. Further research will probably show that it is a 'species complex', consisting of several distinct species with similar external appearances.

**testulalis** Geyer BEAN POD BORER
Cosmopolitan 28 mm. (1·1 in.)
This moth is nocturnal, and seldom seen during the day, although regularly attracted to light. It is common in the warmer parts of all the continents and is occasionally found in Britain. It is probably an extensive migrant, although it may also be spread by the transport of infected beans. It is found commonly all over Africa, S. America, Hawaii, India, and Malaya, although it is possible that several similar looking species are involved in a species complex, a situation similar to the one found recently in the European Corn borer *(Ostrinia nubilalis)*. **56**j

**martia** see **Polystichtis**
**mathewi** see **Atteva**
**matronula** see **Pericallia**
**maxima** see **Macrauzata**
**mayo** see **Papilio**

**Mazuca** Walker (NOCTUIDAE)
The 4 species of this genus of moths are restricted to tropical or subtropical Africa. The hindwings are white and unmarked; the forewings yellow, with black and red markings.

**amoena** Jordan
Congo Basin, Malawi 30 mm. (1·18 in.)
Very few specimens of this strikingly marked moth exist in collections. It was inadvertently redescribed, but aptly renamed by the South African entomologist, Janse, as *M. elegantissima*. **390**c

**haemagrapha** Hampson
W. Africa from Sierra Leone to Ghana 30 mm. (1·18 in.)
This is another very colourful and unusually marked Noctuid. It is rather similar in colour pattern to the illustrated *M. amoena*. The chief diagnostic features are the circle at the outer margin of the forewing which encloses an area of orange-red, and the lines at the inner margin which are wholly dark. Nothing is known about the life history of this species.

**strigicincta** Walker
W. & E. Africa 30–40 mm. (1·18–1·57 in.)
This is a fairly common species, similar in colour to the illustrated *M. amoena* but with a much more broken pattern of lines which give the impression that the wings have been decorated with hieroglyphics. **391**a

MEADOW
ARGUS see **Junonia villida**
BLUE, VIOLET see **Polyommatus icarus**
BROWN see **Maniola jurtina**
MEAL MOTHS see **Ephestia, Plodia, Pyralis**
**meander** see **Prepona**

**Mechanitis** Fabricius (NYMPHALIDAE)
About 20 species of butterflies are placed in this C. American and tropical S. American genus of the subfamily Ithomiinae. Several species of this genus are members of mimetic associations with other species of Ithomiinae and with species of Heliconiinae, Danaiinae and the moth family Hypsidae.

**californica** see **M. isthmia**

**egaensis** Butler
Amazon Basin 70–75 mm. (2·76–2·95 in.)
Several species of Ithomiinae and other groups of butterflies are associated with *M. egaensis* in a mimetic association (see *Tithorea harmonia)*. Some males of this species are very dark in colour, with the orange parts replaced by orange-brown. **131**m

**isthmia** Bates: syn. *M. californica*
C. & tropical S. America 60–80 mm. (2·36–3·15 in.)
In cloudy weather, this butterfly keeps to the shelter of the forest, but in sunny conditions is a frequent visitor to flowers, especially to *Eupatorium*. It is somewhat variable in the shape and colour of the yellow stripes on the forewing. Some colour-forms are probably members of a mimetic association with *Melinaea lilis;* one form is mimetically associated with at least 5 other species of Ithomiinae, a species of Danaiinae, a species of Heliconiinae and a Hypsid moth *Chetone angulosa*. The anomalous name *M. californica* Reakirt is a synonym, but California is not part of the normal range of this tropical butterfly. Its caterpillar feeds on species of *Solanum* (Solanaceae).

**polymnia** Linnaeus
Tropical S. America 70–75 mm. (2·76–2·95 in.)
In Guyana, a form of this species is a member of a large, mimetic association of species of Ithomiinae and Heliconiinae and others (see *Melinaea)*.

**medinae** see **Catocalopsis**

**mediocris** see **Aphthonetus**
**mediopectinellus** see **Ochsenheimeria**
MEDITERRANEAN
BROCADE see **Spodoptera littoralis**
FLOUR MOTH see **Ephestia kuehniella**
**medon** see **Euphaedra**
**medus** see **Orsotriaena**
**megaera** see **Euchloron**
**megalastra** see **Atteva**

**Megalopalpus** Röber (LYCAENIDAE)
The few species of this genus of butterflies are found in Africa S. of the Sahara. All of them have large, asymmetrical palps. Their caterpillars are associated with ants; those of *M. zymna* are carnivorous.

**zymna** Westwood
W. Africa 30–42 mm. (1·18–1·65 in.)
The caterpillars of this species are predaceous on the nymphs and adults of plant-bugs. The latter are fairly active insects and the caterpillars have to move quickly to capture their prey which are despatched with a bite behind the head. Associated with the caterpillars are ants which feed on the secretions of the plant-bugs. The adult butterflies are white above, with a dark brown apex to the forewing like a small Pierid and a faint brown marginal band on the hindwing.

**Megalopyge** Hübner (MEGALOPYGIDAE)
FLANNEL MOTHS
The caterpillars have 6 pairs of prolegs – most others have 4 pairs. The wings not only have scales but long hairs near the base of the wing (hair-like scales). The bodies of the moths are very hairy and the whole appearance is of a very 'woolly' moth.

**crispata** Packard
CRINKLED or WHITE FLANNEL MOTH 25–35 mm.
U.S. (0·98–1·38 in.)
Whitish or yellow wings with curly black and brown hairs, particularly along the front margins of the forewings. The fleshy caterpillar is covered with long brown silk hairs which form a crest along the middle of its back. It feeds on *Rubus villosius*.

**lanata** Stoll
C. & S. America 50 mm. (1·97 in.)
This species is constant in colour and pattern, with little difference between sexes. Several similar looking species are known, including one, as yet undescribed, where the line round the margin of the hindwing of *M. lanata* is absent. **48**d

**opercularis** Smith & Abbott PUSS CATERPILLAR
U.S. – Virginia to 25–35 mm.
W. Texas (0·98–1·38 in.)
The caterpillar, which is about 25 mm. (0·98 in.) long, has little curled tufts on each side of the tail and is covered with long reddish silky hairs. It feeds on hackberry *(Celtio)*, maple and sycamore *(Acer)* and other trees. It has a very severe sting caused by the toxins on the hairs.

**Megalopygidae** FLANNEL MOTHS
A family of moths in the Zygaenoid group found only in America. The mouth parts of the adult are usually reduced and non-functional and the caterpillars are very hairy with powerful stinging hairs. See *Megalopyge, Trosia.*

**Megalorhipida** Amsel (PTEROPHORIDAE)
The moths in this genus are typical of plume-moths having the forewings divided into 2 slender lobes, with the cleft extending to the middle of the wing with pointed tips and the hindwings divided into 3 lobes.

**defectalis** Walker
Worldwide tropics 13 mm. (0·51 in.)
This moth is rather a sober brown colour with some black scales scattered along the wing margin. It was first described from specimens collected in Africa but is widespread in the warmer regions. The host plant of the small spiny caterpillar is *Boerhaavia diffusa* (Nyctaginaceae), but it has been recorded on a number of other plants.

**megarus** see **Graphium**

**Megathymidae** GIANT SKIPPERS
A small family of skippers, which have different life histories from the Hesperiidae (skippers). The caterpillars bore into *Yucca* or *Agave* (Agavaceae). The adults are large, hairy looking butterflies with a strong and rapid flight. See *Megathymus.*

**Megathymus** Scudder (MEGATHYMIDAE)
These are large butterflies with stout abdomens related to Hesperiidae, found only in C. America and the S. part of the U.S. The caterpillars are mostly woodborers, living in the pith and underground roots of species of *Yucca* and related plants. They are put into a popular Mexican drink which is only genuine when the bottle has a caterpillar in it.

**yuccae** Boisduval YUCCA SKIPPER
S. U.S. 45–82 mm. (1·77–3·23 in.)
The males are much smaller than the females. There is one brood a year. This species is variable over its range in S. Carolina to Florida and a number of subspecies have been described. The caterpillar feeds on *Yucca.* **79**b

**megera** see **Lasiommata**
**meges** see **Lamproptera**

**Melanargia** Meigen (NYMPHALIDAE)
Butterflies in this genus are found in Europe and temperate Asia. They belong to the subfamily Satyrinae. The caterpillars, where known, feed on species of grass (Gramineae).

**galathea** Linnaeus MARBLED WHITE
Europe including Britain, W. temperate 46–52 mm.
Asia, N. Africa (1·81–2·05 in.)
Several subspecies of this basically black and white butterfly have been described. The adults fly in June-July in grassy meadows, from sea-level to 1520 m. (5000 ft). The caterpillars which are brown coloured with thin white and brown lines and 2 small tail-like points, feed on different grasses, eg. *Phleum, Triticum.* Several different forms have been named based on variation in the colour pattern of the wings. Related species are found in W. Europe with broadly similar patterns to *M. galathea.* **180**

**larissa** Geyer BALKAN MARBLED WHITE
S.E. Europe 50–60 mm. (1·97–2·36 in.)
Although this species is regarded as confined to Bulgaria, Greece, Albania and Yugoslavia, there are several similar allied species which occur from W. Asia to Iran, their exact relationship with this species is not known. The butterfly is out in June and July on the rocky slopes of hills. **179**

**occitanica** Esper WESTERN MARBLED WHITE
S.W. Europe and N. Africa 50–56 mm. (1·97–2·2 in.)
Broadly similar to the Marbled White but with much less white on the hindwing and with conspicuous black eyespots on the hindwing with reddish centres. The underside hindwing veins are brown, not black as in the Marbled White. The eyespots of the hindwing are large and more distinct than the related *M. ines,* the Spaniard Marbled White. This butterfly has several subspecies and is found in France, Spain, Italy and Portugal, Algeria, Morocco, Sicily. The adult flies in May, June or July, depending on the altitude, in mountainous rocky country and from the lowlands to 1520 m. (5000 ft). The foodplant of the caterpillar is not known but is probably species of grass.

**melaneus** see **Danaus**
**melanippus** see **Danaus**

**Melanitis** Fabricius (NYMPHALIDAE)
This genus of butterflies is in the subfamily Satyrinae and the species in it are found from Africa to Australia. Generally the underside of the wings are blackish and seasonal forms are common. The butterflies are on the wing towards dusk, generally avoiding bright sunlight.

**leda** Linnaeus EVENING BROWN or TWILIGHT BROWN
Africa, Madagascar, India 60–84 mm.
Pakistan and S.E. Asia, New Guinea, (2·36–3·31 in.)
Australia, Fiji, Tahiti
One of the commonest 'browns' over its wide range. The large black eyespot below the apex of the forewing has white spots and is surrounded by orange. There are 1–2 small eyespots near the margin of the hindwing, which generally has a small point on it, similar to a small point below the apex of the forewing on the outer margin. This is a shade-loving butterfly which is well camouflaged when it settles amongst dead leaves. In S.E. Asia the caterpillars feed on rice and sugar cane and at times reach pest proportion. In Africa it feeds on *Cynodon setaria* and other grasses. **174**j

**libya** Distant
Africa 70–85 mm. (2·76–3·35 in.)
This species is widespread from W. Africa, into E. and Central Africa, as far S. as Rhodesia. It is similar to *M. leda* with which it often flies but differs in the lack of the orange-brown round the 2 white spots in the forewing, of which the lower white spot is nearer the wing margin and both are larger than *M. leda.* The hindwings are generally a darker brown. It behaves in the same way as *M. leda* settling on the ground between dead leaves. There are dry and wet season forms but they are not well known.

**melanochroia** see **Calodesma**
**melanochroma** see **Heteropsyche**
**melanospila** see **Epicoma**
**melaxantha** see **Ephestris**
**melia** see **Dismorphia**
**melicerta** see **Neptis**

**Melinaea** Hübner (NYMPHALIDAE)
There are about 25 species in this tropical American genus of the subfamily Ithomiinae. Several species of this genus are members of mimetic associations with similarly patterned species of Heliconiinae, Danainae and other species of Ithomiinae, and with day-flying moths of the family Hypsidae. The colour pattern of these supposedly distasteful species is mimicked by species of Nemeobiidae, Pieridae, Nymphalinae and the moth family Castniidae. Males of *Melinaea* can be distinguished from similarly patterned males of *Mechanitis* by the normally developed forelegs.

**lilis** Doubleday
C. & tropical S. America 70–76 mm. (2·76–2·99 in.)
The C. American subspecies *M. lilis imitata* is a member of a large mimetic complex involving both unpalatable species and their presumably palatable mimics. Involved in this complex are 2 species of *Mechanitis,* 2 species of *Ceratinia* and a species of *Tithorea* (Ithomiinae), a species of *Lycorea* (Danainae), a species of *Heliconius* and a species of *Eueides* (Heliconiinae), and a day-flying Hypsid moth. Mimics of this group include a species of *Protogonius* and a species of *Eresia* (Nymphalinae), 2 Pierids of the genera *Dismorphia* and *Perrhybris.* Many of these species fly together and sit on the same flowerheads flaunting their warning colour pattern, if they are distasteful species, or false-warning colour pattern in the case of the deceptive mimics. Other subspecies of *M. lilis* take part in similar mimetic associations. **134**a, b

**mneme** Linnaeus
Tropical S. America, chiefly the 70–80 mm.
Guianas & Amazon Basin (2·76–3·15 in.)
This butterfly is one of the commonest species in a mimetic complex of at least 19 others which include several species of Ithomiinae *(Melinaea, Mechanitis, Ceratinia)* and of Heliconiinae *(Heliconius, Eueides)*, a Nymphaline *(Eresia)* species, 2 Danaine *(Lycorea)* species and a species of Nemeobiidae *(Stalachtis)*. The Nymphaline and Nemeobiid are probably palatable to predators and are mimics of the remaining species. **131**c

**scylax** Salvin
Costa Rica, Panama 80–88 mm. (3·15–3·46 in.)
This species has brownish orange hindwings and is unmarked except for a dark marginal band. The forewing is similar to that of *M. lilis.* There is little variation in colour pattern between specimens of *M. scylax.*

**melinus** see **Strymon**
**melissa** see **Danaus hamata**

**Melissopus** Riley (Tortricidae)
The micro-moths in this genus have extensive metallic scaling, more so than in species of *Cydia* to which they are related.

**latiferreanus** Walsingham FILBERT WORM
U.S., Canada 12–18 mm. (0·47–0·71 in.)
Reddish brown forewings have a dark grey middle band and coppery coloured bands across them; the hindwings are black. The apex of the forewings have delicate angled marks along the margin. This small Tortricid moth has at least 2 broods a year in the S. U.S. The caterpillar feeds on acorns and is common over much of its range. The species was first described from specimens collected in Oregon in 1875 by an English collector, Lord Walsingham and the specimens he collected are now in the British Museum (Natural History).

**Melitaea** Fabricius (Nymphalidae)
A large genus with some 40 species of butterflies in Europe and Asia but with only one species, *M. minuta,* in N. America. The species in it are related to the Argynnid fritillaries. (See also under *Chlorosyne*).

**arduinna** Freyer FREYER'S FRITILLARY
S.E. Europe, Turkey, Iran & Asia 42–46 mm. (1·65–1·81 in.)
Similar to *M. cinxia* but larger. The butterflies are out in May and June with only a single generation in Europe, possibly more elsewhere. It flies over meadows and banks in hilly country. The foodplant of the caterpillar is unknown although suggested as *Centaurea* (Compositae).

**cinxia** Linnaeus GLANVILLE FRITILLARY
Europe including Britain, N. Africa, temperate Asia 32–45 mm. (1·26–1·77 in.)
This species has 2 broods over most of its range, the second brood adults are usually smaller than the first. The butterfly is out in May-June and again in August-September, flying in grass meadows and lowland areas up to 1830 m. (6000 ft). It is widespread throughout S. and C. Europe, but in Britain is now found only in the Isle of Wight, although formerly it was more widespread. The caterpillar which is black with white spots and rows of short black spines, lives gregariously in silken tents on its foodplant, plantains *(Plantago)* or *Centaurea* (Compositae). It is named after Lady Glanville, an 18th century butterfly collector. Amongst the stories about Lady Glanville is the famous one about her Will. It was disputed by her heirs, who wanted it set aside on the ground of her supposed insanity, proof of which they said, was shown by the fact that she collected butterflies! **191**

**didyma** Esper SPOTTED FRITILLARY
S. & W. Europe, Russia, N. Africa, temperate Asia 36–42 mm. (1·42–1·65 in.)
This butterfly occurs in a wide variety of habitats. Several subspecies have been described but the differences between them are not always clear because this species tends to form local populations, differing slightly from nearby ones, but usually with some intermediates between the extremes. It can be found from sea-level to 1520 m. (5000 ft). The caterpillar feeds on toadflax *(Linaria)* and plantains *(Plantago)*. The butterfly is about from May onwards in N.W. France, Belgian, Switzerland, Germany and Austria, with subspecies in C. and S. Europe, Spain, Portugal, Italy and N. Africa. **190**

**minuta** Edwards
U.S. 28–32 mm. (1·1–1·26 in.)
This is the only known representative of the genus in N. America where it occurs in Colorado, Arizona and Texas. There are 2 generations in suitable localities, one only in the higher ranges up to 3050 m. (10,000 ft). The caterpillar feeds on *Penstemon* (Scrophulariaceae). Externally this is very similar to the widespread *M. cinxia* but differs in some aspects of its morphology. Several forms, differing on pattern, have been named from different parts of its range.

**Melittia** Hübner (Sesiidae) BEE MOTHS or SQUASH BORERS
Worldwide genus of micro-moths including some of the largest species in the family. All *Melittia* species resemble bees or other Hymenopterous insects and are generally characterized by very hairy hind legs. One of the Squash borers in this genus is a serious pest of garden Squashes (Cucurbitaceae). Species of *Melittia* are unlike many other moths in that they are day-flying and so realistic in their imitation of wasps that they are left severely alone. Although they look like wasps, they cannot sting.

**cucurbitae** Harris SQUASH BORER
N. America, Mexico 28–32 mm. (1·1–1·26 in.)
This is a serious pest of squashes (Cucurbitaceae). It is a rapid flyer like others in the family and sometimes occurs in large numbers, swarming over the fields of squashes in the bright sunshine. The caterpillar tunnels into the veins and can cause serious damage or death of the plant. The caterpillar leaves the vine and burrows into the ground to pupate. The species is common in most of N. America except on the Pacific coast. The moth is a typical clearwing-shape, with dark olive green forewings, an orange-red abdomen with black spots along the back and red legs with long black scales. The hindwings are transparent, with thin black veins across.

**gloriosa** Edwards MANROOT or BIG-ROOT BORER
N. America – Oregon, California, Arizona, New Mexico, Utah, Colorado, Texas, Kansas and N. Mexico 40–50 mm. (1·57–1·97 in.)
This species is the biggest of all the American species of clearwings (Sesiidae). The caterpillar feeds on *Cucurbita foetidissima* and on underground tubers of *Marah fabaceus* (Cucurbitaceae) known as Manroot or Big-root. The pupal cases have special structures to help the moth to emerge from the soil, which in some cases may be dry and compacted. There are several colour varieties of this moth in N. America. **28**b

**Mellicta** Billberg (Nymphalidae)
A genus of butterflies found in Europe and temperate Asia. They are generally strong fliers but not quite as powerful as the larger fritillary butterflies to which they are related. The species tend to vary considerably over their range and as a result of this many different forms and subspecies have been named. Although there are probably less than a dozen species in the genus, there are some 200 names listed in catalogues of the genus, which have been applied to the 12 species either as names of subspecific forms or aberrations.

**athalia** Rottemburg HEATH FRITILLARY
Europe including Britain, temperate Asia including Japan 34–46 mm. (1·34–1·81 in.)
One of the commonest and most widespread butterflies which has been divided into a number of subspecies over its range. The butterflies are very variable in pattern and fly in May-June, with a second brood August-September in meadows and open pastures. The caterpillars feed on cow-wheat *(Melampyrum)* or plantain *(Plantago)*. **187**

**britomartis** Assmann ASSMANN'S FRITILLARY
C. Europe, temperate Asia to Korea 34–36 mm. (1·34–1·42 in.)
A rather dark fritillary, broadly similar to the Heath Fritillary, *M. athalia,* but a more patterned underside hindwing with a smaller yellow-brown median area and much darker on the upperside, with the black lines heavier and more continuous. The butterfly is out in May and August in S. Europe, with a single brood in the more N. part of its range. It flies over heaths and similar areas, generally at low altitudes. The caterpillars feed on species of plantain *(Plantago)* and *Veronica* (Scrophulariaceae).

**melonella** see **Galleria**
**melpomene** see **Heliconius**
MELSHEIMER'S SACK-BEARER see **Cicinnus melsheimeri**
**melsheimeri** see **Cicinnus**
**membliaria** see **Euplocia**
**memnon** see **Papilio**
**menalcus** see **Nymphidium**

**Menander** Hemming (Nemeobiidae)
There are about 10 species in the C. American and tropical S. American genus of fast-flying butterflies. They were known under the name *Tharops* Hübner until 1939 when it was found necessary to replace this name which had been used already in zoological nomenclature by an author earlier than Hübner.

**hebrus** Cramer
Tropical S. America 25–32 mm. (0·98–1·26 in.)
The less colourful female (not illustrated) of this species is brown above with a slight violet iridescence and with brown and black markings and an orange patch at the forewing apex. The caterpillar is not known. **263**v

**menander** Stoll BLUE THAROPS
Panama, tropical S. America 34 mm. (1·34 in.)
When not feeding on flowers, often those of *Eupatorium* species, *M. menander* usually settles on the under surface of a leaf. The female is not as colourful as the male (illustrated) the blue being confined to speckles on a chiefly brown background. **263**u

**superba** Bates BLACK-CHEQUERED BLUE
Tropical S. America 38–42 mm. (1·5–1·65 in.)
This rare and beautiful butterfly differs from *M. menander* mainly in the angled, almost tailed hindwings, which are strongly spotted with black, and the more greenish colour of the central part of the iridescent forewings.

**menander** see **Menander**
**menapia** see **Neophasia**
**mencia** see **Epicopeia**
**mendica** see **Diaphora**
**mendocino** see **Saturnia**
**menelaus** see **Morpho**
**meneria** see **Amarynthis**

**Meneris** Westwood: syn. *Aeropetes* (Nymphalidae)
This genus of butterflies has only 2 known species which are both common and widespread African insects.

**tulbaghia** Linnaeus MOUNTAIN BEAUTY or MOUNTAIN PRIDE
S. Africa 80–100 mm. (3·15–3·94 in.)
This is one of the largest species of butterfly in the subfamily Satyrinae. It is a fast flier and occurs on the higher mountain ranges of Africa S. of the Zambesi. The butterfly readily settles on brightly coloured flowers and the caterpillar has been reared on *Hebenstretia* (Scrophulariceae), *Hypparhenia* and other species of Gramineae. The species was named by Linnaeus after an 18th century governor of Cape Colony, called Tulbagh. **174**k

**meolans** see **Erebia**
**mercurea** see **Eudonia**
**mercurella** see **Eudonia**

**Meridarchis** Zeller (Carposinidae)
Some 30–40 species are known from this genus of micro-moths which is mainly Indo-Australian. A collection of moths made in the Philippines recently was found to contain specimens of several new species, which have just been scientifically described. Study of collections of micro-moths made by specialist collectors continually add to the detailed knowledge of these small, but often important species. Studies on the biology of micro-moths is a field in which anyone can make observations which will be of value in building up a picture of the lives of these small insects. No large apparatus is needed for this, but considerable patience and careful observations are required.

**trapeziella** Zeller
India, China, Burma 24–34 mm. (0·94–1·34 in.)
The species is typical in appearance of many in this genus. The specimen illustrated was collected in Burma. Nothing is known of its biology. **47**c

**meridionalis** see **Ornithoptera**

**Mesene** Doubleday (Nemeobiidae)
There are nearly 20 C. American and tropical S. American species in this genus of butterflies. Most of

them are orange above, with black margins; some have white markings on the forewing. Several species of the day-flying Geometrid genus *Eudule* closely resemble *Mesene* in colour pattern. *Mesene* species may be models in this instance, as at least one species, *Mesene phareus,* is known to feed on a poisonous plant and may be capable of storing plant toxins in its tissue in the manner of distasteful Danainae and others. Another possible mimetic partner is the Hypsid genus *Mesenochroa.*

**margaretta** White
C. America including Mexico, N.W. S. America 24–28 mm. (0·94–1·1 in.)
The form illustrated, *semiradiata*, is closely matched by a day-flying Hypsid moth *Mesenochroa guatemalteca.* **266**c

**phareus** Cramer
Tropical C. & S. America 20–25 mm. (0·79–0·98 in.)
This is a red or reddish orange butterfly, with black borders to both pairs of wings. The caterpillar is known to feed on the foliage of *Pallinia pinnata,* a species of Sapindaceae, which has been described as 'very poisonous'. It is a fairly common species.

**Mesenochroa** Felder (HYPSIDAE)
There are 4 known species in this C. and N.W. S. American genus of moths. They are apparently mimicked by similarly patterned orange, black and white species of other groups of moths and butterflies.

**guatemalteca** Felder
Guatemala, Panama 27–35 mm. (1·06–1·38 in.)
This is possibly a model for the Nemeobiid butterfly species *Mesene margaretta* and a species of Zygaenidae *Malthaca radialis.* **381**b

**Mesenopsis** Godman & Salvin (NEMEOBIIDAE)
About 5 species are placed in this C. American and tropical S. American genus of butterflies. Most species are rare in collections. They are mimics of day-flying moths of the family Dioptidae.

**albivitta** Lathy
Brazil 28–35 mm. (1·1–1·38 in.)
This strikingly coloured species is considered to be a mimic of the Dioptid moth *Scea steinbachi* which it closely matches in colour pattern, especially in the female. Some males have a narrow yellow streak on the hindwing, extending outwards from the base of the wing. **266**a

**bryaxis** Hewitson
C. America to Bolivia 31–35 mm. (1·22–1·38 in.)
*M. bryaxis* is probably a mimic of the Dioptid *Josia ligata* (see illustration of the similarly patterned *J. lativitta* **362**). Its wings are dark brown, with an elongate orange-yellow band extending from the base to the outer margin of each wing.

**Mesoacidalia** Reuss (NYMPHALIDAE)
A genus of butterflies which only recently was regarded as separate from the Argynnid fritillaries, to which it is related. They are mainly found in Europe and temperate Asia.

**aglaja** Linnaeus DARK GREEN FRITILLARY
Europe including Britain, N. Africa, temperate Asia including Japan 48–60 mm. (1·89–2·36 in.)
A number of distinct subspecies of this species have been described. The forewings on the upperside are the typical Argynnid pattern but the underside ground-colour of the hindwings are yellowish buff with green overlay and silver spots. The butterfly which is out in June and July is common in meadows and heathlands where violets *(Viola),* the foodplant of the caterpillar, grows. It is a fast-flying butterfly and is more often seen flying rapidly by than at rest.

**Mesosemia** Hübner (NEMEOBIIDAE)
Most species of this genus of over 60 C. and S. American species of butterflies have either one or 2 eyespots on the middle of the forewing. They seldom fly far, and jump rather than fly from one leaf or flower to another. Their distribution is generally rather localized.

**mevania** Hewitson
N. S. America 43–46 mm. (1·69–1·81 in.)
This is a forest species, occurring up to at least 2970 m. (9750 ft). There is considerable variation in the width and clarity of the white transverse band on the forewing. Several other species resemble *M. mevania* in colour pattern type, for example *M. loruhama* Hewitson and *M. messeis* Hewitson, both of which are found in N. S. America. Both pairs of wings have eyespots on the pale brown under surface. Eyespots in species of Saturniidae have been shown by A. D. Blest, previously of London University, to frighten small birds and are probably protective in function. **263**z

**Mestra** Hübner (NYMPHALIDAE)
A genus of butterflies with species in C. and S. America, some of them reaching into the S. U.S. Some of the butterflies in this genus have been bred upon species of *Tragia* (Euphorbiaceae).

**amymone** Ménétries AMYMONE BUTTERFLY
C. America into S. U.S. 35–44 mm. (1·38–1·73 in.)
The pale grey forewing has a white band through the middle; the hindwing has a broad border of orange-brown The caterpillar feeds on species of *Tragia* (Euphorbiaceae). This butterfly is common in S. Texas but only strays into Kansas, where it is unable to survive the winter; each summer it moves N. to repopulate the area.

**Metachanda** Meyrick (METACHANDIDAE)
A small genus of micro-moths typical of the family in their appearance.

**citrodesma** Meyrick
Africa 7–12 mm. (0·28–0·47 in.)
This species has been collected widely over Africa, from Sierra Leone, in the W., to Malawi in the S. It also occurs on the islands in the Gulf of Guinea. A number of similar looking, related species occur throughout Africa. Nothing is known at present of its biology. **43**f

**Metachandidae**
A small family of micro-moths related to the Gelechiidae mostly from Africa, with species also in India. The moths are small, rather inconspicuous. Little is known of their biology. See *Metachanda.*

METALLIC CERULEAN see **Jamides alecto**
**metallica** see **Tascina**
METALMARK BUTTERFLIES see **Apodemia, Calephelis, Lasaia, Nemeobiidae**

**Metamorpha** Hübner: syn. *Victorina* (NYMPHALIDAE)
A small genus of butterflies from C. and S. America with species which reach into the U.S.

**epaphus** Linnaeus
C. & S. America 70–75 mm. (2·76–2·95 in.)
This species is common in the mountainous wet tropical forests where the butterfly keeps fairly near to the ground, generally in the vicinity of the foodplant of the caterpillar *(Ruellia,* Acanthaceae). The butterflies feed on mouldy leaf litter and on flowers. The males are smaller than the females. The eggs are laid in a loose cluster on apical leafbuds of *Ruellia tubiflora.* The caterpillar, in the later stage has a shiny black head with velvet maroon coloured body and long, bright yellow, branched, spines. This species is sometimes placed in the genus *Siproeta.* **231**b

**stelenes** Linnaeus MALACHITE or BAMBOO PAGE BUTTERFLY
S. U.S. 70–80 mm. (2·76–3·15 in.)
A widespread species, with powerful flight which takes it on migration into the S. U.S. and West Indian islands. The pale green median patches and bands contrast with the darker borders of the wings. The males and females are similarly patterned, generally flying in the sunshine and often congregating in numbers. They have a distinctive flight with sudden swift movements. The eggs are laid on *Blechum* and *Ruellia* (Acanthaceae). The fully grown caterpillar is large, some 50 mm. (1·97 in.) long. The adult butterflies feed on juices of a variety of ripe fruits and they will drink at damp patches of soil. **210**

**Metarbela** Holland (METARBELIDAE)
These are very Cossid-like moths, the majority of the species occurring in Africa.

**triguttata** Aurivillius
Africa 22 mm. (0·87 in.)
This species is found in Uganda and the Sudan. It has a typically long body which projects well beyond the hindwings. The biology of this species is not known. **11**d

**Metarbelidae**
A small family of moths related to the Cossidae, found mostly in Africa. Little is known of their biology and they have been little studied. See *Metarbela.*

**metelkana** see **Diacrisia**
**meteus** see **Automolis**
**metharme** see **Heliconius**
**meticulodina** see **Uropyia**
**meticulosa** see **Phlogophora**
**metis** see **Metisella**

**Metisella** Hemming (HESPERIIDAE)
A genus of skipper butterflies with some 20 known species, all from Africa.

**metis** Linnaeus GOLD-SPOTTED SYLPH
South Africa 26 mm. (1·02 in.)
This small skipper is brown with orange or red spots on the fore- and hindwings, the underside is unmarked brown, It is mainly a forest species found in Cape Province, Natal and Transvaal. The caterpillar feeds on species of Gramineae.

**orientalis** Aurivillius
Africa 24 mm. (0·94 in.)
This skipper is found in Uganda and Tanzania. There are a number of similar looking species with the orange patches on the wings. It has the typical fast flight of a skipper but nothing is known of its biology. Evans. **89**

**meton** see **Rekoa**
**mevania** see **Mesosemia**
MIAMI BLUE see **Hemiargus thomasi**
**micale** see **Narathura**

**Micragone** Walker: syn. *Cyrtogone* (SATURNIIDAE)
There are about 17 described species in this genus of moths. It is found in tropical Africa S. to N. South Africa. The wings of most species are finely striate or granulate in appearance and most have a pink area near the front edge of the hindwing.

**ansorgei** Rothschild ANSORGE'S PRINCE MOTH
Africa S. of the Sahara to Rhodesia 65–84 mm. (2·56–3·31 in.)
The female of this species differs from the male (illustrated) in the evenly rounded outer margin of the forewing and in the paler coloration. The caterpillar is chiefly black above and green below, with long white hair and black-haired, red tubercles; it feeds gregariously on species of *Brachystegia* and *Julbernardia.* **331**b

MICROPSYCHID MOTHS see **Solenobia**

**Micropterigidae**
This is a family of small, generally day-flying micro-moths. They are often considered to be the most primitive moths living today. However the family is one whose systematic position is still in dispute amongst the specialists, some regarding the Micropterigids as an Order distinct from the Lepidoptera. Thus for some, these are not moths, but as different from the Lepidoptera as, for example, the Caddis-flies (Trichoptera). Amongst the more unusual features in the Micropterigidae are the presence of functional mandibles (jaws) in the adult moth which they use for chewing. Most moths, when they feed, have a long proboscis with which they suck nectar or plant juices. The Micropterigids use their mandibles to chew pollen

from flowers on which they feed. Species in the family are found in Europe, N. and S. America, Australia and New Zealand. They have been found fossilized in pieces of amber which are known to be 40 to 50 million years old. See *Micropterix*.

**Micropterix** Hübner (MICROPTERIGIDAE)
Worldwide. This genus is often regarded, with the others in the family, as not being true Lepidoptera, and is considered as separate from the Lepidoptera as the Trichoptera or Caddis-flies. Apart from the primitive type of venation, this genus is peculiar in lacking the proboscis or sucking tube used by most moths and butterflies for feeding. The adults feed on pollen of various flowers using mandibles (jaws) which in other Lepidoptera are reduced or absent.

**auricrinella** Walsingham
U.S., Canada 9 mm. (0·35 in.)
This beautiful moth has a golden head and purple, iridescent wings. The grey underside of the wings is also iridescent. The fore- and hindwing venation is broadly similar. The adult feeds on the pollen of flowers, but its complete life history is unknown. It is common throughout the N.E. U.S. and Canada.

**calthella** Linnaeus
Europe including Britain 7–9 mm. (0·28–0·35 in.)
This species is generally common in marsh places in May and June. The caterpillar feeds on the moss, *Hypnum*. The adult, like the others in the genus, has mandibles instead of a proboscis and feeds on pollen. **2**

**schaefferi** Herrich-Schäffer
S. Europe 9–10 mm. (0·35–0·39 in.)
This species is similar to the others in the genus and has been recorded in France, Switzerland and Italy. The adult moths have mandibles for chewing pollen. **1**a

**thunbergella** Fabricius
Europe including Britain 8–9 mm. (0·31–0·35 in.)
The head of this moth is brownish but the forewings are a pale shining gold-bronze colour. The larvae live in moss and the adults, as with others of the genus, feed on pollen of different flowers.

**Microsphinx** Rothschild & Jordan (SPHINGIDAE)
There is a single known species in this hawk-moth genus.

**pumilum** Boisduval BEDSTRAW HAWKLET
South Africa 26 mm. (1·02 in.)
This is one of the smallest species of hawk-moths. The caterpillar is green, marked with yellow and purple longitudinal lines and black dashes; it feeds on bedstraws *(Galium)*. **344**e

**Mictopsichia** Hübner (GLYPHIPTERIGIDAE)
This genus of micro-moth is widespread except for Europe and Asia. Species of this genus all show the iridescent scales which are particularly common in the Glyphipterigids.

**durranti** Walsingham
C. & S. America 20 mm. (0·79 in.)
Little is known about this species. The specimen illustrated is one of the original specimens collected in Peru in 1884. **34**d

**ornatissima** Dognin
Peru 20–22 mm. (0·79–0·87 in.)
First described in 1905 from specimens collected in Peru, this species is found in several parts of S. America. Nothing is known at present about its life history or foodplants of the caterpillar. The colour and patterns, with the iridescence make this a striking looking moth. Related species have orange coloured hindwings.

**midamia** see **Cyclosia**

**Midila** Walker (PYRALIDAE)
This genus of Pyralid moths has about 20 known species occurring from C. America to S. Brazil. The genus was recently placed in its own subfamily, Midilinae, to show it is considered distinct from other Pyralid subfamilies. Some of the species in the genus look rather like noctuid moths, others like geometrid moths; few have the typical, rather delicate, pyralid shape. The life history of most of the species is unknown. One species has caterpillars which feed on the roots of *Colocasia* (Araceae).

**quadrifenestrata** Herrich-Schäffer
S. America: Venezuela to Peru, Guyanas, S. Brazil 50–70 mm. (1·97–2·76 in.)
There are 4 subspecies, each from different parts of S. America. Nothing is known of their life-histories. This species is typical in appearance of most of the Midilinae. **63**c

MIGRANT BUTTERFLIES see **Catopsilia, Phoebus**
**milberti** see **Nymphalis**
MILBERT'S TORTOISESHELL see **Nymphalis milberti**
**milca** see **Vanessula**

**Miletus** Hübner (LYCAENIDAE)
There are probably about 20 species in this genus of butterflies. Its range extends through S.E. Asia to New Guinea. Compared with most species of Lycaenidae, the species of *Miletus* are unusually drab in colour. The males are brown above, with a white patch on the forewing, and pale brown below, with brown markings. Females are paler in colour. *Miletus* species are found mainly in low-elevation rain-forests.

**boisduvali** Moore
Sikkim, N. India, Burma, S.E. Asia to New Guinea 32–38 mm. (1·26–1·5 in.)
This is a fairly common species except in the extreme W. of its range. The males are brown above, with a single, large, round spot on the middle of the forewing. The female is similar to the male, but the forewing spot is poorly marked and the outer margin of the hindwing is evenly rounded, not angulate at the middle. Several subspecies have been described, chiefly from the islands of Malaysia and Indonesia. The eggs are laid on stems and shoots infested with aphids (greenflies and their allies) which are the food of the caterpillars of *M. boisduvali*. Two species of ants attend these aphids and feed on their honeydew secretions, but seem to ignore the competing caterpillars.

**symethus** Cramer
N.E. India to the Philippines, Indonesia E. to Lombok 33–40 mm. (1·3–1·57 in.)
*M. symethus* is typical in colour pattern to many others of its genus. The male is dark brown above, variously marked with one or more patches of white in the middle of the forewing. The hindwings are lighter in colour than the forewing and are nearly white in some specimens. Females are invariably lighter than the males and are usually nearly white. The caterpillars, are often found in company with nymphs (early stages) of Hemiptera (plant-bugs) of the families Fulgoridae and Jassidae, but the nature of this association is uncertain.

**Milionia** Walker (GEOMETRIDAE)
This genus of moths has probably more remarkably brightly metallic coloured species than any other moth genus. Basically the iridescence of greens and blues overlay combinations of patterns in velvety black. Little is known of their life histories but individually some of the adults are well known, being popular subjects for coloured illustrations. Any collection of coloured pictures of moths normally includes at least one species of *Milionia*. The genus has 40 or more species, many of these have formed local races in different islands which differ in details of colour or pattern. A lot of work is needed to elucidate the relationships of the colour forms. Most of the species occur on New Guinea or surrounding islands. A few are found in Indonesia and one reaches as far N. as Japan.

**aglaia** Rothschild & Jordan
New Guinea 40–55 mm. (1·57–2·17 in.)
The metallic iridescent abdomen contrasts with the forewings. Nothing is known of the biology of this species. The pattern is rather more geometric than some of the other species. The underside has yellow, instead of red bands, on the wing. **296**k

**aroensis** Rothschild
New Guinea 48 mm. (1·89 in.)
The males have a red band on the forewing; the females yellow. The underside of the wings is broadly similar to the upperside, but the iridescence at the base of the wings is green instead of blue as on the upperside. **296**j

**basalis** Walker
New Guinea, Indonesia 42–52 mm. (1·65–2·05 in.)
The striking colour forms of this species are so different that many have been named as new species. The specimen illustrated is the form named *M. basalis butleri* Druce. **296**d

**callima** Rothschild & Jordan
New Guinea 38–40 mm. (1·5–1·57 in.)
Several forms with slightly different iridescence known from different areas. This is basically a dark blue species with a triangular green iridescent patch on the forewings. The legs have strongly iridescent patches of scales. The underside is broadly similar but the green iridescent patch is near the middle of the wing. **296**m

**coeruleonitens** Rothschild
Indonesia, W. Irian 38–40 mm. (1·5–1·57 in.)
All the specimens of this species have been collected in mountainous areas between 1520–3050 m. (5000–10,000 ft). The underside is similar to the upperside but has a patch of yellow on the hindwing. **296**p

**eichhorni** Rothschild & Jordan
New Guinea 60 mm. (2·36 in.)
The species has more iridescence on the underside than the upperside of the wing. This would probably be visible if the insect rests with the wings raised in the manner of many Geometrid moths. The forewings are iridescent black with a red band at an angle across the forewing. The abdomen is electric blue above.

**elegans** Rothschild & Jordan
Fergusson Island, Goodenough Island 51–65 mm. (2·01–2·56 in.)
The colour and angled margin on the hindwing are distinctive of this species. This is a rather variable species with quite strikingly different colour patterns. The underside is dark, with a white patch in the middle of the forewing and black spots. **296**c

**euchromozona** Rothschild
New Ireland 50 mm. (1·97 in.)
The iridescent scales are over the legs as well as the wings and body. The hindwing is of a uniform colour with a striking iridescence. The abdomen is an electric blue. The underside has an orange patch and iridescent green on the underside, contrasting with the blue iridescence of the upperside. **296**e

**exultans** Rothschild
New Britain (Bismarck Archipelago) 40–50 mm. (1·57–1·97 in.)
The red mark on the forewing varies from a small patch to a large band, contrasting sharply with the velvet iridescence of deep blue. The underside is black without the red forewing patch and has a strong blue iridescence. **296**h

**grandis** Druce
New Guinea, Fergusson Island 70 mm. (2·76 in.)
The males differ in the shape of the hindwings which are rather pointed, those of the females being rounded as illustrated. Several closely allied species have this sexual dimorphism. **296**b

**lamprima** Rothschild & Jordan
New Guinea 36–40 mm. (1·42–1·57 in.)
Several of the smaller species are similar to this but they all have different coloured bands on the forewings. **296**q

**leucomelas** Montrouzier
Indonesia; Buru, Goodenough Island, Woodlark Island 53–68 mm. (2·09–2·68 in.)
This species varies in colour and pattern on different islands. Specimens from Buru are larger and have more blue. The colour and shape misled the early 19th century entomologists into describing this as a species of Lithosiid. The forewings are blue black with a very broad red band. The hindwing has a black apex and

bright yellow-red area in the middle. This species has rather elongate wings.

**obiensis** Rothschild
Obi 60 mm. (2·36 in.)
Only known from this island which is in Indonesia between Celebes and W. Irian. This is one of the many species described by Lord Rothschild whose private collection at Tring, England, was the largest in the world. Given to the British Museum (Natural History) in 1939, it is an immensely valuable source of scientific specimens, specialists from all over the world coming to study it. **296**f

**ovata** Rothschild & Jordan
New Guinea 75 mm. (2·95 in.)
Although less iridescent than many species in the genus, the colours are quite distinctive. The moth has been collected at altitudes of 1520–2130 m. (5000–7000 ft). **296**l

**paradisea** Jordan
New Guinea 37–45 mm. (1·46–1·77 in.)
This species has been used more often as an example of the genus *Milionia* than the others. In spite of being fairly common in parts of New Guinea, nothing is published of its life history. **296**g

**pericallis** Rothschild & Jordan
New Guinea 50–55 mm. (1·97–2·17 in.)
Similar to *M. elegans* but the forewing has a green area and the iridescence is green instead of blue. There are several different colour forms of this species. **296**a

**plesiobapta** Prout
New Ireland 38–50 mm. (1·5–1·97 in.)
One of the most striking things about this species is the iridescent green on the body. The iridescence dances about as the light angle alters. **296**n

**weiskei** Rothschild
New Guinea 48–60 mm. (1·89–2·36 in.)
The bright yellow on the fore- and hindwings of this species are characteristic. The tip of the apex of the hindwing is yellow, the rest is blackish-blue. On the hindwing the yellow band has a reddish inner edge. Nothing is known of its biology.

MILKWEED BUTTERFLY see **Danaus plexippus**
**milliaria** see **Callioratis**

**Miltochrista** Hübner (ARCTIIDAE)
Over 100 species of moths are placed in this genus of the subfamily Lithosiinae. It is represented in Europe, temperate and tropical Asia and Australia.

**miniata** Forster ROSY FOOTMAN
Europe including Britain, through temperate Asia to Japan 24–30 mm. (0·94–1·18 in.)
The distinctively marked *M. miniata* emerges as an adult in June or July. It flies in wooded areas and on heaths, the daily flight period starting around dusk. Its hairy, brown caterpillar feeds on algae (primitive green plants) and lichens on the branches and trunks of various trees in August and September and, after hibernation, during the following spring to May. **382**c

**Mimacraea** Butler (LYCAENIDAE)
This is a small genus of tropical and subtropical butterflies found in Africa S. of the Sahara. They are large in size compared with most other Lycaenids, and are mimics of butterflies of the Nymphalid subfamilies Danainae and Acraeinae.

**marshalli** Trimen MARSHALL'S ACRAEA MIMIC
Kenya, Tanzania, Malawi Rhodesia, Mozambique 45–55 mm. (1·77–2·17 in.)
This distinctive butterfly is a mimic of the common, Danaine butterfly *Danaus chrysippus*. In spite of its smaller size, the resemblance to *D. chrysippus* is extremely close both in the colour pattern and in the leisurely, gliding flight. The paler under surface of the wings, especially the hindwing which is spotted with black basally, is *Acraea*-like in pattern. The foodplants of the caterpillars are lichens, an unusual source of food for a Lycaenid. **244**b

**neurata** Holland
W. Africa 50–75 mm. (1·97–2·95 in.)
Both sexes of this large Lycaenid species closely resemble the males of *Bemastistes epaea*, a species of the Nymphalid subfamily Acraeinae.

**Mimallonidae**
This is a small family of moths, usually considered to be related to the ancestors of the Saturniidae and Bombycidae and somewhat similar in general appearance to species of the latter family. It is chiefly tropical American in distribution, but 4 species are represented in N. America. Its caterpillars are case-dwellers The species of this family are also widely known under the later and invalid name Lacosomidae. See *Cicinnus, Lacosoma*.

**Mimas** Hübner (SPHINGIDAE)
There is a single known species in this genus of hawk-moths.

**tiliae** Linnaeus LIME HAWK-MOTH
Europe, through Asia to Siberia and Japan 60–75 mm. (2·36–2·95 in.)
This is one of the most effectively camouflaged hawk-moths when at rest. The forewings do not entirely conceal the hindwings so that the general shape is more like a bunch of dead leaves than a moth. It is occasionally more reddish or greenish brown than in the illustration. Elm *(Ulmus)* and lime *(Tilia)* foliage are the chief food-source, but birch *(Betula)* and other trees and shrubs have also been recorded as foodplants of the caterpillar. **346**

MIME BUTTERFLIES see **Papilio**
MIMIC see **Limenitis archippus**

**Mimoniades** Hübner (HESPERIIDAE)
A S. American genus of skipper butterflies with some 7–8 species at present known in it.

**versicolor** Latreille: syn. *M. montana*
S. America 50 mm. (1·97 in.)
A brightly coloured skipper butterfly with yellow spots on the forewing and rather iridescent blue bands on the hindwing. The pattern is variable in this species but all have yellow-orange stripes along the sides of the thorax. The underside is equally colourful with more blue on the hindwing and the red forewing band with yellow towards the back. This butterfly is widespread in parts of Brazil, Peru, Bolivia and Colombia. It is found generally above 2000 m. (6560 ft) in colder areas where it is often abundant, its bright colours making it conspicuous. It does not have the same rapid flight so characteristic of many skipper butterflies but is more leisurely in its progress. **90**

**mimosae** see **Argema**
**mineus** see **Mycalesis**
**miniata** see **Miltochrista**
**minimus** see **Cupido**

**Miniophyllodes** de Joan (NOCTUIDAE)
One species is known. Its closest relative is probably in the African genus *Miniodes*.

**aurora** de Joan
Madagascar 55–65 mm. (2·17–2·56 in.)
This rare species exhibits both cryptic coloration and flash-coloration. At rest, the long stalk-like palps and the remarkably leaf-like forewings are highly cryptic. The hindwings, in contrast, are a brilliant orange, black and yellow; they are hidden by the forewings when the moth is at rest, but are exposed if danger is imminent to produce a sudden flash of colour which may disconcert an enemy and allow the moth to escape. **392**c

**ministra** see **Datana**
**ministrana** see **Eulia**

**Minois** Hübner (NYMPHALIDAE)
A genus of butterflies related to *Satyrus* (Satyrinae).

**dryas** Scopoli DRYAD BUTTERFLY
N. Spain, C. & S. Europe, temperate Asia including Japan 54–58 mm. (2·13–2·28 in.)
The male is a dark black-brown, unpatterned on the upperside except for 2 black eyespots with bluish white centres, one in front of the other, below the apex of the forewing. The outer margin of the hindwing is scalloped. The underside is paler, still a dark brown but the 2 eyespots are larger and have an orange ring round the black. The female is a paler brown, with very large eyespots on the upperside forewing and the underside even paler, with yellow round the large eyespots and some white on the hindwing. Several subspecies have been described, including one in Japan. The adult flies over grassy slopes from lowlands to 910 m. (3000 ft), in July and August. The caterpillars feed on various grasses.

**minuta** see **Melitaea**
**minuticornis** see **Calpe**
**mionina** see **Callicore**
**mirabilis** see **Cechenena, Charagia**
**miraculosa** see **Calliprogonos**
**mirifica** see **Euliphyra**

**Mirina** Staudinger (ENDROMIDAE)
Two temperate Asiatic species belong in this genus of moths; *M. christophi* (illustrated) and the Chinese *M. fenzeli* Mell. Although this genus was placed tentatively in the family Saturniidae by its author, and later in the Endromidae, there is little doubt that its placement in either family is incorrect.

**christophi** Staudinger
E. U.S.S.R. (Ussuri and Amur river valleys) 29–44 mm. (1·14–1·73 in.)
The strange caterpillar of this species feeds on honeysuckle *(Lonicera)*. It is spiny and black when young but green in the later stages with bluish green, eversible, fleshy processes similar to the osmateria of Papilionidae, and it overwinters as a chrysalis. **318**p

**misippus** see **Hypolimnas**
MISTLETOE BROWNTAIL see **Euproctis edwardsi**

**Mithras** Hübner (LYCAENIDAE)
Mithras, the name of the ancient Persian god of light and the sun, is an especially apt name for this group of iridescent butterflies. The few species of this genus are found in tropical America.

**hemon** Cramer BLACK-BACKED BLUE
Tropical S. America 28–33 mm. (1·1–1·3 in.)
This relatively common butterfly flies in forest clearings and along the edges of forests. Nothing is known about its life history. The illustrated specimen is a male. **263**a

**lisus** Stoll FOREST BLUE
C. America including Mexico, and tropical S. America 34–38 mm. (1·34–1·5 in.)
The male of this species is a bright blue above, the female dark brown except for some pale blue at the base of the wings. Similar in general appearance to the illustrated *M. hemon*, but is a more brilliant blue in the male and darker in the female. This is a rare species in collections.

**mittrei** see **Argema**
**mneme** see **Melinaea**

**Mnesarcha** Meyrick (MNESARCHAEIDAE)
These micro-moths are found only in New Zealand. They are primitive moths of the suborder Dacnonypha which have a single genital opening in the female instead of the more usual double one as in most other families of Lepidoptera.

**loxoscia** Meyrick
New Zealand 10–12 mm. (0·39–0·47 in.)
Very little is known of the biology of this primitive species. There are very few related species and these are all in New Zealand. It is a species which has interested morphologists studying the more primitive groups and although little is known of the live moth, its morphology has been studied in some detail. **1**e

**Mnesarchaeidae**
A family of micro-moths known only from New Zealand. They are regarded as a primitive group and a study of their biology will provide interesting information which may be useful in interpreting the relationships of these primitive moths. See *Mnesarcha*.

**mniochlora** see **Oleuthreutes**
**mniszechi** see under **Euploea eleusina**

MOCKER SWALLOWTAIL see **Papilio dardanus**
**modesta** see **Euploea, Pachysphinx**
**molesta** see **Cydia**
**moma** see **Viviennea**

**Momphidae**
This family of micro-moths includes several subfamilies (eg. Cosmopteryginae). Momphids are mostly small moths, found all over the world. They generally have long slender, pointed wings with reduced venation. The wing patterns often have metallic colours in them. See *Aphthonatus, Cosmopterix, Glyphipteryx, Hyposmocoma, Sathrobrota, Sorhagenia, Stilbosis, Trachydora.*

**monacha** see **Lymantria**
MONARCH BUTTERFLIES see **Danaus**
**monerythra** see **Atteva**
**moneta** see **Dione**
MONKEY-SLUG see **Phobetron pithecium**
**monocentra** see **Cyclotorna**
**monodactyla** see **Emmelina**

**Monopis** Hübner (TINEIDAE)
A cosmopolitan genus of micro-moths, the caterpillars feeding on dried vegetable produce. At times they can reach pest proportions. They are liable to be found in dried vegetables, when a few caterpillars will cause rejection of the product by the purchaser!

**crocicapitella** Clemens
Worldwide 11–17 mm. (0·43–0·67 in.)
The white head and thorax contrast with the dark purple brown wings which have traces of yellow along the outer wing margin. The caterpillar is a general scavenger of seeds and refuse and is often associated with old bird's-nests or bat droppings in caves.

**montana** see **Alucita, Mimoniades**
**montrouzieri** see **Papilio**
**monuste** see **Ascia**
MOON-MOTHS see **Actias, Argema**
MOORLAND CLOUDED YELLOW see **Colias palaeno**

**Moresa** Walker (NOTODONTIDAE)
This is a very small genus of distinctive species with green forewings and yellow or white hindwings. They are confined to tropical S. America.

**magniplaga** Schaus 45–60 mm. (1·77–2·36 in.)
This is a rare species in collections. Nothing seems to be known about its biology. Females differ from the illustrated male in the yellow hindwings. **357**c

**morgani** see **Xanthopan**
**mori** see **Bombyx**
**moritili** see **Platylesches**
MORMON BUTTERFLIES see **Papilio**
MOROCCO ORANGE-TIP see **Anthocharis belia**

**Morpho** Fabricius (NYMPHALIDAE)
This genus contains some of the most brilliantly iridescent butterflies in the world. They generally fly in bright sunshine. The brilliant gloss of the male is due to the physical structure of the scales reflecting the light, not to a pigment. The caterpillars mostly live in groups or nests in forest trees. The head has a pair of horizontal conical processes pointing forward. They are usually brightly coloured, often yellow or red stripes with long tufts of bristles. Some of the adults fly slowly near the ground by the edges of woods, others soar around the tops of tall trees. The genus extends from Mexico to S. Brazil and N. Argentine. The Amazon is the main area for them particularly to the west where the rivers run down the E. Cordilleras of Ecuador and Peru. Recent work has separated the various species in the genus *Morpho* into a number of closely allied groups and subgeneric names have been given to these. With the use of the butterflies for jewellery and decoration, vast numbers are collected and this has made some of the S. American governments, concerned about their disappearance, pass laws for their protection. Aeroplanes flying over the areas where morphos are common have reported seeing the flashing of light from the iridescence of the morphos far below them.

**aega** Hübner
Brazil 80–90 mm. (3·15–3·54 in.)
The brilliant blue male is common in S. Brazil. Many morphos are widely used for decorative purposes, in jewellery etc., and it is estimated that some 6,000,000 of this species are harvested each year. In spite of this it appears that the population is not suffering. No females are taken (they lack the attractive metallic blue of the male), and the bamboo foodplant of the caterpillar is abundant. The females are yellow-brown with a brown margin to the wings with yellowish patches in it. **158**

**cypris** Westwood BLUE MORPHO
S. & C. America 130–140 mm. (5·12–5·51 in.)
Although fairly widespread, nothing is known of the life history of this butterfly. It is only found high up in forest clearings from Nicaragua, Costa Rica and Panama down to Colombia, N.W. Venezuela and into Peru. **161**c

**deidamia** Hübner
S. America 120–140 mm. (4·72–5·51 in.)
This butterfly has a steady and rather rapid flight in open paths on clearings in forests. The species has been separated into several subspecies in different parts of its range, which is mainly in the tropics of S. America from the Guyanas through the Amazon Basin. **162**e

**didius** Hopffer
S. America 150–164 mm. (5·91–6·46 in.)
This species is found mostly in Ecuador and Peru. The females are generally larger than the males. The intense electric blue of the male is present on the females at the base of the wings. The margins are brown with large white spindle-shaped marks. There is some variation in this pattern. The undersides are brown in both sexes with eyespots along the paler wing margins. From below as they fly the wings have a thin, almost transparent look. **162**d

**laertes** Druce
S. America 105 mm. (4·13 in.)
A distinctive silvery white butterfly which is only found in Brazil. The caterpillars are found in nests in forest trees. The butterfly is out from January to March and has a most delicate mother-of-pearl look, quite different from the better known blue morphos. **162**b

**menelaus** Linnaeus BLUE MORPHO
S. America 130–140 mm. (5·12–5·51 in.)
This butterfly is about all the year round from early morning until the sun is hot at midday. It can be found in clearings in forests or forest edges. This is one of the typical 'jungle' morphos. The female is black on the wing margin with the rest of the wing mostly white and brown with a trace of blue suffusion near the base. There are small whitish spots on the broad black margins. Several subspecies are known of this *Morpho* which ranges from Venezuela and the Guianas to the S. province of Brazil. The life history is unknown. This butterfly was used by the natives on the Rio Negro in the early part of this century as the model for their dance masks. It is common round Rio de Janeiro in certain localities. The butterflies fly in the sun, chasing one another and will readily chase a blue cloth waved in the air and so this is widely used when catching them, the males only being attracted to this. **162**c

**peleides** Kollar
S. & C. America, W. Indies 95–120 mm. (3·74–4·72 in.)
This beautiful Morpho is common over much of its range. The male has more blue on the upperside than the female and the underside is strongly patterned. Several subspecies have been described and recent work has shown that there is local host specificity and life-cycle differences between them. The caterpillars feed on a variety of Leguminosae, including *Machaerium seemannii* and species of *Lonchocarpus.* They feed round the edges of the leaves and when disturbed produce a strong smell. This is produced from a gland between the forelegs and is a volatile compound described as 'smelling like rancid butter'. The adults feed on rotting fruit but are also common on fermenting fungi in wounds in living and fallen trees. **164, 166**

**perseus** Cramer
C. & S. America, the Guianas, Venezuela, Brazil 120–140 mm. (4·72–5·51 in.)
The female has a broad, dark brown border, brown on the front edge of the forewing and silvery blue or pale at the base of the forewing. The underside is brown with lighter marks and 2 eyespots edged with yellow in the forewing. The illustrated subspecies *M. perseus metellus* from French Guiana has less black and yellow on the forewing. Neither sex has the electric blue coloured wings shown by many species of *Morpho.* **162**f

**portis** Hübner: syn. *M. cyntheris* SKY-BLUE MORPHO
Brazil, Uruguay 90–98 mm. (3·54–3·86 in.)
This butterfly is found locally in forest regions, especially where bamboos are common, in S. Brazil and into Uruguay. The life history is unknown. The butterfly generally is about in mid afternoon when it flies low over the ground, flapping slowly. The female is similar but with a broader brown border. Several subspecies have been described, even though the whole range of the species is relatively restricted. **161**a

**rhetenor** Cramer BLUE MORPHO
Amazon forests, Venezuela, the Guianas S. to Peru 130–150 mm. (5·12–5·91 in.)
This magnificent butterfly lives in the jungles, flying high round the trees and can be seen in most months of the year. In collecting specimens, the males are generally lured to a bait of rotting fruit. The female does not have any blue, but is shades of yellow, brown and orange-brown on the upperside. The males are said to have a strong sulphur smell. Several subspecies have been described in different parts of its range, including one from Peru which has a broad white band on the blue hindwing of the male, continued with a few patches onto the forewing. **162**a

**zephyrites** Butler
S. America 80–85 mm. (3·15–3·35 in.)
This occurs from Colombia to Peru and Ecuador with several subspecies described. Figured is the nominate subspecies, *M. zephyrites zephyrites* Butler. **161**b

MORPHO,
BLUE see **Morpho menelaus, M. rhetenor**
SKY-BLUE see **Morpho portis**
**morphotenaris** see **Taenaris**
**morsei** see **Leptidea**
MORT BLEU,
COCOA see **Caligo teucer**
PURPLE see **Eryphanis polyxena**
MOTH-BUTTERFLY see **Liphyra brassolis**
MOTHER-OF-PEARL BUTTERFLIES see **Rekoa, Salamis**
MOTHER-OF-PEARL MOTH see **Pleurotypa**
MOTTLED
BEAUTY see **Alcis repandata**
UMBER see **Erannis defoliaria**
WILLOW, SMALL see **Spodoptera exigua**
MOUNTAIN, BLUE see **Papilio ulysses**
MOUNTAIN
BEAUTY see **Meneris tulbaghia**
CLOUDED YELLOW see **Colias phicomone**
PRIDE see **Meneris tulbaghia**
**moupinensis** see **Chionaema**
MOURNING
CLOAK see **Nymphalis antiopa**
EMPEROR see **Ubaena fulleborniana**
MOUSE-COLOURED EUCHAETIAS see **Euchaetias murina**
MUD PUDDLE see **Colias philodice**
**mufindiae** see **Aethiopsestis**
**mulciber** see **Euploea**
**multicaudata** see **Papilio**
**mundella** see **Tirathaba**
**munionga** see **Oreisplanus**
**muricata** see **Idaea**
**murina** see **Euchaetias**

**Muschampia** Tutt (HESPERIIDAE)
This genus of skipper butterflies, with species in Europe and temperate Asia, has often been regarded as congeneric with *Syrichtus* Boisduval and some species may be found under that generic name in the literature.

**tessellum** Hübner TESSELLATED SKIPPER
Greece, Yugoslavia, S. Russia, Iran, temperate Asia to E. Russia 32–36 mm. (1·26–1·42 in.)
Somewhat similar to the Grizzled Skipper, *Pyrgus malvae,* but more restricted in its distribution. The butterfly is out in May and June and flies in meadows from lowlands to over 910 m. (3000 ft). The life history is not known. **88**

**muscosa** see **Epia**
MUSLIN MOTH see **Diaphora mendica**
MUSTARD WHITE see **Pieris napi**

**Musurgina** Jordan (AGARISTIDAE)
One species of moth is placed in this genus.

**laeta** Jordan
Madagascar 32 mm. (1·26 in.)
This is one of the few species of Agaristidae able to produce sound and, as in the others, it is only the male which has this capability. The sound is produced by the movement of the raised, terminal segments of the hind-legs against a ridged area under the middle of the forewing. This method contrasts with that of another Agaristid *Platagarista tetrapleura,* where only one ridged 'file' is used. **402**e

**mutabilis** see **Crambus**
**muticum** see **Calephelis**
**muzina** see under **Ecpantheria**

**Mycalesis** Hübner: syn. *Bicyclus* (NYMPHALIDAE)
These butterflies are found from Africa through India to Australia. They are frequently common and often show some seasonal variation. The caterpillars feed on various species of Gramineae. The eyes, when examined under a lens, can be seen to be covered with fine hairs. Each hair is between the facets. The genus is in the subfamily Satyrinae. Some of the species in the genus will be transferred to *Culapa* Moore when more information is available.

**adolphei** Guerin RED-EYE BUSH-BROWN
N. India 48–56 mm. (1·89–2·2 in.)
A dark chocolate brown butterfly with a small white spot below the apex of the forewing and another behind it, surrounded by black with a slight reddish ring. The hindwing has 3–4 eyespots, usually with black centres surrounded by orange-brown. This butterfly is found from May to September from 1220–2130 m. (4000–7000 ft) in N. India but the foodplant of the caterpillar is not known.

**anaxias** Hewitson WHITE-BAR BUSH-BROWN
India, Pakistan, Burma 48–55 mm. (1·89–2·17 in.)
The brown upperside has a white bar at an angle below the apex. The underside sometimes has small eyespots. It flies in evergreen forests up to 1830 m. (6000 ft) in N. India. Like most *Mycalesis* it is a weak flier, settling in the grass. It is more active in the mornings and evenings.

**anynana** Butler SQUINTING BUSH-BROWN
Africa 18–25 mm. (0·71–0·98 in.)
This brown species, which has one black eyespot on the margin of the forewing, is characterized by the eyespots on the underside of the forewing which, especially in the dry season form, have the white in the black eyespot off-centre, hence the popular name. This species is found from E. Africa to South Africa in bush or forest areas.

**janardana** Moore COMMON BUSH-BROWN
Burma, Indonesia 44 mm. (1·73 in.)
This species is common over most of its range from Burma to the Moluccas. The upperside is brown; underneath there is a row of eyespots below the wing margin on the fore- and hindwing. The butterfly flies in the forest undergrowth, gardens and cleared areas of bush. The caterpillar feeds on species of Gramineae but the details of the life history are not well-known. This species will probably be transferred to the genus *Culapa* Moore.

**mineus** Linnaeus DARK-BRANDED BUSH-BROWN
Sri Lanka, India, Pakistan, Burma 40–50 mm. (1·57–1·97 in.)
A very common butterfly found in hills and plains. The brown upperside has eyespots along the margin of fore- and hindwings, generally 2 in the forewing, 4 or more in the hindwing, each surrounded by a dark yellow ring.

**rama** Moore CINGALESE BUSH-BROWN
Sri Lanka, S. India 48–58 mm. (1·89–2·28 in.)
Generally a brown butterfly with a large black, white centred, eyespot on the forewing, halfway from the apex to the hind margins of the forewing, which is surrounded by a ring of yellow-orange. The underside is paler, with 2 eyespots on the forewing on the paler margin and 4 large ones on the hindwing, often with smaller ones on the margin. The eyespots have a distinct yellow line along their inner edge on fore- and hindwing. The underside eyespots vary in almost every specimen, especially between wet and dry season forms. It flies near bamboo jungle in wet lowlands of Sri Lanka, the caterpillar feeding on bamboo. This butterfly which was formerly only found in Sri Lanka and is rare even there, has also now been recorded in S. India.

**rhacotis** Hewitson
Africa 40–50 mm. (1·57–1·97 in.)
This is the commonest species in the genus in Africa and is widespread in forests S. of Sahara. Seasonal forms have been described of this rather dark brown, relatively unpatterned butterfly. It has 1–2 eyespots near the margin of the forewing, just behind the middle; these vary in size and may be surrounded by orange. The underside has a dark brown base to the wing with a paler apical half. The caterpillar feeds on *Ehrharta erecta* (Gramineae). The butterfly is about in all months of the year and can be found in most tree-shaded spots.

**terminus** Fabricius ORANGE BUSH-BROWN
Australia, New Guinea, Bismarck Archipelago, the Moluccas 42 mm. (1·65 in.)
This species has seasonal forms, and several subspecies from different localities. Both sexes are similar but the female is generally larger and paler than the male. The upperside is orange-brown with a black apex and outer margin, there is a small white spot near the middle of the wings. The hindwing is orange-brown with a row of spots between the veins. The greyish underside has a prominent line down the middle and is paler towards the margin with one eyespot near the margin on the forewing and a row along the hindwing, ringed with orange. The dry season forms are usually paler above and below. The caterpillar feeds on *Imperata* (Gramineae) and the butterfly is common over much of its range. **174**b

**Myelois** Hübner (PYRALIDAE) KNOT-HORN MOTHS
A large genus of Pyralid moths at present containing many unrelated species from Europe and Asia. Formerly it was believed to occur in N. America but it has been shown that the N. American species belongs to a different genus, *Apomyelois*. The popular name is the one for most of the subfamily (Phycitinae) to which this genus belongs and refers to the swelling at the base of the horns (antennae) on the head of the males. (See also under *Ectomyelois*.)

**cribella** Hübner: syn. *M. cribumella* THISTLE-ERMINE KNOT-HORN 25–30 mm. (0·98–1·18 in.)
Europe including Britain, temperate Asia including Japan
The forewings are glossy white with black specks, the hindwings are grey. Superficially it looks rather like species of Yponomeutid moths but differs in many structural details which can be seen on close examination. When at rest on the chalk grass lands and downs the moths are well camouflaged by their colouring. The moth rests by day on the leaves, the colour of its wings suggesting the reflection of light from the glossy leaves of the thistles. It is locally common in S. England and is similarly distributed throughout Europe and Asia. The young caterpillar feeds on the flowerheads of *Cardium* (Compositae) and other genera of thistle. In the autumn it burrows into the stems eating out the pith where is hibernates until the following spring. When fully grown it cuts a circular hole in the stem then pupates in the mine in a net-like cocoon.

**mylitta** see **Dynamine**

**Mylothris** Hübner (PIERIDAE)
Widespread over Africa S. of the Sahara. Most of the butterflies in this genus have a white or yellow ground-colour, rarely grey, and are generally rather slow fliers. Over 50 species are known, all restricted to Africa.

**agathina** Cramer
Congo to South Africa, E. Africa into Ethiopia 55–60 mm. (2·17–2·36 in.)
Probably one of the commonest of the genus and found in forests. The caterpillar has red and black transverse bands, sprinkled with yellow spots and white along the side. It feeds on *Loranthus* (Loranthaceae). The caterpillars are noted for their habit of migrating in columns similar to those better known species of *Thaumetopaea* and have the popular name of Processionary caterpillars. The females have less orange on the forewings than the male.

**poppea** Cramer TWIN DOTTED BORDER
E. Cape through Mozambique, Rhodesia to the tropics 50–64 mm. (1·97–2·52 in.)
A large number of subspecies of *M. poppea* have been described. The male is white with the apex and the margin of the wing spotted and a large orange patch on the base of the forewing. The hindwing is white, with orange along the front edge and a band into the wing; the margins are spotted with black. The underside forewing is similar to the upperside. The female is similar but has wider orange areas than the male. This is a common species in the bush. The caterpillar is believed to feed on species of Loranthaceae.

**trimenia** Butler TRIMEN'S DOTTED BORDER
South Africa: Natal, Transvaal, E. Cape 50–62 mm. (1·97–2·44 in.)
The whitish forewings with black apex and black dots on the margin are contrasted with yellow hindwings, also with black dots along the margin. There are several related species, mostly differing in the size and number of the black spots on the border. This butterfly is found in forests.

**myopaeformis** see **Conopia**

**Myrina** Fabricius (LYCAENIDAE)
This is a small genus of iridescent blue butterflies found in Africa S. of the Sahara. All the species have a single, broad, hindwing tail.

**silenus** Fabricius FIGTREE BLUE
Tropical & subtropical Africa 30–40 mm. (1·18–1·57 in.)
As the popular name of this species implies, its caterpillar feeds on the foliage of fig trees *(Ficus)*. The caterpillars are green with white spots, and are well camouflaged amongst terminal shoots of the fig which are green and often spotted with drops of milky, white sap. The chrysalis too is cryptically coloured and is attached to the stem of the fig where it resembles a small fruit. Adult females have a much broader orange-brown border to the wings. The under surface of both sexes is orange-brown, with a single transverse line on the hindwing. **244**q

**myrina** see **Cethosia**
**myrsusalis** see **Banisia**

**Myscelia** Doubleday: syn. *Sagaritis* (NYMPHALIDAE)
A small genus of butterflies from C. and S. America of which 2 species reach into the S. U.S.

**cyaniris** Doubleday & Hewitson
C. America, S. U.S. 60–65 mm. (2·36–2·56 in.)
The female of this species has very little iridescence and is browner in colour. The life cycle is not known. **217**f

**orsis** Doubleday
S. America 50–55 mm. (1·97–2·17 in.)
The vivid blue male is illustrated. The female is quite different and is brown with white spots on the fore- and hindwings arranged in lines across the wing. There is a white streak in the cell on both wings. The hooked apex of the forewing has a red-brown spot, similar to the one in the male. The underside is brown, lightly

patterned. The butterfly is common from C. America to Brazil. **217j**

**Mythimna** Ochsenheimer (NOCTUIDAE)
This is a large genus of moths found in N., Central and S. America, Hawaii, Galapagos, Europe and Asia, but best represented in the New World. Several of the species are pests, for example the 2 species which follow and the Oriental Army-worm *M. separata* Walker, the most destructive pest of cereals in China. The generic name *Pseudaletia* is used by several authors for many species placed by others in *Mythimna*.

**convecta** Walker
Australia, Tasmania 30–39 mm. (1·18–1·54 in.)
The caterpillar of *M. convecta* is an army-worm pest of grasses and cereals in Australia. Adults are similar in pattern to the illustrated *M. unipucta*, but the dark marking on the forewing is seldom associated with a clearly marked white speck and there is more contrast between the dark brown outer part of the hindwing and the pale brown basal part.

**unipuncta** Haworth ARMYWORM, WHITE-SPECK or AMERICAN WAINSCOT MOTH
N. America, C. America, S. America, Hawaii, Galapagos, Madeira, Europe 35–42 mm. (1·38–1·65 in.)
This is a rare species in Europe, but an extremely common and destructive pest in the U.S. where it was recorded as a pest as early as 1743 in New England. It has become more common recently in Israel, an indication that its range may be extending eastwards. In Britain, *M. unipuncta* occurs rarely, as a migrant, but has bred occasionally in S. Devon. Its caterpillars feed at night on grasses, grain and other crops; after having ravaged one crop, masses of caterpillars will move as an 'army' to new feeding grounds. This roving habit is apparently induced by crowded conditions which affect not only the behaviour of the caterpillars but also their coloration. **400**

**Nacna** Fletcher: syn. *Canna* (NOCTUIDAE)
*Canna*, the original name for this small genus of Indian and S.E. Asian moths is unusable for nomenclatural reasons and was replaced by its anagram *Nacna*.

**malachitis** Oberthür
India, Sikkim, China, Japan 25–35 mm. (0·98–1·38 in.)
Little is recorded in Western literature about this beautifully marked species. The ground colour of the wings is similar to that of the mineral malachite. *N. pulchripicta* from Sikkim and N. India is similar in colour pattern, but the forewing marking is angled and V-shaped. **391n**

**pulchripicta** see under **N. malachitis**

**nais** see **Apodemia, Symphaedra**
NAIS METALMARK see **Apodemia nais**
**nama** see **Hestinalis**

**Napata** Walker (CTENUCHIDAE)
There are currently about 50 tropical American species in this genus. Many of them are quite unlike one another and almost certainly not correctly classified.

**atricincta** Hampson
Brazil 27 mm. (1·06 in.)
This is a rare species related to *N. walkeri* and differing from it chiefly in the dark margins to the fore- and hindwings.

**walkeri** Druce BARRED-TIP YELLOW
Mexico, C. America 30–35 mm. (1·18–1·38 in.)
At least 3 other species match *N. walkeri* in colour pattern: a Nemeobiid butterfly *Mesene semiradiata*, a Zygaenid moth *Malthaca radialis* and a Pyralid moth *Mapeta xanthomelas*. It is not yet known whether any mimetic associations are involved. **364g**

**Napeogenes** Bates (NYMPHALIDAE)
There are about 60 species in this genus of the subfamily Ithomiinae. Most are restricted to tropical S. America, others are at least partly C. American in distribution. Many species form mimetic associations with species of other genera of Ithomiinae (as *Ceratinia* Hübner), Heliconiinae, the moth family Hypsidae and with species of supposedly palatable Nymphaline, Nemeobiid and Pierid butterflies.

**corena** Hewitson
Tropical S. America 38–42 mm. (1·5–1·65 in.)
This is a member of a mimetic complex with species of *Leucothyris* and *Hypoleria* (Ithomiinae), and probably the day-flying Hypsid *Hyalurga osiba*. These distasteful species are mimicked by the supposedly palatable *Dismorphia erythroe* (Pieridae), and by *Stalachtis lineata* and *Ithomeis corena* both species of Nemeobiidae. **131n**

**larilla** Hewitson
Ecuador 63–68 mm. (2·48–2·68 in.)
This is a typical *Napeogenes* species, one of many transparent-winged 'ghost' butterflies of the Ecuadorian forests. Nothing is known about the life history. **131j**

**napi** see **Pieris**
**napoleon** see **Dynastor**

**Narathura** Moore (LYCAENIDAE) OAK BLUE
This is a genus of over 150 species of butterflies found in Sri Lanka, India, China, Japan and S.E. Asia to New Guinea, N. Australia and the Solomons. The upper surface in both sexes of most species is a brilliant, iridescent blue or violet; the under surface of the wings is cryptically coloured in a broken pattern of mostly browns and greys. The caterpillars and the chrysalids are attended by green tree-ants of the genus *Oecophylla*.

**amantes** Hewitson LARGE OAK BLUE
Sri Lanka, India, Malay Archipelago to Timor 45–57 mm. (1·77–2·24 in.)
This is a fairly common butterfly, especially in the Himalayas at elevations below 1520 m. (5000 ft), and is one of the largest and most colourful species of its genus. The male is similar in appearance to the illustrated *N. micale*, but has a green iridescence at the base of the wings and broader, black marginal bands. Females are mostly brown. Both sexes are attracted by the scent of nutmeg *(Myristica)* and cinnamon *(Cinnamomum)* trees on which they sometimes congregate in large numbers. The several subspecies of *N. amantes* differ from each other in the hue of the ground-colour of the wings.

**eumolphus** Cramer GREEN OAK BLUE
N. India, Sikkim, Nepal, Burma, E. to the Philippines, New Guinea 44–50 mm. (1·73–1·97 in.)
This is often a common species at low altitudes. The dark, iridescent olive-green colour of the male upper surface of Indian representatives of this species is unusual in this genus. Males from other parts of the range vary in colour from green to a light blue. The female is a dark reddish brown, with some violet scales at the base of the wings.

**madytus** Fruhstorfer BRIGHT OAK BLUE
Australia, New Guinea and associated islands 35–42 mm. (1·38–1·65 in.)
This species is similar in general appearance to *N. micale*, but is distinguished by the purple or bluish purple upper surface of the wings and the paler brown under surface. It flies often in the same localities as *N. micale*. The caterpillars feed at night and rest during the day in a tent of leaves and silk.

**micale** Boisduval COMMON OAK BLUE
Moluccas to New Guinea and N. Australia 40–43 mm. (1·57–1·69 in.)
There are 16 recognized subspecies of this brilliantly blue butterfly. They differ from each other in details of the pattern and in the shade of blue of the upper surface of the wings. Females differ from the illustrated male in the broad, dark band along the margins of the wings. The caterpillars have been recorded from trees of several different families. **244g**

**Naroma** Walker (LYMANTRIIDAE)
The 4 species of this genus of yellowish white moths are restricted to Africa, S. of the Sahara, including South Africa.

**signifera** Walker
Africa, S. of the Sahara 30–43 mm. (1·18–1·69 in.)
The caterpillars of *N. signifera* live on trees in association with those of the Lycaenid butterfly *Teratoneura isabellae* and are invariably attended by ants. The caterpillar of *T. isabellae* is distinctively Lymantriid-like. The forewings are yellowish white, with a dark yellow mark at the end of the cell and a row of black dots extending from this marking to the base of the wing; the hindwing is white.

NARROW BLUE-BANDED SWALLOWTAIL see **Papilio nireus**
**narycia** see **Pseudonympha**
**nastes** see **Colias**
NATAL BARRED BLUE see **Spindasis natalensis**
**natalensis** see **Spindasis**
**nattereri** see **Heliconius**
NAVAL COMMODORE see **Precis touhilimasa**
NAVAL ORANGEWORM see **Paramyelois transitella**
NAWAB BUTTERFLIES see **Polyura**
**nawi** see **Epipomponia**

**Neadeloides** Bryk (PYRALIDAE)
This genus of Pyralid moths has only one included species. This is characterized by the remarkable long antennae.

**cinerealis** Moore
N. India 41 mm. (1·61 in.)
Although the species is common in collection from N. India nothing seems to be known about its biology. It is a silvery brown species with white hindwings and long slender legs. The most striking features are the long slender antennae; these are over twice the length of the body. **63l**

**nearctica** see **Nomophila**
**nefte** see **Parathyma**

**Nemapogon** Schrank (TINEIDAE) CORN MOTHS
Micro-moths, widespread, usually small species and often important economically as pests of stored foods. With several related species it is also a pest in wine cellars where the caterpillar bores into the corks of wine bottles.

**granella** Linnaeus EUROPEAN GRAIN MOTH
Worldwide 10–15 mm. (0·39–0·59 in.)
This moth has been introduced artificially to most countries of the world. The yellowish white caterpillar feeds on corn and grain, but it will also attack a wide variety of substances including almonds, nuts, cigars, wheat, clover seed, dried cherries. In Europe an allied species, *N. cloacella* Haworth, with similar appearance and habits is common. Both species also feed out of doors on various fungi. The Grain Moth has also been recorded as a pest in the wine industry where it damages the corks! The caterpillar bores into the corks of wines in stores, with obvious disastrous results. **5g**

**Nemeobiidae**: syn. *Riodinidae* METALMARKS, ERYCIDS, NEMEOBIIDS, RIODINIDS
This group of butterflies is usually treated as a separate family, but is probably better treated as a subfamily of the Lycaenidae. Its range is essentially tropical American but it is also represented in tropical Africa and Asia. One genus and species, *Hamearis lucina*, is found in Europe. Nemeobiids are remarkably variable in coloration: almost every colour of the rainbow is represented in one species or another, and there is a wide variety of colour patterns. Some species are accepted as mimics of distasteful species of butterflies and day-flying moths, but at least one species (see *Mesene*) feeds on a poisonous plant and may be itself distasteful. Some of the most striking mimics are in the genera *Ithomeis, Stalachtis, Esthmopsis, Uraneis* and *Mesenopsis*. Adults rest typically on the under surface of a leaf with their wings partly spread out. Few species are common over a wide area although some are locally common. The caterpillars of some species are very hairy, not a characteristic of most butterfly species. See *Abisara, Alesa, Amarynthis, Ancycluris, Apodemia, Argyrogrammana, Astraeodes, Audre, Caria, Calephelis, Chamaelimnas, Chorinea, Cremna, Dodona, Echenais, Esthemopsis, Euselasia, Hamearis, Helicopis, Hermathena, Hyphilaria, Ithomeis, Lasaia, Laxita, Lucillella, Lymnas, Lypropteryx, Menander, Mesene, Mesenopsis,*

*Mesosemia, Nymphidium, Polystichtis, Praetaxila, Rhetus, Semomesia, Stalachtis, Styx, Symmachia, Syrmatia, Taxila, Theope, Uraneis.*

**Nemophora** Illiger & Hofmannsegg (INCURVARIIDAE)
An almost worldwide genus with most species in Europe, Asia and Africa. It is not found in New Zealand. The species are all characterized by the long antennae in both sexes, and the metallic coloured wings. The moths fly in bright sunlight.

**degeerella** Linnaeus LONG-HORNED MOTH
Europe including Britain, to Turkey 15–19 mm. (0·59–0·75 in.)
This beautiful little moth is locally common over its whole range. Its wings are a shining golden yellow with brown marks along the veins. The adult moth has very long antennae, even longer in the males than the females. These are held out in front of the moth as it flutters slowly in the sunshine over bushes. The caterpillar feeds on various herbaceous plants. **4**

**scabiosella** Scopoli
Europe including Britain 15–20 mm. (0·59–0·79 in.)
The forewings are a golden-bronze, usually with a coppery tinge. The caterpillars feed on the seed heads of *Scabiosa* (Dipsacaceae). The moths fly in June and July.

**Neocastnia** Hampson (CASTNIIDAE)
The only other genera of Castniidae to occur outside the New World are *Tascina* and *Synemon*. There is a single known species. Like *Tascina* the adults lack a functional proboscis.

**nicevillei** Hampson
Burma 80 mm. (3·15 in.)
This species is known only from southern, peninsular Burma. The figured specimen, the type, is possibly the only specimen to exist at present in collections. Its large eyes indicate that like other Castniids this is a day-flying species. Nothing is known about the life history. **45**n

**neocypris** see under **Phoebus cipris**

**Neola** Walker (NOTODONTIDAE)
Both species of this genus are restricted to Australia.

**semiaurata** Walker
E. Australia 55–60 mm. (2·17–2·36 in.)
The adult of this species is a typical, cryptically coloured Notodontid; the grey and red *Acacia*-feeding caterpillar, in contrast, is a striking creature with clubbed hairs, and a pair of eyespots at its rear, surmounted by a dorsal horn.

**Neomyrina** Distant (LYCAENIDAE)
Only one species is known.

**hiemalis** Godman & Salvin WHITE IMPERIAL
Burma, Thailand, Malaya, Sumatra 35–45 mm. (1·38–1·77 in.)
This forest butterfly is quite unlike a typical Lycaenid in appearance. It is a large species, chiefly white, with dark blue or black outer borders to the forewing, and is much like some species of the family Pieridae. The under surface of both sexes is white, with grey transverse lines and iridescent blue spots at the base of the tail. The caterpillar is unknown. **244**r

**Neophasia** Behr (PIERIDAE)
These butterflies range from C. America into S. U.S. The caterpillars tend to be gregarious, feeding on *Pinus ponderosa* and other species and at time become numerous and causing damage to the trees. Only 2 species are known at present in this genus.

**menapia** Felder PINE WHITE
Mexico, W. U.S. to S. Canada 42–50 mm. (1·65–1·97 in.)
This white butterfly has a black line along the forewing margin which curves backward over the apex of the cell. The apex of the wing is black with white spots, the black extending only to the middle of the outer margin of the forewing. The hindwings are white. The wing markings on the underside sometimes have a pinkish tinge. The caterpillar feeds on species of pine *(Pinus)* and Balsam Fir *(Abies balsamea)*. The caterpillar lowers itself on long silk threads from the tree to pupate on the vegetation below. It can be very destructive in pine woods.

**neophilus** see **Parides**
**neophron** see **Euphaedra**

**Neopseustidae**
A family of micro-moths with only 3 described species. They are small, little known species which are placed in a primitive suborder where they are related to Eriocranids and Mnesarchiids. The species are found in India and Formosa. They have similar venation in the fore- and hindwings instead of the differences which are found in the majority of families. See *Neopseustis.*

**Neopseustis** Meyrick (NEOPSEUSTIDAE)
The only genus in the family, these micro-moths are found in India, Burma and Formosa. They are small and inconspicuous and very few specimens are known. They are scientifically important for their primitive structures which are studied to help understand the origins and evolution of the lepidoptera.

**calliglauca** Meyrick
India 16 mm. (0·63 in.)
This moth has a grey mottled base to the forewing with a white patch in the middle, the apical white part is heavily mottled with black scales. The wings are rather oval shaped, somewhat pointed at the outer margin rather than the elongate triangular shape of many species.

**Neossiosynoeca** Turner (OECOPHORIDAE)
A genus of micro-moths found in Australia. Only one species is known; this has a specialized habitat and has been found in nests of parrots.

**scatophaga** Turner
Australia 29–35 mm. (1·14–1·38 in.)
The caterpillar of this moth is a scavenger, as are many other Oecophorid moths. However, this one has specialized slightly more and lives in the nest of a species of parrot. The caterpillars feed in these nests on the debris and excreta – thus helping to keep the nest clean. It is said that they even clean the feet and feathers of the young birds in search for food! The moths are dark brown, with whitish hindwings and a black spot in the middle of the forewing. They have been called 'the Parrot's Nanny'.

**Nephantis** Meyrick (XYLORYCTIDAE)
A small genus of micro-moths, some of which are pests of crops, mostly in the Oriental region but with a few African species.

**serinopa** Meyrick
Sri Lanka, India, Burma 16–24 mm. (0·63–0·94 in.)
This is a grey-brown moth with small black spots along the front edge of the wing; the hindwings are paler. The caterpillar is often a pest of coconuts *(Cocos)*, feeding on the leaves. When present in large numbers it can be damaging to the young palm. In recent years there has been some evidence that this moth is spreading to other coconut growing areas, in which case it could become a major pest though at present it is regarded as a minor one.

**Nephele** Hübner (SPHINGIDAE)
There are about 20 species in this genus of hawk-moths. It is mostly tropical African in distribution but is also represented in S.E. Asia and Australia. Many species have a silvery white marking on the middle of the forewing.

**argentifera** Walker SILVER-BARRED NEPHELE
Tropical E. Africa to South Africa 68–80 mm. (2·68–3·15 in.)
The head and thorax of the adult are green; the abdomen is green with a row of black patches on each side. The hindwings are brown. Its most conspicuous features are a lustrous, silvery white, crescent-shaped marking on the middle of the green forewing and 2 transverse, white lines. It is commonest in coastal areas where the caterpillar feeds on *Carissa* (Apocynaceae). Caterpillars can be either green or brown, and have a long, nearly straight, terminal horn.

**vau** Walker V-NEPHELE HAWK-MOTH
Tropical & subtropical Africa 68 mm. (2·68 in.)
This species is named after the lustrous white or yellow 'V' on the forewing of this otherwise brown or greenish brown species. Its caterpillar feeds on *Carissa* (Apocynaceae).

NEPHELE, SILVER-BARRED see **Nephele argentifera**

**Nepheronia** Butler (PIERIDAE)
A mainly African genus of butterflies with many described aberrations and subspecies of the few species in the genus. One species is known from the Indian subcontinent.

**argia** Fabricius LARGE VAGRANT
Tropical to South Africa 62–82 mm. (2·44–3·23 in.)
The male is yellowish white with a red apex to the forewing, spotted with black. This forest species is common from South Africa through Mozambique to E. and Equatorial Africa. This is the largest 'white' butterfly in S. Africa. It is a rapid flyer and at times is abundant over much of its range.

**thalassina** Boisduval CAMBRIDGE VAGRANT
Tropical to South Africa 56–70 mm. (2·2–2·76 in.)
The fore- and hindwings of this butterfly are a pale sky blue with blue-black borders. The underside has a silky white sheen but is unpatterned. The female is white with black border, sometimes with yellow hindwing but in some subspecies (e.g. in Cameroons) this colour pattern is reversed and the females have yellowish forewings. This species flies in the more open bush country or at the edges of forests.

**nephrosema** see **Cryptophasa**

**Nepticula** Heyden (NEPTICULIDAE)
These minute micro-moths include the smallest known species. They are to be found all over the world. As far as is known, all the caterpillars of this genus are leaf-miners, producing characteristically shaped mines which include the serpentine scrawls or blotches found on many leaves. Generally the caterpillars mine along the layers near the upperside of the leaf, but some species will mine the lower surface. All developments of the caterpillar take place between the upper and lower surfaces of the leaf. Some mine different sides of the leaf at different stages in their life (still between the upper and lower surfaces). Usually the caterpillar comes out of the mine to pupate in the leaf-litter on the ground. These minute moths often have a metallic coloured appearance to the wings, with delicate patterns.

**altella** Braun
U.S. 6–7 mm. (0·24–0·28 in.)
The forewings are purple-brown near the base and brown with a purple iridescence towards the apex of the wing giving a deep purple appearance to this tiny moth. It has a narrow silvery line across the forewing. This species has been bred from Pin Oak *(Quercus palustis)* in Ohio. The tiny caterpillar mines on the underside of the leaf in a narrow twisting line. It is yellow with a row of dark brown marks along the underside. There are many other similar species in the U.S. but this species can be separated from the others by the colour pattern, purple to the base of the wing and brown beyond. In others where there is a similar colour change along the wing the base is brown.

**aurella** Fabricius
Europe including Britain, N. Africa 5–7 mm. (0·2–0·28 in.)
This tiny moth is common over most of its range. The light yellow coloured caterpillars live in long irregular mines in the leaves of blackberry species, *Rubus* (Rosaceae). The eggs are laid on the under surface of the leaf, although there are records of them on the upper side. The mine in the leaf is irregular, wavy, at first greenish white with a thin black line (frass or excrement) in it. The caterpillar when fully grown comes out of the leaf and pupates either attached to a

leaf or twig or in the leaf-litter below. The moth emerges early in the year and while the weather is suitable will have several broods. **1c**

**Nepticulidae**
These are small micro-moths, worldwide in distribution, whose caterpillars mine in leaves. These leaf-mines are often of characteristic shape and can help in the identification of the species. In the adult moth the wing veins are simplified and reduced compared with many of the larger moths. The adult nepticulids generally do not feed. Some of the smallest moths known belong to this family. See *Nepticula, Stigmella*

**Neptis** Fabricius (NYMPHALIDAE) SAILERS or GLIDERS
A huge genus of butterflies in Africa, Europe and Asia and in the Orient, extending through Indonesia to New Guinea and Australia where there is one species. The genus is related to *Pantoporia*. The popular name is derived from their leisurely gliding or sailing flight. All are blackish species with white bands and spots on the wing.

**alta** Overlaet OLD SAILER
Africa 48–60 mm. (1·89–2·36 in.)
This is a black species with a white band in the middle of the hindwing and a white curved band on the forewing which is divided into three main parts with black between them. The wings have a series of parallel lines with some white running along the margin. This is a typical African bush species from East Africa south to Mozambique, Rhodesia and Transvaal. The life history of this species is unknown.

**melicerta** Drury
Africa, S. of the Sahara 35–45 mm. (1·38–1·77 ins.)
This is a widespread and fairly common African species. The specimen illustrated was photographed in Uganda. Nothing is known of its life history. **218**

**rivularis** Scopoli HUNGARIAN GLIDER
N. Italy, E. Europe, temperate Asia including Japan 50–54 mm. (1·97–2·13 in.)
This can be separated from the similar species *N. sappho* by the single band of white on the hindwing and the lack of black spots on the underside of the wing. There is a distinct subspecies in Japan. The caterpillar feeds on *Spiraea* (Rosaceae).

**sappho** Pallas: syn. *N. hylas* COMMON GLIDER or COMMON SAILER
E. Europe, Asia including Japan, Malaysia, Indonesia 44–48 mm. (1·73–1·89 in.)
This widespread butterfly is out in May and June in the temperate regions with a second generation in August-September. There are a number of similar looking species in Europe and Asia. *N. sappho* has a golden brown underside. In Europe the caterpillar feeds on *Lathyrus* (Leguminosae); elsewhere they are recorded on a variety of plants. **216b**

**trigonophora** Butler BROKEN BARRED SAILER
South Africa N. to the tropics 46–56 mm. (1·81–2·2 in.)
A black and white species with 3 white patches on the forewing and a very broad white band through the middle of the hindwing. The underside has white streaks at the base of the hindwing and the forewing has a white streak on it. This is mainly a forest species with a wide distribution from the tropics to South Africa. It is commoner from Mozambique N.

**neptunus** see **Parides**
**nerii** see **Daphnis**
**Nerissa** see **Cepora**
**nero** see **Appias**
**nesaea** see **Elymnias**

**Nessaea** Hübner (NYMPHALIDAE)
A small genus of butterflies with species mainly found in the tropics of S. America.

**hewitsonii** Felder
S. America 65–75 mm. (2·56–2·95 in.)
The female is brown, with the blue band on the forewing but not on the hindwing. The underside is green. **209c**

**obrinus** Linnaeus
C. & S. America 65 mm. (2·56 in.)
This butterfly is common in the N. half of S. America. It flies in and out of woodlands and likes to sit with its wings spread. In flight it shows the unexpected colour combination contrasting the green on one side with upperside. The female is brown, not black, with red spots in the forewing near the middle of the front margin. The upper and underside are illustrated. **209d, f**

**nessus** see **Anaea**
NEST CATERPILLAR, UGLY see **Archips cerasivorana**
NETTLE
GRUB, BLUE-STRIPED see **Parasa lepida**
PUG SEE **Eupithecia venosata**
NETTLE-TREE BUTTERFLY see **Libythea celtis**
**neurata** see **Mimacraea**

**Neurosigma** Butler (NYMPHALIDAE)
Only one species, which is found from India to Burma is known in this butterfly genus.

**doubledayi** Westwood
N. India, Pakistan, Burma 65–90 mm. (2·56–3·54 in.)
The female has a white ground-colour except at the base of the wing, otherwise the pattern is similar to the male, although the females are larger. Little seems to be known of the biology of this species. **231f**

**neustria** see **Malacosoma**
**ni** see **Trichoplusia**
NI MOTH see **Trichoplusia ni**
**niavius** see **Amauris**
**nicevillei** see **Neocastnia**
**nicothoe** see **Anthela**
NIGHT BUTTERFLY see **Cepheuptychia, Magneuptychia, Taygetis**
NIGHT FIGHTER, PALM TREE see **Zophopetes dysmephila**
**nigricana** see **Cydia**
**nigrofasciatum** see **Capillamentum**
**nigrolineata** see **Grammodora**
**nimbosa** see **Sorhagenia**
**nina** see **Leptosia**
**ninana** see **Cydia**
**ninus** see **Delias**
**niobe** see **Fabriciana**
NIOBE FRITILLARY see **Fabriciana niobe**
**nireus** see **Papilio**
**nivea** see **Acentria, Cyrestis, Leucodonta**
NO-BRAND GRASS YELLOW see **Eurema brigitta**
**nobilis** see **Papilio**
**nobilitella** see **Cydosia**
**noblei** see **Pterothysanus**

**Noctua** Linnaeus (NOCTUIDAE)
This is a small genus of European moths. The caterpillars are cutworms; they overwinter in a dormant state partly grown, and re-start feeding in the following spring.

**pronuba** Linnaeus LARGE YELLOW UNDERWING
Europe including Britain, N. Africa to C. & W. Asia 50–58 mm. (1·97–2·28 in.)
This extremely common species is on the wing in Europe in June and July, both during the day and at night. Moths seen later in the year in the north are probably migrants. The forewing coloration can vary considerably between individuals, and is a rich purple-brown in some specimens. The yellow part of the hindwing rarely has a black, crescent-shaped marking in the centre. The caterpillars feed from August to May of the following year on grasses and low-growing plants and are an occasional pest in gardens; they are either green or brown in ground-colour, marked with brown and black. **399e**

**noctuella** see **Nomophila**

**Noctuidae** NOCTUID, OWLET and UNDERWING MOTHS
There are probably as many as 20,000 species in this almost cosmopolitan family of moths. They are chiefly brown, unspectacular, mostly nocturnal moths which attract little attention; their caterpillars, however, amongst which are the cutworms and armyworms, are the most notorious pests of food-crops throughout the world. Most of the species are between 20 and 40 mm. (0·79–1·57 in.) in wingspan, but the family also includes extremely small species and one of the largest known moths *Thysania agrippina* which can measure up to 305 mm. (12·01 in.) from wingtip to wingtip. The adult moths have tympana (ears) placed one on each side of the thorax with which they are able to sense the squeaks of bats, their main nocturnal enemies. Reaction to bat sounds takes the form of somewhat haphazard evasive flight. See *Acontia, Acronicta, Agrotis, Anomis, Amathes, Amphipyra, Anarta, Annaphila, Apamea, Argyritis, Ascalaphia, Asparasa, Autographa, Baorisa, Bellura, Bena, Busseola, Calamia, Calpe, Capillamentum, Catocala, Cerapteryx, Cucullia, Cydosia, Daphoenura, Diphthera, Dypterygia, Earias, Egira, Emmelia, Eosphoropteryx, Epicausis, Eublemma, Euclidia, Eudaphnaeura, Eugraphe, Exyra, Feltia, Gonodonta, Grammodes, Harrisimemna, Heliothis, Hypena, Lobocraspis, Mamestra, Muzuca, Miniophyllodes, Mythimna, Nacna, Noctua, Nonagria, Nycteola, Ochropleura, Othreis, Panolis, Peridroma, Phlogophora, Phyllodes, Plusia, Pseudoips, Psychomorpha, Schinia, Scoliopteryx, Sesamia, Spodoptera, Thysania, Trichoplusia, Xanthia, Zale.*

**Nola** Leach (NOLIDAE)
This is an almost cosmopolitan genus of over 300 species of moths, most of which have been combined at times with the generic bames *Roeselia* Hübner and *Celama* Walker which are now treated as synonyms of *Nola*.

**albula** Denis & Schiffermüller KENT BLACK ARCHES
Europe including Britain, temperate Asia to Japan 18–23 mm. (0·71–0·91 in.)
This is a typical, sombrely coloured species of the genus *Nola*. The forewing markings of some specimens are much fainter than in the illustration and nearly absent in others. Moths are on the wing in July. The caterpillar is white or pink, with black and grey markings; it feeds on dewberry, blackberry and raspberry (*Rubus*), cinquefoil (*Potentilla*) and other species of the family Rosaceae. *N. albula* overwinters as a caterpillar. The cocoon is often attached to a grass stem. **390a**

**sorghiella** Riley SORGHUM WEBWORM
N., C. & S. America 9–16 mm. (0·35–0·63 in.)
This is one of the most serious pests of grain sorghums in the U.S. The moths are minute white insects, with light brown and black markings; they fly from April to September in the N. of their range, but breed and fly throughout the year in the tropics and subtropics. The caterpillars are gregarious and spin webs on the flowerheads of various wild grasses and sorghums.

**Nolidae**
This small family of moths was long considered to be a subfamily of the Arctiidae but is now thought to be more closely related to the Noctuid subfamily Sarrothripinae. Most species are very small, averaging little more than 10 mm. (0·39 in.) in wingspan and are sombrely coloured. The caterpillars of some species are pests of foodcrops. See *Nola*.

**nomius** see **Graphium**

**Nomophila** Hübner (PYRALIDAE)
A worldwide genus of Pyralid moths, one of whose species is a well-known migrant and has been recorded in almost every country of the world. However, recent research has shown this 'worldwide' species to consist of a number of distinct species with very precise distributions. All are broadly similar externally, hence the confusion. All the species mentioned here in this genus were previously known as *N. noctuella*.

**australica** Munroe
Australia, Tasmania, Christmas Island 24–32 mm. (0·94–1·26 in.)
This species, which has only been recognized since 1973 is darker and slightly more heavily marked than *N. noctuella* with which it was formerly confused.

**nearctica** Munroe CELERY WEBWORM MOTH
U.S., Canada, Mexico, Guatemala, Bermuda, Bahamas, Haiti, Jamaica, Costa Rica 27–35 mm. (1·06–1·38 in.)
This American species was separated from *N. noctuella* in 1973. It is widespread in the New World where it is a defoliator of sweet clover *(Trifolium)*,grasses, etc. and can be a pest on celery *(Apium)*.

**noctuella** Denis & Schiffermüller CELERY WEBWORM or RUSH VENEER
Europe including Britain, Asia, Africa, Madagascar, Hawaii 23–31 mm. (0·91–1·22 in.)
Most of the records of moths in the genus were included under this name, which is known from every continent. However, in future, specimens will have to be examined with care since specimens under this name were found to represent more than 8 different species. The moth is well known as a migrant and northerly populations which do not overwinter are replaced by immigrants which survived further south. One individual of this species was collected in England by a well-known entomologist and on examination found to have traces of radio-active material embedded in it which was regarded as an indication that it had flown past one of the Sahara atomic bomb tests. **56w**

**Nonagria** Ochsenheimer (NOCTUIDAE)
This genus of moths is widely distributed in the temperate and tropical zones of the Old World and is represented by a single S. American species. The caterpillars are at first leaf-miners, then live within the stems of reed-like semiaquatic and marsh plants. The finely striate adult moths are well camouflaged when at rest on a leaf or stem.

**oblonga** Grote LARGE NONAGRIA
E. Canada, U.S., Jamaica 34–52 mm. (1·34–2·05 in.)
This moth is similar in colour pattern to the illustrated *N. typhae*. The caterpillar, like that of *N. typhae*, also feeds on reedmace (cattail) *(Typha)*, at first as a leaf-miner, later as a stem-borer. Before turning into a chrysalis, the caterpillar bores a hole in the stem of the foodplant as an escape exit for the adult moth which will eventually emerge between July and September.

**typhae** Thunberg BULRUSH WAINSCOT
Europe, including Britain 45–52 mm. (1·77–2·05 in.)
The moth, which appears in July and August, can sometimes be seen resting on the stems and leaves of reedmace *(Typha)*, usually head downwards. Females are able to slit open the reedmace stem and deposit their eggs inside. The eggs overwinter there and hatch in the following spring. The caterpillar feeds until July or August inside the stem of the foodplant. **404**

NONAGRIA, LARGE see **Nonagria oblonga**
NONPAREIL MOTH, CLIFDEN see **Catocala fraxini**
**norax** see **Hibrildes**
**Noropsis** see **Diphthera**
NORTHERN
CLOUDED YELLOW see **Colias hecla**
GRIZZLED SKIPPER see **Pyrgus centaureae**
JUNGLE QUEEN see **Stichophthalma camadeva**
METALMARK see under **Calephelis muticum**
SWIFT see under **Hepialus fusconebulosa**
WINTER MOTH see **Operophtera fagata**

**Nothus** Billberg: syn. *Sematura* (SEMATURIDAE)
There are currently 4 species in this genus of brown, long-tailed moths. They are found in Mexico and much of C. America, the Antilles and in tropical S. America.

**lunus** Linnaeus
Mexico to Brazil 75–100 mm. (2·95–3·94 in.)
This species is typical of its genus in colour pattern and wing shape. There is apparently no experimental evidence as to whether the hindwing eyespots have a protective function (see *Automeris*, whose eyespots are said to frighten small birds). **314g**

**Notodontidae** PROMINENTS, PUSS-MOTHS and PROCESSIONARY MOTHS
There are between 2500 and 3000 described species in this family, but many more species await description. It is a member of the superfamily Noctuoidea. Most species are tropical in range but many are found in temperate regions of the earth. They are mostly brown, grey and green in coloration and cryptically patterned, but a few are brightly coloured. Their general appearance is Noctuid-like but the abdomen is typically much longer than in most species of Noctuidae. The forewing colour pattern of many species is continued on the front part of the hindwing which is not covered by the forewing when the moth is at rest. The rear margin of the forewing bears a short process in several species. The tympana (ears) are placed one on each side of the thorax. Caterpillars of the subfamily Thaumetopoeinae are the well known processionary caterpillars; these and some others (for example *Anaphe*) are protected with irritant, barbed hairs. Many species are pests, particularly of forest trees. The caterpillars of *Stauropus* species have been recorded from cocoa *(Theobroma)*. See *Anaphe, Cerura, Chliara, Danima, Datana, Diloba, Disphragis, Epicoma, Furcula, Galona, Leucodonta, Moresa, Neola, Ochrogaster, Phalera, Pheosia, Pterostoma, Ptilodon, Spatalia, Stauropus, Tarsolepis, Thaumetopoea, Uropyia*.

**Notophyson** see **Anthomyza**
NOVEMBER MOTH see **Epirrita dilutata**
**nox** see **Parides**
**nubilalis** see **Ostrinia**
**nuda** see **Perina**
**Nudaurelia** see under **Imbrasia**
**numatus** see **Heliconius**
**numilia** see **Catonephele**
**nupta** see **Catocala**
**Nyctalemon** see **Lyssa**
**nycteis** see **Chlorosyne**

**Nyctemera** Hübner (ARCTIIDAE)
This is one of the largest genera of its family. Most species have black, or dark brown, and white wings and yellow and black abdomen. They are found in India, E. to Japan, through S.E. Asia to Australia and the Solomons. The similarity in colour pattern of a large number of *Nyctemera* species suggests that this is an assemblage of moths which will prove to be distasteful to birds, as are many other genera of Arctiidae. Unpalatable groups often adopt a common colour pattern, which simplifies the predator's task when it is learning to associate unpalatability with particular colour patterns.

**amica** White
E. Australia 35–44 mm. (1·38–1·73 in.)
*N. amica* has dark brown wings, with 2 irregularly shaped, creamy white patches on the forewing and a single patch on the hindwing. The abdomen is banded with black and yellow. Its caterpillar feeds on *Senecio* and *Cineraria* (Compositae).

**apicalis** Oberthür: syn. *N. antinorii* Oberthür
Africa, S. of the Sahara 45–52 mm. (1·77–2·05 in.)
The extensive range of *N. apicalis* includes Sierra Leone and much of W. Africa, the Congo Basin, E. Africa to Mozambique and S. to South Africa. The forewings of the adult are dark brown, with an oblique, yellow bar towards the apex and 2 short white dashes at the outer margin; the hindwing is white, with a dark brown outer marginal band. The abdomen is brown above, with yellow and black sides. The caterpillar produces a frothy secretion which hardens into yellow spheres rather like cocoons of a parasitic Braconid wasp – these are distributed in the structure of the cocoon by the caterpillar and may have a particular protective function against Braconid wasps which tend to avoid cocoons which are already parasitised, or apparently parasitised.

**Nycteola** Hübner (NOCTUIDAE)
The distinctively shaped wings of these small moths are similar to those of the microlepidopterous family Tortricidae.

**frigidana** Walker
Canada, N. U.S. 25 mm. (0·98 in.)
This is a relatively rare species in collections. It is rather similar in general appearance to the European *N. revayana* with which it has been confused. The caterpillar feeds on poplar *(Populus)* and willow *(Salix)*.

**revayana** Scopoli OAK NYCTEOLINE
Europe, including Britain, Asia E. to Japan 18–25 mm. (0·71–0·98 in.)
This is an extremely variable moth, but is normally greenish or brownish grey with black or brown spots and lines. Several names have been applied to the various colour forms. It apparently overwinters as an adult and is on the wing from August to April. The green and white caterpillar feeds in June and July on oak *(Quercus)* and sometimes on willow *(Salix)*.

NYCTEOLINE, OAK see **Nycteola revayana**
NYMPH BUTTERFLIES see **Asterope, Aterica, Cremna, Paiwarria**

**Nymphalidae** BRUSH-FOOTED BUTTERFLIES
This is a large family of butterflies with many colourful species. Most species have the forelegs of the adult reduced and not functioning as legs. Frequently they have dense tufts of scales on them, hence the popular name of the family. The family is separated into a number of subfamilies in this work following some of the recent classifications, although there is still some dispute about the exact status of these subfamilies. Many of them have in the past, and in some recent publications, been treated as separate families. Some of the widespread subfamilies in the Nymphalidae include the Satyrinae (the browns), whose caterpillars are mainly grass feeders and have a number of species which fly at dusk; the Danainae which generally feed on species of Apocynaceae, Asclepiadaceae and Moraceae and the butterflies are usually distasteful to predators; the Charaxinae and Acraeinae which are mainly African subfamilies; the Nymphalinae, a huge, worldwide, subfamily; the Morphinae, with huge iridescent blue species; the Brassolinae, Ithomiinae, and Heliconiinae which are all American, and the Amathusiinae which are Old World butterflies.

ACRAEINAE: *Acraea, Actinote, Bemastis, Pseudergolis, Telchinia.*
AMATHUSIINAE: *Amathusia, Amathuxidia, Discophora, Hyantis, Stichophthalma, Taenaris, Thaumantis, Thauria, Zeuxidia.*
BRASSOLINAE: *Caligo.*
CHARAXIINAE: *Agrias, Aneae, Charaxes, Coenoplebia, Consul, Euxanthe, Hadrodontes, Hypna, Palla, Polyura, Prepona, Siderone, Zingha.*
DANAINAE: *Amauris, Danaus, Diogas, Euploea, Idea, Ideopsis, Ituna, Lycorea.*
HELICONIINAE: *Agraulis, Dione, Dryadula, Dryas, Eueides, Heliconius, Philaethria, Podotricha.*
ITHOMIINAE: *Aeria, Callithomia, Elzunia, Greta, Hyalyris, Hypoleria, Hypothyris, Ithomia, Mechanitis, Melinaea, Napeogenes, Oleria, Olyras, Scada, Tellervo, Thyridia, Tithorea.*
MORPHINAE: *Morpho.*
NYMPHALINAE (including some less well-defined subfamilies): *Aglais, Amnosia, Anartia, Antanartia, Apatura, Argynnis, Asterocampa, Asterope, Aterica, Baeotus, Bassaris, Batesia, Betearia, Biblis, Boloria, Brenthis, Callicore, Callithea, Caracroptera, Catanophele, Cethosia, Chazara, Chersonesia, Chlorosyne, Clossiana, Cymothoe, Cynandra, Cynthia, Cyrestis, Diaethria, Doleschallia, Doxocopa, Dynamine, Dynastor, Epiphile, Eryphanis, Eunica, Euphaedra, Euphydras, Eurytela, Euthalia, Fabriciana, Faunis, Hamadryas, Hestinalis, Hypolimnas, Inachis, Issoria, Kallima, Ladoga, Laringa, Mandarina, Marpesia, Melitea, Mellicta, Mesoacidalia, Mestria, Metamorpha, Myscelia, Neptis, Nessaea, Neurosigma, Nymphalis, Panacea, Pandorina, Pantopora, Parathyma, Parthenois, Perisama, Phyciodes, Polygonia, Precis, Proclossiana, Pseudacraea, Pseudargynnis, Pseudergolis, Pseudonympha, Pyronia, Rhinopalpa, Salamia, Sasakia, Sephisa, Speyeria, Stilbochiona, Symbrethia, Symphaedra, Tanaecia, Temeris, Terinos, Vagrans, Vanessa, Vanessola, Vindula, Yoma.*
SATYRINAE: *Aphysoneura, Aphantopus, Brintesia, Coenonympha, Cepheuptychia, Cithaeris, Elymnias, Enodia, Erebia, Elimniopsis, Eyptychia, Henotesia, Hipparchia, Junonia, Lassommata, Magneuptychia, Maniola, Melanitis, Minois, Mycalesis, Oeneis, Orinoma, Orsotriaena, Pararge, Pierella, Pyronia, Satyrodes, Satyrus, Tarsocera, Taygetis, Ypthima, Zipaetis.*

**Nymphalis** Kluk (NYMPHALIDAE)
A genus of butterflies related to *Vanessa* which are mainly European and Asian but with a N. American representative and one N. Indian species.

**antiopa** Linnaeus MOURNING CLOAK, WHITE-BORDERED, or CAMBERWELL BEAUTY
Europe including Britain, N. America, temperate Asia 60–64 mm. (2·36–2·52 in.)
This butterfly, which is widespread over the N. Hemisphere does not vary much over the whole range. Specimens from Canada are similar to those from Sweden, while those from further S. in Europe are similar to the U.S. specimens, the differences being only in size and the width of the outer border on the wings. In Japan a subspecies has been described which differs slightly from the European-American subspecies. It flies in open country and in N. America is one of the more familiar species, being one of the first spring butterflies to appear. The butterfly is a rare visitor to Britain where its name Camberwell Beauty comes from the earliest record in 1748 when a few were taken in Cool Arbour Lane near Camberwell, London. The caterpillar is glossy black, spotted with white, a row of red dots along the back, with several rows of spines along the body. It feeds on elm *(Ulmus)*, poplar *(Populus)*, and hackberry *(Celtis)*. **197**

**io** see under **Inachis**

**milberti** Latreille MILBERT'S TORTOISESHELL
Canada S. to W. Virginia 43–45 mm. (1·69–1·77 in.)
This butterfly at times occurs in large numbers. It ranges from Newfoundland to C. and S. Canada in the more mountainous parts to Virginia. It is a typical tortoiseshell in appearance, differing from the one illustrated in having the dark brown on the forewing more extensive, leaving a couple of patches of red-brown near the front wing margin and in having a pale yellow band down both wings along the edge of the darker basal area. The spiny caterpillar lives in colonies on nettle *(Urtica)* with 2–3 broods a year, the first being out in early April.

**polychloros** Linnaeus LARGE TORTOISESHELL
N. Africa, Europe including Britain, Turkey to the Himalayas 50–64 mm. (1·97–2·52 in.)
This butterfly is out in June or July, but it has a longer life than many and flies throughout the summer months, hibernating in the winter, to reappear again in the early spring. The caterpillar feeds on elm *(Ulmus)*, willow *(Salix)* and other trees where it lives in large communities. Several subspecies of *N. polychloros* have been described, including one from N. Africa which flies in the mountainous areas over 1520 m. (5000 ft). The nominate subspecies in Europe flies from sea-level up to this altitude on mountains. **202**

**vau-album** Denis & Schiffermüller FALSE COMMA or COMPTON TORTOISESHELL
E. Europe, temperate Asia, including Japan, U.S., S. Canada 60–66 mm. (2·36–2·6 in.)
A woodland, or woodland-edge species which is richly patterned. Like all tortoiseshells it is a conspicuous and fast flier but when it folds its wings at rest the dark underside make it practically invisible against the bark of trees or dead leaves. It is, like other tortoiseshells, fond of feeding on sap from plants, rotting fruit and visits puddles to drink. The caterpillar, which is light green speckled and striped with paler shades with black spines all over, feeds gregariously on various forest trees including elm *(Ulmus)*, sallow *(Salix)* and birch *(Betula)*. There is one generation a year, the butterflies emerging in July and hibernating over the winter to re-emerge in the following spring. In America it is noted for the changes in abundance in different years, some years being common, others rare. **163h**

**xanthomelas** Denis & Schiffermüller YELLOW-LEGGED TORTOISESHELL
E. Europe, temperate Asia including Japan 60–64 mm. (2·36–2·52 in.)
The Japanese specimens are considered as a distinct subspecies where it is known as the Large Tortoiseshell butterfly. This butterfly is like *N. polychloros* but can be separated on small differences in pattern and by the colour of the middle and hind legs and palps which are brown or buff – these are very dark brown in *polychloros*. The butterfly flies from July to September and, after hibernation, in the spring. The caterpillars live in silken nests on willow *(Salix)*.

**nympheata** see **Nymphula**

**Nymphidium** Fabricius (NEMEOBIIDAE)
These are forest butterflies which like many other Nemeobiidae fly chiefly early in the morning. They settle on the under surface of leaves during periods of rest, but on top of a leaf during short rests between flight periods. There are about 50 species in this C. and tropical S. American genus.

**menalcus** Stoll
Tropical S. America 25–35 mm. (0·98–1·38 in.)
There are several colour forms of this common species, most of them not associated with a particular geographical area. The general plan of the colour pattern is similar to that of the illustrated *N. onaeum*, but the white areas of the wings are various shades of yellow or orange.

**onaeum** Hewitson YELLOW-BANDED NYMPHIDIUM
Tropical C. & S. America from Honduras to Venezuela 26–29 mm. (1·02–1·14 in.)
The colour pattern of this species is typical of its genus. It is a generally rare species, but occurs in loosely knit colonies and can be locally common. **268m**

NYMPHIDIUM, YELLOW-BANDED see **Nymphidium onaeum**

**Nymphula** Schrank: syn. *Hydrocampa* (PYRALIDAE) CHINA MARK MOTH
A genus of delicate long-legged Pyralid moths. They have, in many cases, the appearance of being made from fine porcelain. The caterpillars feed on a variety of water plants and are completely aquatic in habit. There are several species in Europe and in the temperate parts of Asia and one species in N. America.

**ekthlipsis** Grote
U.S., Canada 21–25 mm. (0·83–0·98 in.)
A strongly marked Nymphuline separated from other American species by the pattern of brown bordered yellow bands on white wings. Nothing is known of its life history. The species is common in the N.E. States of the U.S. and in the S.E. part of Canada.

**nympheata** Linnaeus: syn. *N. potamogata* BROWN CHINA MARK 23–30 mm. (0·91–1·18 in.)
Europe including Britain, temperate Asia
When the caterpillar of this moth first hatches it mines into the leaf of the water plant, later on making a flat oval case of cut-out fragments of leaves of *Potamogeton*, *Hydrocharis* and other water plants. It lives in the case on the surface of the water feeding on floating leaves of water plants. This insect is found in still water, avoiding fast flowing parts. The adult moth is out from June to August and flies after dusk. The moth is common over much of its range and is often attracted to light. Recently the caterpillars have been found making holes in polythene liners of artificial ponds. With the advent of polythene liners for large reservoirs, these insects could present problems. **68**

OAK MOTH, SCALLOPED see **Crocallis elinguaria**
OAK
- BEAUTY see **Biston strataria**
- BLUE BUTTERFLIES see **Narathura**
- EGGAR see **Lasiocampa quercus**
- LEAF BUTTERFLIES see **Kallima**
- NYCTEOLINE see **Nycteola revayana**
- SILK-MOTHS see **Antheraea**
- TORTRIX, GREEN see **Tortrix viridana**

**obiensis** see **Milionia**
**oblinata** see **Acronicta**
**obliteralis** see **Synclita**
**oblonga** see **Nonagria**
**obrinus** see **Nessaea**
**obscura** see **Lathroteles** and under **Lactura suffosa**
**obscuralis** see **Parapoynx**
**occidentalis** see **Crambus**
**occitanica** see **Melanargia, Zygaena**
**ocellata** see **Smerinthus**

**Ochlodes** Scudder (HESPERIIDAE)
A large genus of skipper butterflies with species throughout Europe, Asia, in N. America and India but not known from Africa S. of Sahara.

**venatus** Bremer & Grey LARGE SKIPPER
Europe including Britain, temperate Asia including Japan 28–34 mm. (1·1–1·34 in.)
This species is common throughout most of its range, although its occurrence in Ireland still needs to be proved and it is absent from most Mediterranean islands. Several subspecies have been described over its range, including a distinct one in Japan. The butterfly is out in meadowlands from June to August. It usually has a single brood in the N. parts but 2–3 broods in the warmer S. parts of its range. The caterpillar feeds on grasses, including *Festuca*, *Poa*. **87**

OCHREOUS PLUME MOTH see **Platyptilia ochrodactyla**
**ochrodactyla** see **Platyptilia**

**Ochrogaster** Herrich-Schäffer (NOTODONTIDAE)
This is a small genus of the subfamily Thaumetopoeinae, Processionary moths, found in Australia and associated islands. Like several species in other genera of this group, the caterpillars of *Ochrogaster* move in procession to and from their feeding area. The hairs of the caterpillar produce an irritant secretion.

**contraria** Walker
E. Australia 38–70 mm. (1·5–2·76 in.)
The wings of this species are dull brown, with a single, white marking on the middle of each wing. Females have an enormous, yellowish white anal tuft of closely packed scales and hair-scales and a covering of particularly long hair-scales on the thorax. The caterpillars of *O. contraria* rest in silken sacs in *Acacia* and other trees during the day, leaving their refuge at night, in procession, to feed on the foliage of the host tree.

**Ochropleura** Hübner (NOCTUIDAE)
About 100 species are placed in this almost cosmopolitan genus of moths.

**praecox** Linnaeus PORTLAND MOTH
Europe including Britain, to Japan 34–44 mm. (1·34–1·73 in.)
The cryptically coloured forewings of this species are usually green with black and white markings and a single pre-marginal, reddish brown band, but the forewings of occasional specimens are entirely reddish brown in ground-colour; the hindwings are brownish white. Towards evening, the caterpillar emerges from its vertical hole in the soil to feed on nearby grass, dwarf sallow *(Salix)*, chickweed *(Stellaria)* and other plants. **391b**

**Ochsenheimeria** Hübner (OCHSENHEIMERIIDAE)
A small genus of micro-moths with species mostly in Europe. Where the life history is known the caterpillars are usually found feeding on grasses (Gramineae). The moths fly during the day. They are mostly small, rather dark-coloured, species.

**mediopectinellus** Haworth: syn. *O. birdella*
Europe including Britain 11–12 mm. (0·43–0·47 in.)
The small, yellowish white, caterpillar feeds in the stems of various grasses including *Poa*, *Dactylis* and *Bromus*. The moth flies in July and August, usually around mid-day. **5j**

**Ochsenheimeriidae**
A small family of micro-moths, related to the Tineidae. The species are known mostly from Europe.

**ochus** see **Manduca**

**Ocnogyna** Lederer (ARCTIIDAE)
This is a genus of 30 species of small moths. It is represented in Europe, N. Africa, temperate Asia E. in China, and in Peru (*O. jelski* Oberthür). The females are typically brachypterous; the hindwings, in particular, are much reduced and hardly visible. Some of the species now placed in *Phragmatobia* probably belong in *Ocnogyna*.

**corsicum** Rambur
Corsica, Sardinia (male) 20–27 mm. (0·79–1·06 in.)
Females have vestigial wings and are incapable of flight. The moths emerge between March and May. The forewings of the males are brown, marked with a reticulate pattern of yellow lines; the hindwings are yellow, with brown spots on the outer part of the wing. Various grasses form the food of the caterpillar; it is brownish red in colour, with black, longitudinal lines.

**leprieuri** Oberthür
Algeria (male) 21–24 mm. (0·83–0·94 in.)
The female of this moth is a small hairy creature with very short, vestigial wings. The moths emerge in March and April, and can be quite common. Its caterpillar is dark brown, with brownish red hairs in the middle and black hairs at either end, and is similar in general appearance to that of *Phragmatobia fuliginosa*; it feeds at night on borage *(Borago)* and other herbaceous plants. **382**d, e

**parasitum** Esper
S. & C. Europe, Turkey (male) 28–34 mm. (1·1–1·34 in.)
The wings of both sexes are greyish brown, with black markings on the forewing; those of the female are narrow and greatly reduced in length, especially the hindwing.

**octavia** see **Precis**
**octomaculata** see **Alypia**
**ocularis** see **Tethea**

**Odezia** Boisduval (GEOMETRIDAE)
This is a small genus of moths found in Europe and temperate Asia.

**atrata** Linnaeus CHIMNEY-SWEEPER
Europe including Britain, Turkey, temperate Asia including Japan 23–26 mm. (0·91–1·02 in.)
This black moth is locally plentiful throughout its range. The caterpillar is green, paler on the sides with 3 darker lines along the back, a reddish mark at the base and a white line along the sides. It feeds on Earthnut *(Conopodium)*. The moth flies in the daytime and can be seen visiting flowers. **292**

**odorata** see **Ascalapha**

**Oecophoridae**
A worldwide family of micro-moths with many species. The caterpillar feeds on plant detritus, deadwood, lichens. Some are household pests. They include a few species with very reduced wings which live on subantarctic islands. Oecophorids have long curved labial palps and a proboscis with scales on it. Although world wide, it is in Australia where they reach their peak of development with the largest number of species. Many species live in the eucalyptus forests where they feed on and break down the very dry litter. The genus *Hypertropha* included here is often placed in a separated family, Hypertrophidae. See *Alabonia, Endrosis, Hoffmannophila, Hypertropha, Neossiosynoeca, Sphaerelictis, Thalamarchella, Tinearupa*.

**oedippus** see **Horama**
**oehlmanniella** see **Lampronia**

**Oeneis** Hübner (NYMPHALIDAE)
A large genus of butterflies with species in Asia, N. America and with a few of the Asian species extending into E. Europe. This genus belongs to the sub-family Satyrinae but differs from many of them in not having a swollen vein at the base of the forewing. Instead species in the genus only have the front margin of the forewing thickened. The species are found in the higher mountains and in the Arctic.

**bore** Schneider ARCTIC GRAYLING
N. Europe, N. Asia, N. America 44–50 mm. (1·73–1·97 in.)
This butterfly is out in June-July and flies in stony slopes in hilly country from sea-level to 610 m. (2000 ft). The adult quickly gets worn after it emerges, becoming almost transparent. The butterfly is widespread in Canada and Alaska, and Norway to Finland. The caterpillar feeds on various grasses. There are a number of similar, related species but generally they have more pattern on the upper and underside of the wing than this rather plain butterfly. The butterfly is a pale grey brown on the upperside, unrelieved by eyespots and with scarcely any pattern, although there is a trace of a lighter yellow brown near the wing margins. The underside forewing is equally unmarked grey brown and only the underside hindwing shows a white, irregular band on a slightly darker brown. The fringes on the wings have alternate lighter and darker marks. Altogether a very undistinguished grey-brown butterfly.

**Ogyris** Westwood (LYCAENIDAE) AZURES
This genus is found only in New Guinea and Australia. Most of the species are a vivid blue, purple or orange above, but have a cryptically coloured under surface. They are essentially butterflies of the tree-canopy. Their caterpillars feed mainly at night on various mistletoes (Loranthaceae) and rest during the day in crevices of bark or in ants' nests at ground level. The chrysalis is able to produce clicking sounds – possibly a protective device. Other groups of butterflies which feed on species of Loranthaceae seem to be equally successful, measured by their numbers of species: they include *Iolaus*, a Lycaenid genus, *Delias* and *Catasticta* (Pieridae).

**abrota** Westwood DARK PURPLE AZURE
S.E. Australia 38 mm. (1·5 in.)
The deep rich violet of the male upper surface is in sharp contrast to the yellow and nearly black coloration of the female. Three species of ants have been recorded as attendants on the caterpillars. The foodplants, parasitic on *Eucalyptus*, are various species of mistletoe *Loranthus* and *Banksia*. *O. abrota* differs from the similarly coloured, but larger, *O. genoveva* by the evenly rounded, though crenulate, margin of the hindwing and the distinctive pattern of the under surface of the wings.

**aenone** Waterhouse
N.E. Australia 35–45 mm. (1·38–1·77 in.)
Coastal swamps where there are paperbark trees *(Melaleuca)* and dry open woodland with *Casuarina* are the localities where adults of this species are found. Its caterpillars feed at night on mistletoes and shelter during the day in bark crevices. Caterpillars feeding on the mistletoe *Dendrophthoe* are attended by brown ants of the genus *Pheidole* which colonize epiphytic ant-plants *(Myrmecodia)* on the same tree as the mistletoe; caterpillars which feed on the mistletoes *Lysiana* and *Amyema* are attended by black ants belonging to the genus *Iridomyrmex*. **244**u

**genoveva** Hewitson
E., S.E. Australia 45 mm. (1·77 in.)
The male of this colourful species is violet above; the female greenish blue with a pale yellow pattern on the forewing. The caterpillars and chrysalids are readily obtainable from ants' nests at the base of *Eucalyptus* trees parasitised by mistletoes. It seems likely that the attendant ants (species of *Camponotus)* collect caterpillars from nearby trees and carry them to their nests. Fully grown caterpillars are guided by the ants to their foodplants, mistletoes of *Amyema* and other genera, which may be up to 30 m. (100 ft) from the ants' nests.

**ianthis** Waterhouse SYDNEY AZURE
E. Australia 32–35 mm. (1·26–1·38 in.)
There are striking differences in colour pattern between the sexes of this rare butterfly. The male (not figured) is a typical blue above, with dark margins to the wings; the female (illustrated) is hardly recognizable as the same species – even the under surfaces of the wings are quite differently coloured. Its caterpillars feed on various mistletoes and are attended by black and brown ants. **244**l

**Oinophila** Stephens (HIEROXESTIDAE)
A small genus of micro-moths, previously included in different families according to various authors (Tineidae, Lyonetidae). Most of the species in the genus have narrow, pointed forewings. The moths are known from Europe, Africa and Asia.

**v-flava** Haworth
Europe including Britain 8–12 mm. (0·31–0·47 in.)
The yellowish white caterpillar of this moth feeds in silken galleries amongst fungi growing on trees and walls. It is also known to feed on cork and can cause problems when it attacks corks in wine bottles. The moth is out in Europe in May and June. On the continent it is known from a number of countries including France, Holland and Germany, but generally is a rather local species. **6**f

OLD
SAILER see **Neptis alta**
WORLD BOLLWORM see **Heliothis armigera**
**oleae** see **Prays**
OLEANDER BUTTERFLY see **Euploea core**
OLEANDER HAWK-MOTH see **Daphnis nerii**
**olena** see under **Aeria eurimedia**

**Oleria** Hübner: syn. *Leucothyris* (NYMPHALIDAE)
This is a small genus of C. American and tropical S. American butterflies of the subfamily Ithomiinae. Many of the species of this genus were known under the generic name *Leucothyris*.

**synnova** Hewitson
Brazil 38–41 mm. (1·5–1·61 in.)
Few specimens exist in collections of this beautiful species. It has been captured only in the upper reaches of the Amazon river. Nothing is known about its life history. **131**l

**Oleuthreutes** Hübner (TORTRICIDAE)
This genus of micro-moths is in the subfamily Oleuthreutinae. They are worldwide in distribution with some species of economic importance. Most have a typical 'tortricid' look when at rest (see Tortricidae). The pests include many which attack fruit such as apples or plums. At least one species has been used in a biological control experiment. In Australia many species have caterpillars which live in the rather dry litter of Eucalyptus forests. These caterpillars feed on the dead leaves, breaking them up and helping to form the compost which in moister forest areas is done mainly by fungi, worms and beetles.

**mniochlora** Meyrick
India 22 mm. (0·87 in.)
This species is found in the hill country in S. India. The related species are mostly brown, often heavily patterned moths. The green colour of this species is unusual in the genus. Nothing is known of its biology and only a few specimens have ever been seen. **11**c

OLIVE MOTH see **Prays oleae**
OLIVE
KERNEL-BORER see **Prays oleae**
SKIPPER see **Pyrgus serratulae**
**olympia** see **Comocritis, Euchloe**
OLYMPIA BUTTERFLY see **Euchloe olympia**

**Olyras** Doubleday (NYMPHALIDAE)
The 4 species of this genus of the subfamily Ithomiinae are rare in collections. They are inhabitants chiefly of mountainous regions in C. America and tropical S. America.

**insignis** Salvin
C. America, Colombia 82–95 mm. (3·23–3·74 in.)
Similar in pattern to the Daniine butterfly *Ituna ilione*, this lacks a dark, transverse band on the basal half of the hindwing which is a transparent, pale yellow, and not entirely colourless. Some C. American specimens have reddish brown scales at the base of the rear margin of the hindwing. **131**d

**onaeum** see **Nymphidium**

**Oncocera** Stephens (PYRALIDAE)
A small genus of Pyralid moths, formerly included in the genus *Laodamia* whose exact generic limits are not known with certainty. One species is common in Europe and temperate Asia.

**semirubella** Scopoli: syn. *O. carnella*
Europe including Britain, 24–27 mm.
Asia including Japan (0·94–1·06 in.)
The pinkish forewing, usually with a white streak along the front edge, makes this a very conspicuous species. The hindwings are grey. The bronze black larva, with pale blue lines along the body, makes webs on leaves of *Lotus* (Leguminosae). **56**v

**opaca** see **Phaeosphecia**
OPAL COPPER see **Poecilmitis thysbe**
OPAL-CENTERED COPPER see **Poecilmitis thysbe**
**opercularis** see **Megalopyge**
**operculella** see **Phthorimaea**

**Operophtera** Hübner (GEOMETRIDAE) WINTER MOTHS
The genus of moths reaches across Europe and N. America with a few occurring in the Oriental region. The moths are attracted to light during milder winter evenings.

**brumata** Linnaeus WINTER MOTH
Europe including Britain, 25–28 mm.
Asia including Japan, Canada (0·98–1·1 in.)
The caterpillars feed on oak *(Quercus)*, hawthorn *(Crataegus)* and apple *(Malus)*, at times becoming common. The moth is out in October at any time during the winter until February. It may appear during milder spells. As a pest of orchards it has been the object of much work on control measures. One of the commonest is the use of sticky bands round orchard trees which are used to stop the wingless females (illustrated) from climbing up the trees to lay their eggs. In N. America it is restricted to Nova Scotia, New Brunswick and Prince Edward Island where it is believed to have been accidentally introduced from Europe. **282**

**fagata** Scharfenberg: syn. *O. boreata*
NORTHERN WINTER MOTH 30–33 mm. (1·18–1·3 in.)
N. Europe including Britain
This species is usually larger than the Winter moth, *O. brumata,* and the forewings are paler and more glossy. In the female the wings are reduced as in the Winter moth. The caterpillar is greenish with grey stripes along the back with a yellowish line along the side. It feeds on birch *(Betula)* during May and June, the moth is out in October and November and is readily attracted to light on milder evenings. The male is illustrated. **283**

**ophiaula** see **Gelechia**
**ophidicephalus** see **Papilio**

**Ophthalmophora** Guenée (GEOMETRIDAE)
Rather angular winged moths usually with prominent eyespots on the hindwings. They are found from Peru to C. America.

**branickiaria** Oberthür
Peru 35–45 mm. (1·38–1·77 in.)
This species has darker forewings than the others in the genus, with more orange in the hindwings. Several others in the genus have a broadly similar pattern. Nothing is known of their biology. **294**e

**preambilis** Oberthür
Brazil 35 mm. (1·38 in.)
This species was first collected in the Amazon jungle. There are several related species differing in details of the intricate patterns, which produces striking visual effects. The biology is unknown. The white forewings have a broad brown line across the middle, leaving white in the front and hind part of the wing. The hindwings are silvery grey with gold eyespots and a silver, yellow edged line round the margin. Very few specimens of this species have been collected.

**opinatus** see **Charaxes**
**opis** see **Cynandra**

**Opisthograptis** Hübner (GEOMETRIDAE)
The species of moths in this genus are widespread in Europe and temperate Asia. The caterpillars are typical of the family.

**luteolata** Linnaeus BRIMSTONE MOTH
Europe including Britain, 29–38 mm.
Asia, N. Africa (1·14–1·5 in.)
The pale lemon yellow coloured wings have brown marks near the apex and a brown 'O' further along the margin towards the body. These patterns make this moth easily recognized. The twig-like caterpillars are brown tinged with green with a double pointed hump on the back with a smaller one further from the head. It feeds on sloe *(Prunus)* and other Rosaceae. The caterpillar, like many Geometrids, when disturbed, sits still and straight, its twig-like appearance a wonderful camouflage. **309**

**Opogona** Zeller (HIEROXESTIDAE)
A genus of micro-moths with many species all over the world. They were formerly placed with species of *Lyonetia* in the Lyonetidae but recently the genus *Opogona* and a few others were removed to a separate family. However until more work is done on the species in the genus, the exact limits of it are not known.

**sacchari** Bojer: syn. *O. subcervinella*
Mauritius, Seychelles, St Helena, 24–28 mm.
Madeira, Europe, including Britain (0·95–1·1 in.)
This species has been found on a variety of host plants including sugarcane *(Saccharum)*, banana *(Musa)*, sweet potato *(Ipomoea)* and grape vines *(Vitis)*. It has been imported, accidentally, into Europe and in Britain has been bred from caterpillars feeding on banana skins. The moth has brown, rather narrow wings with little marking on them. **34**a

**oporana** see **Archips**
**Oporinia** see **Epirrita**

**Opostega** Zeller (OPOSTEGIDAE)
A worldwide genus of moths. The caterpillars are generally legless and live in mines in leaves, passing their caterpillar stages between the upper and lower surfaces of a single leaf.

**salaciella** Treitschke
Europe including Britain 9–12 mm. (0·35–0·47 in.)
The moth is common in June and July over most of its range often being attracted to light. The caterpillar feeds on species of *Rumex* (Polygonaceae). The specimen figured is fairly typical of the appearance of other species in the genus. **1**b

**Opostegidae**
Small micro-moths, leaf-miners where the life history is known. The moths have a very reduced wing venation. The adults are usually white with darker patterns. It is regarded as rather a primitive family in the Lepidoptera. The family has been associated at different times with different families and its true relationships are still a matter of some dispute. The moths in this family are common in the tropics but extend into most temperate regions. See *Opostega.*

**oppelii** see **Perisama**
**optilete** see **Vacciniina**
ORANGE MOTH see **Angerona prunaria**
ORANGE
ADMIRAL see **Antanartia delius**
ALBATROSS BUTTERFLY see **Appias nero**
BUSH BROWN see **Mycalesis terminus**
DOG BUTTERFLY see **Papilio anchisiades** and **P. cresphontes**
FORESTER see **Euphaedra eleus**
OAK LEAF see **Kallima inachus**
PATCH WHITE see **Colotis halimede**
SULPHUR see **Colias eurytheme**
SWALLOWTAIL see **Papilio thoas**
THEOPE see **Theope endocina**
TORTRIX see **Argyrotaenia citrana**
WORM, NAVEL see **Paramyelois transitella**
ORANGE AND BROWN BUTTERFLY see **Vanessula milca**
ORANGE AND LEMON VAGRANT see **Eronia leda**
ORANGE-BANDED SHOEMAKER see **Catonephele acontius**
ORANGE-BARRED SULPHUR see **Phoebus philea**
ORANGE-TIP BUTTERFLIES see **Anthocharis, Colotis, Hebomoia, Ixias, Paramidea**
ORANGE-TIPPED UNDERLEAF see **Lymnas xarifa**
**orbicularis** see **Aganais**
**orbiculella** see **Antispila**
**orbitulus** see **Albulina**
ORCHARD BUTTERFLY see **Papilio aegeus**
**oreas** see **Hypoleria**

**Oreisplanus** Waterhouse & Lyell (HESPERIIDAE)
A genus of skipper butterflies which is found in S.E. Australia and Tasmania, with only 2 known species.

**munionga** Olliff ALPINE SKIPPER
Australia, Tasmania 28–30 mm. (1·1–1·18 in.)
The brown male has orange, squarish patches and an orange and brown fringe to the forewing. The hindwing is similar but with one orange patch near the middle. The underside, particularly on the hindwing has yellowish brown and black in patches round the veins. The underside of the forewing is similar to the upperside but with a broad orange margin and fewer orange spots. The caterpillar feeds on species of *Carex* (Cyperaceae). The adults are about in February and March and are particularly attracted to the yellow coloured Everlasting flower and other daisies (Compositae), being particularly well camouflaged when at rest on these flowerheads.

**orellana** see **Parides**

**Oreta** Walker (DREPANIDAE)
About 40 species are placed in this genus of hook-tip moths. They are found in E. and S.E. Asia, including most of China and Japan and as far as Australia and the Solomons. There is also a single species in the U.S. and Canada.

**rosea** Walker
N. America 25–35 mm. (0·98–1·38 in.)
This shares many characters with its relative *O. pulchripes*, A Russo-Japanese species whose caterpillars feed on the same foodplant, *Viburnum*. This close relationship reflects the existence in the past of a warmer climate in Alaska and adjoining Siberia across what is now the Bering Straits but which at one time was probably a land connection between the continents and a passageway for the dispersal of insects and other animals. **272**m

**singapura** Swinhoe
S.E. Asia 38–49 mm. (1·5–1·93 in.)
One of the most variable species in its family, specimens can vary from unmarked yellow to a mottled brownish pink and dark brown. Both males and females vary in colour, and led one early entomologist astray to the extent that he described this species at least four times under different names. A collection of specimens looks very much like a batch of dead leaves. **272**k

**Orgyia** Ochsenheimer (LYMANTRIIDAE)
*Orgyia* is a widely distributed genus of about 30 species of moths. It is found throughout the Old World from Europe and Africa to Japan and Australia, but in the New World is restricted to N. America. The females of most species have reduced wings, but there are all stages from fully winged to wingless, and also legless in *O. dubia*. Caterpillars have 4 dorsal 'toothbrush' tufts, but no tail tuft; many are forest, orchard and plantation pests.

**anartoides** Walker PAINTED APPLE
S.E. Australia 24–28 mm. (0·94–1·1 in.)
The wingless female, like the female Vapourer, lays its eggs on the cocoon from which it emerged. The foodplants of the caterpillar include species of the native *Acacia* but also apple *(Malus)* and other introduced trees and a variety of garden plants. The name *anartoides* refers to the general similarity in appearance of this species to that of the Noctuid genus *Anarta* whose species also exhibit 'flash-coloration' of the hindwings. The male forewings are cryptically patterned in brown, whereas the conspicuous hindwings are a bright yellow, with a dark brown band close to the outer margin.

**antiqua** Linnaeus VAPOURER OR RUSTY TUSSOCK
N. Africa, Europe, temperate Asia to Siberia, Canada, U.S. 30–34 mm. (1·18–1·38 in.)
This is a common moth both in urban London and New York and is widespread elsewhere in its range wherever there are trees and bushes to provide food for the caterpillar. The foodplants include not only deciduous trees but several conifers, and in Canada the commonest host-tree of more than 50 species is the Balsam Fir *(Abies)*. In orchards there is sometimes damage to the fruit as well as to the foliage of fruit trees. The unillustrated female, has vestigial wings, moves very little and lays her eggs on the outside of the cocoon where they overwinter. **359**

ORIENTAL
FRUIT MOTH see **Cydia molesta**
LEAFWORM see **Spodoptera litura**
**orientalis** see **Metisella, Tascina**

**Orinoma** Gray (NYMPHALIDAE)
These butterflies have a striking pattern which is quite atypical of most of the subfamily Satyrinae to which this genus belongs. The 2 species known are found in N. India, Pakistan, China, S.E. Asia and Burma.

**damaris** Gray
N. India, Pakistan, China, Vietnam 70 mm. (2·75 in.)
There is some variation in this species which may have yellower orange-brown at the base of the forewing. Basically it is a yellow butterfly with broad brown veins. **174**q

**orinoptera** see **Ludia**
**orithea** see **Coronidia**
**orithya** see **Junonia**
**orizaba** see **Rothschildia**
ORIZABA SILK-MOTH see **Rothschildia orizaba**

**Ormetica** Clemens (ARCTIIDAE)
C. and tropical S. America
There are several groups of similarly coloured, distasteful species in this typically black, yellow and iridescent blue or green genus of tropical American moths. When provoked, *Ormetica* species raise their brightly coloured abdomen as a final advertisement of their noxious qualities.

**tanialoides** Rothschild
Venezuela 32–36 mm. (1·26–1·42 in.)
This is typical of its genus in the yellow and black coloration of the wings and the partly iridescent blue of the abdomen. It has not been recorded from outside Venezuela, but it probably also occurs in Colombia. **368**m

**ornatissima** see **Mictopsichia**
**ornatrix** see **Utetheisa**

**Ornithoptera** Boisduval (PAPILIONIDAE)
BIRDWING BUTTERFLIES
This is a genus of 12 species restricted to the Moluccas, New Guinea, the Solomons and N. Australia. There are 2 groups of species: the *priamus* group, the species of which have a sex-patch on the male forewing, and the *paradisea* group which lacks a sex-patch. The imposing males of *Ornithoptera* are orange, yellow, blue or green and often iridescent; the females are larger but less colourful and are mostly black or dark brown, marked with white on the forewing and white and yellow on the hindwing. Specimens of this genus are amongst the most highly prized of all butterflies and frequently command large sums at auction (see, for example *O. allottei*). The caterpillars of most species feed on the foliage of Aristolochiaceae species which are generally considered to be toxic to birds and other vertebrates. It seems likely that both the caterpillars and the adult butterflies will prove to be unpalatable to predators and that their bright colours, like those of other *Aristolochia*-feeding Papilionidae of the tribe Troidini, function as warning signals to their enemies. The chrysalis of *O. priamus* has been shown to store a toxic aristolochic acid in its tissues.
*paradisea* group: *O. goliath, tithonus, chimaera, rothschildi, paradisea, meridionalis.*
*priamus* group: *O. priamus, croesus, aesacus, victoriae, alexandrae, allottei*

**aesacus** Ney
Obi Island 140–185 mm. (5.51–7.28 in.)
The pattern of black and turquoise on the male wings is similar in general features to that of *O. priamus* but there are no black spots on the hindwings. The white spotted, dark brown forewings and the yellowish grey marked hindwings of the female are typical of several other species of *Ornithoptera*. **107**d

**alexandrae** Rothschild QUEEN ALEXANDRA'S BIRDWING
New Guinea 170–280 mm. (6·69–11·02 in.)
This rare and superb birdwing is the largest known species of butterfly. Some females attain a wingspan of up to 280 mm. (11 in.). Deforestation and the enthusiasm of collectors (probably chiefly the former) have made this species much more rare than in the past, but it is now a protected species. The caterpillar feeds on vine-like species of *Aristolochia*. The adult butterfly usually flies at between 15 and 30 m. (50–100 ft) above the ground. Males, at least, have been known to live for up to 12 weeks in natural conditions. The first specimen of this butterfly was ignominiously felled by gun-shot. The female (not illustrated) is brown with yellowish white markings. **104**b

**allottei** Rothschild
Solomon Islands 138–170 mm. (5·43–6·69 in.)
This extremely rare birdwing is known from less than a dozen specimens. Its rarity attracted the highest known price for a butterfly when a male of this species fetched 10,600 francs in a European sale in October 1966. There is some evidence that *O. allottei* is a hybrid between *O. priamus* and another *Ornithoptera* species. **104**a

**chimaera** Rothschild
New Guinea 120–180 mm. (4·72–7·09 in.)
Most of the original specimens of this species were caught on the same flowering tree by A. S. Meek, a collector who supplied Rothschild who named the species. The colour pattern of *O. chimaera* is similar to that of *O. goliath* but the iridescent green and gold areas on the male wings are less extensive, especially on the forewing.

**croesus** Wallace
The Moluccan Islands Bátjan, Ternate, Tidore, Halmahera 140–170 mm. (5·51–6·69 in.)
Many superlatives have been used to describe the beauty of this famous birdwing species. The orange areas of the male forewing become a rich green when viewed obliquely. The first captures, as well as the original description of this species, were made by Alfred Russell Wallace, contemporary of Charles Darwin and co-founder of the theory of Natural Selection. The caterpillars feed on species of *Aristolochia* found in swampy areas. It is not at present a rare species. **107**c

**goliath** Oberthür GOLIATH BIRDWING
Ceram, Moluccas, New Guinea 137–210 mm. (5·39–8·27 in.)
This superbly coloured member of the *paradisea* group flies in dense rain-forests. Very little appears to be known about its life history or behaviour. The male forewing is black, with green markings; the hindwing is orange-yellow, with black-edged, green spots and black wing margins. The female is an enormous butterfly; it lacks green on the forewing, which is dark brown with whitish markings and has a large, black-spotted, brownish orange area on the hindwing. **107**e

**meridionalis** Rothschild
New Guinea 90–150 mm. (3·54–5·9 in.)
The male of this very rare butterfly is similar in colour pattern to *O. paradisea*, but the gold and green streak along the front of the forewing is much narrower and the similarly coloured, triangular area at the rear edge of this wing is broader than in *O. paradisea*. The females differ from those of *O. paradisea* in the purer white markings.

**paradisea** Staudinger PARADISE OR TAILED BIRDWING
New Guinea 127–150 mm. (5–5·91 in.)
The finely tailed hindwings of the male distinguish this species from all birdwings except the smaller *O. meridionalis*. Like several others of its genus, it is on the list of species protected by law. Forested valleys at about 460 m. (1500 ft) are localities favoured by adults of both sexes. They usually fly at tree-top level, but will visit flowers in the early morning and late afternoon. Three subspecies are known; each from a particular region of New Guinea. None of the females would be recognized at first as representatives of the same species as the males – they have larger, greyish forewings with white spots, and much less colourful, yellow, bluish grey and greyish white hindwings. The caterpillar feeds on species of *Aristolochia*. **107**b

**priamus** Linnaeus
The Moluccas to New Guinea, N. Australia and the Solomons 108–130 mm. (4·25–5·12 in.)
Fourteen subspecies of this brilliantly coloured species have been described and it has various common names mostly based on geographical origin or colour. It is quite common and is probably the best known species of birdwing. The males are green (or blue), orange and black; the huge females are black with white and yellow markings. The illustrated 'green' *O. priamus poseidon* and the blue *O. priamus urvillianus* are 2 of the better known subspecies. Flight is usually at tree-canopy level and is languid in character as might be expected from the long, glider-like wings. A warning display similar to that exhibited by some Saturniid moths has been observed in this species: the abdomen is curved downwards like that of a hornet, and the butterfly becomes immobile. Rain-forest species of *Aristolochia* are the caterpillars foodplants. **91**

**rothschildi** Kenrick
New Guinea 110–150 mm. (4·33–5·9 in.)
The forewings have a less precise colour pattern in the male than those of *O. goliath*, and on the hindwing there is a large area of green surrounding the black spots.

**tithorus** de Haan
New Guinea, Waigeu Island, Misol Island 117–190 mm. (4·61–7·48 in.)
This is similar in general features to the colour pattern of *O. goliath*, but differs in details, especially the presence of 3 black spots on the male hindwing. Four subspecies are recognized.

**victoriae** Gray QUEEN VICTORIA'S BIRDWING
Solomon Islands 152–175 mm. (5·98–6·89 in.)
The huge female of this species is black with white and yellow markings; much less colourful than the black, green, yellow and orange male. The first specimen of this butterfly was shot down from amongst the trecs during a voyage of discovery by the Royal Navy to the Solomon Islands during the reign of Queen Victoria. The caterpillar is brown, with long, fleshy red tubercles. Like other birdwing species, this is now on the list of officially protected butterflies. **107**a

**orsis** see **Myscelia**

**Orsotriaena** Wallengren (NYMPHALIDAE)
This genus of butterflies is related to *Mycalesis* and is in the subfamily Satyrinae. The eyes of species of *Orsotriaena* are smooth whereas those of *Mycalesis* are hairy. Species of *Orsotriaena* are found from India, through S.E. Asia and Indonesia to Australia.

**medus** Fabricius
India, Pakistan, S.E. Asia to New Guinea, Australia 28–45 mm. (1·1–1·77 in.)
Distinct subspecies of this are found in different parts of its wide range. The male is brown above with a more strongly patterned underside. This has a prominent white line, from front to back, across each wing and 2 eyespots on the forewing and usually 4–5 on the hindwing, surrounded by a small orange circle. Often common on parts of its range, where the caterpillars feed on various grasses, including Rice *(Oryza sativa)*. There are seasonal forms over parts of its range.

**Orybina** Snellen (PYRALIDAE)
This is a Pyralid genus of Oriental moths with about 12 known species. All are brightly coloured members of the subfamily Pyralinae. Nothing is known about their biology.

**flaviplaga** Walker
Burma, India, Pakistan, China 35–42 mm. (1·38–1·65 in.)
Several subspecies of this have been described. The specimen illustrated is from Burma. Nothing is known of the biology of these insects. **63h**

**osacesalis** see **Herdonia**
**osiba** see **Hyalurga**
**oslari** see **Eacles** and see under **Ecpantheria**
**ostenta** see **Holomelina**

**Osteosema** Warren (GEOMETRIDAE)
A large genus of moths, some of which are probably unrelated. The antennae are short and pectinate (feathery) in both sexes. These short antennae are characteristic of this genus. Most Geometrid moths have rather long antennae. There has been little work on the genus in recent years and as a result a number of species have been placed in here which more correctly should be in other genera.

**sanguilineata** Moore
N. India 34–44 mm. (1·34–1·73 in.)
Several related species show this striking pattern. The disruptive effect of the wavy brown lines across the wings enables this species to hide from predators. This sort of camouflage which breaks up the outline shape, is imitated by man in wartime camouflage. **294g**

**ostia** see **Parasa**
**ostrina** see **Eublemma**

**Ostrinia** Hübner (PYRALIDAE) CORN-BORER
Twenty species are at present recognized in this genus of Pyralid moths as the result of a recent careful study. This genus has species all over the world and contains some of the more important pests of corn *(Zea)*.

**furnacalis** Guenée
China, Japan, Korea, Australia 20–30 mm. (0·79–1·18 in.)
This species was confused until recently with the well-known European Corn-borer, *O. nubilalis*, and many early records of '*nubilalis*' in the Oriental and Australasian region should probably be refered here. Externally it is similar to *O. nubilalis* but is distinguished by details of its morphology.

**nubilalis** Hübner EUROPEAN CORN-BORER
Europe, U.S., Canada 26–34 mm. (1·02–1·34 in.)
This species is a very serious pest of corn. Every year it causes huge losses to corn growers. Much research work, and many publications have been devoted to the biology of this moth and methods of control. With the ever increasing need for greater food production, control of this pest becomes even more vital. In the U.S. the estimated loss of corn due to the activities of the corn-borer in 1969 was valued at 183 million dollars. The caterpillar often feeds on tomatoes in the fields and apart from the trouble they cause to fresh food, often get 'canned' with the tomatoes. Subsequently the housewife opening the tin is less than pleased at finding tomatoes with the added caterpillar! It has been found that sounds produced by bats scare these moths off the corn fields and this has been used in an attempt to keep them off crops as they react sharply to recordings of the bats.

**Othreis** Hübner (NOCTUIDAE)
There are about 20 recognized species in this pantropical genus of large, fruit-piercing moths. Some are pests of *Citrus* in Australia.

**fullonia** Clerck TROPICAL FRUIT-PIERCER
Australia, tropical Africa and Asia 80–102 mm. (3·15–4·02 in.)
The adult moth is able to pierce the skin of ripe *Citrus*, mango *(Mangifera)* and other fruits with its proboscis and suck their juices. The caterpillars have conspicuous eyespot markings that probably deter predators; they feed on climbing plants of the family Menispermaceae. The greenish brown forewing is doubtless cryptic in pattern and the hindwing an example of 'flash-coloration' (see *Phyllodes floralis*). **392d**

**otis** see **Zizina**

**otrere** see **Anaea**
**oulita** see **Hypothyris**

**Ourapteryx** Leach (GEOMETRIDAE)
A small genus of moths, mainly Indo-Malayan with one species in Europe and several in Japan. They are mostly pale lemon coloured moths with a characteristic small tail on each hindwing and with a small spot at the base of each tail. All are rapid fliers, appearing about dusk flying rapidly and strongly about the edge of woods and over open ground.

**sambucaria** Linnaeus SWALLOW-TAIL MOTH
Europe including Britain, Asia 42–54 mm. (1·65–2·13 in.)
This species is widespread and common in Europe and W. and C. Asia in June and July. The moth has a rapid flight and with its pale colour flashes about at dusk. The caterpillar is brown with light and dark stripes. It feeds on honeysuckle *(Lonicera)*, lime *(Tilia)*, elder *(Sambucus)*, and hawthorn *(Crataegus)* from autumn until spring. In Britain it is common in England but rarer further N. in Scotland. **306**

**ovata** see **Milionia**
OWL BUTTERFLIES see **Caligo**
OWL MOTH see **Thysania zenobia**
OWLET MOTHS see **Cucullia, Noctuidae, Scoliopteryx**
**oxyacanthae** see **Phyllonorycter**

**Oxychirota** Meyrick (OXYCHIROTIDAE)
A small genus of moths with species in India and as far S. as Australia. Very little is known about the species. The proboscis does not have scales on it, as do all Pyralidae, and the fore- and hindwings are long and slender with some reduction in the wing veins compared with most in the Pyraloid complex. Species in related genera have caterpillars which tunnel into the seeds of *Avicennia marina* (Avicenniaceae).

**paradoxa** Meyrick
Australia 10 mm. (0·39 in.)
The wings are very slender in this species with almost parallel margins. Little is known of their biology and very few specimens have been collected. The wings are reddish brown with long fringes and whiter patches on the wing. There are some black scales scattered along the wing giving a slightly granular look. The insect is very small and slender and is very similar externally to species of Gracillariidae.

**Oxychirotidae**
A small family of moths at present placed in the Pyraloid complex although there is some dispute of their actual relationship. See *Oxychirota*.

**Oxytenidae**
About 36 species are placed currently in this family of the superfamily Bombycoidea. Its range extends from Mexico S. into tropical S. America. Included are white species similar in appearance to some Uraniidae (Microniinae) and others which resemble species of Eupterotidae and Cercophanidae. See *Asthenidia*, *Oxytenis*.

**Oxytenis** Hübner (OXYTENIDAE)
There are about 18 species in this C. American (including Mexico) and tropical S. American genus of chiefly brown or yellow moths. The forewing of both sexes is produced and variously hooked apically. The male hindwing has a short, outwardly directed tail in some species.

**peregrina** Cramer
Tropical S. America 60–65 mm. (2·36–2·56 in.)
Females (not illustrated) lack the short tail on the hindwing and have an evenly convex margin to the forewing. **318a**

**ozomene** see **Actinote**

**Pachliopta** Reakirt (PAPILIONIDAE)
This is a genus of 13 Indo-Australian species of butterflies whose caterpillars, like those of other genera of the tribe Troidini, feed on species of the poisonous plant family Aristolochiaceae. Most adults of this genus have conspicuous red or yellow markings on the body or head as a warning to predators of their distasteful qualities. Aristolochic acid has been isolated as one of the vertebrate poisons present in these butterflies. The name of this genus has been misspelt by some authors as '*Pachlioptera*'.

**aristolochiae** Fabricius COMMON ROSE
Sri Lanka, India, Sikkim, Burma, S. China, Thailand to Malaya and the Lesser Sundas 80–110 mm. (3·15–4·33 in.)
This striking, red, black and white species has at least one mimic – a female form of *Papilio polytes*. It is a very common species and is frequently seen in gardens, but it seldom flies at elevations above 1220 m. (4000 ft). Nocturnal roosts of several thousand specimens have been reported, comparable with those of Danainae and Heliconiinae which are also known to roost gregariously in this way. Individual and geographical variants of the hindwing colour pattern include a complete loss of the white markings and, more rarely, of the red markings. **111d**

**hector** Linnaeus COMMON ROSE
Sri Lanka, India, Maldive Islands 70–110 mm. (2·76–4·33 in.)
This is generally a common species at low elevations, but also occurs, more rarely, up to 2440 m. (8000 ft). Nocturnal roosts have been recorded for this species (see *P. aristolochiae*). Like many other butterflies, this species is attracted to blossoms of *Lantana*. It is mimicked by a female form of *Papilio polytes*. The wings of the female (not illustrated) are dark brown, not black above, and the spots are usually orange and very weakly marked except at the margin of the wing. The under surface of the wings is similar to the upper surface in both sexes, but the orange or red spots are much larger. The caterpillar feeds on a species of *Aristolochia*. **111e**

**polydorus** Linnaeus RED-BODIED SWALLOWTAIL
Moluccas E. to New Guinea, N.E. Australia and the Solomons 65–80 mm. (2·56–3·15 in.)
In colour pattern this species is like a pale *P. aristolochiae*, but unlike the latter it lacks a tail to the hindwing. It is a common species in parts of its range where there are areas of rain-forest. Both sexes are attracted to flowers at the margins of forests. Their flight is unhurried and precise as in most unpalatable species. The caterpillar feeds on species of *Aristolochia*. The chrysalis has been compared in shape to a cluster of flowerbuds.

**polyphontes** see under **Papilio memnon**

**Pachlioptera** see under **Pachiopta**

**Pachypodistis** Hampson (PYRALIDAE)
Two species only are known in this S. American genus of Pyralid moths.

**goeldii** Hampson
Venezuela, Brazil, Peru 28–48 mm. (1·1–1·89 in.)
The females are much larger than the males. Striking dark red fore- and hindwings have two lines across, and the moth has a flat body, very different looking from the typical, rather delicate Pyralid. The pupal cases are curious and look like mussel shells *(Mytilus* sp). No details are known of its life history.

**Pachysphinx** Rothschild & Jordan (SPHINGIDAE)
The range of this genus of hawk-moths includes S. Canada, the U.S. and higher elevations in N. Mexico. There are 2 species.

**modesta** Harris BIG POLAR SPHINX
Canada, U.S. 100–115 mm. (3·94–4·53 in.)
This is a generally common species in the U.S. whether in the temperate Atlantic States or in subtropical Florida, but it occurs only sporadically in much of Canada. The larval foodplants are poplar *(Populus)* and willow *(Salix)*. The rather larger, similarly patterned *P. occidentalis* Edwards replaces *P. modesta* in the S.W. U.S. – both are similar in pattern to the Old World *Laothoe populi* but are much larger and have the front half of the hindwings reddish purple, unlike those of *L. populi* which are brownish orange basally.

**occidentalis** see under **P. modesta**

**pactolicus** see **Amphicallia**
**padma** see **Satyrus**
PAGE BUTTERFLIES see **Battus, Eurytides, Marpesia, Metamorpha, Papilio, Philaethria**
**pagenstecheri** see **Heliconisa**
PAINTED
APPLE MOTH see **Orgyia anartoides**
BEAUTY BUTTERFLY see **Cynthia virginiensis**
LADY BUTTERFLIES see **Cynthia**

**Paiwarria** Kaye (LYCAENIDAE)
There is a single species in this S. American genus.

**venulius** Kaye BRIGHT BLUE NYMPH
Tropical S. America 29–33 mm. (1·14–1·3 in.)
The male of this species is one of the most intensely blue species of Lycaenidae, whereas the female is a much less colourful greyish blue. The undersurface of the wings resembles that of a 'brown' (Satyrinae) with which it sometimes flies; the flight too is said to be similar.

**palaemon** see **Carterocephalus**
**palaeno** see **Colias**
PALAENO SULPHUR YELLOW see **Colias palaeno**

**Palaeochrysophanus** Verity (LYCAENIDAE)
Two species are placed in this genus of butterflies. Its range includes much of Europe, the Middle E. and central Asia E. to Siberia.

**hippothoe** Linnaeus PURPLE-EDGED COPPER
Europe, temperate Asia to Siberia 28–30 mm. (1·1–1·18 in.)
The various subspecies of this widespread butterfly have different habitat preferences. Some are restricted to meadows and bogs at low or moderate elevations; others fly only above 1520 m. (5000 ft) in Alpine meadows. Although *P. hippothoe* is not now regarded as a British species, it was recorded as such in the year 1666 by Merrett, a contemporary of Samuel Pepys, and may be yet another copper to have become extinct in Britain during the last few hundred years (see *Lycaena dispar)*. The female is orange-brown above, with several dark spots on the forewing and a marginal row of dark spots on the hindwing. **246**

**Palaeoses** Turner (PALAEOSETIDAE)
A genus of moths related to Hepialids, found only in Australia. They do not have a proboscis or mandibles and thus cannot feed in the adult stage. The forewing venation is slightly reduced.

**scholastica** Turner
Australia 16–18 mm. (0·63–0·71 in.)
Only known from the rain-forests of S. Queensland this species has been collected by sweeping the foliage with a net. The life history is unknown. The moth is a yellow brown, with some white spots near the apex, and grey brown hindwings.

**Palaeosetidae**
This small family of micro-moths is related to the Swift moths (Hepialidae). Three genera are at present known from India, Formosa and Australia. The hindwing lacks the frenulum found in the more advanced species, but they have the fore- and hindwings linked by a lobe from the forewing (jugum) as the Hepialidae, also regarded as a primitive family. See *Palaeoses*.

**palamedes** see **Papilio**
PALAMEDES SWALLOWTAIL see **Papilio palamedes**
PALE
ARCTIC CLOUDED YELLOW see **Colias nastes**
CLOUDED YELLOW see **Colias hyale**
CLOUDED YELLOW, EASTERN see **Colias erate**
GRASS-BLUE see **Zizeeria maha**
GREEN TRIANGLE see **Graphium eurypylus**
PROMINENT see **Pterostoma palpina**
SAPPHIRE see **Iolaus lalos**
TUSSOCK MOTH see **Dasychira pudibunda, Halisidota tessellaris**
**pales** see **Boloria**

**Palla** Hübner (NYMPHALIDAE)
A small genus of butterflies related to *Charaxes*. They are fast and powerful fliers. Most species have short tails on the hindwings, similar to species of *Charaxes*. The species in the genus have frequently been listed under the generic name *Charaxes*.

**ussheri** Butler
Africa 65–80 mm. (2·56–3·15 in.)
The male is blackish on the forewing with a white band through the middle which is broader towards the hind margin. This band continues as a short white mark on the hindwing but changes to deep reddish orange band across the wing to the hind margin and on the tails of the hindwing. The rest of the hindwing is brownish. It has 2–3 small eyespots on the hindwing margin. The female is orange-brown with a whitish band across both wings. It flies in the forests of the Congo, W. Africa, W. Uganda and W. Tanzania throughout the year. The caterpillar feeds on *Porana densiflora* (Convolvulaceae) and *Toddalia aculeata* (Rutaceae).

**pallene** see **Iolaus**
**pallicincta** see **Schinia**
PALLID NAWAB see **Polyura arja**
**pallidactyla** see **Platyptilia**
**palinurus** see **Papilio**
PALM MOTHS see **Tirathaba**
PALM
KING BUTTERFLY see **Amathusia phidippus**
TREE NIGHT FIGHTER see **Zophopetes dysmephila**
PALMFLY BUTTERFLIES see **Elymnias**
**palmii** see **Paranthrene**
**palpina** see **Pterostoma**

**Palpita** Hübner: syn. *Margaronia* (PYRALIDAE)
A large genus of Pyralid moths. Many of the species are white with beautiful lustrous or translucent wings which look like very delicate porcelain.

**unionalis** Hübner
Europe including Britain, Asia, Africa 25–29 mm. (0·98–1·14 in.)
A white moth with yellow edges on the forewing. The moths have been recorded in many countries but in the Orient there are several species which may have been confused with this. Common in the warmer parts of Europe and N. Africa this moth does not survive the winter in N. Europe where it is an occasional immigrant. **56**c

**paludum** see **Buckleria**

**Palustra** Bar (ARCTIIDAE) AQUATIC TIGER-MOTHS
About 12 species of mostly brown moths are placed in this tropical and subtropical S. American genus. The caterpillars of at least some species are adapted to life under water where they feed on water plants. The densely packed hairs of the caterpillar are capable of trapping enough air to allow long periods of submergence.

**laboulbeni** Bar
Tropical S. America 29–45 mm. (1·14–1·77 in.)
This is a sombrely coloured moth with yellowish brown forewings and white male, or pale brownish yellow female, hindwings. The remarkable caterpillars feed on water-plants of the genus *Mayaca*, and swim like a worm, coming rarely to the surface. They walk with difficulty out of the water and frequently lose their balance, but they are capable of living without water if necessary. The cocoon floats on the surface of the water and is often joined to other cocoons.

**Pampa** Walker: syn. *Harrisina* (ZYGAENIDAE)
A genus of moths with species in N. and S. America.

**americana** Guérin EASTERN GRAPELEAF SKELETONISER
U.S. 18–25 mm. (0·71–0·98 in.)
The wings are blackish-green, iridescent, with an orange collar. This is the eastern species which feeds on the leaves of grapes. The caterpillars, which are whitish with black marks at the base of the hair tufts, live in groups under the leaves and are covered with stinging hairs, a useful protection against possible predators. The moth flies from May to August.

**brillians** Barnes & McDonnough WESTERN GRAPE LEAF SKELETONISER
U.S., Mexico 17–25 mm. (0·67–0·98 in.)
The metallic blue to greenish blue, long, rather narrow wings makes this day-flying moth conspicuous. The yellow caterpillar has black bands on the thorax and each body segment has long tufts of stinging hairs. The caterpillars are about 13 mm. (0·51 in.) long and gregarious, feeding in groups on the underside of leaves of wild and cultivated grapes *(Vitis)*. Generally there are 3 broods in a year. The species is found from S. California through Utah and into Mexico.

**pamphilus** see **Coenonympha**

**Panacea** Godman & Salvin (NYMPHALIDAE)
The butterflies in this genus are found from the tropics of C. America into S. America, being confined mostly to the tropical areas. They usually have green bands with a strong metallic iridescence on the upper surface of the wings. The undersides are dark, camouflaging, browns and blacks.

**procilla** Hewitson
S. America 60–68 mm. (2·36–2·68 in.)
This species is relatively common in Brazil and Colombia. The females are larger than the male, lighter coloured below and are rare in collections. Little is known about the habits of this species, the females perhaps keeping to areas of denser vegetation than the males and avoiding collectors. **209**h

**prola** Doubleday & Hewitson
S. America 80 mm. (3·15 in.)
This species is illustrated to show the striking difference in pattern between the upperside and the underside. It is widespread in Brazil and Ecuador. Nothing is known of its biology. **217**a, d

**Panacra** Walker (SPHINGIDAE)
This is a small genus of about a dozen European and Asiatic species of hawk-moths.

**pulchella** Rothschild & Jordan
New Guinea 50–70 mm. (1·97–2·76 in.)
The forewing of this moth is cryptically patterned in black and green, with 2 yellowish white lines extending, respectively from the apex towards the base and the anal angle of the wing. The hindwing is a bright orange with a dark brown outer margin. It is covered by the forewing when the moth is at rest, but is suddenly revealed if the moth is disturbed and takes flight. The disconcerting effect of this 'flash-coloration' on a potential predator is recognized as a defence mechanism in many insects.

**panariste** see **Anaea**
**pandarus** see **Hypolimnas**
**pandione** see **Appias**
**pandora** see **Coloradia, Pandoriana**
PANDORA MOTH see **Coloradia pandora**

**Pandoriana** Warren (NYMPHALIDAE)
This butterfly genus is related to the Argynnid fritillaries. The species are found in Europe and temperate Asia and N. Africa.

**pandora** Denis & Schiffermüller CARDINAL BUTTERFLY
S. Europe, N. Africa, S. Russia to Iran and W. Pakistan 64–80 mm. (2·52–3·15 in.)
Broadly similar to *Argynnis paphia* but with the underside of the forewing rose-red with greenish apex. Hindwings pale green on the underside. The butterflies are out in June and July in Europe, but in N. Africa there are 2 broods, May-June and August-September. The caterpillar feeds on violets, especially *Viola tricolor*. The Cardinal is widely distributed in E. Europe and is common in Morocco, Algeria and Tunisia, occurs on the Canary Islands and has been reported from Tenerife. **192**

**pandorus** see **Eumorphia**
PANDORUS SPHINX see **Eumorphia pandorus**

**Panolis** Hübner (NOCTUIDAE)
There are 3 species in this genus of moths. It is represented in Europe and temperate Asia, from the Atlantic to the Pacific.

**flammea** Denis & Schiffermüller PINE BEAUTY
Europe, W. Asia 31–34 mm. (1·22–1·34 in.)
The longitudinally striped caterpillars of this species are well camouflaged amongst the needles of pine trees (*Pinus sylvestris* in Britain) their sole food source. The adult is cryptically, yet beautifully patterned. **405**

PANSY BUTTERFLIES see **Junonia, Precis**
**panteucha** see **Hednota**

**Panthiades** Hübner (LYCAENIDAE)
This is a large, tropical American genus of tailed hairstreak butterflies.

**m-album** Boisduval & Leconte WHITE-M HAIRSTREAK 25–32 mm. (0·98–1·26 in.)
E. N. America (especially S.) to the mountains of Mexico & Guatemala
The male of this species is a brilliant, iridescent blue above, with black margins. There is a characteristic, white 'M' mark on the under surface of the hindwing. A marking similar to this has been looked at from a different viewpoint in the European and Asiatic Lycaenid *Strymonidia w-album*. The range of this species is sharply disjunct; the E. N. American subspecies, *P. m-album m-album*, is not found S. of Texas; the C. American subspecies, *P. m-album moctezuma* Clench, ranges from C. Mexico S. to Guatemala at elevations of between 1520–2440 m. (5000–8000 ft). The foodplant is probably oak (*Quercus*).

**Pantoporia** Hübner (NYMPHALIDAE)
An Oriental genus of butterflies with species in India, S.E. Asia and Indonesia. A few species occur in China. These butterflies tend to fly rather like the Sailers (*Neptis*) but more powerfully, with a few rapid wing movements followed by a short glide. Their colours and pattern tend to camouflage them in the chequered sunlight of the forests where they are generally found.

**eulimene** Godart
Indonesia 88 mm. (3·46 in.)
This species has been collected on some of the Indonesian islands including Ceram and Buru. There are several related, similar looking, species and a number of subspecies of *P. eulimene* have been described. The sexes are similar. The orange stripes of the upperside are replaced on the underside of the wing by a yellowish stripe. The biology does not appear to be known. **216c**

**hordonia** Stoll BURMESE LASCAR
India, Pakistan, S.E. Asia, Formosa, China 45–55 mm. (1·77–2·17 in.)
This butterfly is a rich orange above with a broad black margin. The underside is a paler yellowish colour marked with brown lines. Several subspecies have been described. Common in forest paths in hilly country. The caterpillar feeds on species of *Acacia* (Leguminosae).

**paphia** see **Argynnis**

**Papilio** Linnaeus (PAPILIONIDAE) SWALLOWTAILS
This is an almost cosmopolitan, but mostly tropical genus of over 200 species. Many of the species have a tailed hindwing, but the most distinctive common feature of these handsome butterflies is the concave shape of the inner margin of the hindwing. Several species are known to mimic species of Danainae and Papilionid species of the tribe Troidini (see, for example, *Parides*). However, at least one species of *Papilio* is itself chemically protected (see *P. antimachus*). The caterpillars feed on a variety of plants, but chiefly on species of Rutaceae (including *Citrus*), Lauraceae and Umbelliferae. Those which feed on *Citrus* are sometimes orchard pests, especially in South Africa. This is a group of butterflies popular with biologists, collectors, interior decorators and artists. A close look at a Dutch still-life painting will often reveal an elegantly poised Swallowtail.

**aegeus** Donovan ORCHARD SWALLOWTAIL
Australia, New Guinea and associated tropical islands 78–90 mm. (3·07–3·54 in.)
This species is often common in populated areas of Australia where the adult butterflies feed from garden flowers. The caterpillars feed on *Citrus* foliage, but are seldom common enough to be considered pests. Early stage caterpillars resemble bird-droppings like the caterpillars of many other species of *Papilio*, a characteristic which probably affords some degree of protection from predators. There are 2 forms of the fully grown caterpillar: one with dark brown lateral stripes, the other with brownish white lateral stripes. The chrysalis varies in colour and often matches the colour of its surroundings. The adult female of *P. aegeus* has several distinct colour forms. **106d**

**agestor** Gray
N.W. India to S. China, Formosa, Malaya 80–96 mm. (3·15–3·78 in.)
This is one of several mimics of the species *Danaus sita* and its flight and resting position are also similar. It flies in open woods where the male is one of many 'territorial' butterflies which patrol a particular tract of territory and chase off intruders, unless they are females of its own species. The caterpillars will feed on the foliage of *Machilus* (Lauraceae).

**alexanor** Esper SOUTHERN SWALLOWTAIL
Mediterranean France to the Balkans, Middle-E., Iran, Turkestan (U.S.S.R and China) 44–75 mm. (1·73–2·95 in.)
The French Riviera resort of Nice has the distinction of being the source of the specimens on which the original description of this species was based. In Europe, it is primarily a butterfly of the Alpine foothills where it occurs up to about 1220 m. (4000 ft) and often can be seen feeding at thistle flowers. It is most likely to be confused with *P. machaon* from which it differs in the yellow base to the forewing and the dark band near the base of the hindwing. The caterpillar feeds on species of Umbelliferae.

**anactus** Macleay DINGY SWALLOWTAIL
Australia 55–60 mm. (2·17–2·36 in.)
This rather small species of *Papilio* is entirely restricted to E. Australia. Though 'dingy' by name, it is by no means an unattractive butterfly. Males of this species will patrol a tract of sunlit territory, and return to it after having been disturbed. The flight of this species is normally leisurely in character. The caterpillars feed on the foliage of cultivated, imported *Citrus* (Rutaceae) and on several native species of the same family. The colour of the chrysalis tends to match the colour of its background. **103a**

**anchisiades** Fabricius ORANGE DOG
C. & tropical S. America U.S. 65–95 mm. (2·56–3·74 in.)
One subspecies of this tail-less butterfly is established along the Rio Grande in Texas and occasionally strays to the mid-western States of the U.S. Both sexes mimic species of *Parides*. Males are black, with a red or purple lobate, central marking and white, marginal markings on the hindwing as in many species of *Parides*. Females are similar to the males, but have pink markings on the hindwing and white spots on the forewing. The caterpillars are known popularly as Orange Dogs, a name also given to caterpillars of *P. cresphontes*. They are often common in *Citrus* orchards.

**androgeus** Cramer QUEEN PAGE
C. & tropical S. America, Mexico 85–105 mm. (3·35–4·13 in.)
The tailed males of this species are black, with a broad yellow band across both pairs of wings. Females, in contrast, are somewhat Heliconiine in appearance and lack a tail to the hindwing; they are chiefly dark, iridescent green or blue, with a white or yellow patch near the middle of the forewing. The habits of the sexes also differ; males frequently circle at tree-canopy level and will venture down to drink from muddy patches, whereas females seldom emerge from the forests. Caterpillars will feed on the foliage of *Citrus* trees, especially orange trees.

**antimachus** Drury AFRICAN GIANT SWALLOWTAIL
Tropical Africa from Sierra Leone to W. Uganda 150–250 mm. (5·91–9·84 in.)
This enormous butterfly, described in 1782, is the largest of African butterflies and nearly as large as the Asian birdwings (*Ornithoptera*). It is a species of dense rain-forests. Males are not rare and can be seen congregating on wet mud in large numbers; females are seldom seen and are probably restricted to tree-canopy level. Unlike most species of butterflies, the females are smaller than the males. It has been shown recently that adults contain cardenolides (vertebrate toxins) in their tissues and that *P. antimachus* is therefore a warningly-coloured species like the similarly patterned but much smaller species of *Acraea*. One male specimen was found to have five times the concentration of cardenolides as a typical specimen of *Danaus plexippus*. **100b**

**arcturus** Westwood BLUE PEACOCK
Kashmir, N. India, Sikkim, Burma, China 110–130 mm. (4·33–5·12 in.)
This brightly coloured species is a butterfly of cool-temperate forests, upwards from 1520–3350 m. (5000–11,000 ft) in the Himalayas. It is similar in colour pattern to the tropical lowland species *P. paris* but differs chiefly on the hindwing which has the blue (violet blue or greenish blue) patch restricted to the front half of the wing and has one or more black-centred, purple spots at the outer margin. Adults are attracted by flowers of *Hibiscus* and *Lantana*. The caterpillars feed on species of Rutaceae.

**aristodemus** Esper
Antilles and S.E. U.S. 80–114 mm. (3·15–4·49 in.)
The N. American subspecies, Schaus's Swallowtail *P. aristodemus ponceanus*, is very rare after having been severely reduced in numbers by the 1938 hurricane and possibly by overenthusiastic collectors. It is similar in colour pattern to *P. cresphontes* but the broad, basal, yellow band on the hindwing has less clearly defined edges and there is a rather larger area of blue at the rear angle of the same wing. The caterpillar feeds on torchwood (*Amyris*).

**ascolius** Felder
C. & tropical S. America 85–100 mm. (3·35–3·94 in.)
Like *P. zagreus*, this species is another remarkable mimic of species of Ithomiinae but it differs in the replacement of the orange markings by yellow except in the rear part of the central patch on the hindwing. Nothing is known about the caterpillar.

**bianor** Cramer
N.E. India, Burma, Indo-China, China, Japan 85–130 mm. (3·35–5·12 in.)
The female (not illustrated) of this large butterfly is a pale, olive-green above on the forewing; the hindwing is similar to that of the male, but the sheen is olive-green. The illustrated specimen is a male from Japan of the subspecies *P. bianor dehaani* Felder. Specimens of *P. bianor* have been caught at elevations of up to 1830 m. (6000 ft) in the Himalayas. The caterpillars are green, with 2 dark transverse bands; they feed on species of the family Rutaceae. **103c**

**blumei** Boisduval
Sulawesi (Celebes) 90–120 mm. (3·54–4·72 in.)
The most striking feature of this richly coloured butterfly is the iridescent, greenish blue tail of the hindwing. Females differ little from the illustrated male but are usually slightly lighter in ground colour. Most specimens of this species have been captured in N. Sulawesi. **103f**

**bolina** see under **Euploea core**

**bootes** Westwood
N.E. India to China Formosa 90–115 mm. (3·54–4·53 in.)
This is a mimic of *Parides philoxenus*, at least in part of its range. The Uraniid moth *Epicopeia polydora* is also part of this mimetic complex. Formosan specimens of *P. bootes* have purple, not the usual red markings on the hindwings.

**castor** see under **Euploea core**

**clytia** Linnaeus COMMON MIME
Sri Lanka, India, N. China, Malaya, Philippines, Lesser Sundas 85–100 mm. (3·35–3·94 in.)
Both the male and the female of this species are represented by 2 quite dissimilar colour forms, each of which mimics a species of Danainae in colour pattern and mode of flight. One of these mimics *Euploea crameri* and *E. core*, the other mimics the blue-grey *Danaus hamata* and *D. limniace*. Commonly, only the female of a palatable swallowtail species will mimic a distasteful model, while the male retains what is probably an ancestral colour pattern; the Common Mime is atypical in this respect. The caterpillar feeds on species of Lauraceae; it resembles a bird-dropping in the early stages, but is conspicuously black, white and red when fully grown. The chrysalis is well disguised to the human eye as a piece of broken twig. **99**g

**cresphontes** Cramer ORANGE DOG, GIANT SWALLOWTAIL 100–138 mm. (3·94–5·43 in.)
Canada (Ontario), U.S. to Costa Rica
The caterpillars of this species are dark brown, with whitish markings and orange osmeteria and are popularly known as Orange Dogs, see also the Orange Dog butterfly, *P. anchisiades*. They are sometimes pests in Florida *Citrus* orchards but will also feed on prickly ash *(Zanthoxylum)*, gas plant *(Dictamnus)* and hop tree *(Ptelea)*. This is the largest species of butterfly to occur regularly in N. America. *P. thoas* Linnaeus, a similarly patterned C. and S. American species which occurs at times in Texas, is the only N. American species to equal it in size. **99**h

**cynorta** Fabricius
W. Africa, the Congo Basin, E. Africa, Ethiopia 70–85 mm. (2·76–3·35 in.)
The female of this forest butterfly has several colour forms, mostly brown with violet-white markings, which mimic corresponding forms of *Bermastistes epaea*, a species of Acraeinae. The male is not mimetic and is similar in colour pattern to *P. zoroastres*, but differs in having a row of white spots close to the outer margin of the hindwing. The caterpillar feeds on *Vepris* (Rutaceae).

**dardanus** Brown MOCKER SWALLOWTAIL
Africa, S. of the Sahara 90–105 mm. (3·54–4·13 in.)
There are several female forms of this remarkably variable and biologically famous butterfly, most of which mimic species of *Danaus* or *Amauris* in wing-shape and colour pattern, including the common *Danaus chrysippus* which has several other mimics. The invariably non-mimetic males have retained the typical *Papilio* tails and are scarcely recognizable as representatives of the same species as the females. *P. dardanus* females seldom leave the shade of the forest, whereas males will fly in bright sunshine along forest trails and margins. The caterpillars feed on the foliage of species of *Clausena*, *Teclea*, *Toddalia* and introduced *Citrus*, all of which are species of the family Rutaceae. A great deal of work has been carried out on the genetics of this species and the evolutionary problems associated with mimicry, especially by C. A. Clark and P. M. Sheppard in Liverpool, England. **105**b, d, f

**demetrius** Cramer
China, Japan 85–128 mm. (3·35–5·04 in.)
This is an essentially Japanese butterfly, but a few specimens have been recorded from China. The male is black above, with a green iridescence; the females paler in colour and usually with a few large, black-centred, red spots along the outer margin of the hindwing. The red-spotted under surface (illustrated) is typical of many species of the genus Parides which are generally considered to be distasteful to predators, and it is probable that *P. demetrius* is a mimic of *Parides philoxenus* in part of its range. **98**

**demodocus** Esper CITRUS SWALLOWTAIL or CHRISTMAS BUTTERFLY 90–115 mm. (3·54–4·53 in.)
Tropical & S. Africa, S. Arabia, Madagascar
Although *Citrus* trees are not native to Africa, they are now the main foodplant of the caterpillar of this species. It is a generally common species, and has become a pest in South Africa where *Citrus* is grown commercially. The caterpillars will also eat the foliage of wild, native species of Rutaceae and Umbelliferae and will occasionally feed on *Cosmos* and other plants which produce chemically similar aromatic compounds. Caterpillars feeding on *Citrus* are usually different in colour from those feeding on species of Umbelliferae. Adult butterflies are often common during the Christmas period. *P. demodocus* is replaced in the E. by the Asiatic and Australian *P. demoleus*.

**demoleus** Linnaeus CHEQUERED or LIME SWALLOWTAIL 80–100 mm. (3·15–3·94 in.)
Iran, India, Sri Lanka, S.E. Asia to New Guinea and N. Australia
There are some unexplained gaps in the distribution of this widespread and possibly migratory butterfly. The range of species with special foodplant needs is dependent on the distribution of that foodplant; but the caterpillars of this species are known to feed on cultivated *Citrus* foliage and also on several species of legumes, so that the apparent absence of this butterfly from Sulawesi (Celebes) and the Moluccas is probably the result of some other factor. This is a common species in some localities, especially where flowering legumes provide a source of nectar for the adults. It is replaced by the similarly patterned *P. demodocus* in Africa. The illustration represents subspecies *P. demoleus demoleus*, **103**e

**dravidarum** Wood-Mason MALABAR RAVEN
India 80–100 mm. (3·15–3·94 in.)
Both sexes of this species are considered to be mimics of the common Indian Crow butterfly, *Euploea core*, especially when in flight. It tends to keep to shaded areas, and occurs most commonly in the lowland forests of the west coast of India and S. India. The caterpillar feeds on a species of *Glycosmis* (Rutaceae).

**euterpinus** Godman & Salvin
Tropical W. S. America 72–94 mm. (2·83–3·7 in.)
The unusual colour pattern of this species is probably most similar to that of *Pereute* a genus of Pieridae, but the general pattern of brown above, with a large orange patch on the forewing is similar to that in a colour form of several species of *Heliconius*, one or more of which may be a model for *P. euterpinus* and provide it with a degree of protection from predators. The female differs from the illustrated male in the yellowish orange, not yellow band on the forewing. **106**f

**fulleborni** Karsch
Tanzania 75–86 mm. (2·95–3·39 in.)
This species, like *P. sjoestedti*, is known only from the mountains of Tanzania. The male resembles *P. zoroastres*; it is black with a broad, white, transverse band on the hindwing, which is continuous with a similar but narrower band across most of the forewing. The females are quite different in pattern and are mimics of *Amauris* species; they are brown, with bluish white spots on the forewing, a row of white spots along the edge of the hindwing, and a large, orange area at the base of the hindwing.

**garamas** see under **P. homerus**

**glaucus** Linnaeus TIGER SWALLOWTAIL
Canada, U.S. 100–165 mm. (3·94–6·5 in.)
This widespread, N. American swallowtail is generally quite common and is often seen feeding on garden flowers, even in the centre of large towns. In the S. of its range, as many as half of its females are brown in ground-colour, not yellow as in the illustration. The caterpillar feeds, often high up, on a variety of trees, including cherry *(Prunus)*, poplar *(Populus)*, ash *(Fraxinus)* and the hop tree *(Ptelea)*. There are 2 orange, black-centred eyespots near the head of the caterpillar which are thought to frighten possible enemies. **101**b

**helenus** Linnaeus RED HELEN
Sri Lanka, India, Burma, Thailand Malaya, Japan 110–130 mm. (4·33–5·12 in.)
The upper surface of the wings is chiefly greenish black in both sexes, but with a large pale yellow patch on the hindwing and, in some specimens, several dark red, crescentic marks near the base of the short hindwing tail. The yellow marking is conspicuous in flight, but is covered by the forewings when the butterfly is at rest. It is possible that this is an example of 'flash-coloration' and that the sudden display of the yellow patch may momentarily disconcert an enemy. Often common in and near forested areas the butterfly will also visit garden flowers and, as a caterpillar, feed on garden *Citrus* trees.

**hesperus** Westwood
Tropical and sub-tropical S. Africa 100–125 mm. (3·94–4·92 in.)
Most specimens of this species have paler yellow markings than in the imposing illustrated specimen, while females have duller yellow markings on a dark brown background, with an orange spot at the rear of the hindwing. The curious name *horribilus* was given by Butler to the Cameroun subspecies of *P. hesperus*. Female butterflies seldom leave the forest and are seen rarely. **106**g

**homerus** Fabricius
Jamaica 125–140 mm. (4·92–5·51 in.)
This huge butterfly is a close relative of the continental American *P. garamas*. The forewings are black above, with an oblique yellow band, a yellow central bar and a few yellow apical spots; the hindwings are also black, but with a broader, transverse yellow band, bordered distally with a row of blue spots, followed by a row of 3–4 red dashes at the base of the hindwing tails – one very short, one of moderate length.

**iswaroides** Fruhstorfer
Malaya, Sumatra 80–90 mm. (3·15–3·54 in.)
The forewings of the male are a dark greenish black, with narrow marginal dashes; the hindwing is black, with a short broad tail, a large creamy white patch at the front edge of the wing and usually with 2 black-centred, dull red spots near the base of the tail. This species is closely related to *P. helenus*, which it matches closely in colour pattern above but differs below in the reduction of the number of red spots at the outer margin of the hindwing.

**laglaizei** Depuiset
Waigeu Island, New Guinea 75–98 mm. (2·95–3·86 in.)
This species is a mimic of the day-flying moth *Alcides agathyrsus*, a member of the family Uraniidae, many of whose species are brightly coloured and probably distasteful to predators. The resemblance between the 2 species is very close, especially on the under surface where the red markings on the inner edge of the butterfly's hindwing match, when at rest, the red under surface of the moth's abdomen. Both species fly together, usually at tree-canopy level. The butterfly is invariably the commoner – an unexpected ratio if *P. laglaizei* is the mimic and *A. agathyrsus* the model. The population level of a mimic is kept low by predators who are able to expose the mimic's deception when the latter's numbers are sufficiently great to allow the predator the chance to learn that distastefulness is not almost invariably associated with a particular colour pattern. **105**g

**leucotaenia** Rothschild CREAM-BANDED SWALLOWTAIL 90–100 mm. (3·54–3·94 in.)
E. Africa
This rare butterfly is restricted in range to areas of montane forests where it is most often encountered along pathways and in clearings. The greenish white, transverse band on the upper surface of the black wings is a distinctive characteristic of this species. Nothing is known of its life history.

**lycophron** Hübner
Tropical S. America 82–95 mm. (3·23–3·74 in.)
This is a minor pest of lime and orange plantations where the caterpillar will feed on these and other *Citrus* trees. Males of this butterfly have black wings, with a broad, oblique yellow band crossing both wings and with a row of yellow crescents along the outer edge of the hindwing. Females are black towards the margins of the wings but are otherwise weakly iridescent greenish blue; they have a row of marginal yellow crescents or spots along the outer margin of both pairs of wings. The male, in particular, is somewhat similar to *P. androgeus*.

**machaon** Linnaeus SWALLOWTAIL
N. Africa, Europe including Britain, temperate Asia, Japan, N. America 75–100 mm. (2·95–3·94 in.)
This is one of the most widely distributed butterflies of the N. temperate regions of the world. It is typically a species of meadowlands and mountain-sides up to about 4270 m. (14,000 ft). *P. machaon* was once fairly common in parts of England but is now restricted as a breeding species to the fenlands of the county of Norfolk and a single locality in Cambridgeshire. In continental Europe, it is commonest in the C. latitudes. It varies geographically in colour pattern to some extent, and there is some seasonal and individual variation. The caterpillar feeds on the feathery foliage of fennel, *Foeniculum*, wild carrot, *Daucus*, and other species of the family Umbelliferae. **92**

**mayo** Hewitson
Andaman Islands 100–130 mm. (3·94–5·12 in.)
The chiefly greenish black female of this species mimics a subspecies of the supposedly unpalatable species *Parides rhodifer*, a relative of the illustrated *Parides coon*. The male is quite different from the female; it lacks a tail on the hindwing and is black above with a broad, pale blue band across the middle of the hindwing.

**memnon** Linnaeus GREAT MORMON
Sri Lanka, India, Japan, S. China to Sulawesi (Celebes), Philippines, Moluccas 120–150 mm. (4·72–5·91 in.)
This fairly common butterfly is unusual for its great variety of female forms. One of these, when in flight, is said to resemble females of the birdwing *Troides brookiana* and another mimics a female form of *Parides coon;* others mimic *Pachliopta aristolochiae* or *Pachliopta polyphontes*. Both tailed and tail-less forms of the female exist. Males never have a tailed hindwing and are normally black, with a slight blue or green sheen and with grey stripes parallel to the wing veins. Females are normally confined to the forest, whereas males will wander into more open country and will visit garden flowers.

**montrouzieri** Boisduval
New Caledonia, Loyalty Islands 70–85 mm. (2·76–3·35 in.)
This butterfly is very similar to the more commonly illustrated and slightly larger *P. ulysses* which it replaces in the above islands. It differs from *P. ulysses* in the less strongly dentate outer margin of the blue area on the forewing. Females differ from the illustrated male in the brown ground-colour of the upper surfaces of the wings and the reduced, less brilliant, blue areas. The under surface, like that of *P. ulysses*, is cryptically patterned in brown. **105**a

**multicaudata** Kirby THREE-TAILED SWALLOWTAIL
W. & S. U.S., W. Canada (rarely), Mexico 98–130 mm. (3·86–5·12 in.)
This is sometimes called the Two-tailed Swallowtail because of the shortness of one of its 3 tails. It is similar in colour pattern to the Tiger Swallowtail, *P. glaucus*, of N. America which has a single, long tail to the hindwing. Its caterpillar feeds on the foliage of a number of deciduous trees, including ash *(Fraxinus)*, oak *(Quercus)*, wild cherry *(Prunus)* and privet *(Ligustrum)*.

**nireus** Linnaeus NARROW BLUE-BANDED SWALLOWTAIL 80–120 mm. (3·15–4·72 in.)
Tropical and subtropical Africa (including Ethiopia) and the Comores
This is a typical 'gloss-Papilio'. Its wings are black above, with a glossy, iridescent, blue and greenish blue (or blue and violet in the subspecies from the Comores Islands) transverse band on both pairs of wings. The hindwing has a very short tail. *P. nireus* flies in both thick scrubland areas and in the forests. Its caterpillars feed on a variety of species of the family Rutaceae, including *Citrus* trees.

**nobilis** Rogenhofer
Kenya, Uganda, Tanzania, E. Congo Basin, S. Sudan 85–110 mm. (3·35–4·33 in.)
This is a butterfly of mountain forests where it flies at elevations of up to about 2440 m. (8000 ft). The unusually coloured, pale, almost patternless wings of this species are a particularly distinctive feature in the male. Females have more extensive brown markings, but are not heavily marked. In the S.W. Ugandan subspecies, the brown markings are replaced by grey in the male. Caterpillars feed on *Wahlenbergia*. **106**b

**ophidicephalus** Oberthür EMPEROR SWALLOWTAIL
Africa S. of the Sahara 90–135 mm. (3·54–5·31 in.)
The illustrated male was captured between 1220–1520 m. (4000–5000 ft) on the slopes of Mount Kilimanjaro. Females are paler in ground-colour above, and the yellow markings are deeper in tone. This is a forest species throughout its extensive range. The caterpillar, like that of the related *P. demodocus*, feeds on the leaves of *Citrus* and other genera of the family Rutaceae. **103**

**palamedes** Drury PALAMEDES SWALLOWTAIL
U.S., Mexico 112–140 mm. (4·41–5·51 in.)
This large, slow-flying ally of *P. troilus* is particularly common in S.E. U.S., in wooded, swampy areas where the caterpillars' foodplants occur (Red bay, *Persea; Sassafras;* and Sweet bay, *Magnolia*). The wings are black in the male, dark brown in the female. Above, the most obvious markings on the forewing are a row of large, yellow patches parallel to the outer margin of the wing; the hindwing has a broad, transverse, yellow band in the middle, yellow crescents along the outer margin, and a black-centered, blue and orange spot at the base of the tail.

**palinurus** Fabricius
Burma to Borneo and the Philippines 75–90 mm. (2·95–3·54 in.)
This is similar to the rather larger *P. blumei*, but the transverse band on the wings is a more yellowish green and the hindwing tail is black unlike that of *P. blumei*. The under surface, like that of *P. blumei* is cryptically patterned in brown. This is typically a woodland species, but will visit garden flowers. Its caterpillar is not known.

**paradoxa** Zinken GREAT BLUE MIME
N. India to Borneo 70–115 mm. (2·76–4·53 in.)
This is a species of open forest, at a wide range of altitudes. There are 2 chief forms of both sexes; one with white patches on the wings and one without, each of which closely mimics a different species of *Euploea*. These 2 colour forms occur throughout the range of this species and in each of the several subspecies. The normal, unhurried flight of Danaines is also mimicked by *P. paradoxa*, but this type of behaviour is abandoned in favour of rapid flight if danger threatens. It is closely related to another mimic, *P. clytia*. **99**j

**paris** Linnaeus PARIS PEACOCK
India, Burma, S. China, Thailand, Laos, Vietnam, Sumatra, Java 85–135 mm. (3·35–5·31 in.)
This species belongs to the group of *Papilio* butterflies known as gloss-papilios, in which there are patches of glossy, metallic blue or green patches on the wings. A chiefly tropical and subtropical species of low altitude forests where it is sometimes very common, it also occurs at higher elevations in S.W. China. The males commonly visit flowers and congregate on wet mud. The illustrated male belongs to the S. Indian subspecies, *P. paris tamile*. Females are more yellowish in colour. The caterpillars of this species feed on a wide variety of plants, including *Citrus* and species of Umbelliferae. **103**d

**phorcas** Cramer GREEN-PATCH SWALLOWTAIL
Africa S. of the Sahara 80–100 mm. (3·15–3·94 in.)
This is a common species in the forests of E. Africa and the Congo. There are 2 common colour forms of the female: one like the male; the other dark brown, not black, with yellow markings. Low-growing flowers attract both sexes. The caterpillar feeds on the foliage of *Teclea*. **113**

**polymnestor** Cramer
Sri Lanka, S. India 100–130 mm. (3·94–5·12 in.)
Although a resident apparently only in Sri Lanka and S. India, stray specimens are occasionally seen in N. India. This is a common species of both forested and more open country and is a regular visitor to garden flowers. The illustrated male belongs to the subspecies *P. polymnestor parinda*. The female is usually dark greyish brown and the blue areas are much paler than the male or are replaced by yellowish white, especially in the forewing. In both males and females there is often a brownish red streak at the base of the forewing, a colour repeated at the base of both wings on their under surface. **103**b

**polytes** Linnaeus COMMON MORMON
Sri Lanka, India, China, Japan, Malay Archipelago to the Philippines, Moluccas 90–100 mm. (3·54–3·94 in.)
Like several other typical butterflies, this species has become adapted to an urban life, where the usual foodplant of the caterpillar is cultivated *Citrus*, especially orange and lime, although it also feeds on wild species of Rutaceae. There are 3 female colour forms of this species and hence the origin of its popular name: one resembles the male, a second mimics *Pachliopta aristolochiae*, and the third mimics *P. hector*. Males are nearly black above, with a creamy white, oblique band extending from the apex of the forewing to the middle of the inner margin of the hindwing.

**polyxenes** Fabricius BLACK SWALLOWTAIL
N., Central and tropical S. America 70–80 mm. (2·76–3·15 in.)
This is an extremely widespread and generally common species, with climatic tolerances which include the subarctic zone of Canada and Alaska and the tropics of the New World. Dark females of this species are considered to be mimics of the distasteful Pipe-vine Swallowtail, *Battus philenor*. The caterpillars, sometimes called Parsleyworms, Carrotworms or Celeryworms, feed on Umbelliferae, including cultivated carrot *(Daucus)*, celery *(Apium)*, and parsley *(Petroselinum)*, and are occasionally common enough to be classed as pests. The pattern of the upper surface of this butterfly is similar in general plan to that of its close ally *P. machaon* except that the central yellow band is reduced to a row of yellow spots on both wings, which are consequently chiefly dark brown or nearly black.

**rex** Oberthür REGAL SWALLOWTAIL
Cameroun, S. Sudan, Ethiopia, E. Africa 100–140 mm. (3·94–5·51 in.)
This large species is restricted to the highland forest zones, including the slopes of Mount Kilimanjaro. The illustrated female form is a mimic of the much smaller but similarly patterned *Danaus formosa*; another form of this species, which is dark reddish brown at the base of the forewing, mimics *Danaus mercedonia*. The caterpillar feeds on Cape Chestnut *(Calodendron)*. **105**e

**sjoestedti** Aurivillius KILIMANJARO SWALLOWTAIL
Tanzania 90–100 mm. (3·54–3·94 in.)
The mountain forests of Mount Kilimanjaro and Mount Meru between 2130–2740 m. (7000–9000 ft) are the only localities where this species is known to occur. The upper surface of the male is nearly black, with a white transverse band across both wings as in *P. zoroastres* but narrower. The female which has a yellow band on the hindwing is a mimic of *Amauris* species.

**slateri** Hewitson
N. India, Bhutan, Sikkim 66–84 mm. (2·6–3·31 in.)
This is a close relative of *P. agestor*. Like the latter it is mimetic of Danaine species. There is little difference in pattern between the sexes. Its caterpillars feed on species of Lauraceae. The under surface of the wings is brown, with whitish, ray-like markings at the outer margin of the hindwing, as well as an orange marking on the forewing corresponding to similar marking on the upper surface. **99**k

**thoas** Linnaeus CITRUS or ORANGE SWALLOWTAIL or KING PAGE 80–110 mm. (3·15–4·33 in.)
S. U.S. through Central and S. America to Argentina
S. Texas is the only recorded locality in N. America for this species. It is very similar in colour pattern to the widespread N. American Giant Swallowtail, *P. cresphontes*, except for small differences in the size of 2 of the yellow markings on the forewing. The caterpillars feed on *Citrus* and *Piper* foliage.

**toboroi** Ribbe
Solomon Islands 80–100 mm. (3·15–3·94 in.)
This close ally of the mimetic *P. laglaizei* is a much less strongly marked version of the latter above, and has an

orange-yellow spot at the inner angle of the tail-less hindwing. It is known only from the island of Bougainville where there is no species of *Alcides* (Uraniidae) comparable to the species which *P. laglaizei* mimics in New Guinea. It seems probable that *P. toboroi* was evolved from *P. laglaizei*-like ancestors and that, in the absence of a distasteful Uraniid model, and the associated advantage to the butterfly in mimicking it, the colour pattern has become modified.

**torquatus** Cramer
C. America including Mexico, tropical S. America 58–66 mm. (2·28–2·6 in.)
Except for the absence of hindwing tails, the female of this forest species is a good match for the female of the unpalatable *Aristolochia*-feeding Cattle Heart, *Parides anchises* Linnaeus. The male is non-mimetic and quite unlike the female in colour pattern. **106c, e**

**troilus** Linnaeus SPICEBUSH SWALLOWTAIL
S. Canada, U.S. 100–126 mm. (3·94–4·96 in.)
The female of this species is thought to be a mimic of the *Aristolochia*-feeding Pipe-vine Swallowtail, *Battus philenor*. The caterpillar feeds on Spicebush *(Benzoin)*, *Sassafras*, Sweet Bay *(Magnolia)* and *Zanthoxylum;* it has 2 small eyespots and 2 larger eyespots behind its head, giving it a fearsome appearance which may well deter predators. When provoked the caterpillar rears up in a snake-like attitude. **101a**

**ulysses** Linnaeus ULYSSES OR BLUE MOUNTAIN BUTTERFLY 100–135 mm. (3·94–5·31 in.)
Moluccas, New Guinea, Australia, Solomon Islands
The males of this richly coloured butterfly, a favourite of collectors, can be attracted to almost any blue object placed at the edge of a low altitude rain-forest area. The unfailing attraction of blue lures is apparently effective up to 30 m. (100 ft) away. *Hibiscus* flowers are also particularly attractive to this species. It is similar in pattern to the smaller *P. montrouzieri* but differs in the normally more sharply dentate edge to the outer margin of the blue area on the forewing. There are over a dozen described subspecies and they differ externally in details of the colour pattern. The caterpillar feeds on *Evodia* and *Citrus*.

**xuthus** Linnaeus
C. and N. China, Korea, E. Russia, Japan 80–110 mm. (3·15–4·33 in.)
Similar in colour pattern to *P. machaon*, but the dark markings are more extensive, especially those on the wing veins, and there are 3 black streaks parallel to the front edge of the forewing in its basal half. This is one of the commonest butterflies in Japan during July and August and is especially common in and around towns, where hedges of *Poncirus* (Rutaceae) provide food for its caterpillars.

**zagreus** Doubleday
Tropical S. America 110–115 mm. (4·33–4·53 in.)
This is a presumably palatable member of a mimetic association, which includes species of the Ithomiine genus *Tithorea*. It is closely related to *P. ascolius*. The caterpillar is unknown. **106a**

**zalmoxis** Hewitson GIANT BLUE SWALLOWTAIL
Tropical Africa 140–170 mm. (5·51–6·69 in.)
This huge and a distinctively coloured butterfly is an inhabitant of the rain-forests of the Congo Basin and W. Africa. There is some variation in the coloration of the wings which are usually blue but can be more greenish than in the illustrated specimen or, rarely, bronze with a green sheen. The female which has some yellow scaling on the forewing, is seen much less frequently than the male. The under surface of both sexes is chiefly greyish white. The foodplant of the caterpillar is not known. **100a**

**zoroastres** Druce
Sudan, tropical Africa 60–85 mm. (2·36–3·35 in.)
The unillustrated female of this forest species differs greatly from the black and white male shown in flight and closely mimics species of *Amauris;* its forewings are dark brown with whitish violet spots, its hindwings similar but with a large yellowish white or white basal area. *P. zoroastres* flies throughout the year, like many tropical butterflies. It is closely related to *P. fulleborni* and *P. sjoestedti*. **114**

**papilionaria** see **Geometra**
**papilionaris** see **Amauta**

**Papilionidae** APOLLO, BIRDWING, SWALLOWTAIL and SWORDTAIL BUTTERFLIES
This worldwide family contains some of the largest *(Papilio antimachus)* and most beautifully coloured *(Ornithoptera)* species of butterfly. It is best represented in the tropics where most of the 530 or so species are found, chiefly in forested regions, but there are also several species in the temperate zones of both the N. and S. Hemispheres. Even Britain can boast of one indigenous species, *Papilio machaon*. The hindwing of many species bears a tail and hence the name Swallowtail and Swordtail. Each of the 3 pairs of legs are functional in the adult, in contrast with most of the Nymphalidae. Three subfamilies are currently recognized: the primitive Baroniinae *(Baronia)*, the Parnassiinae *(Parnassius* and its allies) and the Papilioninae *(Graphium, Ornithoptera, Papilio, Troides* and others). The foodplant of the first subfamily is unknown; Parnassiinae caterpillars feed on species of Aristolochiaceae; those of Papilioninae on species of Annonaceae, Lauraceae, Rutaceae, Umbelliferae or Aristolochiaceae. The caterpillars of all the known species possess an eversible forked process on the top of the thorax. This organ, the osmeterium, produces a pungent scent and is considered to be a protective device chiefly against parasitic insects. The caterpillars of those species which feed on members of the largely poisonous or distasteful family Aristolochiaceae are able to pass on plant toxins to the adult butterfly which is then chemically protected against predators. The noxious qualities of these butterflies is advertised by the use of bright colours or striking patterns, especially of red or yellow – colours which are readily seen by predatory birds. See *Allancastria, Archon, Baronia, Battus, Bhutanitis, Cressida, Dabasa, Euryades, Eurytides, Graphium, Hypermnestra, Iphiclides, Lamproptera, Luehdorfia, Ornithoptera, Pachliopta, Papilio, Parides, Parnalius, Parnassius, Protographium, Serecinus, Teinopalpus, Troides.*

PARADISE
BIRDWING see **Ornithoptera paradisea**
SKIPPER see **Abantis paradisea**
**paradisea** see **Abantis, Callipia, Milionia, Ornithoptera**
**paradoxa** see **Oxychirota, Papilio**
**paralellaria** see **Epione**

**Paraleucoptera** Heinrich (LYONETIIDAE)
A small genus with narrow pointed wings, generally of shining white. These micro-moths occur from Europe, N. Africa and Japan.

**sinuella** Reutti
Europe, N. Africa, Japan 6 mm. (0·24 in.)
The head and thorax are shining white with similarly coloured fore- and hindwings. There is a pale yellow apical patch occupying part of the base of the wing. The species is widespread and the caterpillar feeds on *Populus* (Salicaceae). The Japanese specimen differs slightly from the European ones but they are regarded by the specialists as the same species. The distribution of this species in temperate Asia is not known. It could possibly be an introduced species in Japan.

**Parametriotes** Kuznetzov (PARAMETRIOTIDAE)
The single species in this genus is little known and the exact taxonomic position is uncertain. The genus may be placed in the Gelechiidae when more is known about it. The species in it is known from E. Europe and Turkey.

**theae** Kuznetzov
U.S.S.R., Turkey 8–10 mm. (0·31–0·39 in.)
This small greyish brown species has one or two larger black spots on the wing. In shape it broadly resembles a small specimen of *Phthorimaea operculella*. The caterpillar has been bred from tea bushes *(Theobroma)* where it was damaging the young shoots.

**Parametriotidae**
Only one species is known in this family. It is a small moth, similar to some of the Momphidae and was in fact originally described in the Cosmopteryginae, a subfamily of Momphids. The family is known from specimens from the U.S.S.R. and Turkey. See *Parametriotes.*

**Paramidea** Kuznetzov (PIERIDAE)
A small genus of butterflies with species in temperate Asia, including Japan, Korea and a few species in N. America. It is related to *Anthocharis*, the Orange-tip butterflies of the old world.

**genutia** Fabricius FALCATE ORANGE-TIP
U.S. 33–38 mm. (1·3–1·5 in.)
This is a widespread and locally common species found in or around woodlands. The orange patch at the apex of the forewings and their hooked tip in the male make this species easily recognized, even in flight. The females lack the bright orange tip of the male and has black marks round the edge of the forewing, with a slightly hooked tip. She also has a conspicuous black spot on the forewing. In the U.S. this butterfly is found from southern New England to Texas. The caterpillar feeds on Hedge Mustard *(Sisymbrium)* and other Cruciferae.

**Paramyelois** Heinrich (PYRALIDAE)
This is a small American genus of Pyralid moths differing from *Myelois* mainly in the shape of the labial palps on the head. It is in the subfamily Phycitinae.

**transitella** Walker NAVEL ORANGE WORM
U.S. 15–28 mm. (0·59–1·1 in.)
This species has greyish brown forewings faintly sprinkled with black scales and a large black spot, often white spots on the veins. The hindwings are clear white. *P. transitella* is a minor orchard pest in the S.W. U.S. where the caterpillars feed on the nuts in the seed pods or on the fruit of many trees. They seem to prefer fallen, injured or 'mummified' fruit and dry seeds. It is rare that they attack sound oranges. The caterpillar can be distinguished from the Codlin moth caterpillar by 2 crescent-shaped black areas on the back of the thorax which are absent in the Codlin moth.

**Paranthrene** Hübner (SESIIDAE) CLEARWINGS
A worldwide genus of micro-moths which all probably mimic different species of wasps. The similarity is presumably a protection against predators. There are several N. American species and several in Europe, including Britain.

**palmii** Edwards
U.S. 31–40 mm. (1·22–1·57 in.)
This is a wasp-like moth, with yellow marks between the wings on the sides of the thorax. The abdomen is black at the base near the thorax but the rest is bright yellow, often with a thick black edge to each segment. The forewings are mostly black but have clear patches on them. The hindwings are mostly transparent, with black veins. The caterpillar lives in the bark of White Oak *(Quercus alba)*. They form blisters on the trunk which are visible externally. The moth is out in April and May in the S. and June and July further N. in N. America.

**Parapoynx** Hübner (PYRALIDAE) CHINA MARK
This genus of Pyralid moths has one African species, one in Hawaii, and numerous others in Europe and N. America. The caterpillars of this moth have gills and live in the water, feeding on aquatic plants. Eight groups of up to 6 thin gills can be seen on the side of the caterpillar which lives in a web that it spins on the leaves and stems of the water plants. The caterpillar, if watched closely, can be seen undulating every now and then and this is believed to assist in its respiration under water.

**obscuralis** Grote
U.S., Canada and introduced to Britain 14–22 mm. (0·55–0·87 in.)
The male is smaller than the female, with a similar pattern. It has a buff forewing with white and brown bands along the margins of the hindwings. The caterpillar feeds on various water plants including Eelgrass *(Vallisnaria)* and Pondweed *(Potamogeton)*. The moth lays its eggs at night, settling on a floating leaf and inserting the eggs round the edge of the leaf

onto the underside. The caterpillar feeds underwater on the plants. The moths are common from Nova Scotia through Quebec and Wisconsin and S. to Florida. Recently the moth was found in glasshouse ponds in Britain where it had been introduced on waterweed imported from N. America.

**stagnata** Donovan BEAUTIFUL CHINA MARK
Europe including Britain, Asia including Japan 19–23 mm. (0·75–0·91 in.)
This attractive moth, broadly similar to *Nymphula nympheata* but with more white on the wings, is common throughout its range. It is generally found near the margins of rivers, ponds or lakes where it rests during the day on the plants at the water's edge. The caterpillar lives in the water feeding submerged on water plants, particularly Water-Lily *(Nuphar* sp.) and Burreed *(Sparganium* sp.). The young caterpillar bores into the stem but in the spring it spins pieces of leaves together to form a case in which it lives. **67**

**stratiotata** Linnaeus RINGED CHINA MARK
Europe including Britain 20–25 mm. (0·79–0·98 in.)
The forewings vary in colour from white to pale brownish white. The moth flies in July and August after dusk. The caterpillar feeds under water on Pondweed *(Potamogeton)*, Water Thyme *(Elodea)*, Hornwort *(Ceratophyllum)* and other water plants. It spins together leaves and stems and lives in the open web formed. The caterpillar is whitish green with a dark line on the back, often with a few spots, and has gills along the sides of the body; these are groups of up to 6 filaments. The caterpillar undulates the body at regular intervals and this is believed to be connected with respiration. It pupates in a cocoon attached to the stem of a water plant under water. There is some dispute as to whether the moth emerges under water, or when the water level drops in summer and the cocoon is out of water. There is no reason to suppose that the moth cannot emerge underwater. The development of gills is a specialization in several species with aquatic caterpillars. **70**

**Pararge** Hübner (NYMPHALIDAE)
A large genus of butterflies with most species in Europe and temperate Asia. There are a few which can be found in S. Asia extending into N. India and Pakistan. The genus belongs to the subfamily Satyrinae and is related to *Lasiommata*. The latter generic name has often been discarded and *Pararge* used for all the species in it as well, but more recent work separates these 2 as distinct genera. The butterflies are frequently woodland species often with differences between the spring and summer broods.

**aegeria** Linnaeus SPECKLED WOOD
Europe including Britain, Turkey, Syria, W. and central Asia 36–44 mm. (1·42–1·73 in.)
The butterfly is out in March or April with several broods following during the summer. Often the broods are slightly different in pattern and colour. The bright green caterpillar which has darker lines along the side, edged with yellowish, feeds on grasses including *Agropyron* and *Triticum* (Graminae). The butterfly is widespread in shady places in woodlands over most of its range. **182**

**Parasa** Moore (LIMACODIDAE)
This genus has the typical slug-like caterpillar, with poorly developed legs, that is characteristic of the family. Often the caterpillars have quite a powerful sting; this is from the toxin on the hairs which causes a reaction in man, rather like nettle rash. However, in some cases, where the person is more allergic, the stings can cause serious illness.

**hilarata** Staudinger
China 28 mm. (1·1 in.)
The beautiful green colour on the upper surface of the wing is also found in a number of related species. The underside of the wings are unmarked without any trace of green. Nothing is known of the biology of this species. **54**e

**lepida** Cramer BLUE-STRIPED NETTLE GRUB
Africa 30–40 mm. (1·18–1·57 in.)
The green slug-like larva is a serious pest of cocoa on the Ivory Coast. It also attacks tea, coffee and other crops and can cause a serious reduction of crop yield with a resultant financial loss. The caterpillars are gregarious at first, living more solitarily later on but they pupate in a mass in rough cocoons on the bark of the trees. The life cycle from egg to adult is 10–11 weeks but the pupae may be dormant during periods of drought. The hairs on the caterpillar inflict a painful sting if touched; some people are particularly sensitive to this.

**ostia** Swinhoe
India, Burma 45–74 mm. (1·77–2·91 in.)
The males are much smaller than the females. The forewing is green like *P. hilarata* with a brown base, the hindwings are yellowish brown with a brown fringe and the thorax is green. The antennae of the males are strongly pectinate (feathery). The moth is widely distributed through N. India and Burma. The caterpillar is not known but is probably a slug-like one with stinging hairs which are so typical of other species in the genus.

**Parasemia** Hübner (ARCTIIDAE)
Two species of moths are placed in this genus: *P. plantaginis* and the W. Chinese *P. stotzneri* Bang-Haas.

**plantaginis** Linnaeus WOOD TIGER-MOTH
Europe including Britain, temperate Asia to Japan 30–38 mm. (1·18–1·5 in.)
This is one of the most variable species of Arctiidae, particularly in the ground-colour of the hindwing which may vary from white to red through every gradation of yellow and orange. The white male form of the species is restricted in Britain to high elevations, or northerly latitudes. Males fly chiefly during the day, in search of females, often in bright sunshine. They are on the wing in August and September. The caterpillar feeds on a variety of low-growing plants; it overwinters and completes its growth the following spring in April and May. **375**

**parasitum** see **Ocnogyna**

**Parathyma** Moore (NYMPHALIDAE)
A large genus of butterflies related to *Limenitis* and *Pantoporia* with many species in the Oriental region.

**nefte** Cramer COLOUR SERGEANT
India, Pakistan, Burma 55–70 mm. (2·17–2·76 in.)
The forewings are brown with bluish white patches. One patch is below the apex of the forewing, another is a narrow one in the cell and there is a broad one near the middle of the forewing running to the hind margin. There is an orange patch below the apex of the wing. The hindwing has 2 broad bars across the wing, the outer one is orange colour. This species is common in thick forests in wet hilly regions. The caterpillar feeds on *Glochidion* (Euphorbiaceae).

**selenophora** Kollar STAFF SERGEANT
India, Pakistan 55–70 mm. (2·17–2·76 in.)
This species is similar to *P. nefte* but has only one broad white band on the fore- and hindwing on the upperside, otherwise it is broadly similar. This butterfly is common both before and after the rains in the Himalayas from the lower land to 2130 m. (7000 ft). In S. India it occurs in the evergreen forests. It is most likely to be found in forests in the vicinity of streams and rivers. The caterpillar feeds on *Adina cordifolia* (Naucleaceae).

**Parathyris** Hübner (ARCTIIDAE)
Two species are placed in this genus of C. American and tropical S. American moths.

**semivitrea** Joicey
Tropical S. America 54–60 mm. (2·13–2·36 in.)
This is one of many species which resemble dead, partly skeletonized leaves and are believed, at least partly, to be protected in this way from attacks by predators. Many have this combination of partial transparency and irregular, or angulate, margins to the wings. **368q**

**parce** see **Callionima**
**pardalota** see **Heterogymna**
**pardata** see **Cyphura**
**parhassus** see **Salamis**

**Parides** Hübner (PAPILIONIDAE)
Nearly 90 species are currently placed in this genus of butterflies. It is tropical American and Indo-Australian in distribution, but there is a single Madagascan species *P. antenor*. *Parides* is a genus of the tribe Troidini, the caterpillars of which feed, with few known exceptions, on species of Aristolochiaceae. The caterpillars have a pair of usually pink, fleshy tubercles on each segment. Several supposedly palatable species of Papilionidae and other groups of moths and butterflies (eg. Pieridae) mimic these distasteful species of *Parides*, many of which have nearly identical warning patterns. The males of many species have a fluffy, pale-coloured patch of scent-scales along the inner margin of the hindwing. (See also under *Eniscospila marcus)*.

**aeneas** Linnaeus
Tropical S. America 70–90 mm. (2·76–3·54 in.)
There are a number of variations in pattern and colour in the wing-markings, mostly correlated with geographical distribution. For example the red patch on the hindwing is replaced by purple in some specimens of the subspecies *P. aeneas locris* and *P. aeneas bolivar*. Females have brown wings, no markings on the forewing and an orange patch on the outer half of the hindwing. Nothing appears to be known about the life history of this forest butterfly. (See also under *Eurytides harmodius*). **112b**

**aglaope** Gray
Tropical S. America 65–85 mm. (2·56–3·35 in.)
This species is similar in pattern to its relative *P. aeneas*, but in the male the forewing marking is greyish blue with white spots, and in the female there is a large white mark in the middle of the forewing.

**antenor** Drury
Madagascar 110–135 mm. (4·33–5·31 in.)
This is the only species of *Parides* to have been recorded from Madagascar – no species are known from the neighbouring continent of Africa. It is rare in collections. The caterpillars feed on species of the family Combretaceae. **96c**

**arcas** Cramer
Texas, Central, including Mexico, & tropical S. America 60–75 mm. (2·36–2·95 in.)
This is the only species of its genus to have been recorded in N. America. It differs from its close ally *P. zacynthus* in the green, not blue patch on the forewing of the male. The female is dark brown above with a cream patch on the forewing and an approximately triangular, reddish orange patch at the inner margin of the hindwing. Flowers of *Hamelia* and *Impatiens* are often visited by adults of the subspecies *P. arcas mylotes* according to A. M. Young, of the U.S. The caterpillars feed on a species of *Aristolochia;* they are conspicuously marked with white and, like the adults, are probably unpalatable to vertebrate predators.

**ascanius** Cramer ASCANIUS SWALLOWTAIL
Brazil 45–88 mm. (1·77–3·46 in.)
Little is known about the life history of this unusually marked butterfly, but it is probable that this is a typically marshland species and that its caterpillars feed on species of *Aristolochia*. The male like other species of *Parides* has an area of woolly, scent-scales, white in *P. ascanius*, on the hindwing. **112a**

**coon** Fabricius COMMON CLUBTAIL
N.E. India, Burma, Malaya, Sumatra, Java 90–130 mm. (3·54–5·12 in.)
This butterfly inhabits both open forest and, more commonly, rain-forest at a variety of elevations. The male (not illustrated) is similar in colour pattern to the female, but has much more strongly elongate wings. In some parts of its range, the yellow coloration of the abdomen and the hindwing spots is replaced by orange. *Lantana* blossoms attract this species. **111c**

**crassipes** Oberthür
Burma, Indo-China 85–100 mm. (3·35–3·94 in.)
Few specimens of this species exist in collections. It is similar in colour pattern to the closely related *P.*

*philoxenus*, but its hindwings are narrower and have a single red spot at the end of the tail on the upper surface. There are 6 conspicuous red spots on the under surface. It occurs most commonly between 300–760 m. (1000–2500 ft.).

**hahneli** Staudinger
Brazil 80–95 mm. (3·15–3·74 in.)
This unusually shaped and apparently rare butterfly has seldom been captured far away from the banks of the Amazon River. The female is slightly paler in colour than the illustrated male. The caterpillar is black, and bears numerous yellow tubercles. **112**c

**neophilus** Linnaeus SPEAR-WINGED CATTLE HEART
Tropical S. America 55–82 mm. (2·17–3·23 in.)
The male of this species differs from that of *P. aeneas* chiefly in the forewing marking which is less perfectly triangular in shape and is basally turquoise, changing to greyish white at the apex. The caterpillar feeds on the foliage of a species of *Aristolochia*.

**neptunus** Guérin-Ménéville
Burma, Malaya, Sumatra, Borneo 80–95 mm. (3·15–3·74 in.)
The female of this butterfly has a relatively short forewing and a shorter tail to the hindwing than the more elegant illustrated male. The colour of the hindwing patch varies slightly between specimens, and in the female of subspecies *P. neptunus fehri* it is orange-yellow. **112**e

**nox** Swainson
Malaya, Java, Sumatra, Nias Island, Borneo 75–120 mm. (2·95–4·72 in.)
The males are iridescent dark blue and black above, with a large yellowish or greyish white zone of scent-scales along the inner margin of the hindwing. Females are dark brown above, without a blue sheen, and are variably lined with white between the veins, especially towards the apex of the forewing and usually more noticeably so in specimens from Nias Island. The under surface of the wings is basically similar in colour to that of the upper surface.

**orellana** Hewitson
Central Brazil 80–100 mm. (3·15–3·94 in.)
This has been described as the most beautiful species of its group. The male is similar to *P. aeneas*, but lacks a green patch on the forewing and has a more extensive crimson area on the hindwing. The scent-organ on the inner edge of the male hindwing is white. Females differ from those of *P. aeneas* in the absence of a greyish white patch on the forewing.

**philoxenus** Gray COMMON WINDMILL
Kashmir, N. India, Nepal, Bhutan, Sikkim, Burma, S. China, Formosa, Japan 110–140 mm. (4·33–5·51 in.)
This is the model for the Uraniid moth *Epicopeia polydorus* and *Papilio bootes*. The close similarity in shape and colour pattern of *P. philoxenus* and *E. polydorus* is one of the most impressive examples of mimicry between species of distantly related families in the Lepidoptera. *P. philoxenus* occurs up to 3350 m. (11,000 ft) in forested regions, where the males in particular are attracted to blossoms of Horse-chestnut *(Aesculus)*, *Clematis* and *Rhododendron*. Some specimens have less extensive red and white markings than in the illustration and a few lack wing markings entirely. **111**b

**priapus** Boisduval
Java 100–136 mm. (3·94–5·35 in.)
This species was once fairly common up to 1830 m. (6000 ft) in Java where the blossoms of *Lantana* attracted large numbers of adults. The pale patch on the male hindwing (illustrated) is the exposed male scent storage and distributing organ which in the normal position is folded over against the top surface of the hindwing. The female lacks this modification of the hindwing and has a paler ground-colour to the forewings. **111**f

**quadratus** Staudinger
Tropical S. America 80–110 mm. (3·15–4·33 in.)
This apparently rare butterfly is known only from the Amazon Basin. The yellow marking in the middle of the black (male) or brown (female) forewing may be absent in the male and is usually absent in the female. The hindwing invariably has a large pale yellow patch near its middle.

**rhodifer** Butler
Nicobar, Andaman Islands 100–120 mm (3·94–4·72 in.)
In general appearance this species is similar to its close ally *P. coon* but differs in the distinctive, orange-tipped hindwing tail and, the narrower wings, especially in the male whose long slender forewings have an aspect ratio of about 3:1.

**sesostris** Cramer SOUTHERN CATTLE HEART
Tropical S. and Central America 65–90 mm. (2·56–3·54 in.)
This is a forest species, most commonly seen feeding on flowers in clearings and at the margins of the forest. The male is somewhat similar above to *P. aeneas* in colour pattern but the red marking on the hindwing is reduced to a narrow triangle near the inner margin of the wing. Females are dark brown in ground-colour, with a large, cream patch on the forewing and three or more scarlet patches on the hindwing. The caterpillar feeds on a species of Aristolochiaceae.

**tros** Fabricius
Brazil 76–97 mm. (2·99–3·82 in.)
The illustrated male of this species was photographed by Professor K. S. Brown of São Paulo while the butterfly was feeding on flowers of Ipecac *(Caphaelis)*, a plant whose roots have emetic properties. The female is less dark in colour than the male with the forewing marking closer to the middle of the wing and the hindwing marking orange not red and extending as far as the anal margin of the wing. **95**

**zacynthus** Fabricius
Tropical S. America 70–86 mm. (2·76–3·39 in.)
There is some variation in colour of the forewing marking which in some specimens is more extensively violet than in the illustrated male. Females are brown above, usually with 3 large, bluish white spots on the forewing and a longer red band on the hindwing than in the male. **112**d

**paris** see **Papilio**
PARIS PEACOCK see **Papilio paris**
**parisatis** see **Satyrus**

**Parnalius** Rafinesque (PAPILIONIDAE) FESTOON BUTTERFLIES
The generic names *Thais* Fabricius and *Zerynthia* Ochsenheimer have been applied to this genus by authors. *Parnalius* is related to the Apollo butterflies *(Parnassius)* but has a closer ally in the genus *Serecinus*. The caterpillars of the 2 known species (both mentioned below) feed on species of the family Aristolochiaceae.

**polyxena** Denis & Schiffermüller: syn. *P. hypermnestra* SOUTHERN FESTOON
S. Europe, W. Turkey, S.W. Russia 42–60 mm. (1·65–2·36 in.)
This butterfly is an inhabitant of rocky country, up to elevations of about 910 m. (3000 ft). Its colour pattern is similar in general plan to the male of *Allancastria cerisyi*, but the zig-zag marginal lines are distinctive. It rests either with the wings outspread, or closed together over its back; in either position, the red, black and yellow or yellowish white warning coloration is revealed. The caterpillars feed on species of *Aristolochia*. **109**

**rumina** Linnaeus SPANISH FESTOON
S.W. Europe, N. Africa 38–50 mm. (1·5–1·97 in.)
This is similar above in general colour pattern to *P. polyxena*, but the black basal spot on the forewing is replaced by a red spot, and the outer line on the forewing is much less strongly dentate. The under surface of the hindwing has pale yellow spots at its base unlike that of *P. polyxena*. It flies in rocky countryside on hill slopes and mountain sides where it has been captured up to 1520 m. (5000 ft). It is most common, however, in the coastal districts and is practically confined to them in North Africa. The variably coloured caterpillar feeds on species of *Aristolochia*. **94**k

**Parnassius** Latreille (PAPILIONIDAE) APOLLO BUTTERFLIES
There are about 30 recognized species in this genus, although many more names have been applied at times to this group of highly variable butterflies. They are found in W. N. America, Europe, temperate Asia and Japan. Many collectors have concentrated their attention on this genus and there are consequently several extensive collections of these attractive butterflies both in private hands and national museums. The Eisner collection, now in the museum at Leiden, Holland, is probably the largest. Most, but not all species of *Parnassius* are alpine in distribution or inhabit northerly latitudes, but few occur at such high elevations as *P. acco* Gray which has been captured at 5640 m. (18,500 ft) on the slopes of Mount Everest. The caterpillars are typically hairy and spotted with red.

**apollo** Linnaeus APOLLO
Europe (not Britain) to central Asia 50–100 mm. (1·97–3·94 in.)
This is a common species of mountainous regions in most of Europe and central Asia, but of lower elevations in more northerly latitudes. Specimens of this butterfly from different localities vary noticeably in coloration and pattern; not a rare phenomenon, but one which has produced a multitude of names for minor geographical subdivisions of this species. Stonecrops *(Sedum)*, common plants of subalpine and subarctic zones, are the foodplants of the orange and black caterpillars. **102**

**autocrator** Avinoff
Afghanistan 58–65 mm. (2·28–2·56 in.)
The forewings of this rare species are white above with typical grey markings, best marked in the female. The hindwings are whitish grey at the base and at the outer margin along which there are about four blue and black spots; the central part of the wing is grey, with a broad female, or narrower and interrupted, male orange band.

**charltonius** Gray REGAL APOLLO
Pakistan, N.W. India 80–90 mm. (3·15–3·54 in.)
This large, beautifully marked Apollo is rare in collections. It is a high elevation species found between 4420–4880 m. (14,500–16,000 ft) in the Himalayas. Males circle for long periods over mountain peaks and are seldom captured. Females normally fly in alpine meadowland at lower altitudes and are consequently less difficult to capture. **94**f

**clodius** Ménétriés
N.W. U.S., Canada (British Columbia) 64–75 mm. (2·52–2·95 in.)
This is the only species of its genus restricted in distribution to N. America. It is chiefly a mountain species and is found from N. Alaska to California and Wyoming. There is some variation in the amount of black on the forewing and the size of the orange or red spots on the hindwing. The caterpillar feeds on Stonecrops *(Sedum)*, *Viola*, *Saxifraga*, bilberry *(Vaccinium)* and possibly on other low-growing plants. **94**l

**eversmanni** Ménétriés
E. Asia including Japan, N.W. U.S., Canada 58–62 mm. (2·28–2·44 in.)
This butterfly is found only in the forested mountains of E. Asia, Alaska and adjacent Canada. There is some variation in the ground-colour of the upper surface of the wings which can be yellow or nearly white, although females are seldom yellow. The caterpillars feed on species of *Corydalis*. **94**d

**hardwickei** Gray COMMON BLUE APOLLO
N. India, Sikkim, Himalayan China 50–65 mm. (1·97–2·56 in.)
This is a high elevation species, occurring up to 5180 m. (17,000 ft) in the Himalayas and is seldom seen below 2440 m. (8000 ft). It is one of the most strikingly coloured species of its genus. The illustrated wet season form is much more richly coloured and strongly marked than the dry season form, and there is a high degree of variation in details of the colour pattern even between individuals of a local population. There is also considerable geographical variation in colour pattern. Females are normally darker and more heavily marked than the males. **94**e

**phoebus** Fabricius SMALL APOLLO
European Alps, Urals, Siberia, W. Canada & U.S. as far S. as the mountains of New Mexico 56–63 mm. (2·2–2·48 in.)
In Europe this is even more of an alpine species than *P. apollo* and is seldom seen below 1830 m. (6000 ft). The caterpillar feeds on species of stonecrops *(Sedum)*, *Saxifraga* and *Sempervivum* in damp localities. As in other groups of animals and plants, the narrow water-gap between E. Siberia and W. Alaska has no radical effect on the arctic distribution of this species. The red spots just beyond the middle of the female forewing are often absent in the male. **94**n

**parobifera** see **Cecidothyris**

**Parotis** Hübner (PYRALIDAE)
At present the species in this genus of Pyralid moths are not well known. Formerly they were included in the genus incorrectly called *Margaronia* but as the species in that genus are studied, more and more are removed from it and placed in other genera. At present the species in *Parotis* are mostly from the Indo-Australian region but until they are studied in detail, the exact limits of the genus will be uncertain.

**laceritalis** Kenrick
New Guinea, Borneo, Malaya 38–42 mm. (1·5–1·65 in.)
This is a little known species which has several similar related species occurring geographically close together. The holotype is from New Guinea and more work on this group of green species is needed before their exact distribution is known. **63**m

PARROT'S NANNY see **Neossiosynoeca scatophaga**
PARSLEYWORM see under **Papilio polyxenes**

**Parthenos** Hübner (NYMPHALIDAE)
A genus of butterflies with species from India and Sri Lanka to New Guinea. Some of the species range into S. China and the Philippines. They are powerfully built and strong flying.

**cyaneus** Moore
Sri Lanka 96–102 mm. (3·78–4·02 in.)
This species is known only from the island of Sri Lanka (Ceylon) where it has similar habits to the widespread *P. sylvia*. The blue colour separates it from the broadly similarly patterned *P. sylvia*, which is more olive colour. It is a rapid and powerful flier and is often difficult to catch. The caterpillar feeds on *Adenia* (Passifloraceae) and is found from the lowlands up to 1830 m. (6000 ft). It is a common and widespread butterfly in Sri Lanka and is on the wing in most months of the year.

**sylvia** Cramer CLIPPER BUTTERFLY
India, Sri Lanka, S.E. Asia, Indonesia, New Guinea, Solomon Islands 100 mm. (3·94 in.)
This widespread butterfly has many forms described in different parts of its extensive range. It is mainly a forest species, common at the edges of forests and in surrounding areas. Often congregates in large numbers to drink at the edges of streams, it can be found in most months of the year. The caterpillar feeds on *Adenia* (Passifloraceae) and *Tinospora* (Menispermaceae) and also on Cucurbitaceous vines. Males and females are similar. The specimen illustrated is from N. India. The colour of the wings vary from greenish to a blue and even an orange background, although the basic pattern is similar. **216**q

**parthenos** see **Platarctia**
**parthenis** see **Hyphilaria**
**parvulus** see **Trichoptilus**
**pascuellus** see **Crambus**
PASHA BUTTERFLY, TWO-TAILED see **Charaxes jasius**
**pasinuntia** see **Lycorea**
**patagiata** see **Chlanidophora**
**patala** see **Euthalia**
PATCH WHITE, ORANGE see **Colotis helimede**
**patna** see **Elymnias**
**patroclus** see **Lyssa**
**paulina** see **Appias**
**pausanias** see **Eurytides**
**pavana** see **Eueides**
**pavonia** see **Saturnia**
**pavonii** see **Elzunia**
**payeni** see **Dabasa**
PEA MOTHS see **Cydia, Earias**
PEA BLUE see **Lampides boeticus**
PEACH
BLOSSOM PLUSIA see **Eosphoropteryx thyatiroides**
BUD MOTH see **Anarsia lineatella**
TWIG-BORER see **Anarsia lineatella**
PEACH-BLOSSOM MOTH see **Thyatira batis**
PEACOCK BUTTERFLIES see **Anartia, Inachis, Papilio**
PEACOCK MOTH, GREAT see **Saturnia pyri**
PEACOCK
AWL see **Allora daleschalli**
PANSY see **Junonia almana**
ROYAL see **Tajuria cippus**
PEAK WHITE see **Pontia callidice**
PEARL
CHARAXES see **Hadrodontes varanes**
CRESCENT see **Phyciodes tharos**
GRASS-VENEER see **Catoptria pinella**
PEARL-BORDERED FRITILLARIES see **Clossiana**
PEARLY
EYE BUTTERFLY see **Enodia portlandia**
UNDERWING see **Peridroma saucia**
PEBBLE HOOK-TIP see **Drepana falcataria**
PECAN-LEAF CASE-BEARER see **Acrobasis jugulandis**
**pectinataria** see **Colostygia**
**Pectinophora** see **Platyedra**
**pedella** see **Stathmopoda**

**Pediasia** Hübner (PYRALIDAE)
CLOSEWING or GRASS MOTHS
This large genus of Pyralid moths is found in Europe, Africa, N. India and N. America. They are mostly rather similar looking species, typical of the subfamily Crambinae in their appearance and habits. They have a rather triangular shape when they sit with their wings rolled up on grass stems. Generally they are common in meadows or pasture lands where they are easily disturbed during the day. They fly rapidly, in a zig-zag pattern and settle after only a short flight.

**trisecta** Walker LARGER SOD WEBWORM
U.S., Canada 24–30 mm. (0·94–1·18 in.)
Several species go under the name of Sod webworms. The caterpillars eat a number of different species of grasses. The moth is greyish brown with white hindwings. The forewings are often striped with black. When sitting on grass stems it takes the characteristic position for the family of rolling its wings round the body. The caterpillar is a serious pest in pastures and meadows and cornfields. It causes yield reduction in cornfields and by destroying meadows and pasture reduces the amount of grazing for stock. This moth generally has two broods a year, particularly in the warmer parts of the U.S.

**pediasia** see **Lophocorona**
**pelea** see **Tanaecia**
**peleides** see **Morpho**
**pellenea** see **Actinote**
**pellionella** see **Tinea**

**Pempelia** Hübner (PYRALIDAE)
A genus of Pyralid moths with 30–40 species, mostly European, Asian or N. African with one species described from the Sudan. The moths are long-winged, typical in appearance of the subfamily, Phycitinae, to which they belong.

**formosa** Haworth
Europe including Britain, Asia including Japan 19–21 mm. (0·75–0·83 in.)
Richly coloured reddish-brown forewings make this an attractive small moth. The caterpillar feeds on elm *(Ulmus)* and other trees, spinning a web on the upper surface of the leaves. Many of the related species are important crop pests. **56**l

**Pemphigostola** Strand (AGARISTIDAE)
A single species is place in this genus of day flying moth.

**synemonistis** Strand
Madagascar 38–50 mm. (1·5–1·97 in.)
This unusual species was known for many years from a single male specimen. Recent collecting by Viette, Griveau and Ratovoson in Madagascar has produced a good series of both sexes. The male has a radically modified forewing (see illustration) which is swollen at its base into a sound-box. The under surface of this swelling has a long, rasp-like ridge, against which the similarly rasp-like surface of the middle leg is rubbed to produce sound. Females (also illustrated) lack any form of stridulatory organ and it seems likely therefore, that the male's stridulation may have a pre-mating function or possibly be used to warn other males away from its territory. **402**a, b

**penicillana** see **Tamyra**
**pennella** see **Heterogynis**
**pentadactyla** see **Pterophorus**

**Pentila** Westwood (LYCAENIDAE)
The 50 or so species of this genus of butterflies are restricted to Africa S. of the Sahara. Some are mimics of *Acraea* species, others are yellow or white, with black markings, and resemble species of the family Pieridae.

**tachyroides** Dewitz
Africa S. of the Sahara 26–34 mm. (1·02–1·34 in.)
This is a yellowish white, Pierid-like butterfly, with a dark brown apex to the forewing and a series of round spots extending from the apex along the outer margin and continued along the outer margin of the hindwing. The diffuse, yellowish area at the base of the wings varies in clarity between specimens. The life history is unknown.

**tropicalis** Boisduval SPOTTED BUFF
Tropical Africa 25–40 mm. (0·98–1·57 in.)
This unusual Lycaenid is considered to be a mimic of *Pardopsis punctatissima* Boisduval, a species of the generally unpalatable group Acraeinae. The illustrated specimen is an example of the E. African subspecies *P. tropicalis mombasae*. In some examples of *P. tropicalis* the black area at the apex of the forewing is poorly marked or absent. **244**a

PEPPER AND SALT MOTH see **Biston betularia**
PEPPERED MOTH see **Biston betularia**

**Percnia** Guenée (GEOMETRIDAE)
Most of the moths in this genus have a similar basic pattern of black or grey and white often with yellow bodies. Little is known of the biology of species in this genus.

**ductaria** see under **P. felinaria**

**felinaria** Guenée
N. India 80–85 mm. (3·15–3·35 in.)
One of the largest species in the genus, it differs from the related species, *P. ductaria*, in the lack of streaks on the wings, otherwise the 2 species are broadly similar. **272**t

**peregrina** see **Oxytenis**

**Pereute** Herrich-Schäffer (PIERIDAE)
A small genus of butterflies found in C. and S. America.

**leucodrosime** Kollar
S. America 65 mm. (2·56 in.)
This species is widespread in S. America and has been found in Colombia, Peru, Venezuela, Ecuador and Brazil. The butterfly is conspicuous with the orange-red band across the black forewing and the black hindwings with the grey basal areas. The extent of this grey varies in different specimens. The underside of the wings have the orange-red forewing band but the rest of the fore- and hindwings are blackish brown with conspicuous jet black veins. In some specimens the red forewing band is reduced to a small red spot. **119**a

**periander** see **Rhetus**

**Peribatodes** Wehrli (GEOMETRIDAE)
A genus of moths mostly European and temperate Asian in distribution. Species in this genus are often locally common over their range. Most have strongly patterned wings which help to camouflage them when at rest during the day.

**rhomboidaria** Denis & Schiffermüller WILLOW BEAUTY 34–40 mm. (1·34–1·57 in.)
Europe including Britain, temperate Asia
The moth is variable in colour with some dark coloured forms. It is common in Europe where it flies in July and August, sometimes with a second brood in September. The caterpillar is a dull reddish brown mottled with traces of diamond shapes on the back. It feeds on ivy *(Hedera)*, privet *(Ligustrum)*, lilac *(Syringa)*, *Rosa* and many other shrubs. **311**

**Pericallia** Hübner (ARCTIIDAE)
About 49 species are placed in this genus of moths. It is represented in Europe, temperate and tropical Asia as far E. as Sulawesi (Celebes) and the Philippines, in Africa and Madagascar and Mexico. Some of the included species clearly belong elsewhere.

**matronula** Linnaeus LARGE TIGER-MOTH
Central Europe to Japan 50–58 mm. (1·97–2·28 in.)
Well named as a 'large' tiger-moth, *P. matronula* is a normally fairly rare species of hilly, wooded country emerging in June and July. Its forewings are brown with 4, large yellow patches along the front margin and usually a single yellow patch towards the anal angle; the hindwings are orange-yellow with black markings. The abdomen is a bright red with black markings. Males fly both before noon and at night. The caterpillar, which is a 'woolly-bear' like that of *Arctia caja*, feeds on the foliage of several broad-leaved trees, including ash *(Fraxinus)*, hazel *(Corylus)* and elm *(Ulmus)*. It overwinters twice before changing into a chrysalis. **369**a

**pericallis** see **Milionia**
**Pericopis** see **Dysschema**

**Peridroma** Hübner (NOCTUIDAE)
There are about 10 species in this genus. It is found in Europe, W. Asia, N. and S. America, Hawaii, the Galapagos, Falkland Islands and Gough Islands.

**goughi** Fletcher
Gough Island
The wings of the males of this unusual Noctuid are greatly reduced (brachypterous) and useless for flying, a phenomenon found frequently in moths inhabiting wind-swept oceanic islands where the disadvantages of a fairly sedentary existence are outweighed by the decrease in the chances of being blown off the island into oblivion. At least one other Noctuid endemic to Gough Island is similarly brachypterous in the male. Females are not yet known, but brachyptery is often associated solely with the female of a species and it is highly probable that both sexes of *P. goughi* will prove to have reduced wings.

**saucia** Hübner PEARLY UNDERWING or VARIEGATED CUTWORM 42–55 mm.
Europe, N. Africa, Canary Islands, (1·65–2·17 in.)
Madeira, Turkey, N. America
One of many common destructive cutworm species in the New World, it is seldom an important pest elsewhere. The forewings are cryptically patterned in brown and are rather variable in details of the pattern; the hindwings are brownish white. In Britain this is chiefly a migrant, but is established as a breeding species along the south coast of England. The caterpillar feeds on a variety of herbaceous plants but is not a root feeder; it is often a pest of lettuce grown under glass. It hibernates either as a caterpillar or a chrysalis. **391**g

**perimale** see **Cepora**

**Perina** Walker (LYMANTRIIDAE)
At present 3 species have been scientifically described in this genus of Malaysian and W. Indonesian moths and at least 2 new species await description.

**nuda** Fabricius
Sri Lanka, India, Indochina, 27–44 mm.
Burma, China, Formosa (1·06–1·73 in.)
The atypical, partly transparent-winged male (illustrated) of this species resembles a Psychid and, in fact, is very similar in shape and colour pattern to the Psychid species *Acanthopsyche bipars* Walker. Its female, in contrast, is more typical of a Lymantriid, and has yellowish white wings of nearly equal size, with a scattering of brown scales at the base of the hindwing. The caterpillar has 2 dense, black, dorsal tufts of hair and greyish black, lateral tufts; it has been recorded from the shoots of cultivated fig *(Ficus)*. **357**q

**Perisama** Doubleday (NYMPHALIDAE)
This is a huge genus of S. American butterflies which generally have some beautiful iridescent colour in the upperside and a very contrasting underside pattern. The species are often common and provide wonderful colour as they fly about.

**bonplandi** Guérin-Méněville: syn. *P. albipennis*
S. America 48–55 mm. (1·89–2·17 in.)
The upperside is similar to *P. oppelii* and is black with a green iridescence. The hindwings have a greyish patch round the margin instead of the yellow of *P. oppelii*. The species is common from Brazil to Peru. Several varieties have been named where the colour pattern differs. In some the underside of the hindwing is more brown than grey and there is more red near the base of the forewing. **229**q

**oppelii** Latreille
S. America 40–50 mm. (1·57–1·97 in.)
Both the upper and undersides of this species are illustrated and the contrast is striking. It is widespread from Brazil to Venezuela, Colombia and Peru. **229**p, r

**vaninka** Hewitson
S. America 40 mm. (1·57 in.)
This species is found in Ecuador, Bolivia, Venezuela, Peru, Colombia and Brazil and is common over much of its range. Nothing is known of its biology. Several subspecies have been described, differing generally in the patterns on the underside. The upper and undersides are illustrated. **229**j, k

**periscelidactyla** see **Pterophorus**

**Perizoma** Hübner (GEOMETRIDAE)
A large genus with species in Europe, temperate Asia and N. America. The caterpillars are all typical 'loopers' or 'earth-measurers'.

**sagittata** Fabricius MARSH CARPET
Europe including Britain, 24–27 mm.
Asia including Japan (0·94–1·06 in.)
The caterpillar feeds on Meadow-rue *(Thalictrum)* and the moth is out in July. There is some variation in the pattern of the wings. Generally these are brown, with white-edged broad black band across the forewing. The hindwings are a paler brown. It is similar to, but larger than, the Autumn Green Carpet, *Chloroclysta siterata*, which has a similar distribution but feeds on alder *(Alnus)*, birch *(Betula)* and sallow *(Salix)*. *C. siterata* is out later in the year than *P. saggitata*, being on the wing in September, October, overwintering until the following spring. **290**

**perlata** see under **Campaea margaritata**
**perlella** see **Crambus**
**pernyi** see **Antheraea**
**perseus** see **Morpho**
**persimilis** see **Hestinalis**
**perspicilla** see **Arrhenophanes**
**peterseni** see **Lineodes**
**petreus** see **Marpesia**
**petrosoma** see **Prototheora**
**pexa** see **Cecidothyris**
**pfeifferella** see **Antispila**
**phaedusa** see **Stalachtis**

**Phaegorista** Boisduval (HYPSIDAE)
The 13 or so species of this genus of moths occur in Africa S. of the Sahara. Several of the species are apparently mimetic partners of the widespread Danaine butterfly species *Danaus chrysippus*.

**similis** Walker
Congo Basin, W. Africa 61–72 mm. (2·4–2·83 in.)
This is typical in its colour pattern of several species of *Phaegorista* which mimic the common, distasteful butterfly *Danaus chrysippus*. The ground-colour of the wings of *P. similis* varies considerably between specimens, and can be more yellowish or more reddish than in the illustration. **376**k

**Phaeosphecia** Hampson (CTENUCHIDAE)
There is only one species in this S. American genus of moths.

**opaca** Walker
Brazil, Venezuela 40–45 mm. (1·57–1·77 in.)
This is another remarkable wasp-mimicking species as its generic name suggests. Many Sphecoid wasps have similarly dark blue, partly transparent wings and waspish 'waists'. **367**b

**phaetusa** see **Dryadula**

**Phalacropterix** Hübner (PSYCHIDAE)
A small genus of micro-moths with species in Europe and the Middle E. Most of the species have rather smokey wings, similar to the species illustrated.

**apiformis** Rossi
Italy, Spain 18 mm. (0·71 in.)
This species is common locally in its range. The abdomen has large tufts of scales which may be black or yellow. **5**c

**Phalaenoides** Lewin (AGARISTIDAE)
The 3 species of this genus of moths are solely Australian in distribution. Adults are black or dark brown, with yellow wing-markings.

**glycine** Lewin VINE MOTH
E. & S.E. Australia 45–55 mm. (1·77–2·17 in.)
The caterpillars often reach pest proportions attacking grape-vines *(Vitis)* and Virginia creeper *(Parthenocissus)*. There is some variation in the yellow markings on the forewings; some specimens lack the basal yellow bar. **399**q

**Phalera** Hübner (NOTODONTIDAE)
The 30 or so species in this genus of moths are found in Europe, Asia, Australia and Africa.

**bucephala** Linnaeus BUFF-TIP
Europe including Britain, 50–62 mm.
Asia to Siberia but not Japan (1·97–2·44 in.)
The resting moth is a popular example of 'broken-twig' mimicry; the 'buff-tip' representing the broken end of the twig. This is a generally common moth, often especially so in urban areas where lime trees *(Tilia)* have been planted. Apart from lime, the caterpillars will feed on elm *(Ulmus)* and hazel *(Corylus)*; they are gregarious until nearly full-grown. *Phalera* is represented in Japan by *P. fuscescens* Butler, a species similar in general appearance but differing in details of the pattern. **355**

**Phaloesia** Walker (HYPSIDAE)
Two species of moths are placed in this genus: *P. saucia* and the S. American *P. rubriplaga* Dognin.

**saucia** Walker
Central & tropical 38–52 mm.
S. America (1·5–2·05 in.)
This is one of the many species of essentially Central and S. American moths and butterflies whose range includes subtropical, S.E. Texas. The caterpillar is brown, with a bluish black head and tubercles, and with black and white hairs. The chrysalis is light brown, marked with bluish black and is protected by a loosely woven, yellow cocoon. **376**f

**Phaloniidae** see **Cochylidae**
**phaola** see **Appias**
**phareus** see **Mesene**
**phega** see **Elimniopsis lise**
**phegea** see **Syntomis**

**Pheosia** Hübner (NOTODONTIDAE)
There are about 10 species in this genus of moths,

most of them are found only in temperate Europe and Asia, from England to Japan, but a few are found in N. America. The caterpillars have an unusual varnished appearance.

**gnoma** Fabricius LESSER SWALLOW PROMINENT
Europe, temperate Asia to the Pacific 38–42 mm. (1·5–1·65 in.)
This is a very similar moth to the illustrated *P. rimosa,* but is smaller and has a broader, wedge-shaped marking at the outer part of the rear edge of the forewing. The caterpillars feed on birch *(Betula)* foliage. In Britain this species is more common in N. England and Scotland than in the S.

**rimosa** Packard
E. Canada, U.S. S. to about latitude 35°N. 45–60 mm. (1·77–2·36 in.)
The unusual caterpillar of the rather rare *P. rimosa* has a dorsal horn at the rear as in hawk-moth caterpillars. It can be brown, green or grey and feeds on poplar *(Populus),* willow *(Salix),* and birch *(Betula).* The name 'Mirror-back caterpillar' refers to its varnished appearance. **357**l

**tremula** Clerck SWALLOW PROMINENT
Central & N. Europe including Britain, to coastal Russia but not Japan 40–45 mm. (1·57–1·77 in.)
This is extremely close in colour pattern to the illustrated American species, *P. rimosa.* The caterpillars can be either green or brown; they feed usually on poplars and aspen *(Populus)* less commonly on willows and sallows *(Salix).*

**Phiala** Wallengren (EUPTEROTIDAE)
There are about 46 described species in this genus of mainly white moths. They are all found in Africa S. of the Sahara.

**cunina** Stoll
W. Africa 64–82 mm. (2·52–3·23 in.)
This is one of the more unusually marked species of the family Eupterotidae. There is some variation in the colour pattern between individuals, and on some the markings are brown, not black. **317**h

**phicomone** see **Colias**
**phidippus** see **Amathusia**

**Philaethria** Billberg (NYMPHALIDAE)
There are 2 or possibly 3 species in this genus of Heliconiine butterflies. The generic name *Metamorpha* Hübner (Nymphalinae) was misapplied to this genus at one time.

**dido** Linnaeus SCARCE BAMBOO PAGE
Tropical S. & C. America 85–92 mm. (3·35–3·62 in.)
Unlike many species of its subfamily, this well-known butterfly is not averse to exposure in prolonged periods of sunshine, and will often circle above tree-canopy level, especially on hill-tops. It is one of the largest species of its genus. Several subspecies have been described; they differ from each other in details of the pattern or coloration. White, blue and yellow flowers are preferred by this butterfly; red flowers are seldom visited for nectar. The supposed mimicry by the Malachite butterfly *(Victorina steneles)* in Costa Rica has been questioned recently by Young who has pointed out that *P. dido* is chiefly a tree-canopy species unlike its Nymphaline mimic and that selection by predators cannot operate. This is similar in pattern to the illustrated *P. wernickei,* but the green colour of the wings is less intense and the dark areas are dark brown, not nearly black.

**wernickei** Röber
Amazon Basin, S. Brazil, Paraguay 85–100 mm. (3·35–3·94 in.)
This flies together with its less colourful relative *P. dido* in parts of their range, and in Brazil is joined by the Nymphaline mimic *Victorina stelenes* which matches them both in colour pattern and flight characteristics. The illustrated specimen is feeding on the flowerhead of a species of *Wedelia* (Compositae). **149**

**philarchus** see **Kallima horsfieldi**
**philea** see **Phoebus**
**philenor** see **Battus**

**philodice** see **Colias**

**Philotes** Scudder (LYCAENIDAE)
The 9 or so species of this genus are small and often difficult to identify without an examination of the male genitalia. They are apparently restricted in range to the U.S.

**battoides** Behr
W. N. America 18–24 mm. (0·71–0·94 in.)
Males of this species are blue above, with a dark brown outer margin to the wings. Females are brown above except for a brown-spotted band along the outer margin of the hindwing. The caterpillars' foodplant is *Eriogonum.* The illustrated specimen belongs to the subspecies *P. battoides bernardino.* **254**

**sonorensis** Felder SONORA BLUE
N. Mexico, central & S. California 20–22 mm. (0·79–0·87 in.)
This beautifully coloured orange and blue species is unlike any other butterfly in the U.S. The female forewings are similar in colour pattern above to the illustrated male, but the markings are much reduced, and the hindwings usually lack an orange patch. The caterpillar feeds on species of stonecrop *(Sedum).* Felder, an Austrian entomologist, is thought to have based his description of *P. sonorensis* on specimens captured between 1857–59 in Los Angeles, which at that time was part of Sonora, Mexico. **263**l

**philoxenus** see **Parides**

**Philudoria** Kirby (LASIOCAMPIDAE)
About 20 species of moths are placed in this European, temperate Asian and Indian genus.

**potatoria** Linnaeus DRINKER
Europe, temperate Asia to Japan 47–64 mm. (1·85–2·52 in.)
The caterpillar of this widespread species feeds on coarse grasses and reeds, typically in damp areas, and will drink from dew or raindrops on its foodplant, a habit which is the origin of the common name of this species. Eggs are laid in batches on the foodplants in July and August; the caterpillar feeds for a short time, then hibernates until April. The brown chrysalis is usually attached to the stem of the foodplant. Adult males are normally reddish brown in colour and much darker than the illustrated females. **324**

**phlaeas** see **Lycaena**

**Phlogophora** Treitschke (NOCTUIDAE)
The moths of this and related genera rest with their wings in a distorted, folded and wrinkled position. The genus is not large, but is found in both the Old and the New Worlds.

**iris** Guenée
Central and E. Canada, N. central and N.E. U.S. 38–44 mm. (1·5–1·73 in.)
Very similar in pattern to the Old World *P. meticulosa,* and differs chiefly in the more evenly convex outer margin of the forewing. The caterpillar is a cutworm and attacks the roots and stem bases of several species of low-growing plants. There is probably only one brood per year. **391**t

**meticulosa** Linnaeus ANGLE-SHADES
Europe including Britain, W. Asia 45–50 mm. (1·77–1·97 in.)
This is a generally common species and one of the most distinctively patterned European Noctuids. There are 2 generations each year. One flies from August to late autumn, overwinters and appears again in May to lay eggs. The caterpillars resulting from these eggs overwinter, then pupate and produce adults in May and June, so that during May there are 2 generations on the wing. The caterpillar feeds on groundsel *(Senecio),* dock *(Rumex)* and several other low-growing plants including geraniums. Much work has been carried out recently by Dr Martin Birch on the courtship function of the male scent-organs.

**Phobetron** Hübner (LIMACODIDAE)
The caterpillar of the moths in this genus are hairy. These are stinging hairs which are very unpleasant if touched. They provide good protection for the caterpillars against possible predators. The larva weaves some of these stinging hairs into its cocoon which provides added protection for the pupa. The moths occur in N. and S. America.

**pithecium** Smith & Abbot HAG or MONKEY-SLUG MOTH
U.S., Canada 20–25 mm. (0·79–0·98 in.)
These fat-bodied moths with a very woolly looking body and rather triangular forewings are common in the E. and Central States. The caterpillar feeds on various shrubs including species of Rosaceae and Cupuliferae. It is a curious insect, brown, short, with 3 prominent lateral processes and a general woolly look, totally unlike a normal tube-shaped caterpillar. The males are dark grey-black, with rather pointed wings and little pattern, while the females are larger and have very rounded wings with a distinct brown and white pattern. Their hindwings are grey brown, with marks around the margin, and are rounded, not sharply angular as in the male.

**phocides** see **Bindahara**

**Phoebus** Hübner (PIERIDAE) GIANT SULPHURS
These large fast flying butterflies are well known migrants. They are found in N. and central America with some in S. America. Several species have been reared from caterpillars feeding on species of Leguminosae.

**argante** Fabricius APRICOT
America from Texas through Central America, W. Indies, S. to Paraguay 60 mm. (2·36 in.)
This butterfly is bright apricot yellow with a thin black margin to the forewing, some black marks on the edge of the hindwing and reddish veins on the wings. The females have more black on the wings than the males. The butterfly is a rapid flier, frequently settling and feeding on flowers or settling on damp mud of river-edges to drink. This species is often common over much of its range. The caterpillar feeds on *Inga* and *Caesalpina* (Leguminosae).

**avellaneda** Herrich-Schäffer
Cuba 75–85 mm. (2·95–3·35 in.)
The male is a gaudy colour with the amount of red on the forewing varying between specimens. The female illustrated shows the mark of a bird's beak on the right forewing. Clearly this one had a narrow escape from a bird only to be caught later by a collector. Both sexes are illustrated. **127**h, j

**cipris** Fabricius TAILED SULPHUR
C. & S. America, S. U.S. 60 mm. (2·36 in.)
This yellow butterfly has a pointed hindwing giving the appearance of a short tail. There is a strong orange flush to the sulphur yellow on the wings. It is a rare visitor to the S. U.S., but is common in C. and S. America. The species has seasonal forms; the dry season form, which differs from the wet season one, is called *P. cipris* form *neocypris,* and at one time was regarded as a distinct species and known as *P. neocypris. P. cipris* has been collected in most S. American countries.

**philea** Linnaeus ORANGE-BARRED SULPHUR or YELLOW APRICOT
C. & S. America, S. U.S. 75–85 mm. (2·95–3·35 in.)
Bright sulphur yellow over both the fore- and hindwings with a trace of black near the apex of the forewing. An orange patch in the middle of the forewing and orange band along the hindwing margin. This species is common from southern Brazil to Mexico and had been recorded, as an occasional migrant, in Texas, Florida and Georgia. They fly in bright sunshine and are attracted to river banks where they may congregate in large numbers. The adults are also attracted to humans where, if allowed, they drink their sweat. The caterpillars feed on species of *Cassia* (Leguminosae). The female has a row of brown spots on the forewing and lacks the orange patch. The hindwings have a wider band of orange than the males. There are several related species which have orange or red patches on the wing. **127**l

**rurina** Felder
S. & C. America, W. Indies 75–85 mm. (2·95–3·35 in.)
This butterfly is widespread and often common over its range. The female has a black spot in the middle of the pale wings and red on the hind margin. It is typical in appearance of a number of species of the genus. **127**m

**statira** Cramer YELLOW MIGRANT
C. & S. America, W. Indies 50–55 mm. (1·97–2·17 in.)
A common migrant over much of its range, this butterfly can suddenly appear in thousands. It has a pale lemon yellow base and middle to the fore- and hindwings with whitish broad outer parts in the male. The females are generally yellowish green all over, with a broad band of black at the apex of the forewing. In the male this is just a thin strip of black on the edge near the apex. The migration in the W. Indies are often spectacular and the butterfly may be found hundreds of miles out at sea. The butterflies generally fly in bright sunshine, often drinking from muddy puddles or damp sand where they may be seen in large clusters.

**trite** Linnaeus YELLOW LEAF
C. & S. America 58 mm. (2·28 in.)
This is a bright yellow butterfly, with a narrow black edge round the apex of the forewing. On the underside it has a brown line diagonally across the wings looking like a leaf midrib and this distinguishes it from all the other yellow species. The females are variable from a paler to an orange-yellow and have more black on the forewing. This butterfly is found from Central America to the Argentine and is often common.

**phoebus** see **Parnassius**

**Pholisora** Scudder (HESPERIDAE) SOOTY WINGS
This is a genus of small, sooty brown or black, skipper butterflies found in N. America.

**catullus** Fabricius COMMON SOOTY WING
Canada, U.S. 25–28 mm. (0·98–1·1 in.)
This is a common and widespread species from S. Quebec, Ontario and Manitoba to Georgia and Texas in the S. It has not yet been found in Florida. The upperside is blackish brown with a few small white spots near the apex of the forewing and along the margin. This species has a rapid, erratic flight, close to the ground. The number of white spots varies, the female has 2 lines of them. The caterpillar feeds on species of Chenopodiaceae and has been reared on plants in other families (eg. Amaranthaceae).

**pholus** see **Lycomorpha**
**phorbas** see **Hypolycaena**
**phorcas** see **Papilio**

**Phragmatobia** Stephens (ARCTIIDAE)
The 50 or so species in this genus of moths form a very mixed assortment of species, some of which need to be transferred elsewhere (see *Ocnogyna)*. The genus as it stands at present is represented in Europe, Asia, South Africa and N. and S. America.

**fuliginosa** Linnaeus RUBY TIGER-MOTH
Canada, N. U.S., N. Africa, Europe and temperate Asia to Japan 30–35 mm. (1·18–1·38 in.)
This common and widely distributed species is found in both wooded and open country, at a variety of altitudes. There are usually 2 broods per year: the first emerges in April and May (after having overwintered as a caterpillar), the second in July and August. Moths are occasionally seen in flight during the day, but are normally active only at night. The forewings of *P. fuliginosa* are usually reddish brown, but are rarely greyish brown; the hindwings are a rich, rosy red, usually black at the outer margin and towards the front margin. A new N. American species, *P. lineata*, somewhat similar in coloration to *P. fuliginosa*, was described by Donahue and Newman in 1966. **369**b

**Phthorimaea** Meyrick (GELECHIIDAE)
POTATO TUBER MOTHS
Most of the species in this worldwide genus of micro-moths are small and inconspicuous, but in spite of their small size, some have a serious effect upon our food supplies. The notorious Potato Tuber moth described here is one of these. Some countries have legislation controlling import of plants which might carry these pests into areas where they are not at present known. The inspection to prevent Potato Tuber moths entering Britain is one such quarantine measure. There are many European species of this genus.

**operculella** Zeller POTATO TUBER MOTH or TOBACCO SPLITWORM
Worldwide 12–16 mm. (0·47–0·63 in.)
This species breeds continuously in warm climates on potatoes which have been kept in storage. In the U.S. there are more generations of the moth each year in the S. than in the N. The caterpillar is white or pinkish with a black head. In the early stages of its life it is a leaf-miner, later it tunnels in the large veins of the tobacco leaf or the shoots of potatoes. The moth often lays its eggs on potatoes stored in the fields or in sheds. When they hatch the caterpillar then bores into the tubers, tunnelling through, which ruins the tubers. This moth causes serious losses to potatoes in store and tobacco plants in the field and it will also feed on other plants in the same family (Solanaceae). **43**e

**Phyciodes** Hübner (NYMPHALIDAE)
A very large genus of butterflies found in the New World with many species in the U.S. They are often individually very variable over their range and many subspecies have been described.

**ianthe** Fabricius HANDKERCHIEF
S. America into the W. Indies 32–36 mm. (1·26–1·42 in.)
A black and white butterfly, with broad black outer margins and broad white central area to the fore- and hindwing. The forewing area is cut by a black band leaving a white patch near the margin. There is a slight yellow-green tinge near the base of the forewing. The butterfly is out in sunshine and is generally widespread over its range from Venezuela, Colombia to the W. Indies. *P. i. leucodesma*, the Handkerchief butterfly, is now considered to be a subspecies of *P. ianthe*.

**landsdorfi** see under **Heliconius nattereri**

**tharos** Drury PEARL CRESCENT
U.S., Canada, Mexico 33–40 mm. (1·3–1·57 in.)
Probably one of the commonest butterflies in the U.S., this butterfly is often abundant over much of its range. It is found everywhere there are open spaces, near roadsides or in meadows, it is also commonly seen drinking at puddles. When it alights it holds its wings out at the side, then moves them up and down a few times. The butterfly is very constant over its range with little pattern variation. The caterpillar feeds on species of *Aster* and Crown beards *(Verbesina)* and other Compositae. The upperside is a bright orange-brown with black margins to the wings and a row of black spots on the middle of the black margin of the hindwing. There are several other black marks on the forewing. The underside is paler, with more brown marks on the hindwing. **163**j

**phyciodoides** see **Apodemia**

**Phylaria** Karsch (LYCAENIDAE)
There are 2 species in this genus of butterflies; both are found only in tropical Africa.

**cyara** Hewitson
Tropical E. Africa 23–30 mm. (0·91–1·18 in.)
Above, the male of this species is a brilliant, iridescent violet on the forewing, and white with brown borders on the hindwing which also has 2 iridescent blue and green spots at the base of the tail. The female is white above, bordered with brown on both wings. **261**

**heritsia** Hewitson
Tropical Africa 22–28 mm. (0·87–1·1 in.)
The forewing of the male is iridescent violet and brown in ground-colour, edged with brown; the hindwing is white above, bordered with brown. Females differ in the white ground-colour of the forewing. **261**

**Phyle** Herrich-Schäffer (GEOMETRIDAE)
A few species of moths from S. America have been described in this genus. Their biology is unknown.

**arcuosaria** Herrich-Schäffer
S. America 40 mm. (1·57 in.)
Several related species are the same attractive green colour. Presumably these species are well camouflaged when sitting during the day, a time when they are at greatest risk from predators. **294**t

**phyleis** see **Chetone**

**Phyllocnistidae**
A small family of micro-moths in N. and S. America, Africa and Europe. They are now generally regarded as being related to the Tineidae although their exact systematic position is still in dispute. See *Phyllocnistis*.

**Phyllocnistis** Zeller (PHYLLOCNISTIDAE)
These micro-moths are all small, shining white species with a distinctive pattern. They are found throughout the N. Hemisphere and into Australia. The caterpillars mine in leaves, living between the upper and lower surfaces.

**saligna** Zeller
Europe including Britain 6–7 mm. (0·24–0·28 in.)
This moth is very common over its range where the pale green caterpillar makes galleries in the barks and twigs of species of willow *(Salix)*. Eventually they mine in the leaves. The caterpillars are without legs as is usual in this family and they pupate in a cocoon inside their mine. The adult moths are out in September, overwintering, to re-appear in the following April. The long slender wings, coming more or less to a point are characteristic of species of *Phyllocnistis*. **6**b

**Phyllodes** Boisduval (NOCTUIDAE)
The 10 or so species of this genus of moths are found in India, through S.E. Asia to New Guinea. They are all unusually large compared with most other species of their family.

**floralis** Butler
Burma to Borneo 140–150 mm. (5·51–5·91 in.)
The forewings of this species are considered to be leaf-mimics, so that when the moth is at rest amongst foliage the general effect is highly cryptic. The colourful black, orange and white hindwings illustrate the phenomenon of 'flash-coloration' by which the moth may be able to disconcert a potential predator either by suddenly revealing the hindwing pattern prior to flight or concealing it when coming to rest. **392**f

**Phyllonorycter** Hübner: syn. *Lithocolletis* (GRACILLARIIDAE)
A large genus of minute micro-moths, including some of the smallest known species of Lepidoptera with wing spans of half a centimetre. They generally have rather oval forewings with narrow, slender pointed hindwings. Under a magnifying glass a tuft of scales can be seen on the head. The caterpillars in this genus mine in leaves, making characteristic shaped mines which show externally as various spiral and other patterns on the leaf. Even the position of the mine on the leaf varies among the species, some mining only inside the lower side while others stick to the upperside of the leaf, differentiating their living preferences even within the thickness of one leaf.

**cavella** Zeller
N. & C. Europe including Britain 8–10 mm. (0·31–0·39 in.)
This beautiful little moth is so small that normally, as with most of its relatives, it is easily overlooked. The caterpillars feed in leaf-mines on the underside of the leaf of birch *(Betula)*. There are many relatively similar species, all beautifully marked. **6**d

**lautella** Zeller
C., W. Europe including Britain 6–8 mm. (0·24–0·31 in.)
These tiny moths have caterpillars which mine in the leaves of oak *(Quercus)*. The mines can be found on the underside of the leaves. **6**e

**loxozona** Meyrick
Africa 7 mm. (0·28 in.)
The white thorax contrasts with the forewings which

are golden orange or brownish orange, with 2 white bands across the wing. The apex of the wing has a small white triangle on the anterior margin and a larger one on the posterior margin with black scales scattered along the margin. The hindwing is grey. The caterpillar makes a long, narrow gallery mainly along the edge of the leaf. In the latter stages of the leaf-mine a gall-like swelling arises which distorts the leaf. This is unusual in the genus which usually makes leaf mines without showing side effects such as galls. The caterpillar mines the leaves of *Dombeya* species (Sterculiaceae). The moth is known from Uganda and in South Africa from the Transvaal.

**oxyacanthae** Frey
Europe including Britain 7–9 mm. (0·28–0·35 in.)
The exact distribution of this small moth is not known with certainty. It has been confused with a similar, related species which occurs in N. America. Further work on its identity and the relationship with the American species is needed. **26**

**salicifoliella** Clemens
U.S. 7–8 mm. (0·28–0·31 in.)
The white head of this small yellow moth contrasts with the golden or brownish-yellow wings, streaked in white. The caterpillar mines in the underside of leaves of poplar *(Populus)* and willow *(Salix)*.

**sorbicola** Kumata
Japan 6–7 mm. (0·24–0·28 in.)
This tiny species is known only from these islands though similar looking species are found in many parts of Europe. The caterpillars mine in leaves, spending their whole life between the upper and lower surfaces of one leaf before emerging as beautiful but minute moths.

**tremuloidiella** Braun ASPEN BLOTCH MINER
U.S., Canada 7–9 mm. (0·28–0·35 in.)
This tiny dull yellow-brown moth is found in California, Utah, Idaho and British Columbia. The caterpillar makes irregular blotch mines between the upper and lower surfaces of the leaf and causes premature leaf roll in ornamental trees, especially aspen and poplar *(Populus)*.

**Physoptilia** Meyrick (GELECHIIDAE)
The genus has unusual wing venation and its relationships are not clear. It was considered to belong to the Physoptilidae.

**pinguivora** Meyrick
Indonesia 14–18 mm. (0·55–0·71 in.)
This small moth has been bred from caterpillars feeding on the shoots of *Planchonia* (Barringtoniaceae). The specimen illustrated, which is from Java, has the typical appearance of species in this family. **43k**

**piceata** see **Hypochrysops**
**pictalis** see **Epipagis**
PIEDMONT RINGLET see **Erebia meolans**

**Pierella** Westwood (NYMPHALIDAE)
A genus of butterflies in the subfamily Satyrinae. The 40 or 50 species are all found in S. and C. America. Most of the species in the genus are shade-loving butterflies, keeping to the forests and forest edges. Little is known of their life histories.

**hyceta** Hewitson
S. America 66–77 mm. (2·6–3·03 in.)
This species is widespread in Brazil, Peru and Bolivia and occurs in the Guianas. There is some variation in colour of the hindwings between specimens from different localities. Basically this is a brown butterfly with dark lines across the forewing and an orange-brown or orange-yellow patch over most of the hindwing. In many specimens the hindwing is larger then the forewing. The underside is grey-brown in both sexes. The colour on the forewing changes depending on the angle from which it is viewed. This is caused by the diffraction grating which is explained in greater detail in the Introduction. The illustration shows the variation in colour. **188**

PIERID, TIGER see **Dismorphia amphione**

**Pieridae** WHITES or SULPHURS
This is a large, worldwide family with many species which are common and well-known. The word butterfly is probably derived from one of the species in this genus, a butter-coloured-fly. There are a number of species which fit this but possibly the Brimstone butterfly *(Gonepteryx rhamni)* has the best claim. Most species in the family are yellow or white or orange. The colours are due to pigments, which in the case of the Whites is derived from a waste product of the insect's metabolism, based on uric acids, which are not used in other butterflies for pigments. Some of the species in the genus are pests, particularly of various Cruciferae eg. cabbages etc., and a number have been accidentally introduced to continents where they were formerly unknown. Many of these introductions have had serious results, with the introduced species becoming a pest as they are able to multiply in the new area in the absence of their usual parasites and predators. Most of the species are day-flying, particularly active in warm sunshine. See *Anaphaeis, Anteos, Aporia, Appias, Archonias, Ascia, Catasticta, Catopsilia, Cepora, Colias, Colotis, Delias, Dismorpha, Dixeia, Enantia, Eronia, Euchloe, Eurema, Gonepteryx, Hebomoia, Ixias, Kricogonia, Leptidea, Leptosia, Leucodia, Mylothris, Neophasia, Nepheronia, Paramidea, Phoebe, Pieris, Pinacopteryx, Pontia, Prioneris.*

**Pieris** Schrank (PIERIDAE)
A huge genus of many species, of mostly white butterflies. Some are now worldwide, partly or entirely due to accidental introductions. The majority of the species are found in Europe, temperate Asia and N. America.

**brassicae** Linnaeus LARGE or CABBAGE WHITE
Europe including Britain, N. Africa, 56–66 mm.
all Mediterranean countries (2·2–2·6 in.)
This white butterfly is one of the commonest species over most of its range. The male butterflies have black tips to the otherwise unmarked wings, the females have 2 prominent black spots on the forewings. It is a well-known migrant, occasionally moving in large swarms. There are 2 broods each year over most of its range. The caterpillar, which is yellowish-green with darker green bands on either side and black spots, feeds on a variety of Cruciferae, and is particularly a pest of cabbages. Several subspecies have been described in different areas. The chrysalis is often conspicuous, a grey-green colour spotted or streaked with black and can be found fixed upright to walls or fences. The caterpillars are often attacked by parasites which emerge when the caterpillars go to pupate. The result is a rather dried-up caterpillar stuck to a wall surrounded by a conspicuous mass of small golden cocoons which contain the pupae of tiny parasitic wasps. They should be protected since they help to keep down the number of Cabbage White butterflies. **117**

**manni** Mayer SOUTHERN SMALL WHITE
France, Spain, Switzerland, 40–46 mm.
Balkans, N. Africa (1·57–1·81 in.)
This butterfly generally has 2 or more broods each year. It is similar to *P. rapae* but the female generally has more black at the apex of the wing and is generally smaller than that species. The caterpillar feeds on various Cruciferae including *Iberis*.

**napi** Linnaeus GREEN-VEINED or MUSTARD WHITE
Europe including Britain, 36–46 mm.
Asia including Japan, N. America (1·42–1·81 in.)
This white butterfly is easily recognized by the colour of the underside of the hindwings with the dark edging to the veins. It is common over most of its range and a number of subspecies have been described. The caterpillar feeds on various Cruciferae including *Cardamine, Brassica* and others. They are similar to the Small White but the Green Veined White caterpillars generally have dark, rather than yellowish lines on the back. **123**

**rapae** Linnaeus SMALL or GARDEN WHITE CABBAGE BUTTERFLY or IMPORTED CABBAGEWORM
Europe including Britain, N. Africa, 46–54 mm.
Canary Islands, Azores, temperate (1·81–2·13 in.)
Asia including Japan, N. America, Canada, Australia
This butterfly is very common over most of its range and is one of the earliest 'Whites' to appear in the spring. Subspecies have been described from many places, including Japan. There may be several generations a year, which differ slightly from one another. The butterfly often migrates in large numbers in the more northerly parts of its range. The species was introduced into N. America and Australia and is a pest at times on cabbages and other Cruciferae. The males have mostly all white wings with black tips, the females have more black near the base of the forewings and two black spots on them. The caterpillar, which is all green with a thin yellow line along the back and minute black dots on it, feeds on a variety of Cruciferous plants, including Cabbage, Water Cress *(Nasturtium)*, Horseradish *(Cochlearia)*. Where is has become a pest in Australia, attempts are now being made to introduce a parasite in an effort to control the butterfly. This form of biological control has many advantages over other control methods (eg. insecticides). **122**

**virginiensis** Edwards WEST VIRGINIA WHITE
U.S., Canada 36–40 mm. (1·42–1·57 in.)
Local from Ontario New England and New York to Virginia and in the Transition zone – the area where the Canadian and N. species meet those from the S. This species is similar to and has been confused with *P. napi* but the dark markings of *P. virginiensis* are more diffuse brown and underneath it lacks the yellowish tinge. The wings are semi-translucent with the veins faintly marked. It is a woodland species whose yellowish green caterpillar feeds on Toothwort *(Dentaria diphylla)*. The adults are out in early May, even in central New York.

PIERROT BUTTERFLIES see **Castalius**
PIG-WEED CATERPILLAR see **Spodoptera exigua**
**pigmentaria** see **Aphysoneura**
**pilleriana** see **Sparganothis**
PILTDOWN BUTTERFLY see **Gonepteryx eclipsis**

**Pinacopteryx** Wallengren: syn. *Herpaenia* (PIERIDAE)
A small genus of butterfly with species found over most of Africa S. of the Sahara. They generally have unusual patterning on the wings which clearly separate them from the yellow or white of the typical Pierids.

**eriphia** Godart ZEBRA WHITE
Africa S. of the Sahara 55–65 mm. (2·17–2·56 in.)
This is a common and widespread species found in open bush or grasslands and occurs from Ethiopia to South Africa. The caterpillar feeds on *Capparis* and *Maerua* (Capparidaceae). The butterfly is brownish white with yellow bands and spots, both fore- and hindwing have slightly irregular yellow or pale yellow lines. The female is larger than the male and their lines are often less well defined. The dry season form is paler than the wet season. The bands on the wing tend to run from the body towards the outer margins (seen when the wings are spread) rather than from the front to the back of the wing.

**pinastri** see **Sphinx**
PINE
BEAUTY see **Panolis flammea**
CONE MOTHS see **Eucosma**
EMPEROR see **Imbrasia cytherea**
LAPPET MOTH see **Dendrolimus pini**
LOOPER MOTH see **Bupalis piniaria**
HAWK-MOTH see **Sphinx pinastri**
PROCESSIONARY see **Thaumetopoea pityocampa**
WHITE see **Neophasia menapia**
PINE-SHOOT TORTRIX MOTHS see **Rhyacionia**
**pinella** see **Catoptria**
**pinguivora** see **Physoptilia**
**pini** see **Dendrolimus**
**piniaria** see **Bupalis**
PINK
BOLLWORM see **Platyedra gossypiella**
BORER see **Sesamia calamistis**
RICE-BORER see **Sesamia inferens**
PINK-BARRED SALLOW MOTH see **Xanthia flavago**
PIPE-VINE SWALLOWTAIL see **Battus philenor**

PIPER, GOLDEN see **Eurytele dryope**
PIRATE BUTTERFLY see **Catacroptera cloanthe**
PISTOL CASEBEARER MOTH see **Coleophora malivorella**
**pithecium** see **Phobetron**
PITCHER-PLANT MOTH see **Exyra rolandiana**

**Pityeja** Walker (GEOMETRIDAE)
There are several S. American species in this genus of moths. The forewing pattern is probably a form of disruptive camouflage in which the shape of the wing seems to be broken up. Several related genera have this type of pattern, but generally with paler coloured forewing stripes. *Pityeja* is particularly distinguished by the strong pattern.

**histrionaria** Herrich-Schäffer
Costa Rica, Guatemala 42–55 mm. (1·65–2·17 in.)
The biology is unknown. Related species have basically similar pattern, with stripes on the forewing and a relatively plain hindwing. In some species the stripes are broader, but the angle of them on the forewing is similar. **294**r

**pityocampa** see **Thaumetopoea**
**pixe** see **Lymnas**

**Plagerepne** Tams (PYRALIDAE)
The only species in this genus was described recently from Malaya. Most of the genera, which are believed to be related to this one in the subfamily Chrysauginae, are from S. America. A more typical example of the subfamily is shown in the illustration of *Tamyra penicillana*.

**torquata** Tams
Malaya, Burma 47–74 mm. (1·85–2·91 in.)
Only a few specimens of this species are known, probably not more than 6 in collections throughout the world. The underside of the body has long white scales which are conspicuous and beard-like. The females are much larger than the males. Nothing is known of the biology. **63**n

**plagiata** see **Aplocera**

**Plagodis** Hübner (GEOMETRIDAE)
A small genus of moths with species in Europe, temperate Asia and N. America, with a few species reaching into N. India. Most of the species in the genus are rather similar in colour and often in pattern.

**dolabraria** Linnaeus SCORCHED WING
Europe including Britain, Japan 31–34 mm. (1·22–1·34 in.)
The name is derived from the rather crumpled look of the moth, literally as though it has been scorched by fire. Whether this is the real motive behind the pattern or a rather anthropomorphic approach is not clear. The moth certainly is not conspicuous in woods when at rest. The twig-like caterpillar is brownish with greenish or reddish marks and darker marks along the back and has a hump near the back end. It feeds on oak *(Quercus)*, birch *(Betula)*, sallow *(Salix)* from July to September. The moth lives in woods and is on the wing in late May and June. **291**

PLAIN TIGER see **Danaus chrysippus**
PLAINS BLUE ROYAL see **Tajuria jehana**
PLANE BUTTERFLY see **Bindahara phocides**
**plantaginis** see **Parasemia**

**Platagarista** Jordan (AGARISTIDAE)
There is a single known species in this genus of day flying Australian moths.

**tetrapleura** Meyrick
N. & E. Australia 35–42 mm. (1·38–1·65 in.)
The method of sound production in the male of this species involves the movement of a file-like surface on the hind leg over a ridge on the under surface of the hindwing. In another Agaristid moth the hind legs interact with a specialized area of the forewing (see *Musurgina*). **399**m

**Platarctia** Packard (ARCTIIDAE)
Three species are currently placed in this genus of moths: *P. parthenos*, *P. atropurpurea* Bang-Haas from central and N.E. Asia, and *P. souliei* Oberthür from China (Szechwan Province).

**parthenos** Harris ST LAWRENCE TIGER-MOTH
Canada, U.S. 60–62 mm. (2·36–2·44 in.)
This is chiefly a cool temperate, subarctic, and alpine species found N. of the Arctic Circle in Alaska and Popof Island and extending S. as far as Arizona. It occurs in much of Canada, except the far N., as far as Newfoundland and in N.E. U.S. This is not unlike the equally large Old World species *Pericallia matronula* in general appearance except for the abdomen which is mostly black above. There is some variation between specimens in the extent of yellow markings on the forewing and the amount of black on the hindwings. **369**k

**Platyedra** Meyrick: syn. *Pectinophora* (GELECHIIDAE)
The species of micro-moths in this genus occur on all continents but are commoner in the subtropics and tropics where some are very serious crop pests. There are almost as many research publications on the economic species in this genus as on many other families of micro-lepidoptera.

**gossypiella** Saunders PINK BOLL WORM
Worldwide 20 mm. (0·79 in.)
The moth has rather a mottled appearance with greyish brown forewings, 2 or more poorly defined black spots in the middle and black tips to the wings, with a broad patch near the apex. The hindwings are greyish, pointed at the apex. The caterpillar is a pinkish colour with darker bands and yellowish lines. It feeds on cotton and is a serious pest in many cotton growing countries. The caterpillar eats into the flowers and the developing bolls. It is rated as the sixth most serious pest in the world. Losses in the cotton crop caused by this insect run into millions of dollars.

**Platylesches** Holland (HESPERIIDAE) SKIPPERS
These are mostly brown skipper butterflies, with a very uniform pattern on the upperside. The underside of the hindwings are more marked and differ between species. The genus is primarily African with species S. of Sahara.

**moritili** Wallengren COMMON or HONEY HOPPER
Africa 32 mm. (1·26 in.)
Typically greyish brown with white spots on the forewing forming a 'Y'. The hindwing has a faint median group of white marks. The underside of the hindwing has a greyish violet band across the middle with the outer parts dark brown. There are a number of related species, differing mainly on the hindwing underside pattern. This species is common from equatorial Africa to Rhodesia, Mozambique, Natal and Transvaal.

**Platyprepia** Dyar (ARCTIIDAE)
A single species of moth is placed in this genus.

**guttata** Boisduval: syn. *P. virginalis* RANCHMAN'S TIGER-MOTH
W. Canada, U.S. 50–65 mm. (1·97–2·56 in.)
This is a Rocky Mountain species ranging from British Colombia S. to Colorado and California. The colour pattern is similar in some respects to that of the illustrated *Platarctia parthenos* but the ground-colour of the wings is black and the yellow markings are much larger. The hindwings differ in that there is usually seldom any black at the base and that there are additional black markings along the outer margin of the wings. The abdomen is quite distinctive and is alternately banded with black and yellow. Dark varieties have almost entirely black or dark brown hindwings with a few yellow, marginal spots and a black abdomen with a yellow tail-tuft.

**Platyptilia** Hübner (PTEROPHORIDAE)
PLUME MOTHS
A worldwide genus of Plume moths from the tropics to the Arctic. The caterpillar, where known, feeds mostly on species of Compositae, Scrophulariaceae and Labiatae although other families have been recorded. The adults are rather weak fliers and hold their wings horizontally or obliquely to the body and have a characteristic habit of jerking the wings up and down when at rest. The wings are divided into 2 main lobes in the forewing and 3 in the hindwing. All species have long slender legs.

**carduidactyla** Riley ARTICHOKE PLUME MOTH
U.S., Canada, Costa Rica, Guatamala 18–30 mm. (0·71–1·18 in.)
The wings of the moths are brownish-buff with a black triangular mark on the anterior margin of the forewing. The caterpillar has been reared on artichokes in California and on species of *Cirsium* and a number of other species of Compositae. In California the moth has 3 overlapping generations a year. The adults are mainly nocturnal although they are around at dusk.

**ochrodactyla** Denis & Schiffermüller
OCHREOUS PLUME MOTH
Europe including Britain 23–27 mm. (0·91–1·06 in.)
This plume moth is typical of the genus with the forewings each divided into 2 lobes and the hindwings into 3. The long legs make the moth seem to shake on delicate springs as it hangs on leaves. The caterpillar feeds on Tansy, *Tanacetum vulgare* (Compositae). The moth appears in late June and July and is generally rather sluggish during the day. After dark it may be found on the flowers of Tansy. **78**

**pallidactyla** Haworth
Europe including Britain 23–27 mm. (0·91–1·06 in.)
This species is local in occurrence over its range. It is similar to *P. ochrodactyla* and has often been confused with that species though generally *P. pallidactyla* is the darker coloured species but there are differences in the colouring of the legs. In *P. pallidactyla* these are mostly brown, whereas *P. ochrodactyla* there is usually some white in them. The caterpillar feeds on Yarrow *(Achillea)* and sometimes Tansy *(Tanacetum)* and they mine down the stems. It hibernates in the root of the plant and in the following spring attacks the young shoots. The moth is out in June and July and is active as dusk, occasionally flying if disturbed during the day.

PLAYBOY, BROWN see **Deudorix antalus**

**Plebejus** Kluk (LYCAENIDAE)
This is a fairly large genus of butterflies found in the N. temperate zones of the New and Old Worlds.

**argus** Linnaeus SILVER-STUDDED BLUE
Europe including Britain, temperate Asia to Japan 22–26 mm. (0·87–1·02 in.)
This butterfly is found on heathlands and grassy slopes. The wings of the male are purplish blue above, with narrow black margins; the female is mostly brown above with some blue scales at the base of the wings in some specimens. The 'silver studs' are the silvery green markings near the outer edge of the under surface of the hindwing (see illustration). The caterpillar feeds on numerous low-growing plants, mostly species of Leguminosae and Ericaceae; they are often attended by ants which may carry them to foodplants near their nests. **245**

**Plebicula** Higgins (LYCAENIDAE)
The 7 species of this genus of butterflies are found in Europe, the Middle-E., S. Russia and N. Africa.

**atlantica** Elwes ATLAS BLUE
Morocco 26–35 mm. (1·02–1·38 in.)
The only known localities for this species are between 2130–2440 m. (7000–8000 ft) in the Atlas Mountains of Morocco. The male is blue above, the female, in contrast, brown with orange markings. There are 2 broods per year, the second much smaller than the first. Nothing is known about the caterpillar.

**plesiobapta** see **Milionia**

**Pleurotypa** Meyrick: syn. *Sylepta* (PYRALIDAE)
A large and rather mixed genus of Pyralid moths. At present there are a number of species in the genus *Sylepta*, some of which should be in *Pleurotypa* but further study is needed.

**ruralis** Scopoli MOTHER-OF-PEARL
Europe including Britain, N. & C. Asia 30–34 mm. (1·18–1·34 in.)
The popular name is derived from the mother-of-pearl look of the wings. The moth is widely distributed and often common. It can be easily disturbed from dense herbage or nettles in July during the day when it flutters about before looking for somewhere else to hide. The adults have the long-legged look which is typical of Pyralid moths. The light green caterpillar, which has a dark line along the back, feeds on nettle *(Urtica dioica)* in the spring probably having hibernated over winter. It spins leaves together with its silk, wriggling actively forwards and backwards if disturbed. **72**

**plexippus** see **Danaus**
**plicata** see **Callisthenia**

**Plodia** Guenée (PYRALIDAE)
Although worldwide pests of stored food, there are only 2–3 species known in this Pyralid moth genus. The one described below is a serious pest of stored foods. The cost to man of the ravages of this pest every year must be enormous. There are few countries which do not get their share of attention from this pest but it is more serious in warmer countries. The genus is closely allied to *Ephestia* from which it differs in the shape of the labial palp. These project forward in *Plodia* and upwards in *Ephestia*.

**interpunctella** Hübner INDIAN MEAL MOTH
Worldwide 13–18 mm. (0·51–0·71 in.)
The caterpillar feeds on a huge range of foods which may be stored in houses, shops or warehouses. These include dried peaches, currants, chocolate, garlic, beet, flour, maize and a long list of other foods needed by man. It also attacks herbarium specimens. The moth cannot survive out of doors in the temperate climate, but even so is common and widely distributed in these areas. It is commonly spread by the movement of infested cargoes or lorries which were not properly cleaned out. The caterpillars can survive on a relatively small quantity of food. **56**k

**plotzi** see **Epiphora vacuna**
PLUM JUDY BUTTERFLY see **Abisara echerius**
**plumaria** see **Selidosema brunnearia**
PLUME MOTHS see **Agdistis, Buckleria, Capperia, Lantanophaga, Platyptilia, Pterophorus, Trichoptilus**
PLUMED MOTHS, MANY see **Alucita**
**plumogeraria** see **Caniodes**

**Plusia** Ochsenheimer: syn. *Diachrysia* (NOCTUIDAE)
Many of the species once placed in this large genus have now been transferred to other genera of ths subfamily Plusiinae. They are chiefly N. temperate in distribution, but there are representatives in most parts of the world.

**balluca** Geyer
Canada, U.S. 40–50 mm. (1·57–1·97 in.)
This is the largest species of *Plusia* and related genera of Noctuidae to occur in N. America. It is chiefly restricted to the middle latitudes of Canada and to the N.E. U.S. as far S. as New York State. The cryptically pointed forewings are chiefly iridescent green and yellow-green above; the hindwings yellowish brown. Its caterpillar feeds on aspen *(Populus)* and hops *(Humulus)*. **391**u

**chrysitis** Linnaeus BURNISHED BRASS
Europe including Britain, temperate Asia to Japan 30–40 mm. (1·18–1·57 in.)
The highly lustrous, metallic areas on the forewing of this species vary between specimens from greenish yellow to green. The caterpillars feed on stinging nettle *(Urtica)* and, less frequently, on other plants. The moth flies both during the day and at night. **398**

**festucae** Linnaeus GOLD SPOT
Europe including Britain, temperate Asia to Japan 30–36 mm. (1·18–1·42 in.)
The caterpillar of this widely distributed species feeds on various sedges *(Carex)*, *Iris*, coarse grasses and other grass-like plants; it is green with yellow and white markings. The black chrysalis is enclosed in a cocoon which is attached to the underside of a leaf of the foodplant. **396**

PLUSIA MOTH, PEACH BLOSSOM see **Eosphoropteryx thyatiroides**

**Plutella** Schrank (YPONOMEUTIDAE)
This genus is usually placed in a separate subfamily (Plutellinae) in the Yponomeutidae. The genus is worldwide and contains one species which is probably the most widely distributed single species of Lepidoptera. Although this species is known as a migrant it has sometimes been transported artificially. The wings of the moths in this genus are elongated with an oblique outer margin to the forewing.

**xylostella** Linnaeus: syn. *P. maculipennis*
DIAMOND BACK MOTH 11–15 mm.
Worldwide (0·43–0·59 in.)
This is a serious pest of cabbages, cauliflowers and many other cruciferous plants of economic importance. The eggs are laid singly on the leaves of the plant. The fully grown caterpillar is about 8 mm. long, pale-green in colour and eats the leaves, making characteristic holes in them. In Britain it is common and over Europe large migrations of this pest occur in some years. For many years the species was known as *Plutella maculipennis* but recently the name *xylostella* was used because of the rediscovery of the original specimens used by Linnaeus for his original descriptions. It was found that the species he called *P. xylostella* was the same as the one known for years as *P. maculipennis*. Since *P. xylostella* was the older name this has now been given priority over the other. The popular name derives from the diamond-shaped pattern seen when the moth is at rest with the wings folded along the body. **41**j

**pluto** see **Xylophanes**

**Plutodes** Guenée (GEOMETRIDAE)
This is a large and widespread genus ranging from China through Indonesia. Many species have unusual wing patterns. These are made up of lines or stripes in large patches of colour.

**discigera** Butler
India, Singapore, Indonesia, Sumatra, Buru 25–28 mm. (0·98–1·1 in.)
One of the typical species in the genus with large round wing patterns. The biology is unknown. **294**h

POD-BORER, LIMA-BEAN see **Etiella zinckenella**
**podalirinus** see **Iphiclides**
**podalirius** see **Iphiclides**
**podana** see **Archips**

**Podosesia** Möschler (SESIIDAE)
This genus consists of a single species which occurs in N. America from the Rocky Mountains E. in the U.S. and S. Canada. The single included species varies in appearance in different parts of its range presumably as a mimetic response to 'local' species of the wasp genus *Polistes*.

**syringae** Harris ASH BORER MOTH
N. America 25–35 mm. (0·98–1·38 in.)
This species is readily distinguished by its exceptionally long hind legs, due mainly to the elongation of the first tarsal segment. It is a pest of considerable importance attacking ash *(Fraxinus)*, lilac *(Syringa)*, and other members of the family Oleaceae including European introductions, apparently preferring trees and shrubs under cultivation in plantations or planted as ornamentals. The caterpillar bores in the living wood, overwintering in the burrow. A thinly covered exit hole is prepared in the bark through which the adult moth emerges as early as March in the S. portion of the range and June and July in N. areas.

**Podotricha** Michener (NYMPHALIDAE)
This is a tropical S. American genus of the subfamily Heliconiinae. Very little is known about the life history or habits of the 2 included species. Both were placed, at one time, in the genus *Dryas*.

**euchroia** Doubleday
Tropical S. America 67–78 mm. (2·64–3·07 in.)
The various forms of this butterfly form mimetic associations with forms of *P. telesiphe, Heliconius telesiphe telesiphe* and *H. telesiphe sotericus* Salvin. The wing shape of *P. euchroia* is similar to that of the illustrated *P. telesiphe*, but the more posterior of the orange bands on the forewing extends to the base of the wing and the hindwing is dark brown with 2 parallel bands of orange.

**telesiphe** Hewitson
Tropical S. America 68–78 mm. (2·68–3·07 in.)
Two subspecies are known: *P. telesiphe tithraustes* which has a yellow band on the hindwing, and the illustrated *P. telesiphe telesiphe*. Between latitudes 1 °N. and 2 °S. *P. telesiphe telesiphe* flies together with a similarly patterned form of *Heliconius telesiphe*. **142**j

**Poecilmitis** Butler (LYCAENIDAE)
This is a genus of about 20 mostly S. African, dark-spotted 'copper' butterflies.

**thysbe** Linnaeus OPAL OR OPAL-CENTRED COPPER
South Africa 25–30 mm. (0·98–1·18 in.)
This iridescent, pale blue, scarlet and black species is apparently confined to dry hilly areas of South Africa where the caterpillars feed on the foliage of various plants including a species of *Zygophyllum*, a genus of semi-desert and dry grassland species. The adult male, not illustrated, has larger areas of blue above than the female. **244**t

POLAR FRITILLARY see **Clossiana polaris**
**polaris** see **Clossiana**
POLICEMAN BUTTERFLY, STRIPED see **Coeliades forestan**
**polistratus** see **Graphium**
POLKA-DOT MOTH see **Syntomeida epilais**
**pollux** see **Charaxes**
**polychloros** see **Nymphalis**
**polydamas** see **Battus**
POLYDAMAS SWALLOWTAIL see **Battus polydamas**
**polydora** see **Epicopeia**
**polydorus** see **Pachliopta**
**polygonalis** see **Uresiphita limbalis**

**Polygonia** Hübner (NYMPHALIDAE)
A genus of butterflies found mainly in Europe, temperate Asia and N. America. All are powerful fliers and are much attracted to flowers, often being numerous on the flowers of, for example, *Buddleia*. Most are strongly patterned with reds and blacks and have spiny caterpillars which often feed on nettles *(Urtica)*.

**c-album** Linnaeus COMMA
Europe including Britain, N. Africa, temperate Asia including Japan 44–58 mm. (1·73–2·28 in.)
The name is derived from the comma-shaped mark on the underside of the hindwings. The upperside, which is brightly coloured orange-brown with darker markings is conspicuous, when the insect lands. The dark underside with ragged edges to the wing forms a perfect concealment pattern. The caterpillar is black speckled with greyish, the spine on the second and fifth segments are yellowish, those on the back of the other segments are white. On each side there is a reddish line. They feed on nettles *(Urtica)*, hops *(Humulus)* and various trees. The adults fly in meadows, woodland edges and gardens in June and July, with overwintering specimens appearing in March and April. The Japanese specimens tend to be darker on the upperside than the European ones and have been described as a distinct subspecies. **204, 205**

**egea** Cramer SOUTHERN COMMA
S. Europe, Turkey to Iran 44–46 mm. (1·73–1·81 in.)
This is similar to *P. c-album* but has less black on the wings and the underside white mark is more 'Y' shaped than 'C' shaped. The caterpillar feeds on *Parietaria* (Urticaceae) in dry stony valleys and the butterflies are on the wing in May and June with a second brood in August and September. It is widely distributed in S.E. France, Italy and extends into Hungary and Czechoslovakia.

**interrogationis** Fabricius QUESTION MARK
Canada, U.S. 62–66 mm. (2·44–2·6 in.)
This species is common over most of its range from S. Canada throughout the U.S., E. of the Rocky Mountains. It is the largest of the genus in America with a prominent curved line and a dot, hence the name of the species, on the underside of the hindwing. The butterfly has a striking resemblance to a leaf when it is at rest with its wings closed. The reddish brown caterpillar, with pale spots and patches and many branching spines, feeds on elm *(Ulmus)*, hackberry *(Celtis)*, nettle *(Urtica)* and many other plants. There are 2 broods in the N., but up to 5 in the S., in one year. It is similar to the European Comma butterfly, *Polygonia c-album*. **163**n

**polymnestor** see **Papilio**
**polymnia** see **Mechanitis**
**polynice** see **Rhinopalpa**

**Polyommatus** Latreille (LYCAENIDAE)
This is a genus of tailless, usually small species of butterflies. Males are blue or brown above, females are brown. They normally fly in open country, rather weakly and close to the ground. The 50 or so species are found in Europe, Africa, temperate and tropical Asia and Australia.

**icarus** Rottemburg COMMON OR VIOLET MEADOW BLUE 26–36 mm. (1·02–1·42 in.)
Canary Islands, N. Africa, Europe including Britain, temperate Asia
As its popular name suggests, this butterfly is common almost throughout its range. It is found up to 2440 m. (8000 ft) in the Atlas mountains of Morocco and as high as 2740 m. (9000 ft) in Himalayan India. The foodplants include numerous species of the family Leguminosae. Caterpillars are able to convert the toxic cyanide content of Bird's-foot Trefoil *(Lotus)* (one of its foodplants) to harmless thiocyanate. **251**

**polyphemus** see **Antheraea**
POLYPHEMUS MOTH see **Antheraea polyphemus**
**polytes** see **Papilio**

**Polyptychus** Hübner (SPHINGIDAE)
This is a large genus of tropical African and Asian hawk-moths.

**calcareus** Rothschild & Jordan RUBY POLYPTYCHUS HAWK-MOTH 60–80 mm.
Rhodesia, Malawi (2·36–3·15 in.)
The bright red and orange coloration of freshly emerged specimens of this species fades rapidly to a purplish or reddish brown. The only distinct markings on the wings are a dark brown area near the apex of the forewing and a dark brown patch at the posterior or inner angle of the hindwing. The caterpillar of this species is unknown.

POLYPTYCHUS HAWK-MOTH, RUBY see **Polyptychus calcareus**

**Polysoma** Vari (GRACILLARIIDAE)
This genus belongs to the subfamily Lithocolletinae which are micro-moths, generally leaf-miners, in the caterpillar stage. The moths have very narrow fore- and hindwings, often beautifully marked. The African species were recently studied in detail by Dr L. Vari who produced a book on this interesting subfamily.

**aenicta** Vari
Africa 5·5 mm. (0·22 in.)
The forewings of this tiny moth are roughly parallel sided ending in a rounded apex, greyish-brown with patches of white scales making indistinct lines across the wings giving a rather blotchy effect. Many of the other species in the related genera have more slender pointed wings. The caterpillar makes irregular blotch mines on the leaves of *Albizia* (Leguminosae). At present the species is known only from a few specimens collected in Rhodesia.

**Polystichtis** Hübner (NEMEOBIIDAE)
About 15 species are currently placed in this C. American and tropical S. American genus of butterflies. They are variously coloured and differ greatly in coloration between the sexes.

**argenissa** Stoll
Panama, Colombia 36–38 mm. (1·42–1·5 in.)
The female differs radically from the brilliantly iridescent, violet-blue male in the pale brown upper surface and the oblique, white band on the forewing.

**cyanea** Butler
C. America to Ecuador 32–36 mm. (1·26–1·42 in.)
The male of this butterfly is an iridescent blue-black with a white marking on the forewing; the female is a reasonably good match for the Hypsid moth *Sagaropsis brevifasciata* Hering. As this moth also occurs in Ecuador it seems possible that it is the distasteful model, or one of the models, for *P. cyanea*. Both species are partly iridescent blue and have an oblique, orange bar near the apex of the forewing.

**flammula** Bates
Tropical S. America 36–42 mm. (1·42–1·65 in.)
This is another species remarkable for the difference in colour pattern between the males and females. The male is black with a white pre-apical bar and a scarlet patch on the rear edge of the forewing and over much of the central area of the hindwing. The female hindwing is orange and black instead of red and black and has a row of yellow marginal spots; the forewing has a broader, yellow, pre-apical bar and a broader, orange basal patch. The pattern of the female is similar in general plan to several species of Heliconiinae and Ithomiinae.

**martia** Godman 36–40 mm. (1·42–1·57 in.)
The dark markings on the forewing of this rare and striking species are approximately similar in position to those of *P. argenissa*, but the ground colour of the wings is orange except for an extensive violet apical area of the forewing.

**siaka** Hewitson
Tropical S. America 30–40 mm. (1·18–1·57 in.)
The female colour pattern is in marked contrast to that of the colourful male (illustrated); its wings are dull brown with an oblique, distal, orange band on the forewing and a submarginal orange band on the hindwing. The life-history is not known. **268**p

**Polyura** Billberg: syn. *Eulepis* (NYMPHALIDAE)
A genus of tailed butterflies in the Charaxinae which ranges from India through S.E. Asia to Australia, Fiji and is also found in Formosa. The adult butterflies are attracted by rotting fruit on which they feed. The caterpillars are characterised, as in many of the Charaxinae, by horns projecting from the head. Their habits are similar in many ways to the species of *Charaxes*, to which they are related.

**arja** Felder PALLID NAWAB
Pakistan, India, 75–85 mm.
Burma (2·95–3·35 in.)
A brown, 2-tailed *Charaxes*-like butterfly with a broad white band in the middle of the hindwing continuing on the forewing three quarters of the way to the wing margin with a white spot below the apex. There are a series of small bluish spots along the margin of the hindwing. Underneath is similar, but the white band is edged on either side by reddish brown. This is a common butterfly over much of its range flying rapidly round in the thickest jungle, feeding on rotting fruit, plant sap, but not usually found at flowers.

**delphis** Doubleday JEWELLED NAWAB
N. India, 95–100 mm.
Pakistan, Burma (3·74–3·94 in.)
This butterfly is similar in shape to *Charaxes*, with 2 tails. The upperside is pale greenish-yellow with a large black patch on the apex of the forewing, the inner edge of this black is irregular. The veins are brown, clearly marked. The hindwing is similar, greenish-yellow with a blue tinge. A few narrow marks near the margin have blue centres. The underside is pale blue, beautifully marked with the brown veins, with greenish and brown spots between and a series of pale blue, green and brown lines along the hindwing margin. The flight of this butterfly is rapid and generally round the tops of trees. It is said, of a similar looking related species, that they are conspicuous when they bask on leaves in sunshine and that it is not unusual to find wings of those caught by birds lying on the ground. There are many subspecies described of this butterfly which differ in pattern. This is very variable even in specimens from the same locality.

**dolon** Westwood STATELY NAWAB
Pakistan, North India, 80–105 mm.
China (3·15–4·13 in.)
A pale greenish white butterfly with a dark margin and apex with white spots and a small brown mark from the brown front margin of the forewing. The hindwing has two tails, and a yellowish margin with a series of brown spots which almost make a complete line. Underneath is white, with two brown lines across the forewing and one on the hindwing. The margin of the hindwing has a white line edged with brown and with small spots in the white. One of the fast flying *Charaxes*-like butterflies, always considered difficult to catch as they fly round the tops of trees. It is a pugnacious butterfly chasing after others of its kind. Generally it is found between 1220–2440 m. (4000–8000 ft) in May and June, although in parts of its range there is a second brood. **236**h

**eudamippus** Doubleday GREAT NAWAB
India, Pakistan, Burma, China, 100–102 mm.
S.E. Asia, Formosa, Okinawa (3·94–4·02 in.)
This widespread species is variable in pattern and many subspecies have been described in different parts of its range. The female is more heavily marked than the male. The butterfly is generally found in the lower land. Even in N. India it does not reach far up the mountains. **236**d, e

**pyrrhus** Linnaeus TAILED EMPEROR
Moluccas, Lesser Sunda Island, 70–90 mm.
Kai Islands, Tanimbar Island, (2·76–3·54 in.)
New Guinea, Bismarck Archipelago to N. and E. Australia
This species has many subspecies described from different parts of its range. The adults usually stay fairly high up in trees, but are attracted to over-ripe fruit. The caterpillars feed on various species of wattle *(Acacia)* and also on species of Ulmaceae and Lythraceae. They are powerful fliers and difficult to catch, although readily attracted to bait. **236**f

**polyxena** see **Charaxes, Eryphanis, Parnalius**
**polyxenes** see **Papilio**
**pometaria** see **Alsophila**
**pomona** see **Catopsilia**
**pomonella** see **Cydia**

**Pompilopsis** Hampson (CTENUCHIDAE)
There is a single species in this moth genus.

**tarsalis** Walker
Tropical Central and 26–38 mm.
S. America (1·02–1·5 in.)
This is a wasp-mimicking Ctenuchid. Its models are species of the genus *Agenia* (Hymenoptera: Pompilidae). **367**a

**ponderosa** see **Eucosma**

**Pontia** Fabricius (PIERIDAE)
A genus of butterflies with most species in Europe and temperate Asia, but with a few known from Africa. Although in catalogues of the genus there are many names, the majority of them are subspecies or forms of *P. daplidice*.

**callidice** Hübner: syn. *P. protodice* PEAK, CHECKERED OR COMMON WHITE 42–52 mm. (1·65–2·05 in.)
Europe, N. India, Pakistan, Asia, N. America
This widespread butterfly is found in the higher mountains in S. Europe, but not yet recorded in the Italian mountains, and it is found at altitudes of over 1830 m. (6000 ft). In America it is found in the mountains of California and Colorado where a distinct colour form is found, and another colour form which ranges from Quebec to Florida, W. to the Pacific. It is rare N. of New Jersey. The green caterpillar, with tubercles in rows with paler lines, lives on *Erysium* (Cruciferae) and *Reseda* (Resedaceae). **119**e

**chloridice** Hübner SMALL BATH WHITE
S. Europe, Turkey, Iraq, 40–44 mm.
Iran to Mongolia, N. America (1·57–1·73 in.)
Very similar in appearance to *P. daplidice* but smaller. There are several broods each year. The foodplant is not known. The first brood of the butterfly is out in April and May.

**daplidice** Linnaeus BATH WHITE
Europe including Britain, 40–44 mm.
temperate Asia, including Japan (1·57–1·73 in.)
This migratory species occasionally reaches Britain when its arrival causes excitement amongst collectors. The name is derived from the town of Bath in Britain where the portrait of the butterfly is faithfully recorded, embroidered on a tapestry, by a lady of Bath in the 18th century. The earlier name was Vernon's Half-Mourner, after a Mr Vernon who collected one of the earliest known specimens in Britain. The caterpillar feeds on species of Cruciferae and is green with orange-yellow bands on each side completely speckled with black. The Japanese specimens are slightly different from the European ones and are regarded as a distinct subspecies.

**pontificella** see **Epermenia**
**popeanellus** see **Acrolophus**
POPLAR
ADMIRAL see **Ladoga populi**
HAWK-MOTH see **Laothoe populi**
SPHINX, BIG see **Pachysphinx modesta**
**poppea** see **Mylothris**
**populi** see **Ladoga, Laothoe**
PORCELAIN, AFRICAN see **Marpesia camillus**

**Poritia** Moore (LYCAENIDAE) GEM BUTTERFLIES
Found in Burma, eastwards to the Philippines, the species are a brilliant greenish blue, violet-blue or orange. The males of most species are variable in colour and difficult to distinguish from each other. Most species are rare in collections. Their caterpillars resemble those of the moth family Lymantriidae; they are gregarious and processionary (see *Thaumetopoea* – Notodontidae).

**erycinoides** Felder BLUE GEM
India, Burma, 26–36 mm.
Malaysia, Indonesia (1·02–1·42 in.)
This truly gem-like butterfly is apparently a rarity. The female ought to merit perhaps a different popular name; it is orange-yellow above, with a broad, dark brown marginal band on the forewing and an almost entirely brown hindwing. **244**c

**portis** see **Morpho**
PORTLAND MOTH see **Ochropleura praecox**
**portlandia** see **Enodia**
PORTUGUESE DAPPLED WHITE see **Euchloe tages**
POSTMAN BUTTERFLIES see **Heliconius**
POSY, COMMON see **Drupadia ravindra**
**potamogata** see **Nymphula nympheata**
POTATO TUBER MOTHS see **Phthorimaea**
**potatoria** see **Philudoria**
**praecox** see **Ochropleura**
**praeneste** see **Prepona**

**Praetaxila** Fruhstorfer (NEMEOBIIDAE)
This is a genus of about 10 species found in the forested mountains of New Guinea, its associated islands, and in N. Australia, where it is the only member of the family of Nemeobiidae. There is a striking difference in coloration between the males and the females of each species.

**segecia** Hewitson
New Guinea, Aru Islands, 38–46 mm.
N.E. Australia (1·5–1·81 in.)
This rain-forest species is the only butterfly of its family found in Australia. Males are black with 3-4 white spots at the apex of the forewing and a broad, oblique, white band crossing the middle of the forewing. The female forewing is similar to that of the male but is brownish orange basal to the white band; its hindwing is brownish orange with black marginal markings.

**prasina** see **Harpeptila**

**prasinana** see **Bena**
**prattorum** see **Troides**

**Prays** Hübner (YPONOMEUTIDAE)
A worldwide genus of some 3 dozen species of micro-moths some of considerable economic importance.

**citri** Millière
Europe, temperate Asia 12 mm. (0·47 in.)
This species is a serious pest of citrus, particularly in the Mediterranean region. It has been recorded all over the world but recent work has shown that those outside Europe and Asia are not the true *P. citri*. **41**f

**oleae** Bernard OLIVE KERNEL BORER or OLIVE MOTH
S. Europe, Asia, N. Africa, 13–15 mm.
Cyprus, Turkey, S. Russia (0·51–0·59 in.)
This small moth is a serious pest of olives, particularly in the traditional olive growing areas round the Mediterranean. The damage to olives has been known since ancient times. Theophrastus (372–278 BC) records the olive tree being attacked by a worm which devours the leaves, fruit and kernel of the fruit. The Roman writer, Pliny (23–79 AD) mentions a worm 'which is born in the kernel of the live fruit'. The moth is light brown to grey-white in colour with rather pointed fore- and hindwings. It attacks the plant in a sequence of generations. The first generation attacks the leaves; the second the flowers and the third the fruit. At present the olive growing areas in the U.S., India, South Africa and Pakistan are still free of this pest. The moth is similar to the illustrated *P. citri*.

**preambilis** see **Ophthalmophora**

**Precis** Hübner (NYMPHALIDAE) COMMODORE and PANSY BUTTERFLIES
A large genus of butterflies which are found in the New and Old World Tropics. Many of the species show strong seasonal dimorphism. The genus is closely allied to *Junonia* which is sometimes regarded as a subgenus. They are characterized by the presence of eye spots on the outer parts of both wings and are often brightly coloured. They occur in most types of habitats and their strong, rapid, flight close to the ground is also characteristic. They glide repeatedly and generally only make short flights, settling on flowers with their wings half open, alternately opening and closing them like many other Nymphalids. The foodplants include *Barleria, Nelsonia, Justicia* and other species of Acanthaceae.

**octavia** Cramer
Africa 48–60 mm. (1·89–2·36 in.)
This butterfly occurs in 2 very distinct forms which occur in the wet or dry season with each form peculiar to its own season. These forms not only differ in colour but in size and shape as well. The 2 forms in Africa are, in the N.W., *octavia*, in Sierra Leone, Congo, Ethiopia and Somalia and the S. form *sesamus* in Angola, Kenya, S. through Rhodesia to the Cape. There is also a S. form, *P. octavia* form *natalensis*, which is red on both upper and undersides of the wing while *P. octavia* form *sesamus* is blue on the upperside and brown on the underside. It is common in Kenya where it can be found roosting in quarries and under the eaves of houses at night. The caterpillar feeds on *Coleus forskohli* (Labiatae), a hedging plant used by the local farmers. There are 2 generations a year in Kenya, the first in September and November and the second in April. In South Africa the caterpillar feeds on *Plectranthus, Coleus, Iboza* and other species of Labiatae. It has been suggested recently that the different colour pigments are directly caused by the different temperatures at which they are produced and not directly due to the wetness or dryness of the season. Two forms are illustrated. **185**; d, wet season; g, dry season form.

**touhilimasa** Vieillot NAVAL COMMODORE
Central Africa, 50–70 mm.
S. to Rhodesia (1·97–2·76 in.)
A deep royal blue colour with a few eyespots and a brown underside with pale transverse lines characterize this species of *Precis*. This butterfly flies on the edges of thick forest or along forest roads. It is common over much of its range.

**tugela** Trimen EARED COMMODORE or DRY LEAF BUTTERFLY
45–60 mm.
Africa (1·77–2·36 in.)
This butterfly is found throughout the year in the forested areas of W. Uganda, Kenya, and the Congo and S. to Natal and Transvaal. It is a dark brown with a broad orange-red band across each wing. This band has black veins across it with black spots between most of them. The hindwing has a tail while the apex of the forewing, often with a white spot on it, is curved, often being distinctly hooked in the dry season form. The underside is patterned like a dead leaf, no two specimens are exactly the same on the underside but all are very leaf-like. They use their protective colouring by settling on dry leaf litter. The flight is very swift if approached, but otherwise leisurely.

**prema** see **Alesa**

**Premolis** Hampson (ARCTIIDAE)
Six species are at present placed in this C. American and tropical S. American genus of moths.

**semirufa** Walker
Tropical S. America 35–58 mm. (1·38–2·28 in.)
The forewings of this species are yellow, with a partly transparent, roughly triangular area towards the apex. The hindwings and abdomen are a contrasting bright pink, a colour which possibly warns day-time predators that the moth is distasteful (an example of warning coloration). There is no doubt about the unpleasantness of the caterpillar which in Brazil causes a disease known as pararama. Small hairs from the caterpillar's back enter the skin of workers tapping the rubber tree *Hevea brasiliensis* and cause inflammation or, in chronic cases, disability of the fingers. **364**q

**Prepona** Boisduval (NYMPHALIDAE)
A genus of butterflies noted particularly for their very rapid flight. It has been suggested that they are the fastest fliers in the butterfly world. They are found in central and S. America where many species have been described. The caterpillars, where known, feed on species of Annonaceae, and are almost spineless, compared with many other Nymphalid caterpillars which are very spiny. The butterflies make a crackling sound, like thick paper being screwed up, when flying. They commonly feed off sap exuding from trees and drink freely by streams and from damp mud.

**deiphile** Godart
Brazil 88–95 mm. (3·46–3·74 in.)
This butterfly is rare and local in the high mountains of S. Brazil. It generally flies high and is rarely attracted to bait. The vivid blue iridescent upperside is only in the male, the female is brown patterned without the iridescence. **234**

**demophoon** Hübner SILVER KING SHOEMAKER
S. America 100–120 mm. (3·94–4·72 in.)
This species lives in S. America as far S. as Paraguay and Uruguay. When alive the butterfly has a smell of vanilla about it. They live in the forest flying high in the trees and are attracted to rotting fruits, which are often used to lure them away from tree tops for the collector. The caterpillars feed on species of Annonaceae. The sexes are similar, the females usually larger than the male. This is a dark butterfly with a large bluish green iridescent patch on each wing almost from the front of the wing to the back. It has many described subspecies including *P. demophoon antimache* Hübner, which was formerly regarded as a distinct subspecies.

**meander** Cramer BANDED KING SHOEMAKER
S. America, W. Indies, 85–105 mm.
C. America (3·35–4·13 in.)
This butterfly has a very rapid flight but is readily attracted to rotting fruit, otherwise they tend to fly high round tree tops. It is found in Trinidad, but is fairly widespread in Brazil, Peru, Bolivia and Colombia. Many distinct subspecies and forms have been named in different parts of its range, often with small but striking differences in colour or pattern. **236**b

**praeneste** Hewitson
S. America 90 mm. (3·54 in.)
In the illustration the large scent scale patches in the

hindwings can be clearly seen as prominent tufts. Several subspecies have been described. This butterfly is found in Colombia but is generally regarded as rare although some of the subspecies are commoner than others, including those found further S. in Peru. **230**c

**pretiosa** see **Cyanopepla**
PRETTY WIDOW see **Eupithecia venosata**
**priamus** see **Ornithoptera**
**priapus** see **Parides**
PRIDE, MOUNTAIN see **Meneris tulbaghia**
**primella** see **Compsoctena**
PRINCE MOTHS see **Goodia, Ludia, Micragone**
PRINCELING, DEWITZ'S see **Vegetia dewitzi**

**Prioneris** Wallace (PIERIDAE)
A genus of butterflies found from India, through S.E. Asia into Indonesia.

**thestylis** Doubleday SPOTTED SAWTOOTH
India, Pakistan, Burma, S. China 70–90 mm. (2·76–3·54 in.)
A white butterfly, with white streaks or white spotted apex to the rather pointed forewing. The hindwing has a white spotted, black border. The female mimics the Hill Jezebel butterflies *(Delias* sp.) and generally it is found in the same localities. In the male the underside is similar to the Jezebel butterflies with a black hindwing with yellow spots. Since the underside is mimetic of a distasteful species it is not surprising to find that this species settles, closing its wings and displaying the mimetic side. The butterfly is common at flowers and is found up to 1220 m. (4000 ft).

**Prionxystus** Grote: syn. *Xystus* (COSSIDAE)
A small N. American genus of moths with caterpillars that bore into trees.

**robiniae** Peck CARPENTER WORM
U.S., Canada 50–75 mm. (1·97–2·95 in.)
The forewings are nearly white, translucent, irregularly marked with black. The male has bright yellow hindwings with blackish edges. The female has smokey grey underwings. The caterpillar makes tunnels up to 13 mm. in diameter in tree trunks, preferring soft woods, such as a poplar *(Populus)* or locust *(Robinia)*. The moth is widespread throughout the U.S. and S. Canada and is often attracted to light.

PRIVET HAWK-MOTH see **Sphinx ligustri**
**processionea** see **Thaumetopaea**
**procilla** see **Panacea**

**Proclossiana** Reuss (NYMPHALIDAE)
A genus of butterflies, related to the Argynnid fritillaries found in Europe and N. America.

**eunomia** Esper BOG FRITILLARY
N. & W. Europe, N. Asia, N. America 32–46 mm. (1·26–1·81 in.)
This fritillary is found in wet meadows and marshes, and sphagnum bogs from sea level to 1520 m. (5000 ft). The caterpillar feeds on Bistort *(Polygonium bistorta)* and *Vaccinium* (Ericaceae). The butterfly is out in June and early July and has a swift 'flickering' flight. It is very local with only a few colonies in W. Europe but there is another subspecies which is common further N. In America, it ranges S. from the Arctic to Labrador and in the Rocky Mountains down to Colorado.

**procris** see **Limenitis**

**Prodoxus** Riley (INCURVARIIDAE)
The moths in the genus include species whose caterpillars feed on yuccas. They are related to the Yucca Moth, *Tegeticula,* and have a similar life history but they are not associated with pollination of the flowers of Yucca. The species are found in the U.S. and in C. America.

**quinquepunctellus** Chambers
U.S., Mexico 12–19 mm. (0·47–0·75 in.)
The adult moth has a white head and thorax. The whitish forewing may be unmarked or may have a number of black spots on it. The hindwings are darker with a white fringe. The caterpillar bores into the stems of yucca. They are legless are resemble the grub of weevils (Coleoptera). Pupation occurs in the yucca stalks or fruit with a relatively brief pupal period. The pupa is smooth and adult moths emerge from the pupae in April, May and June.

**prola** see **Panacea**
**promethea** see **Callosamia**
PROMETHEA MOTH see **Callosamia promethea**
**prometheus** see **Caligo**
PROMINENT MOTHS see **Disphragis, Pheosia, Pterostoma, Notodontidae**
**pronuba** see **Noctua**
**pronubana** see **Cacoecimorpha**

**Proserpinus** Hübner (SPHINGIDAE)
This is a small group of species restricted to the N. Hemisphere. One species occurs in the Old World, 6 in the New.

**juanita** Strecker
Canada, U.S., Mexico 45–60 mm. (1·77–2·36 in.)
The original material from which this species was described was collected on the banks of the Rio Grande between Mexico and Texas. It also occurs as far W. as Arizona and N. to Alberta. The caterpillar feeds on evening primrose *(Oenothera)* and willowherb *(Epilobium)*; it has no terminal horn – only a raised black area, with a whitish circle in the centre. **344**v

**vega** Dyar
U.S. 50–55 mm. (1·97–2·17 in.)
This is the rarest species of its genus in N. America, having been recorded only from S. Texas and New Mexico, although, it occurs also in Arizona and is more widely distributed in Mexico. It is similar in pattern to *P. juanita* but has broader, less greyish forewings and has a clear green area at the base of the forewing.

**Protambulyx** Rothschild & Jordan (SPHINGIDAE)
The range of this genus encompasses much of tropical C. and S. America. Two species *P. strigilis* and *P. carteri* are represented in S. Florida (U.S.). Unlike other genera in its tribe, Smerithini, the proboscis is fully developed and functional.

**strigilis** Linnaeus
Tropical C. & S. America, S. Florida 95–120 mm. (3·74–4·72 in.)
This species is occasionally recorded from Florida, but is very common in the rest of its range. There are 2 distinctive colour-forms; one with brownish red forewing, the other with yellowish brown forewings. The hindwings are orange or yellowish orange. The caterpillars are known to feed on the foliage of *Anacardium* and other genera of Anacardiaceae.

**proterpia** see **Eurema**
**protesilaus** see **Eurytides**
**proteus** see **Urbanus**

**Protographium** Munroe (PAPILIONIDAE)
There is a single known species in this genus of butterflies. It is related to species of *Graphium* and *Eurytides.*

**leosthenes** Doubleday FOURBAR SWORDTAIL
Australia 45–54 mm. (1·77–2·13 in.)
This species is similar in colour pattern to *Graphium aristeus* which also occurs in Australia, but as its common name implies it has only 4 bars, which are incomplete, on the forewing and differs slightly in the hindwing colour pattern. The butterflies fly in rainforests and scrubland regions. Its caterpillars feed on climbing species of *Rauwenhoffia* (Annonaceae).

**Prototheora** Meyrick (PROTOTHEORIDAE)
These rare micro-moths regarded as very primitive species, are known only from relatively few specimens. This genus has only been found in Africa and there are very few known species.

**petrosoma** Meyrick
S. Africa 21 mm. (0·83 in.)
The fore- and hindwing venations are similar in the species. All the species in the genus, as in the rest of the superfamily Hepialoidea, have this characteristic. Nothing is known of their biology but a study of their life history could provide data useful in the study of the relationship of the genus and species with other Lepidoptera. The wing coupling mechanism, with the lobe (jugum) from the forewing overlapping the hindwing, is characteristic of the more primitive Hepialoidea. **5**h

**Prototheoridae**
A small family of less than a dozen species. These are found in Africa and Australia and are related to the Swift moths (Hepialidae). The moths in this family, which is regarded as a primitive one, are mostly small and inconspicuous with similar venation in the fore- and hindwings. See *Prototheora.*

PROVENCE BURNET see **Zygaena occitanica**
**prunaria** see **Angerona**
**pruni** see **Strymonidia**
**pruniella** see **Argyresthia, Coleophora**
**Psephea** see **Aganais**

**Pseudacraea** Westwood (NYMPHALIDAE)
The species in this genus of butterflies are mostly rather rare and mimic Danaid or Acraeid butterflies. These latter are known to be distasteful and the *Pseudacraea* presumably derive protection from this. They are known from Africa, S. of Sahara, Madagascar and the Comoro Islands. The caterpillars have finely branched spines on each side of the back, with those on the second and eleventh segments extra long. The striking similarity to many species of *Acraea* means that care is needed to identify the *Pseudacraea* specimen in the field.

**boisduvalii** Doubleday TRIMEN'S FALSE ACRAEA
Africa 68–85 mm. (2·68–3·35 in.)
This is strikingly like species of *Acraea* but is not closely related. It is evidently a mimic of the distasteful species and enjoys the protection from birds and small mammals that this gives. The male is bright red with black spots and a black border with an orange bar below the apex and a greyish tinge behind. The female has broader wings but is otherwise similar. This species is found in the wet forest areas, particularly near streams in tropical W. and C. Africa, S. to Mozambique and the E. Cape. The caterpillar feeds on *Chrysophyllum* and *Mimusops* (Sapotaceae). *P. boisduvalii* mimics almost precisely *Acraea egina,* but is larger than that species and it also has local races which themselves correspond to local races of *A. egina.* **231**h

**eurytus** Linnaeus FALSE WANDERER
South Africa to tropical Africa 52–74 mm. (2·05–2·91 in.)
This is a very variable butterfly which mimics species in the subfamily Acraeinae. These are distasteful species and *P. eurytus* forms local populations which mimic the particular local distasteful *Acraea* species. It is generally a forest species and to identify specimens correctly they need to be caught to see whether they are the model or *P. eurytus* the mimic, the patterns are so similar. Generally it is a blackish species with rounded wing and apices with a yellow band on both wings, narrower in the forewing. This species frequently mimics *Bematistes aganice.* The caterpillars feed on species of *Chrysophyllum* (Sapotaceae).

**Pseudaphelia** Kirby (SATURNIIDAE)
This is a genus of about 10 tropical and S. African day-flying or crepuscular moths, somewhat similar in appearance to Apollo *(Parnassius)* butterflies but not at all closely related to them.

**apollinaris** Boisduval APOLLO MOTH
Congo Basin, E. & S. Africa 54–82 mm. (2·13–3·23 in.)
Not only are the shape and colour of this white or yellowish white moth reminiscent of the Apollo butterflies, but also the manner of flight. The black and yellow banded caterpillar, too, is unusual in having a horn at its posterior end like that of most hawk-moths (Sphingidae) caterpillars. The chrysalis is short, with a long, pointed, terminal process. **331**n

**Pseudarbela** Sauber (PSEUDARBELIDAE)
A small genus of micro-moths found in the Philippines and New Guinea. Nothing is known of their biology.

**semperi** Sauber
Philippines 26–30 mm. (1·02–1·18 in.)
This has the long abdomen and slender wings characteristic of the genus. The forewings are iridescent bronze, the female has a green iridescence and lacks the large yellow patch found in the centre of the hindwing of the male. The biology is unknown, but probably a study of this and more knowledge of the biology and morphology of the adult moth would help in deciding the relationship of this moth with others in the Lepidoptera.

**Pseudarbelidae**
This small family of moths was previously thought to be related to the Cossidae. Now it is placed in the Tineoid complex as being near the Psychidae. When the family was recently recognized as distinct from the Cossidae it was estimated that only some 20 specimens of the whole family were known in collections all over the world. See *Pseudarbela*.

**Pseudargynnis** Karsch (NYMPHALIDAE)
This genus of butterflies gets its name from the resemblance it bears to the Argynnid Fritillaries. The reason for the resemblance is not known, whether it derives any benefit from this, or whether, given a certain basic pattern, only a certain number of variations are possible and therefore some unrelated species tend to look similar. The genus is known only from Africa.

**hegemone** Godart FALSE FRITILLARY
Africa 50–55 mm. (1·97–2·17 in.)
Primarily a swamp-dwelling butterfly found throughout the year from W. Africa to the Zambesi River. The wings are orange-brown with numerous black marks and a double row of black spots along the forewing margin. The hindwing has a wavy, incomplete blue line near the margin and a row of black dots inside. Sometimes there are a few more black spots on the orange-brown hindwings. Little seems to be known of the biology of this species.

**Pseudargyrotoza** Obraztsov (TORTRICIDAE)
A genus of micro-moths with 4 species at present known. They are found in Europe and temperate Asia.

**conwagana** Fabricius
Europe including Britain 11–15 mm. (0·43–0·59 in.)
The yellowish white caterpillar feeds on the seeds of ash *(Fraxinus)* and privet *(Ligustrum)*, both species in the Oleaceae. The moth is very common in June over most of its range. Both sexes are similar, although the females are usually larger than the males. **20**

**Pseudatteria** Walsingham (TORTRICIDAE)
This is a genus of micro-moths with large, brightly coloured species, probably some of the most remarkable-looking of all the S. American Tortricids. The patterns on the wing are quite characteristic of the species in this genus. There are no data on the biology or life history of species in this genus. Thus the significance of the bright colours and pattern is not understood although it is assumed to be a warning coloration. There are several other species in different genera which resemble species of *Pseudatteria* and these are presumed to be mimetic resemblances. Since all the species of *Pseudatteria* are brightly coloured and conspicuously patterned it is assumed that they are the distasteful models and the species in other genera, which look like them, the mimics. Field observations on these conspicuous tortricids in S. America would be of great interest and of value if only to see if the speculation about habits from appearances is correct!

**volanica** Butler
Colombia, Peru, C. America 31 mm. (1·22 in.)
There are several very similar species, with the same type of pattern, from S. America. Nothing is known of their biology. *P. volanica* has 2 subspecies, one mainly S. American, the other mostly in C. America. **11**b

**Pseudergolis** Felder (NYMPHALIDAE)
A genus of butterflies with 2 species known. Each species has several subspecies in different parts of their range. These extend from India to Indonesia. The genus is related to the *Acraea* butterflies.

**wedah** Kollar TABBY BUTTERFLY
Pakistan, India, Burma 55–65 mm. (2·17–2·56 in.)
This is a common species, a dark golden-brown with a number of darker lines across the wing, from the front to the back. There are a series of lines near the base of the forewing across the cell, usually 4 short lines. It is almost entirely confined to the edge of rivers and streams from the lowlands to 2440 m. (8000 ft) in the Himalayas. It has a gliding flight, settling with wings spread out on stones by the streams.

**pseudofea** see **Brephidium**

**Pseudoips** Hübner (NOCTUIDAE)
Seven species are placed in this genus of moths. It is represented in Europe and Asia as far E. as China and Japan.

**fagana** Fabricius GREEN SILVER LINES
Europe including Britain, temperate Asia 32–40 mm.
E. to Siberia and Japan (1·26–1·57 in.)
The confusion that existed about the name of this species and the Scarce Silver Lines, *Bena prasinana*, is discussed under the latter. The caterpillar feeds on oak *(Quercus)*, birch *(Betula)*, beech *(Fagus)* and hazel *(Corylus)*. **401**

**Pseudolycaena** Wallengren (LYCAENIDAE)
A small genus of butterflies found in Mexico S. to tropical S. America. The species are large, bright blue or bluish green above, and pale purple beneath. The hindwing has 2 tails. The black-spotted pattern of the under surface resembles that of the genus *Lycaena*.

**marsyas** Linnaeus CAMBRIDGE BLUE
Tropical S. America 42–49 mm. (1·65–1·93 in.)
The common name of this butterfly refers to the light blue of the upper surface of the wings (the colours of Cambridge University). The female is smaller than the male; an unusual phenomenon in the butterflies. The under surface of both sexes is pale grey, marked with black spots.

**pseudonectis** see **Leguminivora tricentra**

**Pseudonympha** Wallengren (NYMPHALIDAE)
This is a small genus of butterflies in the subfamily Satyrinae. They are typical grassland species found from sea-level to over 2740 m. (9000 ft). The species in this genus are confined to the mainland of Africa where they are found over most of the continent S. of Sahara.

**narycia** Wallengren SMALL HILL-SIDE BROWN
South Africa, Rhodesia 28–32 mm. (1·1–1·26 in.)
This brown butterfly has a large squarish or oval orange patch round the eyespot on the apex of the forewing. The hindwings are plain brown without eyespots on the upperside although there are small ones round the edges on the underside. It is generally found in the drier, interior parts of South Africa and the adult flies from November to April in rocky hills, treeless or wooded country. It is rather slow weak, flier. The life history in unknown.

**Pseudopanthera** Hübner (GEOMETRIDAE)
This genus of moths includes species found in Europe and through temperate Asia.

**macularia** Linnaeus SPECKLED YELLOW
Europe including Britain, 25–28 mm.
temperate Asia (0·98–1·1 in.)
The popular name for this moth describes it adequately. The caterpillar, which is pale-green with lines along the back and sides, feeds on wood sage *(Teucrium)*, woundwort *(Stachys)*, deadnettle *(Lamium)* and mignonette *(Reseda)*. This day-flying moth is out in late May or June and generally to be found in woods. **310**

**Pseudosphex** Hübner (CTENUCHIDAE)
As the generic name suggests this is a group of about 30 wasp-like moths, probably mostly day-flying, which have joined the wasps in a mimetic partnership of distasteful (moths) and venomous (wasps) species. They are all found in C. and tropical S. America.

**crabronis** Druce
C. & tropical S. America 30–35 mm. (1·18–1·38 in.)
This is typical of most species of its genus, with black head and body and transparent brown wings. **367**d

**kenedyae** Fleming
Trinidad only 27–33 mm. (1·06–1·3 in.)
Not only is the general appearance of this day-flying species wasp-like, but also the manner of flight and walking, the angle at which the wings are held and the constant motion of the antennae. The similarity between *P. kenedyae* and several Trinidad wasps is extremely close. Birds, lizards and ants reject it as a source of food in the same way that they reject its wasp partners.

**rubripalpis** Hampson
Brazil 34–40 mm. (1·34–1·57 in.)
This dark, purple-black species closely resembles wasps of the genus *Synoeca*. It is similar in general appearance to the illustrated *P. crabronis*, but the wings are a transparent, iridescent dark violet and dark brown.

**pseudospretella** see **Hofmannophila**
**pseudotsugata** see **Hemerocampa**
**psi** see **Acronicta**

**Psyche** Schrank (PSYCHIDAE)
The larvae of these small micro-moths live in portable cases which they make themselves. They feed mostly on lichens.

**casta** Palla
Europe 12–14 mm. (0·47–0·55 in.)
The purplish brown caterpillar of this small moth makes a case which it covers with fragments of grass or lichen. It is widespread over Europe but rarely common. **24**

PSYCHE BUTTERFLY see **Leptosia nina**

**Psychidae** BAGWORM MOTHS
A worldwide family, sometimes of economic importance. The family gets its name from the case or bag made by the caterpillar. Although the males are always winged and can fly, many species have wingless females. Not only are they unable to fly, in some species they do not leave their pupal case and cannot even walk. These females stay inside the pupal case and fertilization by the males occurs without the females ever leaving the cases. The eggs are laid in the cases and caterpillars when they hatch, spread out over the trees. The cases themselves are remarkable for their ornamentation and are some of the most unusual structures made in the animal kingdom. Usually they are made of twigs or leaves often cut to very regular shapes. The shapes of the pieces are often characteristic of the species. See *Phalacropterix*, *Psyche*, *Solenobia*, *Thyridopteryx*.

**Psychomorpha** Harris (AGARISTIDAE)
Both species of this genus of moths are N. American in distribution.

**epimenis** Drury GRAPE-VINE EPIMENIS
U.S. from New England to Texas, 22–28 mm.
also in some mid-W. States (0·87–1·1 in.)
The red, white and black caterpillars of *P. epimenis* which are almost as colourful as the adults and feed in exposed positions on the leaves of grape vines *(Vitis)*. **402**f

**Pterodecta** Butler (CALLIDULIDAE)
There are 2 known species in this genus of day-flying moths. Their range includes Kashmir, N. India, China, E. Russia and Japan.

**anchora** see under **P. felderi**

**felderi** Bremer
China, E. Russia, Japan 27–32 mm. (1·06–1·26 in.)
The flight of this species is said to resemble that of hairstreak butterflies *(Thecla* and its allies). They appear in the spring and can be common in some localities. The species *P. anchora* from Kashmir and Himalayan India differs from *P. felderi* in details of the colour pattern. The ground-colour of the upper surface of the wings varies considerably between individuals and can be uniformly reddish, greenish or yellow-brown or marked with darker bands and speckles; both wings have a normally conspicuous white mark at the end of the cell. **272**u

**Pterolonche** Zeller (PTEROLONCHIDAE)
Less than 12 species are known in this small genus of rather similar looking micro-moths. They are found mostly in the countries around the Mediterranean, but there is one species in America and one in South Africa.

**pulverentella** Zeller
N. Africa, Europe 20–30 mm. (0·79–1·18 in.)
This species is white with yellow-brown wings giving it an overall pale brown appearance. The underside is unmarked and is a plain smokey brown colour. The caterpillar spins a small web on its food plants which are species of Leguminosae. **43**a

**Pterolonchidae**
A small family of micro-moths at present considered to be in the superfamily Gelechioidea. Very few species are known and the family relationships themselves are uncertain. See *Pterolonche.*

**Pterophoridae** PLUME MOTHS
These long legged fragile looking moths get their name from the plume-like, hairy wings. In most species the wings, both fore and hind, are divided into several lobes, generally the hindwing is divided into 3 lobes each of which are fringed with hairs. There is one subfamily (Agdistinae) where the wings are not divided, but they still have the delicate long-legged look of the more typical plume moths. The caterpillars, sometimes mining leaves in the early stages, are generally covered with long fine hairs. The pupal case is often similarly spiny. The adults are weak fliers, easily disturbed from the resting places but they generally only make short flights. Some of the species are attracted to light. The family is worldwide. See *Agdistis, Amiblyptilia, Buckleria, Capperia, Emmelina, Lantanophoga, Leioptilus, Megatoriphida, Platyptilia, Pterophorus, Trichoptilus.*

**Pterostoma** German (NOTODONTIDAE)
This is a small genus of moths found in Europe E. to China and Japan. They are brownish, greyish or yellowish white in wing colour, marked with darker brown especially on the wing-veins. The palps are unusually long.

**palpina** Clerck PALE PROMINENT
Europe including Britain, temperate 38–45 mm.
Asia to E. Russia and Japan (1·5–1·77 in.)
The illustration of this species shows the processes of the inner margin of the forewing which are a noticeable feature of many species of Notodontidae, especially the 'Prominents'. There are often 2 broods in a year except in the north of its range. The colourful caterpillar is bluish green, longitudinally striped with black and yellow lines; it feeds on poplar *(Populus)* and willow *(Salix)*. The chrysalis is protected by a cocoon, usually placed amongst grass roots at the base of the tree on which the caterpillar feeds. **354**

**Pterophorus** Schäffer (PTEROPHORIDAE)
PLUME MOTHS
The plume moths of this genus are worldwide in distribution and their strange, long-legged appearance and 'feathery' wings puzzle many people when they first see them. Although common they are inconspicuous unless attracted to light. They are a little studied but fascinating group with many different and interesting life histories. A few are minor pests, but one species, whose caterpillar feeds on scale insects, has been used in attempts to control these pests.

**pentadactyla** Linnaeus WHITE PLUME MOTH
Europe including Britain 26–29 mm. (1·02–1·14 in.)
This completely white plume moth is probably one of the most conspicuous of plume moths. The caterpillars feed on *Convolvulus*. Similar looking species are found in many countries. Probably this insect, when first seen, arouses more curiosity than any other when it is attracted to light, resting on a window, it presents an odd appearance and it is difficult to persuade the finder of this unusual insect that it is not an 'unknown one' but is a common moth! **55**

**periscelidactyla** Fitch GRAPE PLUME-MOTH
U.S. 18 mm. (0·71 in.)
The forewings, which are divided into 2 lobes, are yellowish brown with brown marks. The hindwings are divided into 3 lobes with the first 2 chocolate brown and the third lobe white. The tiny hairy caterpillar spins a web on the growing tips of the grape, feeding on the tip. The damage is most serious where it occurs early in the year, otherwise the caterpillar only damages shoots which are later pruned away.

**Pterothysanidae**
There are four known genera in this family of day-flying moths. *Pterothysanus* Walker is a small genus from India, the other equally small genera are African *(Hibrildes* Druce and *Pterocerota* Hampson) or Madagascan *(Caloschemia* Mabille). The other genera listed in volume 14 of Seitz's *Macrolepidoptera of the World* are now placed in the Lymantriidae. The family is currently placed in the same superfamily as the Geometridae. The Pterothysanidae are generally regarded as primitive members of the superfamily Geometroidea, but have been united with the Callidulidae another Geometroid family, in a superfamily Calliduloidea by some authors. See *Hibrildes, Pterothysanus.*

**Pterothysanus** Walker
This is a genus of black and white, mottled, day-flying moths found in N. India, Sikkim, W. China, Burma and Thailand. The inner margin of the hindwing bears a fringe of long, fine hairs. Nothing is known about the life history of any of the species; 4 have been described, but at least 2 more await description.

**noblei** Swinhoe
Burma, Thailand 42–58 mm. (1·65–2·28 in.)
The large, pink, marginal spots on both pairs of wings distinguish *P. noblei* from the more widespread *P. laticilia* Walker which is found in India, W. China and Burma. The N.E. Indian *P. astratus* is similar to *P. noblei*, but has much more brown on the wings and lacks pink spots on the front of the forewing. **272**

**Ptilodon** Hübner (NOTODONTIDAE)
Many of the species now included under *Ptilodon* have been combined with the generic name *Lophopteryx*, a later name which is a synonym of *Ptilodon*. There are about 20 species, mostly N. temperate in distribution; including Himalayan India. The caterpillars are glossy and have a pair of tubercles at the posterior end of the abdomen.

**capucina** Linnaeus COXCOMB PROMINENT
N. & C. Europe including Britain, 35–42 mm.
Asia to Siberia and Japan (1·38–1·65 in.)
The forewings of this common species are yellowish or reddish brown, marked with 2 dentate, dark brown lines. The hindwings are yellow brown, marked only at the anal angle which is normally the only part of the wing uncovered by the forewings when the moth is at rest. The name 'Coxcomb' refers to the scaled process at the inner margin of the forewing which projects above the back when the moth is at rest. The caterpillar feeds on birch *(Betula)*, oak *(Quercus)* and other deciduous trees; it is often common in urban areas. There are 2 broods in a year.

PUDDLE BUTTERFLY, MUD see **Colias philodice**
**pudens** see **Euthyatira**
**pudibunda** see **Dasychira**
**pudica** see **Cymbalophora**
PUFFIN BUTTERFLY, SPOT see **Appias pandione**
PUG MOTHS see **Chloroclystis, Eupithecia**
**pulchella** see **Egira, Erasmia, Panacra, Utetheisa**
**pulchellata** see under **Eupithecia linariata**
**pulcherrima** see **Rhipha**
**pulverentella** see **Pterolonche**
**pumila** see **Danaus**
**pumilum** see **Microsphinx**
**punctella** see **Atteva**
**punctidactyla** see **Amblyptilia**
**pura** see **Hyblaea**
PURE WHITE see **Appias drusilla**
PURPLE, RED SPOTTED see **Limenitis arthemis**
PURPLE
AZURE, DARK see **Ogyris abrota**
BANDED BUTTERFLY see **Limenitis arthemis**
EMPEROR BUTTERFLIES see **Apatura iris**
HAIRSTREAK see **Quercusia quercus**
HAIRSTREAK, GREAT see **Atlides halesus**
LESSER FRITILLARY see **Clossiana titania**
LONG HORN see **Adela bella**
MARBLED MOTH see **Eublemma ostrina**
MORT-BLEU see **Eryphanis polyxena**
SPOTTED SWALLOWTAIL see **Graphium weiskei**
THORN see **Selenia tetralunaria**
TIP BUTTERFLIES see **Colotis**
WINGS see **Eunica**
PURPLE-BORDERED GOLD see **Idaea muricata**
PURPLE-EDGED COPPER see **Palaeochrysophanus hippothoe**
PURPLE-SHOT COPPER see **Lycaena alciphron**
PURPLISH COPPER see **Lycaena helloides**
**purpuralis** see **Pyrausta, Zygaena**
**purpurata** see **Diacrisia**
**purpuripulcra** see **Chrysamma**
PURSLANE, GLOVER'S see **Copidryas gloveri**
**pusillidactyla** see **Lantanophaga**
PUSS CATERPILLAR see **Megalopyge opercularis**
PUSS MOTH see **Cerura vinula**
**pustulella** see **Atteva punctella**
**puziloi** see **Luehdorfia**
PYGMY BUTTERFLIES see **Brephidium**
**pylades** see **Graphium**
**pyraliata** see **Eulithis**

**Pyralidae**
A huge family of moths with many species of considerable economic importance. The species are very variable but basically have similar wing venation, long legs and a rather translucent look plus other morphological modifications. The family is worldwide and is divided into a number of subfamilies. The Crambinae is a subfamily with many pest species, including stalk-borers and related pests. Generally at rest they sit on grass stems with their wings rolled round the body. This gives them a characteristic wedge-shaped appearance. The Schoenobiinae are a smaller subfamily with some species of economic importance. Other subfamilies include Galleriinae, the Wax moths or Bee moths, Phycitinae, the Knot-horn moths, again with many species of economic importance, many of which are pests of stored food and grain. The Pyraustinae, Nymphulinae, Scopariinae are large subfamilies with many varied species. The Pyralinae include many species of diverse habit. Finally there are a number of other subfamilies including the Chrysauginae, Epipaschiinae, Linostinae, Cybalominae, Midilinae. The whole family has species which occupy a huge range of habitats, from species with caterpillars aquatic in habit to other species whose cactus-feeding habit has been put to good use by men in biological control work. One species is reputed to have caterpillars which feed only in the hair grooves of Sloths, but recent work has not supported this.

**Pyralis** Linnaeus (PYRALIDAE)
Many species in this worldwide genus feed on hay, straw or stored cereals and can be of some economic importance. The genus at present has several hundred species in it but they are a very mixed and probably unrelated group of moths.

**farinalis** Linnaeus MEAL MOTH
Europe including Britain, N. & C. Asia, 20–28 mm.
Japan, N. America, Australia, (0·79–1·1 in.)
New Zealand
There is some variation in colour of specimens of this

species, often there is a strong purple tinge. The caterpillars feed in grain warehouses, mills, barns and stables on stored cereals and cereal refuse, living in a long tough silken tube. The moth has been collected far out at sea on lightships and probably migrates long distances. **62**

PYRALIS, VINE see **Sparganothis pilleriana**
**Pyrameis tameamea** see under **Vanessa tameamea**
**pyramidoides** see **Amphipyra**

**Pyrausta** Schrank (PYRALIDAE)
This generic name has long been used for new species described in the subfamily Pyraustinae, when the describer was not sure of the correct genus to apply! As a result it is a mixture of many hundreds of species and it will need years of study before the limits of this genus and those wrongly placed in it are transferred to their correct genera. (See also under *Anania*).

**daphalis** Grote
U.S. 16 mm. (0·63 in.)
This is a fairly widespread species which formerly was in the genus *Titanio* Hübner. The eggs of this moth are laid singly on leaves and flowers of *Salvia columbaria* (Labiatae), on which the caterpillar feeds. When fully fed they weave a flattened cocoon on the food plant for pupation. Generally there are 2 broods per year, in spring and autumn. **56**n

**purpuralis** Linnaeus
Europe including Britain 14–20 mm. (0·55–0·79 in.)
The caterpillar of this moth feeds on the leaves of mint *(Mentha)* and thyme *(Thymus)* in June and again as a second brood in the autumn. It spins the leaves together with its silk. The moth is out in May and June with a second brood in July and August. It flies actively in the sunshine and is found in meadows, edges of woods and a variety of habitat where the food plant of the caterpillar occurs. **69**

**pyrella** see **Swammerdamia**

**Pyrgus** Hübner (HESPERIIDAE)
Many similar species of Skipper butterflies in this family are difficult to separate without careful study. The species are found in N. America and throughout Europe and temperate Asia.

**alveus** Hübner LARGE GRIZZLED SKIPPER
N. Africa, Spain, S. & C. Europe to the Causasus and C. Asia 28–32 mm. (1·1–1·26 in.)
Similar colouring to the Grizzled Skipper, *P. malvae,* but larger and with mostly a plain hindwing (spotted with white in the Grizzled Skipper) and with green and white underside to the hindwings. This butterfly is common in meadows, mostly in hilly areas from 910–1830 m. (3000–6000 ft) where the adult flies from the end of June to August. The caterpillar feeds on *Potentilla, Rubus* (Rosaceae) and *Helianthemum* (Cistaceae).

**centaureae** Rambur NORTHERN GRIZZLED SKIPPER 20–30 mm.
Circumpolar (0·79–1·18 in.)
Widely distributed in the N. Hemisphere in America (where this species is known as the Grizzled Skipper, not to be confused with the European species *P. malvae*). This species reaches down the Rockies to Colorado. The upperside is dark grey, with white spots in a prominent pattern and the veins lined with white. It flies in June-July in bogs, tundra and heaths from sea-level to 1100 m. (3600 ft). The caterpillar feeds on *Rubus chamaemorus* (Rosaceae).

**communis** Grote CHECKERED SKIPPER
Canada, S. to Mexico but rare in the N.E. parts of the U.S. & Canada 25–32 mm. (0·98–1·26 in.)
This is a variable coloured species with dark and light forms. The caterpillar feeds on species of Malvaceae, the adults flying in March in the S. and June in the N., with several broods each year. This species must not be confused with the Chequered Skipper of Europe *(Carterocephalus palaemon)*.

**malvae** Linnaeus GRIZZLED SKIPPER
Europe including Britain, temperate Asia 22–26 mm. (0·87–1·02 in.)
This Skipper is found in bogs and meadows from sea-level to 1830 m. (6000 ft). The grey-brown upperside has many white spots on the fore- and hindwings and the underside hindwing is brownish, sometimes with a greenish or yellowish tinge. The caterpillar is a pale green, covered with short whitish hairs with a white edged dark line along the back and similar ones along the side. They feed in 'caves' formed by drawing together the edges of leaves with strands of silk. When resting the caterpillar curls itself sideways, with its head towards its tail. The food plants include *Potentilla, Fragaria,* and other Rosaceae and *Malva* (Malvaceae). The butterfly is out in April to mid June or July and August. At higher altitudes there is only 1 brood per year. Several subspecies of *P. malvae* have been described. **86**

**serratulae** Rambur OLIVE SKIPPER
Spain & C. Europe & across C. Asia 24–28 mm. (0·94–1·1 in.)
This species is similar to the Large Grizzled Skipper *P. alveus,* but has an olive green or grey-green underside hindwing. From the Grizzled Skipper, *P. malvae,* it can be separated by the unmarked upperside to the hindwing. The butterfly is common in mountainous districts over its range from June to August up to 2440 m (8000 ft), although is rare in the lowlands. The caterpillar feeds on *Potentilla* or *Alchemilla* (Rosaceae).

**pyri** see **Saturnia**
**pyrina** see **Zeuzera**
**pyritoides** see **Habrosyne**
**Pyroderces** see **Sathrobrota**

**Pyronia** Hübner (NYMPHALIDAE)
A genus of butterflies found mostly in Europe and temperate Asia. There are many species in the genus, which is in the subfamily Satyrinae. The species are often variable and many subspecies and aberrations of them have been described.

**cecilia** Vallantin SOUTHERN GATEKEEPER
S. Europe to Turkey, N. Africa 30–32 mm. (1·18–1·26 in.)
Very similar to *P. tithonus* but with brown patches on the forewings of male and female, and with more dark colour on the front edge of the hindwings. The underside hindwings also differ, with more white in *P. cecilia.* It flies in rough bushy parts in hot localities where the caterpillar feeds on grasses, including *Aira caespitosa.* The illustration shows the butterfly at rest after being disturbed, when the eyespots, which are otherwise hidden, are flashed out to frighten the attacker. **167**

**tithonus** Linnaeus GATEKEEPER
Europe including Britain, Turkey 34–38 mm. (1·34–1·5 in.)
The female is slightly larger than the male, with more orange on the wings. They fly in July and August around flowering bushes, especially brambles *(Rubus)*. The caterpillars feed on species of grasses including *Poa* and *Milium.* The butterfly is often common and can be found at the edge of fields round hedgerows. **168**

**Pyrrharctia** Packard (ARCTIIDAE)
A single species of moth belongs in this genus.

**isabella** Smith BANDED WOOLLYBEAR
Canada, U.S. 46–58 mm. (1·81–2·28 in.)
The hairy, brown and black 'woolly-bear' caterpillar of *P. isabella* is a familiar sight in the autumn, when they leave their foodplants, various low-growing plants including grasses and plantains, *Plantago,* in the search for a winter refuge. There are 2 broods, the first in June, the second in August. As pointed out by Forbes, several Asiatic species now placed in the genera *Diacrisia* or *Spilosoma* may be better placed with *P. isabella* in *Pyrrharctia.* It has been shown recently by D. C. Dunning, the American biologist, that *P. isabella* adults are generally palatable to bats and that the sounds produced from the tymbal organs at night by the moths are an example of sound mimicry, see Arctiidae. **368**n

**pyrrhus** see **Polyura**
**pythias** see **Hypochrysops**

**quadratus** see **Parides**
**quadrifenestrata** see **Midila**
**quadripunctaria** see **Euplagia**
QUEEN BUTTERFLIES see **Danaus, Euxanthe, Stichophthalma**
QUEEN
ALEXANDRA'S BIRDWING see **Ornithoptera alexandrae**
CRACKER see **Hamadryas arethusia**
PAGE see **Papilio androgeus**
PURPLE TIP see **Colotis regina**
VICTORIA'S BIRDWING see **Ornithoptera victoriae**
QUEEN OF SPAIN FRITILLARY see **Issoria lathonia**

QUEEN VICTORIA'S MOTH
From time to time, the British Museum (Natural History) receives enquiries about the 'moth that saved the life of Queen Victoria'. The story concerns a journey during which the train carrying the Queen came to an abrupt halt. The driver had seen what appeared to be someone running along the track towards him him waving his arms, but which turned out to be a moth fluttering in front of the train's headlamp. As the royal train was about to move off again, it was discovered that the track ahead had been damaged and that the moth had saved Queen Victoria from almost certain death. The moth supposedly ended up on a pin in the Museum. Sadly, however, there is no record in the archives which supports this legend and the type of lamp used in Victorian times was probably incapable of projecting a sufficiently strong beam of light.

**queenslandensis** see **Agathiphaga**
**quenseli** see **Apantesis**
**quercifolia** see **Gastropacha**
**quercus** see **Lasiocampa, Quercusia**

**Quercusia** Verity (LYCAENIDAE)
There is only one, fairly widespread species in this genus of butterflies.

**quercus** Linnaeus PURPLE HAIRSTREAK
Europe, N. Africa and temperate Asia 24–29 mm. (0·94–1·14 in.)
*Q. quercus* inhabits woodlands of ash *(Fraxinus)* and oak *(Quercus)* between sea-level and 1520 m. (5000 ft). The female (not illustrated) differs above from the male in restriction of the blue coloration to the base of the forewing. The under surface of both sexes is pale grey with a white, transverse line on each wing and indistinct marginal markings. **258**

QUESTION MARK see **Polygonia interrogationis**
**quinquemaculata** see **Manduca**
**quinquepunctellus** see **Prodoxus**
**quinta** see **Greta**

**racemosa** see **Imma**
**radians** see **Apsarasa**
**rafflesia** see **Euschemon**
**rajah** see **Eterusia**
RAJAH BUTTERFLIES see **Charaxes**
RAJAH BROOKE'S BIRDWING see **Troides brookiana**
**rama** see **Mycalesis**

**Ramphidium** Geyer (PYRALIDAE)
The long labial palps which project far in front of the head give the Pyralid moths in this genus a curious appearance. Two species are at present known, both from S. America, but there are other new species in museum collections which are yet to be described.

**gigantalis** Jones
Brazil 45–75 mm. (1·77–2·95 in.)
Particularly noticeable in the specimen photographed are the long palps sticking out in front of the head. This species is rather more olive brown on the forewing than the only other species in the genus. The undersides of the wings of these moths are a dull unmarked smoky grey. The females are much larger than the males. Nothing is known of their biology. **63**p

RANCHMAN'S TIGER-MOTH see **Platyprepia guttata**
RANNOCH LOOPER see **Semiothisa brunneata**
**rantalis** see **Loxostege**
**rapae** see **Pieris**

**Rapala** Moore (LYCAENIDAE)
The 40 or so species of this butterfly genus are found in Sri Lanka, India and S.E. Asia to New Guinea and Australia. The males have a scent-distributing mechanism involving a patch of scent-scales on the hindwing which interacts with a brush of hair-scales underneath the forewing. Male scent is stored in the hindwing scales, then released and distributed by the forewing brush prior to mating. The flight of these brightly coloured butterflies is direct and Hesperiid-like. Their caterpillars feed on flowers and young shoots.

**jarbus** Fabricius COMMON RED FLASH
Sri Lanka, India, Sikkim, Burma, S.E. to the Lesser Sunda Islands 35–41 mm. (1·38–1·61 in.)
This is a fairly common species of forested areas up to elevations of 2440 m. (8000 ft). As in many other species of its family, the male of *R. jarbus* is a much more colourful butterfly than the female which is pale brown in this species. The caterpillar feeds on the shoots and flowers of *Nephalium*, *Melostoma*, *Zizyphus* and *Ougeinea*. **244**cc

**sphinx** Fabricius BRILLIANT FLASH
N.E. India, Burma to the Philippines and Borneo 36–38 mm. (1·42–1·5 in.)
Both sexes of this species are a rich, iridescent blue above, but the male is the more colourful. This is one of the largest species of its genus. The under surface of the wings is yellowish brown, with a few darker markings. The silvery green caterpillar feeds on *Melastoma* and *Elaeagnus* foliage.

RARE TIGER BUTTERFLY see **Heliconius ethillus**

**Ratarda** Moore (RATARDIDAE)
Eight species are placed at present in this genus of moths. It is represented in India, China, Japan through S.E. Asia to Borneo. The species have rounded wings, mostly brown or brown and white in colour.

**furvivestia** Hampson
N.E. India 38–50 mm. (1·5–1·97 in.)
This is one of numerous species of Lepidoptera described from the Khasia Hills in Assam. The reticulate or mottled appearance of the wings of this moth is matched in other species of *Ratarda*, but most have some white coloration, at least on the forewing. **318**j

**Ratardidae**
Two genera are placed in this family: *Ratarda* (illustrated) and *Callosiope* Hering. It is a member of the superfamily Bombycoidea. There are about 9 species, which are solely Asiatic in distribution. See *Ratarda*.

**rattrayi** see **Diacrisia**
RAVEN, MALABAR see **Papilio dravidarum**
**ravindra** see **Drupadia**
REAKIRT'S BLUE see **Hemiargus isola**
**reamurella** see **Adela**
**rebeli** see **Spialia**
**rectangulata** see **Chloroclystis**
RED
  ADMIRALS see **Vanessa**
  BELTED CLEARWING see **Conopia myopaeformis**
  BRANCH-BORER see **Zeuzera coffeae**
  FLASH, COMMON see **Rapala jarbus**
  HELEN BUTTERFLY see **Papilio helenus**
  IMPERIAL see **Suasa lisides**
  LACEWING see **Cethosia biblis**
  RIM BUTTERFLY see **Biblis hyperia**
  SPOTTED PURPLE see **Limenitis arthemis**
  TIP BUTTERFLY see **Colotis antevippe**
  UNDERWING see **Catocala nupta**
RED-BANDED
  AEMILIA see **Aemilia ambigua**
  ZEBRA see **Cremna thasus**
RED-BASE JEZEBEL see **Delias aglaia**
RED-BODIED SWALLOWTAIL see **Pachliopta polydorus**
RED-EYE BUSH BROWN see **Mycalesis adolphei**
RED-SPOT JEZEBEL see **Delias descombesi**
RED-WINGED HADENA see **Apamea lateritia**
REDEYE, GIANT see **Gangara thrysis**
REDSPOT DUKE see **Euthalia evelina**
**redtenbacheri** see **Dichogama**
**reducta** see **Limenitis**
REGAL MOTH see **Citheronia regalis**
REGAL
  APOLLO see **Parnassius charltonius**
  SWALLOWTAIL see **Papilio rex**
**regalis** see **Citheronia**, **Evenus**, **Mandarina**
REGENT SKIPPER see **Euschemon rafflesia**
**regificella** see **Elachista**
**regina** see **Colotis**
**reinwardti** see **Ixias**

**Rekoa** Kaye (LYCAENIDAE)
There is a single species of butterfly in this genus.

**meton** Cramer MOTHER-OF-PEARL
C. & tropical S. America 35 mm. (1·38 in.)
The female differs from the illustrated male in the white, not blue upper surface of the wings. They fly typically at the margins of forests, usually settling several feet from the ground. Males are rarely very pale blue in colour, much like Mother-of-pearl in appearance. **263**b

**relicta** see **Catocala**
**repandaria** see **Epione**
**repandata** see **Alcis**
**repleta** see **Eterusia**

**Rethera** Rothschild & Jordan (SPHINGIDAE)
There are about 5 species in this genus of S.W. Asian hawk-moths.

**komarovi** Christoph
S.W. Asia from Iran to Afghanistan 52–67 mm. (2·05–2·64 in.)
The illustrated specimen belongs to subspecies *R. komarovi rjabovi*. Its colour pattern has some characteristics in common with the Australian *Cizara ardeniae*. Nothing is known about the caterpillar of this moth. **344**c

**revayana** see **Nycteola**
**rex** see **Papilio**
**rhacotis** see **Mycalesis**
**rhadama** see **Junonia**
**rhamni** see **Gonepteryx**
**rhamniella** see **Sorhagenia**
**rhetenor** see **Morpho**

**Rhetus** Swainson (NEMEOBIIDAE)
There are about 4 species in this genus of butterflies. Its distribution extends from Mexico to tropical S. America.

**arcius** Linnaeus
Mexico to tropical S. America 26–30 mm. (1·02–1·18 in.)
The female (not illustrated) has little iridescent blue, except on the tail, and is dark brown, rather than black in ground-colour. There are several described subspecies of this butterfly. **268**c

**periander** Cramer BLUE DOCTOR
Central America and tropical S. America 35–40 mm. (1·38–1·57 in.)
The male of this short-tailed species is dark blue above with a black apex and an oblique, white, pre-apical band on the forewing, usually two red patches on the inner margin of the hindwing, and a black, outer marginal band on both pairs of wings. Females are quite different in pattern, with 2 white bands crossing the dark brown forewings. These butterflies are attracted to flowers of the Christmas Bush *(Eupatorium)*, and in the early part of the day to damp sand or mud on the banks of forest streams

**Rheumaptera** Hübner: syn. *Eulype* (GEOMETRIDAE)
This genus of moths includes Asian species which reach to the warmer parts. The genus is also found in Europe and N. America.

**hastata** Linnaeus SPEAR-MARKED or ARGENT AND SABLE MOTH
Europe including Britain, temperate Asia, N. America including Canada 25–33 mm. (0·98–1·3 in.)
This moth is widespread over the N. Continents. The stout caterpillar is a shiny dark olive green, the sides are brownish with a black line along the back. They feed on *Myrica* (Myricaceae) and birch *(Betula)*. In Canada it has been recorded on a number of trees, occasionally in webs with numbers of caterpillars causing defoliation. The moth illustrated is a female. **279**

**undulata** Linnaeus SCALLOP SHELL
Europe including Britain, temperate Asia, N. America including Canada 28–32 mm. (1·1–1·26 in.)
Another very widespread species in the genus, this moth is common over much of the N. Hemisphere. The caterpillar feeds on cherry *(Prunus)*, willow *(Salix)* and a number of other trees. It sometimes spins together the terminal leaves. The stout caterpillar is reddish brown to black with 3–4 paler lines along the back and a creamy stripe along the black spiracles. The moth is out in June and July. In the U.S. it has 2 broods a year. **281**

**Rhinopalpa** Felder (NYMPHALIDAE)
A small genus of butterflies with species from India to Indonesia and one described from the Philippines.

**polynice** Cramer WIZARD
India, Pakistan Burma, Indonesia 70–80 mm. (2·76–3·15 in.)
The butterfly is a rich reddish brown above with broad black borders. The outer margins of the forewing are deeply concave and the hindwing has a wavy edge with a short tail. The hindwing is reddish brown with a black band along the anterior margin, black spots near the margin posterior to the 'tail'. The underside is strongly patterned with tawny and brown with silver blue lines. There are eyespots along the margin. Many subspecies have been described from different parts of its range. The caterpillar feeds on *Poikilospermum suaveolens* (Urticaceae). Both wing surfaces are illustrated. **231**d, g

**Rhipha** Walker (ARCTIIDAE)
About 20 tropical American species of moths are placed currently in this genus of moths. There are at least 3 groups of species within *Rhipha* that clearly should not be associated in one genus, and it is not unlikely that only one species, the black and white *R. strigosa* Walker, will eventually remain in this genus.

**pulcherrima** Rothschild
Brazil 47–60 mm. (1·85–2·36 in.)
The bizarre red, white and black colour pattern of this moth is presumably a warning to predators of unpalatability although there is no evidence as yet that this is so. Females differ from the illustrated male in the yellow ground colour and presence of 3 unbroken, transverse, red bands on the forewing and the reddish orange, not white hindwings. Nothing has been published about the life history of *R. pulcherrima*. **368**h

**rhodesiensis** see **Holocerina**
**rhodifer** see **Parides**

**Rhodinia** Staudinger (SATURNIIDAE)
There are 5 species in this genus of large, robust moths. They are found in India, E. to China, Japan and S. E. U.S.S.R.

**fugax** Butler SQUEAKING SILK-MOTH
Japan, N.E. China E. U.S.S.R. 75–110 mm. (2·95–4·33 in.)
Males differ from the illustrated female in the strongly hooked apex to the forewing and the generally reddish brown coloration. The moths are on the wing in the autumn. The caterpillar is bright green when fully grown, with a blue-dotted, yellow lateral line; it feeds on the foliage of *Phellodendron* (Rutaceae) but in captivity will eat oak *(Quercus)* and willow *(Salix)*. Both caterpillar and chrysalis are capable of producing sound when disturbed. The cocoon is green, an unusual colour for them. **335**

**Rhodogastria** Hübner (Arctiidae)
This is a quite large genus of similarly patterned moths, chiefly tropical African and Madagascan in range, but found also in India and China, through S.E. Asia to New Guinea, Australia and the Solomons. Several species produce a frothy, yellow secretion from the thoracic glands when provoked – this is supposedly a defence mechanism.

**caudipennis** Walker
Sulawesi (Celebes), Aru, New Guinea 49–55 mm. (1·93–2·17 in.)
The tufts of hair-scales in the male (see illustration) are part of its scent-distributing equipment which comes into play during courtship. Some of the original material, from which *R. caudipennis* was described, was collected by Alfred Wallace, the famous naturalist. **368**p

**Rhodometra** Meyrick (Geometridae)
This genus of moths is found in Europe, temperate Asia and Africa and has one species in S. America.

**sacraria** Linnaeus VESTAL MOTH
Europe including Britain, Asia, N. India, N. Africa 22–24 mm. (0·87–0·94 in.)
The moth is commoner in the S. part of Europe and visits the N. and Britain as an occasional migrant. In some years large numbers arrive. These fluctuations in populations are always of interest to biologists but their causes are not generally understood. The caterpillar feeds on various herbaceous plants. Several colour variants of this species are known and are of interest to collectors. **275**

**Rhodoneura** Guenée (Thyrididae)
At present this genus contains a large number of tropical species, many of them probably not related. The moths vary from a delicate silvery, transluscent appearance to a brown camouflaged colour. Some of the latter look very like dead leaves and presumably escape detection in this way.

**limatula** Whalley
Madagascar 29–39 mm. (1·14–1·54 in.)
This curiously patterned species is known only from C. Madagascar where it was collected in forests. The life history is unknown. There are a number of species in Madagascar with this type of pattern but it is rare in moths from elsewhere. **56**t

**zurisana** Whalley
Madagascar 30–36 mm. (1·18–1·42 in.)
Only known from Madagascar where it was discovered by French entomologists in 1960. There are several related species in Africa and India but these are not so vividly patterned. **56**d

**Rhodophthitus** Butler (Geometridae)
An African genus of moths, usually white speckled with black and often with prominent red lines on the wing.

**roseovittata** Butler
Uganda, Malawi, Kenya 42 mm. (1·65 in.)
This species is delicately coloured with red and small black marks on a pale background. There are several related species with broadly similar pattern, often with faint streaks of red across the wings. Some have this type of pattern on a rather yellower background. Little seems to be known of their biology. **294**u

**simplex** Butler
Malawi, Zambia, Rhodesia, N. South Africa 44 mm. (1·73 in.)
Unique in the genus with its plain but beautiful coloration. Many of the related species live in rainforests but the biology of these is not known. **294**f

**Rhodosoma** Butler (Sphingidae)
A single species is placed in this genus of hawk-moths.

**triopus** Westwood
India, Sikkim, Bhutan, S. China 59–70 mm. (2·32–2·76 in.)
Nothing seems to be known of the habits or life history of this richly coloured and distinctive species. **344**k

**rhombella** see **Gelechia**
**rhomboidaria** see **Peribatodes**
**rhodope** see **Gymnautocera**

**Rhyacionia** Hübner: syn. *Evetria* (Tortricidae)
PINE-SHOOT TORTRIX
This genus of micro-moths contains a number of species which are serious pests of pines in the N. Hemisphere. Several species occur both in N. America and Europe. The caterpillars bore into the terminal shoots of pines, this may kill the leader shoot of the pine or cause forking of the tree, others cause the shoot to grow crookedly. They can be serious pests in nurseries and plantations of pines. A few species bore in the cones of spruce. The adults are fairly typical 'tortricid' like moths, see Tortricidae.

**buoliana** Denis & Schiffermüller PINE-SHOOT TORTRIX or EUROPEAN PINE-SHOOT MOTH
Europe including Britain, W. and C. Asia, N. America 17–22 mm. (0·67–0·87 in.)
The forewings of this moth are an orange-brown colour, often suffused with red, with several metallic grey-white lines. The hindwings are light grey. The moth is a typical tortrix shape. This is a common species in Europe where it can be a serious pest of pines causing damage to large trees and serious damage in nurseries. The caterpillar, which is reddish-brown in colour, feeds on the shoots of *Pinus sylvestris*. **18**

**Rhyparioides** see under **Diacrisia**
RICE STEM-BORERS see **Tryporyza**
RICE-BORERS see **Sesamia, Tryporyza**
**ricini** see **Heliconius**
**ridleyanus** see **Graphium**

**Riechia** Oiticica (Castniidae)
*Riechia* was proposed in 1955 to replace the name *Herrichia* which had already been used by Staudinger for a genus of Oecophoridae. Only one species is known as yet.

**acraeoides** Boisduval
Tropical S. America. 50–75 mm. (1·97–2·95 in.)
Although it is wise to be aware of the dangers of replacing confirmatory evidence with 'confident assertion' it is at least a reasonably safe guess that *R. acraeoides* is a mimic or mimetic partner of one or more species of the butterfly genus *Actinote*, a member of the pantropical and generally unpalatable group Acraeinae (Nymphalidae). The caterpillars of *R. acraeoides* feed on the bulbs and roots of a species of *Oncidium*, an epiphytic orchid, and on a *Tillandsia* species, a bromeliad. **45**t

**rileyi** see **Sathrobrota**
**rimosa** see **Pheosia**
RING BUTTERFLIES see **Ypthima**
RINGED
  CARPET MOTH see **Cleora cinctaria**
  CHINA-MARK see **Parapoynx stratiotata**
RINGLET BUTTERFLIES see **Aphantopus, Coenonympha, Erebia, Ypthima**
**Riodinidae** see **Nemeobiidae**
**ripheus** see **Chrysiridia**
**risa** see **Chersonesia**
**rivera** see **Furcula**
**rivularis** see **Neptis**
ROAD PAGE see **Marpesia chiron**
ROBIN MOTH see **Hyalophora cecropia**
**robiniae** see **Prionxystus**
ROCK BROWN, WHITE-EDGED see **Satyrus parisatis**
**Roeselia** see **Nola**
**rogatalis** see **Hellula**
**rolandiana** see **Exyra**
ROOT-BORER, BIG see **Melittia gloriosa**
**rosa** see **Asterope**
**rosacea** see **Acrojana**
ROSE BUTTERFLIES see **Pachliopta**
**rosea** see **Decachorda, Oreta**
**roseana** see **Cochylis**
ROSEATE EMPEROR see **Eochroa trimeni**
**roseilinea** see **Ethopia**
**roseovittata** see **Rhodophthitus**
**rosimon** see **Castalius**
**rosina** see **Lymantria**
ROSS'S ALPINE see **Erebia rossii**
**rossii** see **Erebia**
ROSY
  FOOTMAN see **Miltochrista miniata**
  MARSH see **Eugraphe subrosea**
**rothi** see **Chorodnodes**
**rothschildi** see **Ornithoptera**

**Rothschildia** Grote (Saturniidae)
This genus of large moths is named after Lord Walter Rothschild, who amassed what was probably the largest private collection of Lepidoptera in the world and which now forms part of the British Museum (Natural History) collection. The wings of all the 30 or so species have a large, transparent patch or 'window' in each wing. This genus is basically S. American and C. American in distribution but 3 species are found in the US. Its counterpart in the Old World is *Attacus*, the Atlas moths.

**orizaba** Westwood ORIZABA SILK-MOTH
U.S., Mexico, C. America & tropical S. America 110–145 mm. (4·33–5·71 in.)
This species has been recorded from S. Texas, near the border with Mexico, although authors have cast doubts on some of the records. The foodplant of the caterpillar is unknown. In general appearance the moth is similar to the illustrated specimen of *Attacus atlas* but the transparent patches are relatively much larger.

**zacateca** Westwood
Colombia 70–100 mm. (2·76–3·94 in.)
The female, which is typical of many other species of its genus has relatively larger hindwings than the illustrated male and a straight border to the forewing. The males of most species of *Rothschildia* have a similarly shaped forewing, but none has the extreme elongation of the apex present in *R, zacateca*. **337**c

**roxus** see **Castalius**
ROYAL BUTTERFLIES see **Tajuria**
ROYAL
  ASSYRIAN see **Terinos terpander**
  BLUE THEOPE see **Theope excelsa**
  WALNUT MOTH see **Citheronia regalis**
**rubi** see **Callophrys, Macrothylacia**
**rubida** see **Arctioblepsis**
**rubidata** see **Catarhoe**
**rubripalpis** see **Pseudosphex**
**rubriplaga** see **Darna**
RUBY
  POLYPTYCHUS HAWK-MOTH see **Polyptychus calcareus**
  TIGER-MOTH see **Phragmatobia fuliginosa**
RUDDY DAGGERWING see **Marpesia petreus**
**rufata** see **Chesias**
**rufescens** see **Diacrisia**
**rumia** see **Kallima**
**rumina** see **Parnalius**
RUNG'S COPPER UNDERWING see **Amphipyra berbera**
**rurina** see **Phoebus**
**ruralis** see **Pleurotypa**
RUSH VENEER see **Nomophila noctuella**
RUSSIAN HEATH see **Coenonympha leander**
RUSTIC SPHINX see **Manduca rustica**
**rustica** see **Manduca**
**rusticata** see **Canaea**
RUSTY TUSSOCK see **Orgyia antiqua**

**Sabalia** Walker (Lemoniidae)
*Sabalia* is placed in a family of its own, the Sabaliidae, by some authors, but was transferred to the Lemoniidae by Berger in 1957. It has also been placed, at times, in the Eupterotidae and Brahmaeidae. There are about 8 species, all tropical African in distribution.

**barnsi** Prout
Zambia, S. Congo 58–65 mm. (2·28–2·56 in.)
Nothing is known about the life history of this strikingly patterned species, which was described from a few specimens collected during the early part of this century. Its closest relative is probably the more widely distributed E. African species *S. jacksoni*. **318**d

**sabina** see **Yoma**
**saccharalis** see **Diatraea**

**sacchari** see **Opogona**
SACK-BEARERS see **Cicinnus, Lacosoma**
**sacraria** see **Rhodometra**
SADDLE-BACK see **Sibine stimulea**
SADDLED MOTH see **Disphragis guttivitta**
**saepestriata** see **Japonica**
SAFFRON see **Iolaus pallene**
SAGA FRITILLARY see **Clossiana frigga**
**Sagaritis** see **Myscelia**
SAGEBRUSH SHEEP-MOTH see **Hemileuca hera**

**Sagaropsis** Hering (HYPSIDAE)
Seven species are at present placed in this genus of C. American and N. S. American moths. Most of its members have iridescent blue and black wings, marked with bright yellow on the forewing. Some resemble species of Dioptidae and Ctenuchidae in colour pattern.

**brevifasciata** Hering
Ecuador 38–45 mm. (1·5–1·77 in.)
This is typical in colour pattern of its genus. Nothing is known about its biology. **381**j

**Sagenosoma** Jordan (SPHINGIDAE)
S.W. U.S. is the home of the only known species of this relative of *Isoparce* and *Sphinx*.

**elsa** Strecker
U.S. (N. Arizona, N. New Mexico) 54–72 mm. (2·13–2·83 in.)
No other N. American hawk-moth can be confused with this distinctive species. The whole of the moth is very pale grey, nearly white above with black markings on the thorax and abdomen, and 3 speckled oblique, black lines on the forewing and 2 on the hindwing. The general impression is of a moth which would be well camouflaged on a background lightly covered with snowflakes. It flies in mid May to mid July. The caterpillar is green, with oblique, lateral stripes of brown and yellow. Nothing is known about the foodplants.

**sagittata** see **Perizoma**
**sagulatus** see **Aeolanthus**
SAILER BUTTERFLIES see **Neptis**
ST LAWRENCE TIGER-MOTH see **Platarctica parthenos**
**salaciella** see **Opostega**

**Salamis** Boisduval (NYMPHALIDAE)
This genus of butterflies is related to *Hypolimnas*. They are large, broad-winged species found mostly in Africa, with a few species from the Indian subcontinent. The hindwings always have lobes or tails but the patterns and colours are variable. The caterpillars are similar to *Hypolimnas* with 2 long spines on the head, 2 on each side of segments 2 and 3; and more spines on each segment from 4–10 (often they may have eleven spines on each).

**anacardii** Linnaeus: syn. *S. duprei*
CLOUDED MOTHER-OF-PEARL
Africa, Madagascar, Arabia 75 mm. (2·95 in.)
The underside of the wing is a greyish white, patterned like a leaf. There is some variation in the colour and pattern of this butterfly which always has a general Mother-of-Pearl lustre. Some specimens tend to have more yellow in the wings, while others have a faint mauvish iridescence. This butterfly is found all over Africa south of the Sahara and specimens have been collected in Aden and Yemen. **208**e

**cacta** Fabricius LILAC BEAUTY BUTTERFLY
Tropical Africa to Rhodesia and Mozambique 62–76 mm. (2·44–2·99 in.)
This is a species of dense rain forest. It is variable in size, with a reddish mauve colour and a broad black border to the forewing. The underside is dark grey or orange brown with a dead leaf pattern. Several subspecies have been described.

**parhassus** Drury MOTHER-OF-PEARL BUTTERFLY
Tropical Africa E. to South Africa 76–100 mm. (2·99–3·94 in.)
This is a widespread species and is often common in forest or thick bush. The butterfly is a translucent green with patches of violet or mother-of-pearl, with a black margin. A tail protrudes from the side of the hindwing, which is the same colour as the forewing, but has a prominent red and yellow eyespot. Smaller, less colourful eyespots occur on the forewing. The caterpillar feeds on species of *Asystasia* and *Isoglossa* (Acanthaceae). **223**

**temora** Felder BLUE SALAMIS
Africa 80–90 mm. (3·15–3·54 in.)
This species is found from Nigeria to Kenya and from Ethiopia southwards to Tanzania and Zambia. Generally found in forests it flies in most months of the year. The female is larger than the male but the blue area on the wing is smaller. The life history is unknown. Both wing surfaces are illustrated. **213, 214**

SALAMIS, BLUE see **Salamis temora**
**salicifoliella** see **Phyllonorycter**
**salicis** see **Leucoma**
**saligna** see **Phyllocnistis**
**salmachus** see **Synanthedon**
**salome** see **Eurema**
SALLOW MOTH, PINK-BARRED see **Xanthia flavago**
SALMON ARAB, LARGE see **Colotis fausta**
**salmacis** see **Hypolimnas**
SALOME SULPHUR see **Eurema salome**
**salvinii** see **Callithea**
**saltitans** see **Cydia**
SALTMARSH CATERPILLAR see **Estigmene acrea**

**Samia** Hübner (SATURNIIDAE)
There is a single, variable, widespread species in this genus.

**cynthia** Drury AILANTHUS SILK-MOTH, CYNTHIA MOTH
U.S., Europe, E. Asia to Japan 90–140 mm. (3·54–5·51 in.)
This moth has been introduced into both Europe and the U.S. from Asia. The N. American members of this species feeds, as the common name indicates, on the Tree of Heaven *(Ailanthus)*, the same foodplant as the Chinese subspecies *S. cynthia cynthia* and are considered to represent an introduced component of this subspecies brought into the U.S. as a potential commercially viable producer of silk. Philadelphia, Pennsylvania, is the probable site of introduction of this species in the U.S. The C. European population of *cynthia* probably belongs to the subspecies *S. cynthia walkeri*. **333**b

**sancta** see **Episcea**

**Sandia** Ehrlich & Clench (LYCAENIDAE)
There is a single known species. *Sandia* has been elevated to generic rank in Elliot's recent reclassification of the Lycaenid butterflies.

**mcfarlandi** Ehrlich & Clench
U.S., Mexico 23–26 mm. (0·91–1·02 in.)
This distinctive hairstreak ranges from New Mexico and Texas to Chihuahua, Mexico, and yet was unknown until 1960. Both sexes have greyish brown forewings and tail-less, yellowish brown hindwings. There are two broods per year. The caterpillars feed on beargrass *(Nolina)* flowerheads. The chrysalis overwinters.

**sangaris** see **Cymothoe**
**sanguiflua** see **Erasmia**
**sanguilineata** see **Osteosema**
**sanguinaria** see **Amnemopsyche**
**sanguinea** see **Hyblaea**
**sannio** see **Diacrisia**
**saphirina** see **Argyrogrammana**
**sapho** see **Heliconius**
SAPPHIRE BUTTERFLIES see **Heliophorus, Iolaus**
**sappho** see **Neptis**
**sara** see **Heliconius**
**sarcitrella** see **Endrosis**
**sardanapalpus** see **Lemonia**
**sarpedon** see **Graphium**

**Sasakia** Moore (NYMPHALIDAE)
These butterflies are found in China, Tibet, Japan and Formosa.

**charonda** Hewitson
China, Japan 95–120 mm. (3·74–4·72 in.)
There are several subspecies of this iridescent blue butterfly. The female is brown, without the blue on the wings. This butterfly has been illustrated on Japanese stamps and is the Japanese National butterfly. The adults are out in Hokkaido and N. Honshu in July but in more temperate areas in June. It is a rapid flier, rather like species of *Charaxes*. The caterpillar feeds on *Celtis* (Ulmaceae). **230**f

**Sathrobrota** Hodges: syn. *Pyroderces* (MOMPHIDAE)
Two species are at present known in this genus of micro-moths. Both species have caterpillars which scavenge but each has slightly different habits.

**rileyi** Walsingham
U.S., Jamaica, Australia Philippines, S.E. Asia, Hawaii, Indonesia, India, South Africa, Afghanistan, Turkey 8–11 mm. (0·31–0·43 in.)
The caterpillar of this widespread species is a common scavenger on rotten cotton bolls, maize and various stored grains. It is also a pest on citrus, bananas and on sugar cane. At times the caterpillar can be present in sufficient numbers to create problems in stored foods. **44**h

SATIN, WHITE see **Leucoma salicis**
**saturata** see **Imma**
SATURN BUTTERFLY see **Zeuxida amethystus**

**Saturnia** Schrank (SATURNIIDAE)
This is a genus of 7 species found in N. Africa, Europe, temperate Asia and N. America. The 3 American species are restricted to California. One, *S. pavonia*, species occurs in Britain. *Calosaturnia* Smith is currently combined with *Saturnia*.

**albofasciata** Johnson
U.S. (California) 37–48 mm. (1·46–1·89 in.)
This is the least well known of N. American Saturniidae and is found only at elevations between 1370 m. (4500 ft) and 2130 m. (7000 ft) in the southern part of its range and above 400 m. (1300 ft) in the north. It overwinters as an egg, passes the summer as a chrysalis and emerges as an adult at the end of October or beginning of November. Males start flying during the middle of the afternoon and, if their search is successful, mate in the late afternoon. Females fly only in the late evening and early part of the night. Males of *S. albofasciata* are similar in colour pattern to the illustrated *S. mendocino*, but the females are brownish grey in ground-colour, with a white, marginal band as in the female of the European *S. pavonia*. The caterpillar feeds on the foliage of several chapparal plant species including snowbush *(Ceanothus)* and mountain mahogany *(Cercocarpus)*.

**mendocino** Behrens
U.S. (California) 48–70 mm. (1·89–2·76 in.)
This is probably the best known N. American species of its genus. It gets its name from Mendocino County, California, where the first specimens were captured in a redwood forest of a coastal range of the Sierras. The moths are on the wing between February and May. Both sexes fly during the day. Females are static until mated, after which they take to the wing and normally lay their eggs the same day. The caterpillars are a pale yellowish green, with orange-red tubercles; they feed on manzanita *(Arctostaphylos)* and possibly on *Rhus*, *Ceanothus* and *Arbutus*. **331**h

**pavonia** Linnaeus EMPEROR MOTH
C. & S. Europe, N. & C. Asia, E. to Siberia 52–70 mm. (2·05–2·76 in.)
The day-flying males of this species are most often seen between 2 pm. and 6 pm. on warm, sunny days in April and May. Females fly almost invariably only at night. In Britain the Emperor occurs both on heathland, where the caterpillar feeds normally on ling heather *(Calluna)*, and in river estuaries where blackthorn *(Prunus)* and bramble *(Rubus)* are the foodplants. In Germany the chief foodplant at lower elevations is sloe *(Prunus)*, and bilberry *(Vaccinium)* in the mountains. The cocoon has an 'exit-only' trapdoor which allows passage of the emerging moth but prevents the entry of parasites. Female moths differ from the il-

lustrated male in the absence of brownish orange on the hindwings which are chiefly grey and white. **334**

**pyri** Denis & Schiffermüller GREAT PEACOCK MOTH, VIENNESE EMPEROR
S. Europe, W. Asia 100–150 mm. (3·94–5·91 in.)
This is the largest native species of either moth or butterfly to occur in Europe where it is often common in parts of its range. The moths are on the wing between April and June. The caterpillars feed in July and August on the foliage of apple *(Malus)*, sloe *(Prunus)*, elm *(Ulmus)* and other trees; it is mainly green, with numerous blue tubercles from which long, clubbed hairs arise. **340c**

**Saturniidae** GIANT SILKWORM or EMPEROR MOTHS
This family includes some of the largest moths in the world in terms of wingspan. *Attacus atlas* is the joint largest species, together with the Noctuid *Thysania agrippina*. Most species are tropical in range, but several occur in temperate zones (eg *Automeris*). Many species have eyespots or transparent patches (eg. *Rothschildia*) on the wings and some have a long tail to the hindwing (eg. *Eudaemonia*). The males of most species have strongly pectinate antennae. The proboscis of both sexes is absent or vestigial and the moths are unable to feed. The caterpillars are large, and usually armed with hairs or spines which may contain irritating (eg. *Hylesia*) or lethal (see *Lonomia*) secretions. The anal tuft of the adult moth may also contain irritant hairs. Some species are commercially important as producers of silk (see *Antheraea*). See *Actias, Aglia, Antheraea, Antistathmoptera, Argema, Athletes, Attacus, Automeris, Bunaea, Bunaeopsis, Callosamia, Cinabra, Citheronia, Coloradia, Coscinocera, Decachorda, Eacles, Eochroa, Epiphora, Eubergia, Eudaemonia, Goodia, Graellsia, Heliconisa, Hemileuca, Holocerina, Hyalophora, Hylesia, Imbrasia, Lonomia, Loxolomia, Ludia, Micragone, Pseudaphelia, Rhodinia, Rothschildia, Samia, Saturnia, Ubaena, Urota, Vegetia*

**saturnioides** see **Carthaea**
SATYR BUTTERFLIES see **Satyrus**

**Satyrinae** (NYMPHALIDAE)
A group of butterflies, formerly considered a distinct family, but now regarded as a subfamily of the Nymphalidae. Generally known as 'Browns', the vast majority of species are brown with a prominent pattern on the underside. The species are common, often abundant, and are found from the Arctic to the Antarctic. In America they are found in all parts from the N. to the highest mountains and tropical forests S. to Tierra de Fuego. One of the characteristics of the 'Browns' is a swollen basal part to the veins of the forewing. The shape of the wings are variable and, although many different colours are found in species of the subfamily, the predominant colours are brown. The caterpillars tend to feed on grass, bamboo, sugar cane and palms. Many of the tropical species fly at dusk. Some of the species add to the effect of their cryptic (concealing) patterns by leaning towards the sun when at rest and thus reducing their shadows, which might otherwise be conspicuous.

**Satyrodes** Scudder (NYMPHALIDAE)
A small genus of butterflies, in the subfamily Satyrinae, found in N. America.

**canthus** Linnaeus: syn. *S. euridice* EYED BROWN
Canada, U.S. 45 mm. (1·77 in.)
This butterfly is found from Quebec, Ontario and Manitoba S. to the N. part of Florida. It occurs in marshy meadows where the caterpillar feeds on various species of grass. The upperside is brown with a lighter patch below the apex of the forewing. There are small black spots along the wing margin on the forewing, and large ones, surrounded by orange brown, along the hindwing margin. The underside is paler with very clear eyespots in the fore- and hindwings. The anterior eyespot in the hindwing has several rings round it on the underside (as do the others) but is just a black spot on the upperside.

**Satyrus** Latreille: syn. *Aulocera* (NYMPHALIDAE)
A typical genus of brown butterflies in the Satyrinae. Many subgenera have been described but the detailed classification of the genus has not been settled. For example, a few of the genera used in this book are considered to be subgenera of *Satyrus* by some specialists but not by others. See also *Chazara*.

**actea** Esper BLACK SATYR
S.W. Europe, Turkey, Syria, Iran 48–56 mm. (1·89–2·2 in.)
A dark brown butterfly, with a single black eyespot with a white centre below the apex of the forewing. The hindwings are more reddish brown but dark brown at the base. The underside forewing is more patterned, with the eyespot large, with a narrow orange surround. The hindwings have dark transverse marks on the wing but 2 thin irregular white lines from the front to the back of the hindwing. The adults are common and widely distributed over dry stony slopes from 910–1830 m. (3000–6000 ft), in Spain, Portugal, France and across S.W. Europe. The caterpillar feeds on various species of grass. The butterfly, which is out in July and August is very variable with many local races.

**padma** Kollar GREAT SATYR
N. India 70–98 mm. (2·76–3·86 in.)
The dark brown upperside has a line of white marks forming a band on the forewing, continuing as a solid line on the hindwing. The underside is variegated with grey white and brown. These are strong flying butterflies with a graceful flight like some of the larger Nymphalinae. They occur above 1220 m. (4000 ft) and are out towards the end of April, with a second brood later on. They fly in open places on ridges and hilltops above 2290 m. (7500 ft).

**parisatis** Kollar WHITE-EDGED ROCK BROWN
N. India, Iran Afghanistan 65–70 mm. (2·56–2·76 in.)
A blackish brown, velvet looking species with bluish white border, and 3 white spots below the apex of the forewing. The hindwing border is wider than the forewing. The underside of the wing is pale, prominently streaked. This is one of the larger satyrid butterflies. It is found above 1500 m. (5000 ft) on open mountainsides in N. India from May to July and again in October. It has a very rapid flight.

**swaha** Kollar COMMON SATYR
N. India, Pakistan E. Asia, Afghanistan 60–70 mm. (2·36–2·76 in.)
This is one of the common butterflies of the lower parts of the Himalayas, over 1500 m. (5000 ft). It has a similar pattern to the Great Satyr, but is smaller. The wings have a bronze sheen with a broad white or yellow band on the hindwing, which narrows posteriorly. This is continued on the forewing but is divided by the dark wing veins into white patches, usually with several white patches near the apex. This species flies more over the open ground than in forests.

**saucia** see **Peridroma**, **Phaloesia**
**saundersii** see **Chlorosyne**
SAWTOOTH, SPOTTED see **Prioneris thestylis**
**scabiosella** see **Nemophora**
**scabriuscula** see **Dypterygia**

**Scada** Kirby (NYMPHALIDAE)
A single species of butterfly is placed in this genus of the subfamily Ithomiinae.

**karschina** Herbst
Brazil 40–50 mm. (1·57–1·97 in.)
This beautiful little species is rather unusual in coloration but is matched fairly closely by another Brazilian Ithomiine *Pteronymia euritea hemixanthe* Felder. The wings of the male (not illustrated) are more transparent than those of the female. **131e**

SCALLOP SHELL MOTH see **Rheumaptera undulata**
SCALLOPED
OAK MOTH see **Crocallis elinguaria**
OWLET see **Scoliopteryx libatrix**
SACK-BEARER see **Lacosoma chiridota**
SCARCE
BAMBOO PAGE see **Philaethria dido**
BORDERED STRAW see **Heliothis armigera**
BRINDLE MOTH see **Apamea lateritia**
CORNELIAN BUTTERFLY see **Deudorix hypargyria**
FOOTMAN see **Eilema complana**
GREEN-STRIPED WHITE see **Euchloe falloui**
SHOT SILVERLINE see **Spindasis elima**
SILVERLINES MOTH see **Bena prasinana**
SILVER-SPOTTED FLAMBEAU see **Dione juno**
SWALLOWTAIL see **Iphiclides podalirius**
SCARLET TIGER-MOTH see **Callimorpha dominula**
SCARLETT BUTTERFLY see **Axiocerses bambana**
**scatophaga** see **Neossiosynoeca**

**Scea** Walker (DIOPTIDAE)
The 12 or so species in this genus of moths are tropical S. American in distribution. Most species are similar in pattern to *S. steinbachi*.

**steinbachi** Prout
Argentina 34–40 mm. (1·34–1·57 in.)
Related and similarly patterned species of *Scea* found further north in S. America form mimetic associations with almost identically coloured species of other families (see, for example, the illustrated Nemeobiid butterfly *Mesenopsis albivitta*). Specimens of *S. steinbachi* have been captured at elevations up to 1200 m. (4000 ft). **362b**

**schaenia** see **Antanartia**
**schakra** see **Lasiommata**
SCHAUS'S SWALLOWTAIL see under **Papilio aristodemus**
**schenckii** see **Danaus**

**Schinia** Hübner (NOCTUIDAE)
This genus of moths is best represented in N. America where there are over 100 species, but it also occurs in C. and S. America, Europe and Asia. The caterpillars, like those of related genera, feed on the flowers and fruits of their foodplants.

**marginata** Haworth
E. Canada, U.S. E. of the Rockies 25–30 mm. (0·98–1·18 in.)
The forewings of this species are dark brown, with a pale brown medial band and outer marginal band; hindwings are light brown, becoming darker towards the outer margin. Florida specimens of *S. marginata* are smaller than in the rest of its range and Arizona populations are generally paler. The mostly yellow and brown caterpillar feeds on species of *Ambrosia*.

**pallicincta** Smith
U.S. 24 mm. (0·94 in.)
This is both a night and day-flying species of the deserts of California. Both species are attracted to light, at night, but females lay their eggs by day in the half open flower-heads of desert marigolds *(Baileya* species) in which the caterpillar lives and feeds. The forewings are chiefly olive-brown, with 2 whitish, transverse lines; the hindwings are brown.

**scholastica** see **Paleoses**
**schonbergi** see **Taenaris**

**Schreckensteinia** Hübner (SCHRECKENSTEINIIDAE)
A genus of micro-moths with species in Europe, N. & C. America. Few species are known in this genus. The caterpillar forms an open network cocoon in which it pupates.

**festaliella** Hübner
Europe including Britain, N. America 10–12 mm. (0·39–0·47 in.)
The pale green caterpillar, which has a dark line along the back, feeds on the underside of the leaves of raspberry and other species of Rosaceae. The moth is out in May and June, with a second brood in August in Europe; in N. America it is out earlier in the year. Generally it is found in the E. states and at times can be common on raspberry plants. **43d**

**Schreckensteiniidae**
A small family of micro-moths related to the Heliodinidae and at present placed near the Epermeniidae. These small moths are found in Europe and N. America.

**scierops** see **Strepsimanes**
**scintillans** see **Chionaema**

**Scirpophaga** Treitschke (PYRALIDAE)
Most species in this genus of Pyralid moths are stem-borers feeding inside the stems of sedges although they may go on to wheat and rice. Generally they are white moths, sometimes yellowish with few marks on the wings. The females are larger than the males and usually have rather more pointed wings. The abdomen of the male is generally rather pointed while the female has a tuft of scales which often makes the abdomen broad at the end. The males are more heavily patterned than the females. Species of *Scirpophaga* occur in India, Thailand and throughout the E. Indies with one species in Europe. Similar looking species occur in N. America but they belong to related genera.

**chrysorrhoa** Zeller
Australia, India, Pakistan, Sri Lanka, Thailand, Borneo, Java 20–45 mm. (0·79–1·77 in.)
The female of this moth may be twice the size of the male, it has shining white forewings and white hindwings. The male has brownish forewings often with a grey suffusion. The moth has been found in New South Wales and along the E. coast to the Northern Territory. The caterpillar feeds on different species of sedges (Cyperaceae) where it tunnels in the stem.

**scitiscripta** see **Cerura**

**Scoliopteryx** Germar (NOCTUIDAE)
One species is known and is restricted to the N. Hemisphere.

**libatrix** Linnaeus HERALD, SCALLOPED or EUROPEAN FRUIT-PIERCING MOTH
Canada, U.S., Europe (including Britain), N. Africa, temperate Asia to Japan 35–45 mm. (1·37–1·77 in.)
This attractive and well known species is not only widely distributed from E. to W., but is tolerant of both subarctic Alaskan temperatures and the semi-desert climate of New Mexico, U.S. It is often seen in early spring when it emerges from overwintering as an adult, a genuine herald of spring. Species of willow *(Salix)* are the chief foodplants of the caterpillar. The proboscis of this moth is armed with erectile barbs as in *Calpe* species and is able to penetrate the skin of soft fruit and allow the moth to suck the juice of the fruit. **395**

**Scoparia** Haworth (PYRALIDAE)
A worldwide genus of Pyralid moths, usually grey and black, which resemble lichens on stones. The caterpillars of only a few of the species are known. These feed on the root of various plants including species of *Senecio*, *Tussilago* and *Picris* (Compositae). The moths are attracted to light. All the species in Europe are similarly patterned and coloured. They can be recognized as species of *Scoparia* but distinguishing the individual species requires careful study. In the mountains of New Guinea there are a large number of species of *Scoparia* which have become rather different in pattern from the more typical ones. See also *Eudonia*.

**ambigualis** Treitschke
Europe including Britain 17–22 mm. (0·67–0·87 in.)
Rather inconspicuous Pyralid moth belonging to the subfamily Scopariinae. The moths are out in May and June. The caterpillars are said to feed on mosses.

**apachealis** Munroe
U.S. (Arizona, Utah) 14–20 mm. (0·55–0·79 in.)
This is a variable species, with generally a rather grey colour with dark marks on the wing and often a trace of black X on the forewing. The species was first collected at Greer, White Mountain, Apache Country, Arizona in 1962 and the description of the moth with its name was first published in 1972. There are three subspecies of this species all described together in 1972. They differ in size and locality in which they occur.

**arundinata** Thunberg: syn. *S. dubitalis*
Europe including Britain 16–19 mm. (0·63–0·75 in.)
This is a common species, typical in appearance of others in the genus. Considerable variation in colour exists, with the markings almost absent where the insect is living on calcareous soils. Little is known of the biology. In some instances they are reported as feeding on the roots of ragwort *(Senecio)*, others suggest it feeds on mosses. For a very common, if small, moth a study of its biology would be interesting. **65**

**scopigera** see **Bembecia**
SCORCHED WING see **Plagodis dolabraria**
SCOTCH ARGUS see **Erebia aethiops**

**scotina** see under **Lasaia sessilis**
**scotti** see under **Callidula, Charagia**
**scribaiella** see **Cosmopterix**
**scribonia** see **Ecpantheria**
**scripta** see **Habrosyne**

**Scythridae**
A family of micro-moths related to the Gelechiidae. They are worldwide in their distribution. The caterpillars have tufts of hairs arising from small warts over the body. Little is known of the biology of most of the species in the family. See *Scythris*

**Scythris** Hübner (SCYTHRIDAE)
A very large, worldwide genus of micro-moths. Species occur on all the main Continents. They are generally more numerous in the warmer parts, particularly in S. Europe. Many of the species are rather small dark coloured moths. Some of the caterpillars are leaf-miners, others spin up leaves and feed in long silken galleries along the stems of species of *Erica*.

**cuspidella** Denis & Schiffermüller
Europe 12–17 mm. (0·47–0·67 in.)
The yellow mark near the apex and the yellow near the base of the forewing separates this from related similar species. This species is common in S. Europe. The caterpillar feeds on species of Leguminosae. **44j**

**grandipennis** Haworth
Europe including Britain 14–18 mm. (0·55–0·71 in.)
A fairly typical example of the genus. The caterpillar feeds on gorse *(Ulex)* and often its webs, made by large numbers of caterpillars, are quite conspicuous. The caterpillars are green-grey, with a pale line along the back with black spots, and a darker line along the side. The moth is out in June and although widespread in C. Europe is local in Britain. **43d**

**scylax** see **Melinaea, Zipaetis**
**segecia** see **Praetaxila**
**segetum** see **Agrotis**
**Seirarctia echo** see under **Eumaeus atala**
**seitzi** see under **Acerbia**
**selene** see **Actias, Clossiana**

**Selenia** Hübner (GEOMETRIDAE)
A large number of species of moth are known in this genus which is spread over Europe through temperate Asia to Japan with a few species described from the U.S. They are often yellowish or brownish-yellow with a wavy outer wing margin. The caterpillars feed on the leaves of a variety of trees. Most of the caterpillars, which are popularly known as 'loopers' or 'Earth Measurers', look very like twigs when they stretch out motionlessly. This is their camouflage to avoid predators. The caterpillars instead of having 3 pairs of legs in the front, 4 in the middle and 2 claspers at the back, lack the 4 middle pairs. This is characteristic of Geometrid moths. Their method of movement is to move the back up to the front, so arching the middle of the body, then to stretch out the front. This gives rise to the popular name of 'looper' caterpillar.

**dentaria** Fabricius: syn. *S. bilunaria* EARLY THORN
Europe including Britain, N. Asia 34–41 mm. (1·34–1·61 in.)
The yellow brown forewings have 3 darker lines across them and a dark apex. The hindwing is similarly coloured, with more indistinct lines. These are characteristic of this group of moths. There are related, rather similar species in Europe and N. America. The caterpillars feed on sallow (*Salix* sp.), blackthorn *(Prunus spinosa)*, hawthorn *(Crataegus* sp.*)*. The moths fly in March and April and, with a new generation, in July and August. There is quite a lot of variation in colour of the wings and several colour forms have been named.

**tetralunaria** Hufnagel PURPLE THORN
Europe including Britain, temperate Asia including Japan 32–44 mm. (1·26–1·73 in.)
This moth is widespread and it is interesting that it is the one species which occurs right across Europe and Asia to Japan without much variation or subspeciation. The moth flies in April and May with a second generation later in the year. The spring generation being usually darker than the Autumn one. The caterpillar, which is reddish brown, mottled with darker brown and grey, feeds on alder *(Alnus)*, sallow *(Salix)*, cherry *(Prunus)* and other trees. **313**

**selenophora** see **Parathyma**

**Selidosema** Hübner (GEOMETRIDAE)
A small genus of moths found in Europe and temperate Asia.

**brunnearia** Villiers: syn. *S. plumaria* BORDERED GREY
Europe including Britain 34–36 mm. (1·34–1·42 in.)
The moth is on the wing in July and August and flies at dusk, generally over the heathlands where the food of its caterpillar is common (Ling, *Callumna*). The caterpillar is grey brown with darker brown irregular lines on the back and pale lines on the side. **307**

**Sematura** see **Nothus**

**Sematuridae**
This family of moths is a member of the superfamily Geometroidea where its nearest relatives are possibly the Uraniidae and Epicopeidae. The 40 or so species are found in C. America (including Mexico) and tropical S. America. The hindwings have a short tail in most species but a long tail in *Nothus*. See *Coronidia, Homidiana, Nothus*.

**semele** see **Hipparchia**
**semiaurata** see **Neola**

**Semiothisa** Hübner (GEOMETRIDAE)
A huge genus of moths with many hundreds of species known, mostly from the N. Hemisphere. Where the life history is known the moths lay their eggs in clusters at the base of new growth of conifers.

**bisignata** Walker
U.S., Canada 24–28 mm. (0·94–1·1 in.)
The male is pale yellowish-brown with a single, small black spot on the anterior margin of the wing and some indistinct lines across. The female is whiter and has several black spots on the forewing near the anterior margins, and a larger mark near the apex of the wing. The wings are angular in shape, the hindwing has a distinct pointed edge, not rounded as in many Geometrids. The caterpillar feeds on *Pinus* and the adults are out in May and August.

**brunneata** Thunberg: syn. *S. fulvaria*
RANNOCH LOOPER 24–26 mm. (0·94–1·02 in.)
N. & C. Europe including Britain, N. Asia, N. America
This moth has reddish-yellow wings with a few darker markings, the female having darker transverse lines. Often there are marks round the margin of the hindwings. The moth is widely distributed in Europe but in Britain it is only found in N. Scotland. The caterpillar is reddish-brown with an irregular black-edged, green line along the back and thin white lines on either side. It feeds on bilberry *(Vaccinium)* on the leaves and looks very like a twig of this plant. In N. America the moth is out in July, in Europe in June and July. It is common from Newfoundland to Massachusetts.

**clathrata** Linnaeus LATTICED HEATH
Europe including Britain, W. & C. Asia to Japan 22–26 mm. (0·87–1·02 in.)
This species varies in background wing colour forms, nearly white to brown. The pattern of brown or black cross lines on the wings gives a distinctive appearance with a lattice effect, hence the popular name. The caterpillars feed on clover and trefoils (Leguminosae) and are green with white lines along the back and side. The moth has two broods a year over most of its range and can be seen in April and May and July and August. It is common in clover fields and chalk slopes and

generally flies by day. Several subspecies have been described, including one in Ireland.

**wauaria** Linnaeus V-MOTH
Europe including Britain 25–30 mm. (0·98–1·18 in.)
On the forewings of this moth there are dark marks, often tinged with violet, the middle ones on the wings in the shape of a V. The rest of the forewings are greyish white, spotted in grey. The moth is common over most of Europe in June and July. The caterpillar which feeds on currant and gooseberry *(Ribes)* is green-grey with a black spot and a bright yellow stripe on the side. The moth has been recorded in N. America but this is probably a misidentification of a similar local species.

**semirubella** see **Oncocera**
**semirufa** see **Premolis**
**semivitrea** see **Parathyris**

**Semomesia** Westwood (NEMEOBIIDAE)
There are 7 described species in this tropical S. American genus of butterflies.

**capanea** Cramer
Tropical S. America 30–42 mm. (1·18–1·65 in.)
The pattern of the under surface of the wings is largely repeated on the upper surface of the female wings which are brown with light brown bands and a white marginal band on the forewing. The illustrated male is brilliant blue and violet above with black eyespots. **263**y

**semperi** see **Pseudarbela**
**senegalensis** see **Bebearia**

**Sephisa** Moore (NYMPHALIDAE)
A small genus of butterflies with species in India, Formosa, W. China, Korea N. to the Amur peninsula.

**dichroa** Kollar WESTERN COURTIER
India, Pakistan, China, Korea, U.S.S.R. (Amur) 60–75 mm. (2·36–2·95 in.)
The brown wings have conspicuous yellow-brown patches while the hindwings have more yellow-brown veins. The brown on the veins spread out on the hindwing margin to leave just small patches of yellow-brown on the margin between the veins. This colourful butterfly can be found commonly between 910–2440 m. (3000–8000 ft) in oak forest in N. India.

**sequitiertia** see **Stenoma**

**Serecinus** Westwood (PAPILIONIDAE)
There is a single known species in the genus. It is closely related to *Parnalius*.

**telamon** Donavan
China, Korea, S.E. Russia 50–65 mm. (1·97–2·56 in.)
The female (not illustrated) differs above from the male in the much more strongly developed markings on the wings: both wings are heavily marked with dark grey bands, there is a complete transverse red band near the outer margin of the hindwing, and the tail is almost entirely black. There is some variation in pattern between the various subspecies and between the spring and summer broods of *S. telamon*. This is frequently a common species where its foodplant, *Aristolochia*, occurs. The caterpillar is black with brown markings. **94**g

**serena** see **Galona**
SERGEANT BUTTERFLIES see **Parathyma**
**sericina** see **Esthemopsis**
**serinopa** see **Nephantis**
**serpentaria** see **Loxolomia**
**serratulae** see **Pyrgus**

**Sesamia** Guenée (NOCTUIDAE)
The caterpillars are stem-borers in wild grasses or cultivated cereals. Some reach pest proportions and seriously affect the yield from crops like maize *(Zea)* and *Sorghum* and sugar-cane *(Saccharum)*. The moths are sombrely coloured in greys and browns. The distribution is African and Asian and to a lesser extent European.

**calamistis** Hampson PINK BORER
Africa (S. of the Sahara), Madagascar, Mauritius 22–30 mm. (0·87–1·18 in.)
Foodplants of this species include some minor cereal crops, and wild grasses, but also maize *(Zea)*, rice *(Oryza)* and sugar-cane *(Saccharum)* in Mauritius. It prefers damp, marshy areas. The wing coloration is slightly variable, especially in W. African and Madagascan populations.

**inferens** Walker PINK RICE-BORER
China, Japan, India, S.E. Asia to the Philippines 25 mm. (0·98 in.)
The caterpillars feed on a wide range of cereal crops: maize *(Zea)*, *Sorghum*, wheat *(Triticum)*, sugar-cane *(Saccharum)*, rice *(Oryza)* and a type of millet. It is a particularly serious pest of rice in India and Malaya.

**Sesia** Fabricius (SESIIDAE) CLEARWINGS
This genus of micro-moths imitates the appearance of wasps. This imitation (mimicry) is so good that the moths are frequently mistaken for wasps. Presumably this also confuses the predators that feed on them and the moths are left alone.

**apiformis** Clerck HORNET MOTH
Europe including Britain, temperate Asia, N. America 33–46 mm. (1·3–1·81 in.)
As its name implies this moth for its own protection mimics the Hornet Wasp but lacks the sting of that insect. The caterpillar is yellowish white with a yellow plate near the front and a brown head. It feeds on the roots and lower parts of the trunks of poplar *(Populus)*. The moth emerges from its chrysalis in the early morning from a cocoon of wood chippings woven together with silk by the caterpillar. **28**a

**Sesiidae:** syn. *Aegeriidae* CLEARWINGS
A fairly large family of micro-moths with most scientifically described species in the N. hemisphere but with many in the S. continents. The family is remarkable for their mimicry of different species of bees and wasps. Not only the appearances are similar but many even make a buzzing noise. This resemblance gives them a degree of protection, particularly from birds and small mammals which tend to avoid insects they think can sting. Although the moths imitate the bees and wasps in so many ways, the one thing they cannot do is sting. The Sesiids are usually day-fliers and are around the flowers with their wasp models. They have a rapid flight and are not easy to see in the field. Many species have larvae which tunnel into the stems of trees and shrubs, some are serious pests of fruit trees. Some of the mimicry of the Sesiids of wasps or bees is remarkably accurate, in other cases they merely have a yellow and black pattern and clear wings but the detailed resemblance is not so close. See *Alcathoe, Bembesia, Conopia, Melittia, Paxanthrene, Podosesia, Sesia, Synanthedon.*

**sesostris** see **Parides**
**sessilis** see **Lasaia**
SETACEOUS HEBREW CHARACTER see **Amathes c-nigrum**

**Setina** Schrank: syn. *Endrosa* (ARCTIIDAE)
About 4 species belong in this genus of European and W. Asian moths. The male is typically both day and night flying.

**irrorella** Linnaeus DEW MOTH
Europe including Britain, W. Asia 25–31 mm. (0·98–1·22 in.)
Moths emerge as early as May, but are most common in June and July. Males fly by day as well as during darkness. The tymbal organ of *S. irrorella*, especially of the male, is very large and is probably capable of producing the audible sound which has been recorded frequently for *Setina aurita*, a related European species with a similarly large tymbal organ. There is a great deal of variation between individuals both in the ground-colour of the wings and in the pattern of black spots. The black, yellow and white caterpillar feeds during the day on lichens growing on rocks or the bark of trees. Rocky hillsides are the best localities for this species. **364**j

**Setiostoma** Zeller (STENOMIDAE)
This genus is comprised of 17 species of small, brightly coloured micro-moths which are diurnal in habit and very similar in appearance. The similarity in colour pattern, and to a lesser extent venation, of *Setiostoma* to some groups in the family Glyphipterigidae resulted in the genus being associated with that family for many years. The genus is restricted to the W. Hemisphere and is primarily Neotropical in its distribution. Virtually no information is available on the life histories of the tropical members of the genus; however, the two species which occur in N. America feed on the foliage of trees in the family Fagaceae, primarily species of oaks *(Quercus)*.

**dietzi** Duckworth
Venezuela, Colombia, Costa Rica 11–15 mm. (0·43–0·59 in.)
This species bears the striking and consistent pattern of the forewings found in all members of the genus, featuring a light basal area and a dark brown outer area profusely patterned with iridescent scales. Recently a specimen of *S. dietzi* was observed and photographed visiting the flowers of *Cordia inermis* (Boraginaceae) in Costa Rica, representing the first observation of adult activity patterns for any of the tropical species.

**sexta** see **Manduca**
**shasta** see **Icaricia**
SHASTA see **Icaricia shasta**
SHEEP-MOTH, SAGE BRUSH see **Hemileuca hera**
SHELL MOTH, SCALLOP see **Rheumaptera undulata**
SHEPHERD'S FRITILLARY see **Boloria pales**
SHIELD BEARERS see **Heliozelidae**
SHOEMAKER BUTTERFLIES see **Catonephele, Prepona**
SHOT SILVERLINE, SCARCE see **Spindasis elima**
**siaka** see **Polystichtis**

**Sibine** Herrich-Schäffer (LIMACODIDAE)
Mostly S. American with a few species in Mexico. Their caterpillars are covered in stinging hairs.

**stimulea** Clemens SADDLE-BACK MOTH
E. & S. U.S. 30 mm. (1·18 in.)
The moth is a velvety reddish-brown with 2 white spots near the apex of the forewing. The hindwings are lightly coloured. The caterpillar which feeds on apple *(Malus)*, cherry *(Prunus)* and many ornamental trees, is brown with horns at each end and with a characteristic green saddle with brown centre, from which the popular name is derived. It is covered with stiff hairs which are mildly poisonous producing a severe stinging sensation, more severe than nettles.

**Siderone** Hübner (NYMPHALIDAE)
A small genus of butterflies with species found from Mexico to S. Brazil. The undersides of the wings are generally very leaf-like and camouflage the butterflies well when they are at rest. The genus is related to *Anaea* but little is known of the biology of the species, although a few have been reared from caterpillars feeding on *Viviania* (Vivianiaceae).

**galanthis** Cramer
S. America, W. Indies 62 mm. (2·44 in.)
The female has orange-brown hindwings with a large patch of orange on the forewing. The history of this species is somewhat confused. It was originally described as 'from Surinam' then it was suggested as Colombian, but the exact identity of this species still seems to be in some doubt. Did Cramer, when he originally figured the butterfly in about 1775, have a specimen similar to the one illustrated, or was his a different species? **163**k

**marthesia** Cramer
S. America 55–65 mm. (2·17–2·56 in.)
This is a very local species in Brazil. It is also one of the largest in the genus. The butterfly is strongly attracted to rotting fruit or market garbage. **235**

**Siga** Hübner (PYRALIDAE)
A small S. American genus of large Pyralid moths with similarly patterned species.

**liris** Cramer
Peru, Brazil, Bolivia, Guyana 65–84 mm. (2·56–3·31 in.)
This species is unique in its colouring. The other species in the genus lack the green and are generally brown or grey brown. Its biology is not known. **63j**

**signata** see **Wagimo**
**signella** see **Symmoca**
**signifera** see **Naroma**
**silenus** see **Myrina**
SILK-MOTHS see **Antheraea, Bombyx, Hyalophora, Rhodinia, Rothschildia, Samia**
SILKWORM see **Bombyx mori**
SILVER
HAIRSTREAK see **Chrysozephyrus syla**
KING SHOEMAKER see **Prepona demophoon**
SPOTTED SKIPPER see **Epargyreus clarus**
SILVER-BARRED NEPHELE see **Nephele argentifera**
SILVER-BORDERED FRITILLARY see **Clossiana selene**
SILVER-SPOTTED
FLAMBEAU see **Agraulis vanillae**
FLAMBEAU, SCARCE see **Dione juno**
TIGER-MOTH see **Halisidota argentata**
SILVER-STRIPED HAWK-MOTH see **Hippotion celerio**
SILVER-STUDDED BLUE see **Plebejus argus**
SILVER-WASHED FRITILLARY see **Argynnis paphia**
SILVER-Y MOTH see **Autographa gamma**
SILVERLINE BUTTERFLIES see **Spindasis**
SILVERLINES MOTH see **Bena, Pseudoips**
SILVERY
BLUE see **Glaucopsyche lygdamus**
CHECKERSPOT see **Chlorosyne nycteis**
**similana** see **Epinotia stroemiana**
**similis** see **Chrysopoloma, Euproctis, Phaegorista**
**simonsii** see **Henotesia**
**simplex** see **Rhodophthitus**
**simplicia** see **Liptena**
**Similais** see **Loxostege**
**sinapis** see **Leptidea**

**Sindris** Boisduval (PYRALIDAE)
One of the genera of Pyraline moths which include brightly coloured African species.

**magnifica** Jordan
Africa 45 mm. (1·77 in.)
This beautiful insect is known only from 2 specimens collected in Angola. Nothing is known of its biology and none have been collected in recent years. The specimen illustrated is the holotype. **63a**

**singapura** see **Oreta**
**sinha** see **Issora**
**sinope** see **Urota**
**sinuata** see **Cadarema**
**sinuella** see **Paraleucoptera**
**Siproeta epaphus** see **Metamorpha epaphus**
SIREN BUTTERFLIES see **Hestinalis**
**sita** see **Danaus**

**Sitotroga** Heinemann (GELECHIIDAE)
This genus of micro-moth is found mostly in Europe and Asia, apart from one widespread pest species. There are 2 or 3 species of this genus in Africa.

**cerealella** Oliver ANGOUMOIS GRAIN MOTH
Cosmopolitan 11–16 mm. (0·43–0·63 in.)
The caterpillar of this moth feeds on grain of all kinds in stores. It is known from N. and S. America, Europe, including Britain, and India through to Australia. It is always present in the warmer countries, sometimes as a fairly minor pest but occasionally as a major one, causing heavy loss of food in store. The caterpillar spends its whole life in one grain and in warm climates the life cycle only takes four weeks. The forewings are pointed pale greyish brown, sprinkled with brown spots, with long fringes on the hind margin of both wings, although not at the base of the forewing. The hindwings, which narrow sharply towards the apex, are grey. The adult moth is described at rest as 'looking like a fragment of chaff'.

SIX CONTINENT see **Hypolimnas misippus**
SIX-BELTED CLEARWING see **Bembecia scopigera**
SIX-SPOT BURNET see **Zygaena occitanica**
SIX-TAILED HELICOPIS see **Helicopis endymion**
**sjoestedti** see **Papilio**
SKELETONISERS see **Pampa**
SKIPPER BUTTERFLIES see **Abantis, Carterocephalus, Epargyreus, Erynnis, Euschemon, Hesperia, Leucochitonea, Megathymus, Muschampia, Ochlodes, Platylesches, Pyrgus, Thymelicus, Urbanus, Zophopetes**
SKIRTED CALICO, YELLOW see **Hamadryas fornax**
SKY-BLUE MORPHO see **Morpho portis**
**slateri** see **Papilio**
**sloanus** see **Urania**
SMALL
APOLLO see **Parnassius phoebus**
BATH WHITE see **Pontia chloridice**
BLUE GRECIAN see **Heliconius sara**
CHINA-MARK see **Cataclysta lemnata**
COPPER see **Lycaena phlaeas**
DARK YELLOW UNDERWING see **Anarta cordigera**
FLAMBEAU see **Eueides aliphera**
HEATH see **Coenonympha pampilus**
HILL-SIDE BROWN see **Pseudonympha narycia**
LACEWING see **Actinote pellenea**
MOTTLED WILLOW see **Spodoptera exigua**
MAGPIE MOTH see **Eurrhypara hortulata**
POSTMAN see **Heliconius erato**
PEARL-BORDERED FRITILLARY see **Clossiana selene**
TORTOISESHELL see **Aglais urticeae**
WHITES see **Dixeia, Pieris**
SMALLER
LANTANA see **Strymon bazochii**
WOOD-NYMPH see **Ideopsis gaura**
**smaragdalis** see **Charaxes**
**smaragditis** see **Tinostoma**
SMEARED DAGGER see **Acronicta oblinata**
**smeathmanniana** see **Aethes**

**Smerinthus** Latreille (SPHINGIDAE)
EYED HAWK-MOTHS
About 10 species are placed in this genus. Their distribution includes Europe (including Britain), temperate Asia as far E. as Japan, N. America and Mexico.

**jamaicensis** Drury TWIN-SPOT SPHINX MOTH
Canada, U.S. 50–82 mm. (1·97–3·23 in.)
The scientific name of this moth is misleading, as no species of *Smerinthus* occurs in Jamaica. The material from which it was described by Drury was thought to be from Jamaica, but was probably from New York State. It occurs in much of Canada and the U.S., but is rare in most of the S.E. The blue and black eyespot on the hindwing varies a great deal in shape, a fact which has given rise to numerous names for individual specimens. The eyespot is however almost invariably divided by a black bar unlike that of its close relative *S. ocellata* (illustrated). S. specimens are generally lighter in colour than the illustrated N. example. As in many other species of the tribe Smerinthini, the proboscis is poorly developed and the moths do not feed. The caterpillar will feed on the foliage of several trees including willow *(Salix)*, Aspen *(Populus)* cultivated apple *(Malus)*, peach and plum *(Prunus)*.

**ocellata** Linnaeus EYED HAWK-MOTH
Europe, temperate Asia (not Japan) 70–85 mm. (2·76–3·35 in.)
Apart from the E. Asiatic *S. planus,* this is the only species of its genus in which the eye spot on the hindwing is not divided by a black bar. There is experimental evidence to show that the eye spots are exposed when a moth is provoked and that birds are frightened by them. Similar eye spots in the family Saturniidae (eg. *Automeris*) produce a similar effect on small birds. The caterpillars feed on sallows and willows *(Salix)*, apple *(Malus)*, privet *(Ligustrum)* and poplar *(Populus)*. The normal countershading of many caterpillars is reversed in *S. ocellata* which feeds upside-down and is lighter in colour on its back. The shadow cast by light from above is nullified by the pale back and the three-dimensional caterpillar is apparently flattened and less conspicuous against the flat leaves on which it feeds. **347**

**smithi** see **Epicausis**
SMOKY ORANGE TIP see **Colotis evippe**
SNOUT BUTTERFLIES see **Libythea**
SNOUT MOTHS see **Hypena**
SNOW FLAT, WATER see **Tagiades litigiosa**
SNOWFLAKE see **Leucidia brephos**
SNOWY EUPSEUDOSOMA see **Eupseudosoma involuta**
**sociella** see **Aphomia**
SOD WEBWORM, LARGER see **Pediasia trisecta**
SOD WEBWORM, STRIPED see **Crambus mutabilis**
SOLDIER BUTTERFLY, LITTLE see **Chlorosyne saundersii**

**Solenobia** Duponchel (PSYCHIDAE)
MICROPSYCHID MOTHS
These tiny micro-moths have caterpillars which make cases round themselves which they then carry about. They generally feed on lichens. Species of the genus are known only from Europe. Males are very rare in the genus, most of the eggs are laid by unimpregnated females for several generations (parthenogenesis). When males are produced they have grey forewings with rather rounded wings.

**inconspicuella** Stainton
Europe including Britain 9–13 mm. (0·35–0·51 in.)
The caterpillar feeds in algae and lichens growing on fence posts and tree trunks, carrying their cases with them. The case which is covered with particles of debris, is 5–6 mm. long, black and often powdered with green algae. Recently they have been getting into houses and causing some concern. They are quite harmless and it seems that they have been feeding on algae in chimneys (unused because of central heating). The females are unusual in having their wings reduced in size often just to small filaments and as a result are unable to fly.

**Somabrachyidae**
A small family of micro-moths related to the Psychidae. The species are found in N. Africa and the Near E. The caterpillars are colourful, looking rather like those of Zygaenid moths. The females are wingless, the males have normal, rather rounded, wings. See *Somabrachys.*

**Somabrachys** Kirby (SOMABRACHYIDAE)
A small genus of micro-moths with colourful, polyphagous larvae, feeding on a wide range of trees. The species are restricted to the S. and E. sides of the Mediterranean Sea.

**albinervis** Oberthür
Algeria 25 mm. (0·98 in.)
At present this species is known only from Algeria but related, similar, species are found all along the N. African coast. Some of the species reach into Syria. The moths, which have wingless females, look very like the Psychids to which they are related. The eggs are laid in groups round twigs of trees and the caterpillars feed on a wide range of food plants. **5a**

**sommeri** see **Tarsolepis**
**sondaica** see **Discophora**
SONORA BLUE see **Philotes sonorensis**
**sonorensis** see **Philotes**
SOOTY BLUE see **Zizeeria knysna**
SOOTYWING BUTTERFLIES see **Pholisora**
**sophia** see **Junonia, Synemon**
**sophonisba** see **Eunica**
**sorbicola** see **Phyllonorycter**
**sorenseri** see **Tinearupa**
**sorghiella** see **Nola**
SORGHUM WEBWORM see **Nola sorghiella**

**Sorhagenia** Spuler (MOMPHIDAE)
A genus of small micro-moths with elongated rather pointed forewings. They are related to the *Gelechia.* Species of the genus are found in Europe and N. America. Few of the caterpillars are known, those that have been found feed on buckthorn *(Rhamnus)*.

**nimbosa** Braun
U.S. 9–12 mm. (0·35–0·47 in.)
The forewings are dark reddish-brown, with paler markings. The moth has been found in several of the States including California, Washington and Pennsylvania. The caterpillar has been reared from species of *Rhamnus*, where it feeds inside a large inflated gall-like chamber formed from the two halves of the leaf.

**rhamniella** Zeller
Europe including Britain 9–11 mm. (0·35–0·43 in.)
This small moth has a similar colouring to the related American species but has grey hindwings. The moth flies during June and July, but is rather local, particularly in Britain. When the young caterpillar is feeding it starts in the shoots of buckthorn *(Rhamnus)*, then as it gets larger it moves to the leaves, which it rolls together tying them with silk.

**sorona** see **Callicore**
SOUTH AFRICAN CARNATION WORM MOTH see **Epichoristodes acerbella**
SOUTHERN
CATTLE HEART see **Parides sesostris**
CLOUDYWING see **Thorybes bathyllus**
COMMA see **Polygonia egea**
FESTOON see **Parnalius polyxena**
GATEKEEPER see **Pyronia cecilia**
SMALL WHITE see **Pieris manni**
SWALLOWTAIL see **Papilio alexanor**
WHITE ADMIRAL see **Limenitis reducta**
WHITE, GREAT see **Ascia monuste**
WHITE PAGE see **Eurytides telesilaus**
**southeyae** see **Tarsocera**

**Spalgis** Moore (LYCAENIDAE)
*Spalgis* is placed in the same group of genera as *Liphyra* and in a separate family, the Liphyridae, by Clench and others. Its range includes Africa, Sri Lanka, India, and S.E. Asia to Celebes.

**epeus** Westwood APEFLY
Sri Lanka, India, Burma, Malaya 20–30 mm.
E. to Sulawezi (Celebes) (0·79–1·18 in.)
The popular name for this butterfly refers to the shape of the chrysalis which somewhat resembles a monkey's head. The upper surface of the adult butterfly is brown, with a whitish central patch on the forewing; beneath, the wings are pale brown, with several darker transverse lines. The butterflies have a characteristic leisurely flight; they are confined to forested areas. The caterpillars are predacious on minute scale-insects and mealy-bugs (Hemiptera, Coccidae). Eggs are laid amongst the future prey so that the caterpillar is at once able to burrow into a nearby Coccid after emerging from the egg.

SPAN WORM MOTH, WALNUT see **Coniodes plumigeraria**
SPANGLED FRITILLARY, GREAT see **Speyeria cybele**
SPANISH FESTOON see **Parnalius rumina**

**Sparganothis** Hübner (TORTRICIDAE)
Many American species have been described in this genus of micro-moths. It includes many species which are of economic importance and cause losses in agriculture or horticulture. A few species are known from Europe and temperate Asia.

**pilleriana** Denis & Schiffermüller VINE PYRALIS OR VINE TORTRIX MOTH
Europe including Britain, Asia 16–25 mm.
including Japan, N. America (0·63–0·98 in.)
Forewing often with a golden metallic tint with a yellowish-brown band across the wing and pale lines near the edge of the wing. The caterpillar feeds on a variety of plants but can be particularly destructive to grape vines *(Vitis)*. The leaf-rolling caterpillar has appeared in plague proportions in the past. In the middle of the 16th century the Vine-Pyralis is recorded as destructive and prayers were offered on the orders of the Bishop of Paris for the 'diminution of these insects'. In a ten year period (1828–37) the value of grapevines lost was calculated at 1,500,000 francs and some 3,500,000 francs were lost in trade connected with this in just two Departments of France. The moth is not a Pyralid, in spite of its old name, Vine-Pyralis.

**sparmannella** see **Eriocrania**

**Spatalia** Hübner (NOTODONTIDAE)
There are about 20 species in this genus of moths. It is distributed throughout much of the Old World, except for Africa, from Europe to Japan and south-eastwards to New Guinea. Many of the species have metallic silver or gold markings on the forewings.

**argentina** Denis & Schiffermüller
Central and S. Europe 30–38 mm. (1·18–1·5 in.)
The scientific name of this beautiful species refers to the metallic, silvery markings on the forewing. Its caterpillar feeds on the foliage of oak *(Quercus)*. **357j**

**spatiose** see **Euphaedra**
SPEAR-MARKED MOTH see **Rheumaptera hastata**
SPEAR-WINGED CATTLE HEART see **Parides neophilus**
SPECKLED BLACK NYMPH see **Cremna actoris**
SPECKLED MOTH, CRIMSON see **Utetheisa pulchella**
SPECKLED WOOD see **Pararge aegeria**
SPECKLED YELLOW see **Pseudopanthera maculaia**
SPECTRE BUTTERFLY see **Bebearia senegalensis**

**Speyeria** Scudder (NYMPHALIDAE)
These American butterflies are related to the European *Argynnis* and both genera are popularly known as fritillaries. They are found mostly in N. America. The caterpillars feed on species of violets *(Viola)*. The adults are rapid fliers, frequently feeding at flowers. A large number of subspecies and aberrations of the N. American species have been described.

**cybele** Fabricius GREAT SPANGLED FRITILLARY
U.S. 68–80 mm. (2·68–3·15 in.)
This species occurs widely in the Atlantic states and reaches across N. America to the W. States. The adult is typical of the large group of fritillary butterflies which have a broadly similar external appearance. The caterpillar feeds on violets (*Viola* sp.). A number of forms and subspecies have been described. **163f**

**Sphaerelictis** Meyrick (OECOPHORIDAE)
A small genus of micro-moths with only four known species, two in Australia, one in Sri Lanka and one in India. Little is known of their biology.

**dorothea** Meyrick
India 18 mm. (0·71 in.)
This is a fairly common species in parts of India but details of its life history are not known, although the specimen illustrated has a label indicating that it was bred! The available data on life histories, even of quite common species, is fragmentary and time spent studying and breeding them rather than just adding to a collection would be very rewarding. **44a**

**Sphecodina** Blanchard (SPHINGIDAE)
Both known species of this genus of hawk-moths are found only in the N. Hemisphere; one occurs in the New World, the other in eastern Asia. The abdomen of the adult is short and broad, unlike most other hawk-moths.

**abbotii** Swainson ABBOT'S SPHINX MOTH
Canada, U.S. 60–72 mm. (2·36–2·83 in.)
Eastern N. America is the home of this attractive species where it ranges from S. Canada to N. Florida. It flies at dusk and will visit honeysuckle *(Lonicera)* flowers and the sap of willows *(Salix)*. The caterpillars feed on the foliage of Virginia creeper *(Parthenocissus)* and grape-vine *(Vitis)*. **344s**

**Sphecosoma** Butler (CTENUCHIDAE)
There are about 33 species in this C. and S. American genus. Most of them are similar in general appearance to moths of the genus *Pseudosphex*.

**trinitatis** Rothschild
Trinidad 25–28 mm. (0·98–1·1 in.)
This is not greatly different in coloration and pattern to the illustrated species of *Pseudosphex*. It is probably part of a mimetic association with parasitic wasps of the family Ichneumonidae.

**Sphingidae** HAWK or SPHINX MOTHS
This is possibly the best known and most popular of all families of moths. The streamlined, long-winged adults are favourites with many collectors, and the huge, almost hairless caterpillars are no less popular with breeders. About 800 species are known, most of them tropical in range, but many are found in temperate regions. There are one and a half times as many species in the New World as in the Old, but the Old World fauna is more diverse and has twice as many genera. Hawk-moths are amongst the fastest insects and also are adept at hovering; speeds of up to 30 miles per hour (50 k.p.h.) have been recorded. Some species fly during the day and visit flowers for nectar (see *Macroglossum* and *Hemaris*). The hovering flight of these day-flying species is much like that of various bees and wasps which some are considered to mimic. Many, but not all species, have developed an enormously long proboscis which enables the moth to probe into the nectaries of tubular flowers like orchids and honeysuckle *(Lonicera)* (see *Xanthopan*). There is a typanum (ear) on the palps of the species of some genera. Sound production is not a common phenomenon in this family, but the males of some adults are able to stridulate with their genitalia, and the Death's-head Hawk-moth can produce a rhythmic, muted squeak by forcing air through its proboscis. The chrysalis of many species produces a rustling sound when disturbed, which may deter predators. There is little chemical evidence for chemical defence (but see *Acanthosphinx*) even in *Nephele* some of whose species feed on members of the poisonous plant family Asclepiadaceae. The caterpillars of nearly all species have a characteristic, curved horn at the rear end, and many have one or more pairs of eye-spots towards the front which are displayed in 'threatening' display postures which are frequently snake-like in effect. The caterpillars of a few species are serious pests of crops (see for example *Manduca*). See *Acanthosphinx, Acherontia, Agrius, Antinephele, Arctonotus, Callionima, Cechenena, Chaerocina, Cizara, Cocytius, Daphnis, Deilephila, Erinnyis, Euchloron, Eumorpha, Hemaris, Hippotion, Hyles, Laothoe, Leptoclanis, Macroglossum, Manduca, Microsphinx, Mimas, Nephele, Pachysphinx, Panacra, Polyptychus, Proserpinus, Protambulyx, Rethera, Rhodosoma, Sagenosoma, Smerinthus, Sphecodina, Sphinx, Tinostoma, Xanthopan, Xenosphingia, Xylophanes*.

**Sphinx** Linnaeus (SPHINGIDAE)
This is a large, mostly New World genus of hawk-moths, best represented in N. America, but with a few European and temperate Asian species. The adults are chiefly grey or brownish grey; the caterpillar has 7 pairs of oblique stripes on each side.

**dolli** Neumoegen
S.W. U.S. 45–55 mm. (1·77–2·17 in.)
This is the smallest N. American species of its genus and not likely to be confused with other than *S. sequoiae* which generally is more bluish and has the black line on the forewing interrupted at the middle. The head, thorax, forewings and abdomen of *S. dolli* are pale grey; the hindwings are brown. There is a black oblique line on the forewing extending from the base of the wing to its apex, and a black mid-dorsal line and black, lateral patches on the abdomen. Juniper *(Juniperus)* is the only recorded foodplant of the caterpillar.

**ligustri** Linnaeus PRIVET HAWK-MOTH
Most of Europe, through Asia to China, 80–110 mm.
Japan and E. U.S.S.R. (3·15–4·33 in.)
In S. England this is probably the commonest species of hawk-moth, no doubt partly the result of the wide downland and urban distribution of the caterpillars' main foodplant, privet *(Ligustrum)*. The purple and white striped, pale green caterpillars are probably more often seen than the adult moth, especially in the autumn when they leave their foodplant to burrow in soil or dead leaves before building a flimsy cocoon inside which the chrysalis forms. The chrysalis stage usually lasts about nine months, but may extend rarely to three years. The only species likely to be confused

with *S. ligustri* in England is *Agrius convolvuli*, which frequently migrates across the English Channel. The sphinx-like, larval feeding posture is the origin of the generic name for this species and its allies. The family name Sphingidae has the same derivation. **344j**

**pinastri** Linnaeus PINE HAWK-MOTH
Europe including Britain, through Asia to China and Japan 75–90 mm. (2·95–3·54 in.)
*S. pinastri* has been recorded on frequent occasions from Canada and the U.S. as the result of accidental introductions. In Britain it is confined to the south in regions where the caterpillars' main foodplants, pines *(Pinus)* occur. The moth often visits honeysuckle *(Lonicera)* blooms at dusk. It flies from May to July. The caterpillars are at first striped longitudinally and are well camouflaged amongst the pine needles; later stages, in contrast, disruptively patterned and not striped. **350**

**sphinx** see **Rapala**
SPHINX MOTHS see **Acanthosphinx, Cocytius, Eumorphia, Hyles, Manduca, Pachysphinx, Smerinthus, Sphecodina, Tinostoma**

**Spialia** Swinhoe (HESPERIIDAE)
A genus of skipper butterflies with species found in Africa, Europe and Asia.

**rebeli** Higgins
Africa 21 mm. (0·83 in.)
This skipper is found throughout W. Africa to Angola and in Kenya and Uganda. There are a number of related, and very similar looking species in Africa. The biology is unknown. **82**

SPICEBUSH SWALLOWTAIL see **Papilio troilus**

**Spilosoma** Curtis (ARCTIIDAE)
*Spilosoma* is best developed in the Old World tropics but is also represented in the temperate zones and by a few New World species. Many of these moths are white or yellow with black markings.

**congrua** see under **S. virginica**
**latipennis** see under **S. virginica**

**lubricipeda** Linnaeus WHITE ERMINE MOTH
Europe including Britain, through temperate Asia to Japan 38–43 mm. (1·5–1·69 in.)
Although usually white, the wings of some specimens of the species are pale yellow. It has been suggested recently that *S. lubricipeda* which is distasteful to birds is mimicked by the probably palatable *S. luteum* – it follows the rule that a model should be on the wing earlier in the season than its mimic, so the latter can benefit from unpleasant experiences with the model. Histamines and acetylcholines have been identified from the tissues. **377**

**luteum** Hufnagel BUFF ERMINE MOTH
Europe (including Britain, through temperate Asia to China 32–41 mm. (1·26–1·61 in.)
This is a very common species even in urban areas where its brown hairy, caterpillars will feed on varieties of garden plants and occasionally on the foliage of trees. There is considerable variation between adults in the colour pattern, and melanic (black) forms sometimes occur. *S. luteum* is considered by Rothschild to be a mimic of the unpalatable White Ermine, *S. lubricipeda*. The chrysalis over-winters and is encased in a loosely woven cocoon placed in debris at ground-level.

**prima** see under **S. virginica**

**virginica** Fabricius YELLOW WOOLLY-BEAR or VIRGINIAN TIGER-MOTH 35–45 mm. (1·38–1·77 in.)
Canada (rarely), U.S., Mexico and Central America
This is similar in general appearance to the illustrated Old World *S. lubricipeda*, but with fewer black markings on the wings, especially the forewing. Other N. American species of *Spilosoma (prima, latipennis, congrua)* differ only in details of the coloration and pattern from *S. virginica*. There are two broods per year; adults emerge in May and June, and again in August and September. The caterpillar, a 'woolly bear', is rather variable in colour, but is often yellow or black with yellow hair; it feeds on the flowers, young stems and leaves of many species of low-growing plants and it sometimes is a minor pest of corn *(Zea)* and water-melons *(Citrullus)*; it has been recorded also on willow *(Salix)* and aspen *(Populus)*. *S. virginica* has been introduced into Hungary and is apparently extending its range from there.

**Spindasis** Wallengren (LYCAENIDAE)
BARRED BLUE or SILVERLINE BUTTERFLIES
Africa, India, China, Japan, Malaya to the Philippines is the range of this genus. The species of *Spindasis* are butterflies of both moist and semi-desert areas. The upper surface of these butterflies is dark brown, usually with a dark blue iridescence in the male, and banded with orange in some species. Beneath, the wings are marked with metallic streaks and bars. Burnt grass and the ashes of wood fires have an attraction for the butterflies. The caterpillars of most species are attended by ants.

**elima** Moore SCARCE SHOT SILVERLINE
India, Sri Lanka 28–42 mm. (1·1–1·65 in.)
This is a generally uncommon species, found up to elevations of 1520 m. (5,000 ft). The male forewing is brown above, with a purple iridescence at the base and an orange patch centrally; the hindwing is chiefly iridescent purply brown. The under surface is pale brown, with metallic gold, transverse bands. The female lacks any purple iridescence on the upper surfaces of the wings.

**natalensis** Westwood NATAL BARRED BLUE
Mozambique, Rhodesia, South Africa 26–40 mm. (1·02–1·57 in.)
This is a generally common forest and scrubland species. The wings are pale blue with pale brown borders, with an orange patch on the hindwing at the base of the two tails and usually with three orange bars apically on the forewing. The under surface of the wings is pale yellow, with silver and black spots and bands. The caterpillar feeds on *Mundulea* and *Vigna*.

SPINY BOLLWORM see **Earias biplaga**

**Spiramiopsis** Hampson (LEMONIIDAE)
Only one species is placed in this genus of moths. It has been placed at times in the families Brahmaeidae and Eupterotidae. The caterpillars have several characteristics in common with those of *Brahmaea wallichii*.

**comma** Hampson
South Africa 50–60 mm. (1·97–2·36 in.)
The name 'comma' refers to the large comma-shaped marking on the forewing of this moth. Its caterpillars are spiny from head to tail. Both the second and third thoracic segments of the caterpillar bear a pair of extraordinary, spinose, spiral, dorsal processes which are over a centimetre in length. There is a similar process at the rear end of the abdomen, similar to the 'horn' of most hawk-moth (Sphingidae) caterpillars. **318c**

**Spiris** Hübner (ARCTIIDAE)
A single species is placed in this genus of moths. It has been combined with *Coscinia* at times.

**striata** Linnaeus FEATHERED FOOTMAN
Europe including Britain, Middle-East, temperate Asia to E. China and Russia 35–50 m. (1·38–1·97 in.)
This is a rare, migrant species in Britain; but often common elsewhere in its range. The characteristic striations on the forewing may be nearly absent in some specimens, while in others these are much thicker than in the illustrated specimen. The caterpillar is brown, marked with orange and white; it feeds chiefly on grasses (Gramineae), ling heather *(Calluna)* and plantain *(Plantago)*. *S. striata* overwinters as a caterpillar, feeds until May and then turns into a chrysalis. The moths emerge in June and July. **364p**

**splendens** see **Aulacodes, Chamaelimnas, Citheronia, Eudaphnaeura**
**splendida** see **Acrojana, Chrysochlorosia** and under **Esthemopsis sericina**
SPLITWORM, TOBACCO see **Phthorimaea operculella**

**Spodoptera** Guenée (NOCTUIDAE)
This is a small but almost cosmopolitan genus of economically important pest-species, containing some of the most notorious pests of food-crops. The forewings of most species are rather inconspicuously marked with brown; the hindwings are broad (see illustration of *S. litura*) and chiefly white.

**exempta** Walker COMMON or AFRICAN ARMYWORM
Africa, tropical Asia to Australia 25–32 mm. (0·98–1·26 in.)
This is one of the important leaf and stalk-feeding pests of maize *(Zea)*, grasses, *Sorghum*, rice *(Oryza)*, wheat *(Triticum)* and other cereal crops in Africa. In Australia it is sometimes a pest of sugar-cane *(Saccharum)*. Immense swarms of caterpillars are frequently reported from cultivated areas. It is similar in general appearance to the larger *S. litura* (illustrated) but the forewing pattern is less complicated and less clearly marked.

**exigua** Hübner BEET or LESSER ARMYWORM, PIG-WEED CATERPILLAR, or SMALL MOTTLED WILLOW MOTH
Almost cosmopolitan 25–28 mm. (0·98–1·1 in.)
Nearly every crop cultivated by man is attacked by the caterpillar of this notorious species: corn *(Zea)*, and indigo *(Indigofera)* in Asia; cotton *(Gossypium)* and various legumes in South Africa; tobacco *(Nicotiana)* in Rhodesia; and several other plants including alfalfa *(Medicago)*, beet *(Beta)*, cabbage *(Brassica)*, tomato *(Lycopersicon)* and potato *(Solanum)*. The adults are strongly migratory in habits; probably one of the reasons for its success as a species. Various methods of control have been used in efforts to combat this moth. These include the introduction of *Bacillus thuringensis* which causes disease in caterpillars, gamma radiation of adults which upsets the genetic structure of the sex-cells, and the use of 'hexalure', a simulated female sex-scent which when applied over a large area misleads males who under normal conditions locate females by following scent-gradients originating from their potential mates. One of the mynah birds *(Gracula)* is a natural enemy of *S. exigua* in Hawaii. **390j**

**frugiperda** Smith FALL ARMYWORM
America 35–40 mm. (1·38–1·57 in.)
This well-known pest-species can overwinter in the Gulf States of the U.S., but is an essentially tropical American moth. There are several broods per year in the tropics; the later broods move north into N. America during the late summer, often resulting in severe damage there by caterpillars in the autumn. A great variety of wild and cultivated plants are attacked, including corn *(Zea)*, various other grains and grasses, peanuts *(Arachis)*, potato *(Solanum)*, tomato *(Lycopersicon)* and cotton *(Gossypium)*. The moths are similar in general appearance to the illustrated *S. litura* but are darker and much less well-marked; the pattern is almost completely absent in some specimens.

**littoralis** Boisduval EGYPTIAN COTTON LEAFWORM or MEDITERRANEAN BROCADE MOTH
Africa, Middle East to India 34–42 mm. (1·34–1·65 in.)
The caterpillar of *S. littoralis* is a major pest of cotton *(Gossypium)* in the United Arab Republic and Israel, and will also attack sugar-beet *(Beta)*, tomato *(Lycopersicon)*, potato *(Solanum)*, alfalfa *(Medicago)* and apple *(Malus)*. Adults are difficult to distinguish in colour-pattern from the illustrated *S. litura*. One of the control measures employed has involved the use of chemicals to sterilize adults which are then released and subsequently engage in wasteful (to the species) mating with untreated moths. **397**

**litura** Fabricius ORIENTAL LEAFWORM
India, through S.E. Asia to Australia and the S. Pacific islands 35–42 mm. (1·38–1·65 in.)
This is a common pest of cultivated vegetables, cotton *(Gossypium)* and tobacco *(Nicotiana)*. In New Guinea the caterpillars are a pest of cocoa *(Theobroma)*. The wings are very variable in coloration and, to some extent, in pattern, but are usually clearly marked. **390h**

SPOT, GOLD see **Plusia festucae**
SPOT
PUFFIN see **Appias pandione**
SWORDTAIL see **Graphium nomius**
SPOTLESS GRASS YELLOW see **Eurema herla**
SPOTTED
ACRAEA, LARGE see **Acraea zetes**
BOLLWORM, COTTON see **Earias insulana**
BUFF see **Pentila tropicalis**
CUTWORM see **Amathes c-nigrum**
FRITILLARY see **Melitaea didyma**
PURPLE, RED see **Limenitis arthemis**
SAWTOOTH see **Prioneris thestylis**
SKIPPER, SILVER see **Epargyreus clarus**
SULPHUR see **Emmelia trabealis**
SWALLOWTAIL, PURPLE see **Graphium weiskei**
ZEBRA see **Graphium megarus**
SPRING AZURE see **Celastrina argiolus**
SPRUCE BUDWORM see **Choristoneura fumiferana**
SPURGE HAWK-MOTH see **Hyles euphorbiae**
SQUASH-BORERS see **Conopia, Melittia**
SQUEAKING SILK-MOTH see **Rhodinia fugax**
SQUINTING BUSH BROWN see **Mycalesis anynana**
**stacyi** see **Zelotypia**
STAFF SERGEANT see **Parathyma selenophora**
**stagnata** see **Parapoynx**

**Stalachtis** Hübner (NEMEOBIIDAE)
The species of this small genus of under 10 Central American and tropical S. American butterflies are all mimics, both in colour pattern and manner of flight, of either Acraeinae, Heliconiinae or Ithomiinae. A characteristic of this group is that its species regularly feign death if captured.

**calliope** Linnaeus
N. and N.W. S. America 56–60 mm. (2·2–2·36 in.)
There are several, mostly geographical, variations on the illustrated pattern, but each is a fairly exact copy of species of Ithomiinae and Heliconiinae. This is a common species in much of its range. **265**; a, from Brazil; b, from French Guiana.

**euterpe** Linnaeus
Tropical S. America 46–50 mm. (1·81–1·97 in.)
Some forms of this species resemble typical orange and brown species of *Actinote* (Acraeinae) and are probably their mimics. Northern specimens usually have a transverse, orange band on both wings; Amazonian specimens normally lack an orange band on the hindwing. **268**g

**phaedusa** Hübner
Tropical S. America 37–45 mm. (1·46–1·77 in.)
*S. phaedusa* is one of several *Stalachtis* species which mimic transparent-winged species of Ithomiine butterflies and day-flying Hypsid moths. Few specimens are average in size: they are mostly either small or large – possibly an indication that the species mimics two or more differently sized models. **266**b

STALK-BORER, MAIZE see **Busseola fusca**

**Stamnodes** Eversmann (GEOMETRIDAE)
A genus of moths found in the temperate part of Asia, N. India and America. Many of the species settle with the wings held over the back like butterflies.

**danilovi** Erschoff
S. Siberia and N. China 30 mm. (1·18 in.)
Nothing is known of the biology of this brightly coloured species. Relatively few specimens are to be found in collections in museums throughout the world. **294**n

STATELY NAWAB see **Polyura dolon**

**Stathmopoda** Herrich-Schäffer (STATHMOPODIDAE)
A large genus of micro-moths, many of which occur in Australia. The abdomen of the moths have small spines on the dorsal side. The caterpillars have a wide range of food. Some tunnel in galls, some are predatory on scale insects and others feed on spiders egg sacs.

**pedella** Linnaeus
Europe 10–14 mm.
including Britain (0·39–0·55 in.)
The caterpillar feeds on the fruits of alder *(Alnus)*. The moths are widespread but not common over much of their range. The adult flies in July. **44**e

**Stathmopodidae**
The caterpillars of the micro-moths in this family feed on a wide range of food plants. Some are predatory on scale insects (Coccidae). The moths can be recognized by the hind legs which have hairs arranged in whirls at the apex of each tarsal segment. Most species rest with the hind legs raised. Although many are nocturnal, a few species are day-flying. This family has also been known as the Tinaegeriidae. See *Stathmopoda.*

**statices** see **Adscita**
**statira** see **Phoebus**
**staudingeri** see under **Lasiocampidae**

**Stauropus** Germar (NOTODONTIDAE)
This is a moderately large Old World genus which has been subdivided by taxonomists to some extent during the last few years. The forewings of most species are cryptically patterned in brown, grey and green. The caterpillar of *S. alternatus* Walker is a pest of cocoa trees *(Theobroma)* and tea *(Thea)* plants in Sri Lanka and Java.

**fagi** Linnaeus LOBSTER MOTH
Europe including Britain, 55–68 mm.
and temperate Asia to Japan (2·17–2·68 in.)
The moth of this species is a sombre greyish brown. The caterpillar, in contrast, is an extraordinary scorpion- or lobster-like creature. When disturbed it raises the swollen rear end of the abdomen over its back and vibrates one of the abnormally long, spider-like second and third pairs of thoracic legs. Further provocation may produce an ejection of a formic acid secretion from the thoracic glands of the caterpillar. Its first meal is its egg-shell, or most of it, after which it will feed on beech *(Fagus)*, oak *(Quercus)* and other deciduous trees, including fruit trees. **353**

**steinbachi** see **Scea**
STEINDACHNER'S EMPEROR MOTH see **Athletes steindachneri**
**steindachneri** see **Athletes**
**stelenes** see **Metamorpha**
**stellatarum** see **Macroglossum**
**stellifera** see **Anisozyga**
STEM-BORERS see **Castniidae, Cossidae, Noctuidae Pyralidae, Sesiidae**
STEM-TIP BORER see **Earias biplaga**

**Stenoma** Zeller (STENOMIDAE)
A large and mixed genus of micro-moths containing at present many which are probably not related to one another. The genus has species all over S. and C. America. It tends to have species which are characterized by rather broad and rounded wings. Stenomids are common forest moths of Guyana. Several species resemble bird droppings when at rest. They sit exposed on leaves, with wings extended pressed to the leaf surface. In this way the moth is unrecognized by its potential enemies. Some of the species are pests on cocoa in S. America.

**sequitiertia** Zeller
S. America 36–50 mm. (1·42–1·97 in.)
This species was first described in 1854 by a German entomologist. It is a particularly strikingly patterned species with rounded wings and 4 large spots on the forewing. Many of the other species in the genus are rather sombre coloured. As with so many species of exotic moths, little is known of its biology. The species has been collected in Brazil, Peru and the Guianas. **42**d

**Stenomidae**
This is a family of micro-moths with species mostly in the tropics. In America there are species in the N. and S., but it is in S. America that most species are found. A number of the adults are particularly well camouflaged species. Much of the biology is unknown. Some of the species have caterpillars which mine in leaves, others bore into trees and shrubs. Generally the males have ciliate (hairy) antennae with differences in the wing venations and genitalia which are used to classify them. The relationship of this family to the Xyloryctidae is close, the latter could be described as Old World Stenomids. See *Antaeotricha, Loxotoma, Setiostoma, Stenoma.*

**stenosoma** see **Dracaeunura**

**Stericta** Lederer (PYRALIDAE)
A large genus of Pyralid moths with a wide variety of species at present included, which are probably unrelated. It belongs to the subfamily Epipaschiinae. Species of the genus occur in Australia, India and a few in China. They are often rather soberly coloured. Characteristically they all have very long palps sticking out in front of the head from below, and raised scales on the forewings. Many species in the genus have a rather greenish tinge to the wings.

**flammealis** Kenrick
New Guinea 26–30 mm. (1·02–1·18 in.)
This is the exception in an otherwise relatively sober coloured genus. The bright colours are very conspicuous in specimens in collections but perhaps, unless it is a warning colour, they are probably well camouflaged in the field. Some related species have caterpillars which live in webs and form cocoons in clusters. **56**bb

**Stibochiona** Butler (NYMPHALIDAE)
A small genus of butterflies with less than 6 species, found only in the Orient.

**coresia** Hübner: syn. *S. kannegieteri*
Indonesia 55 mm. (2·17 mm.)
A mostly black species with a few small white spots round the forewing but with a large oval blue-purple iridescent patch on the hindwing. In some specimens this may be a pale Cambridge blue, in others, almost purple. The females are generally brown rather than black and the hindwing colour is paler. The undersides are all brown with white edges and a few white marginal spots. The life history is unknown. **208**g

**Stichophthalma** Felder (NYMPHALIDAE)
This genus is in the subfamily Amathusiinae. It has long slender antennae which gradually thicken towards the tips. Generally the species in the genus, whose food plants are known, have caterpillars which feed on palms or bamboos. The adult butterfly feeds at sap flowing from injured trees or rotting fruit, cattle droppings, but are rarely seen on flowers. They are large, powerful fliers.

**camadeva** Westwood NORTHERN JUNGLE QUEEN
N. India, Pakistan 120–130 mm.
to N. Burma (4·72–5·12 in.)
The pale blue forewing, with darker edges, and dark bluish hindwing with a white band below the margin which is divided by the blue veins, make a striking pattern. The underside is green with black lines near the base of the wing, a white band along the fore- and hindwings followed by a row of round brown spots, usually with a small white centre, between each vein from the front to the back of each wing. The outer part of the underside is yellowish brown. They fly in thick jungles, usually near the ground. There are 2 broods a year in N. India where the adults are said to fly between 610–910 m. (2000–3000 ft), often in clearings, from June to August. The males are particularly active fliers.

**sticticalis** see **Margaritia**

**Stigmella** Schrank (NEPTICULIDAE)
These micro-moths are amongst the smallest known species. The caterpillars are leaf-miners. *Stigmella* is a large genus with species in Europe, Asia and a few African and Indian species. The moths in the genus have not been studied in recent years.

**ulmivora** Heinmann
Europe including Britain 3–5 mm. (0·12–0·2 in.)
Probably one of the smallest moths, this is local over much of its range in W. Europe. The caterpillar lives

in irregularly shaped galleries inside the leaves of elm *(Ulmus)*. The adult is out in May and August. Its shape (see illustration) is typical of many in the family. The egg is generally laid on the midrib of the elm leaf and the mine starts as a slender track. The caterpillar is green (separating it from a related species on elm which is yellow). The mine forms a loose spiral pattern on the leaf. In most areas this species has 2 broods a year. **1d**

**Stilbosis** Clemens (Momphidae)
These small micro-moths which are found in N. and Central America are not well known. Most are relatively inconspicuous. They are related to Gelechid moths.

**tesquella** Clemens
U.S. 9 mm. (0·35 in.)
This species is common and widespread from New York and N. Carolina to Minnesota. It has been collected in Tennessee and Utah. The forewings are shiny dark lead-colour. The caterpillar feeds externally on species of *Amphicarpaea* and *Lespedeza* (Leguminosae).

**Stilpnotia** see under **Leucoma**
**stimulea** see **Sibine**
**stolida** see **Grammodes**
STRAIGHT PIERROT BUTTERFLY see **Castalius roxus**
**stratiotata** see **Parapoynx**
STRAW, SCARCE BORDERED see **Heliothis armigera**
STRAWBERRY TORTRIX see **Acleris comariana**
**Streblote dorsalis** see under **Lasiocampidae**

**Strepsimanes** Meyrick (Strepsimanidae)
A little known genus of micro-moths with only one species which has peculiar wing venation. Very little is known about this genus and few specimens of the species in it are known.

**scierops** Meyrick
India 13–14 mm. (0·51–0·55 in.)
The curved edge to the forewing, with the peculiar venation, were the characters used in the original description to describe this moth. In fact the moths, although of interest for their peculiar structures, did not inspire the original author whose description ran 'the toneless shadowlike colouring and entire absence of marking afford no clue to the origins or habits of this singular insect'. In all probability, when this species is re-examined, it will be found that it is not a distinct family but can be placed in one of the existing ones.

**Strepsimanidae**
A family of micro-moths with only one genus and one species known at present. They are small Gelechid-like moths and only a few specimens are known. The actual relationship of this family with others in the Gelechoid group is still not known. Dr R. Hodges, U.S. Department of Agriculture, considers that the species should be transferred to the Hypeninae (Noctuidae). See *Strepsimanes*.

**stroemiana** see **Epinotia**
**striata** see **Spiris**
**strigicincta** see **Mazuca**
**strigilis** see **Protambulyx**
**strigillata** see **Graphelysia**
STRIPED
BLACK CROW BUTTERFLY see **Euploea alcathoe**
BLUE CROW BUTTERFLY see **Euploea mulciber**
HAWK-MOTH see **Hyles lineata**
POLICEMAN see **Coeliades forestan**
SOD WEBWORM see **Crambus mutabilis**
**streckerianus** see **Battus**

**Strophidia** Hübner (Uraniidae)
Two species are placed in this genus of the subfamily Microniinae: *S. fasciata* (illustrated) and *S. directaria* Walker from Papua and the Moluccas.

**fasciata** Cramer
Sri Lanka, India, Burma, Malaya, Borneo 47–62 mm. (1·85–2·44 in.)
The striate colour patterns varies in detail between specimens, and in some the markings are very faint. Nothing is known about the life history of this species. **317c**

**Strymon** Hübner (Lycaenidae)
This is a large, chiefly tropical genus of butterflies found only in the New World. The species are tailed or tailless. One species *S. melinus*, is widely distributed in the U.S.

**avalona** Wright
U.S. (California) 24–28 mm. (0·94–1·1 in.)
This has one of the smallest ranges of any species of butterfly. It is known only from the island of Santa Catalina off the coast of S. California. The caterpillar feeds on a species of *Lotus*.

**bazochii** Godart SMALLER LANTANA BUTTERFLY
Texas to tropical S. America, Hawaii 22–26 mm. (0·87–1·02 in.)
This species, together with *Tmolus echion* Linnaeus, was introduced into Hawaii in an effort to control *Lantana*, a pest plant species in these islands. The caterpillars of both feed on the flowerheads of *Lantana* and successfully inhibit the production of seeds. Basil *(Ocimum)* and *Hyptis* are also attacked, by the caterpillars, but *S. bazochii* has proved to be a much more satisfactory biological control insect than *T. echion* whose caterpillars have shown an addiction to several garden crops, including potato and egg-plant *(Solanum)* and chili pepper *(Capsicum)*. *S. bazochii* is a tailless butterfly with brown forewings and blue hindwings.

**melinus** Hübner COTTON SQUARE BORER, GREY HAIRSTREAK or HOP VINE THECLA BUTTERFLY
S. Canada, U.S. C. and N.W. S. America 25–28 mm. (0·98–1·1 in.)
The pale blue-grey colour of the upper surface characterizes this common hairstreak butterfly. The caterpillar feeds on the leaves and inside the fruit of a variety of herbaceous plants and on hawthorn *(Crataegus)*, apple *(Malus)*, and lemon *(Citrus)*. It is a minor pest of cotton *(Gossypium)*, cow-peas *(Vigna)*, lima beans *(Phaseolus)* and hops *(Humulus)*.

**Strymonidia** Tutt (Lycaenidae)
The few species of this genus are found in Europe, C. Asia, China, Japan and Formosa.

**pruni** Linnaeus BLACK HAIRSTREAK
Europe and temperate Asia to the Pacific 29–33 mm. (1·14–1·3 in.)
The male is black above with orange, marginal lunules on the hindwing which is single-tailed; the female is similar but often has an orange suffusion in the middle of the forewing. The dark brown under surface has a broad, orange, marginal band enclosing black spots on the hindwing and usually on the posterior part of the forewing. The caterpillar of this species feed on the leaves of various species of *Prunus*. Adults are seldom found far away from areas where the food-plants occur.

**w-album** Knoch WHITE LETTER HAIRSTREAK
N.W. and C. Europe including Britain, to Japan 24–38 mm. (0·94–1·5 in.)
A distinctive white mark in the shape of a large 'W' on the under surface of the hindwing is the origin of both the scientific and common names of this butterfly. The upper surface of the wings is almost entirely brown; dark in the male, paler in the female. The caterpillar feeds on the leaves of elm *(Ulmus)* and other trees near which the adult butterfly is also found flying often at a great height. W. F. Kirby recorded 'myriads that hovered over every flower and bramble blossom' at Ripley, near London in 1827. **259**

**stygia** see **Junonia**

**Styx** Staudinger (Nemeobiidae)
There is only one known species in this genus of butterflies. *Styx* is currently placed in a subfamily of its own, the *Styginae*, in the Nemeobiidae, although has been classified at times with the Lycaenidae.

**infernalis** Staudinger 40–50 mm. (1·57–1·97 in.)
This strange, semitransparent, rather moth-like butterfly, is an inhabitant of elevations of up to 2440 m. (8000 ft) in the Andes of Peru. There is little variation in colour between specimens. **268r**

**Suasa** de Nicéville (Lycaenidae)
Only one species is known in this S.E. Asian butterfly genus.

**lisides** Hewitson RED IMPERIAL
N. India to Indochina, Malaya, the Philippines 27–30 mm. (1·06–1·18 in.)
Little seems to be known about this rare little butterfly. The female (not figured) lacks blue scales at the base of the forewing and has brown hindwings with an area of white at the base of the tails. Males of one subspecies lack an orange patch on the forewings. The under surface of the wings is white with orange-brown markings on the forewing and black lines and dots on the hindwing. **244w**

**subargentana** see under **Eana argentana**
**subrosea** see **Eugraphe** and see under **Lymantria rosina**
**succadanea** see **Agathia**
**suffusa** see **Lactura**
SUGAR-CANE BORERS see **Castnia, Diatraea**
**sula** see under **Lasaia sessilis** and **L. maria**
SULPHUR BUTTERFLIES see **Anteos, Colias, Dercas, Eurema, Phoebus**
SULPHUR MOTH, SPOTTED see **Emmelia trabealis**
SUNBEAM BUTTERFLIES see **Curetis**
**superalis** see **Diaphania**
**superba** see **Menanda**
**swaha** see **Satyrus**

**Swammerdamia** Hübner (Yponomeutidae)
A genus of micro-moths with species in N. America, Europe and Asia.

**caesiella** see under **S. heroldella**

**heroldella** Hübner
Europe including Britain, N. America 11–12 mm. (0·43–0·47 in.)
The white head and thorax of this moth are conspicuous. The wings are greyish white with brown scales scattered over them, the hindwings are grey. The caterpillar feeds on *Betula*, *Alnus* and *Castanea* and the moth is out in May and June with another generation in August. The moth is variable in colour and was only recently recognized in N. America where it was probably introduced. It is more widespread in Canada and the U.S. than *S. pyrella*. The species is also known as *S. caesiella* Hübner.

**pyrella** de Villers
Europe including Britain, N. America 11–13 mm. (0·43–0·51 in.)
The rather narrow forewings of this small moth are a light, shining, gold-bronze colour; the hindwings are yellowish. The caterpillar feeds on *Pyrus*, *Malus*, *Prunus* and other species of Rosaceae as well as on *Betula* (Betulaceae). The moth is generally local in Europe, flying in May and June. In N. America this species has been found in Washington and British Columbia where it is believed to have been introduced on plant material from Europe.

SWAMP
METALMARK see **Calephelis muticum**
TIGER-MOTH see **Diacrisia metelkana**
SWALLOW PROMINENT, LESSER see **Pheosia gnoma**
SWALLOW-TAIL MOTH see **Ourapteryx sambucaria**
SWALLOWTAIL BUTTERFLIES see **Papilionidae**
SWEET POTATO HORNWORM see **Agrius cingulatus**
SWEET-OIL, GREEN see **Aeria eurimedia**
SWIFT MOTHS see **Hepialus, Zelotypia**
SWORD-GRASS MOTH, DARK see **Agrotis ipsilon**
SWORDTAIL BUTTERFLIES see **Eurytides, Graphium, Protographium**
SYCAMORE TUSSOCK MOTH see **Halisidota harrisii**
SYDNEY AZURE see **Ogyris ianthis**
**syla** see **Chrysozephyrus**
SYLPH, GOLD-SPOTTED see **Metisella metis**
**sylphina** see **Chorinea**
**sylvata** see **Abraxas**
**sylvester** see under **Euploea coreta**
**sylvia** see **Parthenos**

**Symbrenthia** Hübner (NYMPHALIDAE)
A genus of butterflies with species from India and Indonesia to Formosa and Japan.

**hypselis** Godart HIMALAYAN JESTER BUTTERFLY
India, Pakistan, Burma, Java 40–50 mm. (1·57–1·97 in.)
This butterfly is common over much of its range up to 2290 m. (7500 ft). It is out in all months of the year, generally in shady parts by water. When it flies in the sunshine it looks like a black and yellow *Neptis* and behaves in a similar manner. The upperside is brown with a slightly irregular red band, narrowing towards the body, and broader towards the apex of the wing. Below the apex is another red band, with a third one on the hind margin of the forewing continued with another short one on the hindwing. There is another red band nearer the margin. The bands tend to go along, rather than across the wing. The underside is a strikingly bright yellow, strongly patterned, with 4–5 blue patches, edged with brown along the margin of the hindwing. **185**h, k

**symethus** see **Miletus**

**Symmachia** Hübner: syn. *Cricosoma* (NEMEOBIIDAE)
About 50 species are placed in this genus of butterflies. It is represented in both C. America and tropical S. America.

**asclepia** Hewitson
C. & tropical S. America 17–24 mm. (0·67–0·94 in.)
This is a bright orange species, marked with black spots along the front and outer margins of the forewing. No comment can be found in the literature concerning the caterpillar's foodplants which the scientific name suggests might have been known to Hewitson as a species of the generally poisonous family Asclepiadaceae.

**Symmoca** Hübner (SYMMOCIDAE)
A genus of micro-moths related to *Gelechia* with species in Europe, including Britain, and N. Africa. Some of the species are detritus-feeders, others feed on lichens on the bark of trees.

**signella** Hübner
Europe 16–31 mm. (0·63–1·22 in.)
This species is out in June and July in S. and C. Europe. The caterpillar feeds on lichen on the barks of various trees. The moths are variable in size, the females usually being larger than the males. They are relatively common over most of their range. **43**h

**Symmocidae**
A small family of micro-moths related to the Gelechiidae. Most of the species in this family are found in Europe but there are a few in N. America.

**Symphaedra** Hübner (NYMPHALIDAE)
A small Indian genus of butterflies related to *Euthalia*.

**nais** Forster BARONET BUTTERFLY
India, Pakistan, Sri Lanka 60–70 mm. (2·36–2·76 in.)
This is quite unlike the other species in the genus and is bright tawny orange above with black marks and dark veins. The black hindwing margin is wider than the forewing margin and there are a row of black spots below the margin on the hindwing. The butterfly generally flies low over the ground, settling with the wings spread flat. It lives in the drier forest (other *Symphaedra* tend to be rain forest species) below 910 m. (3000 ft). The caterpillar feeds on Ebony *(Diospyros melanoxylon)* and species of *Shorea* (Dipterocarpaceae). The insect is about in bright sunshine, often being attracted to sap or over-ripe fruit. **231**e

**Synanthedon** Hübner (SESIIDAE) CLEARWINGS
These wasp-like micro-moths are worldwide. The N. American species look similar to the European counterpart.

**salmachus** Linnaeus: syn. *S. tipuliformis*
CURRANT BORER or CURRANT CLEARWING
Europe including Britain, temperate Asia, N. America, Australia, New Zealand 18–20 mm. (0·71–0·79 in.)
This moth has been accidentally introduced into a number of countries. The caterpillar bores into the branches of currants *(Ribes)*. The forewings are clear with brown apex and a large brown mark below the apex. The hindwings are clear. The body is black with yellow or orange rings. The caterpillar is a serious problem for fruit growers in many countries. It feeds on the pith in the stem, working its way down the stem. It is fully grown in March when it gnaws its way towards the surface of the stem but it does not penetrate it. Instead it leaves a very thin skin over the hole and pupates below this so that, when the moth emerges in June or July, it is able to force its way out of the stem.

**vespiformis** Linnaeus YELLOW-LEGGED CLEARWING
Europe including Britain 17–25 mm. (0·67–0·98 in.)
The caterpillar feeds under the bark of oaks *(Quercus)* or elms *(Ulmus)*. The crossbar on the forewing is orange-red and the body of the male has two yellow spots on the thorax and four yellow stripes on the abdomen. The legs, as its English name suggests, are yellow. The markings of the females are slightly different. The moth is out in July and is found in trees even in the middle of large towns.

**Synclita** Lederer (PYRALIDAE)
Dark coloured Pyralid moths with rounded wings and a rather obscure pattern. The legs are long and delicate looking. The caterpillar, which lives in the water, feeds on aquatic plants. The genus is mainly American with species from Canada to Argentina.

**obliteralis** Walker
U.S., introduced to Hawaii and Britain 9–18 mm. (0·35–0·71 in.)
This species is common in N. America from Nova Scotia to Manitoba and westwards to British Columbia. It has been introduced, presumably on imported water plant, to Hawaii. A few years ago the first ones were introduced accidentally to Britain. The caterpillar is a pest in greenhouses on water plants. They feed on a wide variety of plants, including Duckweed *(Lemna)*. They live in flattened oval cases made of leaves or parts of leaves. The adult males, which are smaller than the females, are brownish with a few darker marks and a small white mark on the forewing. The females are paler, usually greyish-brown or orange-brown.

**Synemon** Doubleday (CASTNIIDAE)
This is a solely Australian group of about 30 species of moths and is the only Castniid genus to be represented there. The species fly by day, usually close to the ground, and regularly visit flowers for nectar in open country or forest clearings. Their flight is very similar in character to that of many Skipper butterflies (Hesperiidae). The hindwings of most species are brightly coloured but are concealed by the forewings when the moth is at rest. The antenna is strongly clubbed.

**sophia** White
S. and N.W. Australia 30–45 mm. (1·18–1·77 in.)
As in other species of *Synemon* the females (see illustration) have a long, pointed ovipositor. At rest, the brightly coloured hindwings are covered by the cryptically coloured forewings. The caterpillars feed on the roots of a sedge *(Lepidosperma)*. **45**p

**synemonistes** see **Pemphegostola**

**Syngamia** Guenée (PYRALIDAE)
This is a large, worldwide genus of moths with at least 100 known species. Unfortunately the species in the genus have not recently been studied and the whole genus consists of rather a mixture of unrelated species. The one discussed below has often been placed in the genus *Anania* Hübner, and other will probably be moved to new genera when more is known about them.

**florella** Cramer
N., S. & C. America, West Indies 16–18 mm. (0·63–0·71 in.)
For a common insect, very little is known about this species. The moth has been recorded from Mexico to Brazil and in Florida in the U.S. The wing patches may be yellow or orange in colour. Like many of its relatives the caterpillar of this moth is probably a leaf-tier, feeding on the leaves and hiding in the tied up piece, but nothing seems to be recorded about it. **75**

**synnova** see **Oleria**

**Synploca** Hodges (MOMPHIDAE)
This genus was originally placed in the family Walshiidae, but is part of a complex of genera which have variously been placed in families or subfamilies in the Gelechoid complex. The Walshiidae is regarded as a subfamily of the Momphidae but more recent work has suggested that the name is a synonym of an older name. This genus of micro-moths is known only from the U.S.

**gumia** Hodges
U.S. 6–9 mm. (0·24–0·35 in.)
This small moth has been collected only in Arizona. Its life history is unknown. Further collecting will probably show it to be more widely distributed. These tiny moths are not only difficult to collect but difficult to handle for study. Their collection is a highly skilled operation. **44**d

**Syntaracus** Butler (LYCAENIDAE)
The 10 or so species of this genus of 'blue' butterflies are mostly rather similar to each other in colour pattern and are often difficult to identify. Their distribution extends from southern Europe through much of Africa, Madagascar and the islands of the Indian Ocean. One species, *S. manusi*, has been described from the island of Manus in the Admiralty Islands, to the N.E. of New Guinea.

**telicanus** Lang COMMON BLUE
Africa 24–32 mm. (0·94–1·26 in.)
This is a common butterfly in most parts of Africa. Its caterpillar feeds on the flowers, buds and young shoots of a variety of plants including *Burkea, Medicago, Melilotus, Crataegus* and *Plumbago*. The male is a bright violet blue above; the female is brown, with a blue area in the middle of the forewing and with some blue scales at the base of the hindwings and 1–2 black spots at the base of the single tail. The under surface of both sexes is white, spotted with brown; at the base of the tail there are 2 black spots, encircled first with iridescent green, then orange.

**Syntomeida** Harris (CTENUCHIDAE)
There are currently under 10 described species in this genus of moths. Their range includes S. U.S. through C. America to tropical S. America.

**epilais** Walker POLKA-DOT
U.S. (Florida) to Guatemala and Honduras 41–47 mm. (1·61–1·85 in.)
The ground-colour of both wings is black, with a slight bluish sheen; the abdomen is iridescent blue and black with a dull crimson anal area. There is a small white spot at the base of the forewing, usually 4 large, white spots on the rest of the forewing, a single, large, white spot in the middle of the hindwing, and 4 especially conspicuous white spots (2 large, 2 small) on the anterior end of the abdomen. *S. syntomoides* Boisduval which is also found in the same regions, differs in usually having five large, white spots on the forewing. The caterpillar of *S. epilais* is red, with black tufts; it feeds on the foliage of oleander *(Nerium)*, a genus of poisonous plants.

**Syntomis** Ochsenheimer (CTENUCHIDAE)
This is a large genus of moths found in temperate and tropical Asia, Africa and Europe. Many of the species are dark brown or black, with transparent patches in the wings. See also *Amata*.

**phegea** Linnaeus
Europe including Britain, Middle East and central Asia 28–38 mm. (1·1–1·5 in.)

Like many other moths of its family, *S. phegea* flies during the day and can be seen on the blossoms of thyme *(Thymus)* and lavender *(Lavandula officinalis)*. It is thought to be a mimetic partner of the Zygaenid moth *Zygaena ephialtes* Linnaeus which has a yellow banded form almost identical in pattern and colour. **365**

**Syrichtus** see **Muschampia**
**syringae** see **Podosesia**

**Syrmatia** Hübner (NEMEOBIIDAE)
There are about 5 long-tailed species in this genus of C. American and tropical S. American butterflies.

**aethiops** Staudinger TADPOLE BUTTERFLY
N. part of tropical S. America 17–20 mm. (0·67–0·79 in.)
This is a smaller version of *S. dorilas* (illustrated) but without markings on the uniformly black forewings. Like *S. dorilas* it is a rare butterfly of the forests.

**dorilas** Cramer WHITE-SPOTTED TADPOLE BUTTERFLY
Brazil, Venezuela 17–20 mm. (0·67–0·79 in.)
Flight is slow in this strange species, but the wing-beat is fast and the butterfly is consequently difficult to distinguish as such, when in the company of wasps and flies. Females (not illustrated) have an orange patch at the base of the forewing. **260**c

TABBY BUTTERFLY see **Pseudergolis wedah**
**tachyroides** see **Pentila**
TADPOLE BUTTERFLIES see **Syrmatia**

**Taenaris** Hübner: syn. *Morphotenaris* (NYMPHALIDAE)
This genus of butterflies is in the subfamily Amathusiinae. It has a large number of species in New Guinea and the surrounding islands. Many of the species are white, with prominent rounded eyespots on the hindwing. Species of the genus occur in Malaysia and Indonesia but the majority are in New Guinea.

**schoenbergi** Fruhstorfer
New Guinea 95–140 mm. (3·74–5·51 in.)
Several species and subspecies of this butterfly have been described from different parts of its range in New Guinea. They differ mostly in the pattern of the eyespots on the hindwings. Nothing is known of the biology of this species. **160**a

**tages** see **Erynnis, Euchloe**

**Tagiades** Hübner (HESPERIIDAE)
This is a small genus of Skipper butterflies with 3 species known from Africa and a few from the Oriental region.

**flesus** Fabricius CLOUDED FLAT BUTTERFLY
Tropical Africa to South Africa 44–50 mm. (1·73–1·97 in.)
This is a typical, fairly large Skipper with broad, dark brown wings with a bluish grey suffusion in the outer part of the wing. There are 5 larger hyaline spots and 4 smaller ones near the apex of the forewing. Underside the hindwing has a row of black spots on the brown front margin, the rest of the wing is white. This species is a fast, erratic flier, common in bush and forest over much of its range. The caterpillar feeds on *Dioscorea* (Dioscoreaceae).

**litigiosa** Möschler WATER SNOW FLAT BUTTERFLY
India, Burma, Pakistan 37–44 mm. (1·46–1·73 in.)
The forewings are dark brown to grey brown, with a few small white spots near the apex of the wing and over the cell. The apical part of the hindwing is dark, the rest white, with some black spots just below the margin. This Skipper butterfly is common over much of its range in the hilly forest regions. It flies fast, keeping to the shade in the forests. They settle on the underside of the leaves. There is a related species which is broadly similar in pattern (*T. menaka* Moore) differing in having one or two more black spots on the white hindwings, inside the main row of spots, and lacking the white spots over the forewing cell. It tends to fly at higher altitudes than *T. litigiosa*.

TAILED
ADMIRAL, LONG see **Antanartia schaenia**
BIRDWING see **Ornithoptera paradisea**
BLUE see **Lampides boeticus**
BLUE, EASTERN see **Everes comyntas**
EMPEROR see **Polyura pyrrhus, Urota sinope**
FLAMBEAU see **Marpesia petreus**
JAY see **Graphium agamemnon**
SULPHUR see **Dercas verhuelli, Phoebus cipris**

**Tajuria** Moore (LYCAENIDAE) ROYAL BUTTERFLIES
The 30 or so species of this genus are found in S.E. Asia from China to the Philippines and the Lesser Sundas. Both sexes of this genus are normally blue above and there are 2 small, slender tails to the hindwing. The foodplants are mistletoes of the genera *Viscum* and *Loranthus*. Both sexes, but especially the females, tend to fly high in the branches of trees bearing these parasitic mistletoes.

**cippus** Fabricius PEACOCK ROYAL BUTTERFLY
Sri Lanka, India, Burma, S. China, Malay Archipelago to Borneo and Lombok 31–45 mm. (1·22–1·77 in.)
The female (not figured) of this fairly common species is mostly greyish blue above. It occurs in forested areas chiefly at low elevations, but has been found up to 2130 m. (7000 ft) where *Loranthus* is available as food for the caterpillar. Blossoms of *Lantana* and *Poinsettia* are often visited by the adult butterfly. Several subspecies have been described – they differ from one another in small details of the colour pattern. **244**v

**jehana** Moore PLAINS BLUE ROYAL BUTTERFLY
Sri Lanka, India 30–37 mm. (1·18–1·46 in.)
The males differ from those of the Peacock Royal in the conspicuous white fringes to the wings. Although known popularly as a plains butterfly, it occurs as high as 2130 m. (7000 ft) in N.W. India. The foodplant of the caterpillar is *Loranthus*.

**tameanea** see **Vanessa**
**tamile** see **Papilio paris**
**tamu** see **Heliophorus**

**Tamyra** Herrich-Schäffer (PYRALIDAE)
This is a small genus of some half a dozen species of Pyralid moths. They all have a striking outline but presumably when they are at rest in their natural surroundings they fold their wings and show the more camouflaged undersides. They occur from C. America to tropical S. America and on some W. Indian islands.

**penicillana** Herrich-Schäffer
Trinidad, C. & S. America 34–50 mm. (1·34–1·97 in.)
Nothing is known of the biology of this species which is in the subfamily Chysauginae. The very long palps which stick out well in front of the head are characteristic of the subfamily. **63**q

**Tanaecia** Butler (NYMPHALIDAE)
A large genus of butterflies found from India to the E. Indies in the warmer parts, as far N. as the Philippines.

**pelea** Fabricius MALAY VISCOUNT BUTTERFLY
N. India, Pakistan, Malaya, Borneo 56–68 mm. (2·2–2·68 in.)
A very variable species with many subspecies described. The underside has a similar pattern but is paler. The butterfly is common over much of its range. The life history is not known. **216**k

**tangens** see **Homidiana**
**tanialoides** see **Ormetica**
TAPESTRY MOTH see **Trichophaga tapetzella**
**tapetzella** see **Trichophaga**
**tarquinius** see **Feniseca**
**tarsalis** see **Pompilopsis**

**Tarsocera** Butler (NYMPHALIDAE)
This small genus of butterflies in the subfamily Satyrinae is found in South Africa.

**southeyae** Dickson
South Africa 48–50 mm. (1·89–1·97 in.)
This butterfly was described in 1969 after specimens had been collected in South Africa by Mrs R. J. Southey at a roadside. It proved a little difficult to capture since it was flying back and forward through a wire fence. However, Mrs Southey persisted and collected a small series. These, when they were examined, proved to be a species new to science and were named after the collector. **174**m

**Tarsolepis** Butler (NOTODONTIDAE)
Six species are currently placed in this genus of moths. They are found in India and Japan, and through S.E. Asia to New Guinea. Their general appearance is remarkably like that of hawk-moths (Sphingidae), with long pointed wings and an elongate abdomen. The posterior end of the abdomen bears a tuft of unusually long, clubbed scales and there are distinctive metallic silver markings on the forewings.

**sommeri** Hübner
India, Burma, Malaya, Sumatra, Java, Borneo, the Philippines 50–80 mm. (1·97–3·15 in.)
The red structures visible at the base of the hindwing in the plate are scent-distributing organs. These are present only in the male, and are probably therefore sex-scent or aphrodisiac in function; their conspicuous colour is unusual in an organ of this kind, however, and it is possible that it may be associated with defence or have a dual sexual and defensive function. **357**a

**Tascina** Westwood (CASTNIIDAE)
This is one of the 3 genera of Castniidae, which contains about 30 genera in all, found outside the New World. Only 2 species are known; both are rare in collections. The proboscis is vestigial, and it is unlikely that the adults are able to feed.

**metallica** Pagenstecher
Borneo, the Philippines 80 mm. (3·15 in.)
This resembles *Neocastnia nicevillei* but the forewing band is narrower, and the iridescent blue area on the hindwing is restricted to that part of the wing adjacent to the abdomen.

**orientalis** Westwood
Singapore 80 mm. (3·15 in.)
The forewing is dark brown with a cream oblique shape, as in *T. metallica* but the hindwing is orange basally, with a broad, dark brown marginal band.

**tatila** see **Eunica**
**tau** see **Aglia**
TAU EMPEROR see **Aglia tau**
TAWNY
COSTA see **Telchinia violae**
EMPEROR see **Asterocampa clyton**
RAJAH see **Charaxes polyxena**

**Taxila** Doubleday (NEMEOBIIDAE)
There are about 6 species in this butterfly genus. As suggested by the anagram these butterflies are rather similar in appearance to *Laxita*. It is found from northern India through S.E. Asia to Borneo.

**haquinus** Fabricius HARLEQUIN
N. India, Burma and Malaya to Borneo 45–55 mm. (1·77–2·17 in.)
This fairly common species is found in woodland areas above 760 m. (2500 ft). Males are brown above, with an orange apex to the forewing. Females are orange-brown, marked with brown and white, and with an oblique, white, apical band on the forewing.

**Taygetis** Hübner (NYMPHALIDAE)
These butterflies are in the subfamily Satyrinae. Mostly they are medium to large butterflies with a few smaller species. They are found throughout tropical, C. & S. America. Very few species have been bred, those known feed on bamboo.

**echo** Cramer: syn. *T. velutina* NIGHT BUTTERFLY
Tropical S. America 55–64 mm. (2·17–2·52 in.)
This dark brown Satyrid has a darker patch in the centre of the wing. A line of yellow spots, more conspicuous on the hindwings, are visible on the under-

side. They fly in shady moist places mostly towards evening, often feeding on rotting vegetation. There are a number of rather similar species in the genus, all rather sombre coloured fairly typical of the Satyrine butterflies.

TEAK MOTH see **Hyblaea pura**

**Tebenna** Billberg (GLYPHIPTERIGIDAE)
At present only the 1 species is included in this genus of micro-moths which was separated from the wide-spread related genus *Choreutis*.

**bjerkandrella** Thunberg
Worldwide 9–11 mm. (0·35–0·43 in.)
Probably one of the widest ranged of any moth, this tiny moth has been collected on most continents. The caterpillar is green with yellow on the sides and black spots. It feeds on various thistles and other Compositae. In Australia it is recorded as living in webs on the underside of the Scotch Thistle. **34**e

**Tegeticula** Zeller (INCURVARIIDAE) YUCCA MOTHS
These small moths are found in N. America and Mexico. While not extremely colourful they have evolved a remarkable life history. This involves a close relationship (called symbiosis or living together) with the Yucca plant. The moths themselves are specially modified for this symbiosis. The females have a pair of specialized processes under the head called tentacles used to gather pollen from the *Yucca* flower. Having gathered the pollen she then lays one egg in the female part of the plant (the pistil), afterwards forcing the pollen held by the tentacles onto the stigma of the flower. This pollination of the flower means that the seeds will develop. The *Tegeticula* egg hatches and the caterpillar feeds on the developing seeds. The female moth thus makes certain of a food supply for her off-spring, and as the caterpillars do not usually eat all the developing seeds, both plant and animal benefit.

**yuccasella** Riley YUCCA MOTH
N. America, Mexico 15–27 mm. (0·59–1·06 in.)
This is the most widespread species in the genus. The remarkable life history was first described in 1872 and recent work has shown the original account, which was disputed at first, to be substantially correct. **5**f

**Teinopalpus** Hope (PAPILIONIDAE)
There are only 3 species in this genus of butterflies. It is represented in N. India, Sikkim, Bhutan, Burma and S. China. The genera *Lamproptera* and *Eurytides* are its closest relatives. The caterpillars feed on *Daphne*, a laurel-like shrub used for making paper.

**aureus** Mell
S. China 80 mm. (3·15 in.)
There are very few specimens of this species in collections. It differs from the larger *T. imperialis* in the more greyish colour of the transverse band on the forewing and a greater amount of yellow on the hindwing.

**imperialis** Hope KAISER-I-HIND BUTTERFLY
N. India, Sikkim, Nepal, Bhutan, Burma, S. China 75–120 mm. (2·95–4·72 in.)
This rare, exquisitely coloured species occurs between 1830–3050 m. (6000–10,000 ft) in thickly forested regions where it is seldom seen except in sunlit clearings and invariably before midday. Males (illustrated) have a single, long tail to the hindwing; females have 2 long tails and a rather different colour pattern though similar in coloration. Unlike most other species of Papilionidae it is not attracted to flowers, but the male will visit wet patches on the forest floor. **96**a

**telamon** see **Serecinus**

**Telchinia** Hübner (NYMPHALIDAE)
This genus of butterflies belongs to the subfamily Acraeinae. The butterflies are often included in the genus *Acraea* and the exact status of the name *Telchinia* is uncertain. The caterpillars feed on poisonous plants and are able to use the plant poisons which they metabolize into their body and utilize for their own protection. Most species are brightly coloured and are distasteful to predators.

**violae** Fabricius TAWNY COSTA
India, Pakistan, Sri Lanka 50–65 mm. (1·97–2·56 in.)
This is a common species over most of its range. The caterpillars feed on members of Cucurbitaceae and Passifloraceae. The butterfly is a reddish brown above with darker wing margins and a few black spots on the forewing. The hindwing is similar, with smaller spots but with a broad black margin with white spots. Little seems to be known of its biology.

**telea** see **Chlorostrymon maesites**
**telesilaus** see **Eurytides**
**telesiphe** see **Heliconius, Podotricha**
**telicanus** see **Syntaracus**

**Tellervo** Kirby (NYMPHALIDAE)
This is the only genus of the subfamily Ithomiinae to occur outside the New World tropics. There are about 6 species; they are found in Celebes and the Moluccas of S.E. Asia, eastwards to New Guinea, N. Australia and the Solomons.

**zoilus** Fabricius
Celebes to New Guinea, N.E. Australia Solomon Is 36–48 mm. (1·42–1·89 in.)
This species is the model for a species of Nymphalinae *Neptis praslini*. Each of the several subspecies of this butterfly is restricted to rain-forest regions; they differ from each other in the size and shape of the white markings. **131**h

**Temenis** Hübner (NYMPHALIDAE)
A small genus of butterflies with species in C. & S. America.

**laothoe** Cramer TOMATO BUTTERFLY
C. & S. America and into the W. Indies 42 mm. (1·65 in.)
The popular name refers to the colour of the butterfly which has red forewings, with a broad, black apex and black base. There is an almost round patch of red on the front of the hindwings surrounded by a broad black border. This species is relatively common in some parts of its range, extending S. into Brazil. It flies particularly along the edge of forests or forest tracks where there are plenty of fallen fruit on which it feeds.

**temora** see **Salamis**
TENT CATERPILLARS see **Malacosoma, Lasiocampidae**

**Teratoneura** Dudgeon (LYCAENIDAE)
There is a single known species of butterfly in this genus.

**isabellae** Dudgeon
C. and W. tropical Africa 32–34 mm. (1·26–1·34 in.)
The remarkable caterpillars of this species live in association with ants, and caterpillars of the Lymantriid moth *Naroma signifera*. The appearance of the caterpillar is quite atypical for a Lycaenid and is closely similar to that of *Naroma*. The Lycaenid caterpillar is probably a mimic of the Lymantriid (most Lymantriid caterpillars are avoided by birds), whereas the adult is non-mimetic. A second possibility is that both caterpillars are advertising their unpalatability to mutual advantage (the hairs of *T. isabellae* are possibly as unpleasant to predators as those of *N. signifera*). The adult moth is orange above, with a narrow black, marginal band on the hindwing, and a similar but broader band on the forewing which is also black in a broad band along the front margin.

**Terias** see **Eurema**

**Terinos** Boisduval (NYMPHALIDAE)
A small genus of butterflies with species found from New Guinea through Indonesia to the Philippines.

**clarissa** Boisduval
Malaysia, Thailand, Indonesia, the Philippines 70–80 mm. (2·76–3·15 in.)
This species is very variable and many subspecies have been described. The specimen illustrated is from the Philippines. **185**l

**terpander** Hewitson ROYAL ASSYRIAN BUTTERFLY
Malaysia 70 mm. (2·76 in.)
This butterfly is blue with a large black patch on the fore- and hindwings. The blue extends to the margin and covers most of the hindwing. On the margin of the hindwings there are 2 large pale triangular patches. The female has less blue on the wings. This is a forest butterfly which frequently drinks by the road-sides on damp muddy patches. The greenish caterpillar has longitudinal stripes, black spines and yellow head. It feeds on *Antidesma* (Stilaginaceae).

**terminus** see **Mycalesis**
**terpander** see **Terinos**
**tesquella** see **Stilbosis**
**tessellaris** see **Halisidota**
TESSELLATED SKIPPER see **Muschampia tessellum**
**tessellum** see **Muschampia**
**testulalis** see **Maruca**

**Tethea** Ochsenheimer (THYATIRIDAE)
About 20 species of moths are placed in this genus. It is represented in Europe, temperate Asia (including Japan) and in India, Sikkim, Burma and S.E. Asia to Borneo.

**ocularis** Linnaeus FIGURE OF EIGHTY MOTH
Europe including Britain, temperate Asia to E. U.S.S.R. and Japan 32–38 mm. (1·26–1·5 in.)
In Britain, the moths are on the wing during May and June. This is a generally common species even in urban areas. The caterpillar feeds at night during July and August on poplar *(Populus)* foliage and retires during the day to its refuge between leaves. The over-wintering chrysalis is protected by a loosely woven cocoon placed amongst debris at the base of the food-plant.

**tetracha** see **Copromorpha**
**tetralunaria** see **Selenia**
**tetrapleura** see **Platagarista**
**teucer** see **Caligo**
**textor** see **Hyphantria**
**thaidina** see **Bhutanitis**

**Thalaina** Walker (GEOMETRIDAE)
A small genus of moths from Australia and Tasmania with silky white wings and geometric patterns. Their caterpillars feed on *Acacia*.

**angulosa** Walker
Australia 38–42 mm. (1·5–1·65 in.)
Several related species occur in Australia. This species has been reared but no information on the life history has been published. The moth is typical in appearance of the species in the genus. The hindwings are white, unpatterned. **305**

**Thalamarchella** Fletcher (OECOPHORIDAE)
At one time this micro-moth genus was placed in a separate family but recent work has shown that this is incorrect and that the family (Thalamarchellidae) was a mixture of genera from 2 different families. This is typical of the problems of studying these smaller, and little known species. When one considers the total number of Lepidoptera already described (in excess of 150,000) many of the species have never been seen since they were originally collected and even more have not been studied since the first person described them as 'new to science'.

**alveola** Felder & Rogenhofer
Australia 26 mm. (1·02 in.)
This moth is found in Western Australia. There are one or two similar species with dark hindwings. The caterpillars tie together the leaves of the plant on which they feed. **39**b

**thalassina** see **Nepheronia**
**thalictri** see **Calpe**
THAROPS, BLUE see **Menander menander**
**tharos** see **Phyciodes**

**Tharsanthes** Meyrick (PYRALIDAE)
Only 2 species known in this Pyralid moth genus, both from S. America. Both have a beautiful geometric pattern of golden orange on the wings. There are

other, as yet undescribed, species in museum collections.

**aurantia** Jones
Brazil 25–35 mm. (0·98–1·38 in.)
This pattern is very striking and there are several other species like it. It is probably protective colouring to blend with its surroundings, but nothing is known about this. All the ones similarly patterned, (including some in related genera) are from S. America. The life history is unknown.

**thasus** see **Cremna**

**Thaumantis** Hübner (NYMPHALIDAE)
These butterflies belong to the subfamily Amathusiinae. The genus is related to *Zeuxidia*. The species are found from India to Indonesia.

**diores** Westwood JUNGLE GLORY BUTTERFLY
N. India and Pakistan to S. China, S.E. Asia to Hainan 95–115 mm. (3·74–4·53 in.)
This butterfly is said to fly only from dusk onwards. It has 2 broods in N. India in May and September. The male is a dark blackish butterfly with large blue patches in each wing, the patches in the hindwings being almost circular. The female is larger than the male and the colours are paler. The adults feed on rotting fruit and any decaying matter. The caterpillar is unknown. There are several rather similar related species from Malaysia.

**thaumasta** see **Apocrisias**

**Thaumetopoea** Hübner (NOTODONTIDAE)
This is a small genus of moths found in Europe, Asia and Africa. It is typical of the subfamily Thaumetopoeinae. The adults are grey or brown and white and apparently cryptically patterned.

**pityocampa** Denis & Schiffermüller
PINE PROCESSIONARY MOTH
Central and S. Europe 38–48 mm. (1·5–1·89 in.)
This species is similar in coloration and pattern to the illustrated *T. processionea*. In habits it also matches *T. processionea*, but its caterpillar feeds on the needles of pine *(Pinus)* and other coniferous trees. The famous French naturalist Fabre recorded how he placed a series of *T. pityocampa* caterpillars on the rim of a plant-pot and watched them follow one another around the rim until they were exhausted.

**processionea** Linnaeus
Central and S. Europe 28–37 mm. (1·1–1·46 in.)
This is probably the best known species of the subfamily Thaumetopoeinae in the N. Hemisphere. The caterpillars rest in communal webs during the day, leaving at night to feed (typically on oak, *Quercus*) in a procession or a wedge-shaped formation. Although not a serious pest, the caterpillars are sometimes a nuisance as their hairs have irritant properties. **357**d

**Thauria** Moore (NYMPHALIDAE)
This small genus of butterflies is in the subfamily Amathusiinae. The species are found in S.E. Asia, S. China into Indonesia.

**aliris** Westwood TUFTED JUNGLE KING BUTTERFLY
S.E. Asia 110–128 mm. (4·33–5·04 in.)
This butterfly is found from Burma, through Thailand and Malaysia to Borneo. It flies just before sunset in the thickly forested areas. The adults feed on rotting fruit but the foodplant of the caterpillar is not known, though bamboo or palms have been suggested. The sexes are similar in colour and pattern but the males have a large tuft of scales on each side of the abdomen. There are several related species from India. The male has slightly pointed forewings, brown or reddish brown, with 2 white spots near the apex and a whitish band from near the middle of the forewing, diagonally towards the base of the outer margin. The hindwings, which are brown near the base, have a warm reddish brown colour, split into a smaller anterior part by the black from the base of the wing reaching the margin, and a large hind part.

**lathyi** Frühstorfer JUNGLE KING BUTTERFLY
India, Pakistan, Burma 110–120 mm. (4·33–4·72 in.)
Very similar externally to *T. aliris*. It is found in, and is locally common in, forests where it flies towards evening and again in the early morning.

**theae** see **Parametriates**

**Thecla** Fabricius (LYCAENIDAE)
There are only 2 species in this Old World temperate genus of hairstreak butterflies, *T. betulae* and the eastern Asian *T. betulina* Staudinger. Several other species, especially tropical American, have been placed in *Thecla* by various authors – these belong in other genera, but the task of sorting out the confused classification is not yet complete (see *T. coronata*).

**betulae** Linnaeus BROWN HAIRSTREAK
Europe, temperate Asia to the Pacific Ocean 30–45 mm. (1·18–1·77 in.)
The female (illustrated) of this widespread species is a much more brightly coloured butterfly than the sombre male. It is a woodland species, inhabiting areas where there are deciduous trees such as sloe and plum *(Prunus)* and birch *(Betula)* which provide food for the caterpillar. There is a single brood each year. *T. betulae* is seldom found N. of latitude 62°N; its S. limit is in N. Spain and C. Italy. **257**

**coronata** Hewitson
Tropical S. America 45–60 mm. (1·77–2·36 in.)
This is one of several species currently but incorrectly placed in *Thecla*. It is one of the largest and most brilliantly coloured species of Lycaenidae. The under surface of the wings is mostly dark green and probably affords a high degree of camouflage when the butterfly is at rest. The male (not illustrated) differs from the female chiefly in the much narrower, black marginal bands on the wings and the black, not red, patch at the base of the hindwing tails. **244**e

THECLA, HOP VINE see **Strymon melinus**
**themis** see **Euphaedra**
**themisto** see **Thyridia**
**theon** see **Hypochrysops**

**Theope** Doubleday (NEMEOBIIDAE)
Many of the 50 or so species of this genus of butterflies are blue in colour and resemble species of Lycaenidae, but others are white or orange. In the resting position, the wings are closed together so that the under surface is exposed. Like many other species of Nemeobiidae they fly in a quick, darting fashion, and settle on the under surface of a leaf. Most species are rare and localized in distribution and are found in parts of C. and tropical S. America.

**endocina** Westwood ORANGE THEOPE BUTTERFLY
C. America and tropical S. America 25–35 mm. (0·98–1·38 in.)
The upper surface of the female of this rare butterfly is orange or orange-red. The male is similar, but has a black, apical patch continuous with a broad, black band along the front edge of the forewing. The under surface of the wings is brownish yellow in both sexes. The hairy and mainly green caterpillars feed on the foliage of cocoa *(Theobroma)*, but are not important pests. They live inside a rolled leaf and are attended by ants.

**excelsa** Bates ROYAL BLUE THEOPE BUTTERFLY
Brazil and the Guianas 32–33 mm. (1·26–1·3 in.)
The wings of this species are an unusual, gold colour below; the upper surface of the wings is a contrasting, rich violet-blue, with a dark brown front and outer margin to the forewing and a dark brown front margin to the hindwing. This is the largest known species of *Theope*.

**pieridoides** Felder WHITE THEOPE BUTTERFLY
Amazon Basin and N. S. America 26–30 mm. (1·02–1·18 in.)
Both surfaces of the wings of this unusually coloured butterfly are white. In Bahia, Brazil, it flies together with another white and Pierid-like species of the genus *Leucochitonea* (Hesperiidae). Whether operational mimicry exists is unknown. **268**q

THEOPE BUTTERFLIES see **Theope**

**Theorema** Hewitson (LYCAENIDAE)
Two species are known, *T. eumenia* and the Colombian *T. dysmenia* Draudt.

**eumenia** Hewitson
S. Mexico S. to Ecuador 46–50 mm. (1·81–1·97 in.)
The under surface of both sexes of this superbly coloured species is dark brown, with silvery green, lunulate markings on the hindwing and a blue patch and green marginal band on the forewing. **243**b, d

**theresiae** see **Epicmelia**

**Thera** Stephens (GEOMETRIDAE)
A genus of moths the species of which are widespread over the N. Hemisphere.

**juniperata** Linnaeus JUNIPER CARPET MOTH
Europe including Britain, temperate Asia 23–26 mm. (0·90–1·02 in.)
This moth gets its popular and scientific names from the foodplant of the caterpillar (Juniper). The yellowish green caterpillar has 3 lines along the back and a white one on the side. It feeds on the leaves of the plant in July and August. The moth is out in October and November. Apart from accidental or dispersed flights the moth is only found in the vicinity of Juniper trees. It is quite local in Britain. **286**

**thestylis** see **Prioneris**
**thetis** see **Curetis, Lycaena**
THISTLE-ERMINE KNOT-HORN see **Myelois cribrella**
**thoas** see **Papilio**
**thoe** see **Lycaena**
**thomasi** see **Hemiargus**
THORN MOTHS see **Ennomos, Selenia**

**Thorybes** Scudder (HESPERIIDAE)
A large genus of Skipper butterflies with species in N. America and in tropical America. Most are rather dark brown in colour.

**bathyllus** Smith SOUTHERN CLOUDY WING
U.S. (Florida to C. Texas, N. to Massachusetts, Wisconsin, Nebraska) 32–42 mm. (1·26–1·65 in.)
This species is common in open spaces and roadsides where it can be seen in its fast erratic flight. It is brown on the upperside with some conspicuous white marks across the wing, below the apex. The white spots generally extend from vein to vein, in related species they are small dots. The caterpillar feeds on species of Leguminosae. The adults are on the wing in June with 1 brood in the N. of its range and 2 in the S.

THREE-SPOT, HARRIS'S see **Harrisimemna trisignata**
THREE-TAILED SWALLOWTAIL see **Papilio multicaudata**
**thunbergella** see **Micropterix**
**thyastes** see **Eurytides**

**Thyatira** Ochsenheimer (THYATIRIDAE)
About 12 species belong in this widespread genus of moths. It is represented in Europe, the Middle East, much of temperate India, Nepal, Sikkim, Burma, and Indo-China through S.E. Asia to Bali and the Philippines, and also in the S. U.S., C. America and tropical S. America.

**batis** Linnaeus PEACH-BLOSSOM
Europe including Britain, temperate Asia to E. U.S.S.R. and Japan 32–38 mm. (1·26–1·5 in.)
This beautiful moth is matched fairly closely in colour pattern by the N. American species *Euthyatira pudens*. Fairly common in wooded districts, adults are on the wing in June and July. Caterpillars feed at night on raspberries and bramble *(Rubus)* during August and September. The chrysalis overwinters. **271**

**thyatira** see **Esthemopsis**

**Thyatiridae**: syn. *Cymatophoridae*
This is a small family of mostly sombrely coloured, medium sized moths which resemble species of the family Noctuidae in wing shape and general appear-

ance but are members of the superfamily Geometroidea and are probably most closely related to the Drepanidae and the Cyclidiidae.

**thyatiroides** see **Eosphoropteryx**
**thyodamus** see **Cyrestis**

**Thymelicus** Hübner (HESPERIIDAE)
A genus of Skipper butterflies found throughout Europe and temperate Asia and into N. America.

**acteon** Rottemburg LULWORTH SKIPPER
S. & C. Europe, including Britain, 22–26 mm.
N. Africa, Canary Islands, Cyprus, (0·87–1·02 in.)
Lebanon and Turkey
This butterfly is on the wing in May but generally has only 1 brood a year. In Britain it is known only from the S., including Lulworth, from where it received its name. It flies in meadows and grassy banks from sea-level to 1520 m. (5000 ft). The caterpillar feeds on Brome grass *(Bromus)*. Several subspecies have been described from the more southerly part of its range. **83**

**lineola** Ochsenheimer ESSEX OF EUROPEAN SKIPPER
Europe including Britain, 24–28 mm.
temperate Asia, N. America (0·94–1·1 in.)
This species was introduced into Canada in or before 1910. It has now spread into the U.S. and is extending its range, its caterpillar proving to be a pest of hay. The butterfly is out from May until August, as a single brood. It flies in meadows from sea-level to 1830 m. (6000 ft). The caterpillars feed on grasses including *Holcus, Agropyrum*. It is local in Britain, where it is found mainly along the coasts of Essex, Suffolk and Kent. The female is illustrated. **84**

**thyra** see **Macroneme**

**Thyretidae**
This family was erected a few years ago by Kiriakoff to accommodate a group of African genera previously placed in the Ctenuchidae. Its affinities lie with both the latter family and the Notodontidae. All 3 families are members of the superfamily Noctuoidea. See *Automolis*.

**Thyridia** Hübner (NYMPHALIDAE)
About 7 species are placed in this genus of the subfamily Ithomiinae. They are found chiefly at low elevations, from N. S. America to Argentina. Their wings are partly transparent, with dark margins and markings as in mimicking species of *Ituna* (Danaiinae); *Philaethria* (Heliconiinae); Hypsidae (eg. *Anthomyza)*, and Castniidae. Transparency of the wings in this genus is achieved by the modification of the scales to slender hairs.

**confusa** Butler
Tropical S. America 60–92 mm. (2·36–3·62 in.)
Several other species of butterflies and moths closely resemble *T. confusa* in colour pattern. They include other species of Ithomiinae, Danainae (eg. *Ituna* species), Pieridae *(Dismorphia* species), and day-flying moths of the families Hypsidae and Castniidae. Some of these, at least, fly together with *T. confusa* and are members of the same mimetic association. (See also under *Gazera linus*). **131**g

**themisto** Hübner
C. Brazil to Argentina 70–80 mm. (2·76–3·15 in.)
This is like a small *T. confusa*, but with a dark band along the front margin of the forewing. Its caterpillars feed on the foliage of *Brunfelsia*, a genus of Solanaceae. (See also under *Gazera linus*).

**Thyrididae**
A family of mainly tropical moths, with one species in Europe (none in Britain) and probably not more than 25 species in the whole of Europe and temperate Asia. The species occur in the tropics of the New and Old World but are rarely abundant. Little is known of their biology but in many respects they are similar to the Pyralidae, with which they are generally considered to be related. The family is probably not homogeneous and consists of 2 or 3 unrelated subfamilies which have a different evolutionary origin. Many of the species have patterns in a reticulate form and are often extremely leaf-like in appearance. The family has species of extremely different appearance and often brightly coloured but the vast majority are rather uniformly coloured, reticulately patterned species. The host plants, where known, include species of *Terminalia* (Combretaceae), rice *(Oryza)*, where the caterpillars are leaf-rollers. A few species are gall-makers and 1 species in S. America has a caterpillar which bores in fruit trees. Recently 1 specimen of a tropical thyridid was collected alive in Holland but probably accidentally introduced. See *Banisia, Belenoptera, Canaea, Cecidothyris, Hepialodes, Herdonia, Kalenga, Rhodoneura, Thyris.*

**Thyridopteryx** Stephens (PSYCHIDAE) BAGWORMS
Micro-moths in this genus usually have transparent wings and the female, which is wingless, never leaves the case.

**ephemeraeformis** Haworth BAGWORM MOTH
U.S., Britain 25 mm. (0·98 in.)
The male has pale brownish wings and is typically moth-like. The female is wingless and remains in the 'bag'. The caterpillar, after hatching, makes a bag which it carries about. After pupation, which takes place in the bag, the males emerge, but the females only partially emerge from the case which they never leave completely. The male is attracted by the scent of the female and after mating, the female lays her eggs inside the bag. After this the female drops out to the ground and dies. The tiny caterpillars get blown about after they hatch from the bag and feed on most trees that they happen to land upon. The moth is widespread throughout much of N. America, but its status in Britain is uncertain. It is even possible that the British specimens were actually collected in America and at some stage wrongly labelled.

**Thyris** Laspeyres (THYRIDIDAE)
In Europe and N. America the species in this genus look very similar. The transparent spot on the fore- and hindwings are characteristic of the genus. They are stout bodied moths which are day fliers. There are very few species in the genus, all looking rather similar.

**lugubralis** Boisduval
U.S., widespread N.Y. 15–23 mm.
southwards (0·59–0·91 in.)
Black, with translucent white markings. The caterpillar feeds on grapes where it can be found in June. This species is very similar to the European *Thyris fenestrella* which while widespread in the warmer parts of Europe, is not common. **56**cc

**thyrsis** see **Gangara**

**Thysania** Dalman (NOCTUIDAE)
This small genus of large moths is essentially tropical American in distribution, but is represented by a single species in N. America. *Thysania* includes one of the 2 largest species of moths in the world.

**agrippina** Cramer
C. America and tropical 230–305 mm.
S. America (9·06–12·01 in.)
This magnificent moth shares with the Saturniid *Attacus atlas* the distinction of being the joint largest moth in the world in terms of wingspan. Specimens of *T. agrippina* sometimes reach S. U.S. **389**

**zenobia** Cramer OWL MOTH
S. U.S., C. America and 100–150 mm.
tropical S. America (3·94–5·91 in.)
No other N. American Noctuid exceeds *T. zenobia* in wingspan. In most of its range it is a common species. The colour pattern is similar in character to that of the much larger *T. agrippina* (illustrated), but the lines are less regular in shape and there is often a dark, oblique shade across the forewing from its base to the apex.

**thysbe** see **Hemaris, Poecilmitis**
**tiberius** see **Euxanthe**
TIGER BUTTERFLIES see **Danaus, Heliconius Lycorea, Tithorea**
TIGER
PALMFLY see **Elymnias nesaea**
PIERID see **Dismorphia amphione**
SWALLOWTAIL see **Papilio glaucus**
TIGER-MOTHS see **Arctiidae**
**tiliae** see **Mimas**

**Timyra** Walker (LECITHOCERIDAE)
This genus of micro-moths is typical of the family. Most species in the genus are found in Sri Lanka.

**cingalensis** Walker
Sri Lanka 22–28 mm. (0·87–1·1 in.)
The large scale tufts on the legs and the long curved labial palps on the head are distinctive. The female has brown hindwings. Related species occur in India but nothing is known of their biology. **39**d

**Timyridae** see **Lecithoceridae**

**Tinea** Linnaeus (TINEIDAE) CLOTHES MOTH
The genus is worldwide but is particularly common in the N. Hemisphere. This is a genus with many small species, some of which are of importance to man because of the damage they cause to clothing. Most are small rather dull-coloured moths of 8–25 mm (0·31–0·98 in.) wingspan but when examined closely through a magnifying glass are very beautifully marked insects.

**pellionella** Linnaeus CASE-BEARING or CASEMAKING CLOTHES MOTH 10–14 mm.
Asia including Japan, N. America, (0·39–0·55 in.)
Europe including Britain
This moth gets its name from the habit of its caterpillar which carries a small silken case round with it. This small case is spun by the caterpillar whose head and foreparts protrude from the case. The caterpillars feed mainly on substances of animal origin and are particularly fond of woollen garments. The moths tend to keep out of the brighter places in houses, lurking in the darker corners. They lay between 100 and 300 small white eggs. These should not be confused with the pieces of frass (excreta), small dark pellets often abundant on moth-damaged wool. The eggs hatch in 4–8 days and the caterpillar crawls away, spinning a silken thread to start making a portable case which it drags around. It is this caterpillar stage which is destructive, causing the damage to clothing and soft furnishings. Although there are various proofed fabrics and insecticides for treatment of clothes, it is still useful to hang them in the sunlight before packing them away for the winter. Wrapping in polythene will help to protect the clothes but is not entirely satisfactory since the caterpillars can bite their way through it.

**Tinearupa** Salmon & Bradley (OECOPHORIDAE)
This genus of micro-moths, with reduced wings in the adult state, is only found on subantarctic islands off New Zealand.

**sorenseri** Salmon & Bradley
Campbell Island, 10–14 mm.
Auckland Island (0·39–0·55 in.)
This small, grey, virtually wingless, moth has formed subspecies on 2 subantarctic islands. A flying insect is probably at some disadvantage on these storm-swept islands as it could easily be blown into the sea. The caterpillars feed on lichen or mosses on the islands. Since it is unlikely to have got to the island in its present wingless state (unless due to some prehistoric land connection) it is likely that its ancestors were winged but that as evolution progressed, those with shorter wings were favoured and had a greater survival rate.

**Tineidae**
A huge family of micro-moths, most of which are small, often rather dull coloured species. It includes many of economic importance, including well-known species like clothes moths and grain moths. Species of Tineidae are found all over the world, even in the colder parts. Frequently the caterpillar has a case round itself which it carries about. The caterpillars feed on animal fibres (wool), grain and in fact almost any dried animal or plant remains. Many of the species are known from only a few specimens, although others

are cosmopolitan pests. See *Monopis, Nemapogon, Tinea, Tineola, Trichophaga.*

**Tineodes** Guenée (TINEODIDAE)
This is an Australian genus of micro-moths. They have rather narrow wings, rather similar to the Agdistid plume moths (q.v.). Several related genera are found in Australia.

**adactylalis** Guenée
Australia 16 mm. (0·63 in.)
This species is common in parts of Australia. It is a rather grey-brown species roughly similar in shape to *Lineodes.*

**Tineodidae**
A small family of moths related to the Pyralids found from India to New Guinea and Australia. They all have narrow, long wings and long legs and superficially resemble species of *Agdistis.* The caterpillars tie leaves together. Species of this family are not common; there are, for example, some 10 species known from Australia. They differ from the Agdistid plumes in having 4 segments to the maxillary palps under the head. See *Tineodes.*

**Tineola** Herrich-Schäffer (TINEIDAE)
CLOTHES MOTH
This is a worldwide genus of micro-moths but with few species. It is related to the genus *Tinea.* Probably it has been spread by artificial means since the caterpillars, as the popular name implies, feed on clothes. The forewings are rather long and narrow.

**bisselliella** Hummel COMMON or WEBBING CLOTHES MOTH 11–15 mm.
Worldwide (0·43–0·59 in.)
This is a shining golden coloured moth between 6–8 mm. (0·24–0·31 in.) long. It lays 40–70 eggs over a period of 24 days. The females do not readily fly – those clothes moths that are seen flying are either males or females that have laid all their eggs. The length of life of the caterpillar depends on the temperature. Generally they are found on raw or soiled wool rather than clean manufactured material. Although they feed mainly on substances of animal origin they are also found on other materials such as man-made fibres. Moth-proofing is helpful in reducing the problem but clothes should be dry-cleaned before storing and preferably sealed in an airtight container. Polythene bags provide some protection but the caterpillars are capable of chewing through them to get at the clothes underneath. **5d**

**Tinostoma** Rothschild & Jordan (SPHINGIDAE)
This genus was erected to accommodate a single species, the famous green hawk-moth of Hawaii.

**smaragditis** Meyrick GREEN SPHINX OF KAUAI
Hawaii 85 mm. (3·35 in.)
Early this century, an American collector of hawk-moths, Preston Clark, offered 100 dollars for a specimen of this elusive species, then known from a single specimen, and sent a collector to Hawaii to search for it. Since then, 2 more specimens have been found: one in 1961 sitting on the lid of a garbage container clearly unaware of the inappropriateness of its perch, the other in 1969. **344r**

TINY ACRAEA see **Acraea uvui**
**tipuliformis** see **Synanthedon salmachus**

**Tirathaba** Walker (PYRALIDAE) PALM MOTHS
A large number of species of these Pyralid moths are found in the Oriental and Australasian regions. Many species are of economic importance as pests of palm trees. Most of the species in the genus have rather blunt, rounded, wings and are generally brown or reddish brown. The veins on the wings are often red, giving quite a strong pattern in some species. The genus is in the subfamily Galleriinae.

**mundella** Walker PALM MOTH
India, Pakistan, Indonesia, 18–25 mm.
Malaysia (0·71–0·98 in.)
This species is one of the common pests of palm trees from India through to New Guinea. When a large number of caterpillars attack Oil Palms or Coconut Palms, particularly if the trees are young, they will completely destroy them.

**Tischeria** Zeller (TISCHERIIDAE)
A widespread genus of micro-moths whose caterpillars are generally leaf-miners. The genus is primarily European and temperate Asian. The adults have rather pointed wings.

**ekebladella** Bjerkander: syn. *T. complanella*
Europe including Britain, 8–11 mm.
Turkey, N. Africa (0·31–0·43 in.)
The caterpillar of this species lives inside the leaves of oak *(Quercus)* where it forms whitish blotches on the leaves. **6a**

**Tischeriidae**
A small family of micro-moths which has variously been associated with the Nepticulidae or with the Incurvariidae. Species are known from Europe, India, Africa and N. America. See *Tischeria.*

TIT BUTTERFLIES see **Hypolycaena, Zeltus**
**titania** see **Clossiana**
TITANIA'S FRITILLARY see **Clossiana titania**
**Titanio daphalis** see under **Pyrausta daphalis**
**tithonus** see **Pyronia**

**Tithorea** Doubleday (NYMPHALIDAE)
There are about 15 species of butterfly in this genus of the subfamily Ithomiinae. They resemble many species of *Mechanitis* and *Melinaea* (Ithomiinae) in colour pattern. Their range is tropical American.

**harmonia** Cramer TIGER BUTTERFLY
C. and tropical 57–63 mm.
S. America (2·24–2·48 in.)
This is a generally common species in virgin forest, areas of secondary growth and in cocoa *(Theobroma)* plantations. It is a member of a mimetic complex which includes at least 3 other species of Ithomiinae in the genera *Melinaea, Mechanitis,* and *Ceratinia,* a species of *Lycorea* (Danaiinae) and a species of *Heliconius* (Heliconiinae). All of these belong to unpalatable groups of butterflies and have evolved the same colour pattern as an efficient group-advertisement of their noxious qualities. The group is mimicked by the Nymphaline species *Protogonius pardalis* Bates and the Pierid *Dismorphia egaena,* which are probably palatable to predators. The caterpillars of *T. harmonia* feed on species of *Echites,* a genus of Apocynaceae. **131a**

**tithorus** see **Ornithoptera**
**tityrus** see **Epargyreus clarus**
TOADFLAX PUG see **Eupithecia linariata**
TOBACCO MOTH see **Ephestia elutella**
TOBACCO
BUDWORM see **Heliothis virescens**
HORNWORM see **Manduca sexta**
SPLITWORM see **Phthorimaea operculella**
**tobleri** see **Euploea**
**toboroi** see **Papilio**

**Tolype** Hübner (LASIOCAMPIDAE)
The 100 or so species of *Tolype* are chiefly C. American and tropical S. American in distribution, but there are a few N. American species. The adults are grey, black and white in colour pattern and difficult to identify. The caterpillars are typical 'lappets' in shape.

**glenwoodi** Barnes
S.W. U.S. 25–40 mm. (0·98–1·57 in.)
This is a typical member of its genus in colour pattern; others are similar in colour, much whiter, or with more grey and black. *T. glenwoodi* has 1–2 broods a year. Its caterpillar has been recorded from Gambel's Oak *(Quercus gambelii).* **319**

TOMATO BUTTERFLY see **Temenis laothoe**
TOMATO FRUITWORM see **Heliothis zea**
TOMATO HORNWORM see **Manduca quinquemaculata**
TOOTH-STREAK HOOKED TIP see **Ypsolopha dentella**

**torquata** see **Dasysphinx, Plagerepne**
**torquatus** see **Papilio**
**torrefacta** see **Hygrochora**
TORTOISESHELL BUTTERFLIES see **Aglais, Nymphalis**
TORTRICES see **Tortrix**

**Tortricidae** TORTRICID, FRUIT, BELL or LEAF ROLLER MOTHS
A large family of micro-moths worldwide in distribution, containing many species which are pests of fruit trees. Some of the species which feed on weeds have been used in biological control (eg. *Epinotia lantana* Busck). Many of the species are variable and occur in different colour forms; these are popular with collectors. The family here includes the subfamily Olethreutinae (often put as a separate family), which includes the notorious Apple Codling moth. With the economic importance of the family Tortricidae they have been much studied in recent years with many papers on their biology and distribution. The numbers of this family can be gauged from the figures for N. America with over 600 species and Australia with over 800 known species. Even in Britain there are well over 300 species. The systematic position of the family is still in dispute, some regard it as nearer to the Cossoidea and Tineoidea as shown in the preface classification, while others put it nearer the Pyraloidea. Most of the moths have a rather similar shape to the typical 'tortricid' or 'bell-moth' shape. This is clearly seen when the moths are at rest and the wings are held flat in a bell shape. See *Acleris, Archips, Argyiotaenia, Cacoecimorpha, Cerace, Choristoneura, Chresmarcha, Cnephasia, Croesia, Cydia, Eana, Epiblema, Epichoristodes, Epinotia, Eucosma, Eulia, Leguminivora, Melissopus, Oleuthreutes, Pseudargyrotoza, Pseudatteria, Rhyacionia, Sparganothis, Tortrix, Zacorisa.*

**Tortrix** Linnaeus (TORTRICIDAE) TORTRICES or BELL MOTHS
A large genus of micro-moths found worldwide. The forewing is usually broad but sharply truncated at the end; the hindwing is rounded. The forewings are usually brightly coloured or strongly variegated. When at rest, the moths extend the forewings flat over the hindwings giving the insect the appearance of a bell. The caterpillars very often roll the leaves on which they feed, tying them down with silk. Others feed on seed and several are fruit pests.

**viridana** Linnaeus GREEN OAK TORTRIX
Europe including Britain 16–24 mm. (0·63–0·94 in.)
This is a widespread and abundant species, easily recognized by its shape and green forewings. The caterpillars feed on oak or sallow. They roll the edges of the leaves, tying them down with silk. The caterpillars are very active and if disturbed will drop off the leaf, hanging by a silken thread. Sometimes they are so abundant on oaks that the whole tree may be stripped of its leaves. **12**

TORTRIX MOTHS see **Acleris, Archips, Argyrotaenia, Cacoecimorpha, Rhyacionia, Sparganothis, Tortrix**

**Tortyra** Walker (GLYPHIPTERIGIDAE)
This worldwide genus of micro-moths has species from India through to Australia, in the Solomon and Philippine Islands. Many species have the remarkable metallic appearance shown in the illustration. This green iridescence is particularly striking when several of the moths are seen together. Most have the iridescence in bands across the wings.

**divitiosa** Walker
Assam, Formosa, Indonesia, New Guinea 20 mm.
Malaya (0·79 in.)
Several species in the genus have a similar type of wing pattern. Many are darker than the one illustrated but most have the green iridescence, and some have yellow hindwings. All are striking looking moths and are quite distinct from species in related genera. Nothing is known of their life history. **34m**

**touhilimasa** see **Precis**
**toxopei** see **Zacorisa**

**trabealis** see **Emmelia**

**Trachydora** Meyrick (MOMPHIDAE)
A large genus of micro-moths from Australia and Africa. Many of the species in the genus are known from only one or two specimens and little is known of their biology.

**leucobathra** Meyrick
N. Australia, Queensland 11 mm. (0·43 in.)
The wings are more grey and silver than others in the genus which are generally yellowish in colour. At present there are no records of the foodplant of the caterpillar. **44**c

**trajanus** see **Euxanthe**
**transitella** see **Paramyelois**
**translucidalis** see **Diaphania superalis**
TRANSPARENT BUTTERFLY, BLUE see **Ithomia drymo**
TRANSPARENT BURNET see **Zygaena purpuralis**
**trapeziella** see **Meridarchis**
TREBLE BAR MOTH see **Aplocera plagiata**
TREE
BROWN, BAMBOO see **Lethe europa**
NYMPH BUTTERFLIES see **Idea**
**tremula** see **Pheosia**
**tremuloidiella** see **Phyllonorycter**
TRIANGLE BUTTERFLIES see **Graphium**
TRIANGLE MOTH see **Heterogenea asella**
**tricentra** see **Leguminivora**

**Trichophaga** Ragonot (TINEIDAE)
A small worldwide genus of micro-moths with one important species whose larvae feed on furs, skins, etc. They are related to the genus *Tinea* but are separated from them by differences in wing venation.

**tapetzella** Linnaeus WHITE-TIPPED CLOTHES MOTH, TAPESTRY or CARPET MOTH
Worldwide 16–21 mm. (0·63–0·83 in.)
This is probably less common in recent years, partly no doubt due to the use of man-made fibres instead of animal skins. The caterpillar of this moth can be very destructive to furs and woollen garments, where they make silken galleries, eating large patches of fur or wool. **5**b

**Trichoplusia** McDunnough (NOCTUIDAE)
About 15 species are placed in this almost cosmopolitan genus of moths. It belongs to the same group of genera as *Plusia*.

**ni** Hübner NI MOTH or CABBAGE LOOPER
America, Hawaii, Europe, N. Africa and the Canary Islands 33–38 mm. (1·3–1·5 in.)
This is a serious pest in the U.S., especially in the S. where its caterpillars feed on cabbage *(Brassica)*, and other species of Cruciferae, and also on cotton *(Gossypium)*, potato *(Solanum)*, tomato *(Lycopersicon)*, soya beans *(Glycine)*, *Chrysanthemum* and many other plants. The various control measures include irradiation of the chrysalids and adults, the application of *Bacillus thuringensis* as a bacterial insecticide and electrified traps baited with the synthetic sex-scents 'looplure' or 'hexalure'. The popular name 'Cabbage Looper' refers to the looping, Geometrid-like movements of the caterpillar which has fewer false legs than most other Noctuids. The moths are similar in colour pattern to the illustrated *Autographa gamma*, but differ in the shape of the silvery mark on the forewing.

**Trichoptilus** Walsingham (PTEROPHORIDAE)
PLUME MOTHS
A large genus of Plume moths, all of which have divided fore- and hindwings. The hindwings are divided into 3, the forewings into 2 by a deep cleft.

**parvulus** Barnes & Lindsey
N. America 10 mm. (0·39 in.)
This species is one of the remarkable insects which feed on insect-eating plants. The plant is Sundew *(Drosera)* whose leaves have sticky hairs which catch insects. Caterpillars of this small plume-moth actually feed on the leaves, avoiding the sticky hairs. The caterpillar feeds mainly at night and will even eat the remains of other insects trapped and partially digested by the plant. The caterpillar feeds below the long sticky hairs, but should it accidentally touch one, it is well protected by the long hairs on its own body. This habit of feeding on *Drosera* is shared by the European species *Buckleria paludum*.

**tricolor** see **Dysschema, Hemileuca, Magneuptychia**
TRICOLOR BUCK see **Hemileuca tricolor**
**tridelta** see **Anticrates**
**tridens** see **Calamia** and see under **Acronicta psi**

**Tridrepana** Swinhoe (DREPANIDAE)
With few exceptions, the 35 or so species of this genus of hook-tip moths are yellow in ground-colour, with brown (and in some specimens, metallic white) markings.

**flava** Moore
India, Sikkim and Asia to Sulawesi (Celebes) 42–66 mm. (1·65–2·6 in.)
This is the largest species of its genus and 3 subspecies have been described. Nothing has been published about the life history of this moth. **272**l

**trifoli** see **Zygaena**
**trigona** see **Bertholdia**
**trigonophora** see **Amata, Neptis**
**triguttata** see **Metarbela**
TRIMEN'S
DOTTED BORDER see **Mylothris trimenia**
FALSE ACRAEA see **Pseudacraea boisduvalii**
**trimeni** see **Eochroa**
**trimenia** see **Mylothris**
**trinitatis** see **Sphecosoma**
**triopus** see **Rhodosoma**
**tripartita** see under **Hecatesia**
**trisecta** see **Pediasia**
**trisignata** see **Harrisimemna**

**Trisophista** Meyrick (YPONOMEUTIDAE)
The few species described in this genus of micro-moths are either from Africa or Madagascar.

**doctissima** Meyrick
Zaire, Uganda, Kenya 22–26 mm. (0·87–1·02 in.)
This species lives in the forest in Africa. It has a striking pattern of iridescent metallic blue with black dots. The undersides of the wings are an unmarked bluish-purple. No data is available on its life history or host plants of the caterpillar and only a few specimens of the moth itself have been found. A related species from Madagascar is similarly shaped but has white forewings with black spots. Although very conspicuous, its colour and pattern provide good camouflage in the forest where there may be small areas brightly illuminated and dappled dark patches. **38**e

**tristella** see **Sphecosoma**
**trite** see **Phoebus**
**trivene** see **Limenitis**
**trochylus** see **Freyeria**

**Troides** Hübner (PAPILIONIDAE)
The *Ornithoptera* birdwings are probably the closest relatives of this genus of butterflies which are found in much of S.E. Asia from Sri Lanka and India to New Guinea. There are about 20 species, some are large and colourful and highly prized by collectors; many are black and dark brown with lustrous yellow areas on the hindwings. Many species feed on poisonous *Aristolochia* species as caterpillars and are mimicked by supposedly palatable species of other genera. It seems probable, therefore, that the adults are distasteful to predators and that their yellow and black coloration is a warning pattern. *Troides aeacus* is known to contain poisonous acetylcholine-like substances in its tissues.

**aeacus** Felder GOLDEN BIRDWING BUTTERFLY
Bhutan, Sikkim, N.E. India, S. China, Formosa, Burma, Malaya 125–175 mm. (4·92–6·89 in.)
This species is rather similar in pattern to the generally smaller *T. helena* (illustrated), but is brown, rather than black in ground-colour and has less intensely yellow, more golden hindwings. It is common in some hilly localities, especially between 300–910 m. (1000–3000 ft) in the foothills of the Himalayas. There is some variation in the colour pattern between the various subspecies.

**brookiana** Wallace RAJAH BROOKE'S BIRDWING
Malaya, Sumatra, Borneo 120–175 mm. (4·72–6·89 in.)
This butterfly was first discovered in Borneo by the famous naturalist Alfred Russell Wallace, and named after the British Rajah Brooke of Sarawak. Tropical rain-forests are the usual home of this species where it flies typically at tree-canopy level. Males of the subspecies *T. brookiana albescens* are seen much more often than the less richly coloured females as they are attracted to wet mud along the banks of streams and elsewhere, whereas both sexes of subspecies *T. brookiana trogon* are attracted to flowers and the females are seen as often as the males. It has been discovered recently that the caterpillar of this butterfly feeds on species of *Aristolochia*. The male of *T. brookiana* differs from the illustrated *T. trojana* in the absence of a blue iridescence to the green markings on both pairs of wings and at the base of the hindwing; the more elongate forewing markings, and the broader hindwing band are also diagnostic.

**helena** Linnaeus COMMON BIRDWING
Sri Lanka, India to Hainan, the Malay Archipelago to New Guinea 95–135 mm. (3·74–5·31 in.)
This widely distributed birdwing is in fact common in places as its name implies. Its colour pattern is typical of many species of *Troides*. The sexes fly together in forested areas, but also sometimes in urban areas, usually some height above the ground. The females differ from the illustrated male in the presence of several black spots on the hindwing; a few have an orange, not yellow patch on the hindwing. The caterpillars feed on a species of *Aristolochia*. **107**f

**hypolitus** Cramer
Sulawesi (Celebes), the Moluccas, Philippines 165–200 mm. (6·5–7·87 in.)
This is the largest species of its genus. Its immense size, and the less extensive yellow markings which are absent from the middle of the hindwing, distinguish this species from *T. helena* (illustrated). The male abdomen is red and black above, the female abdomen yellowish green. Four subspecies have been described, each restricted to one or more islands in S.E. Asia.

**prattorum** Joicey & Talbot
Buru Island 115–170 mm. (4·53–6·69 in.)
This is the most colourful species of its genus. When viewed from above the wings are black and yellow; but when viewed obliquely from behind (with the wings in the set position, as in the illustration) the yellow patches change to greenish blue. **100**c

**trojana** Staudinger
Philippines (Palawan) 140–155 mm. (5·51–6·1 in.)
The island of Palawan was described by Jordan in 1910 as 'very unpleasant and dangerous' and inhabited by 'many doubtful characters who have taken refuge there from the Philippines' and is therefore not surprising that few captured specimens date from this era. The female (not illustrated) is brown, not black, and has mostly white markings on the forewing and an additional row of white spots near the outer margin of the hindwing. The colour pattern of *T. trojana* is basically similar to that of the more common *T. brookiana*. Nothing is known of the foodplant of the caterpillar. **104**c

**troilus** see **Papilio**
**trojana** see **Troides**
TROPICAL FRUIT-PIERCER see **Othreis fullonia**
**tropicalis** see **Pentila**

**Trosia** Hübner: syn. *Langucys* (MEGALOPYGIDAE)
This American genus has species through C. & S. America to Peru. There are several related ones with similar patterns. The caterpillars, where known, have stinging hairs which can produce severe reactions in humans.

**bicolor** Möschler
Amazon, Peru, British Guiana 40 mm. (1·57 in.)
Little is known about this species, the adult is a typical

Megalopygid in appearance. Its biology is unknown but it probably has a slug-like caterpillar with stinging hairs, typical of the family. **48**a

**truncataria** see **Isturgia**

**Tryporyza** Common (PYRALIDAE) RICE STEM BORER MOTHS
The white or yellow-brown forewings of the female are rather pointed at the apex. The forewings of the males are brown and grey with much more rounded forewing margins. Many of the species in this genus are serious pests of rice. When their caterpillars bore in the stem of the rice plant, the subsequent yield of rice grain is seriously reduced. This genus belongs to the subfamily Schoenobiinae.

**innotata** Walker WHITE RICE BORER MOTH
E. Malaysia, Indonesia, S. Vietnam, Australia 20–34 mm. (0·79–1·34 in.)
The adult moth is a slender white insect without any spots on the wing. The female has a pinky brown tip to the abdomen. This is an important rice pest and much research work has been done on it to see how it might be controlled. Identification of this particular pest is difficult as many species in the subfamily (Schoenobiinae) to which the White Rice Borer belongs are very similar looking moths but have different life histories. The damage in the field can be seen by the dead heads of rice plants above the level where the caterpillar has bored into the stem.

TUBER MOTHS, POTATO see **Phthorimaea**
TUFTED JUNGLE KING see **Thauria aliris**
**tugela** see **Precis**
**tulbaghia** see **Meneris**
**tullia** see **Coenonympha**
TURNIP MOTH see **Agrotis segetum**
TUSSOCK MOTHS see **Dasychira, Halisidota, Hemerocampa**
TUSSORE SILK-MOTHS see **Antheraea**
TUSSUR SILK-MOTHS see **Antheraea**
TWIG-BORER, PEACH see **Anarsia lineatella**
TWILIGHT
BROWN see **Melanitis leda**
SKIPPER see **Zophopetes dysmephila**
TWIN DOTTED BORDER see **Mylothris poppea**
TWIN-SPOT
CARPET MOTH, DARK-BARRED see **Xanthorhoe ferrugata**
SPHINX see **Smerinthus jamaicensis**
TWO-BRAND CROW BUTTERFLY see under **Euploea coreta**
TWO-TAILED
PASHA see **Charaxes jasius**
SWALLOWTAIL see under **Papilio multicaudata**
TYGER HAWK, BEE see under **Acherontia atropos**
**typhae** see **Nonagria**

**Tyria** Hübner (ARCTIIDAE)
There is a single species of moth in this solely European genus.

**jacobaeae** Linnaeus CINNABAR MOTH
Europe including Britain 35–45 mm. (1·38–1·77 in.)
This is one of the most unpalatable species of moth, especially in the adult stages, and is invariably rejected by vertebrate predators. The noxious qualities, which are not solely dependant on poisonous plant alkaloids, are advertised in the black and yellow coloration of the gregarious caterpillar and the red and black wings of the moths. Adults are capable of ejecting noxious secretions from their thoracic glands. The introduction of this species into New Zealand in an attempt to control accidentally introduced ragwort *(Senecio)*, its chief foodplant in Britain, failed because the native birds surprisingly ate the conspicuous caterpillars, apparently with some degree of relish. In Britain, the caterpillar is distasteful to birds – the gaudy black and yellow coloration of the caterpillars acting as a warning signal. A normal female and an unusual, unmarked red form of this species are shown in the illustration. **370**

**tyrianthina** see **Anaea**

**Ubaena** Karsch (SATURNIIDAE)
This is a genus of only 2 species of E. African mountain moths. The males fly by day in search of females who usually only fly at night.

**fulleborniana** Karsch MOURNING EMPEROR
Zambia, S. Tanzania 90–110 mm. (3·54–4·33 in.)
The hindwings of the female (not figured) may be either red or yellow basally. Nothing is known about the caterpillar of this species. **333**a

**ucubis** see **Uraneis**
UGLY NEST CATERPILLAR MOTH see **Archips cerasivorana**
**ulmivora** see **Stigmella**
**ulysses** see **Papilio**
ULYSSES see **Papilio ulysses**
UMBER, MOTTLED see **Erannis defoliaria**
UNDERLEAF see **Lymnas iarbus**
UNDERLEAF, BLACK see **Alesa amesis**
UNDERLEAF, ORANGE-TIPPED see **Lymnas xarifa**
UNDERWING MOTHS see **Amphipyra, Anarta, Catocala, Noctua, Peridroma**
**undalis** see **Hellula**
**undulata** see **Rheumaptera**
**unica** see **Gonodonta**
**unionalis** see **Palpita**
**unipuncta** see **Mythimna**
**uraneides** see under **Calodesma**

**Uraneis** Bates (NEMEOBIIDAE)
There are 3 species in this tropical S. American genus of butterflies. They mimic species of the day-flying moth family Hypsidae.

**ucubis** Hewitson 36–40 mm. (1·42–1·57 in.)
One of the forms of this butterfly closely resembles the Nemeobiid *Esthemopsis clonia;* another form has a similarly patterned hindwing, but has a pre-apical, white band on the forewing resembling a form of *Calodesma uraneides* or *Hypocrita aletta* both of which are species of Hypsidae. Males (not illustrated) of this species have straighter outer margins to the wings and narrower white markings. **263**w

**Urania** Fabricius (URANIIDAE)
This is the only New World representative of its family. In colour pattern and particularly in coloration it is similar to the African and Madagascan genus *Chrysiridia.* There are about 8 species, all solely tropical American in distribution.

**leilus** Linnaeus
Tropical S. America, the Antilles 58–85 mm. (2·28–3·35 in.)
This is similar in pattern to the illustrated *U. sloanus*, but the markings are mostly green, and without a golden iridescence. The outer margin of the hindwing has a much broader white margin, and the tail is white except for the black, central vein. The caterpillar, like that of *Chrysiridia ripheus*, bears long, clubbed hairs on its body, and feeds on species of *Omphalea*, a genus of poisonous Euphorbiaceae. **315**

**sloanus** Cramer
Jamaica 50–68 mm. (1·97–2·68 in.)
The remarkable similarity in the type of colour pattern and coloration between this species and the species of the African and Madagascan *Chrysiridia*, can be seen from the illustration. Apart from the presence of a single long tail in *U. sloanus*, the only striking difference between this and *C. ripheus* is the transposition of the multicoloured area on the hindwing from one side of the tail to the other. The black, blue and white caterpillar feeds on *Omphalea* (Euphorbiaceae) (see *Chrysiridea ripheus* and *Alcides agathyrsus*). **314**c

**Uraniidae**
This is a small family of moths containing both flimsy, white, nocturnal members (Microniinae) and dramatically coloured, *Papilio*-like, day-flying species (Uraniinae). They are found in tropical America and Africa and in temperate and tropical Asia E. to Australia, New Guinea and the Solomons. One species, *Alcides agathyrsus* is reputedly distasteful to predators and is mimicked by the butterfly species *Papilio laglaizei*. The family Uraniidae is a member of the superfamily Geometroidea. See *Alcides, Chrysiridia, Cyphura, Lyssa, Strophidia, Urania.*

**Uranothauma** Butler (LYCAENIDAE)
The 10 species of this genus of butterflies are found in Africa, S. of the Sahara.

**falkensteinii** Dewitz
Africa, S. of the Sahara 21–25 mm. (0·83–0·98 in.)
The upper surface of the male is dull purple, with a slight iridescence and black scent-scale streaks between the forewing veins. The female is much paler in ground-colour above and is patterned like the under surface; it is much rarer than the male, as in other species of its genus. The illustrated specimen is drinking on wet sand.

**Urbanus** Hübner (HESPERIIDAE) LONG-TAILED SKIPPER
These have elongate hindwings and are found in N., C. and S. America where some 28 species are known. Many of them feed on species of Leguminosae.

**proteus** Linnaeus LONG-TAILED SKIPPER, BEAN-LEAF ROLLER OR ROLLER WORM
U.S., S. & C. America 40–50 mm. (1·57–1·97 in.)
In the U.S. this species reaches to Connecticut and Texas but further S. is often abundant. The caterpillar feeds on cultivated beans, and when abundant can be a serious pest. Apart from beans *(Phaseolus)* it also feeds on *Wisteria* and other Leguminosae. There are several broods each year.

**Uresiphita** Hübner (PYRALIDAE)
A small genus of Pyralid moths with one widespread species.

**limbalis** Denis & Schiffermüller: syn. *U. polygonalis*
Europe including Britain, Asia, Australia, Hawaii 27–34 mm. (1·06–1·34 in.)
A common species in collections of Pyralids from many parts of the world. It is a migrant, occurring casually in Britain and the more northerly parts of Europe. The caterpillars feed gregariously, webbing together the leaves of their host plant. It has been found on gorse *(Ulex)*, broom *(Cytisus)*, *Acacia* and a wide variety of Leguminosae. **56**q

**uricoecheae** see **Catasticta**

**Uropyia** Staudinger (NOTODONTIDAE)
There are 2 known species in this genus of moths: *U. hammamelis* Mell, described from Chinese material, and the following more widely distributed species.

**meticulodina** Oberthür
China, E. U.S.S.R. (and Askold Island), Japan 48–58 mm. (1·89–2·28 in.)
This is a particularly elegantly marked moth of rich browns and brownish green. Charles Oberthür, who first described the species, amassed the largest private collection ever made of Himalayan Chinese butterflies and moths, including many species previously unknown to science. Many of the moths, including the Notodontidae, were bought by the British Museum (Natural History). **357**f

**Urota** Westwood (SATURNIIDAE)
There is a single species in this genus of moths.

**sinope** Westwood TAILED EMPEROR MOTH
Angola, E. Africa from Ethiopia to Natal 82–100 mm. (3·23–3·94 in.)
The males of this species have a short, outwardly directed tail on the hindwing; the female is tailless but has a small lobe on the outer margin of the hindwing. The coloration of the wings is variable and can vary from very pale brown to dark brown; the forewing has 2 white, nearly straight, transverse bands and the generally pinkish brown caterpillar has a single broad, white band. The conspicuous caterpillar is yellowish white with black bands and short hairs.

**urticata** see **Eurrhypara hortulata**
**urticaea** see **Aglais**
**ussheri** see **Palla**

**Utetheisa** Hübner (ARCTIIDAE)
This is a small genus of mostly distinctively patterned moths. It is almost cosmopolitan in distribution. Some of its species are known to be migratory. New World species feed on *Crotalaria* (Leguminosae) and other plants. Old World species include *Heliotropium* and *Tournefortia* (Boraginaceae) amongst their food plants. Each of these plant genera, though of different families contain substances toxic to vertebrates.

**bella** Linnaeus BEAUTIFUL UTETHEISA or BELLA MOTH 33–47 mm. (1·3–1·85 in.)
U.S. & S. Canada (E. of the Rockies), C. & S. America, S. to Chile and Argentina, Bermuda, Bahamas, Cuba, Dominican Republic, Puerto Rico, the Virgin Islands
The N. American subspecies, *U. ornatrix bella* Linnaeus, was long considered to be a separate species – it also occurs in Bermuda and the Bahamas. Specimens of this species occasionally reach England. A substance poisonous to vertebrates (which includes birds and other enemies of moths) has been isolated from the secretions of thoracic glands; this is mixed with the insect's haemolymph (blood and lymph) and and ejected as a yellowish froth with a strange, unpleasant smell, probably as a defensive mechanism. The caterpillars feed on the flowers of golden rod *(Solidago)*, *Crotalaria* and on lupins *(Lupinus)*. **364**r

**pulchella** Linnaeus CRIMSON SPECKLED MOTH
Europe including Britain, Africa, Asia, Australia 35–42 mm. (1·38–1·65 in.)
This is one of the most widespread moths in the Old World, a result, at least partly, of its migratory habits. Specimens are often taken at sea and at lighthouses. In Europe this is probably not a resident species N. of the Mediterranean coast. One possibly pertinent reason for its rejection of Britain as a breeding place is the need for sunshine before the caterpillars will start to feed. Forget-me-not *(Myosotis)* and borage *(Borago)*, both genera of Boraginaceae, are the usual foodplants. **366**

UTETHEISA, BEAUTIFUL see **Utetheisa ornatrix**
**uvui** see **Acraea**

**Uzucha** Walker (XYLORYCTIDAE)
A small genus of micro-moths with species in Australia. The caterpillars have been found on *Eucalyptus* where they tunnel in the stems. The moths are large, rather fat bodied species rather more similar at first sight to macro-moths than to the delicate micro-moths.

**humeralis** Walker
Australia (Queensland) 40–60 mm. (1·57–2·36 in.)
The female has pale coloured forewings; in the male these are dark grey. Both have black abdomens with some yellow. This species has been reared on *Eucalyptus platyphylla* where the caterpillar bores into stems. The moth flies at night but little is known of the biology of this species. **42**a

V MOTH see **Semiothisa wauaria**
V-FLAVA see **Oinophila**
V-NEPHELE HAWK-MOTH see **Nephele vau**

**Vacciniina** Tutt (LYCAENIDAE)
This is a small genus of butterflies restricted to the N. temperate regions of the world.

**optilete** Knoch CRANBERRY BLUE
European Alps, Arctic Europe, Japan, N. America (Alaska to Hudson Bay) 22–25 mm. (0·87–0·98 in.)
The caterpillars of *V. optilete* feed on the flowers and foliage of species of *Vaccinium* and various other moorland, peat-bog and mountain plants. The adult butterflies are a deep violet blue above.

**vacua** see **Burlacena**
**vacuna** see **Epiphora**

**Vagrans** Hemming: syn. *Issoria* (NYMPHALIDAE)
This butterfly genus is found in Oriental regions, across Indonesia to Australia and the S. Pacific. The caterpillar has long branched spines including a pair on the head.

**egista** Cramer AUSTRALIAN VAGRANT BUTTERFLY
India, Pakistan, Hainan, Philippines, Mariana Islands, Indonesia, New Guinea, Solomons, Samoa, Friendly and Society Islands, Australia 55–60 mm. (2·17–2·36 in.)
This wide ranging species has many subspecies. The general colour is orange-brown with a black margin and 2 black spots on the front margin of the forewing. The hindwing is similar and has a short tail on each side. The underside is more patterned with yellow-brown and white with black spots on the fore- and hindwing and a general purplish sheen. The caterpillar feeds on species of *Xylosoma* and *Homalium* (Flacourtiaceae). The adults are common in rocky areas and roadways, particularly through rain-forests. **163**c

VAGRANTS see **Eronia, Issoria, Nepheronia, Vagrans**
**valeria** see **Lacosoma**
VAMPIRE MOTH see **Calpe eustrigata**

**Vanessa** Fabricius (NYMPHALIDAE)
This is a genus of butterflies found in Europe, Asia, including India, Sri Lanka and Indonesia with 1 species which occurs in America. The genus was recently studied and some changes made in the species included in it. There are 5 known species whose caterpillars feed on species of Urticaceae (nettle family) with a few other recorded foodplants. Most of the butterflies are similarly coloured with deep reds and orange-browns. See also under *Cynthia*.

**atalanta** Linnaeus RED ADMIRAL
Europe including Britain, N. Africa, Turkey, Asia, N. India, U.S., Canada, Mexico, Central America, Bermuda, Cuba, introduced Hawaii 50–70 mm. (1·97–2·76 in.)
This widespread and well known butterfly has 2 subspecies. One, *V. atalanta atalanta* Linnaeus is in the Old World, while *V. atalanta rubria* Fruhstorfer is in the New World. The caterpillars, which are variable in colour, are generally black with rows of yellow spots, many raised, with white warts along the side and with black spines with orange bases. The food is generally nettles *(Urtica)* but it will also feed on hops *(Humulus)* and other plants in the family Urticaceae. The butterflies are fast fliers and are about from May to October, hibernating over the winter. In the warmer parts of its range there are probably 2 broods. The survival of hibernating butterflies in the more northerly part of their range is rare and the populations are replaced each year by immigrants which have survived further S. **203**

**canace** Johanssen BLUE ADMIRAL
India, Pakistan, Sri Lanka, Burma 60–75 mm. (2·36–2·95 in.)
This is a common species in wooded country in hilly areas. It is generally found in the vicinity of streams where its fast flight up and down the river can make it difficult to catch. It often settles on stones with its wings partly open and is seen at flowers and rotting fruit. It is found in most months of the year. The wings are a deep blue with pale blue bands below the margin of both wings. These bands have a few darker marks. The outer margin of the forewing is strongly concave and there is a short tail on the hindwing.

**indica** Herbst INDIAN RED ADMIRAL
India, Pakistan, Japan, Taiwan, Philippines, U.S.S.R., Canary Islands, Madeira, S. Portugal and S. Spain 50–74 mm. (1·97–2·91 in.)
After *V. atalanta* this is the next most widespread of the Red Admirals. The subspecies which occurs in parts of S. Europe and the Canary Islands is different from the very widespread subspecies from India to the Philippines. The foodplants of the caterpillars are species of Urticaceae. Two other subspecies are known, one from S. India, the other from S. Sulawesi (Indonesia).

**tameamea** Eschscholtz
Hawaiian Islands 62–80 mm. (2·44–3·15 in.)
This is the largest species in the genus *Vanessa* but is one of the most restricted in distribution. It was formerly included in the genus *Pyrameis*. The forewings are paler coloured than most others with a very broad yellow-orange centrally placed band in the forewings, broader than any others in the genus. The caterpillar feeds on various species of Urticaceae and the butterfly can be found on most of the mountains on the Hawaiian Islands.

**Vanessula** Dewitz (NYMPHALIDAE)
A genus of butterflies with only 1 species which is found in C. Africa.

**milca** Hewitson ORANGE AND BROWN
Africa 45–50 mm. (1·77–1·97 in.)
This is a brown butterfly with a broad orange band across the fore- and hindwings. It is found in the forests of the Congo and W. parts of Kenya and Uganda to W. Africa. It is a slow, rather weak flier, unlike many Nymphalids, and can be seen fluttering along forest paths throughout the year.

**vanillae** see **Agraulis**
**vaninka** see **Perisama**
VAPOURER see **Orgyia antiqua**
**varanes** see **Hadrodontes**
**vardhana** see **Lycaenopsis**
**variegana** see **Acleris**
VARIEGATED CUTWORM see **Peridroma saucia**
**vau** see **Nephele**
**vau-album** see **Nymphalis**
**vega** see **Proserpinus**

**Vegetia** Jordan (SATURNIIDAE)
The 3 species of this genus of moths are confined to S.W. Africa. They are closely related to species of *Ludia*.

**dewitzi** Maasen & Weymer DEWITZ'S PRINCELING
South Africa 34–60 mm. (1·34–2·36 in.)
The most distinctive feature of this small species is the white or yellow C-shaped marking on each wing. Females differ little in colour pattern from the illustrated male. *V. dewitzi* is known only from the N. part of Cape Province. **331**g

VEIN BUTTERFLY, GREAT BLACK see **Aporia agathon**
VEINED SWALLOWTAIL see **Graphium leonidas**
VEINED WHITE, BLACK see **Aporia crataegi**
**velutina** see **Hamadryas, Taygetis echo**
**venatus** see **Ochlodes**
VENEER, RUSH see **Nomophila noctuella**
**venosata** see **Eupithecia**
**venulinus** see **Paiwarria**
**venus** see **Leto**

**Venusia** Curtis (GEOMETRIDAE)
This genus of moths contains several species spread across the N. Hemisphere, some of which are minor pests as defoliators of trees or shrubs.

**cambrica** Curtis WELSH WAVE or CAMBRIC WAVE MOTH 25–28 mm. (0·98–1·1 in.)
Europe including Britain, Asia including Japan, N. America
This moth was first described in 1839 from specimens collected in Wales. The caterpillar is green with reddish marks and a yellow line along the back. It feeds on Mountain Ash *(Sorbus)* in Europe and on this and several other trees including Red Alder, Birch and Black Cottonwood in N. America. In Europe the moth is widespread, in N. America it is found in W. Canada and in the U.S. from Alaska to Massachusetts and N. California. **293**

**venusta** see **Cercophana, Eurema**
**verbasci** see **Cucullia**
VERDANT HAWK-MOTH see **Euchloron megaera**
**verhuelli** see **Dercas**
**versicolor** see **Mimoniades**
**versicolora** see **Endromis**
**versicularia** see **Corymica**
**vespertilio** see **Hyles**
**vespiformis** see **Synanthedon**
**vesta** see **Acraea**
VESTAL see **Rhodometra sacraria**
**vibilia** see **Eueides**
VICEROY see **Limenitis archippus**

**victoriae** see **Ornithoptera**
VICTORIA'S BIRDWING see **Ornithoptera victoriae**
**Victorina** see **Metamorpha**
VIENNESE EMPEROR see **Saturnia pyri**
**villica** see **Arctia**
**villida** see **Junonia**
**vinaceostriga** see **Arycanda**

**Vindula** Hemming (NYMPHALIDAE)
A small genus of butterflies from Indonesia with some external similarities to species of *Vanessa*. Generally they are large, powerful fliers.

**erota** Fabricius CRUISER BUTTERFLY
India, Pakistan, Malaysia, Indonesia, Burma, Thailand 70–95 mm. (2·76–3·74 in.)
This is richly coloured reddish-orange on the upperside with a band of a lighter colour on both wings. This band is broad on the forewing, tapering across the fore- and hindwing. The hindwing has a short tail. The female is similarly patterned but pale greenish grey. Some darker patterning is present along the wing margins with eyespots on the hindwing. The butterfly is common over most of its range in open country and jungle and often visits *Lantana* flowers (Verbenaceae). The caterpillar, which is pale yellow marbled with brown and armed with branched spines with 2 upright horns, feeds on species of *Passiflora* (Passifloraceae). The pupa resembles a dead and partly decayed leaf. This is a very variable species and many subspecies have been named.

VINE PYRALIS see **Sparganothis pilleriana**
VINE THECLA, HOP see **Strymon melinus**
VINE TORTRIX see **Sparganothis pilleriana**
**vinula** see **Cerura**
**violae** see **Telchinia**
VIOLET MEADOW BLUE see **Polyommatus icarus**

**Virachola** Moore (LYCAENIDAE)
This genus occurs in tropical Africa, S.E. Asia and Australia; it is best represented in Africa. The 20 or 30 species are relatively large for Lycaenidae. As in *Rapala*, there is a scent-scale patch on the male hindwing which interacts with a hair-brush under the forewing. The caterpillars feed inside legume pods, pomegranates *(Punica)*, guavas *(Psidium)* and other fruits. *Virachola* has been combined with the genus *Deudorix* by one authority.

**democles** Miskin
N.E. coast of Australia and Prince of Wales Island 28–33 mm. (1·1–1·3 in.)
This is possibly only a subspecies of the Indian *V. perse* (illustrated). The caterpillars feed inside the fruit of a species of *Strychnos*, a climbing plant of the family Loganiaceae, and ensure that the fruits do not drop as a result of their depredations by attaching them to the stem of the plant with silk, a technique also employed by *V. isocrates*.

**isocrates** Fabricius COMMON GUAVA BLUE
India, Sri Lanka, Burma 34–50 mm. (1·34–1·97 in.)
The caterpillar of this common species feeds on the seeds and pods of pomegranates, guavas and other fruits (see *V. democles)*. The chrysalis is formed inside the remains of the fruits. The adult male is chiefly iridescent brown and violet blue above; the female is mainly brown, with an orange patch on the forewing, and the black spot, encircled with orange, at the base of the hindwing tail. It is a widespread species occurring both on the plains and up to 2130 m. (7000 ft) in the Himalayas.

**perse** Hewitson
Sri Lanka, India, Sikkim 32–45 mm. (1·26–1·77 in.)
The female of this species differs from the illustrated male in the larger blue area on the forewing, which has a white patch at its apex and lacks the orange spot. The pinkish caterpillar is blotched with reddish brown and covered with numerous small black hairs. It feeds inside the fruit of *Randia* and other Rubiaceae and binds its food source to the branch so that the fruit does not fall. **244**aa

**virescens** see **Heliothis**
**virgaureana** see **Cnephasia**
**virgatella** see **Coleophora**
VIRGIN TIGER-MOTH see **Apantesis virgo**
**virginalis** see **Platyprepia guttata**
VIRGINIA WHITE, WEST see **Pieris virginiensis**
VIRGINIAN CTENUCHA see **Ctenucha virginica**
VIRGINIAN TIGER-MOTH see **Spilosoma virginica**
**virginica** see **Ctenucha, Spilosoma**
**virginiensis** see **Calephelis, Cynthia, Pieris**
**virgo** see **Apantesis**
**viridana** see **Tortrix**
VISCOUNT, MALAY see **Tanaecia pelea**
**vitrea** see **Ideopsis**

**Viviennea** Watson (ARCTIIDAE)
Twelve species are placed in this genus of tropical American moths. All the species are dark brown and yellow (or orange), with at least the terminal segment of the abdomen a brilliant iridescent green or blue above. A. D. Blest (U.K.) has shown that *V. tegyra* Druce, a species very closely related to the illustrated *V. moma*, is probably unpalatable to predators. It reacts, when disturbed, by alternately raising and lowering the wings and by directing upwards the brightly coloured abdomen as a signal of distastefulness to a predator with previous experience of this or other similarly coloured species of *Viviennea*.

**moma** Schaus
Tropical S. America 30–43 mm. (1·18–1·69 in.)
The colour pattern of *V. moma* is typical of its genus. Modifications of this pattern involve an extra, incomplete yellow band in *V. zonana* Schaus and the almost complete loss of the dark brown markings on the forewings of *V. salma* Druce and *V. superba* Druce. Nothing is known about its life history. **368**k

**volanica** see **Pseudaterria**
**vollenhouii** see **Ixias**
**vulcanus** see **Heliconius**

**w-album** see **Strymonidia**

**Wagimo** Sibatani & Ito (LYCAENIDAE)
There are 4 species in this genus of butterflies. It is restricted to Japan, China, Formosa, Korea and E. Russia.

**signata** Butler
China, Formosa, Japan 26–33 mm. (1·02–1·3 in.)
The under surface of this species is brownish orange, with darker brown and orange markings and 5 white lines. Males differ little from females in coloration either above or below. **253**

WAINSCOT MOTHS see **Mythimna, Nonagria**
WAITER see **Marpesia coresia**
**wakefieldii** see **Euxanthe**
**walkeri** see **Napata**
WALL BUTTERFLIES see **Lasiommata**
**wallacei** see **Heliconius**
**wallichii** see **Brahmaea**
WALNUT MOTH, ROYAL see **Citheronia regalis**
WALNUT SPAN WORM MOTH see **Coniodes plumigeraria**
WANDERER BUTTERFLIES see **Bematistes, Danaus, Pseudacraea**
WASP-MOTH, DOUBLE-TUFTED see **Didasys belae**
WATER SNOW FLAT see **Tagiades litigiosa**
WATER-VENEER, FALSE-CADDIS see **Acentria nivea**
**wauaria** see **Semiothisa**
WAVE MOTHS see **Cyclophora, Venusia**
WAX MOTHS see **Galleria**
**webbianus** see **Leptotes**
WEBBING CLOTHES MOTH see **Tineola bisselliella**
WEBWORM MOTHS see **Hednota, Hellula, Hyphantria, Hypsopygia, Loxostege, Margaritia, Nola, Nomophila, Pediasia**
**wedah** see **Pseudergolis**
**weiskei** see **Graphium, Milionia**
WEIDEMEYER'S ADMIRAL see **Limenitis weidemeyerii**
**weidemeyerii** see **Limenitis**
WELSH WAVE see **Venusia cambrica**
**wernickei** see **Philaethria**
WEST COAST LADY see **Cynthia anabella**
WEST VIRGINIA WHITE see **Pieris virginiensis**
**westermanni** see **Junonia**
WESTERN
COURTIER see **Sephisa dichroa**
GRAPELEAF SKELETONISER see **Pampa brillians**
MARBLED WHITE see **Melanargia occitanica**

**Weymeria** Karsch (AGARISTIDAE)
There is a single species in this moth genus.

**athene** Weymer
E. Africa 48–50 mm. (1·89–1·97 in.)
This is possibly yet another member of the African mimetic complex having as its focus the distasteful butterfly *Danaus chrysippus*. Very similar to *V. athene* in pattern is another African Agaristid *Heraclia poggei*. **399**k

WHISTLING MOTHS see **Hecatesia**
WHITE, BORDERED see **Bupalis piniaria**
WHITE BUTTERFLIES see **Anaphaeis, Aporia, Appias, Ascia, Colotis, Delias, Dixeia, Euchloe, Leptosis, Leptidea, Melanargia, Neophasia, Pieris, Pontia**
WHITE
ADMIRALS see **Ladoga, Limenitis**
DISMORPHIA see **Anantia licina**
EMPEROR see **Helcyra hemina**
ERMINE see **Spilosoma lubricipeda**
ERYCID see **Hermathena candidata**
FLANNEL see **Megalopyge crispata**
IMPERIAL see **Neomyrina hiemalis**
LADY SWALLOWTAIL, ANGOLA see **Graphium pylades**
LETTER HAIRSTREAK see **Strymonidia w-album**
M HAIRSTREAK see **Panthiades m-album**
ORANGE-TIP see **Ixias marianne**
PAGE, SOUTHERN see **Eurytides telesilaus**
PEACOCK see **Anartia jatrophae**
PLUME MOTH see **Pterophorus pentadactyla**
PROMINENT see **Leucodonta bicoloria**
RICE BORER see **Tryporyza innotata**
SATIN see **Leucoma salicis**
THEOPE see **Theope pieridoides**
TIGER see **Danaus melanippus**
UNDERWING see **Catocala relicta**
WHITE-BAR BUSH BROWN see **Mycalesis anaxias**
WHITE-BORDERED BUTTERFLY see **Nymphalis antiopa**
WHITE-CLOAKED SKIPPER see **Leucochitonea levubu**
WHITE-EDGED ROCK BROWN see **Satyrus parisatis**
WHITE-LINED SPHINX see **Hyles lineata**
WHITE-MARKED TUSSOCK see **Hemerocampa leucostigma**
WHITE-SHOULDERED CLOTHES MOTH see **Endrosis sarcitrella**
WHITE-SHOULDERED HOUSE MOTH see **Endrosis sarcitrella**
WHITE-SPECK see **Mythimna unipuncta**
WHITE-SPOTTED TADPOLE see **Syrmatia dorilas**
WHITE-TAILED ERYCID see **Echenais alector**
WHITE-TIPPED CLOTHES MOTH see **Trichophaga tapetzella**
WHITE-TIPPED CLOVER CASE MOTH see **Coleophora frischella**
**widenmanni** see **Dactyloceras**
WIDOW MOTH, PRETTY see **Eupithecia venosata**
WIDOW SPHINX see **Acanthosphinx guessfeldti**
WILLOW MOTH, SMALL MOTTLED see **Spodoptera exigua**
WILLOW BEAUTY see **Peribatodes rhomboidaria**
WINDMILL BUTTERFLY, COMMON see **Parides philoxenus**
WINTER MOTHS see **Alsophila, Operophtera**
**Witlesia** see **Eudonia**
WITCH MOTH, BLACK see **Ascalapha odorata**
**wittfeldi** see **Alypia**
WITTFELD'S FORESTER MOTH see **Alypia wittfeldi**
WIZARD see **Rhinopalpa polynice**
WONDERFUL HAIRSTREAK see **Chrysozephyrus ataxus**
WOOD BUTTERFLY, SPECKLED see **Pararge aegeria**
WOOD
LEOPARD see **Zeuzera pyrina**
SATYR, LITTLE see **Euptychia eurytus**

TIGER-MOTH see **Parasemia plantaginis**
WHITES see **Delias, Leptidea, Leptosis**
WOOD-BORER see **Xyleutes**
WOOD-NYMPH, SMALLER see **Ideopsis gaura**
WOODLAND GRAYLING see **Hipparchia fagi**
WOODLING, BEAUTIFUL see **Egira pulchella**
WOOLLY BEARS see **Arctiidae, Pyrrharctica, Spilosoma** and see under **Arctia caja**
WOUNDED HAWK-MOTH see **Leptoclanis basalis**

**Xanthabraxas** Warren (GEOMETRIDAE)
Only a single species is known in this genus which is related to *Abraxas*, the Currant moths.

**hemionata** Guenée
N. China 44–56 mm. (1·73–2·2 in.)
The bright yellow colour with a striking wing pattern makes this an easily recognized species. The biology is unknown. **294**w

**Xanthia** Ochsenheimer (NOCTUIDAE)
Between 50 and 60 species are placed in this genus of moths. Its range is chiefly European and temperate Asian, but it is also represented in tropical S.E. Asia and in N. America.

**flavago** Fabricius PINK-BARRED SALLOW MOTH
Canada, N. U.S., Europe including Britain, temperate Asia 28–34 mm. (1·1–1·34 in.)
The orange and reddish brown pattern of the forewings of this species is repeated in general plan in many other species of *Xanthia*. The caterpillar is reddish or purplish brown, speckled with dark brown and with 3 pairs of pale lines along its back; it feeds at first inside the catkins of sallows and willows *(Salix)* and later on the leaves. **391**r

**xanthocosma** see **Cerace**
**xanthomelas** see **Nymphalis**

**Xanthopan** Rothschild & Jordan (SPHINGIDAE)
There is a single species in this genus of African and Madagascan hawk-moths.

**morgani** Walker
Afrotropical 100–135 mm. (3·94–5·31 in.)
Charles Darwin, author of the classical *Origin of Species*, postulated that the orchid *Angraecum sesquipedale* could be pollinated only by an insect with a tongue of over 12 inches in length able to penetrate to the bottom of the long, slender nectary of this orchid. Darwin's predicted insect was discovered later and described as *X. praedicta* (a subspecies of *X. morgani)* by Rothschild & Jordan. A further prediction has been made that an orchid similar to the apparently endemic Madagascan species will be discovered in Africa. The forewings are cryptically patterned with yellow, brown and black; the hindwings are brown, with a yellow, basal area bisected by a dark brown bar. Its caterpillars are green, with oblique, purple stripes on each side and a long, curved horn.

**Xanthorhoe** Hübner (GEOMETRIDAE) CARPET MOTHS
The popular name is misleading as the moth seldom comes indoors unless attracted by light, and their caterpillars (unlike those of species of *Hofmannophila* or *Tinea*) do not feed on carpets but on various plants.

**ferrugata** Clerck DARK-BARRED TWIN-SPOT CARPET MOTH
Europe including Britain, Asia 22–24 mm. (0·87–0·94 in.)
The moth is widespread over Europe. There are several named varieties of this species which is similar in appearance to several other Carpet moths. The closely related species in N. America looks very like the European species and was formerly believed to be the same species. The caterpillars are a mottled brown and grey colour with pale diamond shapes and black spots in the middle of the back.

**fluctuata** Linnaeus GARDEN CARPET MOTH
Europe including Britain, W. & C. Asia, including Japan, N. Africa 22–27 mm. (0·87–1·06 in.)
Probably one of the commonest species in Europe. Its pattern is a good example of camouflage. When the moth is at rest on a tree trunk, it blends perfectly with the background. The caterpillar, which is a typical 'looper', feeds on Cruciferae, including cabbage, and is also said to feed on gooseberries and currants. (Grossulariaceae). The records of this species in N. America have proved to be of a closely related and similar looking, but distinct species.

**xarifa** see **Lymnas**

**Xenosphingia** Jordan (SPHINGIDAE)
There is a single known species in this genus of hawk-moths.

**jansei** Jordan JANSE'S SPHINX OF HAWK-MOTH
Rhodesia 54 mm. (2·13 in.)
*X. jansei* is named after a famous S. African entomologist. The forewings of this species are not elongate as in most species of hawk-moths and its general appearance is more like that of a Noctuid species. The forewing ground colour of most specimens is green, with a single white spot in the centre of the wing; the hindwing is greyish green. A few specimens have pinkish brown forewings.

**xuthus** see **Papilio**

**Xyleutes** Hübner (COSSIDAE) WOOD BORERS or CARPENTER MOTHS
A large genus with many species from Australia and Africa. The caterpillars bore into trees and because of their large size the tunnels they produce can be serious problems for foresters. Some caterpillars tunnel in the soil, where they feed on roots of *Acacia*. These include the 'Witchety grubs' which are eaten by the aborigines in Australia. A few species have females with short wings which are incapable of flying.

**affinis** Rothschild
Australia 120–175 mm. (4·72–6·89 in.)
This huge moth, with an abdomen often 70 mm. long, lays large numbers of eggs. The caterpillars bore into the trunks of *Eucalyptus*, taking 2–3 years before they become adult. They seriously damage the tree, killing it if it is a young one. **8**a

**eucalypti** Herrich-Schäffer
Australia 130–200 mm. (5·12–7·87 in.)
As its specific name implies the caterpillars bore into *Eucalyptus*. The moth is typical in appearance of many in the genus. **8**d

**leucomochla** Turner
Australia 170 mm. (6·69 in.)
This is a brown species, similar in shape to *X. affinis*, with the male darker than the female. The caterpillars feed externally on the roots of *Acacia ligulata*, generally feeding on the sap flow rather than on the roots themselves. This is a dry country insect which avoids dessication by living underground. The caterpillars, called 'Makowitjut', are collected by the aborigines and either cooked or eaten raw.

**Xylomiges** see **Egira**

**Xylophanes** Hübner (SPHINGIDAE)
The 50 or so species of this genus of hawk-moths have long, slender, tapered bodies and narrow, pointed wings. Most of the species are C. American or tropical S. American in distribution, but 4 occur in N. America.

**gundlachi** Herrich-Schäffer
Cuba 40–45 mm. (1·57–1·77 in.)
This is a basically green moth, except for the white-edged, brown hindwings. The only forewing markings are a greenish brown, central spot and a brown and white, transverse line near the outer edge of the wing.

**pluto** Fabricius
S. U.S., through C. America to S. Brazil 60–70 mm. (2·36–2·76 in.)
This is a common species in Florida where it is often seen feeding from the flowers of *Verbena*. The caterpillar has been recorded on *Erythroxylon* and milkberry *(Chiococca)*. **344**f

**Xyloryctidae**
This family and the allied one Stenomidae, have often been confused in the past. Recent work however, is giving us a clearer picture of the family and its relationships, together with that of a third family, Oecophoridae, which has many features in common with the other two. Xyloryctid moths are particularly common in Australia, but occur elsewhere in the tropics of the Old World. The caterpillars tunnel in leaves or bark and in a variety of other places. The moths are all night fliers but are readily attracted to light. See *Aeolanthus, Cryptophasa, Cyanocrates, Nephantes, Uzucha.*

**xylostella** see **Plutella**
**xynias** see **Eurytides**
**Xystus** see **Prionxystus**

**yamamai** see **Antheraea**
YAMFLY see **Loxura atymnus**
**yasudai** see **Lyonetia**
YELLOW BUTTERFLIES see **Colias, Eurema, Kricogonia, Napata, Pseudopanthera**
YELLOW
APRICOT BUTTERFLY see **Phoebus philea**
COSTER see **Acraea vesta**
LEAF BUTTERFLY see **Phoebus trite**
MIGRANT see **Phoebus statira**
PANSY see **Junonia hierta**
SKIRTED CALICO see **Hamadryas fornax**
TIGER-MOTH see **Arctia flavia**
UNDERWING, LARGE see **Noctua pronuba**
UNDERWING, SMALL DARK see **Anarta cordigera**
WOOLLY BEAR see **Spilosoma virginica**
YELLOW-BANDED
BEETLE MIMIC see **Correbidia assimilis**
NYMPHIDIUM see **Nymphidium onaeum**
SWORDTAIL see **Graphium illyris**
YELLOW-LEGGED
CLEARWING see **Synanthedon vespiformis**
TORTOISESHELL see **Nymphalis xanthomeles**
YELLOW-NECKED APPLE-WORM or CATERPILLAR see **Datana ministra**
YELLOW-SHELL see **Camptogramma bilineata**
YELLOW-SPOTTED GONATRYX see **Anteos chlorinde**
YELLOW-TAIL see **Euproctis similis**
YELLOWIE, LITTLE see **Eurema venusta**

**Yoma** Doherty (NYMPHALIDAE)
A genus of butterflies with 2 species which range from Burma through Indonesia, N. Australia and the Solomon Islands.

**sabina** Cramer LURCHER
Burma, Philippines, S.E. Asia, Indonesia to Australia 62–80 mm. (2·44–3·15 in.)
The subspecies described include 1 in N. Australia. The butterfly is a rich brown colour with a broad orange band through the wings from the front to the back. The underside is pale brown with a paler middle band and black spots in the margin between the veins. The hindwing has a small point on the margin giving it a characteristic shape. **231**k

**Yponomeuta** Latreille: syn. *Hyponomeuta* (YPONOMEUTIDAE)
A large genus of micro-moths with species all over the world. The name was first given to it in 1796 when it was spelt Yponomeuta. It was subsequently altered to *Hyponomeuta* in 1837 and this was used for many years, since it was thought the original was a misprint. However the spelling with a 'Y' instead of 'Hy' is now generally accepted. The moths are generally small or very small. The caterpillars have a variety of habits, some mine in leaves, others live gregariously in webs below leaves. Species of *Yponomeuta* often have the forewing covered in spots.

**cagnagella** Hübner: syn. *Y. cognatella*
Europe including Britain 20–26 mm. (0·79–1·02 in.)
This European species is common over most of its range. The caterpillars spin webs over the leaves of Spindle *(Euonymus)* on which they feed. Related species, with similar patterns, sometimes cover trees in a spectacular manner with their silk. There are a number of related species, all rather similar, including some where, although the adults are indistinguishable, their caterpillars feed on different host plants.

The caterpillars generally are specific to one host and will not eat others. They are considered as 'biological species' a slightly misleading term (since we hope all species are biological ones!) but this is to distinguish them from ones that can be separated on structural differences ('morphological species'). **31**

**Yponomeutidae:** syn. *Hyponomeutidae*
The older spelling of the name Yponomeutidae has now replaced the spelling which is found in a number of books as Hyponomeutidae. Yponomeutids are generally small or very small moths with caterpillars which tunnel in shoots or feed exposed, often living in a communal webbing. Sometimes this webbing can be so extensive that whole trees or rows of trees are covered with a silken webbing which attracts popular attention. The family is worldwide, containing many pest species, some of which are pests of crops. Many tropical Yponomeutids are brightly coloured species. Generally 4 subfamilies are recognized in the family (Plutellinae, Yponomeutinae, Amphitherinae and Argyresthiinae) but there is still no firm agreement amongst specialists about this. See *Anticrates, Argyresthia, Atteva, Comocritis, Harpetila, Lactura, Plutella, Prays, Swammerdamia, Trisophista, Yponomeuta, Ypsolopha.*

**Ypsolopha** Latreille (YPONOMEUTIDAE)
Many of the micro-moths in this family have the apex of the forewings produced and curved backwards giving a hooked appearance. The genus is mainly European and Asian but has species in N. America. Many have a characteristic appearance when at rest.

**dentella** Fabricius TOOTH-STREAK HOOKED TIP MOTH 17–21 mm.
Europe including Britain, U.S. (0·67–0·83 in.)
The forewings have a strongly hooked apex, reddish-brown, with yellow streak on hind margin, curving into the wing. The hindwings are brown. The caterpillar is pale yellow-green with a broad brown strip on the back. It feeds on honeysuckle *(Lonicera)*. The moth is out in May and June and is common over most of its range. The status of this species in the U.S. is uncertain. Probably the specimens collected there had been accidentally introduced. **32**

**falciferella** Walsingham
U.S. 19–25 mm. (0·75–0·98 in.)
The long narrow wings with strongly hooked tip are conspicuous characters for identifying this moth. It has grey brown wings with darker marks. The moth has been recorded on the E. and W. coasts of the U.S. The caterpillar is believed to feed on Rosaceae. The moth is out in August.

**Ypthima** Hübner (NYMPHALIDAE) RING BUTTERFLIES
These are small weak fliers in the subfamily Satyrinae, with a bouncing flight. Typically they are brown with prominent eyespots, which, below the apex of the forewing have twin white 'pupils'. They are found from India to Australia and all over Africa S. of Sahara.

**asterope** Klug AFRICAN RINGLET
Africa S. of Sahara 35–40 mm. (1·38–1·57 in.)
A widespread and common species throughout its range it can be found in all months of the year. The caterpillar feeds on various species of grass. The butterfly is greyish brown with a large eyespot near the apex of the forewing. This has a yellow ring round a black, white centred spot with a larger ring surrounding these. There is a small eyespot near the apex of the hindwing. There are a number of similar species but this one can be distinguished by the much less conspicuous eyespots on the underside.

**baldus** Fabricius COMMON FIVE RING BUTTERFLY
India, Pakistan, Burma 32–48 mm. (1·26–1·89 in.)
This is a very common Ring butterfly. It is brown above with a prominent eyespot at the apex of the wing with, nearby, 2 small white spots. The hindwing is similar. Underside is paler, with more eyespots on the hindwing. It flies in most open or forest areas throughout India. It has seasonal forms.

YUCCA MOTHS see **Tegeticula**
YUCCA SKIPPER see **Megathymus yuccae**

**yuccae** see **Megathymus**
**yuccasella** see **Tegeticula**

**zacateca** see **Rothschildia**

**Zacorisa** Meyrick (TORTRICIDAE)
A genus of micro-moths with about 12 known species from New Guinea to the Philippines. They are generally very colourful, often with metallic iridescences, yellow and blacks predominating. Nothing is known of their biology but their striking colour and patterns could make them popular with collectors, but they are known only from regions well removed from the main areas visited by collectors.

**holantha** Meyrick
New Guinea 20–26 mm. (0·79–1·02 in.)
This species is typical of many in the genus. It has striking steel-blue coloured wings with orange tips. This moth is rare in collections and nothing is known of its biology. **14a**

**toxopei** Diakonoff
New Guinea 35 mm. (1·38 in.)
Several subspecies of this have been scientifically described. Nothing is known of its biology; the moths were collected flying in remote areas of New Guinea. The specimen illustrated is the nominate subspecies *(Z. toxopei toxopei)*. **14c**

**zacynthus** see **Parides**
**zagraea** see **Zegara**
**zagraeoides** see **Zegara**
**zagreus** see **Papilio**

**Zale** Hübner (NOCTUIDAE)
Except for a minor representation in C. America, *Zale* is unknown outside N. America. Both wings are marked with numerous sinuous, transverse bands in most species. The caterpillars are capable of jumping like those of *Catocala*.

**lunata** Drury
U.S., E. Canada 45–52 mm. (1·77–2·05 in.)
This somewhat variable moth is distributed throughout most of the U.S. where it flies chiefly in September. Its striped, brown caterpillars feed on the foliage of numerous trees and shrubs. Californian species of *Z. lunata* are paler in colour than specimens from elsewhere in its range. **399g**

**zalmoxis** see **Papilio**
**zayla** see **Limenitis**
**zea** see **Heliothis**
ZEBRA see **Heliconius charitonius, Colobura dirce**
ZEBRA, RED-BANDED see **Cremna thasus**
ZEBRA, SPOTTED see **Graphium megarus**
ZEBRA SWALLOWTAIL see **Eurytides marcellus**
ZEBRA, WHITE see **Pinaccopteryx eriphia**
**zebralis** see **Dichocrocis**

**Zegara** Oiticica (CASTNIIDAE)
There are about 15 species in this genus of C. American and tropical S. American moths. They are orange, brown and white in colour and are mimics of Heliconiine species.

**zagraea** Felder
Tropical S. America 88–100 mm. (3·46–3·94 in.)
*Z. zagraea* is similar in pattern to *Z. zagraeoides*, but differs in the colour of the outer row of spots on the forewing which are dull yellow and the less well marked ray design on the hindwing. It may be part of the same mimetic association as *Z. zagraeoides.*

**zagraeoides** Houlbert
Tropical S. America 85–100 mm. (3·35–3·94 in.)
This large species is not common in collections. It is closely similar in colour and pattern to species of the butterfly genus *Heliconius* such as *H. ethillus* Godart. Nothing is known about its life history. **45k**

**Zelotypia** Scott (HEPIALIDAE) SWIFT MOTHS
Only 1 species has been described in this Australian micro-moth genus, although several colour forms have been named.

**stacyi** Scott BENT-WING SWIFT MOTH
Australia 190 mm. (7·48 in.)
The specimen illustrated is from New South Wales. The caterpillar, which is very large, tunnels into *Eucalyptus* trees where it may take 5–6 years to become mature. This can kill the younger trees and seriously damage older ones. It has been said that this moth, when at rest, resembles the head of a reptile, a *Varanus* lizard has been suggested. The eyespot on the wing is particularly conspicuous and it is possible that this is a protective device. **7a**

**Zeltus** de Nicéville (LYCAENIDAE)
Only 1 species is known.

**amasa** Hewitson FLUFFY TIT
Sikkim, N. India and Indo-China to the Philippines and Sumbawa 28–32 mm. (1·1–1·26 in.)
The long, fluffy, hindwing tails are a distinctive feature of this beautiful species. The female (not illustrated) is dark brown above. This is a low elevation, wet forest butterfly which keeps to the shade, fluttering weakly on its strange wings. **244z**

**zenobia** see **Thysania**
**Zeonia** see **Chorinea**
**zephyrites** see **Morpho**
**Zerynthia** see **Parnalius**
**zetes** see **Acraea, Battus**

**Zeuxida** Hübner (NYMPHALIDAE)
The butterflies in this genus belong to the subfamily Amathusiinae. Species in the genus are found from the Philippines through Thailand and Indonesia.

**amethystus** Butler SATURN BUTTERFLY
Thailand, Malaysia, Indonesia 70–100 mm. (2·76–3·94 in.)
The butterfly is common in Borneo and Sumatra. It may be generally found near streams and waterfalls up to about 1370 m. (4500 ft). It flies in the early morning and late afternoon. The life history is unknown. The underside is very leaf-like and perfect camouflage when at rest. **160c**

**Zeuzera** Latreille (COSSIDAE) LEOPARD MOTHS
A worldwide genus of moths with caterpillars which bore into trees.

**coffeae** Nietner RED BRANCH BORER
India, Sri Lanka through Indonesia to New Guinea, Taiwan, China 50–70 mm. (1·97–2·76 in.)
There are several closely related species which have been confused with this species, which has translucent white wings with bluish black spots. Externally it looks like the European Leopard moth. The caterpillar bores into trees. When the eggs hatch, the tiny caterpillars hang from the trees on silk threads and are blown about by the wind. It is a matter of chance whether they are blown on to a suitable host tree. This type of chance distribution of young is usually the sign that many eggs are laid to increase the chance of one caterpillar finding a host. This is the case with *Z. coffeae* where the female produces many eggs. The life cycle is about 14 months. The damage done by the caterpillars is especially serious in young cocoa trees *(Theobroma)*.

**pyrina** Linnaeus WOOD LEOPARD MOTH
C. & S. Europe including Britain, N. Africa, W. & C. Asia and N. America 50–64 mm. (1·97–2·52 in.)
This species is not common, but can be found in most parts of its range. The caterpillar bores into trunks and branches of apples and hawthorns and can seriously damage young trees. It takes several years to reach maturity, all the time boring through the wood of the host tree. The moth is very conspicuous with white forewings regularly patterned with blue black spots. The hindwings are white.

**zinckenella** see **Etiella**

**Zingha** Hemming (NYMPHALIDAE)
A genus with only 1 species in at present, found in Africa. It is closely related to *Charaxes*.

**zingha** Stoll: syn *Z. berenice*
W. Africa to E., 68–80 mm.
S. to Malawi (2·68–3·15 in.)
This species is found as far E. as Uganda but occurs throughout Zaire and is relatively common in W. Africa. The females are paler coloured than the males. The adults fly in most months of the year. The caterpillars have been reared on *Hugonia platysepala* (Linaceae). **237**d

**Zintha** Elliott (LYCAENIDAE)
There is a single species in this recently named genus, which was formerly included in the genus *Castalius*.

**hintza** Trimen HINTZA BLUE
Tropical C. Africa to Mozambique 24–30 mm.
and South Africa (0·94–1·18 in.)
The male of this species is violet-blue above, with brown markings; the female white, with black margins and distal spots and blue at the base of the wings. It lives in forested and more open country and has a localized distribution where the caterpillars foodplant *Zizyphus* occurs.

**Zipaetis** Hewitson (NYMPHALIDAE)
This genus of butterflies is in the subfamily Satyrinae. There are only 3 species in the genus which are found from India, Burma into S. China.

**scylax** Hewitson DARK CATSEYE BUTTERFLY
N. India, Pakistan, Burma 55–62 mm. (2·17–2·44 in.)
A common species from sea-level up to a few thousand feet. It is a species of thick jungle and is out in April to June and later in the year as well. The butterfly is a weak flier and is not often found out of the thick bush cover. The upperside is dark brown, unmarked, with a pale border with 3 black lines along the margin. On the underside these lines are prominent, with a silver sheen. The forewing has several small eyespots, while the hindwing has large eyespots all surrounded by a silvery line. The margin of the eyespot, with the 3 parallel lines round, is characteristic of this species. Nothing is known of its biology.

**Zizeeria** Chapman (LYCAENIDAE)
This is a genus of a few tailless, blue and brown, mostly minute butterflies. It is widely distributed in the Old World.

**knysna** Trimen AFRICAN GRASS-BLUE, DARK GRASS-BLUE or SOOTY BLUE 18–25 mm.
S.W. Europe, Africa, Asia, Australia (0·71–0·98 in.)
In Europe this species is found in damp localities near streams and elsewhere, at low altitudes; elsewhere in its range open areas are preferred. The caterpillar feeds on plants of various families. The male is lilac blue above, with a broad, greyish margin to the wings; the female is mostly greyish brown with a few blue scales at the base of the wings. The under surface is greyish white, with dark brown spots.

**maha** Kollar PALE GRASS-BLUE
India, Burma 26–30 mm. (1·02–1·18 in.)
The male of this attractive little species, as its name suggests, is pale blue above. The female is dark brown, with only a few blue scales at the base of the wings. The caterpillars feed first on the under surface of wood sorrel *(Oxalis)* leaves, later on the whole leaf.

**Zizina** Chapman (LYCAENIDAE) GRASS-BLUES
This is a small genus of widely distributed species found in Africa, S. Asia E. to Japan and New Guinea, New Zealand and the W. Pacific islands. It is united with the genus *Zizeeria* by some authors.

**otis** Fabricius COMMON GRASS-BLUE
Africa, India to Japan, New Guinea, 20–23 mm.
Australia and the islands of the S.W. (0·79–0·91 in.)
Pacific
The male of this very widely distributed and common species is lilac above, with brown outer margins to the wings; the female is similar in coloration but has broader brown margins. Both sexes are pale brown beneath, with a weakly marked pattern of brown dots and lines. In parts of its range, *Z. otis* is a minor pest of lucerne *(Medicago)*, clover *(Trifolium)*, garden peas *(Pisum)* and other herbaceous legumes (Leguminosae). It is the commonest species of butterfly in Australia.

**Zizula** Chapman (LYCAENIDAE)
Two species are placed in this genus of butterflies: 1 Old World in range, the other New World. Both species are brown above, and white with black spots beneath. They are amongst the smallest butterflies in the world.

**cyna** Edwards IMPORTED BLUE
S.W. U.S., C. America and 17–21 mm.
tropical S. America (0·67–0·83 in.)
*Z. cyna* was confused for a time with the Old World *Z. hylax* which it was supposed had been introduced with fodder brought over with camels to the deserts of W. U.S. from Africa for military use by Jefferson Davis. There is a big gap in the range of this butterfly which is found in Arizona and New Mexico S. to Guatemala and then appears again in W. S. America (Ecuador, Bolivia and N.W. Argentina).

**hylax** Fabricius: syn. *Z. gaika*
Africa, India, S.E. Asia to Australia 15–23 mm.
and the S. Pacific Islands (0·59–0·91 in.)
This is very similar in appearance to *Z. cyna*. Observations by the Ehrlichs in Tanzania have shown that the high aspect ratio (length to width) of the wings is a modification which allows access by the butterfly to the nectaries inside the base of tubular flowers such as *Asystasia*. In Africa *Z. hylax* caterpillars feed on the flowers of *Oxalis*.

**zodiaca** see **Alcides**
**zoe** see **Colotis**
**zoegana** see **Agapeta**
**zoilus** see **Tellervo**

**Zophopetes** Mabille (HESPERIIDAE)
Species of this genus of Skipper butterflies are found over most of Africa S. of Sahara. Many of them are dusk flying (crepuscular) insects, generally remaining hidden in the sunnier part of the day.

**dysmephila** Trimen TWILIGHT SKIPPER or PALM TREE NIGHT FIGHTER
E. to South Africa 40–45 mm. (1·57–1·77 in.)
A brown Skipper with white antennae and a white marginal streak below the tip of the forewing and a few white spots in the middle of the forewing. This species flies at dusk round *Phoenix* palms on which the caterpillars feed. It is one of the crepuscular (dusk-flying) butterflies which are attracted to light and frequently get caught in the mercury vapour light traps of moth collectors.

**zorcaon** see **Eueides**
**zoroastres** see **Papilio**
**zuni** see **Arachnis**
ZUNI TIGER-MOTH see **Arachnis zuni**
**zurisana** see **Rhodoneura**

**Zygaena** Fabricius (ZYGAENIDAE)
A genus of day-flying moths of very characteristic appearance and habits. Very common in pasture and meadowlands in summer. The colours are very variable and, being a popular group with collectors, many subspecies and aberrations have been named. The bright colours are a warning to predators that the moths are distasteful. It was recently shown that part of the distasteful nature of the insects is due to the presence of traces of compounds of cyanide.

**filipendulae** Linnaeus SIX-SPOT BURNET MOTH
Europe including Britain, 28–35 mm.
W. Asia (1·1–1·38 in.)
This moth with deep green or blue-green metallic forewings has 6 deep crimson spots from which it gets its name. It is common in meadows throughout Europe. In July and August this brightly-coloured moth can be seen in its whirring flight from grass head to grass head. It attracts attention through its bright colour, but is protected by being distasteful to mammals and birds. The caterpillars feed on *Trifolium*, *Lotus* or *Lathyrus* (Leguminosae) and are also brightly coloured. Many subspecies of the Six-Spot Burnet have been named, mostly differing in pattern. In some cases the red spots on the forewing are replaced by yellow. **50**

**occitanica** de Villers PROVENCE BURNET MOTH
Spain, S. France, Italy 30 mm. (1·18 in.)
This attractive day-flying moth is brightly coloured and very conspicuous. Like other species in the genus it is distasteful to birds and small mammals and therefore advertises itself. The pale green, rather slug-like caterpillar has yellow spots and black dots on its sides and feeds on species of *Dorycnium* (Leguminosae). Many local forms of this species have been described and named. **49**

**purpuralis** Brunnich TRANSPARENT BURNET MOTH
Europe including Britain, 27–31 mm.
temperate Asia (1·06–1·22 in.)
This typical Burnet moth with red and black wings is widespread in Europe but local in Britain. The moth flies in the sunshine in June. The caterpillar is dark green with rows of black and yellow spots and a line on the back. It feeds on Thyme *(Thymus)* and Burnet Saxifrage *(Pimpinella)*.

**trifolii** Esper FIVE-SPOT BURNET MOTH
Europe including Britain, 28–35 mm.
N. Africa (1·1–1·38 in.)
The specimen illustrated is the subspecies *Z. trifolii barcelonensis* from Catalonia, Spain. The species has been separated into over 130 subspecies and aberrations from different localities. They differ in the colour and pattern of the wings. **46**a

**Zygaenidae**
A family of moths, worldwide in distribution, with many beautiful and strikingly patterned species. Many are brightly coloured and most are distasteful to predators; some of them having substances in their body which are toxic to warm-blooded predators (eg. cyanides). Their bright colour, frequently reds and blacks, is believed to advertise the fact that they are distasteful and to warn off predators. The majority of the species in the family are day-fliers, preferring the sunniest times of day. Because of their colours and day-flying habits they are popular with collectors. Many of the colours are metallic and iridescent colours are common. Some of the species have unusually shaped wings, with the extreme forms in the subfamily Himantopinae where the hindwings are reduced to long thin streamers. See *Adscita, Agalope, Campylotes, Cyclosia, Erasmia, Eterusia, Gymnautocera, Himantopterus, Pampa, Zygaena.*

**zymna** see **Megalopalpus**